Conservation and the Genomics of Populations

Conservation and the Genomics of Populations

Third Edition

Fred W. Allendorf
Division of Biological Sciences, University of Montana, USA

W. Chris Funk
Department of Biology, Colorado State University, USA

Sally N. Aitken
Department of Forest and Conservation Sciences, Faculty of Forestry, University of British Columbia, Canada

Margaret Byrne
Biodiversity and Conservation Science, Department of Biodiversity, Conservation and Attractions, Australia

Gordon Luikart
Flathead Lake Biological Station, University of Montana, USA

With illustrations by
Agostinho Antunes
University of Porto, Portugal

OXFORD
UNIVERSITY PRESS

OXFORD
UNIVERSITY PRESS

Great Clarendon Street, Oxford, OX2 6DP,
United Kingdom

Oxford University Press is a department of the University of Oxford.
It furthers the University's objective of excellence in research, scholarship,
and education by publishing worldwide. Oxford is a registered trade mark of
Oxford University Press in the UK and in certain other countries

First Edition published in 2007 as *Conservation and the Genetics of Populations* by Blackwell Publishing
Ltd.

Second Edition published in 2013 as *Conservation and the Genetics of Populations* by John Wiley & Sons
Ltd.

Third Edition published in 2022

Reprinted (with corrections) 2022

Impression: 2

Published in the United States of America by Oxford University Press
198 Madison Avenue, New York, NY 10016, United States of America

British Library Cataloguing in Publication Data

Data available

Library of Congress Control Number: 2021937945

ISBN 978–0–19–885656–6 (hbk)
ISBN 978–0–19–885657–3 (pbk)

DOI: 10.1093/oso/9780198856566.001.0001

Printed and bound by Sheridan, United States of America

Front cover image: Yellow-spotted monitor (*Varanus panoptes*) in Litchfield National Park,
Northern Territory, Australia. The collapse of this species is perhaps the most high-profile loss
caused by the invasion of the cane toad (Guest Box 14). Photo by Ed Kanze.

Back cover image: Old growth Sitka spruce (*Picea sitchensis*) in Carmanah Walbran Provincial Park,
British Columbia, Canada. The climber is sampling the tree from different
locations in order to estimate somatic mutation rates (Section 12.1). Photo by T.J. Watt.

We dedicate this book to Michael E. Soulé, who died while we were working on this edition (Crooks et al. 2020). Michael was instrumental in the founding of the field of conservation biology by inspiring his basic science friends to apply their efforts to conserve biodiversity and by organizing a series of meetings in the late 1970s. He also co-authored the first book that applied the principles of genetics to conservation (Frankel & Soulé 1981).

Contents

Appendix and References are available only online at www.oup.com/companion/AllendorfCGP3e

Preface to the Third Edition

I have always loved, and will always love, wild nature: Plants and animals. Places that are still intact. Though others might avoid the word, I insist that we talk about "love" in conservation, because we only protect what we love.

(Michael E. Soulé 2018)

The field of conservation genetics has changed dramatically since the second edition of this book was published in 2013. One-third of the references in this edition were written after the publication of the second edition. We have changed the title to reflect the growing and profound influence that genomics has had on applying genetics to problems in conservation. We have witnessed an extraordinary explosion of knowledge of the genetics and genomics of natural populations because genomic approaches have become more affordable and accessible. It has been a real challenge to add the new literature while keeping the book to a reasonable size. To accomplish this, we have put the Appendix and the References online. We understand that this is inconvenient, but we wanted to avoid an unwieldy book. Approximately 10% of the second edition was taken up by the References. The References and Appendix can be downloaded from the following companion website: www.oup.com/companion/AllendorfCGP3e.

We are excited to add Margaret Byrne and Chris Funk as coauthors. The five of us met in Missoula in July 2019 to plan our efforts (see Figure P.1). We have added Chapter 24, which deals with the practical considerations of being a conservation geneticist and applying genetics to problems in conservation. We invited Helen R. Taylor to help write this chapter; she is the primary author of Chapter 24.

This edition was written largely in the midst of the COVID-19 pandemic. Millions of people worldwide have died from this tragic event. We send our deepest condolences to those who have lost loved ones from this global pandemic. The disease spillover from wildlife to humans is intimately linked to the topic of this book: conservation of biodiversity. This tragedy demonstrates that human health and well-being are inextricably tied to the health and well-being of the natural world. We hope this book furthers biodiversity conservation for the benefit of nature and humans.

Our guiding principle in writing has been to provide the conceptual basis for understanding the genetics of biological problems in conservation. We have not attempted to review the extensive and ever-growing literature in this area. Rather, we have tried to explain the underlying concepts and to provide examples and key citations for further consideration. We also have strived to provide enough background so that students can read and understand the primary literature.

There is a wide variety of computer programs available to analyze genetic and genomic data to estimate parameters of interest. However, the ease of collecting and analyzing data has led to an unfortunate and potentially dangerous reduction in the emphasis on understanding theory in the training of population and conservation geneticists. Understanding theory remains crucial for correctly interpreting outputs from computer programs and statistical analyses. For example, the most powerful

Figure P.1 The authors (left to right: Chris Funk, Margaret Byrne, Sally Aitken, Fred Allendorf, and Gordon Luikart) on the campus of the University of Montana.

software programs that estimate important parameters, such as effective population size (Chapter 7) and gametic disequilibrium (Chapter 10), can be misleading if their assumptions and limitations are not understood. We are still disturbed when we read statements in the literature that the loci studied are not linked because they are not in linkage (gametic) disequilibrium.

We have striven for a balance of theory, empirical examples, and statistical analysis (see Figure P.2). Population genomics provides unprecedented power to understand genetic variation in natural populations. Nevertheless, application of this information requires sound understanding of population genetics theory. To quote Joe Felsenstein: "We have the same situation in population genomics. People have vast amounts of data and do completely half-ass things with it because they don't know any better. And, I wish there was some way of persuading people that we need to train students in the development and properties of the methods. And that means population genetics."

The molecular tools being used by population geneticists continue to change rapidly. It has been difficult to decide which techniques to include in Chapters 3 and 4. We present some techniques

that are seldom or no longer used (e.g., allozymes) because they are crucial for understanding much of the previous conservation genetics literature.

We also have included a comprehensive Glossary. Words included in the Glossary are **bolded** the first time they are used in each chapter. Many of the disagreements and long-standing controversies in population and conservation genetics result

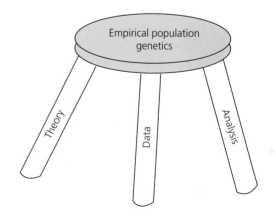

Figure P.2 The application of population genetics to understand genetic variation in natural populations relies upon a combination of understanding theory, collecting data, and understanding analysis.

from people using the same words to mean different things. It is important to define and use words precisely.

Many of our colleagues have written Guest Boxes that present their own work in conservation genetics. Each chapter contains a Guest Box that provides further consideration of the topics from that chapter. These boxes provide the reader with broader voices in conservation genetics from some of the major contributors to the literature in conservation genetics from around the world.

We have lost some special colleagues since the publication of the second edition. We have dedicated this edition to Michel Soulé, who co-authored a Guest Box in the first two editions of this book. We were saddened to learn that Elaina Tuttle, who wrote a Guest Box in the previous edition, passed away in 2016. Fred's good friend and colleague Ian Jamieson passed away in 2015. Ian had an important influence on Fred's understanding of genetic load (Box 17.1), population viability, and rugby.

Fred W. Allendorf
W. Chris Funk
Sally N. Aitken
Margaret Byrne
Gordon Luikart
12 February 2021

Acknowledgments

We are grateful to Ian Sherman, Charlie Bath, and the staff of Oxford University Press for their help in producing this edition. FWA is especially grateful to Ian for his friendship and encouragement over the past eight years.

FWA thanks Diane Haddon for her love, humor, and color vision. WCF would like to thank his wife, Victoria Funk, for her patience and loving support, and his parents, Sue and Bill Funk, for fostering his love of nature and learning. SNA thanks Jack Woods for his endless patience, love, and conversations about topics in this book. She also thanks members of her research group for inspiration and input. MB thanks her family for their support, particularly during weekends of writing and reviewing, and many colleagues for ongoing inspiration in making a difference for conservation of biodiversity. GL thanks his children, Braydon and Madyson, and parents, Nancy and Ed Luikart, for their love and support of his passion for conservation, genetics, and education.

We give special thanks to the following people and organizations for their help: Minitab LLC, for providing statistical software through their Author Support Program; Leif Howard, for help with permissions and much more; Nina Andrascik, for help with the Glossary; John Ashley, Mark Ravinet, Glenn-Peter Sætre, and Diane Whited, for help with figures. Nils Ryman, for his helpful comments on the Appendix; Paul Sunnucks, for his many helpful comments on the first edition of this book; and the following members of the Great Lakes Genetic/Genomic Laboratory, for their detailed comments on the second edition of this book: Amanda Haponski, Carson Pritchard, Matthew Snyder, Shane Yerga-Woolwine, and Carol Stepien.

We thank the following colleagues, friends, and family members who have helped us by providing comments, information, unpublished data, figures, and answers to questions: Darren Abbey, Kea Allendorf, Teri Allendorf, Paulo Alves, Steve Amish, Eric Anderson, Mike Arnold, Jon Ballou, Mark Beaumont, Peter Beerli, Albano Beja-Pereira, Donovan Bell, Steve Beissinger, Kurt Benirschke, Laura Benestan, Louis Bernatchez, Olly Berry, Pierre Berthier, Giorgio Bertorelle, Matt Boyer, Brian Bowen, Jason Bragg, Ron Burton, Chris Cole, Graham Coop, Des Cooper, Rob Cowie, Kirsten Dale, Charlie Daugherty, Sandy Degnan, Pam Diggle, Holly Doremus, Sylvain Dubey, Dawson Dunning, Suzanne Edmands, Norm Ellstrand, Joe Felsenstein, Sarah Fitzpatrick, Brenna Forester, Zac Forsman, Frode Fossøy, Dick Frankham, Ned Friedman, Oscar Gaggiotti, Neil Gemmell, Cameron Ghalambor, John Gillespie, Mary Jo Godt, Dave Goulson, Peter Grant, Noah Greenwald, Ed Guerrant, Brian Hand, Vincent Hanlon, Bengt Hansson, Sue Haig, Kim Hastings, Roxanne Haverkort, Phil Hedrick, Jon Herron, Jon Hess, Kelly Hildner, Rod Hitchmough, Paul Hohenlohe, Denver Holt, Nora Hope, Jeff Hutchings, Brett Ingram, Mike Ivie, Dan Jergens, Mike Johnson, Rebecca Johnson, Rebecca Jordan, Carrie Kappel, Marty Kardos, Kenneth Kidd, Joshua Kohn, Arnie Kotler, Antoinette Kotzé, Siegy Krauss, Bob Lacy, Wes Larson, Pete Lesica, Paul Lewis, Meng-Hua Li, Morten Limborg, Brandon Lind, Curt Lively, Winsor Lowe, Laura Lundquist, Shu-Jin Luo, John McCutcheon, Ian MacLachlan, Siera McLane, Lisa Meffert, Juha Merilä, Don Merton, Mike Miller, Scott Mills, Steve Mussmann, Shawn Narum, Maile Neel, Jeremy Nigon, Rob Ogden, Kathleen O'Malley, Gordon Orians, Sally

Otto, Andy Overall, Jim Patton, Bret Payseur, Rod Peakall, Jill Pecon-Slattery, Eleni Petrou, Robert Pitman, Craig Primmer, Reg Reisenbichler, Bruce Rieman, Pete Ritchie, Bruce Rittenhouse, Bruce Robertson, Rob Robichaux, Marina Rodriguez, Rafael Ribeiro, Beth Roskilly, Norah Saarman, Mike Schwartz, Jim Seeb, Brad Shaffer, Pedro Silva, Stephen Smith, Pia Smets, Doug Soltis, Pam Soltis, Paul Spruell, Dave Tallmon, Mark Tanaka, Barb Taylor, Dave Towns, Kathy Traylor-Holzer, Daryl Trumbo, Hayley Tumas, Susannah Tysor, Dragana Obrecht Vidacovic, Randal Voss, Hartmut Walter, Robin Waples, Tongli Wang, John Wenburg, Andrew Whiteley, Mike Whitlock, Jeannette Whitton, Briana Whitaker, Jack Woods, and Sam Yeaman.

Guest Box Authors

Helen R. Taylor Royal Zoological Society of Scotland, Edinburgh, Scotland, UK (helentaylor23@gmail.com). Primary author of Chapter 24.

Kenneth K. Askelson Biodiversity Research Centre, and Department of Zoology, University of British Columbia, Vancouver, British Columbia, Canada (askelson@zoology.ubc.ca). Chapter 20.

Rachael A. Bay Department of Evolution and Ecology, University of California Davis, Davis, California, USA (rbay@ucdavis.edu). Chapter 16.

Mark A. Beaumont School of Biological Sciences, University of Bristol, Bristol, UK (m.beaumont@bristol.ac.uk). Appendix.

Oliver F. Berry Environomics Future Science Platform, The Commonwealth Scientific and Industrial Research Organisation, Crawley, Western Australia, Australia (oliver.berry@csiro.au). Chapter 13.

Iris Biebach Department of Evolutionary Biology and Environmental Studies, University of Zurich, Zurich, Switzerland (iris.biebach@ieu.uzh.ch). Chapter 18.

Shane C. Campbell-Staton Department of Ecology and Evolutionary Biology and Institute for Society and Genetics, University of California Los Angeles, Los Angeles, California, USA (scampbellstaton@princeton.edu). Chapter 8.

James F. Crow Laboratory of Genetics, University of Wisconsin, Madison, Wisconsin, USA. Deceased (1916–2012). Chapter 5.

Janine E. Deakin Institute for Applied Ecology, University of Canberra, Canberra, Australia (janine.deakin@canberra.edu.au). Chapter 3.

Eleanor E. Dormontt School of Biological Sciences, The University of Adelaide, Adelaide, Australia (eleanor.dormontt@adelaide.edu.au). Chapter 22.

Nicolas Dussex Centre for Palaeogenetics and Department of Bioinformatics and Genetics, Swedish Museum of Natural History, Stockholm, Sweden (nicolas.dussex@gmail.com). Chapter 6.

Holly B. Ernest Wildlife Genomics and Disease Ecology Lab, University of Wyoming, Laramie, Wyoming, USA (holly.ernest@uwyo.edu). Chapter 19.

Peter J.S. Fleming Vertebrate Pest Research Unit, New South Wales Department of Primary Industries, Orange, New South Wales, Australia (peter.fleming@dpi.nsw.gov.au). Chapter 13.

Yasmin Foster Department of Zoology, University of Otago, Dunedin, New Zealand (y.al.foster@gmail.com). Chapter 6.

Armando Geraldes Biodiversity Research Centre and Departments of Zoology and Botany, University of British Columbia, Vancouver, British Columbia, Canada (geraldes@mail.ubc.ca). Chapter 20.

J. Paul Grobler Genetics Department, University of the Free State, Bloemfontein, South Africa (groblerjp@ufs.ac.za). Chapter 23.

Kyle D. Gustafson Department of Biological Sciences, Arkansas State University, Jonesboro, Arkansas, USA (kgustafson@astate.edu). Chapter 19.

Philip W. Hedrick School of Life Sciences, Arizona State University, Tempe, Arizona, USA (philip.hedrick@asu.edu). Chapter 12.

Paul A. Hohenlohe Department of Biological Sciences, Institute for Bioinformatics and

Evolutionary Studies, University of Idaho, Moscow, Idaho, USA (hohenlohe@uidaho.edu). Chapter 4.

Jo Howard-McCombe School of Biological Sciences, University of Bristol, Bristol, UK (j.howard-mccombe@bristol.ac.uk). Appendix.

Darren Irwin Biodiversity Research Centre and Department of Zoology, University of British Columbia, Vancouver, British Columbia, Canada (irwin@zoology.ubc.ca). Chapter 20.

Marty Kardos Northwest Fisheries Science Center, National Marine Fisheries Service, Seattle, Washington, USA.(martin.kardos@noaa.gov). Chapter 17.

Lukas F. Keller Department of Evolutionary Biology and Environmental Studies, University of Zurich, Zurich, Switzerland (lukas.keller@ieu.uzh.ch). Chapter 18.

Antoinette Kotzé Foundational Research and Services, South African National Biodiversity Institute, Pretoria, South Africa (A.Kotze@sanbi.org.za). Chapter 23.

Linda Laikre Division of Population Genetics, Department of Zoology, Stockholm University, Stockholm, Sweden (linda.laikre@popgen.su.se). Chapter 7.

Andrew J. Lowe School of Biological Sciences, The University of Adelaide, Adelaide, Australia (andrew.lowe@adelaide.edu.au). Chapter 22.

Juha Merilä Ecological Genetics Research Unit, Organismal and Evolutionary Biology Research Programme, University of Helsinki, Helsinki, Finland, and Research Division for Ecology and Biodiversity, School of Biological Sciences, The University of Hong Kong, Hong Kong, SAR (merila@hku.hk). Chapter 15.

Paolo Momigliano Ecological Genetics Research Unit, Organismal and Evolutionary Biology Research Programme, University of Helsinki, Helsinki, Finland (Paolo.momigliano@helsinki.fi). Chapter 15.

Sarah P. Otto Biodiversity Research Center, University of British Columbia, Vancouver, British Columbia, Canada (otto@zoology.ubc.ca). Chapter 1.

Sally Potter Australian Museum Research Institute, Australian Museum, Sydney, New South Wales, Australia and Research School of Biology, Australian National University, Acton, Australia (sally.potter@anu.edu.au). Chapter 3.

Uma Ramakrishnan National Centre for Biological Sciences, Tata Institute of Fundamental Research, Bangalore, India (umamakri@ncbs.res.in). Chapter 9.

Bruce C. Robertson Department of Zoology, University of Otago, Dunedin, New Zealand (bruce.robertson@otago.ac.nz). Chapter 6.

Robert H. Robichaux Department of Ecology and Evolutionary Biology, University of Arizona, Tucson, Arizona, USA (robichau@email.arizona.edu). Chapter 21.

Lee A. Rollins Evolution and Ecology Research Centre, Biological, Earth and Environmental Sciences, University of New South Wales. Sydney, New South Wales, Australia (l.rollins@unsw.edu.au). Chapter 14.

Nils Ryman Division of Population Genetics, Department of Zoology, Stockholm University, Stockholm, Sweden (nils.ryman@popgen.su.se). Chapter 7.

Michael K. Schwartz USDA Forest Service, National Genomics Center for Wildlife and Fish Conservation, Missoula Montana, USA (Michael.k.schwartz@usda.gov). Chapter 24.

Richard Shine Department of Biological Sciences, Macquarie University, New South Wales, Australia (rick.shine@mq.edu.au). Chapter 14.

Victoria L. Sork Department of Ecology and Evolutionary Biology and Institute of Environment and Sustainability, University of California Los Angeles, Los Angeles, California, USA (vlsork@ucla.edu). Chapter 11.

Danielle Stephens Zoological Genetics, Inglewood, South Australia, Australia (stephens@zoolgenetics.com). Chapter 13.

Robin S. Waples Northwest Fisheries Science Center, Seattle, Washington, USA (Robin.Waples@noaa.gov). Chapter 10.

Kelly R. Zamudio Department of Ecology and Evolutionary Biology, Cornell University Museum of Vertebrates, Cornell University, Ithaca, New York, USA (krz2@cornell.edu). Chapter 2.

List of Symbols

This list includes mathematical symbols with definitions and references to the primary chapters in which they are used. There is quite a bit of duplication, which reflects the general usage in the population genetics literature. However, the specific meaning should be apparent from the context and chapter.

Symbol	Definition	Chapter
Latin Symbols		
$\hat{x}$	estimate of parameter x	Appendix (A)
A	number of alleles at a locus	3, 4, 5, 6
B	the number of lethal equivalents per gamete	17
CV_A	additive coefficient of variation	11
D	Jost's measure of differentiation	9
D	Nei's genetic distance	9, 20
D	coefficient of gametic disequilibrium	10, 13
D'	standardized measure of gametic disequilibrium	10
D_B	gametic disequilibrium caused by population subdivision	10
D_C	composite measure of gametic disequilibrium	10
E	probability of an event	A
e^2	environmental effect in heritability	11
e_μ	evolvability; the proportional change expected in a trait mean value under a unit strength of selection	11
f	inbreeding coefficient	6
F	realized proportion of genome that is identical by descent	17
F_{ij}	coefficient of coancestry	9
F_{IS}	departure from Hardy–Weinberg proportions within local demes or subpopulations	5, 6, 9, 11, 17, A
F_{IT}	overall departure from Hardy–Weinberg proportions	9
F_k	temporal variance in allele frequencies	A
F_P	pedigree inbreeding coefficient	6, 17, A
F_{SR}	proportion of the total differentiation due to differences among subpopulations within regions	9
F_{ROH}	proportion of the genome that is IBD as estimated by runs of homozygosity	17

Symbol	Definition	Chapter
F_{ST}	proportion of genetic variation due to differences among populations	3, 9, 12, 13, 14, 19, 21, A
$F2_{ST}$	F_{ST} value using the frequency of the most common allele and all other allele frequencies binned together	9
G	generation interval	7, 15, 21
G_{ST}	F_{ST} extended for three or more alleles	9
G'_{ST}	standardized measure of G_{ST}	9
h	gene diversity, computationally equivalent to H_e, especially useful for haploid marker systems	3, 7
h	degree of dominance of an allele	12
h	heterozygosity	6, 7
h^2	narrow sense heritability/proportion of phenotypic variance due to genotypic value	11
H_A	alternative hypothesis	A
H_B	broad sense heritability	11
H_e	expected proportion of heterozygotes	3, 5, 6, 7, 9, 14, 19, 22, A
H_o	observed heterozygosity	3, 6, 9, A
H_0	null hypothesis	A
H_N	narrow sense heritability	11, 15, 18, 21
H_S	mean expected heterozygosity	3, 9, 12, 19
H_T	total genetic variation	9, 12, 14, 19
K	carrying capacity	6, 18
k	number of gametes contributed by an individual to the next generation	7
k	number of populations	20, A
L	number of loci	17
m	proportion of migrants	9, 19, 20, 21
mk	mean kinship	22
mN	number of migrants per generation	9
MP	match probability	22
N	population size	5, 6, 7, 9, 12, A
n	sample size	3, A
n	ploidy level	3
N_b	number of breeders per reproductive cycle	7, 23
N_C	census population size	7, 15, 18, 23, A
N_C	proportion of individuals that reproduce in captivity	21
N_W	proportion of individuals that reproduce in the wild	21
N_e	effective population size	4, 7, 8, 9, 10, 11, 12, 15,18, 21, 23, A
N_{eI}	inbreeding effective population size	7
N_{eV}	variance effective population size	7
N_f	number of females in a population	6, 7, 9
N_m	number of males in a population	6, 7, 9
NS	Wright's neighborhood size	9
P	proportion of loci that are polymorphic	3, 5
P	probability of an event	5, A

Symbol	Definition	Chapter
p	frequency of allele A_1 (or A)	5, 6, 8, 9, 11
p	proportion of patches occupied in a metapopulation	19
PE	probability of paternity exclusion; average probability of excluding (as father) a randomly sampled nonfather	22
PI_{av}	average probability of identity	22, 23
q	frequency of allele A_2 (or a)	5, 6, 8, 9, 11
Q	probability two alleles are identical in state	9
Q	probability of an individual's genotype originating from each population	20
Q_{ST}	proportion of total genetic variation for a phenotypic trait due to genetic differentiation among populations (analogous to F_{ST})	11
r	frequency of allele A_3	5
r	correlation coefficient	A
r	rate of recombination	4, 10
r	intrinsic population growth rate	6, 18
r_A	correlation between two traits	11
R	correlation coefficient between alleles at two loci	10
R	response to selection	11, 15
R	number of recaptured individuals	18
R	rate of adaptation to captivity	21
$R(g)$	allelic richness in a sample of g genes	5, 23
R_{ST}	analog to F_{ST} that accounts for differences length of microsatellite alleles	9, 20
S	self-incompatibility locus	8, 18
S	selection differential	11, 15, 21
S	effects of inbreeding on the probability of survival	17
S	selfing rate	9
s	selection coefficient (intensity of selection)	8, 9
s_x	standard deviation	A
s_x^2	sample variance	A
t	number of generations	6
V_A	proportion of phenotypic variability due to additive genetic differences between individuals	11, 12, 18
V_D	proportion of phenotypic variability due to dominance effects (interactions between alleles)	11
V_E	proportion of phenotypic variability due to environmental differences between individuals	2, 11, 18
V_G	proportion of phenotypic variability due to genetic differences between individuals	2, 11
V_I	proportion of phenotypic variability due to epistatic effects	11
V_k	variance of the number of offspring contributed to the next generation	7
V_m	increase in additive genetic variation per generation due to mutation	12, 18
V_P	total phenotypic variability for a trait	2, 11
V_q	binomial sampling variance	6
W	absolute fitness	8, 11
w	relative fitness	8, 11

Symbol	Definition	Chapter
Greek Symbols		
α	probability of a false positive (Type I) error	A
β	probability of a false negative (Type II error)	A
Δ	change in value from one generation to the next	6, 7, 8
δ	proportional reduction in fitness due to selfing	17
θ	population scaled mutation rate	12
λ	factor by which population size increases each time unit	18, 19
μ	population mean	A
μ	neutral mutation rate	12
π	nucleotide diversity	3, 5
π	probability of an event	A
σ_x^2	population variance	A
Φ_{ST}	analogous to F_{ST} but incorporates genealogical relationships among alleles	9
X^2	chi-square statistic	A
Other Symbols		
x	number of observations or times that an event occurs	A
$\bar{x}$	sample mean	A
$\bar{x}^2$	sample variance (second moment)	A
$\bar{x}^3$	skewness of sample distribution (third moment)	A

List of Abbreviations

Abbreviation	Meaning
ABC	approximate Bayesian computation
AFLP	amplified fragment length polymorphism
AMOVA	analysis of molecular variance
ANOVA	analysis of variance
BAMBI	Baltic Sea Marine Biodiversity
Bd	*Batrachochytrium dendrobatidis*
BIC	Bayesian Information Criterion
BLAST	Basic Local Alignment Search Tool
BLUP	best linear unbiased prediction
BOLD	Barcode of Life Data Systems
bp	base pair
BSC	biological species concept
CBD	Convention on Biological Diversity
CBOL	Consortium for the Barcode of Life
cDNA	complementary DNA
CITES	Convention on International Trade in Endangered Species of Wild Fauna and Flora
CKMR	close-kin mark–recapture
cM	centimorgan
CMS	cytoplasmic male sterility
CMR	capture–mark–recapture
cpDNA	chloroplast DNA
CPSG	Conservation Planning Specialist Group
CRISPR	clustered regularly interspaced short palindromic repeats
CTSG	Conservation Translocation Specialist Group
CU	conservation unit
CWD	chronic wasting disease
DAPC	discriminant analysis of principal components
ddRAD	double digest RAD
df	degrees of freedom
DFTD	devil facial tumor disease
DNA	deoxyribonucleic acid
DPS	distinct population segment
DDT	dichlorodiphenyltrichloroethane
eDNA	environmental DNA
EM	expectation maximization
EST	expressed sequence tag
EPBC Act	Australian Environment Protection and Biodiversity Conservation Act 1999
ESA	United States Endangered Species Act
ESU	evolutionarily significant unit
FAO	Food and Agriculture Organization of the United Nations
FCA	frequency correspondence analysis
FIE	fisheries induced evolution
Gb	gigabase
GBS	genotyping-by-sequencing
GCM	global climate model
GEA	genotype–environment association
GEBV	genomic-estimated breeding value
GEOBON	Group on Earth Observations Biodiversity Observation Networks
GD	gametic disequilibrium
GO	gene ontology
GWAS	genome-wide association study
HDFW	Hawai'i Division of Forestry and Wildlife
HFC	heterozygosity–fitness correlation
HIV	human immunodeficiency virus
HW	Hardy–Weinberg
IAM	infinite allele model
IBD	identical by descent
iBOL	International Barcode of Life Initiative
ICES	International Council for the Exploration of the Sea
IPCC	Intergovernmental Panel on Climate Change
IPBES	Intergovernmental Science-Policy Platform on Biodiversity and Ecosystem Services
ISSR	inter-simple sequence repeat
ITS	internal transcribed spacer
IUCN	International Union for Conservation of Nature
LD	linkage disequilibrium

LDE	language development enzyme	PVA	population viability analysis
LE	lethal equivalent	qPCR	quantitative polymerase chain reaction
LOD	log of odds ratio	QTL	quantitative trait locus
LoF	loss of function	RADs	restriction site-associated DNA markers
MAC	minor allele count	RADseq	restriction site-associated DNA sequencing
MAF	minor allele frequency		
Mb	megabase pairs, equal to million base pairs	RAPD	randomly amplified polymorphic DNA
MCMC	Markov chain Monte Carlo	RCP	representative concentration pathway
MDS	multidimensional scaling	RDA	redundancy analysis
MHC	major histocompatibility complex	rDNA	ribosomal DNA
ML	maximum likelihood	REML	restricted maximum likelihood
MLE	maximum likelihood estimate	RFLP	restriction fragment length polymorphism
MMPA	US Marine Mammal Protection Act	RIL	recombinant inbred line
MOU	memorandum of understanding	RNA	ribonucleic acid
MP	match probability	ROH	runs of homozygosity
MRCA	most recent common ancestor	RONA	risk of nonadaptedness
MSMC	multiple sequentially Markovian coalescent	RRV	raccoon rabies virus
		RZSS	Royal Zoological Society of Scotland
mRNA	messenger RNA	SARA	Species at Risk Act of Canada
mtDNA	mitochondrial DNA	SDM	species distribution model
MU	management unit	siRNA	small interfering RNA
MVP	minimum viable population	SFS	site frequency spectrum
MYA	million years ago	SGS	spatial genetic structure
NCA	nested clade analysis	siRNA	small interfering RNA
NCDE	Northern Continental Divide Ecosystem	SMM	stepwise mutation model
NCPA	nested clade phylogeographic analysis	SNP	single nucleotide polymorphism
NEMBA	South African National Environmental Management: Biodiversity Act	SSR	simple sequence repeat
		STR	short tandem repeat
NGS	next-generation sequencing	*Taq*	DNA polymerase enzyme from *Thermus aquaticus*
NGO	nongovernment organization		
NOAA	US National Oceanic and Atmospheric Administration	TMRCA	time to most recent common ancestor
		UNDRIP	UN Declaration on the Rights of Indigenous Peoples
OTU	operational taxonomic unit		
PAW	Partnership for Action Against Wildlife Crime	UPGMA	unweighted pair group method with arithmetic averages
PCA	principal component analysis	US	United States
PCoA	principal coordinates analysis	USA	United States of America
PCR	polymerase chain reaction	USDA	US Department of Agriculture
PDF	probability density function	USFS	US Forest Service
PE	probability of paternity exclusion	USFWS	US Fish and Wildlife Service
PHR	Pearl and Hermes Reef	VNTR	variable number tandem repeat
PI	probability of identity	VSA	verified subspecies ancestry
PMRN	probabilistic maturation reaction norm	WNS	white-nose syndrome
PSC	phylogenetic species concept	WWF	World Wide Fund for Nature
PSMC	pairwise sequentially Markovian coalescent	YSE	Yellowstone Ecosystem

PART I

Introduction

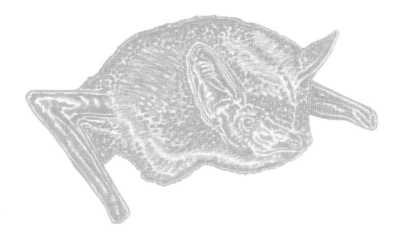

CHAPTER 1

Introduction

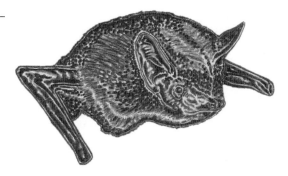

Christmas Island pipistrelle, Example 1.1

The extinction of species, each one a pilgrim of four billion years of evolution, is an irreversible loss. The ending of the lines of so many creatures with whom we have traveled this far is an occasion of profound sorrow and grief. Death can be accepted and to some degree transformed. But the loss of lineages and all their future young is not something to accept. It must be rigorously and intelligently resisted.

(Gary Snyder 1990, p. 176)

The key for conservation genomics will be for the academic and policy spheres to communicate in an effort to maintain a firm grasp on conceptual advances (driven by academic research) and on-site conservation needs (recognized by practitioners).

(Aaron B.A. Shafer et al. 2015, p. 85)

We are living in a time of unprecedented extinctions (Dirzo et al. 2014; Pimm et al. 2014; Humphreys et al. 2019). Current extinction rates have been estimated to be 1,000 times background rates and are increasing (Pimm et al. 2014). Approximately 25% of mammals, 14% of birds, 42% of turtles and tortoises, 40% of amphibians, 34% of conifers, and 35% of selected dicot plant taxa are threatened with extinction (IUCN 2019). Most of these extinction risk projections are based primarily on population declines owing to habitat loss, overharvesting, and pollution. For example, the Christmas Island pipistrelle bat was declared extinct by the International Union for Conservation of Nature (IUCN) in 2017 (Example 1.1). Climate change is anticipated to further increase extinction risks for many species (Urban 2015).

The true picture is likely much worse than this because the conservation status of most of the world's species remains poorly known. In addition, estimates indicate that less than 30% of the world's arthropod species have been described (Hamilton et al. 2010). Only ~6% of the world's described species have been evaluated for the IUCN Red List (Table 1a in IUCN 2019). Few invertebrate species (2%) have been evaluated, and the evaluations that have been done have tended to focus on mollusks and crustaceans. Among the insects, only the swallowtail butterflies, dragonflies, and damselflies have received much attention. A recent analysis has concluded that the number of extinctions of seed plants is more than four times that on the Red List (Humphreys et al. 2019).

Protecting biodiversity poses perhaps the most difficult and important questions ever faced by science (Pimm et al. 2001). The problems are difficult because they are so complex and cannot be approached by the reductionist methods that have

Conservation and the Genomics of Populations, Third Edition. Fred W. Allendorf, *et al.*, Oxford University Press.
© Fred W. Allendorf, W. Chris Funk, Sally N. Aitken, Margaret Byrne, and Gordon Luikart (2022). DOI: 10.1093/oso/9780198856566.003.0001

Example 1.1 Extinction

The nighttime forests of Christmas Island in the Indian Ocean fell silent in 2009 when the last Christmas Island pipistrelle, an echolocating bat, was no longer detected with ultrasound recording devices (Matacic 2017). This species was common until the 1980s, but surveys in the 1990s revealed a drastic decline of unknown cause (Martin et al. 2012). A survey in early 2009 indicated that some 20 bats remained. An attempt to capture these animals for a captive breeding program failed. The last bat evaded capture and was no longer detected with recording devices after 27 August that year. The Christmas Island pipistrelle was officially designated as extinct in 2017 by the IUCN (Matacic 2017).

---

worked so well in other areas of science. Moreover, solutions to these problems require a major readjustment of our social and political systems. An analysis of progress toward international biodiversity targets has concluded that efforts will not result in an improved state for biodiversity in the near future (Tittensor et al. 2014). Biodiversity conservation is arguably the greatest scientific and social challenge currently faced by humanity because biodiversity loss threatens the continued existence of our species and the future of the biosphere itself.

Genetics and genomics have an important role to play in the protection of biodiversity. The earliest applications of genetics to conservation began in the early 1980s at the very beginning of conservation biology (Soulé & Wilcox 1980). "**Genomic**" techniques revolutionized the use of genetics in conservation beginning around 2010 (Allendorf et al. 2010). Applications of genomics to conservation require a fundamental understanding of the theory of population genetics, as well as application of the latest techniques. We have strived to accomplish this goal in this work.

1.1 Genetics and civilization

Genetics has a long history of application to improve human well-being, but also to suppress and discriminate against people (Box 1.1). The domestication of animals and cultivation of plants is thought to have been the key step in the

development of civilization (Diamond 1997). Early peoples directed genetic change in domestic and agricultural species to suit their needs. It has been estimated that the dog was domesticated some 35,000 years ago (Skoglund et al. 2015), followed by goats and sheep around 10,000 years ago (Darlington 1969; Zeder 2008). Wheat and barley were the first crops to be domesticated in the eastern hemisphere ~10,000 years ago; beans, squash, and maize were domesticated in the western hemisphere at about the same time (Kingsbury 2009).

The initial genetic changes brought about by cultivation and domestication were not due to intentional selection but apparently were inadvertent and inherent in cultivation itself. Genetic change under domestication was later accelerated by thousands of years of purposeful selection as animals and crops were selected to be more productive or to be used for new purposes. This process became formalized in the discipline of agricultural genetics after the rediscovery of Mendel's principles at the beginning of the 20th century.

The "success" of these efforts can be seen everywhere. Humans have transformed much of the landscape of our planet into croplands and pasture to support the over 7 billion humans alive today. It has been estimated that 35% of the Earth's ice-free land surface is now occupied by crops and pasture (Foley et al. 2007), and that 24% of the primary terrestrial productivity is used by humans (Haberl et al. 2007). Recently, we have begun to understand the cost at which this success has been achieved. The replacement of wilderness by human-exploited environments is causing the rapidly accelerating loss of species and ecosystems throughout the world. The continued growth of the human population and its direct and indirect effects on environments imperils a large proportion of the wild species that now remain.

Aldo Leopold inspired a generation of ecologists to recognize that the actions of humans are embedded into an ecological network that should not be ignored (Meine 1998):

A thing is right when it tends to preserve the integrity, stability, and beauty of the biotic community. It is wrong when it tends otherwise.

(Leopold 1949, p. 262)

Box 1.1 Eugenics: The dark origins of population genetics and conservation

As conservation geneticists, we recognize the importance of genetic diversity in maintaining healthy natural populations, and in facilitating adaptation to new environmental conditions and challenges. However, both population genetics and the American conservation movement have their roots in the human eugenics movement of a century ago, which viewed genetic diversity among human populations as grounds for discrimination and prejudice. We acknowledge this unfortunate part of the history of both population genetics and conservation, and denounce how it has been used to suppress and disadvantage people.

Many of the early statistical methods that still underlie genetic analysis were developed by devout eugenicists. Francis Galton, a cousin of Charles Darwin, coined the term eugenics in 1883 (Galton 1883, p. 24). Simply put, the field of eugenics viewed human traits as the product of genes, some trait variants more valuable than others, and therefore some human races as better than others (Rohlfs 2020). Galton also developed the concept of linear regression analysis, initially termed "reversion to the mean" or "reversion to mediocrity," which remains widely used in analysis of data of many types. Ronald A. Fisher, who was one of the founders of population genetics, and who developed the

statistical method analysis of variance, was also a staunch eugenicist. Much of *The Genetical Theory of Natural Selection* (Fisher 1930) was devoted to Fisher's concern with the genetic effects of the lower fertility of the English upper class. US President Theodore Roosevelt and his conservation chief Gifford Pinchot, considered fathers of the conservation movement in their country, were both part of the eugenics movement (Wohlforth 2010). The racism that exists in many societies and affects the daily lives of people of color has historical connections to eugenics.

The fields of population genetics and conservations have fortunately progressed a great deal in the past century away from this past. However, they still suffer from the low ethnic and racial diversity typical of ecology and evolutionary biology more broadly (e.g., Graves 2019). The field of conservation genetics will improve further as the diversity of scientists in this field increases, and members of under-represented groups are welcomed warmly and equitably into the community of research and practice. Just as genetic diversity increases the resilience and adaptability of plant and animal populations, the diversity of people in this field will bring new ideas and practices for conserving biodiversity.

The organized actions of humans are controlled by sociopolitical systems that operate into the future on a timescale of a few years at most (e.g., the next election). All too often our systems of conservation are based on the economic interests of humans in the immediate future. We tend to disregard, and often mistreat, elements that lack immediate economic value but that are essential to the stability of the ecosystems upon which our lives and the future of our children depend.

In 1974, Otto Frankel published a landmark paper entitled "Genetic conservation: our evolutionary responsibility," which set out conservation priorities:

First, ... we should get to know much more about the structure and dynamics of natural populations and communities. ... Second, even now the geneticist can play a part in injecting genetic considerations into the planning of reserves of any kind. ... Finally, reinforcing the grounds for nature conservation with an evolutionary perspective

may help to give conservation a permanence which a utilitarian, and even an ecological grounding, fail to provide in men's minds.

(Frankel 1974, p. 63)

Frankel, an agricultural plant geneticist, came to similar conclusions to Leopold, a wildlife biologist, by a very different path. In Frankel's view, we cannot anticipate the future world in which humans will live in a century or two. Therefore, it is our responsibility to "keep evolutionary options open." It is crucial to apply our understanding of genetics and evolution to conserving the natural ecosystems that are threatened by human civilization (Cook & Sgrò 2018).

1.2 Genetics, genomics, and conservation

Darwin (1896, p. 99) was the first to consider the importance of genetics and evolution in the persistence of natural populations. He expressed concern that deer in British nature parks may be subject to loss of vigor because of their small population size and isolation. Voipio (1950) presented the first comprehensive consideration of the application of population genetics to the management of natural populations. He was primarily concerned with the effects of **genetic drift** (Chapter 6) in game populations that were reduced in size by trapping or hunting and fragmented by habitat loss.

The modern concern for genetics in conservation began in the 1970s when Frankel (1970, 1974) began to raise the alarm about the loss of primitive crop varieties and their replacement by genetically uniform cultivars. It is not surprising that these initial considerations of conservation genetics dealt with species that were used directly as resources by humans. Seventy-five percent of crop diversity was lost between 1900 and 2000, and only a few livestock breeds now dominate among domesticated farm animals (FAO 2010). Conserving the genetic resources of wild relatives of agricultural species remains an important area of conservation genetics (Hanotte et al. 2010). For example, the commercial production of sugarcane was saved by the use of germplasm from wild relatives (Soltis & Soltis 2019). However, diversity of crop wild relatives is poorly represented in gene banks (Castañeda-Álvarez et al. 2016). At the same time, geneticists are seeking stress-related genes from wild progenitors of some crops to assist in breeding efforts for new climates (Warschefsky et al. 2014).

The application of genetics and evolution to conservation in a more general context did not blossom until around 1980, when three books established the foundation for applying the principles of genetics and evolution to conservation of biodiversity (Soulé & Wilcox 1980; Frankel & Soulé 1981; Schonewald-Cox et al. 1983). Today conservation genetics is a well-established discipline, with its own journals (e.g., *Conservation Genetics* and *Conservation Genetics Resources*) and two textbooks, including this one and Frankham et al. (2010).

The subject matter of papers published on conservation genetics is extremely broad. However, most articles dealing with conservation and genetics fit into one of the five general categories below:

1. Management and reintroduction of captive populations, and the restoration of biological communities.
2. Description and identification of individuals, genetic population structure, kin relationships, and taxonomic relationships.
3. Detection and prediction of the effects of habitat loss, fragmentation, isolation, and genetic rescue.
4. Detection and prediction of the effects of hybridization and introgression.
5. Understanding the relationships between fitness of individuals or local adaptation of populations and environmental factors.

These topics are listed in order of increasing complexity and decreasing uniformity of agreement among conservation geneticists. Although the appropriateness of captive breeding in conservation has been controversial, procedures for genetic management of captive populations are well developed with relatively little controversy. The relationship between specific genotypes and fitness or adaptation has been a particularly vexing issue in evolutionary and conservation genetics, but new genomic methods have made this more tractable to study. Many recent studies have shown that natural selection can bring about rapid genetic changes in populations that may have important implications for conservation (Homola et al. 2019).

As in other areas of genetics, model organisms have played an important research role in conservation genetics (Frankham 1999). Many important theoretical issues in conservation biology cannot be answered by empirical research on threatened species (e.g., how much gene flow is required to prevent the inbreeding effects of small population size?). Such empirical questions are often best resolved in species that can be raised in captivity in large numbers with a rapid generation interval (e.g., the fruit fly *Drosophila*, the guppy, deer mouse, and the fruit fly analog in plants, *Arabidopsis thaliana*). Such laboratory investigations can also provide excellent training opportunities for students. We have tried to provide examples from both model and threatened species. Where possible we have chosen examples from threatened species or wild populations, even though many of the principles were first demonstrated with model species.

1.2.1 Using genetics to understand basic biology

Molecular genetic descriptions of individuals are also used to understand the basic biology of populations. For example, genetic information can provide valuable insight into the demographic structure of populations (Escudero et al. 2003; Palsbøll et al. 2007). The total population size can be estimated from the number of unique genotypes sampled in a population for species that are difficult to census (Luikart et al. 2010). Moreover, many demographic models assume a single randomly mating population. The distribution of genetic variation over a species range can be used to identify what geographic units can be considered separate demographic units. Consider a population of trout found within a single small lake that might appear to be a demographic unit (Example 9.2). Under some circumstances these trout could actually represent two or more separate reproductive (and demographic) groups with little or no exchange between them (e.g., Ryman et al. 1979).

Genetic analysis can also be used to detect cryptic effects of climate change on the distribution of species. A massive heatwave affected many marine species along the coast of Western Australia in 2011 (Gurgel et al. 2020). The amount of underwater forest cover of two forest-forming seaweeds quickly recovered so that there was no apparent effect of the heatwave. However, genetic analysis of both species before and after the extreme event indicated substantial loss of genetic diversity of both species. Thus, this marine heatwave resulted in a massive and cryptic loss of genetic diversity that may compromise their ability to respond to future environmental change.

Molecular genotyping can also be used to verify the presence of rare species (Chapter 22). For example, wolverines had not been seen in the state of California since 1922 (Moriarty et al. 2009). When photographic evidence suggested the presence of a wolverine in the Sierra Nevada Mountains in 2008, genetic analysis of scat and hair confirmed its presence. In addition, the comparison of the genotype of this individual with samples throughout the west indicated that this individual most likely originated in the Sawtooth Mountains of Idaho, nearly 600 km away.

Genetic analysis has also been used to document some other amazing animal journeys. A cougar originating from the Black Hills of South Dakota left its genetic fingerprints across northern North America from South Dakota to Minnesota to Wisconsin to New York (Hawley et al. 2016). Finally, a cougar killed by a vehicle in Connecticut turned out to be this same animal that had traveled over 2,400 km!

1.2.2 Invasive species and pathogens

Invasive species are recognized as one of the top two threats to global biodiversity (Chapter 14). Studies of genetic diversity and the potential for rapid evolution of invasive species may provide useful insights into what causes species to become invasive (Sakai et al. 2001; Lee & Gelembiuk 2008). More information about the genetics and evolution of invasive species or native species in invaded communities, as well as their interactions, may lead to predictions of the relative susceptibility of ecosystems to invasion, identification of key alien species, and predictions of the subsequent effects of removal (e.g., Roe et al. 2019).

Moreover, genetic descriptions of populations can be used to reconstruct the invasion history and to identify the source populations of biological invasions (Signorile et al. 2016). This information can also be used to detect possible human-mediated translocations and to better manage invasions by identifying transportation pathways that can be targeted for more stringent control.

Similarly, genetic descriptions of population connectivity (e.g., gene flow) can be used to reconstruct pathways of host movement and pathogen spread. Knowing corridors or barriers to movement and spread can help managers monitor, predict, and prevent infectious disease transmission and outbreaks (Blanchong et al. 2008). Study of the genetics of pathogens can also provide valuable information about the genetic population structure and demography of host species (Biek et al. 2006).

1.2.3 Conservation genomics

Recent advances in molecular genetics, including sequencing of the entire genomes of many species, have revolutionized applications of genetics (e.g., medicine, forestry, and agriculture). We currently have complete genome sequences from thousands of species, as well as many individuals within species (Ellegren 2014). The Earth BioGenome Project intends to characterize the genomes of all of Earth's eukaryotic biodiversity over a 10-year period. This coming explosion of information has transformed our understanding of the amount, distribution, and functional significance of genetic variation in natural populations (Allendorf et al. 2010; Shafer et al. 2015).

Now is a crucial time to explore the potential applications of this information revolution for conservation genetics, as well as to recognize limitations in applying genomic tools to conservation issues. The ability to examine hundreds or thousands of genetic markers with relative ease has made it possible to answer many important questions in conservation that have been intractable until now (Figure 1.1).

1.2.3.1 Genetic engineering

Some have proposed that genetic engineering (or genetic modification) should be used to introduce adaptive variants to prevent extinction (e.g., Thomas et al. 2013), but this approach is not likely to be of general utility (Hedrick et al. 2013a). It might be applicable in a very few individual cases, for example, for some long-lived plants, where disease resistance is primarily due to single genes. In general, however, identification of "missing" adaptive single-gene variants in endangered species and increasing their frequency in populations without causing harmful side effects is infeasible in the wild except in a small number of special cases (e.g., where novel pathogens have been introduced; Kardos & Shafer 2018). However, such research is starting to play an important role in agriculture, identifying valuable variants in domesticated species that live in managed environments (Zhu et al. 2020).

Furthermore, when genetically based fitness reductions have been documented in endangered populations, they almost always have been traced not to a lack of adaptive diversity, but to increased frequency of detrimental alleles and increased homozygosity caused by genetic drift and inbreeding (Hedrick et al. 2013a). Those were the factors causing low fitness in Florida panthers and Swedish vipers. In these cases, fitness was increased by genetic rescue, the introduction of unrelated individuals from other populations. The introduction of specific adaptive alleles by genetic engineering in these cases, as proposed by Thomas et al. (2013), will not overcome the genome-wide effects of inbreeding depression.

Nevertheless, there are some situations in which genetic engineering should be considered as a conservation genetics technique (Strauss et al. 2015). Many native trees in the northern temperate zone have been devastated by introduced diseases for which little or no genetic resistance exists (e.g., European and North American elms, and the North American chestnut). Adams et al. (2002) suggested that transfer of resistance genes by genetic modification is perhaps the only available method for preventing the loss of important tree species when no native variation in resistance to introduced diseases exists. Transgenic trees have been developed for both American elm and American chestnut, and are now being tested for stable resistance to Dutch elm disease and chestnut blight (Popkin 2018). The use of genetic engineering to improve agricultural productivity has been controversial but is now widespread for some crops. There will no doubt continue to be a lively debate about the use of these procedures to prevent the extinction of natural populations.

The loss of foundation tree species is likely to affect many other species as well. For example, whitebark pine is currently one of the two most important food resources for grizzly bears in the Yellowstone National Park ecosystem (Mattson & Merrill 2002). However, most of the whitebark pine in this region is projected to be extirpated because of an exotic pathogen, and with predicted geographic shifts in the climatic niche-based habitat of this species in the next century (Warwell et al. 2007; McLane & Aitken 2012).

1.3 What should we conserve?

Conservation can be viewed as an attempt to protect the genetic diversity produced by evolution over the previous 3.5 billion years on our planet. This is an overwhelming task. Over 2 million species have been described and perhaps 100 million species have yet to be described (Soltis & Soltis 2019). Darwin (1859) was the first to represent this diversity in a diagram that he referred to as the "Tree of Life." The first comprehensive Tree of Life for all described species was published in 2015 (Figure 1.2; Hinchcliff et al. 2015).

Genetic diversity is one of three forms of biodiversity recognized by the IUCN as deserving conservation, along with species and ecosystem diversity. Unfortunately, genetics has been generally ignored by the member countries in their National Biodiversity Strategy and Action Plans developed to implement the Convention on Biological Diversity (CBD) (Laikre et al. 2010a; Hoban et al. 2020).

We can consider the implications of the relationship between genetic diversity and conservation at many levels: genes, individuals, populations, varieties, subspecies, species, genera, and so on. Genetic diversity provides a retrospective view of the evolutionary history of taxa (phylogenetics), a snapshot of the current genetic structure within and among populations (population and ecological genetics), and a glimpse ahead to the future evolutionary

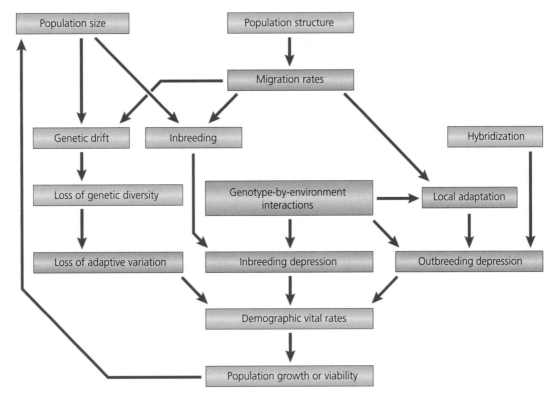

Figure 1.1 Schematic diagram of interacting factors in conservation of natural populations. Traditional conservation genetics, using neutral markers, provides direct estimates of some interacting factors (blue). Conservation genomics can address a wider range of factors (red). It also promises more precise estimates of neutral processes (blue) and understanding of the specific genetic basis of all of these factors. For example, traditional conservation genetics can estimate overall migration rates or inbreeding coefficients, whereas genomic tools can assess gene flow rates that are specific to adaptive loci or founder-specific inbreeding coefficients. From Allendorf et al. (2010).

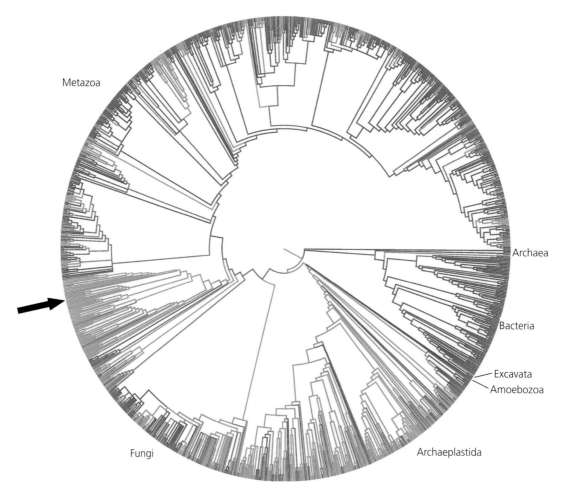

Figure 1.2 Comprehensive Tree of Life showing phylogenetic relationships among over two million described species. The arrow indicates the position of humans in this tree. Redrawn from Soltis & Soltis (2019).

potential of populations and species (evolutionary biology). Genomic tools have provided new insights into all of these areas.

1.3.1 Phylogenetic diversity

While species have historically been prioritized for conservation efforts and resources based on their charismatic appeal and high profile, scientifically justified methods have been developed to identify those species that should be of highest importance from an evolutionary standpoint. The amount of genetic divergence based upon **phylogenetic** relationships is often considered when

setting conservation priorities for different species (Mace et al. 2003; Rosauer & Mooers 2013). For example, the United States Fish and Wildlife Service (USFWS) assigns priority for listing under the Endangered Species Act (ESA) on the basis of "taxonomic distinctiveness" (USFWS 1983). Species of a **monotypic** genus receive the highest priority. The tuatara is an extreme example that raises several important issues about assigning conservation value and allocating our conservation efforts based upon taxonomic distinctiveness (Example 1.2).

Faith (2008) recommended integrating evolutionary processes into conservation decision-making by considering phylogenetic diversity. Faith provides

Example 1.2 The tuatara: a living fossil

The tuatara is a lizard-like reptile that is the remnant of a taxonomic group that flourished over 200 million years ago during the Triassic Period (Figure 1.3). Tuatara are now confined to some 30 small islands off the coast of New Zealand (Daugherty et al. 1990). Three species of tuatara were recognized in the 19th century. One of these species is now extinct. A second species, *Sphenodon guntheri*, was ignored by legislation designed to protect the tuatara, which "lumped" all extant tuatara into a single species, *S. punctatus*.

Daugherty et al. (1990) reported **allozyme** and morphological differences from 24 of the 30 islands on which tuatara are thought to remain. This study supported the status of *S. guntheri* as a distinct species and indicated that fewer than 300 individuals of this species remain on a single island, North Brother Island in Cook Strait. Daugherty et al. (1990) argued that not all tuatara populations are of equal conservation value. As the last remaining population of a distinct species, the tuatara on North Brother Island represent a greater proportion of the genetic diversity remaining in the genus *Sphenodon* and deserved special recognition and protection. However, results with other molecular techniques indicate that the tuatara on North Brother Island do not warrant recognition as a distinct species (Hay et al. 2010; Gemmel et al. 2020).

Figure 1.3 Adult male tuatara on Stephens Island, New Zealand. Photo courtesy of Nicola Nelson.

On a larger taxonomic scale, how should we value the tuatara relative to other species of reptiles? Tuatara species are the last remaining representatives of the Sphenodontida, one of four extant orders of reptiles (tuatara, snakes and lizards, alligators and crocodiles, and tortoises and turtles). In contrast, there are ~5,000 species in the Squamata, the speciose order that contains lizards and snakes.

Continued

Example 1.2 *Continued*

One position is that conservation priorities should regard all species as equally valuable. This position would equate the tuatara with any single species of reptiles. Another position is that we should take phylogenetic diversity into account in assigning conservation priorities. The extreme phylogenetic position is that we should assign equal conservation value to each major sister group in a phylogeny. According to this position, tuatara would be weighed equally with the over 5,000 species of snakes and lizards. Some intermediate between these two positions seems most reasonable.

an approach that goes beyond earlier recommendations that species that are more taxonomically distinct deserve greater conservation priority. He argued that the phylogenetic diversity approach provides two ways to consider maximizing biodiversity. First, considering phylogeny as a product of evolutionary processes enables the interpretation of diversity patterns to maximize biodiversity for future evolutionary change. Second, phylogenetic diversity also provides a way to better infer biodiversity patterns for poorly described taxa when used in conjunction with information about geographic distribution.

Vane-Wright et al. (1991) presented a method for assigning conservation value on the basis of phylogenetic relationships. This system is based on the extent to which branches within a phylogenetic tree are distinct from other lineages. Each extant species is assigned an index of taxonomic distinctiveness that is inversely proportional to the number of branching points to other extant lineages. May (1990) has estimated that the tuatara (Example 1.2) represents between 0.3 and 7% of the taxonomic distinctness, or perhaps we could say genetic information, among reptiles. This is equivalent to saying that the single tuatara species is equivalent to ~10–200 of the "average" reptile species. Crozier & Kusmierski (1994) developed an approach to setting conservation priorities based upon phylogenetic relationships and genetic divergence among taxa. Faith (2002) presented a method for quantifying biodiversity for the purpose of identifying conservation priorities that considers phylogenetic diversity both among and within species.

There is great appeal to placing conservation emphasis on distinct evolutionary lineages with few living relatives. Living fossils, such as the tuatara,

the gymnosperm tree ginkgo (Royer et al. 2003), or the coelacanth fish (Thomson 1991), represent important pieces in the jigsaw puzzle of evolution. Such species are relics that are representatives of taxonomic groups that once flourished. Study of the primitive morphology, physiology, and behavior of living fossils can be extremely important in understanding evolution. For example, tuatara morphology has hardly changed in nearly 150 million years. Among the many primitive features of the tuatara is a rudimentary third, or pineal, eye on the top of the head.

Tuatara represent an important ancestral outgroup for understanding vertebrate evolution. For example, Lowe et al. (2010) used genomic information from tuatara to reconstruct and understand the evolution of 18 human retroposon elements. Most of these elements were inactivated early in the mammalian lineage, and thus study of other mammals provides little insight into these elements in humans. These authors conclude that species with historically low population sizes (such as tuatara) are more likely to maintain ancient mobile elements for long periods of time with little change. Thus, these species are indispensable in understanding the evolutionary origin of functional elements in the human genome.

Winter et al. (2013) pointed out that there is little evidence that consideration of phylogenetic diversity has been applied in conservation. They provide potentially useful approaches to ease uncertainties and bridge gaps between research and conservation with respect to phylogenetic diversity. Pollock et al. (2015) have applied this approach to evaluate how well the nature reserve systems of Victoria, Australia, capture the evolutionary diversity of eucalypts. They conclude that including small but especially crucial areas can have a major effect on the preservation of phylogenetic diversity.

In contrast, others have argued that our conservation strategies and priorities should be more forward- than backward-looking, based primarily upon conserving the evolutionary process rather than preserving only those pieces of the evolutionary puzzle that are of interest to humans (Erwin 1991). Those species that will be valued most highly under the schemes that weigh phylogenetic distinctness are those that may be considered evolutionary failures. Evolution occurs by changes within a single evolutionary lineage (**anagenesis**) and the branching of a single evolutionary lineage into multiple lineages (**cladogenesis**). Groups with large species radiations represent actively speciating lineages. In contrast, conservation of primitive, nonradiating taxa is not likely to be beneficial to the protection of the evolutionary process and the environmental systems that are likely to generate future evolutionary diversity (Erwin 1991).

Figure 1.4 illustrates the phylogenetic relations among six hypothetical species (from Erwin 1991). Species A and B are phylogenetically distinct taxa that are endemic to small geographic areas (e.g., tuatara in New Zealand). Such lineages carry information about past evolutionary events, but they are relatively unlikely to be sources of future evolution. In contrast, the stem resulting in species C, D, E, and F is relatively likely to be a source of future anagenesis and cladogenesis. In addition, species such as C, D, E, and F may be widespread, and therefore are not likely to be the focus of conservation efforts.

The problem is more complex than just identifying species with high conservation value; we must take a broader view and consider the habitats and environments where our conservation efforts could be concentrated. Conservation emphasis on phylogenetically distinct species will lead to protection of environments that are not likely to contribute to future evolution (e.g., small islands along the coast of New Zealand). In contrast, geographic areas that are the center of evolutionary activity for diverse taxonomic groups could be identified and targeted for long-term protection.

Recovery from our current extinction crisis is a central concern of conservation. It is important to maintain the potential for the generation of future

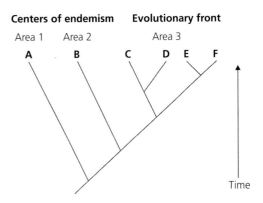

Figure 1.4 Hypothetical phylogeny of six species. Redrawn from Erwin (1991).

biodiversity. We should identify and protect contemporary hotspots of evolutionary radiation and the functional taxonomic group from which tomorrow's biodiversity is likely to originate. In addition, we should protect those phylogenetically distinct species that are of special value for our understanding of biological diversity and the evolutionary process. These species are also potentially valuable for future evolution of biodiversity because of their combination of unusual phenotypic characteristics that may give rise to a future evolutionary radiation. Isaac et al. (2007) have proposed using an index that combines both evolutionary distinctiveness and IUCN Red List categories to set conservation priorities.

1.3.2 Species or ecosystems

A related, and sometimes impassioned, dichotomy is whether to emphasize species conservation or the conservation of habitats or ecosystems (Bowen 1999; Armsworth et al. 2007). Conservation efforts to date have generally emphasized the concerns of individual species. For example, in the USA the ESA has been the legal engine behind many conservation efforts. However, it is frustrating to see enormous resources being spent on a few high-profile species when little is spent on less charismatic taxa or in preventing environmental deterioration that would benefit many species. It is clear that conservation strategies that combine protection of habitat and ecosystems, as well as species, are needed.

Some have advocated a shift from saving things, the products of evolution (species, communities, or ecosystems), to saving the underlying processes of evolution "that underlie a dynamic biodiversity at all levels" (Templeton et al. 2001, p. 5431).

Conservation requires a balanced approach based on habitat protection that also takes into account the natural history and viability of individual species. This is particularly challenging for wide-ranging species. Consider Chinook salmon in the Snake River basin of Idaho, which are listed under the ESA. These fish spend their first 2 years of life in small mountain rivers and streams far from the ocean. They then migrate over 1,500 km downstream through the Snake and Columbia Rivers and enter the Pacific Ocean. There they spend 2 or more years ranging as far north as the coast of Alaska before they return to spawn in their natal freshwater streams. The decline of Chinook salmon populations is implicated in the declining body size, low birth rate, and high mortality rate of the endangered southern resident killer whale populations of the eastern Pacific (Groskreutz et al. 2019). There is no single ecosystem that encompasses these fish, other than the biosphere itself. Protection of this species requires a combination of habitat measures and management actions that take into account the complex life history of these fish.

1.3.3 Populations or species

It has also been argued that more concern should be focused on the extinction of genetically distinct populations, and less on extinction of species (Hughes et al. 1997; Hobbs & Mooney 1998). The conservation of many distinct populations is required to maximize evolutionary potential of a species and to minimize the long-term extinction risks of a species. In addition, a population focus would also help to prevent costly and desperate "last-minute" conservation programs that occur when only one or two small populations of a species remain. The first attempt to estimate the rate of population extinction worldwide was published by Hughes et al. (1997). They estimated that tens of millions of local populations that are genetically distinct go extinct each year. Approximately 16 million of the world's 3 billion genetically distinct natural populations go

extinct each year in tropical forests alone. In a more recent analysis, Ceballos et al. (2017) found that 32% of vertebrate species have decreased in population size and range.

Perhaps more importantly, populations rather than species are the functional unit of ecosystems. The loss of populations throughout most of a species' range could have substantial and widespread environmental effects, even though the species itself is relatively safe from extinction in a part of its former range. Luck et al. (2003) have considered the effects of population diversity on the functioning of ecosystems and so-called **ecosystem services**. They argue that the relationship between biodiversity and human well-being is primarily a function of the diversity of populations within species. They have also proposed an approach for describing population diversity that considers the value of groups of individuals to the services that they provide (Hoban et al. 2020).

One assessment over many taxa concluded the global average rates of population loss to be about 30% per species since 1970 (WWF 2006). Ceballos et al. (2017) found that out of 177 mammals for which they collected detailed data, all have lost 30% or more of their geographic ranges. They also found that more than 40% of these mammal species have experienced a range shrinkage of 80% or more. These estimates, however, all assume that population extirpation and loss of genetic diversity are proportional to loss of range area rather than defining populations using genetic criteria.

The amount of genetic variation within populations may also play an important ecosystem role in the relationships among species in some functional groups and ecosystems. Leigh et al. (2019) have estimated that ~6% of the genetic variation in wild populations has been lost in the past 200 years. Clark (2010) has found that intraspecific genetic variation within forest trees in the southeastern USA allows higher species diversity. Results in **community genetics** suggest that individual alleles within some species can affect community diversity and composition (Crutsinger et al. 2006). For example, alleles at tannin loci in cottonwood trees affect palatability and decay rate of leaves, which in turn influences abundance of soil microbes, fungi, arboreal insects, and birds (Whitham et al.

2008). Genetic variation in the bark characteristics of a **foundation species** (Tasmanian blue gum tree) has been found to affect the abundance and distribution of insects, birds, and marsupials. Loss or restoration of such alleles to populations could thus influence community diversity, ecosystem function, and resilience (Whitham et al. 2008).

1.4 How should we conserve biodiversity?

Extinction is a **demographic** process: the failure of one generation to replace itself with a subsequent generation. Demography is of primary importance in managing populations for conservation (Lacy 1988; Lande 1988). It is possible to project the expected mean and variance of a population's time to extinction if one has an understanding of a population's demography and environment (Lande 1988).

There are two main types of threats causing extinction: **deterministic** and **stochastic** (Caughley 1994). Deterministic threats include habitat destruction, pollution, overexploitation, species translocation, and climate change. Stochastic threats include any random changes in genetic, demographic, or environmental factors. Populations are subject to uncontrollable stochastic demographic factors as they become smaller. Genetic stochasticity includes random genetic changes (drift) and increased inbreeding. Genetic stochasticity leads to loss of genetic variation (including beneficial **alleles**) and can increase the frequency of harmful alleles. One example of demographic stochasticity is random variation in sex ratios; for example, producing only male offspring. Environmental stochasticity is simply random environmental variation, such as the occasional occurrence of several harsh winters in a row. In a sense, the effects of small population size are both deterministic and stochastic. We know that genetic drift in small populations is likely to have harmful effects, and the smaller the population, the greater the probability of such effects. However, the effects of small population size are stochastic because we cannot predict which traits will be affected.

The extinction probability of small populations is influenced by genetic factors. Small populations are subject to genetic stochasticity, which can lead to loss of genetic variation through genetic drift. The "inbreeding effect of small populations" (Box 1.2) is likely to lead to a reduction in the fecundity and viability of individuals in small populations. For example, Frankel & Soulé (1981, p. 68) suggested that a 10% decrease in genetic variation owing to the inbreeding effect of small populations is likely to cause a 10–25% reduction in reproductive performance of a population. This in turn is likely to cause a further reduction in population size, and thereby reduce a population's ability to persist (Gilpin & Soulé 1986). This has come to be known as the **extinction vortex** (Figure 18.2).

In earlier times, some argued that genetic concerns can be ignored when projecting the viability of small populations because they are in much greater danger of extinction caused by purely demographic stochastic effects (Pimm et al. 1988; Caughley 1994; Sarre & Georges 2009). It has been argued that such small populations are not likely to persist long enough to be affected by inbreeding depression, and that efforts to reduce demographic stochasticity will also reduce the loss of genetic variation. The disagreement over whether or not genetics should be considered in demographic predictions of population persistence has been unfortunate and misleading. Extinction is a demographic process that is likely to be influenced by genetic effects under some circumstances. The important issue is to determine under what conditions genetic concerns are likely to influence population persistence (Nunney & Campbell 1993).

Perhaps most importantly, we need to recognize when management recommendations based upon demographic and genetic considerations may be in conflict with each other. For example, small populations face a variety of genetic and demographic effects that threaten their existence. Management plans aim to increase the population size as soon as possible to avoid the problems associated with small populations. However, efforts to maximize growth rate can actually increase the rate of loss of genetic variation if they rely on the exceptional reproductive success of a few individuals.

Box 1.2 What is an "inbred" population?

Population genetics is a complex field. The incorrect, ambiguous, or careless use of words can sometimes result in unnecessary confusion. We have made an effort throughout this book to use words precisely and carefully.

For example, the term "inbred population" is used in the literature to mean very different things (Chapter 17; Templeton & Read 1994). In the conservation literature, "inbred population" is often used to refer to a small population in which mating between related individuals occurs because, after a few generations, all individuals in a small population will be related. Thus, matings between related individuals (inbreeding) will occur in small populations even if they are mating at random (**panmictic**). This has been called the "inbreeding effect of small populations" (Chapter 6).

In formal population genetics, an "inbred" population is one in which there is a tendency for related individuals to mate with one another. In this sense, a small random mating population is not "inbred." However, some extremely large populations of pine trees are inbred because of their spatial structure (Section 9.2.1). Nearby trees tend to be related to one another because of limited seed dispersal, and nearby trees also tend to fertilize each other because of wind pollination. Therefore, a population of pine trees with millions of individuals could still be "inbred."

Population persistence is a multidisciplinary problem that involves many aspects of the biology of the populations involved (Lacy 2000b). A similar statement can be made about most of the issues we are faced with in conservation biology. We can only address this by incorporating both demographics and genetics, as well as other biological factors that are likely to be critical for a particular situation and species (e.g., behavior, physiology, interspecific interactions, as well as habitat loss and environmental change) in conservation plans.

1.5 The future

Genetics is likely to play an even greater role in conservation biology in the future. Genomic knowledge has tremendous potential to play an important role in some aspects of conservation (Chapter 4). For example, description of genome-wide patterns of chromosomal regions that are identical by descent has allowed the estimation of individual inbreeding coefficients and the detection of inbreeding depression as never before possible (Chapter 17). Patterns of population structure, demographics, gene flow, and relatedness can be resolved on a much finer scale from genomic data than from more traditional markers. Identification of adaptive genes with genomics also provides a basis for utilizing genetic diversity to preserve species under

climate change. Nevertheless, most genetic and genomic applications in conservation do not require complete genome sequences, and many questions can still be addressed with relatively few genetic markers. It is more important to use the most appropriate, resource-efficient techniques to address crucial problems in conservation than it is to use the latest fashionable techniques.

There are a variety of efforts around the world to store samples of DNA libraries, frozen cells, gametes, and seeds that could yield DNA (Section 16.7.2; Frozen Ark Project, Millennium Seed Bank Project, Svalbard Global Seed Vault; Ryder et al. 2000). The hope is that species could be resurrected from stored germplasm if extirpated, or at least provide complete genome sequences of species that might become extinct. These sequences could be invaluable for reconstructing evolutionary relationships, understanding how specific genes arose to encode proteins that perform specialized functions, and how the regulation of genes has evolved.

The concept of **de-extinction** has attracted wide attention in the past few years (Box 21.2; Sandler 2014; Shapiro 2015). The possibility of resurrecting individuals from extinct species is truly exciting. This has been proposed to be done by either cloning entire genomes or by inserting gene sequences from extinct species into living relatives using the **CRISPR** (clustered regularly interspaced short palindromic repeats) technique

(Phelps et al. 2020). For example, 14 loci from the extinct wooly mammoth have been used to replace these same 14 loci in the elephant genome (Shapiro 2015).

Although headline making, de-extinction is unlikely to become a practical conservation tool. Even if it does become possible to clone a few individuals of extinct species, there are a host of problems that will make it impossible to recreate a genetically viable population. Moreover, the concept of recreating woolly mammoths by inserting a few of their genes into the elephant genome is as fanciful as creating a Shakespearean masterpiece by inserting a few sonnets penned by Shakespeare into a movie script written by the Coen brothers. Focusing limited conservation resources on cloning, CRISPR, or cryopreservation of gametes could reduce efforts to conserve large viable populations and their habitats (Box 21.2).

This is an exciting time to be interested in the genetics of natural populations. Molecular techniques make it possible to sequence entire genomes of any species of interest, not just those that can be bred and studied in the laboratory (Ellegren 2014). However, interpretation of this explosion of data requires a solid understanding of both population genetics theory and analytical methods (Figure P.2 in the Preface). This book is intended to provide a thorough examination of our understanding of genetic variation in natural populations. Based upon that foundation, we will consider the application of this understanding to the many problems faced by conservation biologists, with the hope that evidence-based actions can make a difference in conservation.

Guest Box 1   Extinction and evolution in a human-altered world
Sarah P. Otto

Every organism alive today descends from ancestors that form a chain of survivors spanning over 4 billion years. Their ancestors survived in the face of calamity, scarcity, disease, and environmental change. In the genomes of living organisms reside many of the secrets of their ancestors' success: the genes that allowed past generations to evade predation, tolerate droughts, and survive famine.

We do not often look at life this way, but every living organism is a treasure trove of genetic innovation—innovation that arose by random mutation but that allowed the ancestors of all living organisms to survive and replicate, while untold others did not.

This treasure is immense. Although we still do not know how many species there are on the planet, and we do not have an estimate of the full evolutionary tree of life, a ballpark estimate for the total length of all branches in the tree relating extant species amounts to over 10^{14} years of evolutionary discovery (Sandell & Otto 2016). By comparison, 10^{14} is about the number of letters ever published in books across human history. This is a tremendous cumulative amount of evolutionary time over which species have, through natural selection, accumulated mutations that allowed their ancestors to survive and reproduce.

Every species that humans drive extinct obliterates a part of the tree of life. Lost with it are the genetic innovations that allowed its ancestors to survive. Antibiotics, temperature regulation, flight, pollution control, neural networks, self-repairing tissues, etc. are not human inventions but evolutionary ones, borrowed from eons of adaptation (Bar-Cohen 2006). While any one species might represent a small twig of this tree, the potential loss of genetic innovation is staggering given the one million species estimated to be at risk of extinction owing to human activities (IPBES 2019).

These losses to the tree of life are not random. In the past 50,000 years, an increasing human presence has led to the extinction of half of the largest bodied terrestrial species. Species with small ranges, as well as plants that are woody, tropical, or dioecious, are more at risk. By contrast, widespread, generalist, dispersive species and human commensals are, on average, less prone to human-caused extinction.

Reducing genetic diversity within and among species also limits the potential for future adaptation. As emphasized in this book, a key ingredient for evolutionary adaptation is variation. Human activities have led to, on average, a 68% decline in population size of wild vertebrate species since 1970 (WWF 2020). As population sizes decline and ranges become more limited, species may lose the genetic variation needed to adapt to a rapidly changing world.

Humans are reshaping the tree of life and altering the course of evolution to a degree and at a speed that is unprecedented for the impact of a single species. In addition to extinction, humans are altering the course of evolution by shifting selective pressures. Selectively favored are those organisms that can tolerate human-altered landscapes and human-caused environmental changes.

As a brief summary, human-altered selection pressures favor species that (Otto 2018):

- can thrive in human-created niches, from house sparrows adapting to our farms to mosquitoes adapting to our subways;
- avoid hunting and harvesting, including elephants with shorter tusks and fish with smaller mouths;
- avoid collisions with buildings, automobiles, and powerlines, which cause the death of >800 million birds in the USA and >60 million birds in Canada annually;
- shift the timing of migration to match a warming climate, with changes occurring at an average rate of 2.3 days per decade.

Why does it matter that we are altering the course of evolution? First, we leave to our children a less natural world, one that has lost diversity through extinction and that has been subject to an overarching new selection pressure: the ability to survive alongside humans. Second, the species that thrive in our human-altered world are often ones that take advantage of our excesses: pests that live on our waste, diseases that can spread in dense human populations, pathogens that thrive on large stands of monoculture crops. Much of human-caused evolutionary change is unintended and undesired. We would be wise to pay more attention. See Otto (2018) for further discussion and references.

CHAPTER 2

Phenotypic Variation in Natural Populations

Western terrestrial garter snake, Section 2.3

Few persons consider how largely and universally all animals are varying. We know however, that in every generation, if we would examine all the individuals of any common species, we should find considerable differences, not only in size and color, but in the form and proportions of all the parts and organs of the body.

(Alfred Russel Wallace 1892, p. 57)

Understanding the genetic and developmental underpinnings of variation in colour promises a fuller understanding of these evolutionary processes, but the path to unravelling these connections can be arduous. The advent of genomic techniques suitable for nonmodel organisms is now beginning to light the way.

(Paul A. Hohenlohe 2014, p. 1529)

Genetics has been defined as the study of differences among individuals (Sturtevant & Beadle 1939). If all of the individuals within a species were identical, we could still study and describe their morphology, physiology, behavior, and so on. However, geneticists would be out of work. Genetics and the study of inheritance are based upon comparing the similarity of parents and their progeny relative to the similarity among unrelated individuals within populations or species.

Variation in species has sometimes been problematic for naturalists and taxonomists. For example,

the king coat color pattern in cheetahs was first described as a cheetah–leopard hybrid (van Aarde & van Dyk 1986). Later, animals with this pattern were recognized as a new species of cheetah. It was then suggested that this coat pattern was a genetic polymorphism within cheetahs. Observations of inheritance of coat patterns with captive cheetahs eventually confirmed that this phenotype results from a recessive allele at a single **autosomal** locus (van Aarde & van Dyk 1986).

Variability among individuals is essential for adaptive evolutionary change. **Natural selection**

Conservation and the Genomics of Populations, Third Edition. Fred W. Allendorf, *et al.*, Oxford University Press.
© Fred W. Allendorf, W. Chris Funk, Sally N. Aitken, Margaret Byrne, and Gordon Luikart (2022). DOI: 10.1093/oso/9780198856566.003.0002

cannot operate unless there are **phenotypic** differences among individuals (Chapter 8). Transformation of variation among individuals within populations into differences between populations, or species, by the process of natural selection is the basis for adaptive evolutionary change described by Charles Darwin over 150 years ago (Darwin 1859). Nevertheless, there is surprisingly little in Darwin's extensive writings about the extent and pattern of differences among individuals in natural populations. Instead, Darwin relied heavily on examples from animal breeding and the success of artificial selection to argue for the potential of evolutionary change by natural selection (Ghiselin 1969).

Alfred Russel Wallace, the co-founder of the principle of natural selection, was perhaps the first biologist to emphasize the extent and importance of variability within natural populations (e.g., Figure 2.1 illustrates his observations of variation in size of red-winged blackbirds). Wallace felt that "Mr. Darwin himself did not fully recognise the enormous amount of variability that actually exists" (Wallace 1923, p. 82). Wallace concluded that for morphological measurements, individuals commonly varied by up to 25% of the mean value; that is, from 5–10% of the individuals within a population differ from the population mean by 10–25% (Wallace 1923, p. 81). This was in opposition to the commonly held view of naturalists in the 19th century that variation among individuals was comparatively rare in nature.

Mendel's classic work was an attempt to understand the similarity of parents and progeny for traits that varied in natural populations. The original motivation for Mendel's work was to test a theory of evolution developed by his botany professor (Unger 1852), who proposed that "variants arise in natural populations which in turn give rise to varieties and subspecies until finally the most distinct of them reach species level" (Mayr 1982, p. 711). The importance of this inspiration can be seen in the following quote from Mendel's original paper (1866): "this appears, however, to be the only right way by which we can finally reach the solution of a question the importance of which cannot

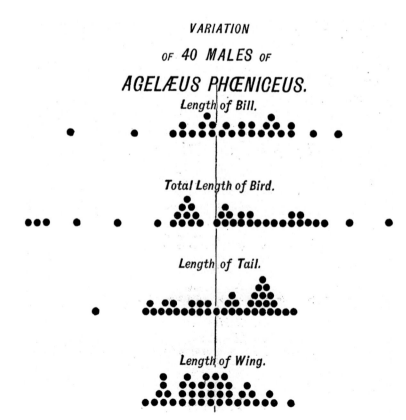

Figure 2.1 Original diagram by Alfred Russel Wallace of variation in body dimensions of 40 red-winged blackbirds in the USA. From Wallace (1923, p. 64).

be overestimated in connection with the history of the evolution of organic forms."

Mendel chose to study the inheritance of seven characters in peas that had clearly distinguishable forms without intermediates: tall versus dwarf plants, violet versus white flowers, green versus yellow pods, and so on. His revolutionary success depended upon the selection of qualitative traits in which the variation could be classified into discrete categories, rather than quantitative traits (Chapter 11) in which individuals vary continuously (e.g., weight, height).

Population genetics was limited to species that could be studied experimentally in the laboratory for most of the 20th century. Experimental population genetics was dominated by studies that dealt with *Drosophila* fruit flies until the mid-1960s because of the difficulty in determining the genetic basis of phenotypic differences among individuals in natural populations (Lewontin 1974). *Drosophila* that differed phenotypically in natural populations could be brought into the laboratory for detailed analysis of the genetic differences underlying phenotypic differences. Similar studies were not possible for species with long generation times that could not be raised in captivity in large numbers. However, population genetics underwent an upheaval in the 1960s when biochemical techniques allowed genetic variation to be studied directly in natural populations of any organism. Lewontin (1974) provided an excellent and highly readable account of the state of population genetics at the beginning of the molecular revolution.

Molecular techniques today make it possible to study variation in the DNA sequence of any species. The complete genomes of thousands of species have been sequenced, and the Earth BioGenome Project proposes to sequence all eukaryotic genomes over a period of 10 years (Lewin et al. 2018). However, even this level of information does not provide understanding of the significance of genetic variation in natural populations (Allendorf et al. 2010). Adaptive evolutionary change within populations consists of changes in morphology, life history, physiology, or behavior. Such traits are usually affected by a combination of many genes and the environment, so that it is difficult to identify single genes that contribute to the genetic differences between individuals for many of the phenotypic traits that are of interest.

In this chapter, we consider the amount and pattern of phenotypic variation in natural populations. We introduce approaches and methodology to understand the genetic basis of phenotypic variation. Finally, we consider the potential importance of phenotypic evolution in conservation.

2.1 Color pattern

Variation in coloration is the most obvious example of qualitative phenotypic variation within species. Three of the seven traits studied by Mendel were discrete differences in coloration. Alfred Russel Wallace (1923, p. 189) argued that color variation in animals must be under natural selection in the wild because color in wild populations is so much less variable than color in domestic species (e.g., dogs, cats, cattle, etc.). Caro (2005) has reviewed the adaptive significance of color polymorphisms in mammals and considered the relative importance of the three classic hypotheses for the function of coloration in mammals: concealment, communication, and regulation of physiological processes. Recent work has shown that color polymorphisms in some species can play an important role in adaptation to changing environmental conditions associated with urbanization and climate change (Kerstes et al. 2019; see Example 2.1).

Discrete color polymorphisms are widespread in plants and animals. Polymorphism in this context is considered to be the occurrence of two or more discrete, genetically based phenotypes in a population in which the frequency of the rarest type is greater than 1% (Hoffman & Blouin 2000). For example, a review of color and pattern polymorphisms in anurans (frogs and toads) cites polymorphisms in 225 species (Hoffman & Blouin 2000). However, surprisingly little work has been done to describe the genetic basis of these polymorphisms or their adaptive significance. Hoffman & Blouin (2000) reported that the mode of inheritance has been described in only 26 species, but conclusively demonstrated in only two! Nevertheless, available results suggest that, in general, color pattern polymorphisms are highly heritable in anurans.

Example 2.1 Evolution of snail shell color in urban heat islands

Extreme environmental conditions in urban centers can bring about selective changes in wild city-dwelling populations inhabiting urban centers (Hendry et al. 2017). For example, Kerstes et al. (2019) used a citizen science approach to study the effects of urban environmental conditions on shell color in the land snail *Cepaea nemoralis* in the Netherlands. This species has been a model organism for ecological genetic studies of adaptation for over 100 years (Clarke 1978).

Shell color in these snails ranges from pale yellow to dark brown with a variable number of up to five black spiral bands (Figure 2.2). The authors employed a simple smartphone app (SnailSnap) that allowed people to take and upload images of snails to a popular Dutch citizen science platform. Analysis of color of almost 10,000 snails photographed throughout the country showed that snails in urban areas that are relatively hot are more likely to be yellow than snails from other areas as predicted based on expectations. In addition, urban yellow snails were also more likely to carry dark bands on the underside of the shell. Previous experiments have demonstrated that yellow snails are better able to survive under high temperatures presumably because of a greater albedo (the proportion of received solar radiation that is reflected rather than absorbed by an object).

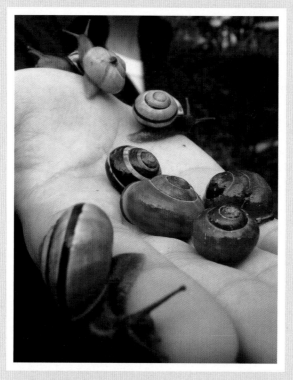

Figure 2.2 *Cepea nemoralis* snails used in a citizen science investigation in the Netherlands of the effects of natural selection in urban environments on shell color polymorphism. Photo courtesy of Menno Schilthuizen.

Example 2.2 Plumage polymorphism in the Arctic skua

A color polymorphism in Arctic skua (or parasitic jaeger) was the subject of a long-term population genetic study by O'Donald (1987). Three color phases (pale, intermediate, and dark) occur throughout the range of the Arctic skua. Some birds have pale plumage with a white neck and body, while other birds have a dark brown head and body. O'Donald & Davis (1959, 1975) classified monogamous breeding adults and their chicks (normally two per brood) on Fair Isle, Scotland, as either pale or melanic (dark or intermediate). The following results were obtained through 1951–1958:

Parental types	Chicks	
	Pale	Melanic
Pale × pale	29	0
Pale × melanic	52	86
Melanic × melanic	25	240
Total	106	326

These results suggested that this color polymorphism was controlled by a single Mendelian locus in which the dark allele is dominant to the pale.

This prediction was confirmed by Mundy et al. (2004), who sequenced melanocortin 1 receptor (*MC1R*) as a candidate locus responsible for this polymorphism. They found that a single amino acid substitution at amino acid 230 ($Arg^{230} \rightarrow His^{230}$) correlates perfectly with this polymorphism (Figure 2.3). All melanic birds were either heterozygous Arg^{230}/His^{230} or homozygous His^{230}/His^{230}; darker birds were more likely to be homozygous. All pale birds were homozygous Arg^{230}/Arg^{230}.

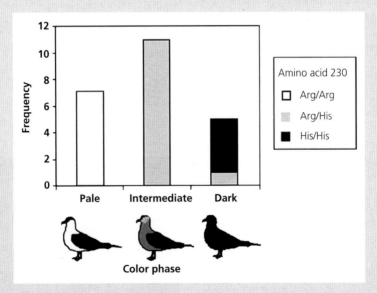

Figure 2.3 Association between genotypes at *MC1R* and color phases of the Arctic skua from Slettnes, Norway. From Mundy et al. (2004).

Sequence analysis of *MC1R* indicates that the ancestral Arctic skuas were pale and the melanic form is the derived trait. Comparison of the amount of sequence divergence between alleles suggests that the His^{230} mutation arose during the Pleistocene ~340,000 years ago.

Continued

Color pattern polymorphisms have been described in many bird species (Example 2.2; Hayward 1987). Galeotti et al. (2003) found that color polymorphisms occur in ~3.5% of all bird species. As a group, owls show the greatest frequency of color polymorphism among birds. For example, red and gray morphs of the eastern screech owl occur throughout its range (VanCamp & Henny 1975). This polymorphism has been recognized since 1874 when it was realized that red and gray birds were conspecific and that the types were independent of age, sex, or season of the year.

The inheritance of this phenotypic polymorphism has been studied by observing progeny produced by different mating types in a population in northern Ohio (VanCamp & Henny 1975). Matings between gray owls produced all gray progeny (Table 2.1). The simplest explanation of this observation is a single locus with two alleles, and the red allele (*R*) is dominant to the gray allele (*r*). Under this model, gray owls are homozygous *rr*, and red owls are either homozygous *RR* or heterozygous *Rr*. Homozygous *RR* red owls are expected to produce all red progeny regardless of the genotype of their mate. One-half of the progeny between heterozygous *Rr* owls and gray birds (*rr*) are expected to be red and one-half are expected to be gray. We cannot predict the expected progeny from matings involving red birds without knowing the frequency of homozygous *RR* and heterozygous *Rr* birds in the population. Progeny frequencies from red parents in Table 2.1 are compatible with most red birds being heterozygous *Rr*; this is expected because the red morph is relatively rare in northern Ohio based upon the number of families with red parents in Table 2.1. We will take another look at these results in Chapter 5 after we have considered estimating genotypic frequencies in natural populations.

Rare color phenotypes often attract wide interest from the public. For example, a photojournalist published a picture of a white-phase black bear near Juneau, Alaska, in the summer of 2002. In response to public concerns, the Alaska Board of Game ordered an emergency closure of hunting on all "white-phase" black bears in the Juneau area during the 2002 hunting season. Similar laws making it illegal to harvest white-phase deer exist in several other states in the USA (e.g., Wisconsin and Illinois).

Table 2.1 Inheritance of color polymorphism in eastern screech owls from northern Ohio (VanCamp & Henny 1975).

		Progeny	
Mating	Number of families	Red	Gray
Red × red	8	23	5
Red × gray	46	68	63
Gray × gray	135	0	439

Figure 2.4 A black bear female with her two cubs in the rainforest of coastal British Columbia. The white phenotype is caused by a recessive allele at a single locus. Photo courtesy of Ian McAllister/Pacific Wild.

The Spirit Bear, or Kermode bear, is a white phase of the black bear caused by a recessive allele at *MC1R* (Hedrick & Ritland 2011). Figure 2.4 shows a female black bear with one black and one white cub. The white morph occurs at low frequencies along the coast of British Columbia and Alaska (Ritland et al. 2001). However, the Kermode bear is at frequencies as high as 40% on some islands off the coast of British Columbia. These bears have been protected from hunting since 1925 (Ritland et al. 2001). Klinka & Reimchen (2009) have reported that white bears are more successful at capturing salmon during the day (see also Reimchen & Klinka 2017). Experiments indicate that salmon were twice as evasive to black as the white morph during the day, but both morphs were equally successful capturing salmon at night. This is an example of **pleiotropy**, which occurs when a single gene affects two or more distinct phenotypic characteristics.

White-phase individuals of many species are maintained in zoos because of public interest in their unusual coloration. For example, wild Bengal white tigers historically were fairly numerous in India (Luo et al. 2019). However, they have not been seen in the wild since 1958. Inheritance studies demonstrated that white coloration is caused by a single, recessive autosomal allele that also causes abnormal vision in several species (Thornton 1978). Genomic analysis supported this result, and has shown that white coloration is caused by a single amino acid substitution in the protein coded by *Slc45a2* (Luo et al. 2019). There are many white tigers held in captivity, but they all derive from a single male white tiger captured in the wild in 1951. White tigers suffer from a variety of health problems, such as premature death, stillbirth, and deformities. However, it is unclear whether these problems are caused by the substitution at *Slc45a2* itself or because of inbreeding resulting from all white tigers being descended from a single male. The maintenance of white tigers in captivity has become somewhat controversial.

A series of papers on flower color polymorphism in the morning glory provides a model system for connecting adaptation with the developmental and molecular basis of phenotypic variation (reviewed in Clegg & Durbin 2000). Flower color variation in this species is determined primarily by allelic variation at four loci that affect flux through the flavonoid biosynthetic pathway. Perhaps the most surprising finding is that almost all of the mutations that determine the color polymorphism are the result of the insertion of mobile DNA elements that are called **transposons**. In addition, the gene that is most clearly subject to natural selection is not a structural gene that encodes a protein, but is rather a regulatory gene that determines the floral distribution of pigmentation (Clegg & Durbin 2003).

Flower color can have a great effect on pollinator visits. Different types of pollinators are attracted to different colors of flowers (see Example 8.3). Bradshaw & Schemske (2003) bred lines of two species of monkeyflowers that had substitutions for a single locus (yellow upper, *YUP*) controlling the presence or absence of yellow carotenoid pigments. They found that a change in color of monkeyflowers caused by a *YUP* allele substitution resulted in a near-wholesale shift in pollinators from bumblebees to hummingbirds.

We have entered an exciting new era where for the first time it has become possible to identify the genes responsible for color polymorphisms, and these genes often play a similar role in multiple species. A series of papers has shown that a single gene, melanocortin-1 receptor (*MC1R*), is responsible for color polymorphism in a variety of birds and mammals, including the Spirit Bear (Majerus & Mundy 2003; Mundy et al. 2004). This same approach has been used to detect color polymorphisms in extinct mammoths and Neandertals (Rompler et al. 2006; Lalueza-Fox et al. 2007). Field studies of natural selection, combined with study of genetic variation, will eventually lead to understanding of the roles of selection in generating similarities and differences between populations and species (Barrett et al. 2019).

2.2 Morphology

Morphological variation is everywhere. Plants and animals within the same population can differ in size, shape, and numbers of body parts. However, there are serious difficulties with using morphological traits to understand patterns of genetic variation. It's more complex than color polymorphisms underlain by a single gene that are relatively unaffected by the environment. The biggest problem is that variation in morphological traits is caused by both genetic and environmental differences among individuals. Therefore, variability in morphological traits in natural environments cannot be used to estimate the amount of genetic variation within populations or the amount of genetic divergence between populations. In fact, we will see in Section 2.7.1 that morphological differences between individuals in different populations in the wild may actually be misleading in terms of genetic differences between populations.

Size traits can be correlated with other important life history traits. For example, in blue-eyed Mary, a small winter annual plant, there is considerable variation among populations in flower size. Populations of small-flowered plants are more likely to be self-pollinating than insect pollinated, with anthers shedding pollen before flowers open (Elle et al. 2010). They also flower earlier, and are found in drier climates than large-flowered populations where the period favorable for growth is shorter.

Most phenotypic differences between individuals within populations have *both* genetic and environmental causes. Geneticists often represent this distinction by partitioning the total phenotypic variability for a trait (V_P) within a population into two components:

$$V_P = V_G + V_E \qquad (2.1)$$

where V_G is the proportion of phenotypic variability due to genetic differences between individuals, and V_E is the proportion due to environmental differences. The **heritability** of a trait is defined as the proportion of the total phenotypic variation that has a genetic basis (V_G/V_P). The greater the heritability of a trait, the more phenotypic differences among individuals within a population are due to genetic differences among individuals. Equation 2.1 is an extreme simplification of complex interactions considered in more detail in Chapter 11.

One of the first attempts to tease apart genetic and environmental influences on morphological variation in a natural population was by Punnett (of Punnett square fame) in 1904. He obtained a number of velvet belly sharks from the coast of Norway to study the development of the limbs in vertebrates. The velvet belly is a small, round-bodied, viviparous shark that is common along the European continental shelf. Punnett counted the total number of vertebrae in 25 adult females and the 2–14 fully developed young they carried (Figure 2.5). He estimated the correlation between vertebrae number in females and their young to test the inheritance of this morphological character. He assumed that the similarity between females and their progeny would be due to inheritance since the females and their young developed in different environments.

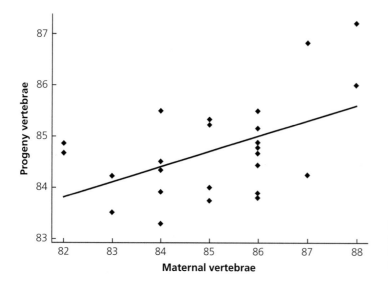

Figure 2.5 Regression of the mean number of total vertebrae in velvet belly sharks before birth on the total number of vertebrae in their mothers ($P < 0.01$). Data from Punnett (1904).

Punnett (1904, p. 346) concluded "the values of these correlations are sufficiently large to prove that the number of units in a primary linear meristic series is not solely due to the individual environment but is a characteristic transmitted from generation to generation." In fact, ~25% of the total variation in progeny vertebrae number can be attributed to the effect of their mothers ($r^2 = 0.254$; $P < 0.01$). We will take another look at these data in Chapter 11 when we consider the genetic basis of morphological variation in more detail.

The use of genomics to understand the genetic basis of morphological variation has shown that some morphological variation results from a few genes with major effects and that some of these genes are responsible for variation in multiple species. For example, a single locus (*IGF1*) is responsible for much of the variation in size in domestic dogs (Sutter et al. 2007). This same gene has been found to be responsible for the presence of either large or small bill size in the black-bellied seed-cracker, an African finch (vonHoldt et al. 2018). The discrete large and small bill phenotypes breed randomly, but differ in diet and feeding performance relative to seed hardness. Inheritance studies demonstrated that this dimorphism appears to be controlled by a single autosomal locus with the large bill allele being dominant (Smith 1993). However, one or few genes of large effects affecting traits like size are the exception rather than the rule with quantitative traits.

Some phenotypic variation cannot be attributed to either genetic or environmental differences among individuals. Bilateral characters of an organism may differ in size, shape, or number. Take, for example, the number of gill rakers in fish species. The left and right branchial arches of the same individual usually have the same number of gill rakers. However, some individuals are asymmetric; that is, they have different numbers of gill rakers on the left and right sides. **Fluctuating asymmetry** of such bilateral traits occurs when most individuals are symmetric and there is no tendency for the left or right side to be greater in asymmetric individuals (Palmer & Strobeck 1986).

What is the source of such fluctuating asymmetry? The cells on the left and right sides are genetically identical, and it seems unreasonable to attribute such variability to environmental differences between the left and right side of the developing embryo. Fluctuating asymmetry is thought to be the result of the inability of individuals to control and integrate development, so that random physiological differences occur during development and result in asymmetry (Palmer & Strobeck 1997; Leamy & Klingenberg 2005). That is, fluctuating asymmetry is a measure of developmental noise— random molecular events (Lewontin 2000). Mather (1953) ascribed the regulation or suppression of these chance physiological differences to the genotypic stabilization of development, and proposed that developmental stability could be measured by fluctuating asymmetry. Thus, increased "noise" or accidents during development (i.e., decreased developmental stability) will result in greater fluctuating bilateral asymmetry. The amount of fluctuating asymmetry in populations may be a useful measure of stress resulting from either genetic or environmental causes in natural populations (Leary & Allendorf 1989; Zakharov 2001; Shadrina & Vol'pert 2016).

2.3 Behavior

Behavior is another aspect of the phenotype and thus will be affected by natural selection and other evolutionary processes, just as any phenotypic characteristic will be. Genetically based differences in behavior are of special interest in conservation because many of these differences are of importance for local adaptation, and because captive breeding programs often result in changes in behavior because of adaptation to captivity (Caro 2007; Moore et al. 2008; Berger-Tal et al. 2016).

Most research in behavioral genetics has used laboratory species such as mice and *Drosophila*. These studies have focused on determining the genetic, neurological, and molecular basis of differences in behavior among individuals. *Drosophila* behavioral geneticists are especially creative in naming genes affecting behavior; for example, they called a gene *couch potato* that is associated with reduced activity in adults (Bellen et al. 1992).

The extent to which genetic factors are involved in differences in bird migratory behavior has been studied systematically over the past 30 years in the

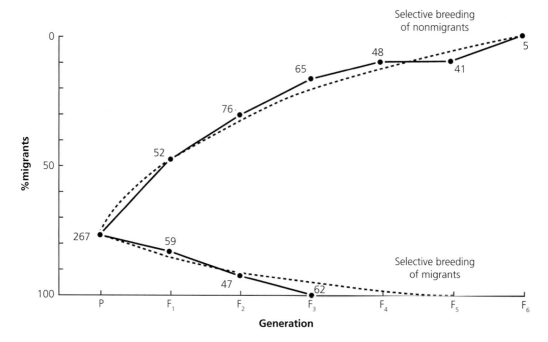

Figure 2.6 Results of two-way selective breeding experiment for migratory behavior with blackcaps from a partially migratory Mediterranean population. From Berthold & Helbig (1992).

blackcap, a common warbler of Western Europe (Berthold 1991; Berthold & Helbig 1992). Selective breeding experiments have shown that the tendency to migrate itself is inherited and is based upon a multilocus system with a threshold for expression (Figure 2.6). Toews et al. (2019) used a genomic approach to conclude that a single gene is associated with migration patterns in golden-winged and blue-winged warblers.

The importance of genetically based differences in behavior for adaptation to local conditions has been shown by an elegant series of experiments with the western terrestrial garter snake (Arnold 1981). This garter snake occurs in a wide variety of habitats throughout the west from Baja to British Columbia, and occurs as far east as South Dakota. Arnold compared the diets of snakes living in the foggy and wet coastal climate of California and the drier, high-elevation inland areas of that state. As hard as it may be to believe, the major prey of coastal snakes is the banana slug; in contrast, banana slugs do not occur at the inland sites.

Arnold captured pregnant females from both locations and raised the young snakes in isolation away from their littermates and mother to remove this possible environmental influence on their behavior. The young snakes were offered a small chunk of freshly thawed banana slug. Native coastal snakes usually ate the slugs; inland snakes did not (Figure 2.7). Hybrid snakes between the coastal and inland sites were intermediate in slug-eating proclivity. These results confirm that the difference between populations in slug-eating behavior has a strong genetic component.

Studies with several salmon and trout species have demonstrated innate differences in migratory behavior that correspond to specializations in movement from spawning and incubation habitat in streams to lakes favorable for feeding and growth (reviewed in Allendorf & Waples 1996; de Leaniz et al. 2007). Fry emerging from lake outlet streams typically migrate upstream upon emergence, and fry from inlet streams typically migrate downstream. Differences in compass orientation behavior of newly emerged sockeye salmon correspond to movements to feeding areas.

There are some very interesting pleiotropic behavioral effects of the *MC1R* gene that we

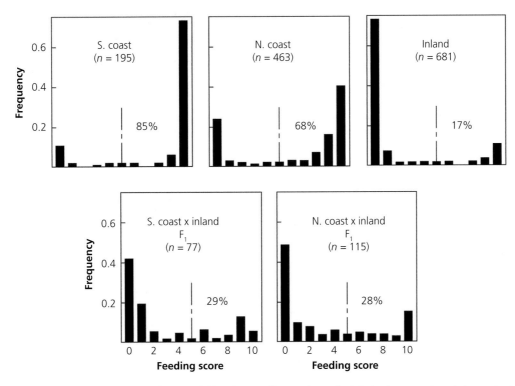

Figure 2.7 Response of newborn garter snakes to availability of pieces of banana slugs as food. Snakes from coastal populations tend to have a high slug feeding score. A score of 10 indicates that a snake ate a piece of slug on each of the 10 days of the experiment. Inland snakes rarely ate a piece of slug. Redrawn from Arnold (1981).

considered in Section 2.1 (Ducrest et al. 2008). Individuals with the darker *MC1R* allele in many species tend to be more aggressive, more sexually active, and more resistant to stress than lighter individuals. This results from effects of *MC1R* within the melanocortin system, which produces a variety of peptide hormones that act as neurocrine and endocrine factors. Pleiotropy can produce some curious associations. For example, people with red hair are less sensitive to the effects of subcutaneous local anesthesia, and they have been found to exhibit greater fear of dental care (Randall et al. 2016).

2.4 Life history

Individual differences in life history (the sequence and timing of events related to survival and reproduction that occur from birth through death) are ubiquitous in plants and animals. Variation in these phenotypes has a major effect on the viability of populations (Section 18.3.1). Variation in life

history traits that have at least a partial genetic basis occurs in virtually all populations of plants and animals. Traditionally, life history variation was thought to be affected by many loci with small effects (Lande 1982). However, somewhat surprisingly, genomic approaches have revealed that a single locus with a major effect is responsible for variation in life history in a variety of species.

A number of papers have found that several different life history characteristics of anadromous salmon and trout are primarily affected by single major genomic regions. For example, Barson et al. (2015) found a single gene (*VGLL3*) that is responsible for 39% of the phenotypic variation in age of sexual maturity in Atlantic salmon. Hess et al. (2016) found a single gene (*GREB1-like*) that explained 46% of the variation in adult migration timing (summer or winter spawning migration) of an anadromous population of rainbow trout (steelhead) in the Columbia River, USA. The summer and winter run types had virtually no genetic divergence at other regions throughout the genome. Prince et al. (2017) reported similar results to those of Hess et al. (2016) in both Chinook salmon and anadromous rainbow trout in the Columbia River. Pearse et al. (2019) discovered a single gene autosomal region that had a major effect of alternative life histories (resident or anadromous) of many populations of rainbow trout.

2.5 Phenology

Many environments fluctuate seasonally between more and less favorable conditions, and for plants and animals living in those environments, timing is critical. The timing of biological events, particularly in relation to climatic or other environmental variation, is called **phenology**. Species have evolved genetic responses to environmental cues to use favorable climatic periods and avoid typically poor conditions for growth and reproduction. Genetically based differences in phenology are of special interest for conservation since these traits are often locally adapted because of the environmental differences among populations. Humans have used phenological events, such as bud break or flowering date, or the arrival of migratory birds, to track seasons and initiate agricultural activities for millennia, and climate change is resulting in major shifts in phenological events in many species (Chapter 16).

Plants need to germinate, grow, flower, and produce seeds while temperatures and moisture levels are favorable, and avoid active growth when conditions are sufficiently cold, hot, or dry to cause damage. **Annual plants** need to synchronize their growth and reproduction with local climatic conditions to complete their lifecycle in a short period of time, and the timing of events is under both genetic and environmental control. The heritability of flowering time in populations of annual field mustard plants in southern California is estimated to be one-third to one-half of the total phenotypic variation in flowering time (Franks et al. 2007). Successive recent droughts resulted in an average advance of the date of flowering by up to a week, showing a genetic response to selection. The capacity for adaptation to new climates will, in part, be determined by the extent of genetic variation for phenological events.

Perennial plants that grow for multiple years in temperate and boreal environments need to cease growth, develop dormancy, shed leaves (if deciduous), and cold-acclimate prior to freezing events. These events are particularly well studied in forest trees. The initiation of growth in spring for many woody species requires a period of low temperatures in winter (chilling requirement), followed by a period of warm temperatures (heat sum requirement) signaling the arrival of spring. The timing of bud break in spring is highly heritable; a given individual will be consistently early or late in bud break from year to year compared with others (Howe et al. 2003), although the average date in any given spring will be dependent upon temperature accumulation that year.

Wang et al. (2018) integrated whole genome resequencing (see Section 4.5) with environmental and phenotypic data from **common garden** experiments (see Section 2.7) to investigate the genomic basis of the timing of bud set in European aspen. They sampled 12 populations along a 10° range of latitude in Sweden. They found very little genetic differentiation among these samples at over 4 million **single nucleotide polymorphisms (SNPs)**. However, a single genomic region containing the *PtFT2* gene was responsible for 65% of the observed genetic variation in timing of bud set. Greenhouse experiments demonstrated that *PtFT2* expression affected the phenotypic variation observed in bud set timing.

Animals also have genetically determined responses to environmental cues indicating seasonality. Key phenological traits include timing of migration, egg laying, fertility, molting, and hibernation. Reproductive synchrony among individuals of the same species is important for reproductive success. Timing of long-distance spring and fall migration is critical to avoid arriving when climatic conditions are unfavorable or when food is unavailable (Coppack & Both 2002). The hibernation period of nonmigratory animals similarly needs to be synchronized with local climate and availability of resources. As in plants, photoperiod plays a key role in many animals in triggering phenological events.

Many organisms have endogenous molecular time-keeping systems that control circadian rhythms, and these are also involved in the sensing of seasons. "Clock genes" have been identified in plants and animals that involve light and cold signaling, and mediate both diurnal and seasonal responses (Resco et al. 2009). O'Malley et al. (2010) found that variation at a clock gene corresponds to latitudinal variation in reproductive timing among species of Pacific salmon along the west coast of North America. Those species (chum and Chinook salmon) that have strong phenotypic latitudinal clines in both spawn timing and age at spawning had the strongest allele frequency clines at *clock*. No allele frequency cline was found in coho salmon, which displays no phenotypic clines in spawn timing or age at spawning. Moderate phenotypic and allelic clines were found in pink salmon.

Genetic variation within animal species exists for many phenological traits. We saw in Section 2.3 that the tendency to migrate and the direction of migration are both under genetic control in the blackcap. In addition, studies have found that the timing of fall migration in the blackcap has a substantial genetic component (Pulido et al. 2001). Selection experiments for later migration in the same captive blackcap population have delayed the onset of migration activity by an average of 1 week over two generations, demonstrating that considerable genetic variation exists for timing of migration.

Differences in reproductive phenology can create reproductive barriers within species. The rockhopper penguin in subantarctic and subtropical waters comprises two geographically and genetically distinct groups, considered subspecies or sibling species. These groups differ by 2 months in breeding phenology, reflecting water temperature differences rather than physical distances between their respective habitats (Jouventin et al. 2006). Similarly, the fragrant orchid in Europe has **sympatric** populations of early- and late-flowering individuals that are considered different subspecies, and their reproductive phenological differences allow little gene flow between them (Soliva & Widmer 1999).

In addition to adapting phenology to the abiotic environment, the timing of lifecycle events needs to correspond to that of conspecific individuals and **mutualist species**. Flowering time needs to be not only synchronous with other individuals of the same species to allow for cross-pollination, but also needs to correspond with the availability of animal pollinators for successful fertilization. The

maturation of fruit should also synchronize with the lifecycles of seed dispersers. Growth phenology (e.g., the timing of bud break and "leafing out" in spring) can also affect the impact of herbivorous insects on host plants. The phenology of forest tree caterpillars is locally synchronized with that of host trees, and is under genetic control in both herbivore and host (van Asch & Visser 2007). The date on which red squirrels in the Yukon, Canada, give birth varies both genetically and with environment, particularly with the cone abundance of white spruce (Berteaux et al. 2004).

Climate change has the potential to disrupt phenological synchrony between plant and animal species and their biotic and abiotic environments (Parmesan 2006). Timing of these events is now being tracked as one measure of the extent of recent climate change (Chapter 16). While phenological traits that are dependent on temperature cues may adjust to new climates without genetic changes, those dependent on photoperiod will need to adapt genetically as photoperiods remain constant but temperatures change. We will explore this more in Chapter 16.

2.6 Disease resistance

Disease is a major cause of decline of populations in many species of conservation concern. Diseases causing species decline may result from endemic pathogens becoming more virulent; for example, fueled by warming due to climate change or exacerbated by other stressors such as pollution. However, diseases threatening species are often caused by the introduction of pathogens into naïve populations that are not adapted to them. The increasing problem of pathogens and parasites in conservation has several causes. Globalization and the associated transport of plants and animals around the world are responsible for the introduction of many diseases. Other pathogens sometimes evolve to switch hosts, infecting new species with little or no resistance. As we write this book, the spread of the coronavirus causing COVID-19 is providing an all-too-real example of this phenomenon in *Homo sapiens*, with this virus emerging from wildlife populations and infecting humans, who have subsequently spread the disease globally.

There are many examples of species in rapid decline due to diseases. For example, amphibians globally are in decline due to chytridiomycosis, a skin disease caused by the fungus *Batrachochytrium dendrobatidis*. This disease has been implicated in the decline or extinction of 43 of Australia's 238 species of amphibians (Scheele et al. 2017; see Guest Box 2). American chestnuts used to be one of the largest and most common trees in Carolinian forests, but they have been functionally **extirpated** from eastern North American forests by the introduced fungus *Cryphonectria parasitica*, which causes chestnut blight, resulting in the death of above-ground stems and branches.

If genetic resistance to or tolerance of a pathogen exists within a population or species, natural selection can increase the frequency of disease-resistant phenotypes over time, and result in **evolutionary rescue** of that population. Similarly, if conservation biologists can identify disease-resistant or tolerant individuals and promote reproduction of those individuals (e.g., in captive breeding programs) they can assist the recovery of those populations, or predict the capacity for recovery resulting from natural selection. In order to do this, phenotypes associated with resistance or tolerance need to be identified, and those phenotypes need to be at least partially under genetic control. Phenotypes for disease response are often simply bivariate: healthy or unhealthy, or alive or dead, underlain by a single gene. However, disease response can also be a quantitative trait, with a range of disease severity impact phenotypes resulting from polygenic resistance.

European ash tree populations are collapsing from ash dieback, a disease caused by the fungus *Hymenoscyphus fraxineus* that is killing over 80% of ash trees, on average, across Europe. Common garden experiments with ash trees have found phenotypic variation and moderate heritability of disease resistance in this species, suggesting breeding for resistance to this disease has a good opportunity for success (Plumb et al. 2020). Natural selection is also likely to be effective if those individuals that have some resistance can survive to reproduction. Whole genome sequencing was used to determine that resistance to this disease is polygenic (Stocks et al. 2019). Selecting resistant individuals from common garden experiments will have a higher success rate than selecting individuals from within forests as the environmental variation in a natural forest is higher than that in a uniform common garden environment, reducing trait heritability (Plumb et al. 2020).

2.7 Variation within and among populations

Populations from different geographical areas are detectably different for many phenotypic attributes in almost all species. Gradual changes in phenotypes across geographic or environmental gradients are found in many species. However, there is no simple way to determine whether such a cline for a particular phenotype results from genetic or environmental differences between populations. The most common way to test for genetic differences between populations is to eliminate environmental differences by raising individuals under identical environmental conditions in a common garden experiment. That is, by making V_E in Equation 2.1 equal to zero, any remaining phenotypic differences must be due to genetic differences between individuals.

The classic common garden experiments were conducted with altitudinal forms of yarrow plants along an altitudinal gradient from the coast of central California to over 3,000 m altitude in the Sierra Nevada Mountains (Clausen et al. 1948). Individual plants were cloned into genetically identical individuals by cutting them into pieces and rooting the cuttings. The clones were then raised at three different altitudes (Figure 2.8). Phenotypic differences among plants from different altitudes persisted when the plants were grown in common locations at each of the altitudes (Figure 2.8). Coastal plants had poor survival at high altitude, but grew much faster than high-altitude plants when grown at sea level. A contemporary example of a similar pattern of phenotypic and genetic divergence along an environmental gradient is presented in Example 2.3.

Transplant and common garden experiments are much more difficult with animals than with plants for several obvious reasons. However, James and her colleagues partitioned clinal variation in size

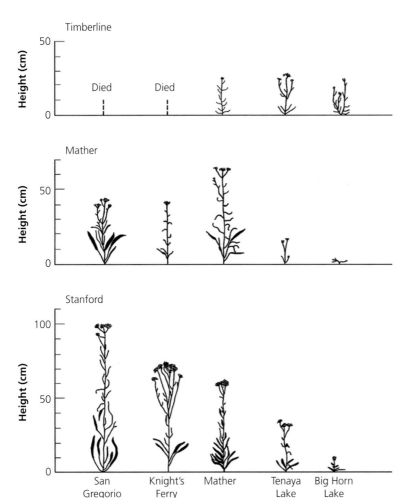

Figure 2.8 Representative clones of yarrow plants originating from five different altitudinal locations grown at three altitudes: 30 m above sea level at Stanford, 1,200 m above sea level at Mather, and 3,000 m above sea level at Timberline. The San Gregorio clone was from a coastal population, and the Big Horn Lake clone was from the highest altitude site (over 3,000 m); the other three clones were from an altitudinal gradient between these two extremes. From Clausen et al. (1948); redrawn from Strickberger (2000).

and shape of the red-winged blackbird into genetic and environmental components by conducting transplant experiments (reviewed in James 1991). Eggs were transplanted between nests in northern and southern Florida, and between nests in Colorado and Minnesota. A surprisingly high proportion of the regional differences in morphology were explained by the locality in which eggs developed (James 1983, 1991).

James (1991) has reviewed experimental studies of geographic variation in bird species. She found a remarkably consistent pattern of intraspecific variation in body size in breeding populations of North American bird species. Individuals from warm humid climates tend to be smaller than birds from increasingly cooler and drier regions. In addition, birds from regions with greater humidity tend to have more darkly colored feathers. The consistent patterns in body size and coloration among many species suggest that these patterns are examples of convergent evolution that have evolved by natural selection in response to similar patterns of differential selection in different environments.

For example, there is some evidence that the color polymorphism in eastern screech owls that we considered earlier affects the survival and reproductive success of individuals. The frequency of red owls increases from north (less than 20% red) to south (~80% red; Pyle 1997). VanCamp & Henny (1975) found evidence that red owls suffered relatively greater mortality than gray owls during severe winter conditions in Ohio, and suggested that this may be due to metabolic differences between red and gray birds (Mosher & Henny 1976). A similar

Example 2.3 Adaptive gradient in Sitka spruce

Sitka spruce has a large geographic range along the Pacific Coast of North America from northern California to Alaska, spanning a wide spectrum of climatic conditions (Figure 2.9). Populations in the southern portion of the range occupy relatively warm, wet habitats with long, favorable growing seasons. There, Sitka spruce trees face interspecific competition from some of the other tallest tree species in the world, including coast redwood and Douglas-fir. In these environments, competition for light results in strong selection for rapid height growth. In the northern portion of the range, temperatures are considerably lower, and the growing season length between late spring and early fall frosts is relatively short. In general, trees cannot withstand temperatures much below freezing without injury during active growth, but can tolerate subfreezing temperatures during the dormant period when tissues are cold-acclimated.

Figure 2.9 (a) Old-growth Sitka spruce trees in the Carmanah Valley, British Columbia. Photo courtesy of T.J. Watt. (b) Phenotypic variation among Sitka spruce populations grown in a common garden experiment in Vancouver, British Columbia, represented by an average tree from each population. The provenance (place of origin) of seed collections and the mean annual temperature of those locations is indicated above each tree. From Aitken & Bemmels (2016).

Continued

Example 2.3 *Continued*

Seedlings grown from seed collected in populations of Sitka spruce across the species range were planted in a common garden experiment in Vancouver, British Columbia (Mimura & Aitken 2007). The results show strong local adaptation, with a trade-off among populations between height growth and adaptation to low temperatures. Trees from the southern portion of the species range did not set bud until late fall, while those from northern Alaska provenances set bud in July or August (Figure 2.10), and achieved much greater cold hardiness. This large difference in growing season length translated into a large difference in total height, with trees from California reaching sizes that were over twice as tall as trees from Alaska (Figures 2.9).

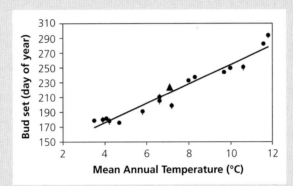

Figure 2.10 Cline of bud set date for Sitka spruce populations along a gradient of mean annual temperature (data from Mimura & Aitken 2007). Seedlings were grown in a common garden in Vancouver, British Columbia. Mean annual temperature for populations of Sitka spruce along the western coast of North America is strongly correlated with latitude as the available frost-free period for growth decreases from south to north. The triangles indicate values for Prince Rupert. This population currently has a mean annual temperature of 7.1°C, but is predicted to warm to 9.8–10.8°C by the 2080s, resulting in maladaptation, assuming the current populations are locally adapted.

Holliday and others have investigated the genomic basis of these population differences in bud set and cold-acclimation phenotypes. Populations from California and Alaska differed in the extent to which genes were expressed during fall cold-acclimation for ~300 of 22,000 genes studied (Holliday et al. 2008). Seedlings in populations from across the species range were genotyped for 768 SNPs in over 200 nuclear genes, and phenotyped for timing of bud set and for cold hardiness in artificial freeze tests (Holliday et al. 2010). A total of 35 SNPs in 28 genes were significantly associated with phenotypes for bud set, cold hardiness, or both. The genes included a homolog to phytochrome A (phyA), which detects photoperiod in plants, and several downstream genes involved in light signal transduction. Fourteen of these SNPs associated with phenotype also had significant genetic clines with climatic variables for population locations.

From this work it is clear that the synchronization of growth and dormancy with local climate in woody plants is genetically complex, involving many genes. It is also now possible to unravel these complex genetics through combining traditional common gardens with genomic methods. For conservation purposes, the greatest contribution of this type of study may be the identification of promising candidate genes that can be focused on for the development of potentially useful adaptive markers for species that cannot be studied in common gardens. For example, there appears to be considerable convergence among conifers in the genomic basis of adaptation to low temperatures (Yeaman et al. 2016).

north–south clinal pattern of red and gray morphs has also been reported in ruffed grouse; Gullion & Marshall (1968) reported that the red morph has lower survival during extreme winter conditions than the gray morph.

2.7.1 Countergradient variation

Countergradient variation is a pattern in which genetic influences counteract environmental influences, so that phenotypic change along an environmental gradient is minimized (Conover & Schultz 1995). For example, Berven et al. (1979) used transplant and common garden experiments in the laboratory to examine the genetic basis of life history traits of green frogs. In the wild, montane tadpoles experience lower temperatures; they grow and develop slowly and are larger at metamorphosis than are lowland tadpoles that develop at higher temperatures. Egg masses collected from high- and low-altitude populations were cultured side-by-side in the laboratory at temperatures that mimic developmental conditions at high and low altitude (18°C and 28°C). The differences observed between low- and high-altitude frogs raised under common conditions in the laboratory for some traits were opposite in direction to the differences observed in nature. That is, at low (montane-like) temperatures, lowland tadpoles grew slower, took longer to complete metamorphosis, and were larger than montane tadpoles.

A reversal of naturally occurring phenotypic differences under common environments may occur when natural selection favors development of a similar phenotype in different environments. Consider the developmental rate in a frog or fish species and assume that there is some optimal developmental rate. Individuals from populations occurring naturally at colder temperatures will be selected for a relatively fast developmental rate to compensate for the reduction in developmental rate caused by lower temperatures. Individuals in the lower temperature environment may still develop more slowly in nature. However, if grown at the same temperature, the individuals from the colder environment will develop more quickly. This will result in countergradient variation (Figure 2.11).

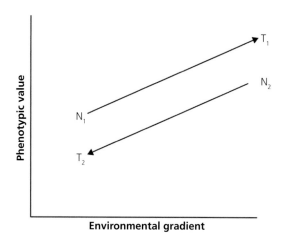

Figure 2.11 Countergradient variation. The end points of the lines represent outcomes of a reciprocal transplant experiment: N_1 and N_2 are the native phenotypes of each population in its home environment. T_1 and T_2 are the phenotypes when transplanted to the other environment. Redrawn from Conover & Schultz (1995).

Therefore, phenotypic differences among populations observed in the wild are not a reliable indicator of genetic differences among populations without additional information. In some cases, all of the phenotypic differences between populations may result from environmental conditions. And, even if genetic differences do exist, they actually may be in the opposite direction of the observed phenotypic differences between populations.

Differences among populations in the amount of total phenotypic variation within populations can also be misleading. Using the relationship represented by Equation 2.1, we would expect a positive association between V_P and V_G. That is, if V_E is constant, then greater total phenotypic variability (V_P) in a population would be indicative of greater genetic variability (V_G). However, assuming that V_E is constant is a very poor assumption, because different populations are subject to different environmental conditions. In addition, the reduction in genetic variation associated with small population size can sometimes decrease developmental stability and thereby increase total phenotypic variability in populations (Leary & Allendorf 1989).

Thus, it is not appropriate to use the amount of total phenotypic variability (V_P) in separate populations to detect differences in the amount of genetic variation between populations. The relationship between V_P and V_G is not straightforward. It differs for different traits within a single population and also depends on the history of the population. The genetic analysis of polygenic phenotypic variation is considered in more detail in Chapter 11.

2.8 Phenotypic variation and conservation

There is accumulating evidence that heritable phenotypic variation plays an important role in the viability of populations (Forsman 2014). Section 18.3 considers how understanding phenotypic variability is important for predicting the viability of populations. Genomic analyses have greatly advanced our understanding of the underlying genetic basis of phenotypic variation and adaptation in the wild (see Chapter 11). These genomic results are now being widely applied to a variety of problems in conservation to conserve biodiversity. Nevertheless, it is often not clear how and when such information can be useful in effective conservation action.

2.8.1 Genetic basis of phenotypic variation

Understanding the underlying genetic basis of phenotypic variation can play an important role in the conservation of populations and species. Appropriate management actions sometimes will differ depending on whether variability underlying an important phenotypic trait is affected by many loci with small effects or a single major locus. If a single major locus affects a trait, then response to selection in that trait depends upon whether or not the locus responsible for the trait is polymorphic in a population. However, if many loci affect a trait, then we would expect a trait to respond to selection, as at least some of those loci will be polymorphic.

For example, we saw in Section 2.4 that major life history differences in anadromous salmon are sometimes affected by a single gene with major phenotypic effects. Two life history types (fall and spring run) of Chinook salmon occurred historically in the Klamath River of California (Langin 2018). Spring run Chinook salmon nearly disappeared because of the construction of dams and water diversions for irrigation. Less than 1% of the remaining fish in some of these populations now carry the allele responsible for the spring migration life history. It has been suggested that if the allele responsible for spring migration is lost, this life history type will likely not evolve again in time frames relevant to conservation (Prince et al. 2017).

2.8.2 Color polymorphism and population viability

It has been proposed that color polymorphisms could play an important role in the ecology and conservation of species and populations (Forsman et al. 2008). Forsman & Åberg (2008) found that species of Australian lizards and snakes with color polymorphisms use a greater diversity of habitats, have larger ranges, and are less likely to be listed as threatened. Betzholtz et al. (2017) found that species of moths in Sweden with variable color patterns had lower extinction risk than those having nonvariable color patterns. Forsman et al. (2008) have predicted that populations with color polymorphisms are less vulnerable to extirpation when facing population declines (see Chapter 18) and are more likely to be successful invasive species (see Chapter 14). In contrast, Bolton et al. (2015) have argued that the presence of color polymorphism might reduce population viability and increase extinction risk (see response by Forsman 2016).

Guest Box 2 The genomic basis of variation in disease resistance
Kelly R. Zamudio

One important phenotype that we know varies among populations is resistance to wildlife diseases. Population-level differences in disease resistance will obviously have important implications for population persistence in the face of emergent infectious diseases, and understanding the genomic basis for that phenotype will be critical for conservation efforts such as genetically informed breeding for reintroductions, genetic rescue of infected populations, and population restoration following declines.

Amphibians, especially frogs, have declined at alarming rates and many species have become extinct because of the emergence of the amphibian-killing fungus *Batrachochytrium dendrobatidis* (*Bd*) (Scheele et al. 2019). Yet, other species are seemingly unaffected by this pathogen. These species-level differences in disease resistance, although interesting, are not ideal for uncovering the genetic basis of this complex phenotype. Species live in different micro-habitats that can alter pathogen exposure and transmission rates, and we also expect that different species will respond to infection using different mechanisms such as behaviors or different immune responses, thus clouding the association between resistance phenotypes and any "resistance genes." The ideal system to query the genome for the basis of variation in disease response is to find a single species where populations differ in their response to infection by the same pathogen.

A few studies have approached the question of the genomic basis for disease resistance by comparing populations within species. Populations of the lowland leopard frog differ dramatically in winter mortality rates across their range (Savage et al. 2011). *Bd* challenge experiments in the lab using individuals from different populations confirm that these differences in susceptibility are in fact genetically based and not the result of different environmental conditions (Savage & Zamudio 2011). Focusing on **major histocompatibility complex** (MHC) genes that are involved in the immune response of frogs to fungal pathogens, we found that within populations, MHC heterozygotes and individuals bearing a specific MHC allele had a significantly reduced risk of death. We also detected a significant signal of positive selection along the evolutionary lineage leading to that protective allele (Savage & Zamudio 2011). Characterizing MHC diversity using field-collected samples across the same populations confirmed that individuals in more resistant populations in fact had higher frequencies of the MHC alleles that were correlated with resistance (Savage & Zamudio 2016).

The studies described above targeted a gene complex with known roles in the immune response to infection. Current studies focused on the genomic basis of disease resistance in wildlife are examining a broader set of candidate genes as well as the differential expressions of those important immune genes in infected frogs (Zamudio et al. 2020). One important potential outcome of studies that examine the link between disease phenotypes and genotypes is the possibility that we find a set of genes that combined explain disease resistance across a broader group of species, allowing us to develop a genetic assay to aid in conservation management and restoration.

CHAPTER 3

Genetic Variation in Natural Populations

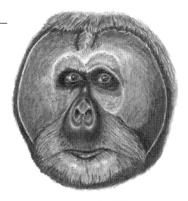

Orangutan, Section 3.1

The empirical study of population genetics has always begun with and centered around the characterization of the genetic variation in populations.

(Richard C. Lewontin 1974, p. 16)

Nevertheless, if populations with unrecognized intraspecific chromosome variation are crossed, progeny fitness losses will range from partial to complete sterility, and reintroductions and population augmentation of rare plants may fail.

(Paul M. Severns & Aaron Liston 2008, p. 1641)

Genetic variation is the raw material of evolution. Change in the genetic composition of populations and species is the basis of evolutionary change. In Chapter 2, we examined phenotypic variation in natural populations. In this chapter, we begin to examine the genetic basis of this phenotypic variation as well as selectively neutral genetic variation by examining genetic differences among individuals and among populations in their chromosomes and DNA sequences at individual loci. In Chapter 4, we will examine the use of genomic techniques to study genetic variation in populations. This sequence, from the chromosomes that are visible under a light microscope down to the study of molecules, reflects the historical sequence of study of natural populations.

In our consideration of conservation, we are concerned with genetic variation at two fundamentally different hierarchical levels:

1. Genetic differences between individuals within local populations.
2. Genetic differences between populations within the same species.

The amount of genetic variation within a population provides insight into the demographic structure and evolutionary history of a population. For example, lack of genetic variation may indicate that a population has gone through a recent dramatic reduction in population size. Genetic divergence among populations is indicative of the amount of **genetic exchange** that has occurred among populations,

Conservation and the Genomics of Populations, Third Edition. Fred W. Allendorf, *et al.*, Oxford University Press.
© Fred W. Allendorf, W. Chris Funk, Sally N. Aitken, Margaret Byrne, and Gordon Luikart (2022). DOI: 10.1093/oso/9780198856566.003.0003

Table 3.1 Historical overview of primary methods used to study genetic variation in natural populations.

Time period	Primary techniques
1900–1970	Laboratory matings and chromosomes
1970s	Protein electrophoresis (allozymes)
1980s	Mitochondrial DNA (mtDNA)
1990s	Nuclear DNA: microsatellites (SSRs)
2000s	Genomics: high throughput sequencing
2010s	High density sampling of mapped or sequenced genomes

and can play an important role in the conservation and management of species. More recently, genetic testing has shown extensive mislabeling of fish species sold in restaurants and grocery stores (Hu et al. 2018).

Population geneticists struggled throughout most of the 20th century to measure genetic variation in natural populations (Table 3.1). Before the advent of biochemical and molecular techniques, genetic variation could only be examined by bringing individuals into the laboratory and using experimental matings. The fruit fly (*Drosophila*) was the workhorse of empirical population genetics during this time because of its short generation time and relative ease of laboratory culture. For example, 41% of the papers (nine of 22) in the first volume of the journal *Evolution* published in 1947 had *Drosophila* in the title; ~5% of the papers (11 of 205) in the volume of *Evolution* published in the year 2019 had *Drosophila* in the title.

Many of the techniques used to detect genetic variation presented in this chapter have been in use for many years. They have been eclipsed by powerful new techniques that we consider in the next chapter that allow direct examination of genetic variation in entire genomes (Figure 4.1). Nevertheless, these "old" tools and the information that they provide remain useful and can complement study of DNA sequences. We are currently experiencing a rejuvenation of chromosomal studies in evolutionary and conservation genetics as new technologies are developed that allow rapid examination of chromosomal differences among individuals (Hoffmann & Rieseberg 2008; Wellenreuther & Bernatchez 2018). The technique of

protein **electrophoresis** has faded away, but a few papers using **allozymes**, allelic variants of enzyme proteins produced by a gene, to detect genetic variation in natural populations continue to be published (Deli et al. 2020). In addition, much of the literature in conservation genetics for many decades relied primarily on the use of allozymes, which provided baseline knowledge of the relative amounts of variation in different species, and the distribution of that variation among and within populations.

The **polymerase chain reaction** (PCR, Box 3.1) has revolutionized our ability to study genetic variation in wild populations beyond our wildest dreams. Who could imagine a few years ago that we would have complete genome sequence information for Neandertals from over 50,000 years ago (e.g., Prüfer et al. 2017)! Moreover, usable DNA samples from **extant** species can be found in a variety of amazing sources: feces, hair left on trees, host blood in ticks, a single pollen grain, and even in the breath of dolphins (Matsuki et al. 2007; Frère et al. 2010).

In this chapter, we will first review genetic variation for chromosomes and discuss its relevance to conservation of wild species. Then we will describe genetic markers used to quantify genetic diversity prior to the advent of the genomics era. Along the way, we will provide an introduction to how data from genetic markers are analyzed and what has been learned about the relative genetic diversity of different taxa.

3.1 Chromosomes

Surprisingly little emphasis has been placed on chromosomal variability in conservation genetics (Benirschke & Kumamoto 1991; Robinson & Elder 1993; Severns & Liston 2008). This is unfortunate because heterozygosity for chromosomal differences often causes reduction in fertility (Nachman & Searle 1995; Rieseberg 2001; Wellenreuther & Bernatchez 2018). For example, the common cross between a female horse with 64 chromosomes and a male donkey with 62 chromosomes produces a sterile mule that has 63 chromosomes. Some captive breeding programs unwittingly have hybridized individuals that are morphologically similar but have distinct chromosomal complements, including orangutans (Ryder & Chemnick 1993), gazelles (Ryder 1987), and dik-diks (Ryder et al. 1989).

Box 3.1 Polymerase chain reaction (PCR)

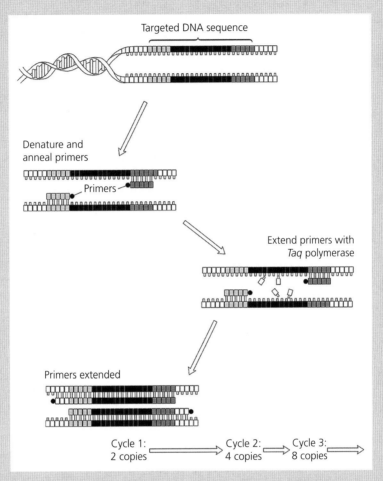

Figure 3.1 The main steps of the polymerase chain reaction (PCR) for amplifying specific DNA sequences: denaturing of the double-stranded template DNA, annealing of primers flanking the target sequence, and extension from each primer by *Taq* polymerase to add nucleotides across the target sequence and generate a double-stranded DNA molecule.

The story of how the polymerase chain reaction (PCR) was discovered illustrates how basic scientific discoveries can lead to huge, unexpected applications—in this case, the revolution in genetic technologies. In 1969, Thomas D. Brock and Hudson Freeze were studying microbes in hotsprings in Yellowstone National Park and identified a thermophilic bacterium they named *Thermus aquaticus*. This bacterium can survive temperatures over 80°C!

The PCR arose from the discovery of *T. aquaticus*. It was developed by Kary Mullis and his colleagues at Cetus Corporation in California in the 1980s. They developed a procedure to use the heat-stable DNA **polymerase** from *Thermus aquaticus* (called "*Taq*") in a process that can generate millions of copies of a specific target DNA sequence in a short time starting with small DNA samples (e.g., one target DNA molecule!). Amplifying many identical copies of

---

Box 3.1 Continued

DNA facilitates sequencing or otherwise analyzing variation.

PCR of a DNA sample involves three steps: (1) denaturing DNA (making double-stranded DNA become single stranded) by heating it to 95°C; (2) cooling the sample to about 60°C to allow hybridization (i.e., **annealing**) with a **primer**—a DNA sequence of ~20 bp designed to match each flanking region of the target sequence; (3) reheating slightly (~72°C) to facilitate extension of the single strand into double-stranded DNA by *Taq* polymerase, which adds nucleotides to the second strand in an order that complements the sequence on the template strand. These three

steps are repeated 30–40 times until millions of copies result (Figure 3.1). The value of *Taq* is that, unlike other polymerases, *Taq* does not denature during the heating phase of these cycles and so does not need to be replaced.

PCR revolutionized modern biology, and continues to have widespread applications in the areas of genomics, population genetics, forensics, medical diagnostics, and gene expression analysis. For example, a temporary shortage of this enzyme resulted from the global demands for testing for COVID-19 in 2020. Mullis was awarded the Nobel Prize in Chemistry in 1993 for his contributions to the development of PCR.

Similarly, translocation or reintroduction programs may cause problems if individuals are translocated among chromosomally distinct groups, as matings of individuals from different groups can cause **outbreeding depression**, a reduction in fitness due to genetic incompatibilities (e.g., Example 3.1). Dobigney et al. (2017) have provided an extremely valuable review of chromosomal polymorphism within mammal species.

Reduced fitness resulting from hybridization between captive individuals from chromosomally distinct populations has been more common than expected because small, isolated populations have a greater rate of chromosomal evolution than common widespread taxa (Lande 1979). Thus, the very demographic characteristics that make a species a likely candidate for captive breeding are the same characteristics that favor the evolution of chromosomal differences between groups. For example, extensive chromosomal variability has been reported in South American primates (Matayoshi et al. 1987), and Banes et al. (2016) found that captive orangutans have been reintroduced into areas where native animals have different chromosomal arrangements (Section 21.8.1).

The direct examination of genetic variation in natural population began with the description of differences in chromosomes between individuals. One of the first reports of differences in the chromosomes of individuals within populations was by

Stevens (1908), who described different numbers of **supernumerary chromosomes**, small extra chromosomes (Section 3.1.4.1), in beetles (White 1973). For many years, study of chromosomal variation in natural populations was dominated by the work of Theodosius Dobzhansky and his colleagues on *Drosophila* (Dobzhansky 1970) because of the presence of giant polytene chromosomes in salivary glands (Painter 1933). However, the study of chromosomes in other species lagged far behind. For example, until 1956 it was thought that humans had 48, rather than 46 chromosomes in each cell. It is amazing that the complete human genome was sequenced within 50 years of the development of the technical ability to even count the number of chromosomes!

3.1.1 Karyotypes

A **karyotype** is the characteristic chromosome complement of a cell, individual, or species. Chromosomes in the karyotype of a species are usually arranged beginning with the largest chromosome (Figure 3.2). The large number of **microchromosomes**, that is, very small chromosomes, in this karyotype of a cardinal is typical for many bird species (Shields 1982). Evidence indicates that bird microchromosomes are essential, unlike the supernumerary chromosomes discussed later in this section (Shields 1982).

The graceful tarplant is a classic example of the importance of chromosomal differentiation between populations for conservation and management. Clausen (1951) described the karyotype of plants from four populations of this species endemic to California. Populations in northern versus southern California can hardly be distinguished morphologically and live in similar habitats. Plants from all of these populations had a **haploid** set of four chromosomes ($n = 4$). However, the size and shape of these four chromosomes differed among populations. Experimental crossings revealed that matings between individuals in different populations either failed to produce F_1 individuals or the F_1 individuals were sterile.

Clausen (1951) concluded that these populations were distinct species because of their chromosomal characteristics and infertility. Nevertheless, he felt that it would be "impractical" to classify them as taxonomic species because of their morphological similarity and lack of ecological distinctness. These chromosomally different populations are now classified as different subspecies. For purposes of conservation, each of these subspecies should be managed separately because of their reproductive isolation. Translocations of individuals among subspecies could cause outbreeding depression through reduction in fertility or the production of sterile hybrids.

Chromosomes of eukaryotic cells consist of DNA and associated proteins. Each chromosome consists of a single highly folded and condensed molecule of DNA. Some large chromosomes would be several centimeters long if they were stretched out—thousands of times longer than a cell nucleus. The DNA in a chromosome is coiled again and again and is tightly packed around histone proteins. Chromosomes are generally thin and difficult to observe, even with a microscope. Before cell division (mitosis or meiosis), however, they condense into thick structures that are readily seen with a light microscope. This is the stage that we usually observe chromosomes (Figure 3.2). The chromosomes right before cell division have already replicated so that each chromosome consists of two identical sister chromatids, joined at the **centromere** (Figure 3.3).

Chromosomes function as the vehicles of inheritance during the processes of mitosis and meiosis. Mitosis involves the separation of the sister chromatids of replicated chromosomes during somatic cell division to produce two genetically identical cells. Meiosis involves the pairing of, crossing over between, and separation of homologous replicated chromosomes during the division of sex cells to produce gametes that vary genetically.

Certain physical characteristics and landmarks are used to describe and differentiate among chromosomes. The first is size. The centromere appears as a constricted region and serves as the attachment point for spindle microtubules, which are

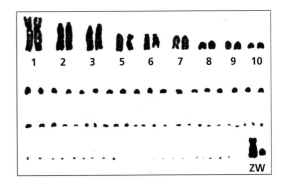

Figure 3.2 Karyotype ($2n = 84$) of a female cardinal. From Bass (1979).

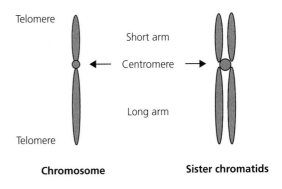

Figure 3.3 Diagram of an unreplicated chromosome (left) and a chromosome that has replicated into two identical sister chromatids that are joined at the centromere (right).

the filaments responsible for chromosomal movement during cell division (Figure 3.3). The centromere divides a chromosome into two arms. Chromosomes in which the centromere occurs approximately in the middle are called **metacentric**. In **acrocentric** chromosomes, the centromere occurs near one end of the chromosome. **Chromosomal satellites** are small chromosomal segments separated from the main body of the chromosome by a secondary constriction. Staining techniques have been developed that differentially stain different regions of a chromosome to help distinguish chromosomes that have similar size and centromere location (Figure 3.4). A process called chromosome painting allows determination of regions of shared common ancestry (**homology**) between populations and species (Ferguson-Smith & Trifonov 2007).

3.1.2 Sex chromosomes

Many groups of animals and some plant taxa have evolved sex-specific chromosomes that are involved in the process of sex determination (see Rice 1996 for an excellent review). In mammals, females are **homogametic** XX and males are **heterogametic** XY (e.g., Figure 3.4). The chromosomes that do not differ between the sexes are called **autosomes**. For example, there are 28 pairs of chromosomes in the karyotype of African elephants ($2n = 56$; Houck et al. 2001); thus, each African elephant has 54 autosomes and two **sex chromosomes**. The heterogametic sex is reversed in birds, butterflies, and moths: males are homogametic ZZ and females are heterogametic ZW (Figure 3.2). Note that the XY and ZW notations are strictly arbitrary and are simply used to indicate which sex is homogametic.

The heterogametic sex varies among species within some taxonomic groups (Charlesworth 1991). Some fish species are XX/XY, some are ZZ/ZW, some do not have detectable sex chromosomes, and a few species even have more than two sex chromosomes (Devlin & Nagahama 2002).

Over 95% of plant species are **hermaphrodites** and therefore do not have sex-determining chromosomes or sex-determining loci within chromosomes (Charlesworth 2002; Heslop-Harrison & Schwarzacher 2011). However, both XX/XY and

ZZ/ZW sex determination systems occur in **dioecious** plant species that have separate male and female individuals. Sex chromosomes appear to have evolved rather recently in plant species. There are no examples of ancient sex chromosomes that are shared among large taxonomic groups of plants, such as the XY system of mammals or the ZW system of birds. While some plant species, like white campion, have two morphologically distinguishable **heteromorphic** sex chromosomes, others like papaya have indistinguishable nonheteromorphic sex chromosomes or chromosomal regions, and some are at earlier stages in the evolution of dioecy with only sex-determining loci (Heslop-Harrison & Schwarzacher 2011). In multiple species of poplar trees, a single gene functions as an on–off switch to determine female versus male development (Müller et al. 2020).

Heteromorphic sex chromosomes can provide useful markers for conservation. The sex of individuals can be determined by karyotypic examination. However, many other easier procedures can be used to sex individuals by their sex chromosome complement. For example, one of the two X chromosomes in females of most mammal species is inactivated and forms a darkly coiling structure (a **Barr body**) that can be readily detected with a light microscope in epithelial cells scraped from the inside of the mouth of females but not males (White 1973). We will see later in this chapter that genetic markers or DNA sequences specific to one of the sex chromosomes can be used in many taxa to identify the sex of individuals (Section 3.3.3).

3.1.3 Polyploidy

Most animal species contain two haploid sets of chromosomes and therefore are **diploid** ($2n$) for most of their lifecycles. The eggs and sperm of animals are haploid, containing only one set of chromosomes ($1n$). However, some species are polyploid because they possess more than two sets of chromosomes: triploids ($3n$), tetraploids ($4n$), pentaploids ($5n$), hexaploids ($6n$), and even greater numbers of chromosome sets. **Polyploidy** is relatively rare in animals, but it does occur in invertebrates, fishes, amphibians, and lizards (White 1973).

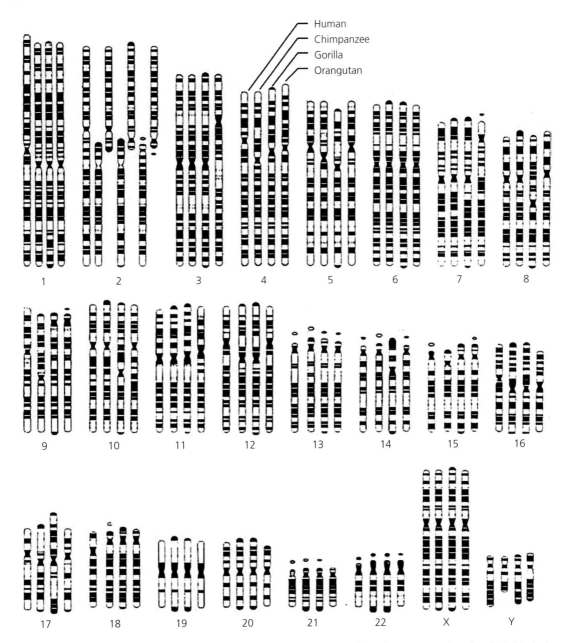

Figure 3.4 Karyotypes and chromosomal banding patterns of humans, chimpanzees, gorillas, and orangutans. Redrawn from Yunis & Prakash (1982).

Plants and some fungi have lifecycles that include both multicellular haploid and diploid phases. Polyploidy is common in plants and is a major mechanism of speciation (Soltis & Soltis 1999; Otto & Whitton 2000). Fifteen percent of angiosperm species and 31% of fern species are thought to have arisen through polyploidy (Wood et al. 2009). In plants, polyploidy can lead to the evolution of asexual reproduction through **apomixis**, the production of seed without meiosis, resulting in embryos that are clones of maternal plants (Whitton et al. 2008). Flow cytometry is a valuable method that can be used to quickly evaluate ploidy level in individual plants (Doležel & Bartoš 2005).

Perhaps the most interesting cases of polyploidy occur when diploid and tetraploid forms of the same taxon exist in sympatry. For example, both diploid (*Hyla chrysoscelis*) and tetraploid (*Hyla versicolor*) forms of gray tree frogs occur throughout the central USA (Ptacek et al. 1994). Reproductive isolation between diploids and tetraploids is maintained by call recognition: the larger cells of the tetraploid males result in a lower calling frequency that is recognized by the females (Bogart 1980). Hybridization between diploids and tetraploids does occur and results in triploid progeny that are not fertile (Gerhardt et al. 1994).

Fireweed provides another example of reproductive isolation between diploids and polyploids. The diploids and tetraploids largely avoid mating through ecological specialization that results in spatial isolation, differences in flowering phenology, and different pollinators (Husband & Sabara 2004). Survival is as high in triploid offspring as it is for diploids and tetraploids; however, pollen production and viability were significantly less in triploids. The overall **relative fitness** of diploids, triploids, and tetraploids is 1, 0.09, and 0.61, respectively.

A thorough treatment of polyploidy is beyond the scope of this chapter. Nevertheless, examination of ploidy levels is an important taxonomic tool when describing units of conservation in some plant taxa.

3.1.4 Numbers of chromosomes

Many closely related species have different numbers of chromosomes, as the horse and mule discussed in Section 3.1. For example, a haploid set of human chromosomes has $n = 23$ chromosomes, while the extant species that are most closely related to humans all have $n = 24$ chromosomes (Figure 3.4). This difference is due to a fusion of two chromosomes to form a single chromosome (human chromosome 2) that occurred sometime in the human evolutionary lineage following its separation from the ancestor of the other species. This is an example

of a **Robertsonian fusion**, which we will consider in Section 3.1.7 and Guest Box 3.

Chromosome numbers have been found to evolve very slowly in some taxa. For example, the ~100 or so species of cetaceans have either $2n = 42$ or 44 chromosomes (Benirschke & Kumamoto 1991; O'Brien et al. 2006). Most species of conifer trees in the pine family (Pinaceae) have $2n = 24$, except for Douglas-fir, which has $2n = 26$ (Krutovsky et al. 2004). In contrast, chromosome numbers have diverged rather rapidly in other taxa. For example, the $2n$ number in horses (genus *Equus*) varies from $2n = 32$ to 66. The Indian muntjac has a karyotype that is extremely divergent from other species in the same genus (Figure 3.5). Some species have surprisingly high numbers of chromosomes. The African baobob, also called the tree of life, has 168 chromosomes (Islam-Faridi et al. 2020)! In the next few sections, we will consider the types of chromosome rearrangements that bring about karyotypic changes among species (Ferguson-Smith & Trifonov 2007).

3.1.4.1 Supernumerary chromosomes

Supernumerary chromosomes (also called **B chromosomes**) are small, lack functional genes (i.e., are **heterochromatic**), and do not pair and segregate during meiosis (White 1973; Jones 1991). Unlike microchromosomes, they are not needed for normal development. The number of supernumerary chromosomes can vary among individuals within a species. In general, the presence or absence of B chromosomes does not affect the phenotype or the fitness of individuals (Battaglia 1964). It is thought that B chromosomes are parasitic genetic elements that do not play a role in adaptation (Jones 1991). B chromosomes have been reported in many species of higher plants (Müntzing 1966). In animals, B chromosomes have been described in many invertebrates, but are rarer in vertebrates. However, Green (1991) has described extensive polymorphism for B chromosomes in populations of the Pacific giant salamander along the west coast of North America.

3.1.5 Chromosomal size

In many species, differences in size between homologous chromosomes have been detected. In most

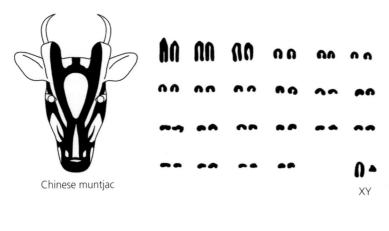

Chinese muntjac

XY

Indian muntjac

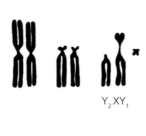

Y₂XY₁

Figure 3.5 The Chinese ($2n = 46$) and Indian muntjac ($2n = 6$ in females and $2n = 7$ in males) and their karyotypes. The Indian muntjac has the lowest known chromosome number of any mammal. First-generation hybrids created in captivity between these species are viable (Wang & Lan 2000). The two Y chromosomes in this species have resulted from a centric fusion between an autosome and the sex chromosomes (White 1973). Redrawn from Strickberger (2000).

of the cases it appears that the "extra" region is due to a heterochromatic segment that does not contain functional genes (White 1973, p. 306). These extra heterochromatic regions resulting in size differences between homologous chromosomes are analogous to supernumerary chromosomes, except that they are inherited in a Mendelian manner. Heterochromatic differences in chromosomal size seem to be extremely common in several species of South American primates (Matayoshi et al. 1987).

3.1.6 Inversions

Inversions are segments of chromosomes that have been turned around so that the DNA sequence has been reversed. They are produced by two chromosomal breaks and a rejoining with the internal piece inverted. There has been renewed interest in inversions because of the ability to detect them with **comparative genomics**, approaches that compare the sequence and spatial arrangement

of DNA and genes in individuals from different species or populations (Hoffmann & Reisberg 2008; Kirkpatrick 2010). For example, comparison of karyotypes identified only nine inversions that distinguish humans and chimpanzees; comparison of their complete genome sequences has revealed over 1,000 inversions (Feuk et al. 2005). Inversions can result in complementary alleles at loci within a chromosomal region being inherited as a coadapted block, as recombination within a heterozygote is suppressed within the inversion as it will be unlikely to produce gametes or spores with a full complement of genes. Lowry & Willis (2010) have detected a widespread chromosomal inversion polymorphism in the yellow monkeyflower that contributes to major life history differences, local adaptation, and reproductive isolation (Example 3.2).

Heterozygosity for inversions is often associated with reduced fertility. **Recombination** (crossing over) within inversions produces aneuploid

Example 3.2 A widespread chromosomal inversion in the yellow monkeyflower

Lowry & Willis (2010) described a widespread chromosomal inversion in the yellow monkeyflower, which has become an important model species for ecological genomic studies. One arrangement of the inverted region is found in an ecotype of this species that lives in habitats characterized by reduced soil water availability in the summer and has an annual life history. The other arrangement lives in habitats with high year-round soil moisture and has a perennial life history. The inversion influences morphological and flowering time differences between these ecotypes throughout its range in western North America.

Yellow monkeyflower plants were bred to reciprocally swap the alternative chromosomal arrangements between the annual and perennial genetic backgrounds, and the offspring were planted in field tests. Late-flowering coastal perennial plants failed to flower before the onset of the hot seasonal summer drought in the inland habitat. Inland annual plants were at a disadvantage in coastal habitat because they invested more resources in reproduction instead of growth and thus failed to take advantage of year-round soil moisture and cool foggy conditions.

Thus, this inversion polymorphism contributes to local adaptation, the annual versus perennial life history polymorphism, and reproductive isolating barriers. These results indicate that adaptation to local environments can drive the spread of chromosomal inversions and promote reproductive isolation. It will be important to discover how common such polymorphisms are in other plant species.

butterflies (Turner 1985), and flower structure loci in *Primula* (Kurian & Richards 1997). Genomic data are revealing more chromosomal regions involved in local adaptation with reduced recombination rates, many of which are due to inversions. For example, Todesco et al. (2020) found numerous such supergenes in wild sunflowers that were associated with locally adaptive phenotypes including flowering time and adaptation to sand dune environments.

An inversion is **paracentric** if both breaks are situated on the same side of the centromere, and **pericentric** if the two breaks are on opposite sides of the centromere. Paracentric inversions are difficult to detect from karyotypes because they do not change the relative position of the centromere on the chromosome. They could only be detected by examination of meiotic pairing or by using some technique that allows visualization of the genic sequence on the chromosome such as in polytenic chromosomes of *Drosophila* and other Dipterans prior to genome sequencing. Several chromosome-staining techniques that reveal banding patterns were discovered in the early 1970s (Figure 3.4; Comings 1978). These techniques have been extremely helpful in identifying homologous chromosomes in karyotypes and for detecting chromosomal rearrangements such as paracentric inversions. However, relatively few species have been studied with these techniques so we know little about the frequency of paracentric inversions in natural populations. Genome sequencing and genetic mapping (Chapter 4) have made detecting inversions and other chromosomal rearrangements much easier (Example 4.1).

Two pericentric inversions have been described in orangutans (*Pongo* spp.). The separate species of orangutans from Borneo and Sumatra are fixed for different forms of an inversion for chromosome 2 (Seuanez 1986; Ryder & Chemnick 1993). These taxa were originally recognized as subspecies, and they sometimes were mixed together in captivity. Today these taxa are recognized as separate species (Banes et al. 2016). Wild captured orangutans have been homozygous for these two chromosomal types while over a third of all captive-born orangutans have been heterozygous (Table 3.2). A pericentric inversion of chromosome 9 is polymorphic

gametes that form inviable zygotes (Figure 3.6). The allelic combinations at different loci within inversion loops will tend to stay together because of the low rate of successful recombination within inversions. In situations where several loci within an inversion affect the same trait, the loci are collectively referred to as a "**supergene**," with allelic combinations across loci within them acting as alleles (Thompson & Jiggins 2014). Early examples of phenotypes controlled by supergenes include shell color and pattern in snails (Ford 1971), mimicry in

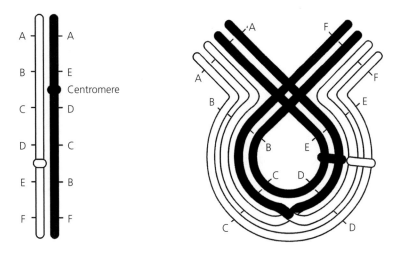

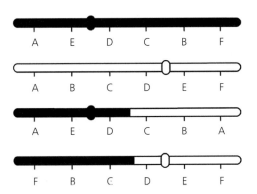

Figure 3.6 Crossing over in a heterozygote for a pericentric inversion. The two chromosomes are shown in the upper left. The pairing configuration and crossing over between two nonsister chromatids is shown in the upper right. The resulting products of meiosis are shown below. Only the two top chromosomes have complete sets of genes; these are noncrossover chromosomes that have the same sequences as the two original chromosomes. From Dobzhansky et al. (1977).

Table 3.2 Chromosomal inversion polymorphisms in the orangutan. The inversion in chromosome 2 distinguishes the Sumatran (*S*) and Bornean (*B*) species. The two inversion types in chromosome 9 (*C* and *R*) are polymorphic in both subspecies. From Ryder & Chemnick (1993).

	Chromosome 2			Chromosome 9		
	BB	**SB**	**SS**	**CC**	**CR**	**RR**
Wild born	51	0	41	67	22	3
Zoo born	90	44	82	71	34	3

in both species. The persistence of the polymorphism in chromosome 9 for this period of time is surprising.

Chromosomal polymorphisms seem to be unusually common in some bird species (Shields 1982; Tuttle et al. 2016). Figure 3.2 shows the karyotype

of a cardinal that is heterozygous for a pericentric inversion of chromosome 5 (Bass 1979). There is evidence that suggests that such chromosomal polymorphisms may be associated with important differences in morphology and behavior among individuals (Guest Box 3). For example, Rising & Shields (1980) have described pericentric inversion polymorphisms in slate-colored juncos that are associated with morphological differences in bill size and appendage size. They suggest that this polymorphism is associated with habitat partitioning during the winter when these birds live and forage in flocks.

3.1.7 Translocations

A **translocation** is a chromosomal rearrangement in which part of a chromosome becomes attached to a different chromosome. Reciprocal or mutual translocations result from a break in each of two nonhomologous chromosomes and an exchange of chromosomal sections. In general, polymorphisms for translocations are rare in natural populations because of infertility problems in heterozygotes.

Robertsonian translocations are a special case of translocations in which the breakpoints occur very close to either the centromere or the **telomere**. A Robertsonian fusion occurs when a break occurs in each of two acrocentric chromosomes near their centromeres, and the two chromosomes join to form a single metacentric chromosome. **Robertsonian fissions** occur when this process is reversed. Robertsonian polymorphisms can be relatively frequent in natural populations because the translocation involves the entire chromosome arm and balanced gametes are usually produced by heterozygotes (Searle 1986; Nachman & Searle 1995).

The Western European house mouse has an exceptionally variable karyotype (Piálek et al. 2005). The rate of evolution of Robertsonian changes in this species is nearly 100 times greater than in most other mammals (Nachman & Searle 1995). Over 40 chromosomal races of this species have been described in Europe and North Africa on the basis of Robertsonian translocations and whole-arm reciprocal translocations (Hauffe & Searle 1998; Piálek et al. 2005). As expected, hybrids between three of these races have reduced fertility. On the average, the litter size of crosses with one hybrid parent is 44% less than crosses between two parents from the AA race (Table 3.3).

3.1.8 Chromosomal variation and conservation

The main concern of chromosomal differences for conservation has been as a possible source of reduced fitness in individuals resulting from crosses between different populations (outbreeding depression; Frankham et al. 2011; Ralls et al. 2020). As we have seen, chromosomal heterozygotes often have reduced fitness. We expect greater rates of chromosomal evolution by genetic drift in small, isolated populations in the case of reduced fitness of heterozygotes (Lande 1979; see Section 8.2.3). Thus, we should be most concerned about outbreeding depression because of chromosomal rearrangements in taxa that tend to have small, fragmented populations (e.g., rodents and primates).

Plants tend to have greater rates of chromosomal rearrangements than animals (Hoffmann & Rieseberg 2008). Chromosomal evolution appears to be highest in annual plants, probably because they are prone to dramatic fluctuations in population size (Harrison et al. 2000). Second, many plant species can reproduce by **selfing** and this greatly increases the probability of chromosomal rearrangements becoming fixed in populations. Differences in ploidy among plant populations are another possible source of reduced fitness of chromosomal heterozygotes resulting from matings between populations.

3.2 Mitochondrial and chloroplast DNA

As well as DNA in the cell nucleus, both animal and plant cells also have DNA in mitochondria, and plants have DNA in **chloroplasts**. These organellar DNA molecules are relatively small, circular, are generally inherited from a single parent, and usually undergo no recombination. The first studies of DNA variation in natural populations examined animal **mitochondrial DNA** (mtDNA) because it is small (~17,000 base pairs (bp) in vertebrates and many other animals), relatively easy to isolate from genomic DNA, and occurs in thousands of copies per cell. These characteristics allowed investigators

Table 3.3 Litter sizes produced by mice heterozygous for Robertsonian translocations characteristic of three different chromosomal races (AA, POS, and UV). From Hauffe & Searle (1998).

Female	Male	No. litters	Litter size
AA	AA (control)	17	6.7 ± 0.8
AA	(AA × POS)	16	4.1 ± 0.4
AA	(AA × UV)	18	2.6 ± 0.3
AA	(UV × POS)	19	3.8 ± 0.3
AA (control)	AA	18	6.8 ± 0.4
(AA × POS)	AA	7	1.0 ± 0
(AA × UV)	AA	10	3.1 ± 0.6
(POS × UV)	AA	11	4.0 ± 0.5

Box 3.2 Restriction enzyme analysis of mtDNA

The discovery of **restriction endonucleases** (also known as restriction enzymes) in 1968 (Meselson & Yuan 1968) marked the beginning of the era of genetic engineering (i.e., the cutting and splicing together of DNA fragments from different chromosomes or organisms). Restriction endonucleases are enzymes in bacteria that provide a protective function by cleaving foreign DNA from intracellular viral pathogens (bacteriophages) harmful to the bacteria. Each restriction enzyme recognizes and cleaves DNA where a specific sequence of nucleotides occurs. The bacterial DNA itself is protected from cleavage because it is **methylated**. Nearly 1,000 different restriction enzymes have been described from bacteria. Restriction enzymes are essential to some of the most commonly used methods of generating population genomic data, as we will see in the next chapter.

The most commonly used restriction endonuclease is *Eco*RI from the bacterium *Escherichia coli*. *Eco*RI cleaves a specific six-base sequence: GAATTC (and the reverse complement CTTAAG). The cleavage is uneven such that each strand is left with an overhang of AATT, where the overhang is in bold and the Xs represent the sequence flanking the restriction site sequence.

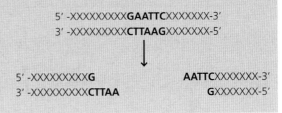

to separate mtDNA from **nuclear DNA** by density gradient ultracentrifugation.

In the 1970s and 1980s, there was an explosion of population genetics research with humans and other animals fueled by mtDNA. In 1979, two independent groups published the first reports of genetic variation in DNA from natural populations. Avise et al. (1979a, 1979b, 1986) used **restriction enzyme analysis** of mtDNA (Box 3.2) to describe sequence variation and the genetic population structure of mice and pocket gophers. Brown & Wright (1979) used the maternal inheritance of mtDNA to determine the parental sex of two lizard species that originally hybridized to produce a **parthenogenetic** species. A paper by Brown et al. (1979) compared the rate of evolution of mtDNA and nuclear DNA in primates. This latter work was done in collaboration with Allan C. Wilson, whose lab became a center for the study of the evolution of mtDNA (Wilson et al. 1985).

Sequence differences between individuals can produce different results when DNA molecules are digested with restriction enzymes. Within the same segment of DNA, some individuals might have only one restriction cut site, while others might have two or three. A circular DNA molecule (such as mtDNA)

with one restriction site will yield one linear DNA fragment after cleavage (Figure 3.7). If two cleavage sites exist, then two linear DNA fragments are produced from the cleavage. We can visualize the number of fragments using gel electrophoresis to separate them by length; short fragments migrate faster than long ones (Figure 3.7).

This is the basis of the **restriction fragment length polymorphism** (RFLP) technique, which was the primary method to detect mtDNA polymorphisms before sequencing became commonly used. Restriction site polymorphisms are usually generated by a single nucleotide substitution in the restriction site (e.g., from GA*A*TTC to GA*T*TC). This causes the loss of the restriction site in the individual because the enzyme will no longer cleave the individual's DNA. As a result, a restriction site polymorphism is detectable as an RFLP following digestion of the molecule with a restriction enzyme and gel electrophoresis.

Figure 3.8 shows RFLP variation in the mtDNA molecules of two subspecies of cutthroat trout digested by two restriction enzymes (*Bgl*I and *Bgl*II) that each recognize six base pair sequences. There are three cut sites for *Bgl*I in the W (westslope cutthroat trout) **haplotype**; there is an

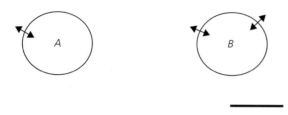

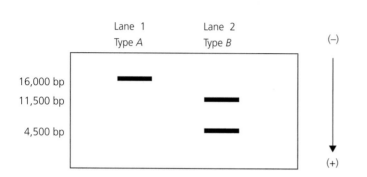

Figure 3.7 Hypothetical examination of sequence differences in mtDNA revealed by restriction enzyme analysis. Type *A* has only one cleavage site (arrow), which produces a single linear fragment of 16,000 bp; type *B* has two cleavage sites that produce two linear fragments of 11,500 and 4,500 bp. Electrophoresis of the digested products results in the gel banding pattern shown. The DNA fragments move in the direction indicated by the arrow, and smaller fragments migrate faster.

additional cut site in the Y (Yellowstone cutthroat trout) haplotype so that the largest fragment in the W haplotype is cut into two smaller pieces. The Y haplotype also has an additional cut site for *Bgl*II, resulting in the second fragment being cut into two smaller fragments.

RFLP analysis is also useful for studies of nuclear genes following PCR amplification of gene fragments. For example, Toews et al. (2016) studied hybridization between the golden-winged and blue-winged wood warblers. Whole genome sequence comparisons found genetic differentiation between these species at only six loci. They then used primers to **amplify** these six regions, and cut each of these with a restriction enzyme for which cut sites differed between species. This RFLP assay allowed them to quickly genotype and assign species or hybrid status to 132 birds based on these six regions.

Several characteristics of animal mtDNA make it especially valuable for certain applications in understanding patterns of genetic variation. mtDNA is haploid and maternally inherited in most species. That is, a progeny generally inherits a single mtDNA genotype from its mother (Figure 3.9). There are thousands of mtDNA

molecules in an egg, but relatively few in sperm. In addition, mitochondria from the sperm are actively destroyed once they are inside the egg. There are many exceptions to strict maternal inheritance. For example, there is evidence of some incorporation of male mitochondria ("paternal leakage") in species that generally show maternal inheritance; for example, mice (Gyllensten et al. 1991), sparrows (Päckert et al. 2019), and humans (Awadalla et al. 1999). In addition, some species (e.g., many mussels) show an unusual doubly uniparental inheritance of mitochondrial DNA, in which female mussels inherit their mtDNA only from their mothers and pass it on to both male and female offspring, while males inherit mtDNA from both their mothers and fathers, but only pass on the paternal mtDNA (Sutherland et al. 1998). Both maternal leakage and mutation can lead to **heteroplasmy** (the presence of more than one organelle genotype within a cell or individual).

Mitochondrial DNA molecules are especially valuable for reconstructing phylogenies because there is generally no recombination between mtDNA molecules. Unlike nuclear DNA, the historical genealogical record of descent is not "shuffled" by recombination between different

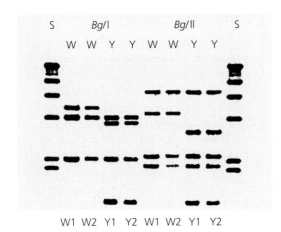

Figure 3.8 Restriction fragment polymorphism of mtDNA in cutthroat trout digested by two restriction enzymes (*Bg*lI and *Bg*lII). The W lanes are two (W1 and W2) westslope cutthroat trout and the Y lanes are two (Y1 and Y2) Yellowstone cutthroat trout. The S lanes are size standards. DNA fragments migrate from top to bottom during electrophoresis. The smallest fragments are at the bottom. From Forbes (1990).

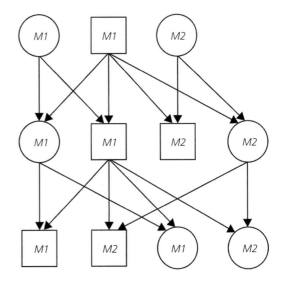

Figure 3.9 Pedigree showing maternal inheritance of two mtDNA genotypes: *M1* and *M2*. By convention in pedigrees, males are represented by squares, and females are represented by circles. Each progeny inherits the mtDNA of its mother.

mtDNA lineages during gamete production, as occurs in nuclear DNA during meiosis. Recombination between lineages is unlikely because mtDNA generally occurs in only one lineage per individual (one haploid genome) as the male gamete does not contribute mtDNA to the zygote. However, there is some evidence for rare recombination events in animal mtDNA (Slate & Gemmell 2004; Ujvari et al. 2007). Thus, the mtDNA of a species can be considered a single nonrecombining genealogical unit with multiple alleles or haplotypes (Avise 2004).

The lack of recombination, which makes mtDNA especially valuable in reconstructing phylogenies, reduces its value for describing genetic population structure within species. The primary problem is that the entire mtDNA genome acts as a single locus because there is no recombination. In addition, there is evidence that mtDNA is affected by natural selection and therefore might not accurately reflect the demographic or evolutionary history and processes within a species (Ballard & Whitlock 2004; Galtier et al. 2009; Balloux 2010). Finally, maternal inheritance makes it especially inappropriate to be used as

the sole source of genetic knowledge for describing units of conservation since patterns of divergence are not influenced by genetic contributions of males (Example 9.3).

Plant mitochondrial DNA has been less useful than animal mtDNA for genetic studies of natural populations primarily because of a lower mutation rate resulting in low levels of variation (Clegg 1990; Powell 1994; Petit et al. 2005). In addition, in some cases rearrangements within the plant mtDNA genome and the transfer of so-called promiscuous DNA between the nuclear and chloroplast genomes make it difficult to use so-called universal primers with plant mtDNA (Kubo & Mikami 2007). There is also some recombination for mitochondrial and chloroplast molecules in many plant species (Mackenzie & Macintosh 1999; McCauley & Ellis 2008).

DNA from plant chloroplasts (cpDNA), however, has proven very useful for phylogeographic studies (Petit et al. 2005), and there are some useful regions of cpDNA that are variable across a wide range of taxa and have been used extensively (Byrne & Hankinson 2012; Shaw et al. 2014; see Section 4.2.1).

Whole chloroplast DNA sequences are now becoming common and provide rich datasets for phylogenetics and other applications (Tonti-Fillipini et al. 2017).

Assessment of cpDNA variation uncovers signals of long-term evolutionary history. Signals of long-term persistence through multiple glacial cycles have been demonstrated using cpDNA sequences, particularly in unglaciated landscapes (Byrne et al. 2014; Sampson et al. 2018). Similarly, analysis of cpDNA in species with distributions of isolated populations on granite outcrops called inselbergs in Brazil and southwestern Australia shows long-term isolation and persistence of populations across inselbergs (Tapper et al. 2014; Hmeljevski et al. 2017).

Most plant species have largely or completely maternal inheritance of cpDNA, although conifers in the pine family have largely paternal inheritance (Petit et al. 2005). Within some plant species, inheritance can occasionally be biparental. For example, McCauley et al. (2005) found primarily maternal inheritance (96% of all offspring) in *Silene vulgaris*, a **gynodioecious** species with some bisexual individuals and some plants producing only female flowers. Despite this high rate of maternal inheritance, the cumulative genetic effects of paternal leakage were evident, as heteroplasmy was detected in over 20% of all individuals.

Chloroplast DNA markers have also played a key role in detecting hybridization and introgression between plant species, and revealing cases in which one species has taken on the chloroplast genotype of another, in a process called chloroplast capture. For example, the European oaks *Quercus petraea* and *Quercus robur* are morphologically distinct, yet share chloroplast haplotypes in **sympatric** populations (locations where both species occur; Petit et al. 1997). Oak seeds are large and do not disperse far. During postglacial recolonization, when a population of one species of oak was invaded by the other, it occurred primarily through pollen from the invader. Even when the original species is genetically swamped by the invader through hybridization followed by multiple generations of backcrossing to the invading species, the chloroplasts of the original colonizer can persist through the maternal lineage.

3.3 Single-copy nuclear loci

Virtually all of the DNA within an individual is found in pairs of chromosomes in the nucleus. For example, there are ~3 billion base pairs in the human nuclear genome compared with only 16,569 base pairs in the human mitochondrial genome. The beginning of population genetics was based upon the study of individual nuclear genes that had observable phenotypic effects (e.g., violet versus white flowers in Mendel's pea plants). A variety of molecular techniques have been developed since the discovery of DNA to directly study genetic differences between alleles at individual nuclear loci.

3.3.1 Protein electrophoresis

The first major advance in our understanding of genetic variation in natural populations began in the mid-1960s with the advent of protein electrophoresis (Powell 1994). The detection of variation in amino acid sequences of proteins by electrophoresis allowed an immediate assessment of genetic variation in a wide variety of species (Lewontin 1974). There is a direct relationship between genes (DNA base pair sequences) and proteins (amino acid sequences). Proteins have an electrical charge and migrate in an electrical field at different rates depending upon their charge, size, and shape. A single amino acid substitution can affect migration rate and thus can be detected by electrophoresis. Moreover, the genomes of all animals (from elephants to *Drosophila*), all plants (from sequoias to Furbish's lousewort), and all microbes (from *E. coli* to HIV) encode proteins, and most share a set of enzymes involved in basic metabolism. Empirical population genetics became universal, and genetic variation has been described in natural populations of thousands of species in the past 50 years. Allozymes are allelic enzymes produced by a single gene, while **isozymes** are alternative isoforms of an enzyme that may be produced by the same or by **paralogous** loci.

There are two fundamental steps to protein electrophoresis. The first is to separate proteins with different electrophoretic mobilities in a supporting medium (usually a gel of starch or polyacrylamide).

However, most tissues contain proteins encoded by hundreds of different genes. The second step of the process, therefore, is to locate the presence of specific proteins. This step is usually accomplished by taking advantage of the specific catalytic activity of different enzymes. Specific enzymes can be located by staining gels with a chemical solution containing the substrate specific for the enzyme to be assayed, and a salt that reacts with the product of the reaction catalyzed by the enzyme and produces a visible product. Figure 3.10 shows variation at the enzyme aconitate dehydrogenase in Chinook salmon from a sample from the Columbia River of North America. Alleles are generally identified by their relative migration distance in the gel. Thus, the *86* allele migrates ~86% as far as the common allele, *100*.

Allozymes were the workhorse for describing the genetic structure of natural populations for 40 years (Lewontin 1991). They were the first genetic markers used to estimate the amount of genetic variation in natural populations, and to compare levels of variation among different species. In addition, they provided initial glimpses into the reproductive behavior of populations of wild species (Avise 2004).

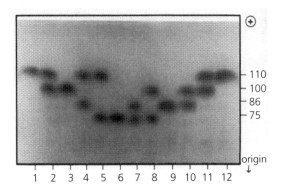

Figure 3.10 Protein electrophoresis of the enzyme aconitate dehydratase in livers of 12 Chinook salmon. The relative mobilities of the allozymes encoded by four alleles at this locus are on the right. The genotypes of these 12 individuals are (1) *110/110*, (2) *100/110*, (3) *100/100*, (4) *110/86*, (5) *110/75*, (6) *75/75*, (7) *86/75*, (8) *100/75*, (9) *86/86*, (10) *100/86*, (11) *110/100*, and (12) *110/110*. From Utter et al. (1987).

3.3.2 Microsatellites

Microsatellites became the most widely used DNA marker in population genetics for genome mapping, molecular ecology, and conservation studies beginning in the 1990s. Microsatellite DNA markers were first discovered in the 1980s (Schlötterer 1998). They are also called SSRs (simple sequence repeats) or VNTRs (variable number of tandem repeats). They consist of tandem repeats of a short sequence motif of one to six nucleotides (e.g., "CGTCGTCGTCGTCGT," which can be represented by $(CGT)^n$ where $n = 5$). The number of repeats at a polymorphic locus generally ranges from ~5 to 100. PCR primers are designed to hybridize to the conserved DNA sequences flanking the variable repeat units and amplify the repetitive region. Alleles are scored based on the length of the amplified region. Microsatellite PCR products are generally between 75 and 300 bp long, depending on the locus and the location of the primers.

Microsatellites are often highly polymorphic, even in small populations and endangered species (e.g., polar bears or cheetahs). This high rate of polymorphism results from a high mutation rate due primarily to slippage during DNA replication (Ellegren 2000a; Chapter 12). A microsatellite mutation usually results in a change in the number of repeats (usually an increase or decrease of one repeat unit). The rate of mutation is typically around one mutation in every 1,000 or 10,000 meioses (i.e., 10^{-3} or 10^{-4} per generation). The high mutation rate of microsatellites (Chapter 12) results in greater heterozygosity and allelic diversity at microsatellites than allozymes, and means that even a small number of microsatellite loci can be quite informative for some purposes.

Alleles for microsatellites are scored on the basis of the length of the repetitive sequence resulting from variation in the number of repeats. For example, Figure 3.11 shows the distribution of alleles for two di-nucleotide microsatellite loci in a wild Argali sheep population in Afghanistan. For the first marker, allele length ranges from 184–198 bp. The longest allele has seven more repeats of the two-bp sequence than the shortest allele.

(a)

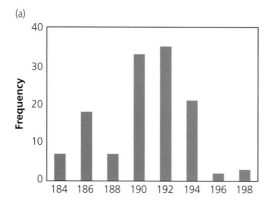

(b)

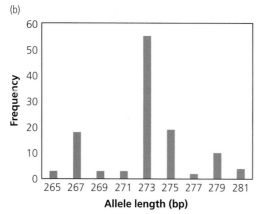

Figure 3.11 Allele length distributions for microsatellite loci from a wild Argali sheep population in the Pamir Mountains, Afghanistan. Allele length is the total length of the PCR product (amplicon). (a) is the *MMP9* locus, and (b) is an MHC locus (major histocompatibility complex gene region). The loci are dinucleotide repeats, and thus most alleles (length classes) differ by two base pairs. Data from Harris et al. (2010). See also Luikart et al. (2011) for information on loci and populations.

3.3.3 Single nucleotide polymorphisms (SNPs)

Single nucleotide polymorphisms (SNPs) are variations at single base pair locations. They are typically bi-allelic, so provide less information than microsatellites per marker. SNPs exist within coding regions of genes (**exons**), in **introns**, in regulatory regions, and in intergenic regions, and so can be used to study both selectively neutral and adaptive variation. They are also **codominant**. For these reasons, and since the potential number of SNPs is orders of magnitude greater than microsatellites, they are rapidly replacing other types of genetic

markers. There are single locus methods for genotyping targeted SNPs and multiple locus techniques for genotyping SNPs at many locations throughout the genome or within targeted genomic regions. In Chapter 4, we discuss SNPs in detail and outline the different approaches for genotyping them for population genomic studies.

3.3.4 Sex-linked markers

Genetic markers in sex-determining regions can be especially valuable in understanding genetic variation in natural populations of animals. For example, markers that are specific to the sex-determining Y or W chromosomes can be used to identify the sex of individuals in species in which it is difficult to identify sex phenotypically, as in many bird species (Ellegren 2000b). In addition, Y chromosome markers, like mtDNA, are especially useful for phylogenetic reconstruction because Y chromosome DNA, like mtDNA, is haploid and nonrecombining in mammals.

Mammals can be sexed using PCR amplification of Y chromosome fragments, or by co-amplification of a homologous sequence on both the Y and X that are subsequently discriminated by size, restriction enzyme cleavage of diagnostic sites, or by sequencing (Fernando & Melnick 2001). Similar molecular sexing techniques exist for birds and other taxa (e.g., amphibians). The W chromosome of birds has conserved sequences not found on the Z, allowing nearly universal avian sexing with PCR (e.g., Huynen et al. 2002). Some plants also have sex-linked sequences (Korpelainen 2002). Robertson and Gemmell (2006) provide a needed caution and guidelines for using these techniques.

3.4 Multiple locus techniques

Multiple locus techniques discussed here, including **amplified fragment length polymorphisms (AFLP)**, randomly amplified polymorphic DNA (**RAPD**), **minisatellites**, and inter-simple sequence repeats (**ISSRs**), assay many anonymous locations across the genome simultaneously with a single PCR reaction (Bruford et al. 1998). These methods have largely been replaced by the genomic approaches for sequencing and genotyping SNPs introduced in Chapter 4, but we review them briefly here. The advantage of these techniques, prior to the genomics revolution, was that many loci could be examined readily with little or no prior information about sequences from the genome.

The major disadvantages are that it is generally difficult to associate individual bands with particular genes, and that alternative alleles are either dominant or recessive. In contrast to the codominant alleles at microsatellite, allozyme, or SNP loci, these methods usually cannot resolve between a heterozygote and the **homozygote** for the dominant type. For example, with a presence–absence polymorphism, the same band will appear if there is one copy of the DNA fragment or two. Consequently, we cannot compute individual (observed) heterozygosity to test for Hardy–Weinberg (HW) proportions. In addition, these markers (like SNPs) are bi-allelic and thus provide less information per locus than the more polymorphic microsatellites. It can require 5–10 times more of these loci to provide the same information as microsatellite loci with multiple alleles (e.g., Waits et al. 2001). However, more loci (10–25) can be analyzed per PCR and per gel lane using some of these techniques, compared with microsatellites (5–10 loci per lane using fluorescent labels).

The RAPD method involved PCR amplification using short (usually 10 nucleotides) arbitrary primer sequences (Welsh & McClelland 1990). RAPD markers were only used for a brief period of time in the 1990s, but they were found to be too unreliable and not sufficiently reproducible to be useful.

3.4.1 Minisatellites

Minisatellites are tandem repeats of a sequence motif that is ~20 to several hundred nucleotides long—much longer than **microsatellite** motifs. Minisatellites were first discovered by Jeffreys et al. (1985), and their use in DNA "fingerprinting" revolutionized human **forensics** cases. They were used in wildlife populations soon after, for example, to study paternity and detect extra-pair copulations in birds thought to be monogamous. However, their use was limited and they are not in use today.

The high polymorphism of minisatellites made them most useful for interindividual studies such as parentage analysis and individual identification before microsatellites became widely available. The minisatellite genotyping technique was the first to be referred to as **DNA fingerprinting** because individuals possess unique minisatellite signatures (Jeffreys et al. 1985). Alleles are identified by the number of tandem repeats of the sequence motif. However, it can be difficult to identify alleles that belong to one locus because most genotyping systems reveal bands (alleles) from many loci together in one gel lane. Also, the repeat motifs are long, so they cannot be studied in samples of partially degraded DNA that generally contain fragments of only 100–300 bases.

3.4.2 AFLPs and ISSRs

The AFLP technique first involves digesting genomic DNA with restriction enzymes to produce many fragments. Adaptors are then ligated to the sticky ends of DNA fragments, and selective PCR is used to amplify a subset of the DNA fragments. This generates DNA fingerprints (i.e., multilocus band profiles) from fragment lengths at many anonymous locations across the genome. This technique was named AFLP because it resembles the RFLP technique (Vos et al. 1995). However, these authors said that AFLP is not an acronym for amplified fragment length polymorphism because it does not detect length polymorphisms. The main advantage of the AFLP method was that many polymorphic markers could be developed quickly for most species, even if no sequence information existed for the species. In addition, the markers generally provide a broad sampling of the genome. Foll et al. (2010) developed a Bayesian statistical treatment of AFLP banding patterns that allows a more accurate use of these markers for understanding genetic population structure. However, this and other methods producing dominant markers were rapidly replaced by microsatellites as they are codominant, and more recently by restriction site-associated DNA sequencing (**RADseq**) and other genomic tools that produce codominant SNP genotypes at many loci (Chapter 4).

ISSR markers generate a large number of DNA fragments from a single PCR. ISSR primers are based upon the SSRs found in microsatellites. Bands are generated by a single-primer PCR reaction where the primer is a repetition of a di-, tri-, or tetranucleotides and the amplified region is a portion of the genome between two identical microsatellite primers with an opposite orientation on the DNA strand. These primer sequences are broadly distributed across the genome. Therefore, the ISSR-PCR technique allows one to quickly screen many parts of the genome without prior DNA sequence knowledge. ISSRs are limited but continue to be used in agriculture and horticulture to confirm the identity of cultivated varieties and clones, as well as in a modest number of population genetic studies of wild species.

3.5 Genetic variation within and among populations

We can use the genetic markers and techniques described in this chapter (and the next) to test directly for genetic variation within populations and genetic differences among populations from different geographical areas. This knowledge is useful for managing populations and developing strategies for conserving genetic diversity. Some of this information can also be used to compare different species for levels of genetic diversity, but these comparisons need to be made cautiously, using the same type of genetic marker, to be valid.

The first comparisons of the amount and geographic pattern of genetic variation in different species were provided by allozymes. With protein electrophoresis, the same or a similar suite of loci were used to estimate genetic variation in different species so that the amount of genetic variation can be compared. In addition, all available allozyme loci were screened, even those that were **monomorphic** in a given species. SNPs are now also providing data for valid species comparisons as long as similar methods are used.

The other types of markers discussed in this chapter are less useful for interspecific comparisons. With microsatellite markers, for example, many loci are typically screened, and then only those that are polymorphic in the target species are chosen for analysis. This selection process results

in an **ascertainment bias** so that the amount of variation found in different species does not provide a meaningful comparison. Assume we select a group of microsatellites for use in a particular species because they are polymorphic. A comparison between the amount of genetic variation at these loci in our target species and a closely related one will be biased because we selected loci that were polymorphic in our target species without considering variability in the other species (Morin et al. 2004). In the remainder of this chapter we will illustrate most comparisons of genetic variation among taxa with allozyme and SNP data that avoid ascertainment bias.

3.5.1 Quantifying genetic variation within natural populations

The most commonly used measure to quantify variation within populations and to compare the amount of genetic variation in different populations is **heterozygosity (_H_)**. At a single locus, heterozygosity is the proportion of individuals that are heterozygous; heterozygosity ranges between zero and one. Two different measures of heterozygosity are used. H_o is the observed proportion of heterozygotes. For example, seven of the 12 individuals in Figure 3.10 are heterozygotes so H_o at this locus in this sample is $7/12 = 0.58$. H_e is the expected proportion of heterozygotes if the population is mating at random, calculated from allele frequencies using HW expectations. The estimation of H_e is discussed in detail in Chapter 5. H_e provides a better measure than H_o to compare the relative amount of variation in different populations as long as the populations are mating at random (Nei 1977). Average heterozygosity is typically calculated across many loci in a population to estimate genetic variation. Expected heterozygosity can also be calculated for SNPs at individual base pairs. This is called **nucleotide diversity (π)**. When nucleotide diversity is averaged over many contiguous base pairs it becomes small because most base pairs (nucleotide sites) are monomorphic (e.g., $\pi = 0.001$; Figure 3.12, see "Mammoth (island)").

Another measure of variation often used is **polymorphism**, the proportion of loci that are genetically variable (_P_). The likelihood of detecting genetic variation at a locus increases as more individuals are sampled from a population. This

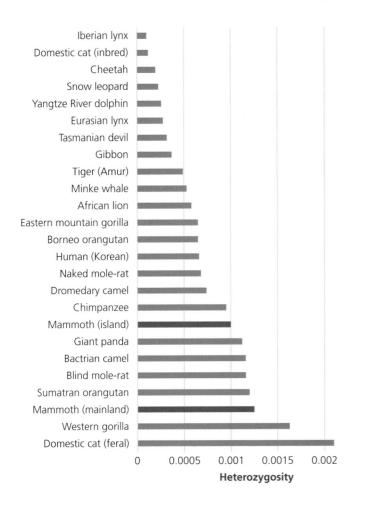

Figure 3.12 Genome-wide single nucleotide polymorphism heterozygosity (nucleotide diversity) in mammals. Modified from Abascal et al. (2016) and updated with the addition of data from a mammoth from the mainland of Asia (Palkopoulou et al. 2015).

dependence on sample size is partially avoided by setting an arbitrary limit for the frequency of the most common allele. We often use the criterion that the most common allele must have a frequency of 0.99 or less for the locus to be considered polymorphic.

3.5.2 Estimates of genetic variation within natural populations

Two protein electrophoresis studies in 1966 were the first to estimate the amount of genetic variation within a species. Harris (1966) described genetic variation in the human population from England and found three of 10 loci to be polymorphic ($P =$ 0.30). On average, individuals were expected to be heterozygous at ~10% of all loci examined ($H_e =$ 0.097). Two back-to-back papers described genetic variation at protein loci in *Drosophila pseudoobscura*

(Hubby & Lewontin 1966; Lewontin & Hubby 1966). A recent discussion of the importance of these two foundational studies for empirical population genetics concluded that these first estimates were surprisingly good, despite the small numbers of genes examined and the insensitivity of standard allozyme analysis compared with sequencing of DNA (Charlesworth et al. 2016).

Nevo et al. (1984) summarized the results of protein electrophoresis surveys of some 1,111 species! Average heterozygosities (H_S) for major taxonomic groups are shown in Figure 3.13. While average heterozygosity varies among taxonomic groups, different species within groups sometimes have enormous differences in the amount of genetic variation they possess (Table 3.4). For example, red pine and ponderosa pine are closely related species with vastly different amounts of genetic variation (Table 3.4). Differences among species in amounts of

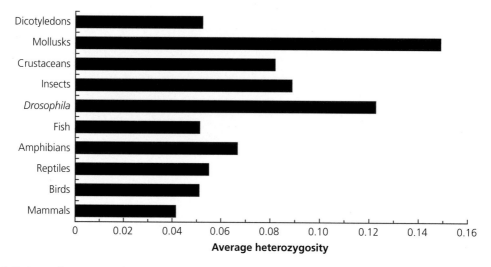

Figure 3.13 Average heterozygosities (H_S) at allozyme loci from major taxa. Modified from Gillespie (1992); data are from Nevo et al. (1984).

Table 3.4 Summary of genetic variation demonstrating range found in different species (Nevo et al. 1984). H_S is the mean expected heterozygosity (H_e) over all loci for all populations examined. P is the percentage of loci that is variable.

Species	No. loci	P	H_S
Alligator	44	7	0.016
American toad	14	34	0.116
Cheetah	52	0	0.000
Humans	107	47	0.125
Moose	23	9	0.018
Polar bear	29	2	0.000
Roundworm	21	29	0.027
Gilia	13	52	0.106
Ponderosa pine	35	83	0.180
Red pine	35	3	0.007
Salsify	21	9	0.026
Yellow evening primrose	20	25	0.028

Table 3.5 Summary of genetic variation in brown bears (also known as grizzly bears) from four regions of North America. The allozyme (34 loci) data are from K.L. Knudsen et al. (unpublished); the microsatellite (eight loci) and mtDNA data are from Waits et al. (1998). The allozyme samples for Alaska/Canada are from the Western Brooks Range in Alaska, and the microsatellite and mtDNA samples for this sample are from Kluane National Park, Canada. H_e is the mean expected heterozygosity (see Section 3.3), $\bar{A}$ is the average number of alleles observed, and h is gene diversity. h is computationally equivalent to H_e, but is termed gene diversity because mtDNA is haploid so that individuals are not heterozygous (Nei 1987, p. 177).

Sample	Allozymes		Microsatellites		mtDNA	
	H_e	$\bar{A}$	H_e	$\bar{A}$	h	A
Alaska/Canada	0.032	1.2	0.763	7.5	0.689	5
Kodiak Island	0.000	1.0	0.265	2.1	0.000	1
NCDE	0.014	1.1	0.702	6.8	0.611	5
YSE	0.008	1.1	0.554	4.4	0.240	3

NCDE, Northern Continental Divide Ecosystem (including Glacier National Park); YSE, Yellowstone Ecosystem (including Yellowstone National Park).

genetic variation can be important for conservation, as evolutionary change cannot occur unless there is genetic variation present.

SNPs also provide an opportunity to compare heterozygosity among species, as long as SNP genotypes are generated in a comparable way; that is, genome wide, or targeting the same regions (e.g., random, intergenic, exons, or introns). SNP heterozygosity (i.e., nucleotide diversity) for a wide range of mammals is shown in

Figure 3.12. In general, wild cats have relatively low heterozygosity, while primates have moderate to high levels. Two ancient DNA samples from mammoths revealed higher heterozygosity in a mainland sample than in an island sample, presumably reflecting population size and isolation. Another interesting comparison is provided by samples of feral and inbred domestic cats—the former has exceptionally high heterozygosity, while the latter has almost no variation.

Example 3.3 Wollemi pine: The botanical find of the century lacks genetic diversity

In 1994, David Noble was exploring a deep, narrow canyon in Wollemi National Park, remote yet just 150 km from Sydney, Australia, and found some trees he didn't recognize. The species turned out to be what is now called Wollemi pine, a species that was thought to have been extinct for over 100 million years (Figure 3.14). There are no other extant species in this genus (Jones et al. 1995). There are currently fewer than 100 mature individuals known to exist within the park, and their location has not been disclosed to protect them (Zimmer et al. 2016).

Figure 3.14 Wollemi pine in Wollemi National Park, New South Wales, showing the habit of coppicing in which multiple stems grow from the base of a single tree. Photo courtesy of Heidi Zimmer.

An initial study of 12 allozyme loci and 800 AFLP fragments failed to reveal any genetic variation (Hogbin et al. 2000). A study of 20 microsatellite loci also failed to detect any genetic variation in this species (Peakall et al. 2003). However, recent sequencing of the chloroplast genome identified six SNPs that make up three different chloroplast genotypes (Greenfield et al. 2016). Whole genome sequencing of this species is underway. The exceptionally low genetic variation, combined with its known susceptibility to exotic fungal pathogens, provides strong justification for current policies of strict control of access and the secrecy of the location of this species.

While heterozygosity estimates will depend on the type of genetic marker used, the relative diversity of populations within a species should be similar for different markers. For example, for four populations of brown bears (Table 3.5), the Kodiak Island population has the lowest genetic diversity for allozymes, microsatellites, and mtDNA; followed by the Yellowstone population and the Northern Continental Divide population. The Alaska/Canada population from the Brooks Range has the highest diversity. However, heterozygosity estimates are far higher for microsatellite markers than for allozymes for all populations.

As we will see in the next few chapters, the local population size and amount of exchange between populations have a major influence on the amount and pattern of genetic variation in natural populations (Example 3.3). In subsequent chapters, we will also see how demographic history, mating systems, and taxonomic differences in mutation rates affect the amount and distribution of genetic variation.

3.5.3 Significance of the amount of variation within populations

Differences in the amount of genetic variation among species need to be carefully interpreted.

Species with lower variation are not necessarily more vulnerable to extinction (Hedrick et al. 1996). For example, initial studies indicated that the cheetah has much less genetic variation than other large cats (O'Brien et al. 1983, 1985). This finding of low allozyme variation led to the conclusion that the cheetah is vulnerable to extinction because of its lack of genetic variation (Table 3.4). However, the equilibrium genetic variation among species is expected to vary, largely because of differences in long-term effective population size. Just because a species has low genetic variation, we can't assume it suffers from reduced fitness or is less able to become adapted to future environmental conditions (Chapter 12).

On the other hand, low genetic variation in a species might indicate a recent reduction in population size, and there are many reasons to expect that the loss of variation and increase in inbreeding associated with such a reduced population size (bottleneck) do potentially indicate vulnerability to extinction. First, a recent bottleneck might indicate demographic instability that is not obvious from contemporary population size alone. Second, a species that has gone through a bottleneck severe enough to erode detectable molecular genetic variation might suffer from fixation of detrimental alleles, resulting in reduced fitness that might increase vulnerability to extinction. Finally, loss of genetic variation caused by the bottleneck may limit the ability of the population to evolve and adapt. The more recent and severe a bottleneck has been, the more we would expect the bottleneck to influence the future of a species.

Returning to the cheetah, more recent studies suggest that the lower genetic variation results from a severe population bottleneck that occurred at the end of the last ice age, some 10,000 years ago (Marker et al. 2008). Since that time, cheetah populations have regained genetic variation at markers with higher mutation rates. As we will see, allozymes recover very slowly from bottlenecks because of their low mutation rate (Section 12.5).

Low genetic variation in itself may not indicate a conservation concern. However, a recent loss of genetic variation is a concern. We will see later in Chapters 6 and 12 that there are ways to detect recent versus historic bottlenecks and losses of genetic variation.

Guest Box 3 Widespread chromosomal diversity across rock-wallabies and implications for conservation
Sally Potter and Janine E. Deakin

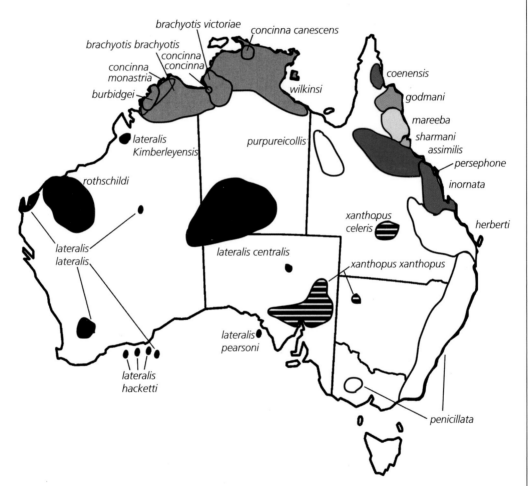

Figure 3.15 Map of the 23 *Petrogale* taxa, including species and subspecies. The colors represent chromosomal groups: *brachyotis* (gray), *xanthopus* (striped), *lateralis* (black), and *penicillata* (colored and white).

The Australian marsupial genus *Petrogale* (rock-wallabies) has the greatest diversity of chromosomal rearrangements in living marsupials (Figure 3.15). With 23 known chromosomally distinct forms, the rock-wallabies are a classic example highlighting the role of chromosomes in divergence and speciation (King 1993; Potter et al. 2017). The chromosomal rearrangements among rock-wallabies range from simple to complex, predominantly involving Robertsonian fusions and

Guest Box 3 Continued

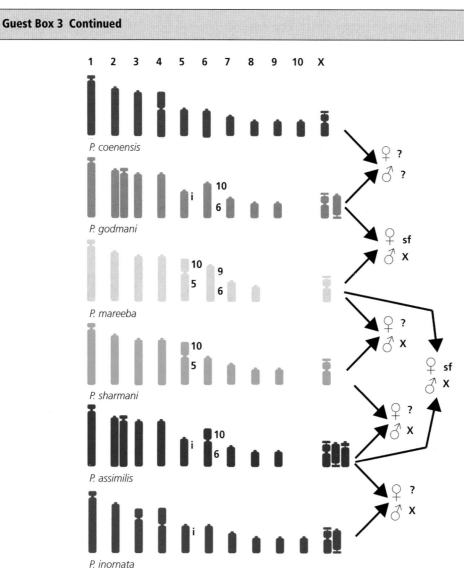

Figure 3.16 Karyotypes for six of the chromosomally complex *penicillata* group of rock-wallabies, including results for offspring from hybrid cross experiments on the right (X = infertile, sf = subfertile, ? = unknown). Chromosomal rearrangements include fusions (5–10, 6–10, 6–9), inversions on chromosome 5, polymorphic chromosomes (X and 2), and centric shifts (3, 4).

centromere repositioning (centric shifts). Fixation of chromosomal rearrangements in rock-wallabies is likely a consequence of their biology and small population sizes. There are currently five Endangered and seven Vulnerable rock-wallaby taxa listed under the Australian Environment Protection and Biodiversity Conservation Act 1999 (EPBC Act). As a result, six species are the subject of conservation management including translocations, reintroductions and captive breeding

(M. Eldridge, personal communication). The level of chromosome diversity between and even within rock-wallaby species makes it imperative that we consider chromosome variation in their conservation.

The importance of looking at chromosome variation and not just variation in genome sequence is highlighted by the description of three morphologically similar yet chromosomally distinct species (*Petrogale sharmani*, *Petrogale mareeba*, and *Petrogale coenensis*; Eldridge & Close 1992). If DNA sequences were analyzed alone, we would have a very different outlook on the taxonomy of this group. With recent divergence and high levels of gene flow estimated from neutral genetic markers, in the absence of chromosome data, we would conclude they are a single species. Hybridization experiments uncovered reproductive isolation between these species (Sharman et al. 1990; Close & Bell 1997), with infertile male offspring and subfertile female offspring resulting. This example highlights how genomics alone could potentially mislead our taxonomic understanding of species, and could result in disastrous consequences for conservation management.

In addition to interspecific chromosomal variation within *Petrogale*, there is also intraspecific karyotype variation for autosomes 1 and 2, and the X chromosome within multiple species (Eldridge & Close 1997; Figure 3.16). One species, the black-footed rock-wallaby (*Petrogale lateralis*) consists of five subspecies with varying chromosomal rearrangements (Eldridge et al. 1991). Three of these subspecies are currently listed as Endangered or Vulnerable and have ongoing conservation management (Pearson 2013). Although these rearrangements appear to include simple fusions and a centric shift, we are still unclear as to the genomic impacts of this structural variation. Hybrid crosses between *Petrogale lateralis centralis* and *Petrogale l. pearsoni* produced fertile offspring (R. L. Close, unpublished data). However, we do not know the long-term viability of these hybrids. These subspecies inhabit different geographic regions within Australia, including different biomes, and thus could exhibit adaptation. We are yet to understand the significance of this intraspecific variation within *P. lateralis* and other *Petrogale* species. Further research linking genomics and chromosome variation is required to establish the fine-scale differences within species to understand the impacts of chromosome rearrangements on genome function, as well as continue to understand how chromosome variation influences adaptation and speciation (Deakin & Potter 2019).

CHAPTER 4

Population Genomics

Bighorn sheep, Section 4.1.3

Just as the polymerase chain reaction leveled the genetic playing field at the end of the 20th century by providing easy access to the genes of all organisms, so the 21st century promises to sweep away the technological privileges of classic model organisms and democratize genomic exploration.

(Camille Bonneaud et al. 2008, p. 587)

The most direct, but unfortunately not the most useful, approach to the phylogeny of recent animals is through their genetics. The stream of heredity makes phylogeny; in a sense, it is phylogeny. Complete genetic analysis would provide the most priceless data for the mapping of this stream.

(George Gaylord Simpson 1945, p. 5)

Population genomics defined broadly is the use of genome-wide data at thousands to hundreds of thousands or even millions of variable loci to infer population demographics and evolutionary processes including selection, genetic drift, gene flow, and mutation. For some applications, genomic approaches have provided incremental improvements to traditional markers (Chapter 3) for quantifying neutral genetic diversity, understanding demography, describing population differences, estimating inbreeding, or inferring patterns of gene flow. For these purposes, genomic data with thousands or more markers provide more precise and nuanced assessments, and in some cases have identified previously undetected but important population structure. More importantly, genomic tools have been revolutionary in opening up a wide range of new population genetic analyses by harnessing information on **linkage** and relationships among loci, and by providing sufficient numbers of informative markers to connect genotypes with phenotypes or environments (Table 4.1; Allendorf et al. 2010).

The explosion in genomic methods, both in sequencing technologies and in **bioinformatic** computing approaches, is having a profound impact on population and conservation genetics. While many people think of genomics in terms of whole genome sequencing, the primary impacts for conservation genetics are the potential for far greater numbers of marker loci, rapid and inexpensive genotyping methods, and the capacity for studying both selectively neutral and adaptive variation. Up to 2013, when the last edition of this book was released, most genomic studies were limited to model species. Since that time, we have seen an explosion in the use of genomic tools to inform conservation

Conservation and the Genomics of Populations, Third Edition. Fred W. Allendorf, *et al.*, Oxford University Press.
© Fred W. Allendorf, W. Chris Funk, Sally N. Aitken, Margaret Byrne, and Gordon Luikart (2022). DOI: 10.1093/oso/9780198856566.003.0004

Table 4.1 Primary genetic problems in conservation and how genomics can contribute to their solution.

Primary problem	Possible genomic solution
Estimating N_e, m, and s	Increasing the number of markers, reconstructing pedigrees, and using haplotype information will provide greater power to estimate and monitor N_e and m, as well as to identify migrants, estimate the direction of migration, and estimate s for individual loci within a population
Reducing the amount of admixture in hybrid populations	Genome scanning of many markers will help to identify individuals with greater amounts of admixture so that they can be removed from the breeding pool
Identification of units of conservation: species, evolutionarily significant units, and management units	Incorporating adaptive genes and gene expression will augment our understanding of conservation units based on neutral markers. Individual-based landscape genetics will more precisely identify boundaries between conservation units
Minimizing adaptation to captivity	Numerous markers throughout the genome could be monitored to detect whether populations are becoming adapted to captivity
Predicting harmful effects of inbreeding depression	Understanding the genetic basis of inbreeding depression will facilitate the prediction of the effectiveness of purging. Genotyping of individuals at loci associated with inbreeding depression will allow the selection of individuals as founders or mates in captive populations. Pedigree reconstruction will allow more powerful tests of inbreeding depression
Predicting occurrence and intensity of outbreeding depression	Understanding divergence of populations at adaptive genes will help to predict effects on fitness when these genes are combined. Detecting chromosomal rearrangements will help to predict outbreeding depression
Predicting the viability of local populations	Incorporating genotypes that affect vital rates and the genetic architecture of inbreeding depression will improve population viability models
Predicting the capacity of populations to adapt to climate change and other anthropogenic challenges	Understanding adaptive genetic variation will help to predict the response to a rapidly changing environment or to harvesting by humans and allow the selection of individuals for assisted migration
Determine genetic basis and environmental drivers of local adaptation of populations	Analyzing genotype–environment associations for genome-wide markers can identify the strength, pattern, and drivers of local adaptation, and inform choice of source populations for reintroductions, genetic rescue, and assisted gene flow
Identifying individuals that have key phenotypic traits, e.g., disease resistance, in conservation breeding programs	Finding genes or genetic markers underlying genetic resistance or other traits critical to population or species survival can enable selective breeding to increase fitness
Determining genetic architecture of adaptive traits	Mapping genes underlying key traits on a linkage map or reference genome can provide information on capacity for adaptation and genomic tradeoffs
Determine relatedness and reconstruct pedigrees	Reconstructing pedigrees and estimating relatedness from genome-wide markers can inform mating designs that reduce inbreeding and increase N_e

and management of threatened species and natural populations.

A major advance provided by genomic data for applications in conservation is the ability to detect the effects of selection using genetic rather than phenotypic data. Loci underlying key phenotypic traits such as disease tolerance can be identified. Local adaptation to environmental conditions can be detected without the necessity of conducting labor-intensive experiments, and the environmental factors driving local adaptation can be identified by interrogating the genomes of individuals from different populations (Hohenlohe et al. 2010; Hoban et al. 2016).

Population genomics also allows for novel analyses that cannot be conducted with individual loci but that require many loci mapped along chromosomes. These genome-wide approaches can detect both selection and demographic events such as population bottlenecks and expansion that have occurred over the past few thousand generations. With genomic data, levels of inbreeding of individuals can be quantified more precisely, and relatedness among individuals managed; for example, in captive breeding populations. Genomic data also provide a much richer understanding of gene flow, introgression, and hybridization by characterizing the pattern, timing, and extent of admixture across the genome, and the extent of recombination in each genomic region between ancestral species or populations. Genomic tools allow geneticists to reconstruct detailed and precise pedigrees and estimate relatedness among individuals, facilitating captive population management.

However, genomic approaches are not a panacea. While larger and more powerful than their predecessors, genomic datasets require careful processing and extensive filtering of data, knowledgeable use of the many analytical tools available, careful attention to the assumptions underlying each analytical method, and thoughtful consideration of possible alternative explanations for results.

In this chapter, we first describe the breakthroughs in DNA sequencing that paved the way for the genomics revolution. We then introduce some of the most widely used methods for generating genome-wide markers, and provide examples of where genomic approaches have informed conservation. Finally, we introduce some novel applications of DNA sequencing in conservation that are outside of the scope of traditional population genetics and genomics, including gene expression, **epigenetics**, and **metagenomics**. As technologies change quickly, we don't provide detailed technical recipes, but rather sketch out the range of possibilities available. Many of the concepts and approaches introduced here will be covered in detail in subsequent chapters. Some of the language may be new and unfamiliar—but it will become part of your vocabulary as you work your way through this book and re-encounter these terms with more context in subsequent chapters.

4.1 High throughput sequencing

4.1.1 History of DNA sequencing technology

Genomic approaches have been made possible by the development of high throughput, low cost DNA sequencing technologies, in parallel with vastly increased computational power. DNA sequencing methods were first developed independently by Walter Gilbert and Frederick Sanger. Gilbert and Sanger, along with Paul Berg, were awarded the Nobel Prize in Chemistry in 1980. Sanger and his colleagues used their own sequencing method to determine the complete nucleotide sequence of the bacteriophage φX174, the first genome ever completely sequenced. The first application of DNA sequencing to the study of genetic variation in natural populations was by Kreitman (1983), who published the DNA sequences of 11 alleles at the alcohol dehydrogenase locus from *Drosophila melanogaster*. Initial studies of DNA variation were technically complex and time consuming, so it was expensive and difficult to sequence and genotype large numbers of individuals from natural populations. However, the advent of the **polymerase chain reaction (PCR)** in the mid-1980s (Box 3.1) removed these obstacles. DNA sequencing started becoming more common as the process became less expensive and more automated around the turn of the century. The complete **mitochondrial DNA (mtDNA)** sequences (each just over 15,000 base pairs (bp)) for 53 humans from diverse origins was published by Ingman et al. in 2000, and the full human nuclear genome sequence (3.2 billion bp) was completed in 2003.

4.1.2 Next-generation sequencing (NGS)

Next-generation sequencing (NGS), also called massively parallel sequencing, has widely replaced Sanger sequencing for population genetics, and has revolutionized the study of genetic variation. NGS involves several steps: (1) fragmenting DNA, commonly with **sonication** (sound energy at ultrasonic frequencies) or restriction enzymes; (2) using PCR to clonally amplify all or targeted DNA fragments and produce **sequencing libraries**; and (3) DNA sequencing of many fragments in these libraries (each ~50–300 bp in length) simultaneously through the addition of nucleotides to complementary strands. As sequencing capacity has increased, prices have dropped rapidly. In 2001, it cost approximately US$10,000 to sequence one megabase (Mb = 1 million bp) of DNA. In 2019, it cost 1 cent US. The cost of sequencing one human genome (3.2 gigabases or billion bp) dropped over that period from US$100,000 to under US$1,000. As a result of decreased costs and increased sequencing capacity globally, large-scale sequencing has become an accessible tool for the study of natural populations.

High throughput, low cost sequencing has facilitated several major steps forward in population genomics. Whole genome sequencing and **assembly** generate **reference genome** sequences for species of interest or their close relatives that can be used to locate genes and markers within the genome. High throughput sequencing of nonmodel organisms allows for the identification of polymorphisms, usually **single nucleotide polymorphisms (SNPs)**, but also insertions, deletions, duplications, inversions, and translocations for population genomic analyses and for linking genotypes with phenotypes or environments (Wellenreuther et al. 2019). Sequencing of **complementary DNA** (cDNA) made by reverse transcribing **RNA** samples provides information about transcribed regions of the genome as well as levels of gene expression. Collectively, NGS has been transformational for understanding the genetics and evolution of populations.

While most high throughput NGS technologies only sequence up to a few hundred base pairs in length, **third-generation sequencing** can produce reads of thousands of base pairs or longer. These long-read technologies can have higher error rates than the short-read approaches, but are improving rapidly. They are particularly useful for assembling genomes (Section 4.3), for distinguishing among **paralogs**—multiple copies of the same gene—and for assembling other highly repetitive regions of genomes.

4.1.3 Single nucleotide polymorphisms (SNPs)

DNA sequence data, usually from multiple individuals or populations, allows for the identification of **SNPs**, the most abundant type of polymorphism in the genome, with one occurring about every 200–500 bp in many wild animal populations (Brumfield et al. 2003; Morin et al. 2004). For example, a G and a C might exist in different individuals at a particular nucleotide position within a population (or within a heterozygous individual). Because the mutation rate at a single base pair is low (about 10^{-8} changes per nucleotide per generation), SNPs usually consist of only two alleles; that is, are bi-allelic. **Transitions** are a replacement of a purine with a purine (G $\Leftrightarrow$ A) or a pyrimidine with a pyrimidine (C $\Leftrightarrow$ T). **Transversions** are a replacement of a purine with a pyrimidine (A or G $\Leftrightarrow$ C or T). Even though there

are twice as many possible transversions as transitions, SNPs in most species tend to be transitions. This is both because of the physical and chemical nature of the mutation process (transition mutations are more common than transversions), and because transversions in coding regions are more likely to cause an amino acid substitution than transitions and be subject to selection. While many SNPs will be located in intergenic regions, introns, or silent sites that do not change amino acid sequences and will likely have no impact on phenotypes, others in coding or regulatory regions may be associated with phenotypes and fitness.

SNPs are useful markers for describing genetic variation in natural populations because of their abundance and genome-wide distribution. An analysis of whole genome sequence data from 2,504 humans from 26 populations around the world found 84.7 million SNPs (Sudmant et al. 2015). Around three-quarters of these SNPs have a minor allele that is rare (<0.5% frequency). Each person is heterozygous at ~5 million SNPs, on average. Two randomly chosen humans will differ at several million base pairs over their entire genomes. SNPs may be even more frequent in other species because humans arose relatively recently in evolutionary terms from relatively few founders, and thus have somewhat limited genome-wide variation. While humans are not the focus of this book, the technological and analytical advances in human genomics have foreshadowed potential applications of genomics in nonmodel species and their natural populations. Millions of SNPs have already been identified for many species of conservation concern. For example, Kardos et al. (2016a) found 4.6 million SNPs in a population sample of 20 semicollared flycatchers and 2.8 million SNPs among three populations of bighorn sheep (Kardos et al. 2015b).

The marker types described in Chapter 3 are largely assumed to be **selectively neutral** (i.e., not affecting fitness), but very useful for determining genetic relationships among individuals, gene flow, population structure, and demographic history. SNPs in noncoding regions of genes (i.e., introns), or in intergenic regions are also likely to be selectively neutral. However, SNPs in coding or regulatory regions may have effects on phenotypes and affect fitness. Within coding regions, SNPs may code for different amino acids, but third

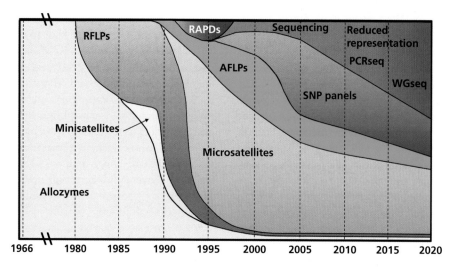

Figure 4.1 Relative popularity of molecular markers in conservation genetics over time. RFLP, restriction fragment length polymorphism; AFLP, amplified fragment length polymorphism; RAPD, randomly amplified polymorphic DNA; SNP, single nucleotide polymorphism. SNP panels include array-based or PCR-based genotyping (by hybridization or Fluidigm microfluidics). Sequencing is subdivided into three approaches: PCRseq (Sanger sequencing of single or few loci), reduced representation sequencing (mainly RADseq and also targeted DNA capture and amplicon sequencing), and whole genome resequencing (WGseq). Modified from Schlötterer (2004).

base pair substitutions in **codons** can be **synonymous** or silent. This means that SNPs are powerful markers for separating the effects of history and population demography from natural selection on a genome-wide basis (Luikart et al. 2003); however, it also means that their analysis can be complex as different evolutionary forces can be acting on different SNPs (Allendorf et al. 2010). Selectively neutral SNPs and those under selection provide different yet complementary information on population structure, demographics, and adaptation, and require different analytical approaches, as we will discuss in later chapters.

SNPs have largely replaced microsatellites as the marker of choice for many applications in conservation (Figure 4.1). One big advantage of SNPs over microsatellites is that it is much easier to standardize the scoring of SNP genotypes when more than one laboratory is studying the same species (Stokstad 2010). SNPs are especially useful for studies involving partially degraded DNA (from noninvasive and ancient DNA samples) because short DNA fragments can still be PCR amplified and sequenced. Another advantage of genomic methods is that the lab work is now easy to outsource, which reduces the need for in-house equipment, technical expertise, and troubleshooting. Both sequencing and SNP genotyping can be outsourced to

high throughput, specialized facilities once DNA is extracted from samples. As a result of all these advances, the management, analysis, and archiving of genomic data through bioinformatic methods has become a greater challenge than the sequencing itself.

4.1.4 Inferences from sequence data

DNA sequence data provides several useful types of information that traditional, unlinked genetic markers scattered across the genome cannot. The distribution of allele frequencies for polymorphisms segregating across regions sequenced, called the **site frequency spectrum (SFS)**, can be used to make inferences about the demographic history of populations and the past occurrence of bottlenecks (Section A10; Charlesworth et al. 2003). For example, a recently bottlenecked population will have a deficit of low-frequency alleles, whereas a growing population will have an excess of low-frequency alleles (e.g., see Appendix Figure A14). Local deviations from the genome-wide SFS can provide information about portions of the genome that are under selection, which can be used to identify potential candidate genes for future study of adaptation (Kreitman 2000). For example, Kardos et al. (2015b) used whole genome resequencing in wild bighorn sheep to identify regions with low variation. A gene previously shown to be associated with horn development in domestic sheep was found in a low-diversity genomic region and appears to be under positive selection for large horn size, as large horns are associated with high **fecundity** in this species.

DNA sequences provide great insight into the demographic history of populations. For example, Robinson et al. (2019) sequenced the whole genomes of wolves from different populations around the world, and assessed heterozygosity of SNPs across all chromosomes (Figure 4.2). Their results nicely illustrate the influence of demographic history and population size on genetic diversity. The wolves from Xinjiang, China, and Minnesota, USA, were from large, outbred populations and had high levels of heterozygosity across all chromosomes. Populations in Tibet and Ethiopia were small and isolated, and had much lower levels of heterozygosity spread across their genomes. Populations of wolves from Mexico, and from Isle Royale, Minnesota, have experienced high levels of recent inbreeding, resulting in a pattern of patchy heterozygosity (or the

converse, long stretches of homozygosity) across the genome. The wolves on Isle Royale have a high frequency of morphological defects and other effects caused by inbreeding. This example shows the power of genome-wide sequence variation for understanding the historical dynamics of populations.

DNA sequence data can also be used to distinguish **cryptic species** that are morphologically indistinguishable but are reproductively isolated. Sequence data can be used in combination with other types of markers and phenotypic trait variation to clarify taxonomic relationships and identify cryptic species. For example, researchers have been cataloging the biodiversity of the ~15,000 species of butterflies and moths in the Area de Conservación Guanacaste in Costa Rica for nearly four decades. Janzen et al. (2017) analyzed DNA sequence data for nearly 15,000 genes in samples of the butterfly *Udranomia kikkawai*, the Costa Rican skipper butterfly, from this area. Their results indicated that this apparent species actually comprises three cryptic species that are phenotypically indistinguishable, but differ in habitat and diet.

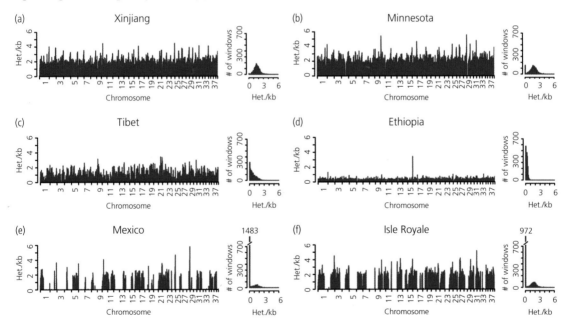

Figure 4.2 Distributions of heterozygosity across the genome of gray wolves from different populations with different demographic histories. In each panel: (left) example barplots showing per-site heterozygosity in nonoverlapping 1-Mb (1 million bp) windows across the autosomal genome; (right) histograms of number of heterozygous sites per 1,000 bp. (a and b) The Xinjiang and Minnesota wolves represent large outbred populations. (c and d) The Tibetan and Ethiopian wolves represent small isolated populations. (e and f) The Mexican and Isle Royale wolves represent populations with recent inbreeding. Het/kb is the number of heterozygous SNPs per 1,000 bp. From Robinson et al. (2019).

One of the challenges of managing captive breeding populations is to reconstruct the pedigrees of individuals for which no pedigree record exists. Genomic quantitative genetic studies now facilitate understanding the complex links between phenotype, genotype, and fitness in wild populations without pedigrees (Gagnaire & Gaggiotti 2016; Gienapp et al. 2017; see Chapter 11).

4.2 Linkage maps and recombination

In population genetics studies with tens of markers, researchers often assumed that the markers were unlinked. This was done to get a genome-wide estimate of diversity, relatedness, and population structure. However, one of the great powers of genomic data is the insights they provide into the physical location of genes, the extent of recombination, and the degree of association of alleles between loci. Stapley et al. (2017) have recently provided an extremely valuable review of variation in recombination rates across eukaryotes.

Linkage maps are **genetic maps** that show the relative locations of genes across the genome based on the recombination frequency between loci, measured in **centimorgans (cM)**. If two loci are 1 cM apart, then the average number of recombination events between them would be 0.01 or 1% (i.e., 1 crossover between the loci observed out of 100 progeny); if two loci are 50 cM apart, they are unlinked (i.e., alleles at the two loci are inherited independently). Linkage maps differ from **physical genome maps** (i.e., assembled reference genome sequences), where the distance between loci is measured in base pairs.

Linkage maps are traditionally generated from estimates of recombination in progeny or in multi-generational pedigrees. To make a linkage map for some model organisms, individuals from two **recombinant inbred lines (RILs)**, homozygous lineages produced through repeated inbreeding or self-pollination in plants, are crossed. Then two of the F_1 (first-generation) progeny are crossed again, and the F_2 (second-generation progeny) are sequenced or genotyped at a large number of loci. The recombination rates in the F_2 generation can be calculated for pairs of loci for which the RILs are homozygous for different alleles. All loci that have

different alleles in the two RILs can be used to make a single genetic map. Recombination rates provide estimates of the genetic distance between pairs of loci, and information on the relative positions of all pairs of loci or chromosomal intervals can be used to determine the order of markers along chromosomes. Mapping can also be done with individuals from full-sib families for all markers where one parent is homozygous and the other is heterozygous—the analysis is just more complex. Box 4.1 illustrates the principle of linkage mapping and how recombination rates are used to estimate genetic distance.

Information on the location of loci in the genome is useful for many reasons, and is one of the factors that differentiates population genetics, using a relatively small number of preferably unlinked markers, from population genomics, where knowledge of the relative positions of markers and genes in the genome is capitalized on. While a fully assembled reference genome provides a complete physical map, for many species with large genomes and highly repetitive content that prevents assembly into chromosomes, a linkage map provides information that a genome assembly cannot. Third-generation sequencing platforms that generate much longer sequences than NGS platforms are, however, facilitating assembly of even very large genomes. For example, the genome of giant sequoia has been assembled into 11 chromosomes by combining long-read sequencing platforms, which revealed chromosomes of up to one gigabase (1 billion bp) long (Scott et al. 2020).

Both physical and genetic linkage maps provide useful information for population genomics research in several ways (Luikart et al. 2018). Knowing the relative position of many loci improves the power to identify those genes associated with phenotypic traits of interest and fitness. Independent loci that are required for some analyses, for example, effective population size, can be identified. A comparison of genetic and physical maps allows for an understanding of variation in recombination rate across the genome. Physical or genetic maps allow for inferences of whether gametic disequilibrium is due to physical linkage or other causes of associations among loci such as demographics or selection (Chapter 10). Genetic parameters (e.g., heterozygosity) can be plotted for loci in order along chromosomes in a **Manhattan plot**, so

Box 4.1 Linkage mapping

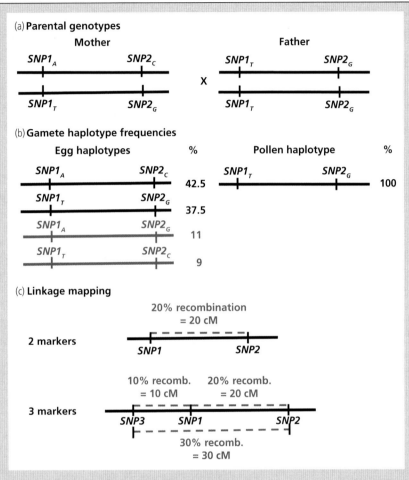

(a) **Parental genotypes**

Mother

$SNP1_A$ $SNP2_C$

$SNP1_T$ $SNP2_G$

X

Father

$SNP1_T$ $SNP2_G$

$SNP1_T$ $SNP2_G$

(b) **Gamete haplotype frequencies**

Egg haplotypes   %   Pollen haplotype   %

$SNP1_A$ $SNP2_C$ 42.5 $SNP1_T$ $SNP2_G$ 100

$SNP1_T$ $SNP2_G$ 37.5

$SNP1_A$ $SNP2_G$ 11

$SNP1_T$ $SNP2_C$ 9

(c) **Linkage mapping**

20% recombination
= 20 cM

2 markers $SNP1$ $SNP2$

10% recomb.   20% recomb.
= 10 cM   = 20 cM

3 markers $SNP3$ $SNP1$ $SNP2$

30% recomb.
= 30 cM

Figure 4.3 Estimation of recombination rates. Each horizontal line indicates a portion of a chromosome. (a) The mother oak is heterozygous at two SNP loci; the father is homozygous at both loci. (b) Since the father is homozygous, there is only one possible pollen haplotype. The more common egg haplotypes (in black) are those with no recombination, while the less common haplotypes (in blue) have been produced through recombination between $SNP1$ and $SNP2$. We assume there are no double crossovers between the two markers. (c) Since 20% of the egg haplotypes (and progeny genotypes) are produced through crossing over, the genetic distance between $SNP1$ and $SNP2$ is 20 cM ($r = 0.20$). $SNP3$ can then be added to the map and the order of the three SNPs determined based on the genetic distance from $SNP3$ to $SNP1$ and $SNP2$.

The genetic distance between any two loci that are heterozygous in an individual can be estimated from the rate of recombination (r) in their offspring (Figure 10.1). Imagine two SNPs for which an oak tree is heterozygous: $SNP1_{AT}/SNP2_{CG}$ (Figure 4.3) Female flowers from this tree are control-pollinated using pollen from another tree with the double homozygous genotype $SNP1_{TT}/SNP2_{GG}$. The haplotype of all of the pollen will be $SNP1_T/SNP2_G$, and the haplotype of the egg can be deduced by subtracting the pollen genotype from the diploid genotype. The genetic

Box 4.1 *Continued*

distance between *SNP1* and *SNP2* can be calculated from the proportion of offspring that have recombinant genotypes inherited from the mother tree. If *SNP1* and *SNP2* are unlinked—either on different chromosomes, or far apart on the same large chromosome, you would expect equal frequency of the four possible two-locus haplotypes (haploid genotypes). In this example, there are two haplotypes that are more common and two haplotypes that are less common (Table 4.2).

As the majority of haplotypes are either $SNP1_A/SNP2_C$ or $SNP1_T/SNP2_G$, these are assumed to be the parental types of the mother tree. That is, the mother tree has one chromosome with the haplotype $SNP1_A/SNP2_C$, inherited from one parent, and the second chromosome has the haplotype $SNP1_T/SNP2_G$, inherited from the other parent.

To calculate the distance between *SNP1* and *SNP2*, we can calculate the proportion of offspring that have recombinant genotypes (*r*). This is (22 + 18)/200 = 20%. This means that one in five meiotic divisions will have a recombination event between *SNP1* and *SNP2*. One cM is equal to a genetic distance with 1% recombination per generation, so the genetic distance between *SNP1* and *SNP2* is 20 cM (*r* = 0.20).

These samples have also been genotyped for a third SNP, creatively named *SNP3*. Ten percent of the samples are recombinants between *SNP1* and *SNP3*, and 30% are recombinants between *SNP2* and *SNP3*. Based on the frequency of recombination, the order of these SNPs on a chromosome must be *SNP3*—*SNP1*—*SNP2*. In this manner, by estimating the genetic distance between pairs of polymorphic SNPs across the genome, and inferring their relative positions, linkage maps can be developed.

Table 4.2 Hypothetical example of linkage mapping (see Figure 4.3).

Genotype of offspring	Haplotype inherited from mother	Type	Expected frequency if unlinked (N = 200)	Observed frequency (N = 200)
$SNP1_{AT}/SNP2_{CG}$	$SNP1_A/SNP2_C$	Parental	50	85
$SNP1_{TT}/SNP2_{GG}$	$SNP1_T/SNP2_G$	Parental	50	75
$SNP1_{AT}/SNP2_{GG}$	$SNP1_A/SNP2_G$	Recombinant	50	22
$SNP1_{TT}/SNP2_{CG}$	$SNP1_T/SNP2_C$	Recombinant	50	18

called because it looks something like a city skyline, with high and low values of the parameter along each chromosome that provide a visualization of how that parameter varies across the genome (e.g., Figure 4.2). Finally, information from linkage or physical maps can help parameterize theoretical models that assess interactions among evolutionary forces (e.g., how does recombination interact with gene flow and selection in adaptation).

Recombination rates can vary greatly between the sexes and across the genome, often with less recombination near centromeres and telomeres. In some regions, recombination hotspots can occur where crossing over occurs more frequently than average (Example 4.1). While the sequence of genes on a chromosome will be the same for a linkage map or a physical map, the relative distances between genes (genetic distance versus physical distance) will vary due to variation in recombination. In the absence of a reference genome sequence—a fully sequenced genome of one individual of a species, ideally assembled into chromosomes—linkage maps provide useful information on the **genomic architecture** of traits, that is, the number and location of associated genes, by indicating whether those genes are scattered across the genome or clustered in genomic islands on chromosomes.

Example 4.1 Linkage mapping in great tit songbirds

van Oers et al. (2014) used a 10,000 SNP array to genotype great tits in a captive population from the Netherlands and a wild population in the United Kingdom. In total, 6,554 SNPs were mapped to 32 linkage groups. The zebra finch is the model passerine bird species for genomics, and the great tit linkage map was compared with the physical map provided by the zebra finch reference genome (Figure 4.4). A comparison of physical distance (Mb) in the zebra finch with linkage map positions in the great tit reveals considerable conserved gene order along chromosomes, but also some chromosomal inversions, and regions of high and low recombination. For example, linkage group 19 shows high concordance of marker sequence and a consistent, positive linear relationship between physical distance in the zebra finch (x-axis) and map distance in the great tit (y-axis). Chromosome 10 shows a small inversion in the great tit genome in the first 5,000 Mb compared with the zebra finch. Chromosome 6 shows high map distance compared with physical distance at the beginning and end of the chromosome, indicating high recombination rates, and a stretch of low recombination in the middle where the map distance is much shorter than the physical distance.

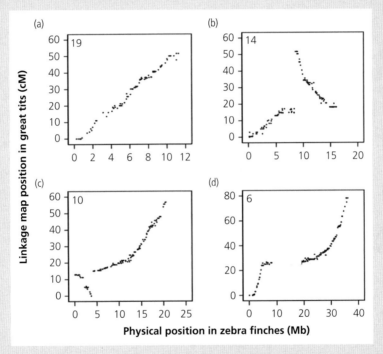

Figure 4.4 Comparative physical (x-axis) and genetic (y-axis) map distances for four great tit chromosomes. Genetic distances are based on recombination rates, while physical positions are based on corresponding chromosomes of the zebra finch reference genome. Regions with a positive slope have the same gene order in the two species, whereas regions with negative slopes indicate chromosomal inversions. Areas with steeper slopes have high recombination rates, while those with flatter slopes have low rates. Figure modified from van Oers et al. (2014).

The authors were also able to compare differences in recombination rates between males and females in the two populations. They found substantial differences in recombination rate between males and females for some chromosomes, consistent with other species with sex-dependent recombination rates, which may reflect differences in the strength of selection between sexes. These differences between sexes were largely conserved between the two populations (i.e., chromosomes that had lower recombination rates in females in the Netherlands population also had lower recombination rates in females in the United Kingdom).

4.3 Whole genome sequencing and reference genomes

A reference whole genome sequence is not a necessity for many population genomic analyses. Inferences can be made about demographic history, inbreeding, relatedness, gene flow, and selection with genomic data without a reference. However, an assembled and **annotated** reference genome for the focal species or for a close relative "provides the ultimate resource for genomic approaches" (Ekblom & Wolf 2014, p. 1027). An assembled reference genome consists of most or all of the nuclear DNA sequences from one individual of a species, assembled into contiguous segments. An annotated reference genome lists predicted protein-coding genes and infers gene function, often based on both experimental evidence and gene sequences from model species. Advances in NGS combined with bioinformatic pipelines for assembling reads from sequenced DNA fragments have made the sequencing and assembly of reference genomes quite straightforward, and there are companies that can provide this service.

The first human genome sequence, completed in 2003, took 13 years to sequence and assemble; less than two decades later, around a million people have been sequenced! As of 2020, genomes had been sequenced and assembled for hundreds of birds, mammals, and plants. The Vertebrate Genome Project has the goal of sequencing the genomes of all vertebrate species globally, and as of May 2021 had 146 genomes assembled with a goal of completing 500 by the end of 2021. Some exceptionally large reference genomes have been assembled, including for the sugar pine (31 Gb; Stevens et al. 2016) and the axolotl (32 Gb; Nowoshilow et al. 2018). The goal of the "Earth BioGenome Project" is sequencing all of Earth's eukaryotic species before 2030 (Lewin et al. 2018). These ambitious projects suggest most species will have a reference genome in the not-too-distant future, providing a tremendous resource for population geneticists who know how to use them. Reference genome sequences and other **genomic resources** are available through the National Institutes of Health National Center for Biotechnology Information website.

Whole genome sequencing involves several steps, detailed by Ekblom & Wolf (2014) and summarized briefly here (Figure 4.5). First, high-quality tissue is sampled from a reference individual, and genomic DNA is extracted and fragmented. DNA fragments within an acceptable range of lengths are size-selected, then amplified using PCR. A sequencing library, comprising all of the amplified products, is then shotgun sequenced. Paired-end sequencing generates reads from both ends of short fragments (a few hundred bp long), while mate-pair reads are sequences from longer fragments (a few thousand bp long). Generating the resulting sequence data is the easy part compared with the bioinformatic challenges that follow!

Individual sequences are tiny puzzle pieces in a massive jigsaw puzzle, but instead of one piece per location in a puzzle, there could be five or 10 or 50, all overlapping slightly (Figure 4.6). Fortunately, there are well developed bioinformatic pipelines to align short reads from the same genomic regions into **contigs**—longer contiguous sequences. Then contigs are joined into longer genomic regions called **scaffolds** using reads that span between contigs. While most next-generation shotgun sequencing produces reads for 250 bp or fewer, the development of third-generation sequencing technologies that produce reads of tens of thousands of bp have greatly facilitated genome assembly by providing links across scaffolds and larger genomic regions (van Dijk et al. 2018). Highly repetitive genomic regions and **paralogous** sequences are difficult to assemble without long reads. Finally, scaffolds are joined into chromosomes, if possible. For some species with large genomes and extensive highly repetitive regions, it has not yet been possible to assemble reference genomes to the chromosomal level, but for many genomic analyses in conservation genetics, this is not necessary.

Once a genome is assembled, it needs to be annotated with biologically relevant information. First, highly repetitive regions such as **retrotransposons** need to be identified and excluded, as these regions have multiple copies of almost identical sequences in different places in the genome and so are difficult or impossible to assemble correctly. Next, coding sequences are predicted. A reference transcriptome

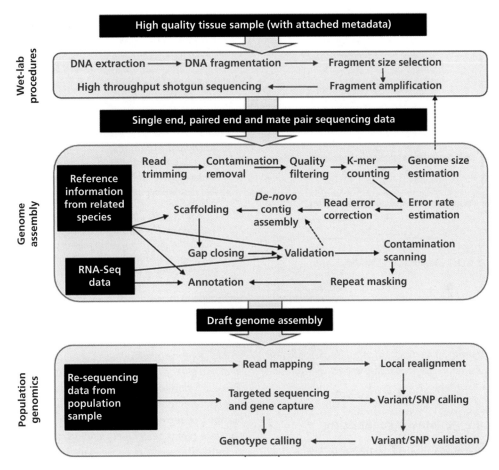

Figure 4.5 Workflow of a typical *de novo* whole genome sequencing project generating data for analysis of populations. Black boxes with white text indicate genomic resources becoming available during the course of the project. From the top: wet-lab procedures, *de novo* assembly bioinformatic pipeline, and post-assembly sequencing and analyses of samples from different populations (population genomics). Modified from Ekblom & Wolf (2014).

(Section 4.8) can be useful for identifying coding regions, intron–exon boundaries, and untranslated regions. In order to annotate the genome to predict gene function, the DNA sequences of putative protein-coding regions are then matched with similar regions from other species using available databases and tools to help identify genes conserved among divergent taxa, for example, from lab mice to moose, or from corn to conifers, comparing model species with known gene function to wild species. Many of the useful comparisons come from experiments with model species that have identified gene functions. Genes of interest can also be classified by putative function using **Gene Ontology**

(GO) terms (Ashburner et al. 2000; The Gene Ontology Consortium 2019).

The availability of genome sequences for model species and associated genetic resources has also greatly facilitated the development of genomic tools for related wild species of conservation concern; for example, use of the dog genome for population studies of wild canids such as wolves and foxes (Gray et al 2009; Box 4.2). While it is very useful to have a reference genome available for a species of interest or a close relative, some genomic analyses can proceed without a reference with **de novo sequence assembly**, where sequence reads are aligned and assembled with each other without

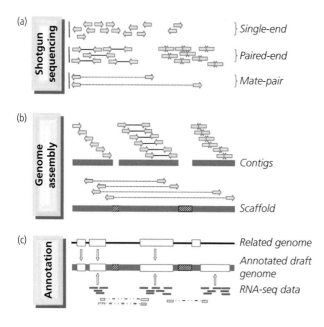

Figure 4.6 Simplified illustration of the genome assembly process and terminology. (a) Shotgun sequencing: short fragments of DNA from the target organism are sequenced at random positions across the genome to a given depth of coverage. Fragments can consist of single reads (typically 50–1,000 bp) or of paired-end reads of varying insert size. Mate-pair **libraries** span larger genomic regions (~2–20 kb inserts) with reads generally facing outwards and can be complemented with fosmid-end libraries (~40 kb inserts). (b) Genome assembly: short-read *de novo* assemblers extend the sequence information from the reads into continuous stretches called contigs. Contigs usually reflect the consensus sequence and do not contain any polymorphisms. Paired-end reads provide additional information on whether a read is supported for a given contig. Read pairs from mate-pair or fosmid-end libraries can be used to order and orient contigs into scaffolds. Gap size between contigs is estimated from the expected length of mate-pairs and marked with Ns (indicated by hatched gray boxes). Long reads from single molecule sequencing provide an alternative. (c) Annotation: gene models can be inferred *in silico* by prediction algorithms, from genomes of related organisms, and by using transcriptome data from the target organism itself. From Ekblom & Wolf (2014).

being informed by a reference genome. If analyses are focused on genic regions, reads can also be aligned to a reference transcriptome if no reference genome is available.

4.4 Whole genome resequencing

With the rapid decline in sequencing costs, it is now financially feasible in some cases, especially for species with small genomes, to resequence the entire genome in multiple individuals from more than one population. **Whole genome resequencing** is the sequencing of multiple individuals at lower coverage (fewer reads, on average, per individual or sequence) than the deep sequencing used to assemble reference genomes. Reads from whole genome shotgun sequences are aligned to a reference genome, and genotypes can then be called for each individual at each SNP. Whole genome resequencing can identify millions of SNPs across the genomes of some species. This provides an enormous amount of information about polymorphisms across the entire genome, evidence of selection by genomic region, population demographic history, regional estimates of recombination, areas of the genome with greater population differentiation than others suggesting a role in local adaptation, and other analyses.

For many applications in **conservation genomics**, whole genome resequencing generates more data than necessary to address the question at hand. There is a tradeoff between the amount of sequence or genotypic data obtained per individual and the number of individuals that can be analyzed, given finite resources. Resequencing a small number of individuals can provide data for developing a genotyping approach that is less costly per sample. For example, Ekblom et al. (2018) first sequenced and assembled a reference genome for one Scandinavian wolverine, and then used whole genome resequencing of 10 additional wolverines from this International Union for the Conservation of Nature (IUCN) Red Listed population. Demographic analysis of these data suggested the effective population size has declined from around 10,000 before the last glaciation to fewer than 500. Using data from the whole genome resequencing, Ekblom et al. (2018) then designed an array with just 96 highly informative SNPs for further population genetic studies.

4.5 Reduced representation sequencing

There are various approaches to partial genome resequencing, called **reduced representation sequencing**. The simplest of these methods

---

Box 4.2 Use of genomic tools from model species for wild populations

Subsets of SNPs in one species can often be useful in closely related species. As most SNPs are bi-allelic, individual SNPs are less informative than individual microsatellite loci with several to many alleles; however, the sheer number of SNPs available, and the development of SNP arrays (also called SNP chips) makes genotyping many loci simultaneously relatively easy. vonHoldt et al. (2011) used a ~48,000 SNP array developed for the domestic dog to assess among- and within-species genetic diversity in several species of wolves and coyotes. The proportion of these SNPs that were polymorphic ranged from 7% in the Ethiopian wolf to 97% in the gray wolf. However, **ascertainment bias** will affect species comparisons as SNPs will have been selected for polymorphism based on the original target species.

The power in numbers of SNPs is well illustrated in a case applying cow SNPs to the endangered European bison (Figure 4.7; Tokarska et al. 2009). This bison went through a severe bottleneck in the last century, was extirpated in the wild, and genetic variation in animals maintained in captivity was low. The lowland population of bison, with a current population size of ~1,800, was founded by only four bulls and three cows, with two animals estimated to be responsible for 80% of the current gene pool. The bovine SNP array containing 52,968 SNPs was used to identify 960 polymorphic SNPs in the European bison. These SNPs were then tested against 17 microsatellite loci for determining paternity

of 92 offspring in a reintroduced Polish bison population. By using 50 of the most polymorphic SNPs, or 90 SNPs chosen at random, fathers could be identified with 95% confidence in at least 50% of the cases. In contrast, the microsatellite loci could only identify fathers with high confidence for two of the offspring.

Figure 4.7 An endangered European bison near Springe, Hanover, Germany. A SNP array designed for use in planning breeding of domestic cattle was used by Tokarska et al. (2009) to determine the parentage of offspring in a reintroduced European bison population in Poland. Photo courtesy of Michael Gäbler.

---

involves sequencing anonymous sites across the genome, while more involved methods target specific regions or involve resequencing exomes, candidate genes, or other specific genomic regions.

4.5.1 Restriction site-associated DNA sequencing (RADseq) methods

An inexpensive and rapid way to discover and genotype thousands of SNPs is through sequencing DNA adjacent to restriction enzyme cut sites, generally called **restriction site-associated DNA sequencing (RADseq)** (see Guest Box 4). DNA is fragmented with restriction enzymes, fragments adjacent to cut sites are amplified using PCR,

and then sequenced to identify SNPs (Figure 4.8). **RAD tags** are anonymous markers, meaning their location within the genome is unknown unless they can be aligned with a reference genome, but unlike amplified fragment length polymorphisms (**AFLPs**), they are co-dominant. There are several different methods of RADseq, including **genotyping-by-sequencing** (GBS) and **double digest RAD (ddRAD)**, that vary in the number of restriction enzymes used (one or two), whether fragments are size-selected, and what **adaptors** are ligated to DNA fragments (Figure 4.8 and Andrews et al. 2016 for details). Adding unique DNA **barcodes** to adaptors allows DNA from different individuals to be **multiplexed** for some stages of library preparation and for sequencing. Restriction

Sequence next to a single restriction enzyme cut site

Sequence flanked by two restriction enzyme cut sites

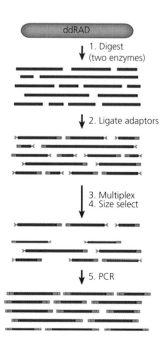

Figure 4.8 Examples of two members of the RADseq family of reduced representation library sequencing protocols. These all begin by digesting high-molecular-weight genomic DNA with one or more restriction enzymes. They vary in whether sequenced regions are flanked by one or two restriction sites, by when and how often adaptor oligonucleotides are added, and whether size selection of DNA fragments occurs. Modified from Andrews et al. (2016).

enzymes are chosen based on genome size, number of loci desired, and intended amount of sequencing. Enzymes with shorter recognition sequences will have more cut sites and produce more fragments than those with longer recognition sequences, but will also require more total sequencing to achieve the same average **read depth** (number of sequenced reads per nucleotide). In some cases, **methylation-sensitive** restriction enzymes are used to enrich sequence data for genic regions. After sequencing, reads are aligned to a reference genome, or assembled *de novo* with each other if no genome sequence is available, after which anonymous SNP loci can be identified and individuals genotyped.

If a reference genome is available, reads are mapped to the reference, which provides SNPs of

known chromosomal locations and generally lower genotyping error rates (Figure 4.9; Hoban et al. 2016; Shafer et al. 2017). This allows use of more informative population genomic analyses including localization of selective sweeps and identification of functional genes, if **annotations** exist for the genome. Mapped loci allow detection of extent of inbreeding, extent and distribution of **hybridity** for dating introgression events, adaptive introgression, and historical gene flow. They also facilitate estimation of the number and density of markers required to interrogate all genomic regions.

For example, a RADseq approach was used to genotype 3,583 birds for over 15,000 SNPs in a Florida scrub jay population that was monitored and sampled over nearly two decades (Chen et al.

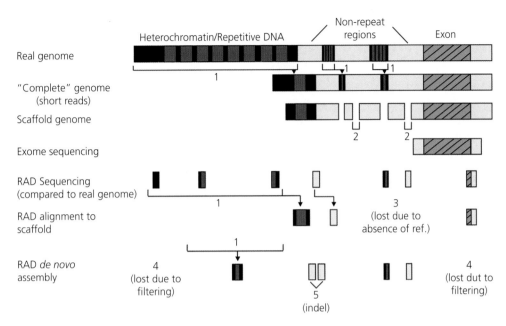

Figure 4.9 Genomic information is lost to varying degrees with different approaches to sequencing. Errors can also accumulate during sequencing, assembly, and alignment. Reference genome assembly: a "complete" reference genome typically excludes highly repetitive regions, due to the difficulty of assembling these regions. A scaffold genome uses short-paired reads to connect shorter contigs into longer scaffolds by indicating that an unknown chunk of sequence exists between the two known linked pieces; however, significant portions of nonrepetitive sequence will remain unassembled in a scaffold genome. Exome sequencing and RADseq: the raw data from exome sequencing will include exons and flanking regions, while RADseq will produce short fragments from throughout the genome. Exome sequencing will miss potentially functional noncoding polymorphisms that are not close to an exon. RADseq can capture some noncoding polymorphisms, but the data are sparse, with variation among samples in which loci are sequenced. In the alignment of RADseq to a scaffold, some sequences may be lost in alignment due to missing homologous sequences in the reference genome. *De novo* assemblies of RADseq data produce only very short, unordered contigs compared with alignment to a reference genome. Modified from Hoban et al. (2016).

2016). The analysis of these genotypes showed that a decline in gene flow into this relatively large and demographically stable population over time resulted in increased inbreeding and decreased fitness. Example 4.2 shows the power of RADseq over microsatellite markers for detecting differentiated populations that potentially warrant separate management as distinct units in cisco fish in Lake Superior (Ackiss et al. 2020).

The advantages of RADseq approaches are that they are relatively low cost and do not require prior genomic knowledge or a reference genome. One major disadvantage is that they often result in large amounts of missing data; that is, different individuals only share a subset of their sequenced DNA regions. During PCR, one allele can be amplified more than the other by chance in a process called **PCR duplication**, producing a bias in genotyping

and allele frequency estimates. **Allelic dropout** can make heterozygous individuals appear homozygous when a lack of a restriction enzyme cut site in one allele results in a fragment too long to pass size selection (Box 5.2). All of these disadvantages have led to a lively debate about the value of RADseq, with some scientists questioning the value of RADseq methods for population genomics, especially for detecting loci involved in local adaptation due to relatively sparse sampling of the genome compared with whole genome resequencing (Lowry et al. 2017). Nonetheless, Catchen et al. (2017) pointed out that these methods have inexpensively generated genotypes for thousands of SNPs for estimating genomic diversity, relatedness, effective population size, population structure, **conservation units**, and levels of introgression.

Example 4.2 RADseq data provide resolution of taxa and population structure that microsatellites could not for cisco fish in Lake Superior

In order to manage and conserve species, distinct populations of individuals need to be identified. When species have recently radiated in new environments, they can be difficult to clearly identify. This is the case of the genus *Coregonus*, which includes ciscoes and whitefish (Ackiss et al. 2020). As a result of postglacial adaptive radiation into lakes, they have evolved distinct phenotypes that occur in sympatry with incomplete reproductive isolation. Within the *Coregonus artedi* complex, at least eight morphologically distinct forms occur in the Laurentian Great Lakes. Stable isotope analysis indicates that these forms have different diets, and feed at different water depths and trophic levels. However, restriction fragment length polymorphism (**RFLP**), mtDNA, and microsatellite data have revealed little or no genetic differentiation among these forms. In the past century, the abundance and diversity of cisco in the Great Lakes has declined due to overfishing, habitat loss, and introduced invasive fish species. There is interest in reintroducing these fishes where they have been lost, but taxa need to be identified to guide this process.

Ackiss et al. (2020) used RADseq to resolve the taxonomic units within *Coregonus artedi* in Lake Superior. They sampled the three most common forms of cisco—*C. artedi*, *C. hoyi*, and *C. kiyi*, and also included two rare forms—*C. zenithicus* and a form resembling *C. nigripinnis*. Briefly, DNA was extracted from fin samples, and digested with the restriction enzyme *SbfI*, then ligated with barcoded adaptors to identify sequences from each sample. The samples were then fragmented to 300–500 bp using sonication, PCR amplified, and sequenced. Sequence reads were assembled *de novo* (without a reference genome) and filtered for quality including missing data (removing loci genotyped in <70% of individuals, and individuals genotyped for <50% of loci) and read depth. Approximately 29,000 SNPs remained after filtering.

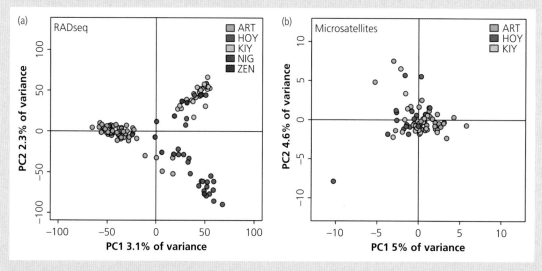

Figure 4.10 A comparison of population structure of samples of cisco fish from the *Coregonus artedi* complex in Lake Superior based on (a) RADseq data for 29,068 SNPs (five putative taxa) and (b) 12 microsatellite markers (three of the five putative taxa). The first two principal components (PCs) are plotted for each marker type. The percentage of variance explained by each PC is labeled on the *x*- and *y*-axes. From Ackiss et al. (2020).

While microsatellite markers failed to detect clear population structure in the *C. artedi* complex, these SNPs identified three clear genetic clusters in Lake Superior (Figure 4.10). Genomic differentiation among these types was widespread across the genome, suggesting genetic drift played a major role, although there were some genomic islands of divergence—clusters of

Continued

Example 4.2 *Continued*

genes that are differentiated among populations—that may have resulted from selection. Population assignment rates of individuals with the SNP data were 93–100%, compared with 62–77% with microsatellites. The genomic markers allowed for the identification of hybrid individuals and provided some evidence of their reduced fitness. The authors concluded that the three major forms within this complex should be managed as separate taxa.

RADseq can be improved by adding a targeted DNA capture step called **Rapture** to reduce the problems of missing data and to reduce sequencing costs by only sequencing informative loci. Rapture allows flexibility of targeting 500 to more than 100,000 loci based on genome position, heterozygosity, adaptive (outlier) gene behavior, and species-diagnostic or population-diagnostic loci for monitoring hybridization and introgression (e.g., Dorant et al. 2019; Kelson et al. 2020; see Guest Box 4).

4.5.2 Targeted sequence capture

Targeted sequence capture is a method for sequencing preselected rather than anonymous genomic regions, concentrating resources on loci of interest and generating high-quality sequence data for multiple individuals and populations with little missing data (Jones & Good 2016). Often the targeted region is the exome (**exome capture**), sequencing exons as well as their flanking regions (Figure 4.9). Targeted capture is a cheaper but efficient alternative to whole genome sequencing, and can be scaled to sequence anywhere from 10s to 10,000s of genes (Hodges et al. 2007). A capture array designed for exons in one species can often be transferred to a related species because of the conservation of coding sequences. For example, Cosart et al. (2011) were able to use capture probes designed for cattle to capture exons and sequence SNPs in American bison.

For targeted capture, the first step is to design and manufacture many DNA probes based on transcriptome or reference genome sequences, with each probe including a short portion of a target sequence (typically 60–120 bp long). These probes are often tiled so that they overlap adjacent probes sequentially along each targeted region. Genomic DNA from samples is sheared into fragments, and

the probes are then hybridized with those fragments. The probes include a **biotinylated** tag. DNA fragments that are hybridized with probes are separated magnetically from nonhybridized fragments using streptavidin magnetic beads that have a high affinity to biotin. These fragments are then used to make libraries and sequenced as per other methods.

RNAseq-based *de novo* transcriptomes can be used to determine sequences for probes for exome capture. Probe success is improved if a reference genome is used to identify intron/exon boundaries, but these boundaries are generally highly conserved so references from related species usually suffice. A shortcut to this process was developed by Puritz & Lotterhos (2018), who created cDNA probes directly from expressed mRNA that captured genomic DNA from exonic regions.

4.5.3 Sequencing of population pools (pool-seq)

Whole genome resequencing and targeted sequence capture of whole exomes remain costly per individual, especially for species with large genome sizes. Pooling DNA from multiple individuals per population prior to sequencing, referred to as **pool-seq** (Schlötterer et al. 2014), can be an effective means of identifying SNPs, estimating allele frequencies, and quantifying genetic diversity of populations. However, it precludes analyses that require individual genotypes such as estimating inbreeding and relatedness of individuals.

For pool-seq, DNA is extracted from individuals, then normalized to a standard concentration before combining equal amounts into a pool for sequencing library preparation. This can be used for whole genome resequencing or for the reduced representation approaches described above. Pool-seq can also be used to pool individuals from a population that are phenotypically similar, for example, pooling healthy individuals together and diseased individuals together, to identify genes associated with phenotypic traits (e.g., Micheletti & Narum 2018) in a method referred to as case-control pooling in medicine. Pool-seq can be used to detect millions of SNPs genome-wide in nonmodel species if a reference genome exists, and bioinformatic pipelines are available to facilitate discovery of loci associated with phenotypes and variation important for conservation (Micheletti & Narum 2018).

4.6 Filtering sequence data

While it is now straightforward to generate vast amounts of sequence data using whole genome resequencing or reduced representation approaches, the resulting sequence data need to be filtered (quality controlled) stringently prior to calling SNP genotypes, especially for RADseq-type methods with species lacking reference genomes. This often results in filtering out a large proportion of sequence data, but a smaller number of high-confidence SNPs will produce more reliable results than a larger set of lower-confidence markers. Filtering can bioinformatically address a number of issues with genomic data. Here we identify a few of the common issues, but see O'Leary et al. (2018) and other best-practice protocols for details on how to filter data.

If read depth, the number of unique sequencing reads that cover a given nucleotide in the reconstructed sequence, for an individual is too low for a given locus, a heterozygous genotype may appear to be a homozygote as the alternate allele may not be sequenced. Sequencing to greater read depth will mitigate this issue (Hendricks et al. 2018). Filtering out loci with inadequate read depth and excess homozygosity will address this problem bioinformatically. If a SNP is observed in only one readout of many (e.g., >10), it is likely a sequencing error.

If reads from paralogous loci or other similar sequences are clustered together erroneously, heterozygosity will be falsely inflated for that apparent locus. For example, if there are two paralogs of a gene that are each fixed for a different nucleotide at a given position within the gene, and these sequences are erroneously combined into a single apparent gene, all individuals will appear to be heterozygous at that locus. Genomes with extensive highly repetitive sequences (e.g., retrotransposons) or large gene families will be particularly prone

to this issue. Filters for excess heterozygosity can identify this problem and eliminate the misclustered reads from analyses. For example, if heterozygosity is higher than expected for a bi-allelic SNP (>0.5), SNP density is excessive along the sequence contig or gene region, and read depth is double the average depth among loci, it is likely that two paralogous loci are being clustered together. Filtering can detect this as a questionable or erroneous locus (e.g., see Box 5.2 and Graffelman et al. 2017).

PCR errors during amplification can generate apparent but false polymorphisms, and PCR bias can amplify one allele more than another. These issues can be minimized with good laboratory procedures, and filters for mapping quality ratios, allele balance, and strand bias can mitigate these errors.

Individuals are pooled into sequencing libraries, and if there are differences among libraries in the DNA fragment size selection step, sequencing coverage, or lanes on a sequencer, it may produce artifacts within results. It is important to randomize biological replicates, for example, individuals from populations or families, among libraries, and can be crucial to re-run a subset of individuals (technical replicates) to detect such problems and quantify potential error rates (e.g., allelic dropouts). Replicate samples can help identify loci that vary among libraries, for example, through principal component analysis, and these are best filtered from datasets.

To get genome-wide estimates of many population genetic parameters, it is best to analyze SNPs that are not located on the same contig or are not closely physically linked. Usually one SNP is retained per contig, but an alternative is to infer multi-SNP haplotypes and analyze these as they will be multi-allelic rather than bi-allelic. This can increase heterozygosity and statistical power for applications such as parentage analysis, and provide genealogical information for **coalescent** approaches to infer historical demography (Sunnucks 2000; see Appendix Section A10). When multiple SNPs exist within a short-read locus (RADtag), **microhaplotypes** can be directly called without **statistical inference** or phasing (Section 10.7.1). For example, if individual sequence reads from a gene reveal three bi-allelic SNPs (with possible alleles A/T, C/G, and C/G), these can be combined into six possible haplotypes (i.e., ACC, ACG, AGC, TCC, TCG, and TGC).

4.7 Other SNP genotyping methods

Species-specific SNP genotyping arrays can be designed for different purposes by selecting SNPs from DNA sequencing data. Markers may be selected for inclusion that are distributed throughout the genome, or selected because they are in or near **candidate genes** of interest based on gene expression patterns, associations with phenotypes, or associations with environments in previous analyses. Arrays contain allele-specific oligonucleotide probes attached to a solid surface. When DNA hybridizes with a probe, an allele-specific dye fluoresces, allowing determination of the alleles present at the focal locus for that individual. SNP arrays can genotype anywhere from tens to hundreds of thousands of SNPs from known genes or contigs, and produce genotypes more easily than raw sequence reads. Unlike sequencing costs, SNP arrays have not dropped much in price in the past decade, but the maximum number of SNPs that can be on an array has increased. Despite their cost, due to their high genotyping quality, lack of missing data, and ease of data analysis, SNP arrays are a useful technology. Some SNP array technologies can be used on historical samples with degraded DNA like fish scales or mammal fecal samples, whereas many NGS technologies cannot (Johnston et al. 2013).

SNP arrays are in widespread use in some breeding and conservation programs. For example, 200,000 SNP arrays have been used in salmon breeding to discover loci associated with growth and disease resistance (Madsen et al. 2020). SNP arrays designed for one species can sometimes be useful for closely related species. For example, vonHoldt et al. (2010) used a SNP genotyping microarray developed for the domestic dog to assay variation at ~48,000 loci in the Great Lakes wolf and red wolf. Results from an analysis across all 38 canid autosomes suggested that these two canids both are admixed varieties, derived from gray wolves and coyotes, respectively. This interspecific admixture could complicate decisions regarding endangered species restoration and protection (Chapter 13).

Amplicon sequencing is a targeted approach for analyzing genetic variation in anywhere from a few to a few thousand selected genomic regions. Oligonucleotide probes are designed to amplify genomic regions of interest. Library preparation is rapid. This is now a common approach for DNA barcoding and metagenomic studies, and for diagnostic tools for detecting disease-causing pathogens and invasive species (Campbell et al. 2015; Johnson et al. 2019; Wilcox et al. 2020).

4.8 RNA sequencing and transcriptome assembly

With the development of NGS technologies, **RNA sequencing (RNAseq)** has become a common practice for sequencing expressed genes, assembling **transcriptomes**, identifying SNPs, and assessing gene expression (Section 4.9). First, mRNA is extracted from one or more tissues, individuals, or experimental treatments. RNA isolation is more challenging than DNA extraction as when cells are lysed, RNases in cells in the sample and the environment immediately start to degrade RNA. cDNA libraries are produced from RNA through reverse transcription. The cDNA is then sequenced, and reads are aligned into exons. A transcriptome includes all of the assembled expressed sequences from an individual, treatment, or species. As not all genes will be expressed in all environments or developmental stages of an organism, transcriptomes from different ages and environments are often combined into a **reference transcriptome**. A transcriptome aligned with a reference genome can be used to design capture probes for exome sequencing, or for targeted sequencing of a subset of candidate genes. Sequencing technologies that produce longer reads are useful for determining alternative splicing of exons. RNAseq can also be used to genotype individuals for SNPs within exons. For example, Rogier et al. (2018) used RNAseq to genotype black poplars for several hundred thousand SNPs, and then used these data to assess population structure.

Prior to NGS, Sanger sequencing was commonly used to generate **expressed sequence tags (ESTs)** from bulk mRNA reverse transcribed into cDNA libraries (Nagaraj et al. 2007). ESTs were often used to develop PCR primers to resequence targeted genes in multiple individuals and populations. These methods have been largely replaced by RNAseq.

4.9 Transcriptomics

Genetic variation exists in regulatory DNA regions as well as in protein-coding sequences. Gene expression also varies with developmental stage, tissue, environment, and health, and is an important factor in phenotypic plasticity. One way to study genetic and environmental variation in gene regulation is by quantifying levels of mRNA present in different tissues, individuals, environments, or treatments. In the recent past, cDNA microarrays and oligonucleotide microarrays were used for this purpose. Thousands or tens of thousands of different short DNA fragments were spotted onto a glass slide or other template, and cDNA from the individuals being studied, labeled with fluorescent dyes or other markers, was hybridized with the array. The intensity of fluorescence provided a quantification of the relative expression levels of targeted genes. Results could be validated with more precise estimates of RNA levels using real-time PCR for a subset of genes. However, RNAseq has largely replaced these methods of quantifying gene expression, in which all of the mRNA from a sample is used to produce cDNA, and the number of reads in a sequencing library for a given gene is used to estimate the relative level of gene expression.

Levels of gene expression can be viewed as phenotypes as they are the joint product of genetic and environmental variation (Hansen 2010). To assess genetic differences in gene expression, individuals need to be reared in a common environment. Information on gene expression differences among populations can be used to complement data on neutral genetic markers and on adaptive traits to group genetically similar individuals into conservation units.

Gene expression differences among populations or between environments can also be used to identify candidate genes that may be involved in adaptation or disease response. Gene expression levels for specific stress-related genes can also be used as biomarkers for health or response to environmental toxins for a range of animal species. While these assays typically investigate environmental rather than genetic sources of variation, they are another example of the use of genomic tools in conservation, and may be used in the future to monitor the

health of individuals in populations. For example, Miller et al. (2011) identified distinct transcriptomic responses to viral infection in sockeye salmon that reduced migratory success, and these responses were used to develop biomarkers for rapid screening of smolts for infection (Miller et al. 2014).

4.10 Epigenetics

Phenotypic plasticity can buffer individuals and populations from environmental changes, and can allow for more rapid responses than adaptation via natural selection on genetic variation. **Epigenetics** refers to mitotically or meiotically heritable changes in gene expression (i.e., passed on to daughter cells through cell division) that are not the result of variation in DNA sequence. Epigenetic effects can generate phenotypic plasticity within an individual, and if passed on to the next generation, can play a role in adaptation. These effects include **DNA methylation**, **histone modification**, and regulation by **small RNAs**. DNA methylation and histone modifications of DNA are referred to as **epigenetic marks**. While the extent of phenotypic plasticity can be under some genetic control, that is, coded in DNA, changes in the epigenetic state of chromatin can alter phenotypes on a much shorter timeframe (hours to months) than genetic changes such as allele frequency shifts between generations (Hu & Barrett 2017). Epigenetic modifications can stop or start gene expression like an on/off switch, or turn expression up or down like a volume control.

While the term epigenetics was coined in broad terms for developmental biology in 1942 by Waddington, epigenetics is a relatively young field of study within ecology and evolutionary biology, and its importance for nonmodel organisms is not well understood. Epigenetic changes within a generation that alter gene expression for different developmental stages, tissues, and environments are well documented for plants (reviewed by Thiebault et al. 2019) and animals (reviewed by Hu & Barrett 2017). However, there is less evidence for, and more debate around, the extent and importance of transgenerational epigenetic modifications passed from parents to offspring (Heard & Martienssen 2014).

If environmentally caused shifts in gene expression are transmitted over multiple generations, they could facilitate adaptation to environmental change such as climate warming (e.g., Horsthemke 2018; Lind & Spagopoulou 2018). Such plastic physiological adaptation might "buy time" for subsequent evolutionary adaptation; however, plasticity could also slow evolutionary adaptation if it impedes natural selection (Nunney 2016). There is some evidence for environmentally induced transgenerational inheritance of epigenetic gene expression changes that influence fitness. There is less evidence that such epigenetic changes persist over multiple generations (Heard & Martienssen 2014). In animals, there is considerable epigenetic erasure and resetting in the germline, leaving less opportunity for transmission of epigenetic conditions. There is more opportunity for epigenetic inheritance in plants.

Environmental factors known to cause transgenerational epigenetic inheritance of phenotypic variation include thermal stresses, drought, salt stress, diet, and exposure to toxins, such as hydrocarbons from plastics, pesticides, and agricultural fungicides (Ben Maamar et al. 2018; Miryeganeh & Saze 2020; Van Cauwenbergh et al. 2020). These stressors have caused transgenerational epigenetic inheritance in humans, fish, birds, plants, and insects. Important research questions for conservation include: How many generations do inherited epigenetic marks persist? Can epi-alleles (e.g., methylation) increase in frequency in a population via transgenerational inheritance like DNA alleles? Are they common enough to be an important source of adaptation compared with DNA sequence changes (Charlesworth et al. 2017)? Complicating factors in natural populations are that epigenetic differences among individuals can be the result of genetic differences affecting epigenetic modification, can be caused by the health of the maternal parent, or can result from shared parent–offspring environmental effects.

DNA methylation of cytosines is the most commonly studied epigenetic mark in nonmodel organisms (Kilvitis et al. 2014). A methyl group is added to the fifth carbon of cytosines within particular sequences or genomic regions, affecting gene expression. Histone modifications are another epigenetic modification to DNA that alter the way DNA is packaged and affect access to it for transcription. Small interfering RNAs (siRNAs) are involved in methylation and histone modification pathways. DNA methylation across the whole genome can be detected through bisulfite sequencing using NGS. A modification of the RADseq method GBS, called epiGBS, is a type of restriction site-associated reduced representation bisulfite sequencing (van Gurp et al. 2016).

DNA methylation has been shown to be heritable in several nonmodel species (Thiebault et al. 2019). For example, higher temperatures during development result in changes in DNA methylation that increase the proportion of males in European sea bass, red-ear slider turtles, and American alligators (Hu & Barrett 2017). Temperature, moisture, and nutrient availability have resulted in methylation differences in several plant species (Thiebault et al. 2019). Plants and animals differ in methylation in some key ways. Animal DNA is typically methylated at sequences with the motif $5'CpG3'$, where C is cytosine, G is guanine, and p is a phosphate group between the two. In plants, most methylation occurs at CG and CHG sequences (where H is A, C, or T). In animals, methylation patterns are usually reset globally immediately after fertilization, whereas in plants there is more opportunity for transgenerational stability of methylation.

4.11 Metagenomics

Metagenomics is the analysis of all of the DNA in an environmental sample. A wide range of environmental conditions and sample types can be used, depending on the objective of the study. NGS has enabled an explosion in metagenomic techniques, and growing numbers of applications to conservation issues. Metagenomic analyses relevant to conservation biology include sampling the gut, skin, foliage, or roots of individuals to assess their health or the presence of symbionts or disease-causing microbes; evaluating habitat quality for conservation or restoration using soil or water samples; detecting the presence of elusive threatened or endangered species of concern from **environmental DNA (eDNA)**; or detecting introduced, invasive species using eDNA.

Example 4.3 Diversity of gut microbiome of howler monkeys reflects habitat

The gut microbiome of animals consists of a diverse array of microbes that affect nutritional efficiency and health. The composition of these microbes in turn reflects diet quality, habitat, and animal health. Fecal samples can be used to noninvasively assess gut microbiome composition, dietary differences, and the presence of pathogenic microbes. Amato et al. (2013) collected fecal samples from black howler monkeys in Mexico inhabiting continuous evergreen rainforest in Palenque National Park, a nearby fragment of evergreen rainforest, a continuous, semi-deciduous forest, and a captive population in a wildlife center. Fresh fecal samples were collected and microbial DNA was extracted. The V1–V3 region of the 16S ribosomal gene was amplified using PCR, and the resulting amplicons were sequenced. After filtering and trimming reads, a 385-bp sequence of all reads was analyzed. The reads were clustered, and sequences with over 97% similarity were considered to belong to the same operational taxonomic unit (OTU). With microbial data, OTUs rather than species are identified and used to assess diversity as many microbes have not been described to the species level. Figure 4.11 shows **rarefaction curves** plotting the number of OTUs versus the depth of sequencing for each animal. Diversity in OTUs needs to be scaled by the depth of sequencing for comparative purposes if depth varies among samples as the diversity detected will generally increase with the depth of sampling.

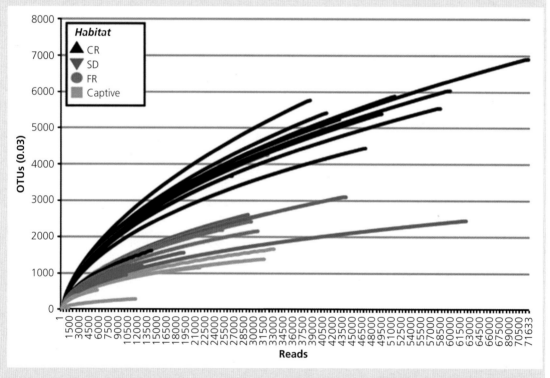

Figure 4.11 Diversity of black howler monkey microbiomes extracted from fecal samples of 32 individuals for the 16S ribosomal RNA gene reflects the host habitat. Rarefaction curves show the number of unique OTUs sharing ≥97% sequence identity per total sequenced reads for each sample. Habitats: CR = continuous evergreen rainforest; SD = continuous semi-deciduous rainforest; FR = evergreen rainforest fragment; captive = captive population of rescued animals in a wildlife center fed an atypical diet. Analysis of variance of the log-linear regression coefficients for each curve confirmed habitat differences ($P < 0.0001$). From Amato et al. (2013).

Howler monkeys inhabiting the continuous rainforest had the highest microbiome diversity, while the captive population had the lowest (Figure 4.11). Although there were fewer sequencing reads from the captive animals, it is clear that at the same

Continued

Example 4.3 *Continued*

depth of sequencing, the diversity is lower for the captive animals, and highest for those animals in continuous ever-green rainforest habitats. Notably, the individuals in fragmented forests had fewer microbes that produce butyrate, a key short-chain fatty acid providing energy to colon cells, and this may be important for animal health, nutrition, and metabolism.

The **microbiome** of an environment is the community of all microorganisms found in that environment, for example, in an animal gut or soil sample. DNA sequences from microbes within a sample can be clustered based on sequence similarity into **operational taxonomic units (OTUs)** or functional groups to describe diversity, or can be searched for the presence of pathogenic microbes. The composition of microbiomes can then be compared among environments, or among individuals or populations sampled. For example, Gellie et al. (2017) compared the bacterial microbiome of remnant, restored, and cleared sites in a eucalypt woodland in Australia by sampling soil eDNA, sequencing the 16S ribosomal DNA, and estimating the relative representation of different OTUs in sites. They found that following vegetation restoration, the soil bacterial community became more similar over time to the undisturbed remnant woodland.

Metatranscriptomics is the analysis of all of the RNA in an environmental sample or tissue. It employs NGS to determine the taxonomic composition and function of complex microbial communities. It has improved understanding of multispecies parasite infections, which could advance conservation of host species like birds infected by multiple blood parasites (Galen et al. 2020).

4.11.1 Host-associated microbial communities

The microbiomes of animals and plants have direct impacts on their physiology and health. Anthropogenic activities such as pollution or land use changes can have large impacts on these microbial communities. A better understanding of the microbiomes of animal host species could improve conservation strategies, and inform captive population management, translocation, and restoration (Trevelline et al. 2019). DNA can be extracted from fresh fecal samples and used to characterize gut microbiome diversity. This approach can provide

insights into diet, animal health, and habitat quality. For example, Amato et al. (2013) found substantial differences in the gut microbiome diversity of black howler monkeys from intact rainforest habitats compared with those from habitat fragments or captive environments (Example 4.3). Habitat fragmentation was also associated with changes in amphibian skin microbial communities (Becker et al. 2017), and agricultural habitats were associated with differences in frog gut microbiomes compared with natural habitats (Chang et al. 2016). Environmental contaminants such as heavy metals and pesticides have been shown to alter the gut microbiomes of fish, amphibians, and insects, and warming ocean temperatures and ocean acidification are associated with changes in the microbial communities of marine sponges and coral (reviewed by Trevelline et al. 2019).

While there is growing interest in the roles of endophytic (internal) microbial communities in agriculture on plant health and productivity, there have been fewer studies of plant microbiomes in the context of conservation of species at risk. There remains considerable potential for using metagenomics to inform conservation activities such as restoration. For example, Singapore has lost over 99% of its original forest cover, and native species need to be conserved in highly modified **ecosystems**. Epiphytic orchids have obligate relationships with mycorrhizal fungi that allow them to establish and grow on tree bark. Izuddin et al. (2019) used a metagenomic approach to assess the extent of epiphytic orchid mycorrhizal fungi on tree bark and on orchid roots to assess the potential for orchid translocation success and establishment in urban environments.

4.11.2 Environmental DNA (eDNA)

eDNA is DNA extracted from an environmental sample such as soil or water (Section 22.1.4). eDNA can be used to establish species presence with diagnostic markers that identify specific taxa. Analysis of eDNA from feces can identify not only the animal that produced the feces, but also other species eaten by that animal, as well as macro-parasites and the microbiome. Along with camera trapping, eDNA is on its way toward revolutionizing endangered species detection and population monitoring. Species-specific PCR primers can be designed to amplify diagnostic sequences called barcodes for one or for many species. Mitochondrial sequences are often the target of these primers. For example, Davy et al. (2015) used eDNA from water samples to detect three endangered and six nonthreatened species of freshwater turtles. They were able to detect all nine species with PCR primers amplifying the mtDNA *COI* gene. The eDNA approach was much less expensive than traditional survey methods. It is also being used to detect invasive species, including disease-causing bacteria, fungi, and viruses.

4.12 Other "omics" and the future

Several other "omic" tools exist to bridge the gap between genomes and phenotypes beyond transcriptomics (Kristensen et al. 2010). The **proteome** includes all of the proteins and the **metabolome** is the entire set of small-molecule metabolites found in a cell, tissue, or organism. These "omics" require highly specialized and instrument-intensive chemical analysis to identify molecules present and quantify their concentrations. These technologies are being used to study genetic diversity and develop tools for plant breeding in some crop plants and their wild progenitors. Proteomics and metabolomics to date have had little use or impact in conservation genetics; however, they may find some useful, specific applications for particular populations or species in the future, or provide biomarkers for monitoring population health and gene-targeted conservation (Kardos & Shafer 2018; Novak et al. 2020; Phelps et al. 2020).

For example, multiple "omic" technologies are being applied to corals to mitigate the massive die-offs of these foundational species from climate warming (Guest Box 16). The goal is to facilitate gene-targeted conservation and industrial-scale production and release of novel lineages of corals resistant to climate warming and pollution. Researchers are using transcriptomics, proteomics, and metabolomics to identify genes and clones that are thermally tolerant to facilitate climate adaptation (Ruiz-Jones & Palumbi 2017; Novak et al. 2020; Seveso et al. 2020). This omics work is paired with stem cell research because stem cells with adaptive genes could be transplanted directly into wild corals to repair damaged reefs. **Gene editing** has been conducted in corals and could be used to deliver genes for heat resilience into coral reefs. Similar work will eventually be conducted with other foundational species.

Guest Box 4 Genomics and conservation of Tasmanian devils in the face of transmissible cancer
Paul A. Hohenlohe

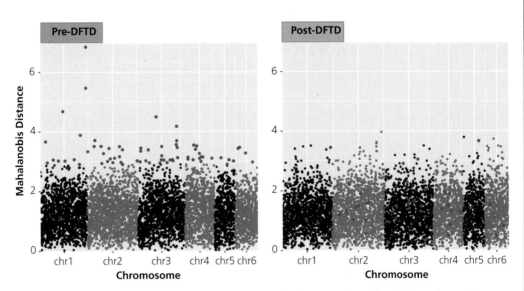

Figure 4.12 Evidence of local adaptation (estimated by a composite metric of selection signals, Mahalanobis distance) across the Tasmanian devil genome. SNP loci showing significant evidence of selection prior to DFTD arrival (left) appear in blue. After DFTD arrival (right), these loci have mostly lost signatures of local adaptation, and the total number of significant SNPs across the genome is reduced. Modified from Fraik et al. (2020).

In 1996, a strange form of cancer was detected in a population of Tasmanian devils (Figure 24.1). Named devil facial tumor disease (DFTD), the cancer was soon discovered to be transmissible; that is, tumor cells are transferred directly from one individual to another, typically through biting, so that DFTD behaves as an infectious disease (Hawkins et al. 2006). A quarter century later, DFTD has spread across nearly the entire species range and caused severe declines in devil populations (Lazenby et al. 2018). Population genomics contributes to understanding the genetic impacts of DFTD on devils and informing management of both captive and wild populations.

Devil populations are evolving in response to DFTD. A study using RAD sequencing identified regions of the genome that exhibit signatures of selection in parallel across populations, indicating that they are unlikely to be due to drift alone (Epstein et al. 2016; Example 23.3). A genome-wide association study using Rapture (RAD plus sequence capture) found that female infection and survival after infection have significant genetic variation, providing the

raw material for an evolutionary response to DFTD (Margres et al. 2018). Mechanisms such as life history shifts (Lazenby et al. 2018) or tumor regression (Margres et al. 2020) may allow devils to persist with DFTD. At the same time, DFTD has affected adaptation of devils to other environmental factors. For example, Fraik et al. (2020) used genotype–environment association analysis to identify SNP loci showing evidence of local adaptation prior to DFTD. Following population declines and selection caused by DFTD, signatures of local adaptation at these loci and across the genome were reduced (Fraik et al. 2020; see Figure 4.12).

A disease-free captive metapopulation of devils has been established. Molecular and pedigree information are used to maintain genetic diversity (Grueber et al. 2019a), and supplementation from captive to wild populations is being implemented in some areas (Grueber et al. 2019b). While this could alleviate the loss of genetic diversity in populations reduced by DFTD, it also risks increasing the severity of the DFTD epizootic by introducing susceptible individuals and impeding ongoing evolution of resistance (Hamede et al. 2020). This illustrates potential tradeoffs between **genetic rescue** (introducing individuals in order to reduce inbreeding depression) and **evolutionary rescue** (the ability of populations to evolve in response to a threat) in evaluating conservation strategies.

Remarkably, a second transmissible cancer, called DFT2, has also arisen recently in Tasmanian devils (Pye et al. 2016). Only a handful of transmissible cancers are known across the animal kingdom, suggesting that devils may be uniquely susceptible to such diseases. Genomic studies are revealing some similarities between DFTD and DFT2 (Patchett et al. 2020), although it is too early to know how devil populations may respond to DFT2. Nonetheless, conservation can benefit from understanding what types of genetic variation are critical for devil populations to persist in the face of current and future transmissible cancers.

PART II

Mechanisms of Evolutionary Change

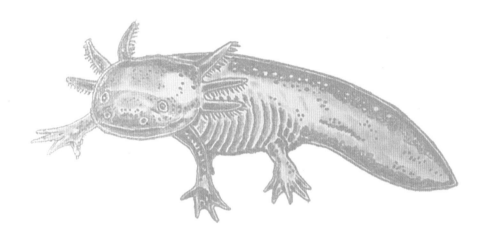

CHAPTER 5

Random mating populations: Hardy–Weinberg Principle

Little spotted kiwi, Example 5.2

Today, the Hardy–Weinberg Law stands as a kind of Newton's First Law (bodies remain in their state of rest or uniform motion in a straight line, except insofar as acted upon by external forces) for evolution: Gene frequencies in a population do not alter from generation to generation in the absence of migration, selection, statistical fluctuation, mutation, etc.

(Robert M. May 2004, p. 790)

In a sexual population, each genotype is unique, never to recur. The life expectancy of a genotype is a single generation. In contrast, the population of genes endures.

(James F. Crow 2001, p. 771)

A description of genetic variation by itself, as in Chapters 3 and 4, will not help us understand the evolution and conservation of populations. We need to develop mathematical models of the effects of Mendelian inheritance in natural populations in order to understand the influence of natural selection, small population size, and other evolutionary factors that affect the persistence of populations and species (Charlesworth 2019; Guest Box 5). The strength of population genetics is the rich foundation of theoretical expectations that allows us to test the predictions of hypotheses to explain the patterns of genetic variation found in natural populations (Provine 2001).

This chapter introduces the structure of the basic models used to understand the genetics of populations. In later chapters, we will explore expected changes in allele and genotype frequencies in the presence of such evolutionary factors as natural

selection or mutation. In this chapter, we will focus on the relationship between allele frequencies and genotype frequencies. In addition, we will examine techniques for estimating allele frequencies and for testing observed genotypic proportions with those expected.

We will use a series of mathematical models to consider the pattern of genetic variation in natural populations and to understand the mechanisms that produce evolutionary change. Models allow us to simplify the complexity of the world around us (Servedio et al. 2014). Models may be either conceptual or mathematical. Conceptual models allow us to simplify the world so that we can represent complex reality with words and in our thoughts. Mathematical models allow us to specify the relationship between empirical quantities that we can measure and parameters that we specify in our biological theory. These models are essential in

Conservation and the Genomics of Populations, Third Edition. Fred W. Allendorf, *et al.*, Oxford University Press.
© Fred W. Allendorf, W. Chris Funk, Sally N. Aitken, Margaret Byrne, and Gordon Luikart (2022). DOI: 10.1093/oso/9780198856566.003.0005

understanding the factors that affect genetic change in natural populations, and in predicting the effects of human actions on natural populations. While these simple models are used in this chapter for one or perhaps two genes at a time, they also apply to genomic data containing tens of thousands of genes or millions of **single nucleotide polymorphisms (SNPs)**.

Models are essential for understanding the genetics of natural populations in a variety of ways:

1. Models make us define the parameters that need to be considered.
2. Models allow us to test hypotheses.
3. Models allow us to generalize results.
4. Models allow us to predict how a system will operate in the future.

The use of models in biology is sometimes criticized because genetic and ecological systems are complex, and simple models ignore many important properties of these systems. This criticism has some validity. Nevertheless, it is impossible to think without using models because reality is too complex (de Brabandere & Iny 2010). Our brains receive information and process this information in order to construct a mental model of how things work. We are often not aware of the models that our brains are using to interpret reality. Construction of mathematical models, such as we will be using in population genetics, forces us to explicitly define the assumptions and parameters that we need to include in order to interpret observed patterns of genetic change in populations.

As a general rule of thumb, models that we develop to understand natural populations should be as simple as possible. That is, a hypothesis or model should not be any more complicated than necessary (**Ockham's razor**). There are several reasons for this. First, hypotheses and models are scientifically useful only if they can be tested and rejected. Simpler models are easier to reject, and, therefore, are more useful. Second, simple models are likely to be more general and therefore more applicable to a wider number of situations.

5.1 Hardy–Weinberg principle

We will begin with the simplest model of population genetics: a random mating population in which no factors are present to cause genetic change from generation to generation. This model is based upon the fundamental framework of **Mendelian segregation** for diploid organisms that are reproducing sexually in combination with fundamental principles of probability (Box 5.1). These same principles apply to virtually all species, from elephants to pine trees to violets. We will make the following assumptions in constructing this model:

1. *Random mating.* "Random mating obviously does not mean promiscuity; it simply means . . . that in the choice of mates . . . there is neither preference for nor aversion to the union of persons similar or dissimilar with respect to a given trait or gene" (Wallace & Dobzhansky 1959, p. 107).

   A population can be random mating with regard to most loci, but be mating nonrandomly with regard to other loci that influence mate choice. For example, snow geese are commonly white, but there is a blue phase that is caused by a dominant allele at the melanocortin 1 receptor (*MC1R*) locus (Section 2.2; Mundy et al. 2004). Snow geese prefer mates that have the same coloration as the parents that reared them (Cooke 1987). Thus, snow geese show **positive assortative mating** with regard to *MC1R*, but they mate at random with regard to the rest of their genome. See Section 9.1.2 for another example of positive assortative mating at *MC1R*.

2. *No mutation.* We assume that the genetic information is transmitted from parent to progeny (i.e., from generation to generation) without change. Mutations provide the genetic variability that is our primary concern in genetics. Nevertheless, mutation rates are generally quite small and are only important in population genetics from a long-term perspective, generally hundreds or thousands of generations. We will not consider the effects of mutations on changes in allele frequencies in detail since in conservation genetics we are more concerned with factors that can influence populations in a more immediate time frame.

3. *Infinite population size.* Many of the theoretical models that we will consider assume an infinite population size. This assumption may effectively be correct in some populations of insects or plants. However, it is obviously not true for

---

Box 5.1 Probability

Genetics is a science of probabilities. Mendelian inheritance itself is based upon probability. We cannot know for certain which allele will be placed into a gamete produced by a heterozygote, but we know that there is a one-half probability that each of the two alleles will be transmitted. This is an example of a random, or **stochastic**, event. There are a few simple rules of probability that we will use to understand the extension of Mendelian genetics to populations.

The **probability** (P) of an event is the number of times the event will occur (a) divided by the total number of possible events (n):

$$P = a/n$$

For example, a die has six faces that are equally likely to land up if the die is tossed. Thus, the probability of throwing any particular number is one-sixth:

$$P = a/n = 1/6$$

We often are interested in combining the probabilities of different events. There are two different rules that we will use to combine probabilities.

The **product rule** states that the probability of two or more independent events occurring simultaneously is equal to the product of their individual probabilities. For example, what is the probability of throwing a total of 12 with a pair of dice? This can only occur by a six landing up on the first die and also on the second die. According to the product rule,

$$P = 1/6 \times 1/6 = 1/36$$

The **sum rule** states that the probability of two or more mutually exclusive events occurring is equal to the sum of their individual probabilities. For example, what is the probability of throwing either a five or six with a die? According to the sum rule:

$$P = 1/6 + 1/6 = 2/6 = 1/3$$

In many situations, we need to use both of these rules to compute a probability. For example, what is the probability of throwing a total of seven with a pair of dice?

Solution: There are six mutually exclusive ways that we can throw seven with two dice: 1 + 6, 2 + 5, 3 + 4, 4 + 3, 5 + 2, and 6 + 1. As we saw in the example for the product rule, each of these combinations has a probability of 1/6 × 1/6 = 1/36 of occurring. They are all mutually exclusive so we can use the sum rule. Therefore, the probability of throwing a seven is:

$$1/36 + 1/36 + 1/36 + 1/36 + 1/36 + 1/36 = 6/36$$

$$= 1/6$$

---

many of the populations of concern in conservation genetics. Nevertheless, we will initially consider the ideal large population in order to develop the basic concepts of population genetics, and we will then consider the effects of small population size in later chapters.

4. *No natural selection.* We will assume that there is no differential survival or reproduction of individuals with different genotypes (i.e., no natural selection). Again, this assumption will not be true at all loci in any real population, but it is necessary that we initially make this assumption in order to develop many of the basic concepts of population genetics. We will consider the effects of natural selection in later chapters.

5. *No immigration.* We will assume that we are dealing with a single isolated population. We will later consider multiple populations in which gene flow between populations is brought about through exchange of individuals.

There are two important consequences of these assumptions. First, the population will not evolve. Mendelian inheritance has no inherent tendency to favor any one allele. Therefore, allele and genotype frequencies will remain constant from generation to generation. This is known as the Hardy–Weinberg (HW) equilibrium. In the next few chapters, we will explore the consequences of relaxing these assumptions on changes in allele frequency from generation to generation. We will not be able to consider all possibilities. However, our goal is to develop an intuitive understanding of the effects of each of these evolutionary factors.

The second important outcome of these assumptions is that genotype frequencies will be in **binomial (HW) proportions**. That is, **genotypic** frequencies after one generation of random mating will be a binomial function of allele frequencies. It is important to distinguish between the two primary ways in which we will describe the genetic characteristics of populations at individual loci: allele frequencies and genotypic frequencies.

The HW principle greatly simplifies the task of describing the genetic characteristics of populations; it allows us to describe a population by the frequencies of the alleles at a locus rather than by the many different genotypes that can occur at a single diploid locus. This simplification becomes especially important when we consider multiple loci. For example, there are 59,049 different genotypes possible at just 10 loci, with each having just two alleles. We can describe this tremendous genotypic variability by specifying only 10 allele frequencies if the populations is in HW proportions.

This principle was first described by G.H. Hardy (1908), a prominent English mathematician, and independently by a German physician Wilhelm Weinberg (1908). The principle was actually first used by an American geneticist W.E. Castle (1903) in a description of the effects of natural selection against recessive alleles. However, this aspect of the paper by Castle was not recognized until nearly 60 years later (Li 1967). A detailed and interesting history of the development of population genetics is provided by Provine (2001). There is great irony in our use of Hardy's name to describe a fundamental principle that has been of great practical value in medical genetics and now in our efforts to conserve biodiversity. Hardy (1967, p.49) saw himself as a "pure" mathematician whose work had no practical relevance: "I have never done anything 'useful'. No discovery of mine has made, or is likely to make, directly or indirectly, for good or ill, the least difference to the amenity of the world."

5.2 HW proportions

We will first consider a single locus with two alleles (A and a) in a population such that the population consists of the following numbers of each genotype:

AA	Aa	aa	Total
N_{11}	N_{12}	N_{22}	N

Each **homozygote** (AA or aa) contains two copies of the same allele while each **heterozygote** (Aa)

contains one copy of each allele. Therefore, the allele frequencies are:

$$p = freq(A) = \frac{(2N_{11} + N_{12})}{2N} \quad (5.1)$$

$$q = freq(a) = \frac{(N_{12} + 2N_{22})}{2N}$$

where $p + q = 1.0$.

Our assumption of random mating will result in random union of **gametes** to form **zygotes**. Thus, the frequency of any particular combination of gametes from the parents will be equal to the product of the frequencies of those gametes, which are the allele frequencies. This is shown graphically in Figure 5.1. Thus, the expected genotypic proportions are predicted by the binomial expansion:

$$(p + q)^2 = p^2 + 2pq + q^2 \quad (5.2)$$
$$ AA \quad Aa \quad aa$$

These proportions will be reached in one generation, providing all of the assumptions are met and allele frequencies are equal in males and females. Additionally, these genotypic frequencies will be maintained forever, as long as these assumptions hold.

The HW principle can be readily extended to more than two alleles with two simple rules:

1. The expected frequency of homozygotes for any allele is the square of the frequency of that allele.
2. The expected frequency of any heterozygote is twice the product of the frequency of the two alleles present in the heterozygote.

In the case of three alleles the following genotypic frequencies are expected:

$$p = freq(A_1)$$
$$q = freq(A_2)$$
$$r = freq(A_3)$$

and:

$$(p + q + r)^2 = p^2 + 2pq + q^2 + 2pr + 2qr + r^2$$
$$ A_1A_1 \ A_1A_2 \ A_2A_2 \ A_1A_3 \ A_2A_3 \ A_3A_3$$
$$(5.3)$$

5.3 Testing for HW proportions

Genotypic frequencies of samples from natural populations can be tested readily to see if they conform to expected HW proportions. This is generally the first step in analyzing population genetic data

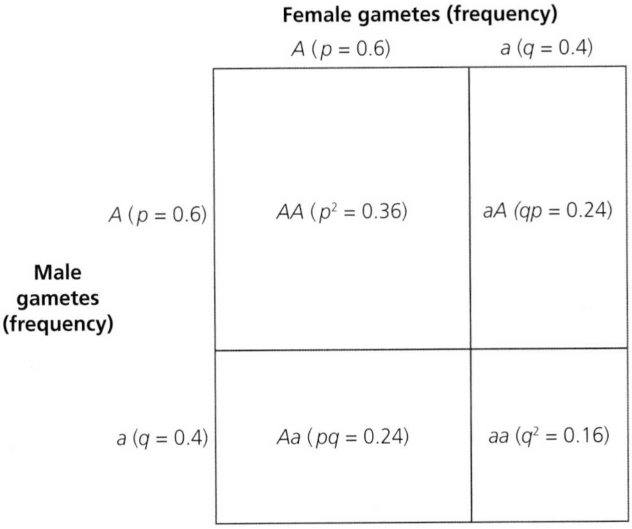

Female gametes (frequency)

Figure 5.1 HW proportions at a locus with two alleles (*A* and *a*) generated by random union of gametes produced by females and males. The area of each rectangle is proportional to the genotypic frequencies.

(Waples et al. 2019). There is a profusion of papers that discuss the sometimes hidden intricacies of testing for goodness-of-fit to HW proportions (Fairbairn & Roff 1980). Lessios (1992) provided an interesting and valuable overview of this topic. Waples (2015) has provided a more recent overview of this topic.

Many papers in the literature refer to these analyses as testing for "Hardy–Weinberg equilibrium," rather than the more precise testing for "Hardy–Weinberg proportions." This is potentially confusing because populations whose genotypic proportions are in HW proportions will not be in HW equilibrium (no mutation, infinite population size, etc.). In general, genotypic proportions will be in expected proportions under the HW model as long as the population is randomly mating. The effects of mutation, small population size, natural selection, and immigration are generally too small to cause genotypic proportions to deviate from HW expectations.

Dinerstein & McCracken (1990) described genetic variation using allozyme electrophoresis at 10 variable loci in a population of one-horned rhinoceros from the Chitwan Valley of Nepal. Both allozymes and SNPs often have two alleles, which makes this example also illustrative for SNP datasets. The following numbers of each genotype were detected at a lactate dehydrogenase (*LDH*) locus with two alleles (*100* and *125*). Do these values differ from what we expect with HW proportions?

100/100	100/125	125/125	Total
$N_{11} = 5$	$N_{12} = 12$	$N_{22} = 6$	$N = 23$

We first need to estimate the allele frequencies in this sample. We do not know the true allele frequencies in this population that consisted of some 400 animals at the time of sampling. However, we can estimate allele frequencies in this population based upon the sample of 23 individuals. The estimate of the allele frequency of the *100* allele obtained from this sample will be designated as $\hat{p}$ (called *p* hat) to designate that it is an estimate rather than the true value.

$$\hat{p} = \frac{2N_{11} + N_{12}}{2N} = \frac{10 + 12}{46} = 0.478$$

and

$$\hat{q} = \frac{N_{12} + 2N_{22}}{2N} = \frac{12 + 12}{46} = 0.522$$

We now can estimate the expected number of each genotype in our sample of 23 individuals' genotype assuming HW proportions:

	100/100	100/125	125/125
Observed	5	12	6
Expected	($\hat{p}^2 N = 5.3$)	($2\hat{p}\hat{q}N = 11.5$)	($\hat{q}^2 N = 6.3$)

The agreement between observed and expected genotypic proportions in this case is very good.

In fact, this is the closest fit possible in a sample of 23 individuals from a population with the estimated allele frequencies. Therefore, we would conclude that there is no indication that the genotype frequencies at this locus are not in HW proportions.

The **chi-square** method provides a statistical test to determine if the deviation between observed genotypic and expected HW proportions is greater than we would expect by chance alone. We first calculate the chi-square value for each of the genotypes and sum them into a single value:

$$X^2 = \sum \frac{(OBSERVED - EXPECTED)^2}{EXPECTED}$$
$$= \frac{(5 - 5.3)^2}{5.3} + \frac{(12 - 11.5)^2}{11.5} + \frac{(6 - 6.3)^2}{6.3}$$
$$= 0.02 + 0.02 + 0.01 = 0.05$$

The X^2 value becomes increasingly greater as the difference between the observed and expected values becomes greater.

The computed X^2 value is then compared to a set of values (Table 5.1) calculated under the assumption that the null hypothesis we are testing is correct; in this case, our null hypothesis is that the population from which the samples was drawn is in HW proportions. We need one additional value to apply the chi-square test, the **degrees of freedom**. In using the chi-square test for HW proportions, the value for degrees of freedom is equal to the number of possible genotypes minus the number of alleles.

Number of alleles	Number of genotypes	Degrees of freedom
2	3	1
3	6	3
4	10	6
5	15	10

By convention, if the probability estimated by a statistical test is less than 0.05, then the difference between the observed and expected values is said to be significant. We can see in Table 5.1 that the chi-square value with one degree of freedom must be greater than 3.84 before we would conclude that the deviation between observed and expected proportions is greater than we would expect

Table 5.1 Critical values of the chi-square distribution for up to five degrees of freedom (v). The proportions in the table (corresponding to $\alpha = 0.05$, 0.01, etc.) represent the area to the right of the critical value of chi-square given in the table, as shown in the figure below. The null hypothesis is usually not rejected unless the probability associated with the calculated chi-square is less than 0.05.

Degrees of freedom	Probability (P)					
	0.90	0.50	0.10	0.05	0.01	0.001
1	0.02	0.46	2.71	3.84	6.64	10.83
2	0.21	1.39	4.60	5.99	9.21	13.82
3	0.58	2.37	6.25	7.82	11.34	16.27
4	1.06	3.86	7.78	9.49	13.28	18.47
5	1.61	14.35	9.24	11.07	15.09	20.52

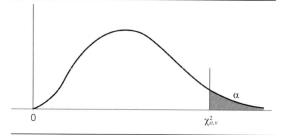

by chance with one degree of freedom. Our estimated X^2 value of 0.05 for the *LDH* locus in the one-horned rhino is much smaller than this. Therefore, we would accept the null hypothesis that the population from which this sample was drawn was in HW proportions at this locus. See Example 5.1 for a situation where the null hypothesis of HW proportions can be rejected. An example of the chi-square test for HW proportions in the case of three alleles is given in Example 5.2.

5.3.1 Small sample sizes

Sample sizes in conservation genetics are often smaller than our statistical advisors recommend because of the limitations imposed by working with rare species. The chi-square test is only an approximation of the actual probability distribution, and the approximation becomes poor when expected numbers are small. The usual rule of thumb is to not use the chi-square test when any expected number is less than five. However, some have argued that this rule is unnecessarily conservative and have suggested using smaller limits on expected values (three by Cochran 1954 and one by Lewontin & Felsenstein 1965).

Example 5.1 Testing for HW proportions

Leary et al. (1993) reported the following genotype frequencies at an allozyme locus (*mIDHP-1*) in a sample of bull trout from the Clark Fork River in Idaho, USA.

Genotype	Observed	Expected	Chi-square
100/100	1	$(\hat{p}^2N = 5.8)$	3.97
100/75	22	$(2\hat{p}\hat{q}N = 12.5)$	7.22
75/75	2	$(\hat{q}^2N = 6.8)$	3.38
Total	25	(25.1)	14.58

Estimated frequency of $100 = \hat{p} = [(2 \times 1) + 22]$

$$/(2 \times 25) = 0.480$$

Estimated frequency of $75 = \hat{q}$

$$= [(22 + 1) + (2 \times 2)]$$

$$/(2 \times 25) = 0.520$$

Degrees of freedom = 1

The calculated X^2 of 14.58 is greater than the critical value for $P < 0.001$ with one degree of freedom of 10.83 (Table 5.1). Therefore, the probability of getting such a large deviation by chance alone is less than 0.001. We would reject the null hypothesis that the sampled population was in HW proportions at this locus.

There is a significant excess of heterozygotes in this sample of bull trout. We will return to this example in the next chapter to see the probable cause of this large deviation from HW proportions (Section 6.6).

Example 5.2 Testing for HW proportions with more than two alleles

The following genotype frequencies were found at a microsatellite locus (*Rowi2*) in a sample of little spotted kiwi from the Zealandia sanctuary in Wellington, New Zealand (Taylor et al. 2017a).

Genotype	Observed	Expected	Chi-square
173/173	27	$(\hat{p}^2N = 29.9)$	0.31
173/177	54	$(2\hat{p}\hat{q}N = 45.9)$	1.43
177/177	12	$(\hat{q}^2N = 17.6)$	1.78
173/179	3	$(2\hat{p}\hat{r}N = 5.4)$	1.07
177/179	7	$(2\hat{q}\hat{r}N = 4.2)$	1.87
179/179	0	$(\hat{r}^2N = 0.2)$	0.20
Total	103	(103.2)	6.66

Estimated frequency of $173 = \hat{p}$

$$= [(2 \times 27) + 54 + 3]$$

$$/(2 \times 103) = 0.539$$

Estimated frequency of $177 = \hat{q}$

$$= [54 + (2 \times 12) + 7]$$

$$/(2 \times 103) = 0.413$$

Estimated frequency of $179 = \hat{r}$

$$= [7 + 3 + (2 \times 0)]$$

$$/(2 \times 103) = 0.049$$

There are six genotypic classes and two independent allele frequencies at a locus with three alleles.

Degrees of freedom = 6 − 3 = 3

The calculated X^2 of 6.66 is less than the critical value with three degrees of freedom of 7.82 (Table 5.1). Therefore, we accept the null hypothesis that the sampled population was in HW proportions at this locus.

In addition, there is a systematic bias in small samples because of the discreteness of the possible numbers of genotypes. Levene (1949) has shown that in a finite sample of N individuals, the heterozygotes are increased by a fraction of $1/(2N − 1)$ and homozygotes are correspondingly decreased (Crow & Kimura 1970, pp. 55–56). For example, if only one copy of a rare allele is detected in a sample then the only genotype containing the rare allele must be heterozygous. However, the simple binomial HW proportions will predict that some fraction of the sample is expected to be homozygous for the rare allele; however, this is impossible because there is only one copy of the allele in the sample. This adjustment will correct for this bias.

Exact tests provide a method to overcome the limitation of small expected numbers with the chi-square test (Fisher 1935). Exact tests are performed by determining the probabilities of all possible samples assuming that the null hypothesis is true

Example 5.3 Exact test for HW proportions

We would reject the null hypothesis of HW proportions in the table below because our calculated chi-square value is greater than 3.84. However, an exact test indicates that we would expect to get a deviation as great or greater than the one we observed some 8% (0.082) of the time (Weir 1996). Therefore, we would not reject the null hypothesis in this case using the exact test.

	Genotypes			
100/100	**100/80**	**80/80**		
21	19	0	$\hat{p} = 0.763$	$X^2 = 3.88$
(23.3)	(14.5)	(2.3)		

There are 10 possible samples of 40 individuals that would provide us with the same allele frequency estimates. We can calculate the exact probabilities for each of these possibilities if the sampled population was in HW proportions using the binomial distribution as shown below:

Possible samples				Cumulative	
100/100	**100/80**	**80/80**	**Probability**	**Probability**	**X^2**
30	1	9	0.0000	0.0000	34.67
29	3	8	0.0000	0.0000	25.15
28	5	7	0.0001	0.0001	17.16
27	7	6	0.0023	0.0024	10.69
26	9	5	0.0205	0.0229	5.74
21	**19**	**0**	**0.0594**	**0.0823**	**3.88**
25	11	4	0.0970	0.1793	2.32
22	17	1	0.2308	0.4101	1.20
24	13	3	0.2488	0.6589	0.42
23	15	2	0.3411	1.0000	0.05

In practice, exact tests are performed using computer programs because calculating the exact binomial probabilities is extremely complicated and time consuming.

5.3.2 Many alleles

Testing for HW proportions at loci with many alleles, such as microsatellite loci, is a problem because many genotypes will have extremely low expected numbers. There are $A(A - 1)/2$ heterozygotes and A homozygotes at a locus with A alleles. Therefore, there are the following possible number of genotypes at a locus in a population with A alleles:

$$\frac{A(A-1)}{2} + A = \frac{A(A+1)}{2} \qquad (5.4)$$

For example, Olsen et al. (2000) found an average of 23 alleles at eight microsatellite loci in pink salmon in comparison with an average of 2.3 alleles at 24 polymorphic allozyme loci in the same population. A total of 279 genotypes are possible with 23 alleles at a single locus (Equation 5.4). Exact tests for HW proportions are possible with more than two alleles (Louis & Dempster 1987; Engels 2009). However, the number of possible genotypes increases very quickly with more than two alleles and computation time becomes prohibitive. Engels (2009) has provided software that uses a likelihood ratio approach for exact HW tests that is much faster than previous approaches.

Permutation tests are useful to analyze data using computer-based randomization in cases where many genotypes will have low expected numbers, as is often true with many alleles. In the case of Example 5.3, a computer program would randomize genotypes by sampling, or creating, 40 diploid individuals from a pool of 61 copies of the *100* allele and 19 copies of the *80* allele. A chi-square value is then calculated for 1,000 or more of these randomized datasets and its value compared with the statistic obtained from the observed dataset. The proportion of chi-square values from the randomized datasets that give a value as large as or larger than the observed provides an unbiased estimation of the probability that the null hypothesis is true.

5.3.3 Multiple simultaneous tests

In most studies of natural populations, multiple loci are examined from several populations resulting in multiple tests for HW proportions (Waples 2015). For example, if we examine 10 loci in 10 population samples, 100 tests of HW proportions

(Example 5.3). The probability of the observed distribution is then added to the sum of all less probable possible sample outcomes. Weir (1996, pp. 98–101) describes the use of the exact test, and Vithayasai (1973) presented tables for applying the exact test with two alleles.

will be performed. If all of these loci are in HW proportions (i.e., our null hypothesis is true at all loci in all populations), we expect to find five significant tests if we use the 5% significance level. Thus, simply applying the statistical procedure presented here would result in rejection of the null hypothesis of HW proportions approximately five times when our null hypothesis is true.

There are a variety of approaches that can be used to treat this problem (Rice 1989). One common approach is to use the so-called **Bonferroni correction** in which the significance level (say 5%) is adjusted by dividing it by the number of tests performed (Cooper 1968). Therefore in the case of 100 tests, we would use the adjusted nominal level of $0.05/100 = 0.0005$. The critical chi-square value for $P = 0.0005$ with one degree of freedom is 12.1. That is, we expect a chi-square value greater than 12.1 with one degree of freedom less than 0.0005 of the time if our null hypothesis is correct. Thus, we would reject the null hypothesis for a particular locus only if our calculated chi-square value was greater than 12.1. This procedure is known to be conservative and results in a loss of statistical power to detect multiple deviations from the null hypothesis. A procedure known as the **sequential Bonferroni** can be used to increase power to detect more than one deviation from the null hypothesis (Rice 1989).

It is also extremely important to examine the data to detect possible patterns for those loci that do not conform to HW proportions. For example, let's say that eight of our 100 tests have probability values less than 5%; this value is not much greater than our expectation of five. If the eight cases are spread fairly evenly among samples and loci and none of the individual probability values is less than 0.0005 obtained from the Bonferroni correction, then it is reasonable to not reject the null hypothesis that these samples are in HW proportions at these loci.

However, we may reach a different conclusion if all eight of the deviations from HW proportions occurred in the same sample, and all the deviations were in the same direction (e.g., a deficit of heterozygotes). This would suggest that this particular sample was taken from a population that was not in HW proportions. Perhaps this sample was collected from a group that consisted of two separate random mating populations (Wahlund effect; Section 9.1.1).

Another possibility is that all eight deviations from HW proportions occurred at the same locus in eight different population samples, and all the deviations were in the same direction (e.g., a deficit of heterozygotes). This would suggest that there is something unusual about this particular locus. For example, the presence of a null allele (Section 5.4.2) would result in a tendency for a deficit of heterozygotes.

5.3.4 Testing large-scale genomic data for HW proportions

Waples (2015) presented an excellent consideration of the primary issues relevant here with a special emphasis on testing for HW proportions with large-scale genomic datasets. In addition, testing for HW proportions is also widely used for quality control in large genomic datasets. Loci deviating from HW proportions might have errors in the sequencing or genotyping process. Next-generation sequencing (NGS) datasets can have high genotyping error rates, especially with low-coverage sequencing (Section 4.4). We can identify questionable loci using tests for HW proportions (Box 5.2).

A major issue with using datasets having thousands of loci is that multiple independent hypothesis tests for HW proportions can lead to a high rate of incorrectly rejecting the null hypothesis, as explained in Section 5.3.3. This can result in a high false positive rate and removal of informative loci from a dataset. If we test 10,000 loci for HW proportions, we expect 500 significant tests by chance alone if we use the 5% significance level. This could lead to the removal of many quality loci that deviate from HW proportions as false positives. A solution to the problem of multiple testing is to apply a Bonferroni or sequential Bonferroni correction to adjust P-values, as discussed in Section 5.3.2.

For example, Figure 5.2a shows the distribution of deviations from HW proportions at nearly 4,000 loci in Channel Island foxes (Funk et al. 2016; Figure 5.3). There is a slight excess of homozygotes in this sample that appears to result from nonrandom mating on the island because of some geographical isolation (Section 9.1). The bottom plot (5.2b) shows simulated values at 3,940 loci for two random mating populations of 100 individuals connected with some

---

Box 5.2 Identifying questionable loci by testing for HW proportions

Tests for deviations from HW proportions are commonly used to assess quality of NGS datasets. For example, NGS data often have many loci with a deficit of heterozygotes when sequencing coverage is low because of allelic dropout. Allelic dropout is detection of only one of the two alleles present at a heterozygous locus (Pompanon et al. 2005). Allelic dropout rates increase with low sequence coverage (<10 sequence reads per locus per sample) because all of the reads from a locus can originate from only one of the two alleles in a diploid individual. This can result from biased PCR amplification of one allele, stochastic sampling error (e.g., only one of two alleles is sampled), or because divergent alleles fail to map (align) to a reference genome during the genotyping process (Hendricks et al. 2018).

To detect allelic dropout problems, we can test for a multi-locus deficit of heterozygotes. We should also test if the heterozygote deficit decreases as depth of sequencing coverage increases. If read depth is low and negatively correlated with heterozygote deficit magnitude, there is likely allelic dropout. To avoid problems from allelic dropout, researchers can get more sequence data or use methods that are based on genotype likelihoods instead of called genotypes (e.g., Korneliussen et al. 2014; Hendricks et al. 2018).

NGS datasets also often have many loci with heterozygote excess. For example, duplicated loci are common in most genomes and can lead to an apparent heterozygote excess during the genotyping process. NGS reads from different loci often align together as a "single locus" because duplicated loci have high sequence similarity. This leads to an apparent heterozygote excess because duplicated loci often have different alleles that when aligned together appear as a single-locus heterozygote. Duplicated loci also often have high sequencing coverage (i.e., many reads) because two (or more) loci contribute sequencing reads (Graffelman et al. 2017).

To identify questionable loci, it is helpful to test for consistent HW deviation at a locus in multiple independent populations (as mentioned in Section 5.3.3). Testing multiple populations for a HW deviation (e.g., excess or deficit) is facilitated by combining the P-values from multiple populations for a locus (Wilson 2019). Combined tests can improve power to identify questionable loci that are too subtle to detect in a single population sample (Wilson 2019).

We should try to understand the cause of departures from HW proportions rather than just discarding loci with significant departures. For example, if a locus has an excess of heterozygotes in multiple independent populations, is not duplicated, and is in a gene known to confer high fitness in heterozygotes, the cause of the heterozygote excess might be selection for heterozygotes. However, power is notoriously low for HW tests for detecting selection (Lachance 2009).

---

gene flow resulting in similar F_{IS} values. There is more scatter in the observed than in the simulated data, especially for negative F_{IS} values. This could result from technical issues in scoring the genotypes at these loci (Box 5.2). It could also result from natural selection or mating system differences in the fox population not included in the simulations.

Corrections for multiple testing with population genomic data are used too seldomly and inconsistently. Sethuraman et al. (2019) reviewed 205 studies published from 2013 to 2018 and reported a lack of use of HW testing along with inconsistent use of corrections for multiple testing. For example, in the same study authors have used different correction methods. Sethuraman et al. (2019) recommends that researchers correct P-values for multiple testing and explicitly report the reasoning behind the correction method they use.

Which multiple testing correction method should we use? There is no consensus on the most appropriate correction. However, guidelines have been proposed (Narum 2006; White et al. 2019). It is crucial that we understand the concepts of **false positives** (sometimes called **Type I errors**), and **false negatives** (sometimes called **Type II errors**) and choose a correction method appropriate for our tolerance of making these errors. For example, if your research question is not affected by using loci that deviate slightly from HW proportions, you might use a less conservative correction and err on the side of including more loci in your downstream analyses. Papers seldom explain reasons they choose a certain correction method or their tolerance for making false positives or false negatives (Waples 2015; Sethuraman et al. 2019).

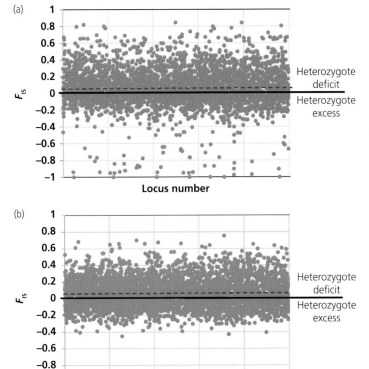

Figure 5.2 (a) Distribution of F_{IS} values at 3,940 SNP loci in a sample of 46 Channel Island foxes from Santa Catalina Island (data from Funk et al. 2016). F_{IS} is the proportional excess (positive values) or deficit (negative values) of homozygotes (see Section 9.1). (b) Distribution of simulated F_{IS} values at 3,940 SNP loci for two random mating populations of 100 individuals each connected with some gene flow resulting in a similar F_{IS} value (see Section 9.1). Dashed line is the mean F_{IS}.

We can use both a conservative and liberal correction method to assess the effects of the correction method on our conclusions. Saarman et al. (2019) used three correction methods to compute P-values. They used the relatively liberal Benjamini–Hochberg correction for multiple testing with $P < 0.02$. They also used the relatively conservative Bonferroni correction $P < 0.02$, and a bootstrapping method with 1,000 bootstrap replicates with $P < 0.02$. Saarman et al. (2019) found almost no differences among the three correction methods, which provided confidence that there is no strong influence of correction method on their conclusions.

5.4 Estimation of allele frequencies

So far we have estimated allele frequencies when the number of copies of each allele in a sample can be counted directly from the genotypic frequencies. However, sometimes we cannot identify the alleles in every individual in a sample. The HW principle can be used to estimate allele frequencies at loci in which there is not a unique relationship between genotypes and phenotypes. We will consider two such situations that are often encountered in analyzing data from natural populations.

5.4.1 Recessive alleles

There are many cases in which heterozygotes cannot be distinguished from one of the homozygotes. For example, color polymorphisms and metabolic disorders in many organisms are caused by recessive alleles. The frequency of recessive alleles can be estimated if we assume HW proportions.

$$\hat{q} = \sqrt{\frac{N_{22}}{N}} \qquad (5.5)$$

Dozier & Allen (1942) described differences in coat color in the muskrat of North America. A dark phase, the so-called blue muskrat, is generally rare relative to the ordinary brown form but occurs at high frequencies along the Atlantic coast between

Figure 5.3 Channel Island fox from Santa Catalina Island. Photo courtesy of Julie King.

New Jersey and North Carolina. Breeding studies (Dozier 1948) have shown that the blue phase is caused by an allele (*b*) that is recessive to the brown allele (*B*). A total of 9,895 adult muskrats were trapped on the Blackwater National Wildlife Refuge, Maryland in 1941. The blue muskrat occurred at a frequency of 0.536 in this sample. If we assume HW proportions (Equation 5.2), then:

$$\hat{q}^2 = 0.536$$

and taking the square root of both sides of this relationship:

$$\hat{q} = \sqrt{0.536} = 0.732$$

and the estimated frequency of the *B* allele is {1 − 0.732} = 0.268.

Example 5.4 demonstrates how the genotypic proportions in a population for a recessive allele can be examined for HW proportions.

5.4.2 Null alleles

In general, null alleles are alleles that produce no detectable product. Null alleles at protein coding loci do not produce a detectable protein product; null alleles at microsatellite loci do not produce a detectable **polymerase chain reaction (PCR)** amplification product. Null alleles at allozyme loci result from alleles that produce either no protein product or a protein product that is enzymatically nonfunctional (Foltz 1986). Null alleles at microsatellite loci result from substitutions that prevent the primers from binding (Brookfield 1996). Heterozygotes for a null allele and another allele appear to be homozygotes. The presence of null alleles results in an apparent excess of homozygotes relative to HW proportions. Brookfield (1996) discusses the estimation of null allele frequencies in the case of more than three alleles. Kalinowski & Taper (2006) have presented a **maximum likelihood** approach to estimate the frequency of null alleles at microsatellite loci.

The familiar ABO blood group locus in humans presents a parallel situation to the case of a null allele in which all genotypes cannot be distinguished. In this case, the I^A and I^B alleles are codominant, but the I^O allele is recessive (i.e., null). This results in the following relationship between genotypes and phenotypes (blood types):

Genotypes	Blood types	Expected frequency	Observed number
$I^A I^A$, $I^A I^O$	A	$p^2 + 2pr$	N_A
$I^B I^B$, $I^B I^O$	B	$q^2 + 2qr$	N_B
$I^A I^B$	AB	$2pq$	N_{AB}
$I^O I^O$	O	r^2	N_O

where p, q, and r are the frequencies of the I^A, I^B, and I^O alleles.

We can estimate allele frequencies at this locus by the **expectation maximization** (EM) algorithm, which finds the allele frequencies that maximize the probability of obtaining the observed data from a sample of a population assumed to be in HW proportions (Dempster et al. 1977). This is an example of a **maximum likelihood estimate** (Section A5), which has many desirable statistical properties (Fu & Li 1993).

We could estimate the frequency p directly, as in Example 5.2, if we knew how many individuals in our sample with blood type A were $I^A I^A$ and how many were $I^A I^O$:

Example 5.4 Color polymorphism in the eastern screech owl

We concluded in Chapter 2 based on the data from Table 2.1 that the red morph of the eastern screech owl is caused by a dominant allele (R) at a single locus with two alleles; gray owls are homozygous for the recessive allele (rr).

Mating	Number of families	Progeny	
		Red	Gray
Red × red	8	23	5
Red × gray	46	68	63
Gray × gray	135	0	439
Total	189	91	507

We can estimate the allele frequency of the r allele by assuming that this population is in HW proportions and the total progeny observed are representative of the entire population:

$$\hat{q}^2 = (507/598) = 0.847$$
$$\hat{q} = \sqrt{0.847} = 0.921$$
$$\text{and}$$
$$\hat{p} = 1 - \hat{q} = 0.079$$

We can also check to see if the progeny produced by matings between two red birds is close to what we would expect if this population was in HW proportions. Remember that red birds may be either homozygous (RR) or heterozygous (Rr). What proportion of gray progeny do we expect to be produced by matings between two red parents?

Three things must occur for a progeny to be gray: (1) the mother must be heterozygous (Rr), (2) the father must be heterozygous (Rr), and (3) the progeny must receive the recessive allele (r) from both parents:

Prob (progeny rr) = Prob (mother Rr) x Prob (father Rr) x 0.25

The proportion of red birds in the population who are expected to be heterozygous is the proportion of heterozygotes divided by the total proportion of red birds:

Prob(parental bird (Rr)) = $(2pq)/(p^2 + 2pq) = 0.959$

Therefore, the expected proportion of gray progeny is:

0.959 x 0.959 x 0.25 = 0.230

This is fairly close to the observed proportion of 0.178 (5/28). Thus, we would conclude that this population appears to be in HW proportions at this locus.

$$\hat{p} = \frac{2N_{AA} + N_{AO} + N_{AB}}{2N} \quad (5.6)$$

where N is the total number of individuals. However, we cannot distinguish the phenotypes of the $I^A I^A$ and $I^A I^O$ genotypes. The EM algorithm solves this ambiguity with a technique known as gene counting. We start with guesses of the allele frequencies, and use them to calculate the expected frequencies of all genotypes (step E of the EM algorithm), assuming HW proportions. Then, we use these genotypic frequencies to obtain new estimates

of the allele frequencies, using maximum likelihood (the step M). We then use these new allele frequency estimates in a new E step, and so forth, in an iterative fashion, until the values converge.

We first guess the three allele frequencies (remember $p + q + r = 1.0$). The next step is to use these guesses to calculate the expected genotype frequencies assuming HW proportions. We next use gene counting to estimate the allele frequencies from these genotypic frequencies. The count of the I^A alleles is twice the number of $I^A I^A$ genotypes plus the number of $I^A I^O$ genotypes. We expect p^2 of the total individuals with blood type A ($p^2 + 2pr$) to be homozygous $I^A I^A$, and $2pr$ of them to be heterozygous $I^A I^O$. These counts are then divided by the total number of genes in the sample ($2N$) to estimate the frequency of the I^A allele, as we did in Equation 5.6. A similar calculation is performed for the I^B allele with the following result:

$$\widehat{p} = \frac{\left[2\left(\dfrac{p^2}{p^2 + 2pr} \right) N_A + \left(\dfrac{2pr}{p^2 + 2pr} \right) N_A + N_{AB} \right]}{2N}$$

$$= \frac{2\left(\dfrac{p+r}{p+2r} \right) N_A + N_{AB}}{2N} \tag{5.7}$$

$$\widehat{q} = \frac{2\left(\dfrac{q+r}{q+2r} \right) N_B + N_{AB}}{2N}$$

$$\widehat{r} = 1 - \widehat{p} - \widehat{q} \tag{5.8}$$

These equations produce new estimates of p, q, and r that can be substituted into the right-hand side of Equations 5.7-5.8 to produce new estimates of p, q, and r. This iterative procedure is continued until the estimates converge. That is, until the estimated values on the left side are nearly equal to the values substituted into the right side.

5.5 Sex-linked loci

We so far have considered only autosomal loci in which there are no differences between males and females. However, the genotypes of genes on sex chromosomes will often differ between males and females. The most familiar situation is genes on the X chromosome of mammals (and *Drosophila*) in which females are homogametic XX and males are heterogametic XY. In this case, genotype frequencies for females conform to the HW proportions. However, the Y chromosome is largely void of genes so that males will have only one gene copy, and the genotype frequency in males will be equal to the allele frequencies. The situation is reversed in bird species: females are heterogametic ZW and males are homogametic ZZ (Ellegren 2000b). In this case, genotype frequencies for the ZZ males conform to the HW proportions, and the genotype frequency in the ZW females will be equal to the allele frequencies (Figure 5.4). Fossøy et al. (2016)

have described a fascinating case in which the genes for egg color in the common cuckoo are located on the W chromosome (Example 5.5).

Phenotypes resulting from rare recessive X-linked alleles will be much more common in males than in females because q^2 will always be less than q. The most familiar case of this is X-linked red–green color blindness in humans in which ~8% of males of northern European origin (including the first author of this book) lack a pigment in the retina of their eye so they do not perceive colors as most people (Deeb 2005). In this case, $q = 0.08$ and therefore we expect the frequency of color blindness in females to be $q^2 = (0.08)^2 = 0.0064$. Thus, we expect more than 10 times more red–green color blind males than females in this case.

A variety of other mechanisms for sex determination occur in other animals and plants (Bull 1983; see Section 3.1.2). Many plant species possess either XY or ZW systems. The use of XY or ZW indicates which sex is heterogametic (Charlesworth 2002). The sex chromosomes are identified as XY in species in which males are heterogametic and ZW in species in which females are heterogametic. Many reptiles have a ZW system (e.g., all snakes; Graves & Shetty 2001). A wide variety of genetic sex determination systems are found in fish species (Devlin & Nagahama 2002) and in invertebrates. Some species have no detectable genetic mechanism for sex determination. For example, sex is determined by the temperature in which eggs are incubated in some reptile species (Graves & Shetty 2001).

5.5.1 Pseudoautosomal inheritance

The classic XY system of mammals and *Drosophila* with the Y chromosome being largely devoid of functional genes taught in introductory genetic classes has been overgeneralized. A broader taxonomic view suggests that mammals and *Drosophila* are exceptions and that both sex chromosomes contain many functional genes across a wide variety of animal taxa (Vicoso 2019). The sex chromosomes of many species have so-called **pseudoautosomal** regions in which functional genes are present on both the X and Y, or Z and W, chromosomes. Morizot et al. (1987) found that functional genes for the creatine kinase enzyme locus are present on

	Z-bearing eggs		W-bearing eggs
	Z^A (p)	Z^a (q)	W
Z^A (p)	$Z^A Z^A$ (p^2)	$Z^A Z^a$ (pq)	$Z^A W$ (p)
Z^a (q)	$Z^A Z^a$ (pq)	$Z^a Z^a$ (q^2)	$Z^a W$ (q)
	Males		Females

Sperm

Figure 5.4 Expected genotypic proportions with random mating for a Z-linked locus with two alleles (*A* and *a*).

Example 5.5 W-linked egg color polymorphism in the common cuckoo

The common cuckoo is a brood parasite that lays its eggs in the nests of a variety of other bird species in Eurasia (Fossøy et al. 2016). This species has evolved many host-specific races that produce eggs which mimic in color and pattern the eggs of different host species (Figure 5.5). More than 15 host-specific egg morphs have been recognized in Europe.

Figure 5.5 Host-specific egg mimicry and variation in egg color among common cuckoo host races that parasitize different species. From left: meadow pipit, brambling, great reed warbler, and the common redstart. The common cuckoo egg is the slightly larger egg in each clutch. Photo courtesy of Frode Fossøy.

Fossøy et al. (2016) described genotypes at 15 nuclear loci and sequenced mitochondrial DNA (mtDNA) in birds from seven host races throughout Europe. They focused on the distinctive immaculate blue eggs, which are relatively rare. They found that birds laying blue eggs belong to a highly divergent maternal lineage at both mtDNA and the W chromosome, but have similar frequencies at nuclear loci from other common cuckoos. Thus, the blue egg polymorphism is maternally inherited and the genes controlling color are apparently on the W chromosome. The maternal inheritance of mtDNA and the W chromosome are completely genetically "linked" in birds even though the genes are not physically "linked." Comparison of sequence divergence in the two mtDNA lineages suggests a divergence time of over 2 million years ago.

---

both the Z and W chromosomes of Harris's hawk. Genome mapping efforts support these early results. For example, three of six genes found to be sex linked in the Siberian jay were present on both the Z and W chromosomes (Jaari et al. 2009). Wright and Richards (1983) found that two of 12 allozyme loci that they mapped in the leopard frog were sex linked and that two functional gene copies of both loci are found in XY males. Functional copies of a peptidase locus are present on both the Z and W chromosomes in the salamander *Pleurodeles waltl* (Dournon et al. 1988).

Differences in allele frequencies between the males and females for genes found on both sex chromosomes will result in an excess of heterozygotes in comparison with expected HW proportions in the heterogametic sex (Clark 1988; Allendorf et al. 1994). This excess of heterozygotes can persist for many generations if the locus is closely linked to the sex-determining locus. These regions happen to be quite small in those species for which we are most familiar with sex-linked inheritance (e.g., humans and *Drosophila*). However, these pseudoautosomal regions comprise a large proportion of the sex chromosomes in many taxa (e.g., salmonid fishes and birds). Detecting such loci in population genetic studies will become much more frequent in the future as more and more loci are examined with genomic techniques. For example, a genetic linkage map of the Siberian jay found that three of six **sex-linked loci** are pseudoautosomal (Jaari et al. 2009). This linkage map included a total of 117 loci, so that nearly 3% of genome-wide loci were pseudoautosomal.

Differences in allele frequency between sex chromosomes will result in an excess of heterozygotes in comparison with expected HW proportions in the heterogametic sex for pseudoautosomal loci (Clark 1988). When possible, the genotypes of the sexes should be examined separately. This is especially important with genomic datasets in which many sex-linked loci are likely to be included (Benestan et al. 2017). For example, Taylor et al. (2017a) detected a significant excess of heterozygotes ($P < 0.05$) at a microsatellite locus (*Rowi20*) with two alleles (*176* and *182*) in the little spotted kiwi for which males are ZZ and females are ZW. Comparing the genotypes of males and females indicated that this locus is sex linked:

		182/182	182/176	176/176
Females	ZW	0	28	0
Males	ZZ	35	0	0
Total		35	28	0
(Expected)		(38.1)	(21.8)	(3.1)

Based on these data, the *182* allele is fixed on the Z chromosome because it is not present in females, and the *176* allele is fixed on the W chromosome. Marshall et al. (2004) provided general methods for estimating allele frequencies on sex chromosomes for pseudoautosomal loci.

5.6 Estimation of genetic variation

We often are interested in comparing the amount of genetic variation in different populations. For example, we saw in Table 3.5 that brown bears from Kodiak Island and Yellowstone National Park had less genetic variation than other populations for allozymes, microsatellites, and mtDNA. In addition, comparisons of the amount of genetic variation in a single population sampled at different times can provide evidence for loss of genetic variation because of population isolation and fragmentation due to habitat loss or other causes. In this section we will consider measures that have been used to compare the amount of genetic variation.

5.6.1 Heterozygosity

The average expected HW heterozygosity at n loci within a population is the best general measure of genetic variation:

$$H_e = 1 - \sum_{i=1}^{n} p_i^2 \tag{5.9}$$

It is easier to calculate one minus the expected homozygosity, as Equation 5.9, than summing over all heterozygotes because there are fewer homozygous than heterozygous genotypes with three or more alleles. Nei (1987) has referred to this measure as gene diversity, and pointed out that it can be thought of as either the average proportion of heterozygotes per locus in a randomly mating population or the expected proportion of heterozygous loci in a randomly chosen individual. Gorman & Renzi (1979) have shown that estimates of H_e are generally insensitive to sample size and that even a few individuals are sufficient for estimating H_e if a large number of loci are examined. In general, comparisons of H_e among populations are not valid unless a large number of loci are examined. As we saw in Section 3.5, expected heterozygosity calculated over a series of contiguous base pairs, including both variable and nonvariable sites, is called **nucleotide diversity (π)**.

There are a variety of characteristics of average heterozygosity that make it valuable for measuring genetic variation. It can be used for genes of different ploidy levels (e.g., haploid organelles) and in organisms with different reproductive systems. We will see in later chapters that there is considerable theory available to predict the effects of reduced population size on heterozygosity (Chapter 6), that average heterozygosity is a good measure of the expected response of a population to natural selection (Chapter 11), and that it can also provide an estimate of individual inbreeding coefficients (Chapter 17).

5.6.2 Allelic richness

The total number of alleles at a locus has also been used as a measure of genetic variation. This is a valuable complementary measure of genetic variation because it is more sensitive to the loss of genetic variation because of small population size than heterozygosity, and it is an important measure of the long-term evolutionary potential of populations (Section 6.4; Allendorf 1986).

The major drawback of the number of alleles is that, unlike heterozygosity, it is highly dependent on sample size. That is, there will be a tendency to detect more alleles at a locus as samples sizes increase. Therefore, comparisons between samples are not meaningful unless samples sizes are similar because of the presence of many low frequency alleles in natural populations. This problem can be avoided by using **allelic richness**, which is a measure of **allelic diversity** that takes into account sample size (El Mousadik & Petit 1996). This measure uses a **rarefaction** method to estimate allelic richness at a locus for a fixed sample size, usually the smallest sample size if a series of populations are sampled (Petit et al. 1998). Allelic richness can be denoted by $R(g)$, where g is the number of gene copies sampled.

The **effective number of alleles** is sometimes used to describe genetic variation at a locus. However, this parameter provides no more information about the number of alleles present at a locus than does heterozygosity. The effective number of alleles is the number of alleles that if equally frequent would result in the observed heterozygosity or homozygosity. It is computed as $A_e = 1/\Sigma p_i^2$ where p_i is the frequency of the ith allele. For example, consider two loci that both have an H_e of 0.50. The first locus has two equally frequent alleles ($p = q = 0.5$), and the second locus has five alleles at frequencies of 0.68, 0.17, 0.05, 0.05, and 0.05. Both of these loci will have the same value of $A_e = 2$.

5.6.3 Proportion of polymorphic loci

The proportion of loci that are polymorphic (P) in a population has been used to compare the amount of variation between populations and species at allozyme loci (Table 3.5). Strictly speaking, a locus is polymorphic if it contains more than one allele. However, generally some standard definition is used to avoid problems associated with comparisons of samples that are different sizes. That is, the larger the sample, the more likely we are to detect a rare allele. A locus is usually considered to be polymorphic if the frequency of the most common allele is less than either 0.95 or 0.99 (Nei 1987). The 0.99 standard has been used most often, but it is not reasonable to use this definition unless all sample sizes are greater than 50 (which is often not the case).

This measure of variation is of limited value. In some circumstances it can provide a useful measure of another aspect of genetic variation that is not provided by heterozygosity or allelic richness. It is potentially valuable in studies of allozyme loci and SNPs with large sample sizes in which many loci are studied, and many of the loci are monomorphic. However, it is of much less value in studies of highly variable loci such as microsatellites because most loci are polymorphic in most populations. In addition, microsatellite loci are often selected to be studied because they are highly polymorphic in preliminary analysis.

Guest Box 5 Is mathematics necessary?
James F. Crow

Much of our understanding of the application of genetics to problems in conservation depends upon the field of population genetics. Population genetics used to consist of two quite different disciplines. One utilized observations of populations in nature or laboratory studies. These were often descriptive and involved no mathematics. This area is epitomized by the early work of Theodosius Dobzhansky, Ernst Mayr, and G. Ledyard Stebbins. At the same time a mathematical theory was being developed by J.B.S. Haldane, R.A. Fisher, and Sewall Wright. One of the earliest bridges was built in 1941 when Dobzhansky and Wright collaborated in a joint experimental paper with lots of theory.

Since that time, most work in population genetics has had some mathematical involvement. Almost every experiment or field observation now utilizes quantitative measurements, and that means statistics. The day is past when one can simply report results with no test of their statistical reliability. Increasingly, experiments are performed or observations are made based on some underlying theory. The person doing the experiments may develop the theory or make use of existing mathematical theory. Finally, there is the development of ever-deeper, more general, and more sophisticated theory. Much of this is being done by people with professional mathematics training.

We cannot all be mathematicians. But we can learn a minimum amount. Every population geneticist must know some mathematics and some statistics. I have done both experimental (usually driven by theory) and theoretical work. But my mathematics is limited and some of the research that I most enjoyed was done in collaboration with better mathematicians, notably Motoo Kimura.

There are two recent changes in the field. Computers have altered everything, and it is hardly necessary for me to mention that you need to know how to use them. It used to be that theoretical work was regularly stymied by insoluble problems. The computer has greatly broadened the range of problems that can be solved, not in the mathematical sense but numerically, which is often what is wanted. At the same time, the mathematical theory itself is advancing as mathematicians enter the field.

The second change is the advent of molecular methods. Population genetics used to have a theory that was too rich for the data. That is no longer true. DNA analysis can yield mountains of data that call for improved, computerized analyses. Even in nonmodel species, datasets are becoming large enough that some sophisticated statistical methods can take days to conduct computations, and some analyses might not be feasible because, for example, computer programs take too long or do not converge.

If you are going to be an experimenter or analyze data with modern statistical tools, you need to know some mathematics and statistics, and be adept at computers. If you are going to develop theory (even if for application to natural populations and conservation), you usually need to be a real mathematician or collaborate with one.

Most readers of this book are primarily interested in understanding, but not contributing to the primary literature in population genetics. Much of the current literature in population genetics employs advanced mathematical methods that are beyond the reach of most biology students. Dobzhansky's method of reading and understanding the papers of Sewall Wright is one possible approach (see the quote at the beginning of the Appendix). Examining the biological assumptions being made is crucial, but not sufficient. However, a healthy amount of skepticism is probably a good thing. There was only one Sewall Wright!

CHAPTER 6

Small Populations and Genetic Drift

Land snail, Example 6.2

> The race is not always to the swift, nor the battle to the strong, for time and chance happens to us all.
>
> **(Ecclesiastes 9:11)**

> . . . the conservationist is faced with the ultimate sampling problem—how to preserve genetic variability and evolutionary flexibility in the face of diminishing space and with very limited economic resources. Inevitably we are concerned with the genetics and evolution of small populations, and with establishing practical guidelines for the practicing conservation biologist.
>
> **(Otto H. Frankel & Michael E. Soulé 1981, p. 31)**

Genetic change will not occur in populations if all the assumptions of the Hardy–Weinberg (HW) equilibrium are met (Section 5.1). However, these assumptions are not met in natural populations, and genetic changes result. In this and the next several chapters, we will see what happens when the assumptions of HW equilibrium are violated. In this chapter, we will examine what happens when we violate the assumption of infinite population size. That is, what is the effect on **allele** and **genotype** frequencies when population size (N) is finite?

All natural populations are finite so **genetic drift** will occur in all natural populations, even large ones. For example, consider a new mutation that increases fitness which occurs in an extremely large population of insects that numbers in the millions. Whether or not the single copy of this advantageous mutation is lost by chance from this population will be determined primarily by the sampling process that determines what alleles are transmitted to the next generation. For example, if the individual with the mutation does not reproduce, the new allele will

be lost immediately. And, even if the individual with the mutation produces two progeny, there is a 25% chance, based on **Mendelian segregation**, that the mutation will be lost. Thus, the fate of a rare allele in an extremely large population will be determined primarily by genetic drift.

Genetic drift is the primary force bringing about allele frequency changes throughout the genome over time. This has been confirmed by genomic approaches that now allow evaluating the relative importance of genetic drift in natural populations by examining allele frequency changes at thousands of **single nucleotide polymorphism (SNP)** loci over time. Chen et al. (2019) used pedigrees in a long-studied natural population of Florida scrub jay to test the relative roles of different evolutionary processes in shaping patterns of genetic change at over 15,000 SNP loci. They concluded that genetic drift is the predominant evolutionary force causing allele frequency change. Funk et al. (2016) and Perrier et al. (2017) came to a similar conclusion in their studies of the island fox and lake trout, respectively.

Conservation and the Genomics of Populations, Third Edition. Fred W. Allendorf, *et al.*, Oxford University Press.
© Fred W. Allendorf, W. Chris Funk, Sally N. Aitken, Margaret Byrne, and Gordon Luikart (2022). DOI: 10.1093/oso/9780198856566.003.0006

Understanding genetic drift and its effects is extremely important for conservation. Fragmentation and isolation due to habitat loss and modification have reduced the population size of many species of plants and animals throughout the world. We will see in future chapters how genetic drift is expected to affect genetic variation in these populations. More importantly, we will consider how genetic drift may reduce the fitness of individuals in these populations and limit the potential of these populations to evolve by natural selection.

6.1 Genetic drift

Genetic drift is random change in allele frequencies from generation to generation because of sampling error. That is, the finite number of genes transmitted to progeny will be an imperfect sample of the allele frequencies in the previous generation (Figure 6.1). The mathematical treatment of genetic drift began with R.A. Fisher (1930) and Sewall Wright (1931), who independently considered the effects of binomial sampling in small populations of constant size N in which the next generation is produced by drawing $2N$ genes at random from a large gamete pool to which all individuals contribute equally. This model is referred to as the **Wright–Fisher model** or Fisher–Wright model. However, Fisher and Wright strongly disagreed on the importance of drift in bringing about evolutionary change (Crow 2010). Genetic drift is sometimes called the "Sewall Wright effect" in recognition that the importance of drift in evolution was largely introduced by Wright's papers.

It is often helpful to consider extreme situations in order to understand the expected effects of relaxing assumptions on models. Consider the example of a plant species capable of self-fertilization with a constant population size of $N = 1$, consisting of a single individual of genotype Aa; the allele frequency in this generation is 0.5. We cannot predict what the allele frequency will be in the next generation because the genotype of the single individual in the next generation will depend upon which alleles are transmitted via the chance elements of Mendelian inheritance. However, we do know that the allele frequency in the next generation will be 0.0, 0.50,

or 1.0 because the only three possible genotypes are AA, Aa, or aa. Based upon Mendelian expectations, there is a 50% probability that the frequency of the A allele will be either zero or one in the next generation.

Genetic drift is an example of a **stochastic** process in which the actual outcome cannot be predicted because it is affected by random elements (chance). Tossing a coin is one example of a stochastic process. One-half of the time, we expect a head to result, and one-half of the time we expect a tail. However, we do not know what the outcome of any specific coin toss will be. We can mimic or simulate the effects of genetic drift by using a series of coin tosses. Consider a population initially consisting of two heterozygous (Aa) individuals, one male and one female. **Heterozygotes** are expected to transmit the A and a alleles with equal probability to each gamete. A coin is tossed to specify which allele is transmitted by heterozygotes; an outcome of a head (H) represents an A allele; and a tail (T) represents an a. No coin toss is needed for homozygous individuals since they will always transmit the same allele.

The results of one such simulation using these rules are shown in Table 6.1 and Figure 6.2. In the first generation, the female transmitted the A allele to both progeny because both coin tosses resulted in heads. The male transmitted an A to his daughter and an a to his son because the coin tosses resulted in a head and then a tail. Thus, the allele frequency (p) changed from 0.5 in the initial generation to 0.75 in the first generation, and the expected heterozygosity in the population changed as well. This process is continued until the seventh generation when both individuals become homozygous for the a allele, and, thus, no further gene frequency changes can occur.

Table 6.1 shows one of many possible outcomes of genetic drift in a population with two individuals. However, we are nearly certain to get a different result if we start over again. In addition, it would be helpful to simulate the effects of genetic drift in larger populations. In principle, this can be done by tossing a coin; however, it quickly becomes extremely time consuming.

A better way to simulate genetic drift is with computer simulations (Section A13). Computational

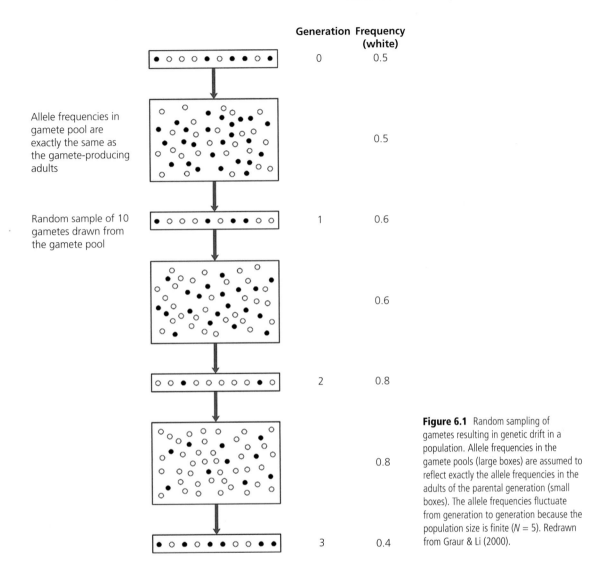

Generation   Frequency
(white)

0   0.5

Allele frequencies in
gamete pool are
exactly the same as
the gamete-producing
adults

0.5

Random sample of 10
gametes drawn from
the gamete pool

1   0.6

0.6

2   0.8

Figure 6.1 Random sampling of
gametes resulting in genetic drift in a
population. Allele frequencies in the
gamete pools (large boxes) are assumed to
reflect exactly the allele frequencies in the
adults of the parental generation (small
boxes). The allele frequencies fluctuate
from generation to generation because the
population size is finite ($N = 5$). Redrawn
from Graur & Li (2000).

0.8

3   0.4

methods are available to produce a random number
that is uniformly distributed between zero and one.
This random number can be used to determine
which allele is transmitted by a heterozygote. For
example, if the random number is in the range of
0.0 to 0.5, we can specify that the A allele is transmitted; similarly, a random number in the range of
0.5 to 1.0 would specify an a allele. Models such as
this are often referred to as Monte Carlo simulations
in reference to the gambling tables in Monte Carlo. Figure 6.3 shows changes in allele frequencies
in three populations of different sizes as simulated
with a computer. The smaller the population size,

the greater are the changes in allele frequency due
to drift (compare N of 10 with 200).

The sampling process that we have examined
here has two primary effects on the genetic composition of small populations:

1. Allele frequencies will change.
2. Genetic variation will be lost.

We can measure genetic changes in small populations either by changes in allele frequencies or
increases in homozygosity caused by inbreeding.
As allele frequencies change because of genetic
drift, heterozygosity is expected to decrease (and

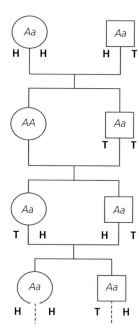

Figure 6.2 Simulation of genetic drift in a population consisting of a single female (circle) and male (square) each generation. A coin is tossed twice to simulate the two gametes produced by each heterozygote. A head (**H**) indicates that the *A* allele is transmitted and a tail (**T**) indicates the *a* allele. Homozygotes always transmit the allele for which they are homozygous.

Table 6.1 Simulation of genetic drift by coin tossing in a population of one female and one male over seven generations. A coin is tossed twice to specify which alleles are transmitted by heterozygotes; an outcome of a head (H) represents an *A* allele; and a tail (T) represents an *a*. The first toss represents the allele transmitted to the female in the next generation and the second toss the male (as shown in Figure 6.2). *p* is the frequency of the *A* allele. The observed and expected heterozygosities (assuming HW proportions) are also shown.

Generation	Mother	Father	*p*	H_o	H_e
0	*Aa* (HH)	*Aa* (HT)	0.50	1.000	0.500
1	*AA*	*Aa* (TT)	0.75	0.500	0.375
2	*Aa* (TH)	*Aa* (HT)	0.50	1.000	0.500
3	*aA* (HH)	*Aa* (TH)	0.50	1.000	0.500
4	*Aa* (TT)	*AA*	0.75	0.500	0.375
5	*aA* (HT)	*aA* (TT)	0.50	1.000	0.500
6	*Aa* (TT)	*aa*	0.25	0.500	0.375
7	*aa*	*aa*	0.00	0.000	0.000

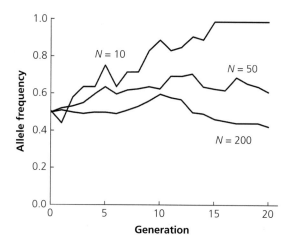

Figure 6.3 Results of computer simulations of changes in allele frequency by genetic drift for each of three population sizes (*N*) with an initial allele frequency of 0.5.

homozygosity increase). For example, heterozygosity became zero in generation 16 with *N* of 10 because only one allele remained in the population. Once such a "**fixation**" of one allele or another occurs, it is permanent; only mutation (Chapter 12) or gene flow (Chapter 9) from another population can introduce new alleles. We will consider the effects of genetic drift on both allele frequencies and genetic variation in the next two sections.

6.2 Changes in allele frequency

We cannot predict the direction of change in allele frequencies from generation to generation because genetic drift is random. The frequency of any allele is equally likely to increase or decrease from one generation to the next because of genetic drift. Although we cannot predict the direction of change, we can describe the expected magnitude of the change in allele frequency. In general, the smaller the population, the greater the change in allele frequency that is expected (Figure 6.3).

The change in allele frequencies from one generation to the next because of genetic drift is a problem in sampling. A finite sample of gametes is drawn from the parental generation to produce the next generation. Both the sampling of gametes and the coin toss can be described by the binomial sampling distribution (Section A3.1). The variance of change in allele frequency from one generation to the next is thus the binomial sampling variance:

$$V_q = \frac{pq}{2N}$$

Given that the current allele frequency is p with a population size of N, there is approximately a 95% probability that the allele frequency in the next generation will be within the interval:

$$p' = p \pm 2\sqrt{\frac{(pq)}{(2N)}} \qquad (6.1)$$

For example, with an allele frequency of 0.50 and an N of 10, the allele frequency in the next generation will be in the interval 0.28 to 0.72 with 95% probability (Equation 6.1). In contrast, with a p of 0.5 and an N of 200, this interval is only 0.45 to 0.55.

6.3 The inbreeding effect of small populations

Genetic drift is expected to cause a loss of genetic variation from generation to generation. **Inbreeding** occurs when related individuals mate with one another. Inbreeding is one consequence of small population size; see Chapter 17 for a detailed consideration of inbreeding. For example, in an animal species with $N = 2$, the parents in each generation will be full sibs (i.e., brother and sister). Matings between relatives will cause an increase in homozygosity. The **inbreeding coefficient** (f) is the probability that the two alleles at a locus within an individual are identical by descent (i.e., identical because they are derived from a common ancestor in a previous generation). We will consider several different inbreeding coefficients that have specialized meaning (e.g., F_{IS}, F_{ST}, etc. in Chapter 9 and F in Chapter 17). We will use the general inbreeding coefficient f in this chapter as defined here, along with its counterpart heterozygosity (h), which is equal to $1 - f$.

In general, the increase in homozygosity due to genetic drift will occur at the following rate per generation:

$$\Delta f = \frac{1}{2N} \qquad (6.2)$$

This effect was first discussed by Gregor Mendel, who pointed out that one-half of the progeny of

Table 6.2 This table appeared in Mendel's original paper in 1866. He was considering the expected genotypic ratios in subsequent generations from a single hybrid (i.e., heterozygous) individual that reproduced by self-fertilization. He assumed that each plant in each generation (Gen) had four offspring. The homozygosity (Homo) and heterozygosity (Het) columns did not appear in the original paper.

Gen	AA	Aa	aa	AA :	Aa :	aa	Homo	Het
				Ratio				
1	1	2	1	1	2	1	0.500	0.500
2	6	4	6	3	2	3	0.750	0.250
3	28	8	28	7	2	7	0.875	0.125
4	120	16	120	15	2	15	0.938	0.062
5	496	32	496	31	2	31	0.969	0.031
n				$2^n - 1$	2	$2^n - 1$	$1 - (1/2)^n$	$(1/2)^n$

a heterozygous self-fertilizing plant will be heterozygous; one-quarter will be homozygous for one allele; and the remaining one-quarter will be homozygous for the other allele (Table 6.2). This is as predicted by Equation 6.2 ($N = 1$, $\Delta f = 0.50$).

We have seen that the expected rate of loss of heterozygosity per generation is $\Delta f = 1/2N$; therefore, after t generations:

$$f_t = 1 - (1 - \frac{1}{2N})^t \qquad (6.3)$$

f_t is the expected increase in homozygosity at generation t and is known by a variety of names (e.g., **autozygosity**, **fixation index**, or the inbreeding coefficient) depending upon the context in which it is used.

It is often more convenient to keep track of the amount of variation remaining in a population using h (heterozygosity), where:

$$f = 1 - h \qquad (6.4)$$

Therefore, the expected decline in h per generation is:

$$\Delta h = -\frac{1}{2N} \qquad (6.5)$$

so that after one generation:

$$h_{t+1} = (1 - \frac{1}{2N})h_t \qquad (6.6)$$

Example 6.1 Bottleneck in the Mauritius kestrel

Kestrels on the Indian Ocean Island of Mauritius went through a bottleneck of one female and one male in 1974 (Nichols et al. 2001). The population had fewer than 10 birds throughout the 1970s, and there were fewer than 50 birds in this population for many years because of the widespread use of pesticides from 1940–1960. However, this population grew to nearly 500 birds by the mid-1990s. Nichols et al. (2001) examined the loss in genetic variation in this population at 10 microsatellite loci by comparing living birds to 26 ancestral birds from museum skins that were up to 170 years old. The heterozygosity of the restored population was 0.099 compared with heterozygosity in the ancestral birds of 0.231. Thus, 43% of the initial heterozygosity was retained.

The amount of heterozygosity expected to remain in Mauritius kestrels after one generation of a bottleneck of $N = 2$ can be estimated with Equation 6.6:

$$(1 - \frac{1}{2N})h_t = (1 - \frac{1}{4})(0.231) = 0.173$$

We can use Equation 6.7 to see that the amount of heterozygosity in the restored population of Mauritius kestrels is approximately the same as we would expect after a bottleneck of two individuals for three generations:

$$(1 - \frac{1}{2N})^t h_o = (0.75)^3 (0.231) = 0.097$$

The actual bottleneck in Mauritius kestrels was almost certainly longer than three generations with more birds than two birds each generation. However, the equations in this chapter all assume **discrete generations** and cannot be applied directly to species such as the Mauritius kestrels that have **overlapping generations**.

The heterozygosity after t generations can be found by:

$$h_t = (1 - \frac{1}{2N})^t h_0 \qquad (6.7)$$

where h_o is the initial heterozygosity. Example 6.1 shows how these expressions can be used to predict the effects of population **bottlenecks**.

Figure 6.4 shows this effect at a locus with two alleles and an initial frequency of 0.5 in a series of computer simulations of eight populations that

consist of 20 individuals each. These 20 diploid individuals possess 40 gene copies at any given locus. Forty gametes must be drawn from these 40 parental gene copies to form the next generation. The genotype of any one selected gamete does not affect the probability of the next gamete that is drawn; this is similar to a coin toss where one outcome does not affect the probability of the next toss.

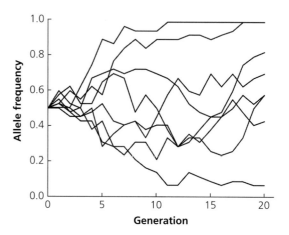

Figure 6.4 Computer simulations of genetic drift at a locus having two alleles with initial frequencies of 0.5 in eight populations of 20 individuals each.

Two of the eight populations simulated in Figure 6.4 became fixed for the A allele. Both of the alleles were retained by six of the populations after 20 generations. The heterozygosity in each of the populations is shown in Figure 6.5. There are large differences among populations in the decline in heterozygosity over time. Nevertheless, the mean decline in heterozygosity for all eight populations is very close to that predicted with Equation 6.6.

The heterozygosity at any single locus with two alleles is equally likely to increase or decrease from one generation to the next (except in the case of maximum heterozygosity when the allele frequencies are at 0.5). This may seem counterintuitive in view of Equation 6.6, which describes a monotonic decline in heterozygosity. Heterozygosity at a locus with two alleles is at a maximum when the two alleles are equally frequent ($p = q = 0.5$; Figure 6.6). The frequency of any particular allele is equally likely to increase or decrease due to genetic drift. Thus, heterozygosity will increase if the allele frequency

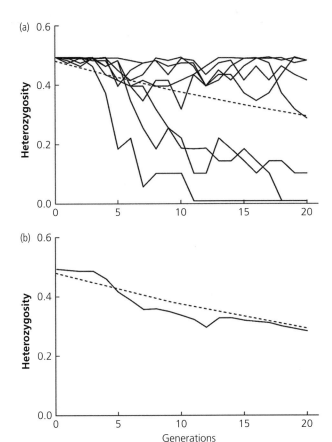

Figure 6.5 (a) Expected heterozygosities (2*pq*) in the eight populations (*N* = 20) undergoing genetic drift as shown in Figure 6.4. The dashed line shows the expected change in heterozygosity using Equation 6.6. (b) Mean heterozygosity for all loci (solid line) and the expected heterozygosity using Equation 6.6 (dashed line).

drifts toward 0.5, and it will decrease if the allele frequency drifts toward 0 or 1. However, the expected net loss is greater than the net gain in heterozygosity in each generation by 1/2*N*.

6.4 Loss of allelic diversity

We have so far measured the loss of genetic variation caused by small population size by the expected reduction in heterozygosity (*h*). There are other ways to measure genetic variation and its loss. A second important measure of genetic variation is the number of alleles present at a locus (*A*). There are advantages and disadvantages to both of these measures.

Heterozygosity has been widely used because it is proportional to the amount of **genetic variance** at a locus, and it lends itself readily to theoretical

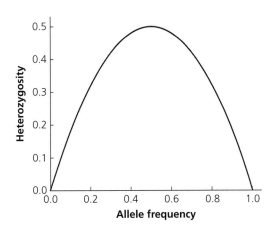

Figure 6.6 Expected heterozygosity (2*pq*) at a locus with two alleles as a function of allele frequency.

considerations of the effect of finite population size on genetic variation. In addition, the expected reduction in heterozygosity because of genetic drift is independent of the number of alleles present. Finally, estimates of heterozygosity from empirical data are relatively insensitive to sample size, whereas estimates of the number of alleles in a population are strongly dependent upon sample size. Therefore, comparisons of heterozygosities in different species or populations are generally more meaningful than comparisons of the number of alleles detected.

Nevertheless, heterozygosity has the disadvantage of being relatively insensitive to the effects of bottlenecks (Allendorf 1986). The difference between heterozygosity and A is greatest with extremely small bottlenecks (Figure 6.7). For example, a population with two individuals is expected to lose only 25% ($1/2N = 25\%$) of its heterozygosity. Thus, 75% of the heterozygosity in a population will be retained even through such an extreme bottleneck. However, two individuals can possess a maximum of four different alleles. Thus, considerably more of the allelic variation may be lost during a bottleneck if there are many alleles present at a locus.

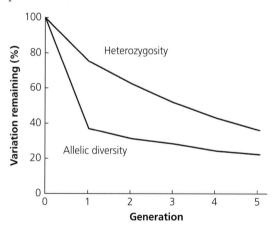

Figure 6.7 Simulated loss of heterozygosity and allelic diversity at eight microsatellite loci during a bottleneck of two individuals for five generations. The initial allele frequencies are from a population of brown bears from the Western Brooks Range of Alaska. Redrawn from Luikart & Cornuet (1998).

The effect of a bottleneck on the number of alleles present is more complicated than the effect on

heterozygosity because it is dependent on both the number and frequencies of alleles present (Allendorf 1986). The probability of an allele being lost during a bottleneck of size N is:

$$(1 - p)^{2N} \qquad (6.8)$$

where p is the frequency of the allele. This is the probability of sampling all of the gametes to create the next generation ($2N$) without selecting at least one copy of the allele in question. Rare alleles (say $p < 0.10$) are especially susceptible to loss during a bottleneck. However, the loss of rare, potentially important, alleles will have little effect on heterozygosity. For example, an allele at a frequency of 0.01 has a 60% chance of being lost following a bottleneck of 25 individuals (Equation 6.8). Figure 6.8 shows the probability of the loss of rare alleles during a bottleneck of N individuals. Greenbaum et al. (2014) provide a method to estimate the expected effects of a bottleneck on allelic diversity.

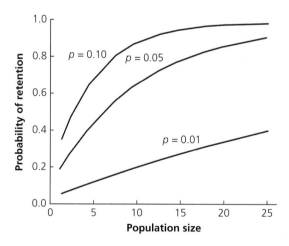

Figure 6.8 Probability of retaining a rare allele ($p = 0.01$, 0.05, or 0.10) after a bottleneck of size N for a single generation (Equation 6.8).

In general, if a population is reduced to N individuals for one generation then the expected total number of alleles (A') remaining is:

$$E(A') = A - \sum_{j=1}^{A}(1 - p_j)^{2N} \qquad (6.9)$$

where A is the initial number of alleles and p_j is the frequency of the jth allele. For example, consider a locus with two alleles at frequencies of 0.9 and 0.1 and a bottleneck of just two individuals. In this case:

$$E(A') = 2 - (1 - 0.9)^4 - (1 - 0.1)^4 = 1.34$$

Thus, on the average, we expect to lose one of these two alleles nearly two-thirds of the time. In contrast, there is a much greater expected probability of retaining both alleles at a locus with two alleles if the two alleles are equally frequent:

$$E(A') = 2 - (1 - 0.5)^4 - (1 - 0.5)^4 = 1.88$$

Thus, the expected loss of alleles during a bottleneck depends upon the number and frequencies of the alleles present. This is in contrast to heterozygosity, which is lost at a rate of $1/2N$, regardless of the current heterozygosity.

The loss of alleles during a bottleneck will have a drastic effect on the overall genotypic diversity of a population. As we saw in Section 5.2, the number of genotypes grows very quickly as the number of alleles increases. For example, the number of possible genotypes at a locus with two, five, and 10 alleles is three, 15, and 55, respectively. Thus, the loss of alleles during a bottleneck will greatly reduce the genotypic diversity in a population.

6.5 Founder effect

The founding of a new population by a small number of individuals will cause abrupt changes in allele frequency and loss of genetic variation (Example 6.2). Such severe bottlenecks in population size are a special case of genetic drift. Perhaps surprisingly, however, even extremely small bottlenecks have relatively little effect on heterozygosity. For example, with sexual species the smallest possible bottleneck is $N = 2$. Even in this extreme case, the population will only lose 25% of its heterozygosity in one generation (Equation 6.5). Stated in another way, just two individuals randomly selected from any population, regardless of size, will contain 75% of the total heterozygosity in the original population. We can also use Equation 6.5 to estimate the size of the founding population if we know how much heterozygosity has been lost through the founding bottleneck.

Table 6.3 Allele frequencies (p) and heterozygosities (h) at two loci in 16 subpopulations of guppies four generations after being founded by a single female and a single male. H_e is the mean heterozygosity at the two loci. Data from Nakajima et al. (1991).

Subpopulation	AAT-1		PGM-1		
	p	h	p	h	H_e
1	0.521	0.499	0.677	0.437	0.468
2	0.738	0.387	0.600	0.480	0.433
3	0.377	0.470	0.131	0.227	0.349
4	0.915	0.156	0.939	0.114	0.135
5	0.645	0.458	0.638	0.461	0.460
6	0.571	0.490	0.548	0.495	0.492
7	0.946	0.102	0.833	0.278	0.190
8	0.174	0.287	0.341	0.449	0.368
9	0.617	0.473	0.500	0.500	0.486
10	0.820	0.295	0.640	0.461	0.378
11	0.667	0.444	0.917	0.152	0.298
12	0.219	0.342	0.531	0.498	0.420
13	1.000	0.000	0.838	0.272	0.136
14	0.250	0.375	0.853	0.251	0.313
15	0.375	0.469	0.740	0.385	0.427
16	0.152	0.258	0.582	0.486	0.372
—	—	—	—	—	—
Average	0.562	0.344	0.644	0.372	0.358
Original Colony	0.581	0.487	0.605	0.478	0.482

A laboratory experiment with guppies clearly demonstrates the effect of reduced heterozygosity as a function of the size of the bottleneck (Nakajima et al. 1991). Sixteen separate **subpopulations** were derived from a large random mating laboratory colony of guppies by mating a female with a single male. After four generations, each of these subpopulations contained more than 500 individuals. Approximately 45 fish were then sampled from each subpopulation and genotyped at two allozyme loci that were polymorphic in the original colony (Table 6.3).

The mean heterozygosity at both loci in these 16 subpopulations was 0.358, in comparison with heterozygosity in the original colony of 0.482. Thus, the mean heterozygosity in the subpopulations, following a bottleneck of two individuals, was 26% lower than in the population from which the subpopulations were founded. This agrees very closely with Equation 6.5, which predicts a 25% reduction following a bottleneck of two individuals. Nevertheless, there are large differences in the amount of heterozygosity lost among subpopulations. For example, subpopulations 4 and 13 lost over 70%

Example 6.2 Effects of founding events on allelic diversity in a snail

The land snail *Theba pisana* was introduced from Europe into Western Australia in the 1890s. A colony was founded in 1925 on Rottnest Island with animals taken from the mainland population near Perth. Johnson (1988) reported the allele frequencies at 25 allozyme loci. Figure 6.9 shows the loss of rare alleles caused by the bottleneck associated with the founding of a population in Perth on the mainland and in the second bottleneck associated with the founding of a population on nearby Rottnest Island. The height of each bar represents the number of alleles in that sample that had the frequency specified on the x-axis. For example, there were seven alleles that had a frequency of less than 0.05 in the founding French population. However, there were no alleles in either of the two Australian populations at a frequency of less than 0.05.

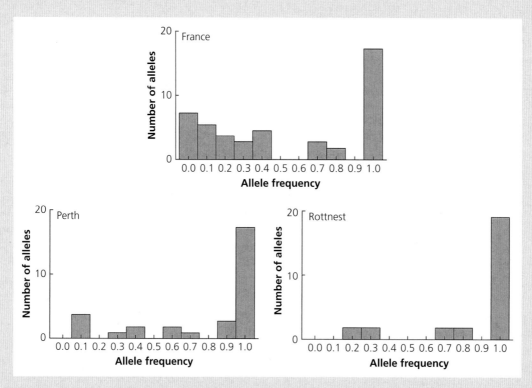

Figure 6.9 Effects of bottlenecks on the number of rare alleles at 25 allozyme loci in the land snail *Theba pisana* that was introduced from Europe into Western Australia in the 1890s. Data from Johnson (1988).

The distribution of allele frequencies, such as plotted in Figure 6.9, can be used to detect bottlenecks even when data are not available from the pre-bottlenecked population (Luikart et al. 1998). Rare alleles (frequency < 0.05) are expected to be common in samples from populations that have not been bottlenecked in their recent history, such as observed in the French sample of these snails. The complete absence of such rare alleles in Australia would have suggested that these samples came from recently bottlenecked populations even if the French sample was not available for comparison.

of their heterozygosity, while subpopulations 6 and 9 actually had increased heterozygosity!

The total amount of heterozygosity lost during a bottleneck depends upon how long it takes the population to return to a "large" size. That is, species such as guppies, in which individual females may produce 50 or so progeny, may quickly attain large enough population sizes following a bottleneck so that little further variation is lost following the initial bottleneck. However, species with lower population growth rates may persist at small population sizes for many generations during which heterozygosity is further eroded.

The growth rate of a population following a bottleneck can be modeled using the so-called logistic growth equation, which describes the size of a population after t generations based upon the initial population size (N_0), the intrinsic growth rate (r), and the equilibrium size of the population (K):

$$N(t) = \frac{K}{1 + be^{-rt}} \qquad (6.10)$$

The constant e is the base of the natural logarithm (~2.72), and b is a constant equal to $(K - N_0)/N_0$.

We can estimate the total expected loss in heterozygosity in the guppy example depending upon the rate of population growth of the subpopulations. The initial size of the subpopulations (N_o) was 2, and we assume the carrying capacity (K) was 500. We can then examine three different intrinsic growth rates (r): 1.0, 0.5, and 0.2. An r of 1.0 indicates that population size is increasing by a factor of 2.72 (e) each generation when population size is far below K. Similarly, r values of 0.5 and 0.2 indicate growth rates of 1.65 and 1.22 at small population sizes, respectively.

Equation 6.10 can be used to predict the expected population size each generation following the bottleneck. We expect heterozygosity to be eroded at a rate of $1/2N$ in each of these generations. Figure 6.10 shows the expected loss in heterozygosity in our guppy example for 10 generations following the bottleneck. As expected, populations having a relatively high growth rate ($r = 1.0$) will lose little heterozygosity following the initial bottleneck. However, heterozygosity is expected to continue

to erode even 10 generations following the bottleneck in populations with the slowest growth rate. In general, bottlenecks will have a greater and more long lasting effect on the loss of genetic variation in species with smaller intrinsic growth rates (e.g., large mammals) than species with high intrinsic growth rates (e.g., insects).

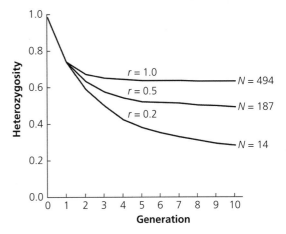

Figure 6.10 Expected heterozygosities in three subpopulations of guppies going through a bottleneck of two individuals and growing at different rates (r) according to the logistic growth equation (Equation 6.10). N is the expected population size for each subpopulation in the 10th generation.

Founder events and population bottlenecks will have a greater effect on the number of alleles in a population than on heterozygosity (Figure 6.7). Loci that are part of the major histocompatibility (MHC) system in vertebrates have been found to have many nearly equally frequent alleles. For example, Knafler et al. (2012) described allelic variation in 50 breeding pairs of the Magellanic penguins at the MHC *DRβ1* gene. They discovered 45 alleles in this sample of 200 gene copies. As we have seen, two birds chosen at random from this population are expected to retain 75% of the heterozygosity. However, two birds can at best possess four of the 45 different *MHC* alleles. Thus, at least 41 of the 45 alleles (91%) will be lost in a bottleneck of two individuals compared with the loss of 25% of the heterozygosity. Thus, bottlenecks of short duration may have little effect on heterozygosity but will severely reduce the number of alleles present at some loci (Example 6.3).

Example 6.3 Founding events in the Laysan finch

The Laysan finch is an endangered Hawaiian honeycreeper found on several islands in the Pacific Ocean (Tarr et al. 1998, Figure 6.11). The species underwent a bottleneck of ~100 birds on Laysan Island after the introduction of rabbits in the early 1900s. The population recovered rapidly after eradication of the rabbits and has fluctuated around a mean of 10,000 birds since 1968. In 1967, the US Fish and Wildlife Service translocated 108 finches to Southeast Island, one of several small islets ~300 km northwest of Laysan that comprise Pearl and Hermes Reef (PHR). The translocated population declined to 30–50 birds and then rapidly increased to some 500 birds on Southeast Island. Several smaller populations have since become established in other islets within PHR. Two birds colonized Grass Island in 1968 and six more finches were moved to this islet in 1970. The population of birds on Grass Island has fluctuated between 20 and 50 birds. In 1973, a pair of finches founded a population on North Island. The population of birds on North Island has fluctuated between 30 and 350 birds.

Figure 6.11 Laysan finch. Photo courtesy of C.R. Kohley.

Tarr et al. (1998) assayed variation at nine microsatellite loci to examine the effects of the founder events and small population sizes in these four populations (Table 6.4). Their empirical results are in close agreement with theoretical expectations. The average heterozygosity on Southeast Island is ~8% less than on Laysan Island; the heterozygosities on the two other islands are ~30% less than the original founding population on Laysan. All three newly founded populations have fewer alleles than the founding population on Laysan. In all cases, the proportion of alleles retained is less than the proportion of heterozygotes retained. The extreme case is Grass Island, where 75% of the heterozygosity was retained, but only 53% of the allelic variation was retained.

It is worth noting that heterozygosities at four of the nine loci are actually greater in the post-bottleneck population on Southeast Island than on Laysan; see the discussion in Section 6.3 and Figure 6.5. This demonstrates that it is important to examine many loci in order to detect and quantify the effects of bottlenecks in populations on heterozygosity.

Continued

Example 6.3 *Continued*

Table 6.4 Numbers of alleles (*A*) and observed heterozygosities (Het) at nine microsatellite loci in Laysan finches in four island populations. The number in parentheses after the name of the island is the sample size. The populations on the other three islands were all founded from birds from Laysan. The bottom row shows the average allelic diversity and heterozygosity in the sample. The value at the bottom of the *A* column is (*A'* − 1)/(*A* − 1), where *A'* is the number of alleles in the sampled population, and *A* is the number of alleles in the Laysan sample (Allendorf 1986). This value ranges from 100%, when all alleles are retained, to zero when all alleles except one are lost. Data from Tarr et al. (1998).

Locus	Laysan (44) A	Het	Southeast (43) A	Het	North (43) A	Het	Grass (36) A	Het
Tc.3A2C	2	0.558	2	0.535	2	0.535	2	0.528
Tc.4A4E	2	0.386	2	0.605	2	0.209	2	0.556
Tc.5A1B	3	0.372	3	0.233	1	0	2	0.583
Tc.5A5A	3	0.409	2	0.071	2	0.372	2	0.278
Tc.1A4D	3	0.659	3	0.744	3	0.698	2	0.528
Tc.11B1C	3	0.636	3	0.674	3	0.628	3	0.194
Tc.11B2E	3	0.614	3	0.488	1	0	2	0.500
Tc.11B4E	4	0.614	4	0.628	2	0.256	3	0.444
Tc.12B5E	5	0.568	4	0.442	3	0.372	1	0
All loci	3.11	0.535	2.89	0.491	2.11	0.341	2.11	0.401
	100%	100%	90%	92%	53%	64%	53%	75%

6.6 Genotypic proportions in small populations

We saw in the guppy example (Table 6.3) that the separation of a large random mating population into a number of subpopulations can cause a reduction in heterozygosity, and a corresponding increase in homozygosity. However, genotypes within each subpopulation will be in HW proportions as long as random mating occurs within the subpopulations. It may seem paradoxical that heterozygosity is decreased in small populations, but the subpopulations themselves remain in HW proportions. The explanation is that the reduction in heterozygosity is caused by changes in allele frequency from one generation to the next, while HW genotypic proportions will occur in any one generation as long as mating is random (Section 5.3).

In fact, there actually is a tendency for an excess of heterozygotes in small populations of animals and plants with separate sexes (Example 6.4). Different allele frequencies in the two sexes will cause an excess in heterozygotes relative to HW proportions

(Robertson 1965; Kirby 1975). An extreme example of this is hybrids produced by males from one strain (or species) and females from another so that all progeny are heterozygous at any loci where the two strains differ. In this case, however, genotypic proportions will return to HW proportions in the next generation.

In small populations, allele frequencies are likely to differ between the sexes just due to chance. On average, the frequency of heterozygotes in the progeny population will exceed HW expectations by a proportion of:

$$\frac{1}{8N_\mathrm{m}} + \frac{1}{8N_\mathrm{f}} \qquad (6.11)$$

where N_m and N_f are the numbers of male and female parents (Robertson 1965). This result holds regardless of the number of alleles at the locus concerned. This reduces to $1/2N$ if there is an equal number of males and females.

Let us consider the extreme case of a population with one female and one male ($N = 2$) and two alleles (Table 6.5). There are six possible types of matings.

Mating between identical **homozygotes** (either *AA* or *aa*) will produce monomorphic progeny. Progeny produced by matings between two heterozygotes will result in the expected HW proportions. However, the other three matings will result in an excess of heterozygotes. The extreme case is a mating between opposite homozygotes, which will produce all heterozygous progeny. On average, there will be a 25% excess of heterozygotes in populations produced by a single male and a single female (Equation 6.11).

With more than two alleles, there will be a deficit of each homozygote and an overall excess of heterozygotes. However, some heterozygous genotypes may be less frequent than expected by HW proportions, despite the overall excess of heterozygotes.

Table 6.5 Expected Mendelian genotypic proportions for all possible matings at a locus with two alleles in a population with a single female and a single male. F_{IS} is a measure of the deficit of heterozygotes observed relative to the expected HW proportions (Section 9.1). A negative F_{IS} indicates an excess of heterozygotes.

Mating	AA	Aa	aa	Freq(A)	F_{IS}
AA × AA	1.00	0	0	1.00	—
AA × Aa	0.50	0.50	0	0.75	−0.33
AA × aa	0	1.00	0	0.50	−1.00
Aa × Aa	0.25	0.50	0.25	0.50	0.00
Aa × aa	0	0.50	0.50	0.25	−0.33
aa × aa	0	0	1.00	0.00	—

Let's revisit the Channel Island fox study (Funk et al. 2016) that we looked at in Section 5.3.4. In

---

Example 6.4 An island population of little spotted kiwi

The population of little spotted kiwi on Long Island, New Zealand, was founded by the introduction of one female in 1982 and one male in 1989 (Taylor et al. 2017a). This population was sampled in 2011–2013 when it comprised the founding pair (alive and still reproductively active), first-generation offspring (27 samples), and later descendants (14 samples).

Table 6.6 shows tests for HW proportions at five of the 15 polymorphic microsatellite loci in this population. The two founders were homozygous for different alleles (*11* and *22*) at the *Aptowe31* locus. All F₁ individuals were heterozygous (*12*) at this locus. A 25% excess of heterozygotes is expected if all of the progeny came from just the two founding individuals (Equation 6.11). On average, there was a 37% excess of heterozygotes at these five polymorphic loci, just slightly greater than the expected value of 25%.

Table 6.6 Observed (and expected) genotypic proportions at five microsatellite loci in little spotted kiwi sampled from Long Island, New Zealand. $\hat{p}(1)$ and $\hat{p}(2)$ are the estimated frequencies of the 1 and 2 allele. F_{IS} equals $[1 − (H_O/H_e)]$ and is a measure of the departure from HW proportions (see Section 9.1). A negative F_{IS} indicates an excess of heterozygotes. From Taylor et al. (2017a) and personal communication. * $P < 0.05$, *** $P < 0.001$.

Locus	Genotype						$\hat{p}(1)$	$\hat{p}(2)$	F_{IS}
	11	12	22	13	23	33			
Aptowe1	27 (28.5)	16 (13.0)	0 (1.5)	—	—	—	0.814	0.186	−0.23
Aptowe29	6 (11.6)	16 (9.9)	0 (2.1)	14 (8.8)	2 (3.8)	0 (1.7)	0.553	0.237	−0.42*
Aptowe31	6 (11.3)	31 (20.4)	4 (9.3)	—	—	—	0.524	0.476	−0.52***
Aptowe35	3 (2.9)	14 (10.5)	3 (9.5)	2 (5.8)	20 (10.5)	0 (2.9)	0.261	0.476	−0.35*
Apt59	18 (20.9)	24 (18.1)	1 (3.9)	—	—	—	0.696	0.304	−0.32*
Mean									−0.37

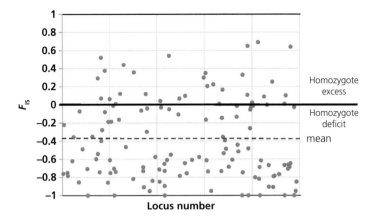

Figure 6.12 Distribution of F_{IS} values at 131 SNP loci in a sample of 46 Channel Island foxes from San Nicolas Island (data from Funk et al. 2016). F_{IS} is the proportional excess (positive values) or deficit (negative values) of homozygotes (Section 9.1).

Figure 5.2, we saw the distribution of F_{IS} values in a sample of 46 foxes from Santa Catalina Island. There was a slight excess of homozygotes in this sample. Very different results were found in an examination of the same loci in 46 foxes from nearby San Nicolas Island. There is a substantial deficit of homozygotes in this sample ($F_{IS} = -0.386$; Figure 6.12). This almost certainly results from the small population size of foxes on this island. An estimate of the genetically effective population size (Chapter 7) of foxes on this island is approximately two. Supporting this observation, only 131 of these SNP loci were polymorphic compared with 3,940 loci in the Santa Catalina sample.

6.7 Effects of genetic drift

We have considered in some detail how genetic drift is expected to affect allele frequencies and reduce the amount of genetic variation in small populations. We will now preview the effects that this loss of genetic variation is expected to have on the population itself (Box 6.1). That is, how will the loss of genetic variation expected in small populations affect the capability of a population to persist and evolve? We will take a more in-depth look at these effects in later chapters.

6.7.1 Changes in allele frequency

Large changes in allele frequency from one generation to the next are likely in small populations due to chance. This effect may cause an increase in frequency of alleles that have harmful

effects. Such **deleterious** alleles are continually introduced by mutation but are kept at low frequencies by natural selection. Moreover, most of these harmful alleles are recessive so that their harmful effects on the phenotype are only expressed in homozygotes.

Let us consider the possible effect of a population bottleneck of two individuals. As we have seen, most rare alleles will be lost in such a small bottleneck. However, any allele for which one of the two founders is heterozygous will be found in the new population at a frequency of 25%. Thus, rare deleterious alleles present in the founders will jump in frequency to 25%. Of course, at most loci the two founders will not carry a harmful allele. However, every individual carries harmful alleles at some loci. Therefore, we cannot predict which particular harmful alleles will increase in frequency following a bottleneck, but we can predict that several harmful alleles that were rare in the original population will be found at much higher frequencies. And if the bottleneck persists for several generations, these harmful alleles may become more frequent in the new population.

This effect is commonly seen in domestic animals such as dogs in which breeds often originated from a small number of founders. Different dog breeds usually have some characteristic genetic abnormality that is much more common within the breed than in the species as a whole (Hutt 1979). For example, dalmatians were originally developed from a few founders that were selected for their running ability and distinctive spotting pattern. Dalmatians are susceptible to kidney stones because they excrete

---

Box 6.1 Population bottlenecks and decreased hatching success in endangered birds

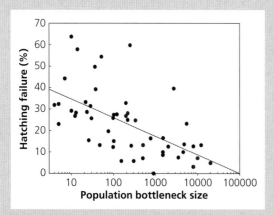

Figure 6.13 Effect of bottleneck size (smallest number of individuals recorded in the population) and percent hatching failure in threatened birds. Hatching failure is plotted on a linear scale and bottleneck size is plotted on a logarithmic scale, although both were log transformed in analyses. Redrawn from Heber & Briskie (2010).

Many populations have undergone severe bottlenecks because of reduced population size and increased fragmentation caused by habitat loss. As we have seen in this chapter, small populations are vulnerable to genetic drift and the inbreeding effect of small populations. Increased hatching failure is a common result of inbreeding in

birds. Heber & Briskie (2010) have shown that bottlenecks in endangered birds have a major effect on hatching success.

Hatching failure is generally less than 10% in noninbred bird populations. Hatching failure is sometimes more than 50% in some inbred populations. Given the increase of inbreeding in smaller populations, Heber and Briskie tested if hatching success decreases in bird populations that have gone through a population bottleneck.

They summarized rates of hatching failure in 51 threatened bird species from 31 families. The size of introductions ranged from 4 to 20,000 individuals. They estimated hatching failure as the proportion of eggs incubated to term that failed to hatch, excluding failure due to desertion, predation, or adverse weather. Under this definition, eggs that failed to hatch were either infertile or died during embryonic development, both of which are thought to increase as inbreeding increases (Jamieson & Ryan 2000).

Heber and Briskie found a substantial increase in hatching failure associated with bottlenecks with smaller population sizes ($P < 0.001$; Figure 6.13). The exact threshold of population size below which inbreeding depression is likely to cause a problem varies among species and traits. Nevertheless, hatching failure was greater than 10% in all species that went through a bottleneck of fewer than 150 individuals.

---

exceptionally high amounts of uric acid in their urine. This difference is due to a recessive allele at a single locus (Trimble & Keeler 1938). Apparently, one of the principal founders of this breed carried this recessive allele, and it subsequently drifted to high frequency in this breed.

Similar effects occur in small wild populations of conservation concern. For example, the Scottish population of red-billed choughs currently has fewer than 60 breeding pairs (Trask et al. 2016). Some of the nestlings in this population are affected by lethal blindness, which is inherited as a Mendelian autosomal recessive allele. The estimated frequency of the blindness allele is 0.126. As we will see in Chapter 8 and Example 21.3, reducing the frequency of such harmful recessive alleles is very

difficult because almost all of the copies are present in phenotypically normal heterozygotes rather than in affected homozygotes.

6.7.2 Loss of allelic diversity

We have seen in Section 6.4 that genetic drift will have a much greater effect on the allelic diversity of a population than on heterozygosity if there are many alleles present at a locus. Evidence from many species indicates that loci associated with disease resistance often have many alleles. The best example of this is the MHC in vertebrates (Edwards & Hedrick 1998). MHC in humans consists of five major tightly linked genes on chromosome 6 (Pierini & Lenz 2018). Many alleles occur at all of these loci; for example, there are 10 or more nearly equally frequent alleles at the *HLA-A* locus and 15 or more at the *HLA-B* locus.

MHC molecules assist in the triggering of the immune response to disease-causing organisms. Individuals heterozygous at MHC loci are relatively more resistant to a wider array of pathogens than are homozygotes (Hedrick 2002). Most vertebrate species that have been studied have been found to harbor many MHC alleles. Thus, the loss of allelic diversity at MHC loci is likely to render small populations of vertebrates much more susceptible to disease epidemics (e.g., Ujvari & Belov 2011; Quigley et al. 2020).

A similar situation holds in Resistance genes (R-genes) that are found in many plants. These genes are involved in pathogen recognition, and they often have many alleles at individual loci (Bergelson et al. 2001). Marden et al. (2017) found that tropical species with smaller local population sizes had lower R-gene allelic diversity and lower recognition-dependent immune responses. This resulted in greater susceptibility to species-specific pathogens that may facilitate disease transmission in species with smaller local populations.

6.7.3 Inbreeding depression

The harmful effects of inbreeding have been known for a long time. Experiments with plants by Darwin and others demonstrated that loss of vigor generally accompanied continued selfing and that crossing different lines maintained by selfing restored the lost vigor. Livestock breeders also generally accepted that continued inbreeding within a herd or flock could lead to a general deterioration that could be restored by outcrossing. The first published experimental report of the effects of inbreeding in animals was with rats (Crampe 1883; Ritzema-Bos 1894).

The implication of these results for wild populations did not go unnoticed by Darwin. It occurred to him that fallow deer kept in British parks might be affected by isolation and "long-continued close interbreeding." He was especially concerned because he was aware that the effects of inbreeding

may go unnoticed because they accumulate slowly. Darwin (1896, p. 99) inquired about this effect and received the following response from an experienced gamekeeper:

. . . the constant breeding in-and-in is sure to tell to the disadvantage of the whole herd, though it may take a long time to prove it; moreover, when we find, as is very constantly the case, that the introduction of fresh blood has been of the greatest use to deer, both by improving their size and appearance, and particularly by being of service in removing the taint of "rickback" if not other diseases, to which deer are sometimes subject when the blood has not been changed, there can, I think, be no doubt but that a judicious cross with a good stock is of the greatest consequence, and is indeed essential, sooner or later, to the prosperity of every well-ordered park.

Despite Darwin's concern and warning, these early lessons from agriculture were largely ignored by those responsible for the management of wild populations of game and by captive breeding programs of zoos for nearly 100 years (see Voipio 1950 for an exception).

A seminal paper in 1979 by Kathy Ralls and her colleagues had a dramatic effect on the application of genetics to the management of wild and captive populations of animals. They used zoo pedigrees of 12 species of mammals to show that individuals from matings between related individuals tended to show reduced survival relative to progeny produced by matings between unrelated parents. The pedigree inbreeding coefficient (F_P) is the expected increase in homozygosity for inbred individuals; it is also the expected decrease in heterozygosity throughout the genome of inbred individuals (Section 17.1.1). One of us (FWA) can clearly remember being excitedly questioned in the hallway by our departmental mammalogist who had just received his weekly issue of *Science* and could not believe the data of Ralls and her colleagues. Subsequent studies (Ralls & Ballou 1983; Ballou 1997) have supported their original conclusions (Figure 6.14).

Inbreeding depression results from both increased homozygosity and reduced heterozygosity (Section 17.4). That is, a greater number of deleterious recessive alleles will be expressed in inbred individuals because of their increased homozygosity. In addition, fitness of inbred

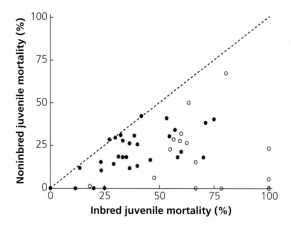

Figure 6.14 Effects of inbreeding on juvenile mortality in 44 captive populations of mammals (16 ungulates, 16 primates, and 12 small mammals). The line shows equal mortality in inbred and noninbred progeny. The preponderance of points below the line (42 of 44, 95%) indicates that inbreeding generally increased juvenile mortality. The open circles indicate populations in which juvenile mortality of inbred and noninbred individuals were significantly different (P < 0.05; exact test). Data from Ralls & Ballou (1983).

individuals will be reduced at loci at which the heterozygotes have a selective advantage over all homozygous types (**heterozygous advantage** or overdominance). Both of these mechanisms are likely to contribute to inbreeding depression, but it is thought that increased expression of deleterious recessive alleles is the more important mechanism (Charlesworth & Charlesworth 1987; Hedrick & Kalinowski 2000). Inbreeding depression is considered in detail in Chapter 17.

Guest Box 6 Detecting bottlenecks in the critically endangered kākāpō
Yasmin Foster, Nicolas Dussex, and Bruce C. Robertson

Reducing the impact of bottlenecks and genetic drift by maintaining or increasing genetic diversity (via genetic rescue) is paramount to species recovery (Bell et al. 2019). Bottlenecks and the subsequent inbreeding and loss of heterozygosity have implications for population persistence and adaptive potential. To understand the role genetics plays in conservation outcomes, whole genome sequencing approaches can be used to detect bottlenecks and resulting losses of genetic diversity, and study the associated impacts of inbreeding depression.

The New Zealand kākāpō is a unique parrot, being flightless, nocturnal, and displaying a polygynous **lek** breeding system where certain males dominate breeding (Figure 6.15). The extant population consists of 211 individuals (June 2020) founded by a small insular population from Stewart Island discovered in the 1970s (61 birds) and one surviving male from the mainland of New Zealand (Powlesland et al. 2006). Kākāpō are unusually long-lived and individuals from the founding population still contribute their genetic potential. Understanding the genetic diversity of these founding individuals, including the differences between the two descendant groups, is crucial for long-term species management. Since the late 1990s, genetic management of kākāpō has aimed to increase genetic diversity by favoring the mating of the mainland lineage, and reducing consanguineous matings via placement of breeders on various offshore islands (Robertson 2006).

Figure 6.15 Male kākāpō, Jamieson, named after the late Professor Ian Jamieson (see Section 18.8). Photo courtesy of Jake Osborne.

Kākāpō genetic management has employed microsatellites, mitochondrial, and immunity-associated DNA-based approaches to assess differences in heterozygosity, allelic diversity, and relatedness between the founding island and mainland kākāpō, to construct their demographic history and to detect bottlenecks (Robertson et al. 2009; Grueber et al. 2015a).

Genetic analyses have informed relatedness and assigned relationships of the founding population, ultimately to maintain diversity by choosing unrelated individuals for mating and artificial insemination (Clout & Merton 1998; Bergner et al. 2014). Significant loss of immunity-associated MHC allelic diversity has been identified between the two descendant groups (Knafler et al. 2014), and **heterozygosity–fitness correlations (HFCs)** were detected for both clutch size and hatching success in female kākāpō (White et al. 2015). Temporal comparisons between historical and modern data reveal the genetic consequences of bottlenecks (Díez-del-Molino et al. 2018); using microsatellite and mitochondrial DNA of museum and modern kākāpō samples indicated severe reduction in genetic diversity (Bergner et al. 2016; Dussex et al. 2018b), consistent with a near-extinction of the species.

Genome-wide SNPs and a chromosome-level genome assembly are currently used in kākāpō management. Genomic relatedness matrices using reduced-representation sequencing now aid in choosing the most unrelated kākāpō for artificial insemination and breeding management. Inbreeding events lead to the accumulation of runs of homozygosity (ROH) across the genome (Section 17.1.2). Historical and modern kākāpō genomes indicate differences in inbreeding and mutational load between extinct and extant populations (Figure 6.16, Dussex et al. 2021). Despite having lower genetic diversity than mainland descendants, the highly inbred Stewart Island population (isolated from the mainland for some 10,000 years) shows reduced mutational load consistent with purging of harmful mutations (via purifying selection). Additionally, individual inbreeding is not associated with the survival of chicks, despite significant differences between descendant groups, further supporting purging of deleterious alleles (Foster et al. 2021).

Guest Box 6 Continued

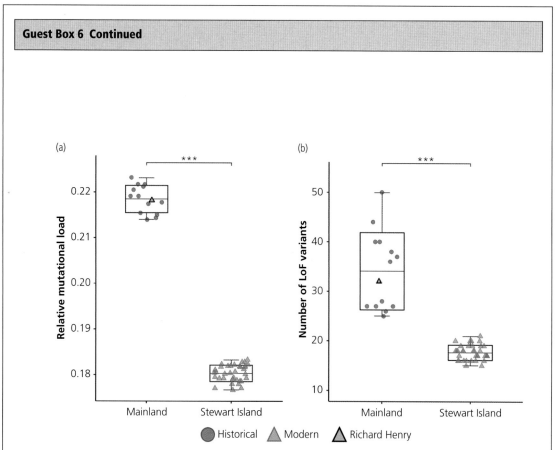

Figure 6.16 Estimates of mutational load in kākāpō (Dussex et al. 2021). (a) Individual relative mutational load measured as the sum of all homozygous and heterozygous derived alleles multiplied by their conservation score over the total number of derived alleles. (b) Number of loss of function (LoF) variants. Modern birds include Stewart Island birds and Richard Henry. Middle thick lines within boxplots and whiskers extending from it represent mean and standard deviation, respectively ($P < 0.001$).

CHAPTER 7

Effective Population Size

Medium ground finch, Example 7.2

Effective population size (N$_e$) is one of the most fundamental evolutionary parameters of biological systems, and it affects many processes that are relevant to biological conservation.

(Robin S. Waples 2002, p. 148)

Use of genomic SNP data could improve N$_e$ estimation precision substantially, but could also cause a bias due to marker linkage (i.e. limited genome size).

(Mark Beaumont & Jinliang Wang 2019, p. 462)

We saw in the previous chapter that we expect heterozygosity to be lost at a rate of $1/2N$ per generation in ideal populations (Equation 6.5). However, this expectation holds only under conditions that rarely apply to real populations. For example, such factors as the number of individuals of reproductive age rather than the total of all ages, the sex ratio, and differences in reproductive success among individuals must be considered. Thus, the actual number of adult individuals in a natural population (census size, N_C) is not sufficient for predicting the rate of genetic drift. We will use the concept of effective population size to deal with the discrepancy between the demographic size and population size relevant to the rate of genetic drift in natural populations.

Perhaps the most important assumption of our model of genetic drift has been the absence of natural selection. That is, we have assumed that the genotypes under consideration do not affect the fitness (survival and reproductive success) of individuals. We would not be concerned with the retention of genetic variation in small populations if the

assumption of genetic neutrality were true for all loci in the genome. However, the assumption of neutrality and the use of neutral loci allow us to predict the effects of finite population size with great generality. In Chapter 8, we will consider the effects of incorporating natural selection into our basic models of genetic drift.

7.1 Concept of effective population size

Our consideration in the previous chapter of genetic drift dealt only with "ideal" populations. **Effective population size** (N_e) is the size of the **ideal** (Wright–Fisher) **population** (N) that will result in the same amount of genetic drift as in the actual population being considered. The basic ideal population consists of "N diploid individuals reconstituted each generation from a random sample of $2N$ gametes" (Wright 1939, p. 298). In an ideal population, individuals produce both female and male gametes (**monoecy**) and self-fertilization is possible. Under these conditions, heterozygosity will decrease by exactly $1/2N$ per generation.

Conservation and the Genomics of Populations, Third Edition. Fred W. Allendorf, *et al.*, Oxford University Press.
© Fred W. Allendorf, W. Chris Funk, Sally N. Aitken, Margaret Byrne, and Gordon Luikart (2022). DOI: 10.1093/oso/9780198856566.003.0007

We can see this by considering an ideal population of N individuals (say 10) in which each individual is heterozygous for two unique alleles for a total of 20 alleles (Figure 7.1). All of these 10 individuals will contribute equally to the gamete pool, which is sampled to create each individual in the next generation. Thus, each allele will be at a frequency of $1/2N = 0.05$ in the gamete pool. A new individual will only be homozygous if the same allele is present in both gametes. For the purposes of our calculations, it does not matter which allele is sampled first because all alleles are equally frequent. Let us say the first gamete chosen is $A15$. This individual will be homozygous only if the next gamete sampled is also $A15$. What is the probability that the next gamete sampled is $A15$? This probability is simply the frequency of the $A15$ allele in the gamete pool, which is $1/2N = 0.05$ because all 20 alleles (2×10) are at equal frequency in the gamete pool (Figure 7.1). Therefore, the expected homozygosity is $1/2N$, and the expected heterozygosity of each individual in the next generation is $1 - (1/2N) = 0.95$.

This conceptual model becomes more complicated if self-fertilization is prevented, or if the population is **dioecious**. In these two cases, the decrease in heterozygosity due to sampling individuals from the gamete pool will skip a generation because both gametes in an individual cannot come from the same parent. Nevertheless, the mean rate of loss per generation over many generations is similar in this case; heterozygosity is lost at a rate more closely approximated by $1/(2N+1)$ (Wright 1931; Crow and Denniston 1988). The difference between these two expectations, $1/2N$ and $1/(2N+1)$, is usually ignored because the difference is small except when N is very small (Luikart et al. 1999).

For our general purposes, the ideal population consists of a constant number of N diploid individuals ($N/2$ females and $N/2$ males) in which all parents have an equal probability of being the parent of any individual progeny. We will consider the following effects of violating the assumptions of such idealized populations on the rate of genetic drift:

1. Equal numbers of males and females.
2. All individuals have an equal probability of contributing an offspring to the next generation.
3. Constant population size.
4. Nonoverlapping (discrete) generations.

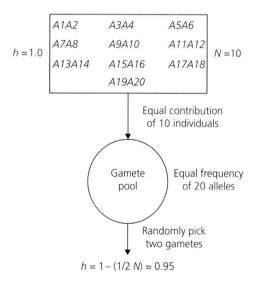

Figure 7.1 Diagram of reduction in heterozygosity (h) in an ideal population consisting of 10 individuals that are each heterozygous for two unique alleles ($h = 1$). Two gametes are picked from the gamete pool to create each individual in the next generation.

We have examined two expected effects of genetic drift: changes in allele frequency (Section 6.2) and a decrease in heterozygosity (Section 6.3). Thus, there are at least two possible measures of the effective population size. First, the "**variance effective number**" (N_{eV}) is whatever must be substituted in Equation 6.1 to predict the expected changes in allele frequency. Second, the "inbreeding effective number" (N_{eI}) is whatever must be substituted in Equation 6.2 to predict the expected reduction in heterozygosity (Example 7.1). See Ryman et al. (2019) for an extensive consideration of a variety of measures of effective population sizes. We will only consider the first two kinds of effective population size (N_{eV} and N_{eI}) because they have the most relevance for understanding the loss of genetic variation in populations.

Example 7.1 Effective population size of grizzly bears

Harris & Allendorf (1989) estimated the effective population size of grizzly bear populations using computer simulations based upon life history characteristics (survival, age at first reproduction, litter size, etc.). They estimated N_{eI} by comparing the loss of heterozygosity in the simulated populations to that expected in an ideal population of $N = 100$ (Figure 7.2). Over a wide range of conditions, the effective population size was ~25% of the actual population size. However, this method, like others, does not account for all factors that might reduce N_e (e.g., high variance in reproductive success).

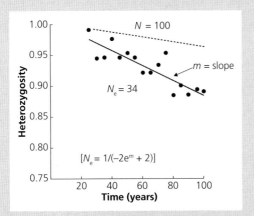

Figure 7.2 Estimation of effective size of grizzly bear populations ($N_C = 100$) by computer simulation. The dashed line shows the expected decline in heterozygosity over 10 generations (100 years) in an ideal population using Equation 6.7. The solid line shows the decline in heterozygosity in a simulated population. The decline in heterozygosity in the simulated population is equal to that expected in an ideal population of 34 bears; thus, $N_e = 34$ (m is the slope of the regression of the log of heterozygosity on time). From Harris & Allendorf (1989).

Crow & Denniston (1988) have clarified the distinction between these two measures of effective population size. In many cases, a population has nearly the same effective population size for either measure. Specifically, their values are identical in constant size populations in which the age and sex distributions are unchanging. We will first consider N_e to be the inbreeding effective population size under different circumstances because this number is most widely used, and we will then consider when these two numbers will differ.

7.2 Unequal sex ratio

Populations often have unequal numbers of males and females contributing to the next generation. The two sexes, however, contribute an equal number of autosomal genes to the next generation regardless of the total of males and females in the population. Therefore, the amount of genetic drift attributable to the two sexes must be considered separately. Consider the extreme case of one male mating with 100 females. In this case, all progeny will be half-sibs because they share the same father. In general, the rarer sex is going to have a much greater effect on genetic drift so that the effective population size will seldom be much greater than twice the size of the rarer sex.

What is the size of the ideal population that will lose heterozygosity at the same rate as the population we are considering which has different numbers of females and males? We saw in Section 6.3 that the increase in homozygosity due to genetic drift is caused by an individual being homozygous because its two gene copies were derived from a common ancestor in a previous generation. The inbreeding effective population size in a **monoecious** population in which selfing is permitted may be defined as the reciprocal of the probability that two uniting gametes come from the same parent. With separate sexes, or if selfing is not permitted, uniting gametes must come from different parents; thus, the effective population size is the probability that two uniting gametes come from the same grandparent.

The probability that the two uniting gametes in an individual came from a male grandparent is 1/4. (One-half of the time uniting gametes will come from a grandmother and a grandfather, and 1/4 of the time both gametes will come from a grandmother.) Given that both gametes come from a grandfather, the probability that both come from the same male is $1/N_m$, where N_m is the number of males in the grandparental generation. Thus, the combined probability that both uniting gametes come from the same grandfather is

$(1/4 \times 1/N_m) = 1/4N_m$. The same probabilities hold for grandmothers. Thus, the combined probability of uniting gametes coming from the same grandparent is then:

$$\frac{1}{N_e} = \frac{1}{4N_f} + \frac{1}{4N_m} \qquad (7.1)$$

This is more commonly represented by solving for N_e with the following result:

$$N_e = \frac{4N_fN_m}{N_f + N_m} \qquad (7.2)$$

As we expect, if there are equal numbers of males and females ($N_f = N_m = 0.5N$), then this expression reduces to $N_e = N$.

In general, a skewed sex ratio will not have a large effect on the N_e/N ratio unless there is a great excess of one sex or the other. Figure 7.3 shows this for a hypothetical population with a total of 100 individuals. N_e is maximum (100) when there is an equal number of males and females, but declines as the sex ratio departs from 50:50. However, small departures from 50:50 have little effect on N_e. The dashed lines in this figure show that the N_e/N ratio will only be reduced by half if the least common sex is less than 15% of the total population. In the most extreme case, the N_e will be approximately four times the rarer sex:

N_f	N_m	N_e
1,000	1	4.0
1,000	2	8.0
1,000	3	12.0
1,000	4	15.9
1,000	5	19.9

Some populations of ungulates in which males are more likely to be hunted can have highly skewed sex ratios. For example, males comprised less than 1% of all adult elk in the Elkhorn Mountains of Montana in 1985 (Lamb 2010). Several authors have suggested that only one male per 25–100 females is sufficient for maintaining population growth and demographic productivity in hunted ungulate populations (e.g., White et al. 2001).

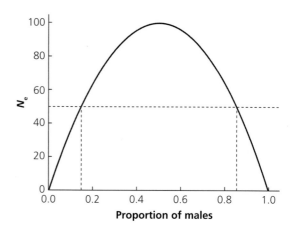

Figure 7.3 The effect of sex ratio on effective population size for a population with a total of 100 males and females using Equation 7.2. The dashed lines indicate the sex ratios at which N_e will be reduced by half because of a skewed sex ratio.

7.3 Nonrandom number of progeny

Our model of an ideal population assumes that all individuals have an equal probability of contributing progeny to the next generation. That is, a random sample of $2N$ gametes is drawn from a population of N diploid individuals. In real populations, parents seldom have an equal chance of contributing progeny because they differ in fertility and in the survival of their progeny. The variation among parents results in a greater proportion of the next generation coming from a smaller number of parents. Thus, the effective population size is reduced.

It is somewhat surprising just how much variation in reproductive success there is even when all individuals have equal probability of reproducing as in the ideal population. Figure 7.4 shows the expected frequency of progeny number in a very large stable population in which the mean number of progeny is two and all individuals have equal probability of reproducing. Take, for example, a stable population of 20 individuals (10 males and 10 females). On the average, each individual will have two progeny. However, ~12% of all individuals will not contribute any progeny! Consider that the probability of any male *not* fathering a particular progeny in this population is 0.90 (9/10). Therefore, the

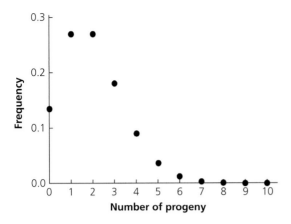

Figure 7.4 Expected frequency of number of progeny per individual in a large stable population in which the mean number of progeny per individual is two and all individuals have equal probability of reproducing.

probability of a male not contributing any of the 20 progeny is $(0.90)^{20}$, or ~12%. The same statistical reasoning applies for females as well. Thus, on average, two or three of the 20 individuals in this population are not expected to contribute any genes to the next generation, while one of the 20 individuals is expected to produce five or more progeny.

We can adjust for nonrandom progeny contribution following Wright (1939). Consider N individuals that contribute varying numbers of gametes (k) to the next generation of the same size (N) so that the mean number of gametes contributed per individual is $\bar{k} = 2$. The variance of the number of gametes contributed to the next generation is:

$$V_k = \frac{\sum_{i=1}^{N} (k_i - 2)^2}{N} \qquad (7.3)$$

The proportion of cases in which two random gametes will come from the same parent is:

$$\frac{\sum_{i=1}^{N} k_i(k_i - 1)}{2N(2N - 1)} = \frac{2 + V_k}{4N - 2} \qquad (7.4)$$

As we saw in the previous section, the effective population size may be defined as the reciprocal of the probability that two gametes come from the same parent. Thus, we may write the effective population

size as:

$$N_e = \frac{4N - 2}{2 + V_k} \qquad (7.5)$$

Random variation of k will produce a distribution that approximates a **Poisson distribution**. A Poisson distribution has a mean equal to the variance; thus, $V_k = \bar{k} = 2$ and $N_e = N$ for the idealized population (Section A3.3). However, as the variability in reproductive success among parents (V_k) increases, the effective population size decreases. An interesting result is that the effective population size will be larger than the actual population size if $V_k < 2$. In the extreme where each parent produces exactly two progeny, $N_e = 2N - 1$. Thus, in captive breeding where we can control reproduction, we may nearly double the effective population size by making sure that all individuals contribute equal numbers of progeny.

This potential near doubling of effective population size occurs because there are two sources of genetic drift: reproductive differences among individuals and Mendelian segregation in heterozygotes. These two sources contribute equally to genetic drift. Thus, eliminating differences in reproductive success will approximately double the effective population size. Unfortunately, there is no way to eliminate the second source of genetic drift (Mendelian segregation), except by nonsexual reproduction (cloning, etc.).

The following example considers three hypothetical populations of constant size $N = 10$ with extreme differences in individual reproductive success (Table 7.1). Each population consists of five pairs of mates. In Population A, only one pair of mates reproduces successfully. In Population B, each of the five pairs produces two offspring so that there is no variance in reproductive success. There is an intermediate amount of variability in reproductive success in Population C.

We can estimate N_e of each of these populations using Equations 7.3 and 7.5, as shown in Table 7.2. Thus, population B is expected to lose only ~3% ($1/2N_e = 0.026$) of its heterozygosity per generation, while populations A and C are expected to lose 24% and 5%, respectively. There are very few examples in natural populations where

Table 7.1 Estimation of effective population size in three hypothetical populations of constant size $N = 10$ with extreme differences in individual reproductive success. Each population consists of five pairs of mates. In Population A, only one pair of mates reproduces successfully. In Population B, each of the five pairs produces two offspring so that there is no variance in reproductive success. There is an intermediate amount of variability in reproductive success in Population C.

	A			B			C		
i	k_i	$k_i - \bar{k}$	$(k_i - \bar{k})^2$	k_i	$k_i - \bar{k}$	$(k_i - \bar{k})^2$	k_i	$k_i - \bar{k}$	$(k_i - \bar{k})^2$
1	10	8	64	2	0	0	0	−2	4
2	10	8	64	2	0	0	0	−2	4
3	0	−2	4	2	0	0	3	1	1
4	0	−2	4	2	0	0	3	1	1
5	0	−2	4	2	0	0	2	0	0
6	0	−2	4	2	0	0	2	0	0
7	0	−2	4	2	0	0	1	−1	1
8	0	−2	4	2	0	0	1	−1	1
9	0	−2	4	2	0	0	4	2	4
10	0	−2	4	2	0	0	4	2	4
			160			0			20

Table 7.2 Estimation of effective population size for three hypothetical populations in Table 7.1 with high, low, and intermediate variability in family size using Equation 7.5.

	$\Sigma(k_i - \bar{k})^2$	V_k	N_e	$1/2N_e$
Population A	160	16	2.11	0.237
Population B	0	0	19.00	0.026
Population C	20	2	9.50	0.053

the lifetime reproductive success of individuals is known so that N_e can be estimated using this approach (Example 7.2).

Equation 7.5 assumes that the variance in progeny number is the same in males and females. However, the variation in progeny number among parents is likely to be different for males and females. For many animal species, the variance of progeny number in males is expected to be larger than that for females. For example, according to the *Guinness Book of World Records*, the greatest number of children produced by a human mother is 69; in great contrast, the last Sharifian Emperor of Morocco is estimated to have fathered some 1,400 children! The current use of sperm donors can also result in males with many progeny. Some sperm donors have apparently fathered hundreds of progeny (Romm 2011).

We can take such differences between the sexes into account as shown:

$$N_e = \frac{8N - 4}{V_{km} + V_{kf} + 4} \quad (7.6)$$

The estimation of effective population size with nonrandom progeny number becomes much more complex if we relax our assumption of constant population size. In the case of separate sexes, the following expression may be used:

$$N_e = \frac{N_{t-2}\bar{k} - 2}{\bar{k} - 1 + \frac{V_k}{\bar{k}}} \quad (7.7)$$

where N_{t-2} is N in the grandparental generation (Crow & Denniston 1988).

7.4 Fluctuating population size

Natural populations sometimes fluctuate greatly in size. The rate of loss of heterozygosity $(1/2N)$ is proportional to the reciprocal of population size $(1/N)$. Thus, generations with small population sizes will dominate the effect on loss of heterozygosity. This is analogous to the sex with the smallest population size dominating the effect on loss of heterozygosity (Section 7.2). Therefore, the average population size over many generations is a poor metric for the loss of heterozygosity over many generations.

For example, consider three generations of a population that goes through a severe bottleneck, say $N_1 = 100$, $N_2 = 2$, and $N_3 = 100$. A very small proportion of the heterozygosity will be lost in generations 1 and 3 $(1/200 = 0.5\%)$; however, 25% of the heterozygosity will be lost in the second generation. The exact heterozygosity remaining after these three generations can be found as shown:

$$h = \left(1 - \frac{1}{200}\right)\left(1 - \frac{1}{4}\right)\left(1 - \frac{1}{200}\right) = 0.743$$

The average population size over these three generations is $(100 + 2 + 100)/3 = 67.3$. Using Equation 6.7 we would expect to lose only ~2% of the heterozygosity over three generations with a population size of 67.3, rather than the 25.7% heterozygosity that is actually lost.

We can estimate the effective population size over these three generations by using the mean of the

Example 7.2 Effective population size of Darwin's finches

Grant & Grant (1992a) estimated the lifetime reproductive success of two species of Darwin's ground finches on Daphne Major, Galápagos: the cactus finch and the medium ground finch. They followed survival and lifetime reproductive success of four cohorts born in the years 1975–1978. Figure 7.5 shows the lifetime reproductive success of the 1975 cohort for both species. The variance in reproductive success for both species was much greater than expected in an ideal population (Figure 7.4). Over one-half of the birds in both species did not produce any recruits to the next generation, and several birds produced eight or more recruits. Eighteen cactus finches produced 33 recruits ($\bar{k} = 1.83$) distributed with a variance (V_k) of 6.74; 65 medium ground finches produced 102 recruits ($\bar{k}=1.57$) distributed with a variance (V_k) of 7.12. The average number of breeding birds (census population sizes) for these years was ~94 cactus finches and 197 medium ground finches. The estimated N_e based on these data is 38 cactus finches and 60 medium ground finches. Thus, the N_e/N_C ratios for these two species are 38/94 = 0.40 and 60/197 = 0.30.

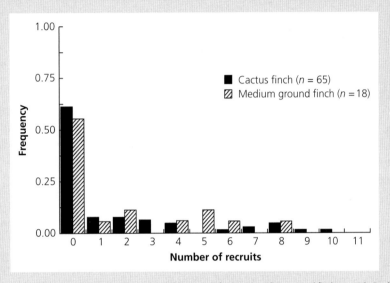

Figure 7.5 Lifetime reproductive success of the 1975 cohort of the cactus finch and medium ground finches on Isla Daphne Major, Galápagos. The x-axis shows the number of recruits (progeny that breed) produced. Thus, over 50% of the breeding birds for both species did not produce any progeny that lived to breed. From Grant & Grant (1992a).

reciprocal of population size ($1/N$) in successive generations, rather than the mean of N itself. This is known as the harmonic mean. Thus,

$$\frac{1}{N_e} = \frac{1}{t}\left(\frac{1}{N_1} + \frac{1}{N_2} + \frac{1}{N_3} + \dots + \frac{1}{N_t}\right) \qquad (7.8)$$

After a little algebra, this becomes:

$$N_e = \frac{t}{\sum\left(\frac{1}{N_i}\right)} \qquad (7.9)$$

Generations with the smallest N have the greatest effect. A single generation of small population size

may cause a large reduction in genetic variation. A rapid expansion in numbers does not affect the previous loss of genetic variation; it merely reduces the current rate of loss. This is known as the "bottleneck" effect as discussed in Section 6.5.

We can use Equation 7.9 to predict the expected loss of heterozygosity in the example that we began this section with:

$$N_e = \frac{3}{\left(\dfrac{1}{100} + \dfrac{1}{2} + \dfrac{1}{100}\right)} = 5.77$$

We expect to lose 23.8% of the heterozygosity in a population where $N_e = 5.77$ over three generations

(Equation 6.7). This is very close to the exact value of 25.7% that we calculated previously.

7.5 Overlapping generations

We so far have considered only populations with discrete generations. However, most species have overlapping generations. Hill (1979) has shown that the effective number in the case of overlapping generations is the same as that for discrete-generation populations having the same variance in lifetime progeny numbers and the same number of individuals entering the population each generation. Thus, the presence of overlapping generations itself does not have a major effect on N_e. However, this result assumes a constant population size and a stable age distribution. Crow & Denniston (1988) concluded that Hill's results are approximately correct for populations that are growing or contracting, as long as the age distribution is fairly stable. Waples et al. (2011) have provided a comprehensive consideration of this problem.

On the other hand, some biological aspects of overlapping generations can have a major effect on N_e (Nunney 2002). For example, N_e is likely to be reduced in polygamous species in which individuals reproduce over many years. In this case, the variance in reproductive success can be greatly increased if the same individuals tend to be relatively successful over many years (Chen et al. 2019; Example 7.3). In contrast, the presence of seed banks or diapausing eggs of freshwater crustaceans can greatly reduce the loss of heterozygosity over time, and thereby increase N_e (Nunney 2002).

Recent efforts to understand the N_e of species with overlapping generations have focused on estimating the effective number of breeders in one reproductive cycle (N_b), rather than on N_e per generation. Surprisingly, Waples et al. (2013) found that N_b for a single reproductive cycle is often larger than N_e per generation, and that an N_e larger than N_C is possible for species with delayed age at maturity. They also found that differences between species in the N_b/N_e ratio are explained largely by just two life history traits: age at sexual maturity and adult lifespan. These important results allow estimating and monitoring N_e per generation based on estimates of the number of breeders per reproductive cycle (N_b).

7.6 Variance versus inbreeding effective population size

The two primary measures of effective population size (N_{eI} and N_{eV}) differ when the population size is changing. In general, the inbreeding effective population size (N_{eI}) is more related to the number of parents since it is based on the probability of two gametes coming from the same parent. The variance effective population size (N_{eV}) is more related to the number of progeny since it is based on the number of gametes contributed rather than the number of parents (Crow & Kimura 1970, p. 361).

Consider the extreme of two parents that have a very large number of progeny. In this case, the allele frequencies in the progeny will be an accurate reflection of the allele frequencies in the parents; therefore, N_{eV} will be nearly infinite. However, all the progeny will be full sibs and thus their progeny will show the reduction in homozygosity expected in matings between full sibs; thus, N_{eI} is very small. In the other extreme, if each parent has exactly one offspring, then there will be no tendency for inbreeding in the populations, and, therefore, N_{eI} will be infinite. However, N_{eV} will be small.

Therefore, if a population is growing, the **inbreeding effective number** is usually less than the variance effective number (Waples 2002). If the population size is decreasing, the reverse is true. In the long run, these two effects will tend to cancel each other and the two effective numbers will be roughly the same (Crow and Kimura 1970; Crow & Denniston 1988).

7.7 Cytoplasmic genes

The effective population size of **cytoplasmic gene** systems (e.g., mitochondria and chloroplasts) is different than the N_e of nuclear genes. We will consider mitochondrial DNA (mtDNA) because so much is known about genetic variation of this molecule. The principles we will consider also apply to genetic variation in chloroplast DNA (cpDNA). However, cpDNA is paternally inherited in some plants (Harris & Ingram 1991).

There are three major differences between mitochondrial and nuclear genes that are relevant for this comparison:

1. Individuals usually possess many mitochondria that share a single predominant mtDNA

Example 7.3 Reduced effective population size in red-winged blackbirds because of the extraordinary reproductive success of one male

Occasionally a truly superior individual graces a population.

(Beletsky & Orians 1989, p. 10)

A long-term study of reproduction of red-winged blackbirds (Figure 7.6) on the Columbia National Wildlife Refuge in central Washington demonstrates the potential for N_e to be greatly reduced in polygamous species with overlapping generations (Beletsky & Orians 1989). Males in this population held breeding territories on average only 2.1 years. Half of all male breeders held territories for just a single year, and annual adult male mortality was ~40%.

Figure 7.6 Male red-winged blackbird. Photo courtesy of John Ashley.

The male known as RYB-AR was banded as a nonterritorial subadult during the 1977 spring breeding season. He first acquired a breeding territory in 1978 and held the same breeding territory through 1988 over 11 consecutive years. The mean annual harem size of RYB-AR was almost double that of other males. Harem size is strongly correlated with reproductive success in this population. Over his lifetime, RYB-AR produced 176 fledged young; this is 17 times greater than the average for males in this population.

RYB-AR fathered 4.2% of the total progeny in this population over the 11 years that he bred. Over these years there were nearly 400 breeding males in this population! As Beletsky and Orians conclude, even if the offspring of RYB-AR are genetically no better than average, he is sure to become a direct ancestor of many individuals in future generations. And, if his exceptional reproductive success is partially inherited by his descendants, his contributions to future generations will be even greater.

sequence. That is, individuals are effectively haploid for a single mtDNA type.

2. Individuals inherit their mtDNA genotype from their mother in most species.
3. There is no recombination between mtDNA molecules.

The effective population size for mtDNA is generally smaller than that for diploid nuclear genes because each individual has only one haplotype (allele) and uniparental inheritance (Birky et al. 1983).

For purposes of comparison, we will use h to compare genetic drift at mtDNA with nuclear genes even though mtDNA is haploid so that individuals are not heterozygous. It might seem inappropriate to use h as a measure of variation for mtDNA since it is haploid and individuals therefore cannot be heterozygous for mtDNA. Nevertheless, h is called gene diversity in this context and is a valuable measure of the variation present within a population (Nei 1987, p. 177). It can be thought of as the probability that two randomly sampled individuals from a population will have different mtDNA genotypes (Nei 1987, p. 177).

The probability of sampling the same mtDNA haplotype in two consecutive gametes is $1/N_f$, where N_f is the number of females in the population. And since $N_f = 0.5N$,

$$\Delta h = -\frac{1}{N_f} = -\frac{1}{0.5N} \tag{7.10}$$

In the case of a 1:1 sex ratio, there are four times as many copies of each nuclear gene as each mitochondrial gene (N_f):

$$\frac{N_e(\text{nuc})}{N_e(\text{mt})} = \frac{2(N_f + N_m)}{N_f} = 4 \tag{7.11}$$

In general, drift is more important and bottlenecks have greater effects for genes in mtDNA than for nuclear genes because of the generally smaller N_e (Example 7.4). Figure 7.7 shows the relative loss of variation during a bottleneck of a single generation for a nuclear and mitochondrial gene based upon Equation 7.11.

Table 7.3 Expected heterozygosity (H_e), diversity (h) at mtDNA, and average number of alleles ($\bar{A}$) per locus in three populations of the southern Australian spotted mountain trout from Tasmania at 22 allozyme loci and mtDNA (Example 7.4). The Allens Creek and Fortescue Creek populations are coastal populations that are connected by substantial exchange of individuals. The Isabella Lagoon population is an isolated landlocked population. From Ovenden & White (1990).

Sample	Nuclear loci		mtDNA	
	$\bar{A}$	H_e	A	h
Allens Creek	1.9	0.123	28	0.946
Fortescue Creek	1.9	0.111	25	0.922
Isabella Lagoon	1.3	0.104	2	0.038

Things become more complicated with an unequal sex ratio. If there are more females than males in a population, then the N_e for mtDNA can actually be greater than the N_e for nuclear genes. If we use Equation 7.2 for N_e for nuclear genes, then the ratio between the effective number of copies of nuclear genes to the effective number of copies of mitochondrial genes is:

$$\frac{N_e(\text{nuc})}{N_e(\text{mt})} = \frac{\dfrac{2\,(4N_f\,N_m)}{(N_f + N_m)}}{N_f}$$

Example 7.4 Effects of a bottleneck in the Australian spotted mountain trout

Ovenden & White (1990) demonstrated that genetic variation at mtDNA is much more sensitive to bottlenecks than nuclear variation in the southern Australian spotted mountain trout from Tasmania. These fish spawn in fresh water, and the larvae are immediately washed to sea where they grow and develop. The juvenile fish re-enter fresh water the following spring where they remain until they spawn. Landlocked populations of spotted mountain trout also occur in isolated lakes that were formed by the retreat of glaciers some 3,000–7,000 years ago.

Ovenden and White found 58 mtDNA genotypes identified by the presence or absence of restriction sites in 150 fish collected from 14 coastal streams. There is evidence of substantial exchange of individuals among the 14 coastal stream populations. In contrast, they found only two mtDNA genotypes in 66 fish collected from landlocked populations in isolated lakes. However, the lake populations and coastal populations had nearly identical heterozygosities at 22 allozyme loci (Table 7.3). As expected, the allelic diversity of the lake populations was lower than the coastal populations.

The reduced genetic variation at mtDNA in the landlocked populations is apparently due to a bottleneck associated with their founding and continued isolation. Oveden and White suggested that the founding bottleneck may have been exacerbated by natural selection for the landlocked life history in these populations. Regardless of the mechanism, the reduced N_e of the landlocked populations has had a dramatic effect on genetic variation at mtDNA but virtually no effect on nuclear heterozygosity.

which, after a bit of algebra, becomes:

$$\frac{8N_m}{(N_f + N_m)} \tag{7.12}$$

This expression will be less than one if there are more than seven times as many females as males. Therefore, the N_e for mtDNA will be less than the N_e for nuclear genes unless there are at least seven times as many females as males.

We have assumed so far in this section that the variance in reproductive success is equal in males and females. As we saw in Section 7.2, this is often

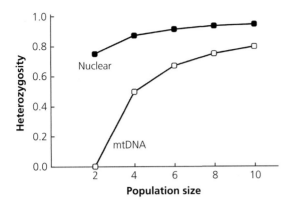

Figure 7.7 Amount of heterozygosity (nuclear) or diversity (mtDNA) remaining after a bottleneck of a single generation for a nuclear and mitochondrial gene with equal numbers of males and females. For example, there is no mitochondrial variation left after a bottleneck of two individuals because only one female is present. In contrast, 75% of the nuclear heterozygosity will remain after a bottleneck of two individuals (see Equation 6.5).

not true. This will decrease the difference in effective population size between nuclear and mitochondrial genes in the many species for which there is much greater variance in reproductive success in males than in females.

7.8 The coalescent

So far we have described genetic changes in populations due to genetic drift by changes in allele frequencies from generation to generation. There is an alternative approach to study the loss of genetic variation in populations that can be seen most easily for the case of mtDNA in which each individual receives the mtDNA haplotype of its mother. We can trace the transmission of mtDNA haplotypes over many generations in the past. That is, we can use a backward-time approach to trace the genealogy of the mtDNA genotype of each individual in a population (Figure 7.8). We can see in the example shown in Figure 7.8 that only one of the original 10 haplotypes remains in a population after just 18 generations due to a process called stochastic **lineage sorting**.

The **gene genealogy** approach also can be applied to nuclear genes, although it is somewhat more complex because of diploidy and recombination. The recent development of the application of genealogical data to the study of population-level

genetic processes is perhaps the major advance in population genetics theory in the past 50 years (Hudson 1990; Wakeley 2009). This development has been based upon two primary advances, one technical and one conceptual. The technological advance is the collection of DNA sequence data that allow tracing and reconstructing gene genealogies. The conceptual advance that has contributed to the theory to interpret these results is called "coalescent theory" (Section A10).

Lineage sorting, as in Figure 7.8, will eventually lead to the condition where all alleles in a population are derived from (i.e., coalesce to) a single common ancestral allele. Therefore, the number of generations to coalescence is expected to be shorter for smaller populations. In fact, the mean time to coalescence is approximately equal to N_e generations for

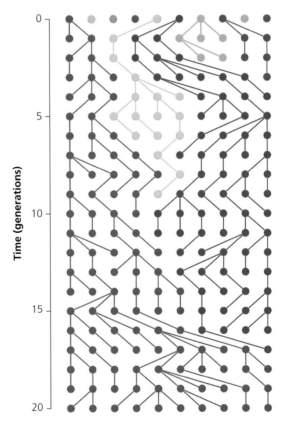

Figure 7.8 Genealogy of mtDNA in a population evolving by genetic drift. Each node represents an individual female and branches lead to daughters. From Revell (2019).

mtDNA, and is four times as long for a nuclear gene (Felsenstein 2019, p. 466). Coalescent theory provides a powerful framework to study the effects of genetic drift, natural selection, mutation, and gene flow in natural populations (Rosenberg & Nordborg 2002; Crandall et al. 2019).

The coalescent approach can be used to study effective population size over relatively long periods of time. Peart et al. (2020) estimated the long-term coalescent N_e for 17 pinniped species (true seals, sea lions, fur seals, and walruses). N_e estimates ranged from ~9,000 to 90,000, and they were strongly correlated with contemporary estimates of N_C ($r^2 = 0.59$, $P < 0.0002$). The N_e/N_C ratios were low (mean, 0.31; median, 0.13) and were strongly associated with demographic history. Residual variation in N_e/N_C, after controlling for past demographic fluctuations, contained information about recent population size changes. Species of conservation concern typically had positive residuals that indicated a smaller contemporary N_C than would be expected from their long-term N_e.

The coalescent is also used to estimate the current or recent effective population size (e.g., Anderson 2005; Luikart et al. 2010). For example, Miller & Waits (2003) estimated N_e for the Yellowstone grizzly bear population to be ~80 from the 1960s to the 1990s using multiple coalescent-based estimators. The coalescent approach enables the extraction of information on genealogical relationships among alleles at a locus, which can improve the estimation of N_e and other parameters (e.g., see Berthier et al. 2002 and Section A10).

7.9 Limitations of effective population size

Effective population size can be used to predict the expected rate of loss of heterozygosity or change in allele frequencies resulting from genetic drift. However, we generally need to know the rate of genetic drift in order to estimate effective population size. Thus, effective population size is perhaps best thought of as a standard, or unit of measure, rather than as a predictor of the loss of heterozygosity. That is, if we know the rate of change in allele frequency or the rate of loss of heterozygosity in a given population, we can use those observed rates to estimate effective population size (Examples 7.1

and 7.4, and Guest Box 7). We will consider the estimation of effective population size in more detail in Chapters 10 and 17.

Perhaps the greatest value of effective population size is heuristic. That is, we can better our understanding of genetic drift by comparing the effects of different violations of the assumptions of ideal populations on N_e (e.g., Figure 7.3). For example, Tanaka et al. (2009) have estimated N_e under different management regimes in order to compare the effects of different measures to control population size in overabundant koala populations. Similarly, in applying the concept of effective population size to managing populations, certain specific effective population sizes are often used as benchmarks. For example, it has been suggested that an N_e of at least 50 is necessary to avoid harmful effects of inbreeding depression in the short term (Franklin 1980; Allendorf & Ryman 2002; Jamieson & Allendorf 2012).

7.9.1 Allelic diversity and N_e

We have considered two measures of the loss of genetic variation in small populations: heterozygosity and allelic diversity. By definition, the inbreeding N_e is an estimate of the rate of loss of heterozygosity, but it is not a good indicator of the loss of allelic diversity within populations. That is, two populations that go through a bottleneck of the same N_e may lose very different amounts of allelic diversity. This difference is greatest when the bottleneck is caused by an extremely skewed sex ratio. Bottlenecks generally have a greater effect on allelic diversity than on heterozygosity. However, a population with an extremely skewed sex ratio may experience a substantial reduction in heterozygosity with little loss of allelic diversity.

The duration of a bottleneck (intense versus diffuse) will also affect heterozygosity and allelic diversity differently (England et al. 2003). Consider two populations that fluctuate in size over several generations with the same N_e, and therefore the same loss of heterozygosity. A brief but very small bottleneck (intense) will cause substantial loss of allelic diversity. However, a diffuse bottleneck spread over several generations can result in the same loss of heterozygosity, but will cause a much smaller reduction in allelic diversity.

In summary, populations that experience the same rate of decline of heterozygosity can experience very different rates of loss of allelic diversity. Therefore, we must consider more than just N_e when considering the rate of loss of genetic variation in populations.

7.9.2 Generation interval

In conservation, we are usually concerned with the loss of genetic variation over some specified number of years in developing policies. For example, according to the International Union for Conservation of Nature (IUCN), species are considered to be "vulnerable" if they have a greater than 10% probability of extinction within 100 years (Table 18.1). The rate of loss of genetic variation through calendar time (e.g., years) depends upon both N_e and mean **generation interval** (G) because $1/(2N_e)$ is the expected rate of loss per generation. Therefore, it is necessary to consider both G and N_e when predicting the expected rate of decline of heterozygosity in natural populations. There are many estimates of N_e in the literature, but very few include estimates of G, which are needed to predict the rate of loss of heterozygosity in calendar time.

There is some confusion in the literature about how to estimate generation interval. The generation interval is the average age of parents (Felsenstein 1971; Hill 1979). Generation interval is not the age of first reproduction, nor is it the average age of reproduction if individuals of different ages produce different numbers of offspring. See Table 7.4 for an example of estimating the generation interval.

It is especially important to estimate the generation interval when comparing the effects of different management schemes on the rate of loss of heterozygosity because conditions that reduce N_e often lengthen the generation interval (Hard et al. 2006). For example, Ryman et al. (1981) found that different harvest regimes for moose in Sweden can have strong effects on both effective population size and generation interval (Figure 7.9). Populations with smaller N_e tended to lose heterozygosity at a slower rate over calendar time because those effects of hunting that reduced N_e (e.g., harvesting young animals) also tended to increase the generation interval. That is, hunted populations with relatively smaller N_e and longer generation interval

Table 7.4 Hypothetical example of estimation of generation interval (G) in a demographically stable population of sockeye salmon, which die after spawning. The mean age of adult females at sexual maturity is 4.680. However, the mean generation interval of females (4.742) is estimated by using the mean number of eggs produced by females of different ages to estimate the proportion of progeny produced by females of different ages. We assume that males of all ages are equally reproductively successful. The generation interval in this population (4.441) is the mean of the generation interval in females (4.742) and males (4.140).

| | **Females** | | | **Male** |
Age	Adults	Eggs	Progeny	Adults
3	0.010	2,500	0.007	0.230
4	0.310	2,825	0.255	0.510
5	0.670	3,712	0.726	0.210
6	0.010	4,000	0.012	0.060
Mean	4.680	—	4.742	4.140

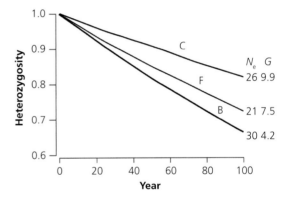

Figure 7.9 Expected decline of heterozygosity under three different sets of hunting regulations in moose from Sweden with a census size of 100 adults following the hunting season. The effective population size and generation interval for each hunting regime is indicated on the right. In hunting regime B, all adults experience identical mortality rates, but calves (less than 1 year old) are protected and are not hunted. In C, only calves are hunted. In F, adult females with calves are protected so that the risk of mortality of an adult female is reduced as a function of the number of calves (0, 1, or 2) with her at the beginning of hunting season. The regime (B) with the largest N_e is expected to lose heterozygosity at nearly twice the rate of the regime (C) with a smaller N_e that has a longer generation interval. Redrawn from Ryman et al. (1981).

would lose genetic variation over calendar time (not generations) more slowly than some populations with large N_e and shorter generation interval.

Generation interval was not considered in the koala example that we previously considered (Tanaka et. al. 2009), and it is possible that some strategies producing larger values of N_e might actually lose heterozygosity at a faster rate over calendar time than strategies resulting in smaller values of N_e. In fact, the strategy recommended to increase N_e was to administer contraception to all female koalas beyond a particular age; this strategy is likely to reduce the generation interval, and actually increase the rate of loss of heterozygosity over calendar time for a given N_e!

The inverse relationship between N_e and generation interval also is often true for differences between species. For example, Keall et al. (2001) estimated the census population size of five species of reptiles on North Brother Island in Cook Strait, New Zealand (Table 7.5). The generation interval for these five species was estimated based upon their life history (age of first reproduction, longevity, etc.; C.H. Daugherty, personal communication). As expected, the species with larger body size (e.g., tuatara) have smaller population sizes and longer generation intervals. The loss of heterozygosity over calendar time is strikingly similar in these five species, although they have very different population sizes. In general, species with larger body size (e.g., elephants) will tend to have smaller population sizes but longer generation intervals than species with smaller body size (e.g., mice), and these effects will tend to counteract each other.

Table 7.5 Expected loss in heterozygosity after 1,000 years for five species of reptiles on North Brother Island, New Zealand. N_e values for each species are assumed to be 20% of the estimated census size (N_C, Keall et al. 2001). The estimated generation interval (G) was then used to calculate the number of generations in 1,000 years (t) in order to predict the proportion of heterozygosity (H_e) remaining after 1,000 years using Equation 6.3.

Species	N_C	N_e	G (years)	t	H_e
Tuatara	350	70	50	20	0.866
Duvaucel's gecko	1,440	288	15	67	0.890
Common gecko	3,738	747	5	200	0.875
Spotted skink	3,400	680	5	200	0.863
Common skink	4,930	986	5	200	0.904

7.9.3 Gene flow

Our consideration of effective population size in this chapter has assumed a single panmictic population that is isolated from other populations. However, most natural populations experience some gene flow from other populations. The different measures of N_e are all similar for a stable isolated population, but these values can differ dramatically in populations experiencing migration. Ryman et al. (2019) show that both N_{eI} and N_{eV} are poor indicators of the rate of genetic drift in subpopulations affected by migration, and that they both consistently overestimate the rate of genetic drift in a subdivided population (i.e., they underestimate the local effective population size). This result has important implications in the consideration of how large populations should be in order to avoid serious problems caused by genetic drift (Section 18.8).

7.10 Effective population size in natural populations

The ratio of effective to census population size (N_e/N_C) in natural populations is of general importance for the conservation of populations (Guest Box 7). Census size is often easier to estimate than N_e. Therefore, establishing a general relationship between N_C and N_e would allow us to predict the rate of loss of genetic variation in a wide variety of species (Waples 2002). The actual values of N_e/N_C in a particular population or species will differ greatly depending upon demography and life history. The ratio of N_e to N_C is expected to be in the range of 0.1–0.5 for many populations (Waples 2016).

The effective size can decline without a decline in N_C. For example, a sudden increase in variance of family size where few families produce the entire next generation could reduce N_e with no reduction in N_C. This can happen in species with high reproductive output such as plants, fish, insects, and amphibians. A cryptic genetic bottleneck can also occur when there are few breeders of one sex due to a skewed sex ratio or a polygynous (or polyandrous) breeding system, as mentioned in Section 7.2. Thus, genetic monitoring is important even when demographic monitoring is possible, especially when the N_e/N_C ratio can fluctuate lower than expected (Mimura et al. 2017).

Frankham (1995) provided the first review of estimates of effective population size in natural populations. He concluded that estimates of N_e/N_C averaged ~10% in natural populations for studies in which the effects of unequal sex ratio, variance in reproductive success, and fluctuations in population size were included. However, Waples (2002) concluded that Frankham (1995) overestimated the contribution of temporal changes by computing the N_e/N_C ratio as a harmonic mean divided by an arithmetic mean. The empirical estimates of N_e that do not include the effect of temporal changes suggest that 20% of the adult population size is perhaps a better general value to use for N_e for many species (Waples 2002).

Palstra & Fraser (2012) provided a critical review of estimates of N_e, N_b, and N_C. Interpreting estimates of N_e and N_b in natural populations is complex. For example, most estimates of contemporary effective population size are based on models that assume N_e is constant over time (Waples 2005a). In real populations, N_e can change dramatically over time. Therefore, it is important to properly match estimates of N_e to the appropriate time periods. In addition, uncertainty in estimates of N_e, N_b, and N_C often have not been considered in estimating N_e/N_C and N_e/N_b ratios.

Palstra & Fraser (2012) presented published N_e/N_b and N_e/N_C estimates including only those studies in which the appropriate time periods were used. Overall, they found a weak and nonsignificant positive regression between both N_e and N_b with N_C. However, they did find strong associations if they used only relatively low values of N_C, less than 3,000 and 600 for N_e and N_b, respectively (Figure 7.10). In general, the effective number of breeders was ~25% of the census size at the time of breeding. There was much greater variability in the estimates of N_e/N_C.

Extremely small values of N_e/N_C have been reported in marine species with high fecundities, high mortalities in early life history stages, and high variance in reproductive success (Hedgecock & Pudovkin 2011). In these species, N_C may be many orders of magnitude greater than N_e (Hauser & Carvalho 2008; Examples 7.5 and 7.6). This suggests that even very large exploited marine fish populations may be in danger of losing crucial genetic variation (Allendorf et al. 2014). However, Waples (2016) has shown that there is a bias in the estimation of N_e when the true N_e is very large so that small estimates of N_e/N_C are expected to occur

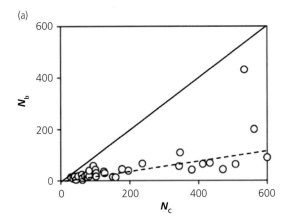

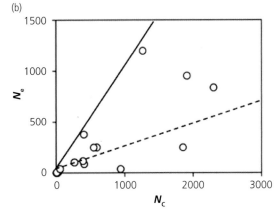

Figure 7.10 Relationships in a variety of species between effective number of breeders (N_b) and census size at the time of breeding (a), and effective population size (N_e) and generational census size (b). The solid line indicates $N_b = N_C$ and $N_e = N_C$; the broken line indicates $N_b/N_C = 0.25$ and $N_e/N_C = 0.25$. Data from Palstra & Fraser (2012).

even when the true N_e/N_C ratio is 0.1 or higher. Therefore, extremely small estimates of N_e/N_C in large populations must be interpreted with caution.

7.11 How can genomics advance understanding of N_e?

Genomics is improving our understanding of the effective population size in natural populations in several ways. First, the use of more loci increases precision and power to detect changes in N_e, and to estimate N_e, which is challenging when N_e is large. A major barrier to understanding N_e in natural

Example 7.5 Effective population size in tiger prawns

A comparison of allele frequencies at eight microsatellite loci in tiger prawns from Moreton Bay, Australia, has shown that the N_e may be nearly three orders of magnitude smaller than N_C in this population (Ovenden et al. 2007). This is an ideal population in which to provide reliable estimates of N_e because it is isolated and does not have overlapping generations. Spawning occurs once a year and only 1% of the individuals live more than 12 months.

Approximately 500 prawns were genotyped in each of three consecutive spawning seasons (2000, 2001, and 2002). Estimates of N_{eV} for the 2001 and 2002 spawning groups were made using the amount of change in allele frequency between consecutive years (i.e., generations) using three different statistical methods. The mean estimated effective populations sizes of the three estimators for the 2001 and 2002 spawning groups was 992 and 1,089. In comparison, the estimated number of adult prawns was 648,898 in 2001 and 464,627 in 2002. Thus, N_e/N_C is ~0.002! These results support the conclusion that the N_e/N_C ratio might be very small in a variety of marine species.

Example 7.6 Effective population size in a seaweed

Fucus serratus (a brown algae) is a key foundation species on rocky intertidal shores of northern Europe. Coyer et al. (2008) estimated the N_e of *F. serratus* in a population from southern Norway sampled in 2000 and 2008 by changes in allele frequencies at 26 microsatellite loci.

Estimates of N_{eV} ranged from 99 to 188, depending on whether the generation interval was assumed to be 1 or 2 years. The estimated census size during this period was 208,000 individuals. Thus, the N_e was approximately nearly three orders of magnitude smaller than N_C in this population. In further support of a small N_e, allelic richness decreased by 14% over this 6-year interval in these samples. Coyer et al. (2008) concluded that this species, and closely related species, are likely to be less resilient to environmental change than generally assumed.

populations has been the low precision of estimators. For example, confidence intervals are often enormous and include infinity. Second, having many loci can provide less biased estimates of N_e by facilitating identification of loci that violate assumptions of N_e estimators (e.g., that loci are independent, no selection, and no genotyping error). For example, Larson et al. (2013) reported imprecise N_e estimates using 39 loci (N_e = 174–infinity) for the population of Chinook salmon from the Tubutulik River. However, N_e estimates became much more precise when using 1,118 loci (N_e = 1,295–3,602).

Genomic approaches also facilitate estimation of N_e at multiple time points in the past to help detect both historical and recent population growth and declines. This is possible thanks to the use of information from linked loci, recombination rates, and runs of homozygosity that can be analyzed using genomic data with mapped loci (e.g., Ceballos et al. 2018; Chapters 10 and 17).

Guest Box 7 Effective population size in brown trout: Lessons for conservation
Linda Laikre and Nils Ryman

The brown trout was one of the first species whose natural population genetic structure was described using the allozyme technique in the 1970s (Allendorf et al. 1976; Ryman 1983). The brown trout is an important model species in conservation genetics. Its status for cultural, sport, and commercial fisheries and the fact that it is often subjected to hatchery breeding and large-scale releases have resulted in many conservation genetics issues being addressed by studying brown trout (Bekkevold et al. 2020).

We initiated a long-term genetic monitoring program of brown trout populations in small mountain lakes and one creek in the Hotagen Nature Reserve in central Sweden in the 1970s. We have genotyped 14 polymorphic allozyme loci continuously for ~100 individuals per population each year from eight sampling locations. We have estimated effective population size in these populations by the amount of allele frequency change between consecutive cohorts (Section 6.2; Jorde 2012). The N_e estimates generally ranged from 20 to 200 (Jorde & Ryman 1996, 2007; Palm et al. 2003).

With such small N_e values, we would expect low genetic variation in these populations. They do not, however, have low genetic variation, and they have maintained genetic diversity over the monitoring period (Charlier et al. 2012; Andersson et al. 2017). Further, the amount of heterozygosity is quite similar in localities with very different local N_e estimates. We explored these patterns further using less frequent monitoring data with SNP genotyping from over 20 additional lakes in the same region. We found the same trend of similar levels of heterozygosity among interconnected lakes in spite of often very low local N_e (Figure 7.11).

These observations suggest that genetic exchange occurs between populations, and that such migration maintains genetic diversity in the separate populations. Ryman et al. (2019) have shown that estimates of local N_e are not good predictors of the rate of loss of genetic variation in populations that are connected to other populations by gene flow. An exception to this pattern of genetic exchange is observed in the one creek locality. This locality appears to be quite isolated. No brown trout occur above a waterfall that constitutes an upstream migration barrier. Similarly, a smaller waterfall below this locality appears to impede migration from downstream populations. This locality exhibits the smallest N_e of only 20, and the amount of heterozygosity is only one-half that of other sites monitored within the same nature reserve.

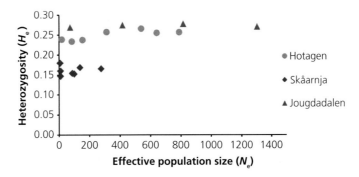

Figure 7.11 Effective population size (N_e) and expected heterozygosity (H_e) estimated from 96 SNPs for populations of naturally occurring brown trout in lakes representing three separate systems of interconnected populations located in different nature reserves in central Sweden (Skåarnja, Hotagen, and Jougadalen). The lack of correlation between N_e and H_e, and the similar amounts of heterozygosity among populations within the different reserves, suggest that gene flow among local populations, rather than N_e, determines the amount of local genetic variation.

We assessed changes in N_e over time in detail in one lake and two creek localities and found large temporal differences in the lake, but more stable N_e in the creek localities. This observation suggests that estimating N_e at only one point in time can provide quite different estimates in some cases, underlining the value of monitoring genetic diversity over longer periods.

We estimated census size (N_C) with mark–recapture techniques in Lake Blanktjärnen, and we estimated the ratio N_e/N_C as 0.11, and as 0.04 in a nearby lake (Charlier et al. 2011). These estimates support the common finding that N_e is much smaller than census size. In Lake Blanktjärnen, we observed a correlation between the effective number of breeders per year (N_b; Charlier et al. 2012) and the number of fish caught per annual sampling effort. This observation suggests that genetic techniques are also useful for monitoring or gaining insights into population demography (Schwartz et al. 2007).

Valuable lessons from our long-term genetic monitoring include: (1) the importance of connectivity, which is crucial for maintaining genetic diversity in metapopulations with small local N_e (Ryman et al. 2019); (2) N_e can vary considerably over time, which can make assessments from single samples unrepresentative; and (3) the N_e/N_C ratio is low in this species.

CHAPTER 8

Natural Selection

Elder-flowered orchid, Example 8.3

Then comes the question, Why do some live rather than others? If all the individuals of each species were exactly alike in every respect, we could only say it is a matter of chance. But they are not alike. We find that they vary in many different ways. Some are stronger, some swifter, some hardier in constitution, some more cunning. **(Alfred Russel Wallace 1923, p. 11)**

Theoretical models predict that the ability to adapt to local conditions despite ongoing gene flow depends on the genetic basis of adaptive traits and overall levels of genetic variability, but we still have a limited understanding of the genomic patterns underlying adaptive divergence in highly connected natural populations, particularly for complex traits with a multigenic basis. **(Aryn P. Wilder et al. 2020, p. 431)**

We have so far assumed that different genotypes have equal probability of surviving and passing on their alleles to future generations. That is, we have assumed that natural selection is not operating. If this assumption were true in real populations, we would not be concerned with genetic variation in conservation because genetic changes would not affect a population's longevity or its evolutionary future. However, as we saw in Chapter 6, there is ample evidence that the genetic changes that occur when a population goes through a bottleneck often result in increased frequencies of alleles that reduce an individual's probability of surviving to reproduce.

In addition, some alleles and genotypes affect survival and reproductive success under different environmental conditions. Remember the white Kermode bear from Section 2.1 that had an advantage in coastal populations of black bears because it is more successful at fishing for salmon.

Genetic differences between locally adapted populations can be important for continued persistence of populations. In addition, individuals that are moved by humans between populations or environments may not be genetically suited for surviving and reproducing in their new surroundings. And, worse from a conservation perspective, gene flow caused by such translocations can reduce the adaptation of local populations (Section 9.7).

For example, many native species of legumes (*Gastrolobium* and *Oxylobium*) in Western Australia naturally synthesize large concentrations of fluoroacetate, which is the active ingredient in 1080 (a poison used to remove nonnative mammalian pests; King et al. 1978). Native marsupials in Western Australia are resistant to 1080 because they have been eating plants that contain fluoroacetate for thousands of years. Therefore, 1080 does not kill native mammals in Western Australia, which means it can be used as a specific poison for introduced foxes

Conservation and the Genomics of Populations, Third Edition. Fred W. Allendorf, *et al.*, Oxford University Press.
© Fred W. Allendorf, W. Chris Funk, Sally N. Aitken, Margaret Byrne, and Gordon Luikart (2022). DOI: 10.1093/oso/9780198856566.003.0008

and feral cats that are a serious problem. However, members of the same 1080-resistant mammal species (e.g., brush-tailed possums) that occur to the east beyond the range of the fluoroacetate-producing legumes are susceptible to 1080 poisoning. Therefore, translocating brush-tailed possums into Western Australia from eastern populations might not be successful because the introduced individuals would not be "adapted" for consuming the local vegetation.

Many of the best traditional examples of local adaptation are from plant species because it is possible to do reciprocal transplantations and measure components of fitness (Joshi et al. 2001; Hall et al. 2010). Nagy & Rice (1997) performed reciprocal transplant experiments with coastal and inland California populations of the native annual *Gilia capitata*. They compared performance for four traits: seedling emergence, early vegetative size (leaf length), probability of surviving to flowering, and number of inflorescences. Native plants significantly outperformed nonnatives for all characters except leaf length. Figure 8.1 shows the results for the proportion of plants that survived to flowering. On average, the native inland plants had over twice the rate of survival compared to nonnative plants grown on the inland site; the native coastal plants had 5–10 times greater survival rates compared with nonnative plants grown on the coastal site.

The ability to use many markers to assign individuals to particular populations (Section 22.4) now makes it possible to compare the fitness of animals from different genetic populations living in a common environment. For example, Mobley et al. (2019) compared the reproductive success of Atlantic salmon spawning in their natal environment versus dispersers from nearby populations in the Teno River drainage in Finland. These salmon had considerable variation in the number of years spent at sea before returning to spawn (sea age). A range of sea ages at maturity, and therefore sizes, was observed in both local and dispersing fish. They found that local individuals had a large and consistent reproductive fitness advantage over dispersers (Figure 8.2). Parentage analysis conducted on adults and juvenile fish showed that local females and males had 9.6 and 2.9 times higher reproductive success than dispersers, respectively.

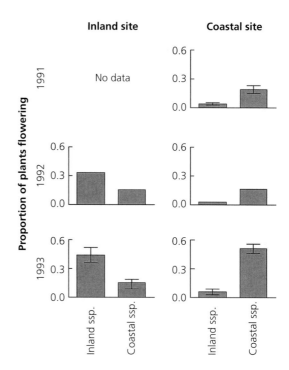

Figure 8.1 Reciprocal transplant experiment with *Gilia capitata* showing local adaptation for the proportion of plants that survived to flowering. The native subspecies had significantly greater survival than the nonnative subspecies in each of the five experiments. For example, ~45% of the seeds from the inland subspecies survived to flowering in 1993 at the inland site; however, only some 15% of the seeds from the nonnative subspecies survived to flowering in the same experiment. From Nagy & Rice (1997).

The adaptive significance of the vast genetic variation detected with molecular techniques has been controversial for over 50 years (King & Jukes 1969; Ohta 1973; Kimura 1983). Initially, the debate focused on whether the surprising amount of genetic variation detected with allozymes was primarily maintained by some form of "**balancing selection**" versus a combination of neutral mutations and genetic drift (Lewontin 1974; Nei 2005). Eventually, the Neutral Theory of Molecular Evolution emerged, which holds that most molecular evolution and most of the molecular variation within species are due to genetic drift of alleles that are selectively or nearly selectively neutral (Kimura 1983).

This view has been challenged recently by authors who concluded that "the neutral theory has been overwhelmingly rejected" (Kern & Hahn

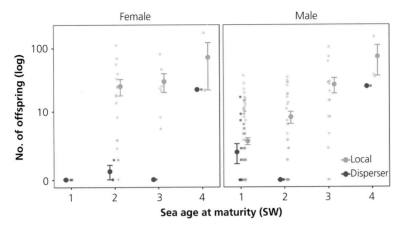

Figure 8.2 Origin (local or disperser) and sea age at maturity (sea winters, SW) with reproductive success (number of offspring) for Atlantic salmon. Large circles with error bars indicate the means ± standard error (SE), and small circles show individual data points. Points are jittered on the x-axis for clarity. There were no very young (1 SW) local females and only one old (3–4 SW) dispersing male. From Mobley et al. (2019).

2018, p. 1369). These authors also concluded that natural selection is the major force determining both between-species divergence and within-species variation. Jensen et al. (2019) have critically examined these conclusions, and rejected Kern & Hahn's arguments and conclusions. Jensen et al. (2019) state that part of the disagreement results from Kern & Hahn (2018) using a definition of the Neutral Theory that is too narrow. Jensen et al. (2019) present a modern version of the Neutral Theory that is based on new genomic evidence. There are aspects of genomic evolution that are not readily explained by the Neutral Theory. Nevertheless, the Neutral Theory is useful as the null model in attempts to detect the effects of natural selection.

Most of the models that we use to interpret data and predict effects in natural populations assume selective neutrality. This is done not because we believe that all genetic variation is neutral. Rather neutrality is assumed because we sometimes have no choice if we want to use the rich theory of population genetics to interpret data and make predictions as most models assume the absence of natural selection. Moreover, allele frequency distributions at neutral loci are far more useful than at adaptive loci in describing effective population sizes and exchange among populations (Section 9.7).

In this chapter, we consider the effects of natural selection on allele and genotype frequencies. Sewall Wright developed powerful theoretical models that allow us to predict the effects of small populations on genetic variability. Most of these models assume

selective neutrality. For example, we have seen that heterozygosity will be lost at a rate of $1/2N$ per generation in the ideal population (Equation 6.5). What is the expected rate of loss of heterozygosity if the genetic variability is affected by natural selection? The answer depends on the pattern and intensity of natural selection in operation. And since it is so difficult to estimate fitness in natural populations, we generally cannot predict the expected rate of loss of heterozygosity, unless we ignore the effects of natural selection. And worse yet, since natural selection acts differently on each locus, there is not one answer, but rather there is a different answer for each of the many loci affected by selection within a population.

8.1 Fitness

Natural selection is the differential success of genotypes in contributing to the next generation. In the simplest conceptual model, there are two major life history components that bring about selective differences between genotypes: **viability** and **fertility**. Viability is the probability of survival to reproductive age, and fertility is the average number of offspring per individual that survive to reproductive maturity for a particular genotype.

The effect of natural selection on genotypes is measured by **fitness**. Fitness is the average number of offspring produced by individuals of a particular genotype. Fitness can be calculated as the product of viability and fertility, as defined above, and we can define fitness for a locus with two alleles as shown:

Genotype	Viability	Fertility	Fitness
AA	v_{11}	f_{11}	$(v_{11})(f_{11}) = W_{11}$
Aa	v_{12}	f_{12}	$(v_{12})(f_{12}) = W_{12}$
aa	v_{22}	f_{22}	$(v_{22})(f_{22}) = W_{22}$

These are absolute fitnesses that are based on the total number of expected progeny from each genotype. It is often convenient to use relative fitnesses to predict genetic changes caused by natural selection. Relative fitnesses are estimated by the ratios of absolute fitnesses. For example, in the data that follow, fitnesses have been standardized by dividing by the fitness of the genotype with the highest fitness (*AA*). Thus, the relative fitness of heterozygotes is 0.67 because, on average, heterozygotes have 0.67 times as many progeny as *AA* individuals (1.80/2.70 = 0.67).

Genotype	Viability	Fertility	Absolute fitness	Relative fitness
AA	0.90	3.00	2.70	1.00
Aa	0.90	2.00	1.80	0.67
aa	0.45	2.00	0.90	0.33

8.2 Single locus with two alleles

We will begin by modeling changes caused by differential survival (viability selection) in the simple case of a single locus with two alleles. Consider a single locus with two alleles having differential reproductive success in a large random mating population in which all of the other assumptions of the Hardy–Weinberg (HW) model are valid. We would expect the following result after one generation of selection:

Genotype	Zygote frequency	Relative fitness	Frequency after selection
AA	p^2	w_{11}	$(p^2 w_{11})/\bar{w}$
Aa	$2pq$	w_{12}	$(2pq w_{12})/\bar{w}$
aa	q^2	w_{22}	$(q^2 w_{22})/\bar{w}$

where $\bar{w}$ is used to normalize the frequencies following selection so that they sum to one. This is the

average fitness of the population, and it is the fitness of each genotype weighted by its frequency.

$$\bar{w} = p^2 w_{11} + 2pq w_{12} + q^2 w_{22} \qquad (8.1)$$

After one generation of selection the frequency of the *A*-allele is:

$$p' = \frac{p^2 w_{11}}{\bar{w}} + \left(\frac{1}{2}\right)\left(\frac{2pq w_{12}}{\bar{w}}\right) = \frac{p^2 w_{11} + pq w_{12}}{\bar{w}} \qquad (8.2)$$

and, similarly, the frequency of the *a*-allele is:

$$q' = \frac{pq w_{12} + q^2 w_{22}}{\bar{w}} \qquad (8.3)$$

It is often convenient to predict the change in allele frequency from generation to generation, Δp, caused by selection. We get the following result if we solve for Δp in the current case:

$$\Delta p = \frac{pq}{\bar{w}} [p(w_{11} - w_{12}) + q(w_{12} - w_{22})] \qquad (8.4)$$

We can see that the magnitude and direction of change in allele frequency are both dependent on the fitnesses of the genotypes and the allele frequency.

Equation 8.4 can be used to predict the expected change in allele frequency after one generation of selection for any array of fitnesses. The allele frequency the following generation will be:

$$p' = p + \Delta p \qquad (8.5)$$

We will use this model to study the dynamics of selection for three basic modes of natural selection with constant fitnesses:

1. Directional selection.
2. Heterozygous advantage (overdominance).
3. Heterozygous disadvantage (underdominance).

8.2.1 Directional selection

Directional selection occurs when one allele is always at a selective advantage. The advantageous allele under directional selection may be dominant, intermediate, or recessive to the alternative allele as shown in the **fitness sets** below:

Dominant	$w_{11} = w_{12} > w_{22}$
Intermediate	$w_{11} > w_{12} > w_{22}$
Recessive	$w_{11} > w_{12} = w_{22}$

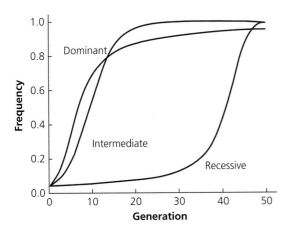

Figure 8.3 Change in allele frequency under directional selection when the homozygote for the favored allele has twice the fitness of the homozygote for the disfavored allele (1.00 vs. 0.50). The heterozygote has the same fitness as the favored allele (1.0, dominant), the same fitness as the disfavored allele (recessive; 0.50), or has intermediate fitness (0.75). The initial frequency of the favored allele is 0.03.

The advantageous allele will increase in frequency and will be ultimately fixed by natural selection under all three modes of directional selection (Figure 8.3). Thus, the eventual or equilibrium outcome is independent of the dominance of the advantageous allele. However, the rate of change of allele frequencies does depend on dominance relationships as well as the intensity of selection. For example, selection on a recessive allele is ineffective when the recessive allele is rare because most of the copies of that allele occur in heterozygotes and are therefore "hidden" from selection.

8.2.2 Heterozygous advantage (overdominance)

Heterozygous advantage occurs when the heterozygote has the greatest fitness:

$$w_{11} \; < \; w_{12} \; > \; w_{22}$$

This mode of selection is expected to maintain both alleles in the population as a **stable equilibrium**. This pattern of selection is often called **overdominance**. In the case of dominance, the phenotype of the heterozygote is equal to the phenotype of one of the homozygotes. In overdominance, the phenotype (i.e., fitness) of the heterozygote is greater than phenotype of either homozygote (Example 8.1).

Example 8.1 Natural selection at an allozyme locus

Patarnello & Battaglia (1992) described an example of heterozygous advantage at a locus encoding the enzyme glucosephosphate isomerase (GPI) in the copepod *Gammarus insensibilis* that lives in the Lagoon of Venice. Individuals were collected in the wild, acclimated in the laboratory at room temperature, and then held at a high temperature (27°C) for 36 hours. Individuals with different genotypes differed significantly in their survival at this temperature (Table 8.1; $P < 0.005$). Heterozygotes survived better than either of the homozygotes.

Table 8.1 Differential survival of *GPI* genotypes in the copepod *Gammarus insensibilis* held in the laboratory for 36 hours at high temperature (27°C). From Patarnello & Battaglia (1992).

	Genotype		
	100/100	*100/80*	*80/80*
Alive	48	90	12
Dead	47	53	27
Total	95	143	39
Relative survival	0.803	1.000	0.490

A persistent problem in measuring fitnesses of individual genotypes is whether any observed differences are due to the locus under investigation or to other loci that are linked to that locus (Eanes 1987). *In vitro* measurements show that heterozygotes at the *GPI* locus in *G. insensibilis* have greater enzyme activity than either homozygote over a wide range of temperatures. In addition, the *80/80* homozygote has the greatest mortality and the lowest enzyme activity. Patarnello & Battaglia (1992) have argued that the observed differences are caused by the *GPI* genotype on the basis of these enzyme kinetic properties and other considerations.

Let us examine the simple case of heterozygous advantage in which the two homozygotes have equal fitness:

$$\begin{array}{cccc} & AA & Aa & aa \\ \text{Fitness} & 1-s & 1.0 & 1-s \end{array}$$

where s (the **selection coefficient**) is greater than zero and less than or equal to one. We can examine the dynamics of this case of selection by plotting the values of Δp as a function of allele frequency (Figure 8.4a). When p is less than 0.5, selection will increase p, and when p is greater than 0.5, selection will decrease p. Thus, 0.5 is a stable equilibrium; that is, when p is perturbed from 0.5, it will return to that value.

Any overdominant fitness set with two alleles will produce a stable intermediate equilibrium allele frequency (p^*). However, the value of p^* depends on the relative fitnesses of the homozygotes. If we solve Equation 8.4 for $\Delta p = 0$ we get the following result:

$$p^* = \frac{w_{12} - w_{22}}{2w_{12} - w_{11} - w_{22}} \qquad (8.6)$$

Thus, the equilibrium allele frequency will be near 0.5 if the two homozygotes have nearly equal fitness. However, if one homozygote has a great advantage over the other, that allele will be much more frequent at equilibrium.

Heterozygous advantage was once thought to be one of the major mechanisms maintaining genetic variation in natural populations (Lewontin 1974). However, we now know that heterozygous advantage in natural populations is relatively rare (Hedrick 2012; but see Examples 8.1 and 8.2). We saw in Box 5.2 that genomic analysis has revealed that duplicated loci are common in many genomes. Haldane (1954) suggested that duplicated loci with fixed "heterozygous advantage" could evolve by selection for alternative alleles at the two genes. This would provide increased fitness through overdominance without the presence of homozygotes with lower fitness (Spofford 1969). Milesi et al. (2017) used an experimental approach with insecticide resistance in mosquitoes to demonstrate that tandemly duplicated genes fixed for alternate alleles would result in permanent overdominance. They also showed that such duplications are selected for over single-copy alleles allowing the fixation of the heterozygous phenotype. By allowing the rapid fixation of divergent alleles, this could contribute to the rarity of overdominance.

8.2.3 Heterozygous disadvantage (underdominance)

Underdominance occurs when the heterozygote is least fit:

$$w_{11} > w_{12} < w_{22}$$

An examination of Δp as a function of p reveals that underdominance will produce what is called an **unstable equilibrium** (Figure 8.4b). The p^* value is found using the same formula as for overdominance (Equation 8.6). However, this equilibrium is unstable because allele frequencies will tend to move away from the equilibrium value once they are perturbed. Underdominance, therefore, is not a mode of selection that will maintain genetic variation in natural populations.

We saw in Chapter 3 that heterozygotes for chromosomal rearrangements often have reduced fertility because they produce unbalanced or aneuploid gametes. Foster et al. (1972) examined the behavior of translocations in population cages of *Drosophila*

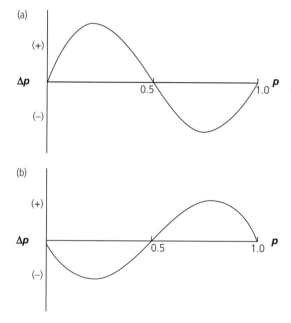

Figure 8.4 Expected change in allele frequency (Δp) as a function of allele frequency (p). (a) Heterozygous advantage when the homozygotes have equal fitness. There is a stable equilibrium at $p = 0.5$. (b) **Heterozygous disadvantage**. There is an unstable equilibrium at $p = 0.5$.

Example 8.2 Heterozygous advantage for a color polymorphism in the common buzzard

The European common buzzard has a plumage polymorphism controlled by a single locus (Figure 8.5; Boerner & Krüger 2009). Observations of 162 offspring and their parents indicated that dark brown individuals and light-colored individuals at this locus

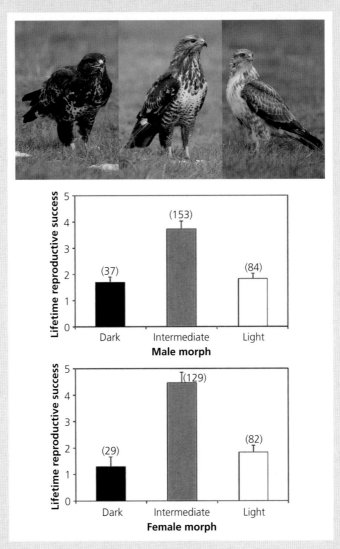

Figure 8.5 Mean lifetime reproductive success (number of progeny) in three color morphs (above) of male and female European common buzzards. Photo courtesy of Oliver Krüger. From Boerner & Krüger (2009).

Continued

Example 8.2 *Continued*

are alternative homozygotes and that heterozygotes are intermediate in color (Krüger et al. 2001). There is assortative mating at this locus in that individuals are more likely to mate with individuals who have the same plumage pattern as their mother.

A long-term study of lifetime reproductive success revealed that heterozygotes tended to show greater annual survival and marked differences in aggression, habitat preference, and parasite load (Boerner & Krüger 2009). Overall, the life-time reproductive success of heterozygotes was nearly twice that of either homozygote averaged over females and males (Figure 8.5).

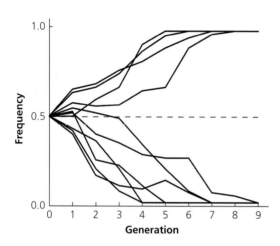

Figure 8.6 Population cage results with *Drosophila melanogaster* showing change in frequency of a chromosomal translocation in 10 populations when the two homozygotes have equal fitness that is approximately twice that of heterozygotes. Populations were founded by 20 individuals and population sizes fluctuated between 100 and 400 flies. Redrawn from Foster et al. (1972).

melanogaster. They set up cages in which homozygotes for the chromosomal rearrangements had equal fitness; in this case, the unstable equilibrium frequency is expected to be 0.5 (Equation 8.7). As predicted by this analysis, the populations quickly went to fixation for whichever chromosomal type became most frequent in the early generations because of genetic drift (Figure 8.6).

8.2.4 Selection and HW proportions

The absence of departures from HW proportions is sometimes taken as evidence that a particular locus is not affected by natural selection. However, this interpretation is incorrect for several reasons

(Lachance 2009). First, differences in fertility will not affect HW proportions. Thus, only differential survival can be detected by testing for HW proportions. Second, even strong differences in survival may not cause departures from HW proportions. For example, Lewontin & Cockerham (1959) have shown that at a locus with two alleles, differential survival will not cause a departure from HW proportions if the product of the fitnesses of the two homozygotes is equal to the square of the fitness of the homozygotes. Finally, the goodness of fit test for HW proportions has little power to detect departures from HW proportions caused by differential survival.

8.3 Multiple alleles

Analysis of the effects of natural selection becomes more complex when there are more than two alleles at a locus because the number of genotypes increases dramatically with a modest increase in the number of alleles; remember, there are 15 possible genotypes with just 5 alleles at a single locus (Equation 5.4). Nevertheless, our model of selection can be readily extended to three alleles (A_1, A_2, and A_3):

Genotype	A_1A_1	A_1A_2	A_2A_2	A_1A_3	A_2A_3	A_3A_3
Fitness	w_{11}	w_{12}	w_{22}	w_{13}	w_{23}	w_{33}
Frequency	p^2	$2pq$	q^2	$2pr$	$2qr$	r^2

The average fitness of the population is:

$$\bar{w} = p^2 w_{11} + 2pq w_{12} + q^2 w_{22} + 2pr w_{13} + 2qr w_{23}$$
$$+ r^2 w_{33} \tag{8.7}$$

And the expected allele frequencies the next generation are:

$$p' = \frac{(p^2 w_{11} + pq w_{12} + pr w_{13})}{\bar{w}}$$

$$q' = \frac{(pq w_{12} + q^2 w_{22} + qr w_{23})}{\bar{w}} \qquad (8.8)$$

$$r' = \frac{(pr w_{13} + qr w_{23} + r^2 w_{33})}{\bar{w}}$$

We can find any equilibria that exist for a particular set of fitnesses by setting $p' = p = p^*$ and solving these equations. The following conditions emerge after a bit of math.

$$p^* = \frac{z_1}{z} \text{ where } z_1 = (w_{12} - w_{22})(w_{13} - w_{33})$$
$$- (w_{12} - w_{23})(w_{13} - w_{23})$$

$$q^* = \frac{z_2}{z} \text{ where } z_2 = (w_{23} - w_{33})(w_{12} - w_{11})$$
$$- (w_{23} - w_{13})(w_{12} - w_{13})$$

$$r^* = \frac{z_3}{z} \text{ where } z_3 = (w_{13} - w_{11})(w_{23} - w_{22})$$
$$- (w_{13} - w_{12})(w_{23} - w_{12})$$

where:

$$z = z_1 + z_2 + z_3 \qquad (8.9)$$

If these equations give negative values for the allele frequencies, that means there is no three-allele equilibrium (i.e., at least one allele will be lost due to selection). The equilibrium will be stable if the equilibrium is a maximum for average fitness (Equation 8.6) and will be unstable if it is a minimum for average fitness. In general, a three-allele equilibrium will be stable if z_1, z_2, and z_3 are greater than zero and:

$$(w_{11} + w_{12}) < (2w_{13}) \qquad (8.10)$$

There are no simple rules for a locus with three alleles as there are for a locus with two alleles. However, the following statements may be helpful.

1. There is at most one stable equilibrium for two or more alleles.
2. A stable equilibrium will be globally stable; that is, it will be reached from any starting point containing all three alleles.

3. If a **stable polymorphism** exists, the mean fitness of the population exceeds that of any homozygote. If such a homozygote existed, it would become fixed in the population.
4. Heterozygous advantage (i.e., all heterozygotes have greater fitness than all homozygotes) is neither necessary nor sufficient for a stable polymorphism.

8.3.1 Heterozygous advantage and multiple alleles

Overdominance was thought to be the major mechanism maintaining genetic variation in natural populations at the time when most empirical evidence suggested that most polymorphic loci had two primary alleles (see discussion in Lewontin 1974, pp. 23–31). However, molecular techniques quickly revealed that many alleles exist at most loci in natural populations. For example, Singh et al. (1976) discovered 37 different alleles at a locus coding for xanthine dehydrogenase in a sample of 73 individuals collected from 12 natural populations of *Drosophila pseudoobscura*.

Can overdominance maintain many alleles at a single locus? This question was approached in a classic paper by Lewontin et al. (1978). They estimated the proportion of randomly chosen fitness sets that would maintain all alleles through overdominance. For a locus with two alleles, heterozygous advantage is both necessary and sufficient to maintain both alleles. If fitnesses are selected at random, the heterozygotes will have the greatest fitness one-third of the time because there are three genotypes (Table 8.2). However, it becomes increasingly unlikely that all heterozygotes will have greater fitness than all homozygotes as the number of alleles increases. In addition, heterozygous advantage (i.e., all heterozygotes have greater fitness than all homozygotes) is neither necessary nor sufficient to maintain an A-allele polymorphism when A is greater than two. In fact, fitness sets capable of maintaining an A-allele polymorphism quickly become extremely unlikely as A increases (Table 8.2). For example, heterozygous advantage is sufficient to always produce a stable polymorphism with two alleles. However, only 34% of all

Table 8.2 Proportion of randomly chosen fitness sets that maintain all *A* alleles in a stable equilibrium (Lewontin et al. 1978). The third column shows the proportion of fitness sets expected to maintain all *A* alleles considering only those fitness sets in which all heterozygotes have greater fitness than all homozygotes.

A	All fitness sets	Heterozygous advantage
2	0.33	1.00
3	0.04	0.71
4	0.0024	0.34
5	0.00006	0.10
6	0	0.01

fitness sets with four alleles in which all heterozygotes have greater fitness than all homozygotes will maintain all four alleles. Thus, overdominance with constant fitness is not an effective mechanism for maintaining many alleles at individual loci in natural populations (Kimura 1983).

Spencer & Marks (1993) have revisited this issue using a different approach. Rather than randomly assigning fitness as done by Lewontin et al. (1978), they simulated evolution by allowing new mutations with randomly assigned fitnesses to occur within a large population and then determined how many alleles could be maintained in the population by viability selection. They found that up to 38 alleles were sometimes maintained by selection in their simulated populations. In general, they found many more alleles could be maintained by this type of selection than predicted by Lewontin et al. (1978).

Spencer & Marks (1993) argued that their approach, which examines how a polymorphism may be constructed by evolution, is a complementary approach to understanding evolutionary dynamics when used along with traditional models that focus only on conditions that maintain equilibrium. Nevertheless, the general conclusions of Lewontin et al. (1978) are still likely to be valid, even if the approach of Spencer & Marks (1993) is more realistic. One major drawback of the results of Spencer & Marks (1993) is that their models do not include genetic drift, and, as we will see in Section 8.5, heterozygous advantage is only effective in maintaining alleles that are relatively common in a population at equilibrium.

Hedrick (2002) has considered the maintenance of many alleles at a single locus by "balancing selection" at the **major histocompatibility complex** (*MHC*) locus. He then assumed resistance to pathogens is conferred by specific alleles and the action of each allele is dominant. He concluded that this model of selection could maintain stable multiple allele polymorphisms, even in the absence of any intrinsic heterozygous advantage, because heterozygotes will have higher fitness in the presence of multiple pathogens.

8.4 Frequency-dependent selection

We have so far assumed that fitnesses are constant. However, fitnesses are not likely to be constant in natural populations (Kojima 1971). Fitnesses are likely to change under different environmental conditions. Fitnesses may also change when allele frequencies change; this is called **frequency-dependent selection**. This type of selection is a potentially powerful mechanism for maintaining genetic variation in natural populations (Clarke & Partridge 1988).

8.4.1 Two alleles

Let us begin with the simple case where the fitness of a genotype is a direct function of its frequency. For example,

$$\begin{array}{cccc} & AA & Aa & aa \\ \text{Fitness} & 1-p^2 & 1-2pq & 1-q^2 \end{array} \qquad (8.11)$$

With this model of selection, a genotype becomes less fit as it becomes more common in a population. The change in allele frequency at any value of p can be calculated with Equation 8.4. We can predict the expected effects of this pattern of selection by examination of the plot of Δp versus allele frequency; we will get the same plot as Figure 8.3. In this case, there is an equilibrium at $p^* = 0.5$ where Δp is zero. Is this equilibrium stable or unstable? When p is less than 0.5, $w_{11} > w_{22}$ and therefore p will increase; when p is greater than 0.5, $w_{11} < w_{22}$ and p will decrease. This is a stable equilibrium.

Note that the homozygotes have a fitness of 0.75, and the heterozygote has a fitness of 0.5 at equilibrium. Therefore, this is a stable polymorphism in which the heterozygote has a disadvantage at equilibrium. We can see that our rules for understanding the effects of selection with constant fitnesses are not likely to be helpful in understanding the effects of frequency-dependent selection. In general, frequency-dependent selection will produce a stable polymorphism whenever the rare phenotype has a selective advantage. However, there is no general rule about the relative fitnesses at equilibrium.

8.4.2 Frequency-dependent selection in nature

Frequency-dependent selection is an important mechanism for maintaining genetic variation in natural populations. You are encouraged to read the review by Clarke (1979); additional references on frequency-dependent selection can be found in a collection of papers edited by Clarke & Partridge (1988). Frequency-dependent selection often results from mechanisms of sexual selection, predation and disease, and ecological competition (Example 8.3).

8.4.3 Self-incompatibility locus in plants

In contrast to heterozygous advantage, frequency-dependent selection can be extremely powerful for maintaining multiple alleles. The self-incompatibility locus (S) of many flowering plants is an extreme example of this (Wright 1965a; Castric & Vekemans 2004; Charlesworth 2010). In the simplest system, pollen grains can only fertilize plants that do not have the same S allele as carried by the pollen. Homozygotes cannot be produced at this locus, and at least three alleles must be present at this locus.

The expected equilibrium with three alleles will be a frequency of 0.33 for each allele because fitnesses are equivalent for all three alleles. At equilibrium, any pollen grain will be able to fertilize one-third of the plants in the population (Table 8.3). However, a fourth allele produced by mutation (S_4) would have a great selective advantage because it will be able to fertilize every plant in the population. Thus, we

Table 8.3 Genotypes possible at the self-incompatibility locus (S) locus in species of flowering plants with three alleles.

Parental genotypes		Progeny frequencies		
Ovule	Pollen	$S_1 S_2$	$S_1 S_3$	$S_2 S_3$
$S_1 S_2$	S_3	0.00	0.50	0.50
$S_1 S_3$	S_2	0.50	0.00	0.50
$S_2 S_3$	S_1	0.50	0.50	0.00

would expect the fourth allele to increase in frequency until it reaches a frequency equal to the other three alleles.

Any new mutation at the S-locus is expected to have an initial selective advantage because of its rarity regardless of the existing number of alleles. However, we also would expect rare alleles to be susceptible to loss through genetic drift. Therefore, the actual number of S-alleles will be an equilibrium between gain through mutation and loss through genetic drift.

Emerson (1940) described 45 nearly equal-frequency S-alleles in the Organ Mountains evening primrose, a narrow endemic plant that occurs in an area of ~50 km^2 in the Organ Mountains, New Mexico. Emerson originally thought that the total population size of this species was ~500 individuals. More recent surveys indicate that the total population size may be as great as 5,000 individuals (Levin et al. 1979). Regardless of the actual population size, this is an enormous amount of variability at a single locus. As expected because of its small population size, this species has very little genetic variation at other loci as measured by protein electrophoresis (Levin et al. 1979).

8.4.4 Complementary sex determination locus in invertebrates

A common method of sex determination in animals also results in frequency-dependent selection that can maintain many alleles. Nearly 15% of all species of invertebrates have a haplodiploid mechanism of sex determination in which females are diploid and males are haploid (e.g., ants, bees, and wasps) (Crozier 1971; Cook & Crozier 1995).

Example 8.3 Frequency-dependent selection in an orchid

Gigord et al. (2001) have presented an elegant example of frequency-dependent selection in the elder-flowered orchid. This species has a dramatic flower color polymorphism; both yellow- and purple-flowered individuals occur throughout the range of the species in Europe (Figure 8.7). Elder-flowered orchids do not provide any reward to insect pollinators, and are usually pollinated by newly emerged insects that are naïve. Laboratory experiments showed that bumblebees tend to sample different color morphs in alternation, because visiting an empty flower increases the probability of switching to a different color morph. This behavior results in rare morphs being proportionately overvisited. This was confirmed in an experiment that demonstrated that whichever color morph is rare has a selective advantage in natural populations (Figure 8.7).

Figure 8.7 Frequency-dependent selection in the elder-flowered orchid. Relative male reproductive success of the yellow morph increases as the frequency of the yellow morph decreases. Male reproductive success was estimated by the average proportion of pollinia (mass of fused pollen produced by many orchids) removed from plants by insect pollinators. The horizontal line corresponds to equal reproductive success between the two morphs. The intersection between the regression line and the horizontal line is the value of predicted morph frequencies at equilibrium (represented by vertical dotted lines). From Gigord et al. (2001). Photo courtesy of Nick Ransdale, CC BY 2.0, Flickr.

Table 8.4 Complementary sex determination system at the *csd* locus found in many haplodiploid species. All females are heterozygous at *csd* and therefore transmit two alleles to their gametes with equal frequency. Females control release of sperm to produce either haploid or diploid progeny. From Hedrick et al. (2006).

Male gametes	Female gametes	
	½ A₁	½ A₂
½ A₁	¼ A_1A_1 (diploid males)	¼ A_1A_2 (females)
½ no fertilization	¼ A_1 (haploid males)	¼ A_2 (haploid males)

Sex is determined in most haplodiploid species by genotypes at the *csd* locus, which results in frequency-dependent selection similar to the *S*-locus in plants (Table 8.4; Hedrick et al. 2006). Heterozygotes at *csd* develop into normal females, and haploid **hemizygotes** develop into normal males. However, diploid homozygotes generally are either inviable or develop into sterile males (although see Elias et al. 2009). Thus, rare alleles are advantageous because they are less likely to produce homozygous inviable or sterile males.

Large populations commonly have 10–20 *csd* alleles, and, therefore, produce very few diploid males (Zayed & Packer 2005). However, genetic drift in small populations reduces allelic diversity at *csd* and increases the proportion of diploid males produced (Figure 8.8). The increase in diploid males reduces the number of females produced in the population and can decrease the population growth rate (Hedrick et al. 2006). We will see in Chapter 18 that this can increase the probability of extinction in small populations.

8.5 Adaptive significance of cytoplasmic genomes

Phylogeographic studies in animals and plants have focused on variation in the genomes of mitochondria and chloroplasts for over 40 years (Avise et al. 1979b; Scowcroft 1979). These studies have generally assumed that sequence polymorphism in these genomes is selectively neutral. There is now accumulating evidence that variation in these genomes sometimes affects the function of these organelles and has adaptive significance (Burton et al. 2013; Bock et al. 2014; Aw et al. 2018).

8.5.1 Plants

Bock et al. (2014) have reviewed the adaptive significance of cytoplasmic genomes in plants. They conclude that there is substantial evidence in favor of the long-held view of selective neutrality. However, there is also recent empirical evidence for the view that sequence variation in plant cytoplasmic genomes is sometimes adaptive. The strongest evidence for the possible adaptive significance of cytoplasmic genomes in plants comes from studies of cytonuclear interactions (Greiner & Bock 2013; Roux et al. 2016).

8.5.2 Animals

A variety of diseases in humans are now known to be caused by sequence variation in mitochondria (Lightowlers et al. 2015). Such phenotypic effects of mitochondria have also affected the pattern of genetic divergence in **mitochondrial DNA (mtDNA)** among human populations (Mishmar et al. 2003). Shtolz & Mishmar (2019) have recently reviewed evidence for the action of natural selection on human mtDNA variation.

There is also evidence that mutations in mtDNA might decrease the viability of small populations (Gemmell et al. 2004). Mitochondria are generally transmitted maternally so that deleterious mutations that affect only males will not be subject to natural selection (Dowling et al. 2008), and empirical evidence has supported this expectation (Innocenti et al. 2011). Sperm are powered by a group of mitochondria at the base of the flagellum, and even a modest reduction in power output may reduce male fertility yet have little effect on females. A study of human fertility has found that mtDNA genotypes are associated with sperm function and male fertility (Ruiz-Pesini et al. 2000).

A few studies have shown phenotypic effects of mtDNA. For example, Derr et al. (2012) demonstrated that individuals with cattle mtDNA were smaller than individuals with bison DNA in two bison herds that had substantial introgression

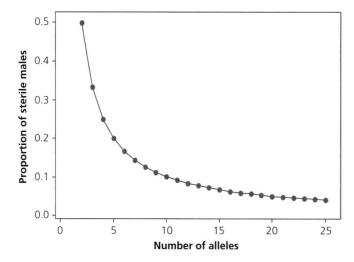

Figure 8.8 Proportion of sterile or inviable males that are produced as a function of the number of equally frequent alleles at the *csd* locus with haplodiploid sex determination. Redrawn from Cook & Crozier (1995).

from cattle. Dobelman et al. (2019) found reduced fitness associated with mtDNA genotype in the invasive common wasp in New Zealand.

8.6 Natural selection in small populations

What happens when we combine the effects of genetic drift and natural selection? More specifically, what are the effects of finite population size on the models of natural selection that we have just considered? There are two general effects of adding genetic drift to these models. First, natural selection becomes less effective because the random changes caused by drift can swamp the effects of increased survival or fertility. Second, the effects of natural selection become less predictable.

As a general rule of thumb, changes in allele frequency are determined primarily by genetic drift rather than by natural selection when the product of the effective population size and the selection coefficient ($N_e s$) is less than one (Li 1978). Thus, a deleterious allele that reduces fitness by 5% will act as if it were selectively neutral in a population with an N_e of 20 ($20 \times 0.05 = 1.00$) or less. The results of our models of selection are deterministic so that we always get the same result if we begin with the same fitnesses and the same initial allele frequency. However, the stochasticity due to genetic drift makes it

more difficult to predict what the effects of natural selection will be on allele frequencies.

8.6.1 Directional selection

Genetic drift will make directional selection less effective. This may be harmful in small populations in two ways. First, the effects of random genetic drift can outweigh the effects of natural selection so that alleles that have a selective advantage may be lost in small populations. Second, alleles that are at a selective disadvantage may go to fixation in small populations through genetic drift. This will increase the **genetic load** of a population (Section 17.6).

Wright (1931, p. 157) first suggested that small populations would continue to decline in vigor slowly over time because of the accumulation of deleterious mutations that natural selection would not be effective in removing because of the overpowering effects of genetic drift. A number of theoretical papers have considered the expected rate and importance of this effect for population persistence (Lynch & Gabriel 1990; Gabriel & Bürger 1994; Lande 1995). As deleterious mutations accumulate, population size may decrease further and thereby accelerate the rate of accumulation of deleterious mutations. This feedback process has been termed **mutational meltdown** (Section 18.6; Lynch & Gabriel 1990).

8.6.2 Underdominance and drift

Most chromosomal rearrangements (translocations, inversions, etc.) cause reduced fertility in heterozygotes because of the production of aneuploid gametes. Homozygotes for such chromosomal mutations, however, may have increased fitness. Thus, chromosomal mutations generally fit a pattern of underdominance and will always be initially selected against, regardless of their selective advantage when homozygous. However, we know that chromosomal rearrangements are sometimes incorporated into populations and species. In fact, rearrangements are thought to be an important factor in reproductive isolation and speciation.

How can we reconcile our theory with our knowledge from natural populations? That is, how can chromosomal rearrangements be incorporated into a population when they will always be initially selected against? The answer is, of course, genetic drift. If random changes in allele frequency perturb the population across the threshold of the unstable p^*, then natural selection will act to "fix" the chromosomal rearrangement. Thus, we would expect faster rates of chromosomal evolution in species with small local deme sizes.

In fact, it has been proposed that the rapid rate of chromosomal evolution and speciation in mammals is due to their social structuring and reduced local deme sizes (Wilson et al. 1975). A paper by Lande (1979) examined the theoretical relationship between local deme sizes and rates of chromosomal evolution. As discussed in Chapter 3, chromosomal variability is of special importance for conservation because the demographic characteristics that make a species a likely candidate for being threatened are the same characteristics that favor the evolution of chromosomal differences between groups. Therefore, reintroduction or translocation programs may reduce the average fitness of a population if individuals are exchanged among chromosomally distinct groups.

8.6.3 Heterozygous advantage and drift

We have seen that heterozygous advantage in a two-allele system will always produce a stable polymorphism with infinite population size. However, overdominance may actually accelerate the loss of genetic variation in finite populations if the equilibrium allele frequency is near zero or one (Robertson 1962).

Consider the following fitness set:

$$\begin{array}{cccc} & AA & Aa & aa \\ \text{Fitness} & 1 - s_1 & 1 & 1 - s_2 \end{array}$$

The following equilibrium allele frequency results if we substitute these fitness values into Equation 8.6:

$$p^* = \frac{s_2}{s_1 + s_2} \qquad (8.12)$$

$N_e(s_1 + s_2)$ is used as a measure of the effectiveness of selection here; the effectiveness of selection increases as effective population size (N_e) increases and the intensity of selection ($s_1 + s_2$) increases. For example, $N_e(s_1 + s_2)$ will equal 60 when ($s_1 + s_2$) = 0.2 and N_e = 300 or ($s_1 + s_2$) = 0.4 and N_e = 150. When the equilibrium allele frequency (p^*) is less than 0.2 or greater than 0.8, this mode of selection will actually lose genetic variation more quickly than the neutral case ($s_1 = s_2 = 0$), unless selection is very strong or the population size is very large (Figure 8.9). Thus, heterozygous advantage is only effective at maintaining fairly common alleles (frequency > 0.2). This is similar to our conclusion in Section 8.3.1.

8.7 Detection of natural selection

Evidence for natural selection on phenotypes is widespread in natural populations (Endler 1986). Nachman et al. (2003) have presented an elegant example of the action of natural selection on an individual locus resulting in local adaptation. Rock pocket mice are generally light-colored and match the color of the rocks on which they live. However, mice that live on dark lava are dark colored (melanic), and this concealing coloration provides protection from predation (Figure 8.10). These authors examined several candidate loci that were known to result in changes in pigmentation in other species. They found mutations in the melanocortin-1-receptor gene that were responsible for the dark coloration in one population of lava-dwelling mice that were melanic. However, they found no evidence of mutations at this locus in another melanic

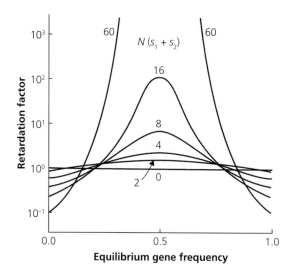

Figure 8.9 Relative effectiveness of heterozygous advantage to maintain polymorphism. Retardation factor is the reciprocal of the rate of decay of genetic variation relative to the neutral case, and N is N_e. Values less than one (10^0) indicate a more rapid rate of loss of genetic variation than expected with selective neutrality. Thus, even strong natural selection (e.g., $N(s_1 + s_2) = 60$) is not effective at maintaining a polymorphism if the equilibrium allele frequency is less than 0.20 or greater than 0.80. Redrawn from Robertson (1962).

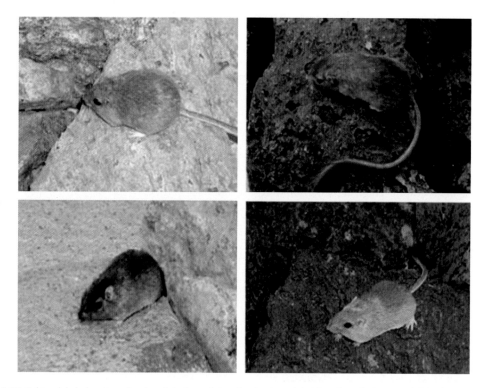

Figure 8.10 Light and dark phenotype of rock pocket mice on light-colored rocks and dark lava. From Nachman et al. (2003).

population. Thus, the similar adaptation of dark coloration apparently has evolved by different genetic mechanisms in different populations.

Many local adaptations are difficult to detect because they will only be manifest during periodic episodes of extreme environmental conditions, such as winter storms (Example 8.4; Guest Box 8), drought, hurricanes (Huey & Grant 2020), disease outbreaks (Quéméré et al. 2020), or fire (Gutschick & BassiriRad 2003). Wiens (1977) has argued that short-term studies of fitness and other population characteristics are of limited value because of the importance of "**ecological crunches**" in variable environments. For example, Rieman & Clayton (1997) suggested that the complex life histories (e.g., mixed migratory behaviors) of bull trout are adaptations to periodic disturbances such as fire that may occur only every 25–100 years.

8.8 Natural selection and conservation

An understanding of natural selection is important for the management and conservation of populations. Local adaptation can play an important role in the long-term viability of populations. We saw in Section 8.6 that natural selection is expected to be less effective in small populations. Leimu and Fischer (2008) found with meta-analysis that local adaptation in plant populations is much more common for large plant populations than for small populations; local adaptation was actually very rare in small populations. The importance of large population size for the ability to evolve raises considerable doubt on the ability of small populations to evolve with changing environments.

Natural selection plays an important role in the ability of populations to respond to anthropogenic changes to the environment. We will consider this in detail in Chapters 13–16 with regard to hybridization, invasive species, exploited populations, and climate change.

There is substantial evidence for natural selection acting on loci associated with disease resistance (e.g., Guest Box 2). For example, MHC loci have been shown to be associated with disease resistance in many species (Garrigan & Hedrick 2003). Nevertheless, how this relationship should

be applied in a conservation perspective has been controversial. Hughes (1991, p. 251) recommended that "all captive breeding programs for endangered vertebrate species should be designed with the preservation of MHC allelic diversity as their main goal." There are a variety of potential problems associated with following this recommendation (Vrijenhoek & Leberg 1991). The primary problem is that "selecting" individuals on the basis of their MHC genotype could reduce genetic variation throughout the rest of the genome (Lacy 2000a). And, there are many other loci important for disease resistance throughout the genome (Quéméré et al. 2020). We will revisit these issues in later chapters when we consider the identification of units of conservation (Chapter 20) and captive breeding (Chapter 21).

Frequency-dependent selection has special importance for conservation because of the many functionally distinct alleles that are maintained by frequency-dependent selection at some loci. We have seen that allelic diversity is much more affected by bottlenecks than is heterozygosity (Section 6.4). Reinartz & Les (1994) concluded that some one-third of the remaining 14 natural populations of *Aster furactus* in Wisconsin, USA, had reduced seed sets because of a diminished number of S-alleles. Young et al. (2000) have considered the effect of loss of allelic variation at the S-locus on the viability of small populations. In addition, frequency-dependent selection probably contributes to the large number of alleles present at some loci associated with disease resistance. Thus, the loss of allelic diversity caused by bottlenecks is likely to make small populations more susceptible to epidemics (Hedrick 2003).

Reduced allelic diversity in R genes has been shown to have harmful effects on seedling survival in small populations of tropical trees (Marden et al. 2017). R genes in plants detect pathogens and determine allele-specific activation of defenses (Meyers et al. 2005; Karasov et al. 2014). Marden et al. (2017) concluded that smaller local tree populations with limited connectivity to other populations have reduced R gene diversity. This results in lower recognition-dependent immune responses, along with greater susceptibility to species-specific

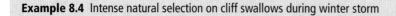

Example 8.4 Intense natural selection on cliff swallows during winter storm

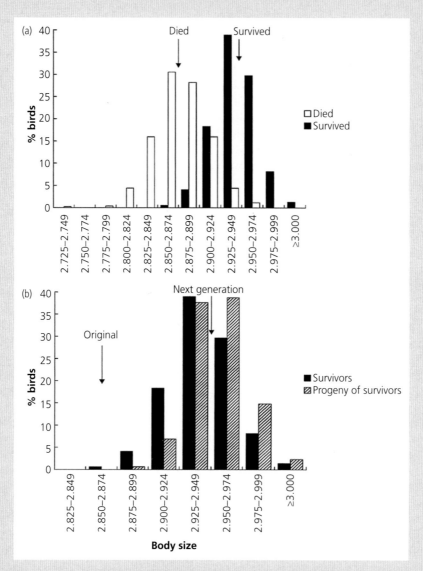

Figure 8.11 Intense natural selection on cliff swallows during a harsh winter storm. Panel (a) shows that larger birds were much more likely to survive the storm than smaller birds. Body size is a multivariate measure that includes wing length, tail length, tarsus length, and culmen length and width. Panel (b) shows that adult progeny of the next generation were much larger than the mean of the population before the storm event. Thus, natural selection increased the size of individuals in this population. Arrows indicate means. From Brown & Brown (1998).

Continued

Example 8.4 *Continued*

Brown & Brown (1998) reported dramatic selective effects of body size on survival of the cliff swallows in a population from the Great Plains of North America (Figure 8.11). Cliff swallows in these areas are sometimes exposed to periods of cold weather in late spring that reduce the availability of food. Substantial mortality generally results if the cold spell lasts 4 or more days. A once in a 100-year 6-day cold spell occurred in 1996 that killed ~50% of the cliff swallows in southwestern Nebraska. Comparison of survivors and dead birds revealed that larger birds were much more likely to survive (Figure 8.11). Mortality patterns did not differ between males and females, but older birds were less likely to survive. Morphology did not differ with age. Nonsurvivors were not in poorer condition before the storm, suggesting that selection acted on size and not condition. Larger birds apparently were favored in extreme cold weather due to the thermal advantage of larger size and the ability to store more body fat.

Examination of the adult progeny of the survivors indicated that mean body size of the population responded to the selective event caused by the storm. The body size of progeny after the storm was significantly greater than the body size of the population before the storm (Figure 8.11). Thus, body size had relatively high heritability (Section 11.1).

pathogens that may facilitate disease transmission and reduce seeding survival.

There has been much excitement in the recent literature about the identification of adaptive molecular variation using population genomics. However, perhaps the most useful contribution of genomics to conservation will be improving inferences about population demography and evolutionary history. Only neutral loci should be used to estimate crucial parameters such as N_e, migrations rates, and genetic similarity among populations.

Guest Box 8 Winter storms drive rapid phenotypic, regulatory, and genomic shifts in the green anole lizard
Shane C. Campbell-Staton

Extreme weather events—such as heat waves, droughts, hurricanes, and winter storms—offer unique opportunities to observe the effects of natural selection in real time. Indeed, extreme weather anomalies allowed some of the earliest empirical studies of natural selection operating in the wild (Bumpus 1898). Despite their brief nature, extreme weather perturbations can account for a significant proportion of the total selection experienced by a population

(O'Donald 1973). However, studying such events can prove challenging due to their fleeting and unpredictable nature. Consequently, there are still surprisingly few empirical examples of biological response to intense weather events, and fewer still have identified the regulatory and genetic targets of selection during extreme weather events (e.g., Lamichhaney et al. 2015).

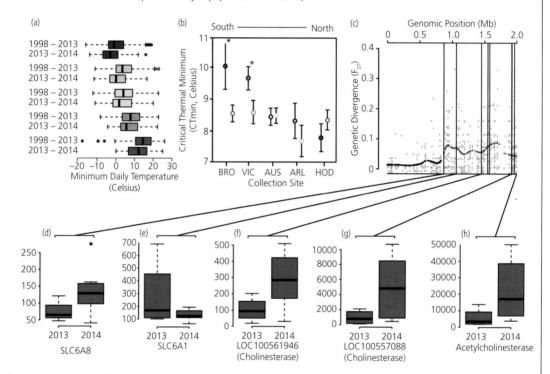

Figure 8.12 Signatures of winter storm selection in green anole populations. (a) Boxplot of daily minimum winter temperatures experienced by five green anole populations along a latitudinal gradient from south to north: Brownsville, Texas (BRO, red); Victoria, Texas (VIC, orange); Austin, Texas (AUS, yellow); Arlington, Texas (ARL, light blue); and Hodgen, Oklahoma (HOD, dark blue). The winter of 2013–2014 was significantly colder than previous winters at each location. (b) Southern populations (BRO and VIC) displayed a significant increase in cold tolerance after the winter storm. Circles and lines represent mean cold tolerance (CT$_{min}$ ± 1 standard error of the mean (SEM)). Filled and open dots represent samples collected before and after the extreme weather event, respectively. (c) One region of the genome displayed significantly elevated genetic divergence (F_{ST}) between samples collected before and after the extreme winter of 2013–2014. Gray dots represent F_{ST} values for individual single nucleotide polymorphisms identified from RNA sequence data. Black dots indicate nonsignificant F_{ST} values within 5-Mb windows ($P \geq 0.05$) in pre- versus post-storm comparisons. Red dots indicate regions of significantly elevated F_{ST} between time points ($P < 0.01$). Black lines indicate genomic positions of differentially expressed genes within F_{ST} outlier peaks. (d–h) This region of the genome contains several genes differentially expressed between time points related to neurological function. From Campbell-Staton et al. (2017).

Guest Box 8 Continued

Anole lizards provide a useful example for understanding how populations respond to extreme weather events and identifying the genomic targets of such selection. Green anoles originate in subtropical climates on the island of Cuba, but established in peninsular Florida before the Pliocene (~6.8–17.8 MYA), eventually spreading across the southern USA to occupy temperate regions in Oklahoma, Tennessee, and North Carolina (Campbell-Staton et al. 2012). Green anole populations display patterns of local adaptation associated with local winter conditions; northern populations occupying colder winter regions display greater cold tolerance than their more subtropical counterparts (Campbell-Staton et al. 2016).

During the winter of 2013–2014, green anole populations across Texas and southeastern Oklahoma were subject to uncommonly severe winter storms, which produced the coldest winter temperatures experienced across the region for at least a decade (Figure 8.12a, Campbell-Staton et al. 2017). Because cold tolerance in the green anole displays evidence of local adaptation, this event provided an opportunity to investigate physiological and genomic response to natural selection resulting from a single extreme weather event. We sequenced liver transcriptomes from 48 lizards collected across this range before and after the extreme winter. Combining the gene expression and sequence data from these samples with physiological experiments measuring individual variation in cold tolerance, we asked two important questions: (1) Has there been selection for cold tolerance in the survivors of this extreme weather event? and (2) Which genes may be the target of such selection?

First, we found that survivors at the southern edge of the species' range displayed greater cold tolerance than their pre-storm counterparts, evidence for temperature-dependent selection (Figure 8.12b). Next, we searched for regions of the species' genome with extreme differences in allele frequencies after the winter storm event. We identified several genes of the genome displaying genetic divergence in the southern survivors (Figure 8.12c) that also show significant differences in gene expression after the storm (Figure 8.12d–h). The combination of genetic and functional expression divergence provides strong support for these genes as targets of storm-induced selection contributing to individual differences in cold tolerance among green anoles.

By comparing time series data on temperature-dependent performance, gene expression, and allele frequency, we were able to provide a particularly detailed account of natural selection associated with a single weather event. In doing so, we identified several genes that may play an important role in cold resilience and winter survival. This type of integrated genomic approach can be applied to ask questions that are important for the conservation of wild populations of many species: What aspects of form and function (phenotypes) are important for the survival of populations facing specific environmental challenges? Which gene(s) are responsible for producing individual variation in those phenotypes? Which alleles within those genes might confer increased resilience to predicted patterns of climate changes faced by a population? The answers to these questions can provide valuable information for breeding and assisted migration programs seeking to increase the resilience of vulnerable populations.

CHAPTER 9

Population Subdivision

Grevillea barklyana, Example 9.1

There is abundant geographical variation in both morphology and gene frequency in most species. The extent of geographic variation results from a balance of forces tending to produce local genetic differentiation and forces tending to produce genetic homogeneity.

(Montgomery Slatkin 1987, p. 787)

Many studies have found that genomic data provide increased resolution for delineating population structure compared to last generation markers, such as microsatellites.

(Amanda S. Ackiss et al. 2020, p. 1049)

So far we have considered only random mating (i.e., panmictic) populations. However, natural populations of most species are subdivided or "structured" into separate local random mating units that are called **demes**. The subdivision of a species into separate subpopulations means that genetic variation within species exists at two primary levels:

1. Genetic variation within local populations.
2. Genetic diversity between local populations.

In contrast, individuals in some species are distributed continuously across large landscapes (e.g., coniferous tree species across boreal forests) and are not subdivided into discrete subpopulations by barriers to **gene flow**. Nevertheless, genetic differences accumulate between areas so that individuals that are further apart are more genetically different (i.e., **isolation by distance**).

There are large differences between species in the proportion of total genetic variation that is due to differences among populations. For example, Schwartz et al. (2002) found little genetic divergence among populations at nine microsatellite loci among 17 Canada lynx population samples collected from northern Alaska to central Montana (over 3,100 km). However populations of other species of vertebrates, including carnivores, can be highly genetically differentiated over a relatively short geographic distance (Figure 9.1). Spruell et al. (2003) found 20 times as much genetic divergence among bull trout populations within the Pacific Northwest of the USA as were found among lynx populations. Even separate spawning populations of bull trout just a few kilometers apart within a small tributary of Lake Pend Oreille in Idaho had twice the amount of genetic divergence as the widespread population samples of lynx (Spruell et al. 1999).

Conservation and the Genomics of Populations, Third Edition. Fred W. Allendorf, *et al.*, Oxford University Press.
© Fred W. Allendorf, W. Chris Funk, Sally N. Aitken, Margaret Byrne, and Gordon Luikart (2022). DOI: 10.1093/oso/9780198856566.003.0009

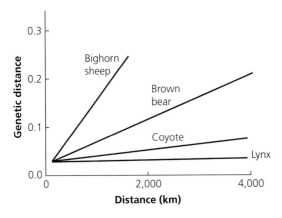

Figure 9.1 General relationship between geographic distance and genetic distance at microsatellite loci for four species of mammals. Lynx and coyotes show little genetic differentiation over thousands of kilometers; wolves (not shown) are similar to coyotes in this respect. However, less mobile species have significant differences in allele frequencies between populations over only a few hundreds of kilometers. Bighorn sheep, for example, live on mountain tops and forests, which often tend not to disperse across deep valleys and forests, which often separate mountain ranges. Modified from Forbes & Hogg (1999) with additional unpublished data from M.K. Schwartz.

Understanding the patterns and extent of genetic divergence among populations is crucial for protecting species and developing effective conservation plans (Guest Box 9). For example, translocation of animals or plants to supplement imperiled populations may have harmful effects if the translocated individuals are genetically different from the recipient population (Frankham et al. 2011). In addition, developing priorities for the conservation of a species often requires an understanding of adaptive genetic differentiation among populations. Perhaps most importantly, an understanding of genetic population structure is essential for identifying units to be conserved. For example, distinct populations can be listed under the US Endangered Species Act (ESA) and receive the same protection as biological species. Similar provisions also exist in endangered species legislation in Australia, Canada, South Africa, and other countries (Box 20.1).

The various uses of the terms **migration**, **dispersal**, and **gene flow** can be somewhat confusing. The classic population genetics literature uses migration and gene flow synonymously

to refer to the movement of individuals or gametes (e.g., pollen) from one genetic population to another (i.e., genetic exchange among breeding groups). This exchange is generally referred to as dispersal in the ecology literature. Migration in the ecology literature refers to movement of individuals during their lifetime from one geographic region to another (e.g., during seasonal migrations to winter ranges or breeding grounds). Dispersal is the same as migration and gene flow in the genetics literature if the dispersers reproduce in their new location.

In this chapter, we will consider populations that are subdivided into a series of partially isolated **subpopulations** (demes) that are connected by some amount of genetic exchange (migration). We will first consider how genetic variation is distributed at neutral loci within subdivided populations because of the effects of two opposing processes: gene flow and genetic drift. We will next consider the effects of natural selection on the distribution of genetic variation within species. Finally, we will consider the application of these analyses to the observed distribution of genetic variation in natural populations.

9.1 *F-statistics*

The oldest and most widely used metrics of genetic differentiation (i.e., divergence) are *F*-statistics. Sewall Wright (1931, 1951) developed a conceptual and mathematical framework to describe the distribution of genetic variation within a species that used a series of inbreeding coefficients: F_{IS}, F_{ST}, and F_{IT}. Holsinger & Weir (2009) provide an insightful review of the application of *F*-statistics to describe and understand genetic population structure.

F_{IS} is a measure of departure from **Hardy–Weinberg (HW) proportions** within local demes or subpopulations. The term "subpopulation" as used in this context is the same as the common use of the term local population. F_{ST} is a measure of allele frequency divergence among demes or subpopulations, and F_{IT} is a measure of the overall departure from HW proportions in the entire base population (or species) due to both nonrandom mating within local subpopulations (F_{IS}) and allele frequency divergence among subpopulations (F_{ST}).

In general, **inbreeding** is the tendency for mates to be more closely related than two individuals drawn at random from the population. It is crucial to define inbreeding relative to some clearly specified base population, which may be the species as a whole, or some specific geographical collection of subpopulations forming a "population." For example, using the entire species as the base population, a mating between two individuals within a local population will produce apparently "inbred" progeny because individuals from the same local populations are likely to have shared a more recent common ancestor than two individuals chosen at random from throughout the range of a species. As we will see, F_{ST} is a measure of this type of inbreeding.

These parameters were initially defined by Wright for loci with just two alleles. They were extended to three or more alleles by Nei (1977), who used the parameters G_{IS}, G_{ST}, and G_{IT} in what he termed the analysis of gene diversity. F- and G- are often used interchangeably in the literature. See Chakraborty & Leimar (1987) for a comprehensive discussion of F- and G-statistics.

F-statistics are a measure of the deficit of heterozygotes relative to expected HW proportions in the specified base population. That is, F is the proportion by which heterozygosity is reduced relative to heterozygosity in a random mating population with the same allele frequencies:

$$F = 1 - (H_o/H_e) \qquad (9.1)$$

where H_o is the observed proportion of heterozygotes and H_e is the expected HW proportion of heterozygotes.

F_{IS} is a measure of departure from HW proportions within local subpopulations:

$$F_{IS} = 1 - (H_o/H_S) \qquad (9.2)$$

where H_o is the observed heterozygosity averaged over all subpopulations, and H_S is the expected heterozygosity averaged over all subpopulations. F_{IS} will be positive if there is a deficit of heterozygotes and negative if there is an excess of heterozygotes. Inbreeding within local populations, such as in self-pollinating plants, will cause a deficit of heterozygotes (Example 9.1). As we saw in Chapter 6, a small effective population size can cause an excess of heterozygotes and result in negative F_{IS} values.

Example 9.1 Selfing in an Australian shrub

Ayre et al. (1994) studied genetic variation in the rare Australian shrub *Grevillea barklyana*, which reproduces by both selfing and outcrossing. They found a significant ($P <$ 0.001) deficit of heterozygotes at the *Gpi* locus in a sample of progeny from 10 maternal plants in one of their four populations:

	Genotypes				
	A1/A1	A1/A2	A2/A2		
Observed	112	43	31	$\hat{p}=$ 0.718	$\hat{F}_{IS}=$ 0.429
Expected	(95.9)	(75.3)	(14.8)		

We can estimate the proportion of selfing that can explain these results by solving for S in Equation 9.7:

$$S = \frac{2F_{IS}}{(1 + F_{IS})}$$

This results in an estimated 60% of the progeny in this population being produced by selfing and the remaining 40% by random mating, if we assume that the deficit of heterozygotes is caused entirely by selfing. Similar estimates of selfing were found at both of the loci studied in these four populations. Inbreeding due to selfing and **biparental inbreeding** (mating between relatives) cannot be distinguished based on single-locus genotypes, but multilocus analysis can distinguish between selfing and biparental inbreeding as causes of an excess of homozygotes (Ritland 2002).

F_{ST} is a measure of genetic divergence among subpopulations:

$$F_{ST} = 1 - (H_S/H_T) \qquad (9.3)$$

where H_T is the expected HW heterozygosity if the entire base population were panmictic (Example 9.2). H_S is the expected HW proportion of heterozygotes calculated for each subpopulation separately and then averaged over all subpopulations. F_{ST} ranges from zero, when all subpopulations have equal allele frequencies, to one, when all the subpopulations are fixed for different alleles. F_{ST} is sometimes called a "**fixation index**."

F_{IT} is a measure of the total departure from HW proportions that includes departures from HW proportions within local populations and divergence among populations:

$$F_{IT} = 1 - (H_o/H_T)$$

These three F-statistics are related by the expression:

$$F_{IT} = F_{IS} + F_{ST} - (F_{IS})(F_{ST}) \qquad (9.4)$$

This approach will be used in this chapter to describe the effects of population subdivision on the genetic structure of populations.

9.1.1 The Wahlund effect

The deficit of heterozygotes relative to HW proportions caused by the subdivision of a population into separate demes is often referred to as the "**Wahlund principle**" (Wahlund 1928). For example, a large deficit of heterozygotes was found at many loci when brown trout captured in Lake Bunnersjöarna (Example 9.2) were initially analyzed without knowledge of the two separate subpopulations (Ryman et al. 1979). Wahlund was a Swedish geneticist who first described this effect in 1928. He analyzed the excess of homozygotes and deficit of heterozygotes in terms of the variance of allele frequencies among S subpopulations:

$$Var(q) = \frac{1}{S} \sum (q_i - \bar{q})^2 \qquad (9.5)$$

When $Var(q) = 0$, all subpopulations have the same allele frequencies and the population is in HW proportions. As $Var(q)$ increases, the allele frequency differences among subpopulations increase and the deficit of heterozygotes increases. In fact,

$$F_{ST} = \frac{Var(q)}{pq} \qquad (9.6)$$

so that we can express the genotypic array of the population in terms of either F_{ST} or $Var(q)$:

Genotype	HW	Wright	Wahlund
AA	p^2	$p^2 + pqF_{ST}$	$p^2 + Var(q)$
Aa	$2pq$	$2pq - 2pqF_{ST}$	$2pq - 2Var(q)$
aa	q^2	$q^2 + pqF_{ST}$	$q^2 + Var(q)$

Example 9.2 The Wahlund effect in a lake population of brown trout

The approach of Nei (1977) can be used to compute F-statistics with genotypic data from natural populations. For example, two nearly equal size demes of brown trout occurred in Lake Bunnersjöarna in northern Sweden (Ryman et al. 1979). One deme spawned in the inlet, and the other deme spawned in the outlet. The fish spent almost all of their life in the lake itself rather than in the inlet and outlet streams. These two demes were nearly fixed for two different alleles (*100* and *null*) at the *LDH-A2* locus. Genotype frequencies for a hypothetical sample taken from the lake itself of 100 individuals made up of exactly 50 individuals from each deme are shown:

	100/100	100/null	null/null	Total	$\hat{p}$	2pq
Inlet deme	50	0	0	50	1.000	0.000
Outlet deme	1	13	36	50	0.150	0.255
Lake sample	51	13	36	100	0.575	0.489
(expected)	(33.1)	(48.9)	(18.1)			

The mean expected heterozygosity within these two demes combined (H_S) is 0.128 (the mean of 0.000 and 0.255). Thus, the value of F_{ST} for this population at this locus is 0.738:

$$F_{ST} = 1 - (H_S/H_T) = 1 - (0.128/0.489) = 0.728$$

That is, the heterozygosity of the sample of fish from the lake is ~73% lower than we would expect if this population was panmictic.

These two approaches for describing the genotypic effects of population subdivision (Wright and Wahlund) are analogous to the two ways we modeled genetic drift in Chapter 6: either as an increase in homozygosity or a change in allele frequency.

The Wahlund effect can readily be extended to more than two alleles (Nei 1965). However, the variance in frequencies will generally differ for different alleles. The frequency of particular heterozygotes may be greater or less than expected with HW proportions. Nevertheless, there will always be an overall deficit of heterozygotes due to the Wahlund effect.

9.1.2 When is F_{IS} not zero?

Generally, the first step in analyzing genotypic data from a natural population is to test for HW proportions (Waples 2015). As we have seen, F_{IS} is a measure of departure from expected HW proportions. A positive value indicates an excess of homozygotes, and a negative value indicates a deficit of homozygotes. Interpreting the causes of an observed excess or deficit of homozygotes can be difficult (Chapter 5). Graffelman et al. (2017) address these issues when using genomic data.

9.1.2.1 Excess of homozygotes: $F_{IS} > 0$

The most general causes of an excess of homozygotes are nonrandom mating or population subdivision. In the case of the Wahlund effect, the presence of multiple demes within a single population sample will produce an excess of homozygotes at all loci for which the demes differ in allele frequency.

Inbreeding within a single local population will produce a similar genotypic effect. That is, the tendency for related individuals to mate will also produce an excess of homozygotes. Perhaps the simplest example of this is a plant that reproduces by both self-pollination and outcrossing. Assume that a proportion S (i.e., the selfing rate) of the matings in a population are the result of selfing and the remainder $(1 - S)$ result from random mating. The equilibrium value of F_{IS} in this case will be:

$$F_{IS}^* = \frac{S}{(2 - S)} \qquad (9.7)$$

For example, consider a population in which half of the progeny are produced by selfing and half by outcrossing ($S = 0.5$). In this case, F_{IS} will be 0.33. See Example 9.1.

Assortative mating can also cause an excess or deficit of heterozygotes. For example, the white Spirit Bear that we considered in Chapter 2 displays positive assortative mating. There is an average deficit of heterozygotes of 36% at the *MC1R* locus responsible for this color polymorphism in black bears in three populations ($F_{IS} = 0.360$; Hedrick & Ritland 2011). In contrast, 10 microsatellite loci were all in HW proportions in these populations. This excess of homozygotes at least partially results from

the tendency for bears to mate with individuals having the same phenotype as their mother, resulting in positive assortative mating.

Null alleles that cannot be detected using a particular assay are another possible source of an excess of homozygotes (see Section 5.4.2 for a description of null alleles at allozyme and microsatellite loci). Heterozygotes for a null allele and another allele appear to be homozygotes on a gel and thus will result in an apparent excess of homozygotes.

Perhaps the best way to discriminate between nonrandom mating (either inbreeding within a deme or unknowingly including multiple populations in a single sample) and a null allele to explain an excess of homozygotes is to examine if the effect appears to be locus-specific or population-specific. All loci that differ in allele frequency between demes will have a tendency to show an excess of homozygotes when analyzed together. Assume you examine 10 loci in 10 different population samples ($10 \times 10 = 100$ total tests), and that you detect a significant ($P < 0.05$) excess of homozygotes for 12 tests. If eight of the 12 deviations are in a single population, this would suggest that this population sample consisted of more than one deme. In contrast, a homozygote excess due to a null allele should be locus-specific. In the same example, if eight of the deviations were at just one of the 10 loci, this would suggest that a null allele at appreciable frequency was present at that locus.

It may also be possible to discriminate between inbreeding versus including multiple populations in a single sample (the Wahlund effect) to explain an observed excess of homozygotes caused by nonrandom mating. Inbreeding will reduce the frequency of all heterozygotes equally (e.g., Equation 9.7). However, as discussed in the previous section, some heterozygotes will be in excess and some will be in deficit in the case of more than two alleles when more than two subpopulations are unknowingly sampled.

Next-generation sequencing (NGS) datasets often have **allelic dropout** and an excess of homozygotes (Section 5.3.4). Allelic dropout is similar to null alleles (e.g., at microsatellite loci) except that dropout often occurs at many loci in NGS datasets because of low sequencing coverage genome-wide. This problem can be detected using tests for a negative correlation between read depth and homozygote excess, and corrected with adequate filtering for read depth (Hendricks et al. 2018).

9.1.2.2 Excess of heterozygotes: $F_{IS} < 0$

A deficit of homozygotes (excess of heterozygotes, $F_{IS} < 0$) may also occur under some circumstances. We saw in Section 6.6 that we expect an excess of heterozygotes in small randomly mating populations. This effect can be substantial in populations with a very small effective population size due to differences in allele frequencies between sexes (Figure 6.12). Certain social structures, inbreeding avoidance, and sex-biased dispersal can also lead to an excess of heterozygotes (Parreira & Chiki 2015). Natural selection may also cause an excess of heterozygotes if they have a greater probability of surviving than homozygotes (Section 8.2 and Table 8.1). However, the differential advantage of heterozygotes has to be very great to have a detectable effect on genotypic proportions (Lachance 2009).

Sex linkage can cause an excess of heterozygotes if males and females have different allele frequencies at sex-linked loci. Sex linkage affects only loci near the sex-determining genomic region, unlike the heterozygote excess at many loci caused by small population size. The effects of sex linkage can be complicated and can cause an excess of heterozygotes in either or both sexes for sex-linked loci (Graffelman et al. 2017). To help identify loci that have an excess (or deficit) of heterozygotes due to sex linkage, we can test for HW proportions in each sex separately (Section 5.5). We can also test a locus simultaneously for HW proportions and allele frequency differences between sexes to understand causes of HW deviations (Graffelman et al. 2017).

9.2 Spatial patterns of relatedness within local populations

Genotypes are sometimes distributed nonrandomly spatially within local populations. This will be especially true for plant species and for animal species that are not mobile (e.g., many marine invertebrates) or species that are philopatric or living in social groups (e.g., some marmots, shrews, and monkeys). In these cases, understanding genetic population structure requires an understanding of the distribution of genotypes on a small geographic scale. For example, if genotypes within a local population are spatially structured, then within-population variation may be underestimated if samples are not spatially well distributed across the population.

9.2.1 Effects of dispersal distance and population density

The **spatial genetic structure** (SGS) within a local population depends on both the average dispersal distance of genes and population density. To illustrate this, imagine two mature populations of a tree species (Figure 9.2). One population has relatively low density (few, widely spaced individuals; Figure 9.2a), while the other has relatively high density (Figure 9.2b). Most gravity-dispersed tree seed falls within approximately two tree heights of the maternal parent. In the lower-density stand, the seed shadows of individual mother trees will overlap relatively little, so there will be a relatively high probability that two seedlings that germinate and grow within that seed shadow are related. In contrast, in the high-density population, many seed shadows will overlap, and the probability of adjacent seedlings being related is lower. At pairwise distances exceeding the seed shadow width, the probability of relatedness is low. If dispersal increases (i.e., seed shadows increase in diameter), then the SGS will decrease. This simplified example considers only offspring sharing a mother, but restricted pollen dispersal (e.g., via insect pollination) can also create SGS.

The extent of SGS is usually quantified using statistics that estimate spatial autocorrelation among genotypes at varying distances apart. The probability that two alleles are identical in state (Q) generally decreases with the spatial distance between them (r); thus, SGS is characterized by the function $Q(r)$. Individuals sampled across a population are genotyped for a set of neutral markers, and the pairwise physical distance between all possible pairs of individuals is estimated. For two individuals i and j, the probability that a random allele from i is identical to a random allele from j is the **coancestry** coefficient, F_{ij} (Vekemans &

(a)   **Lower-density population**

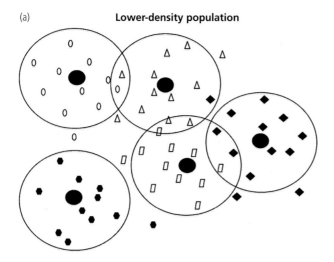

(b)   **Higher-density population**

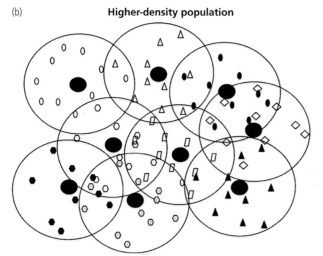

Figure 9.2 Effect of population density on spatial genetic structure (SGS) in tree populations. Larger filled circles indicate tree locations; open circles around them delineate seed shadows within which most seed will fall. Small symbols indicate seedlings from each maternal parent. The lower-density population (top) has less overlap in seed shadows and therefore a higher probability of neighboring seedlings inheriting identical alleles from the same mother tree.

Hardy 2004):

$$F_{ij} = \frac{Q_{ij} - \bar{Q}}{1 - \bar{Q}} \qquad (9.8)$$

Pairs of individuals are pooled into pairwise distance classes, and the average coancestry for each distance class is plotted in a spatial correlogram (Figure 9.3). A full review of the statistical methods of assessing SGS is beyond the scope of this section. Vekemans & Hardy (2004) review several statistical approaches for estimating patterns of relatedness for quantifying SGS.

Population density and SGS can vary substantially among populations if dispersal distance stays relatively constant. Gapare & Aitken (2005) found strong SGS and higher inbreeding levels in lower-density Sitka spruce populations at the northern and southern range margins, and low SGS in high-density central populations. Figure 9.3 illustrates the coancestry of individuals in different pairwise distance categories in one peripheral, geographically isolated population (Kodiak Island, Alaska), and in one population in the center of the species range (Port McNeill, British Columbia). The positive coancestry coefficient in the Kodiak Island population up to a distance of nearly 500 m indicates that individuals within 500 m of each other are more genetically similar to each other than

(a) Kodiak Island, Alaska

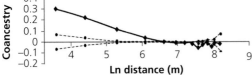

(b) Port McNeill, British Columbia

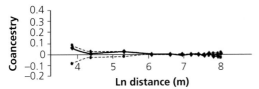

Figure 9.3 Spatial correlograms showing spatial genetic structure (SGS) within populations of Sitka spruce. The solid line is the coancestry coefficient, and the dashed lines show the 95% confidence interval under the null hypothesis that genotypes are randomly distributed. The pairwise distance classes range from tens to thousands of meters (4 = 55 m, 6 = 403 m, and 8 = 2,980 m). The positive coancestry coefficient in the Kodiak Island population (a) up to a distance of nearly 500 m indicates that individuals within 500 m of each other are more genetically similar to each other than are pairs of individuals taken at random from the population. From Gapare & Aitken (2005).

are pairs of individuals taken at random from the population. While these populations now have a similar density of mature trees, the Kodiak Island population was founded only ~500 years ago, and the majority of trees appear to have descended from relatively few founders that persist as very large, old trees in the population. As densities were low for the initial colonizing generations, the SGS persists and is strong (Gapare & Aitken 2005). A similar SGS exists in populations at the southern periphery of the species range, not due to colonization dynamics as that population has likely reached a quasiequilibrium between density and dispersal, but more likely due to low population density.

9.2.2 Effects of spatial distribution of relatives on inbreeding probability

The average relatedness of neighboring individuals and the population density impact the mating system and probability of inbreeding in a population. For example, geographically isolated Sitka

spruce populations at the northern and southern range peripheries receive pollen from just two effective donors, on average, while those in the high-density populations in the center of the species range have tens of effective pollen donors in the pollen cloud they sample (Mimura & Aitken 2007). The peripheral, isolated populations have an effective self-pollination rate of 15–35% (see Example 9.1 for how selfing rate is calculated). In contrast, high-density populations with no significant SGS in the range center have an effective selfing rate of less than 5%. Note that this effective selfing can be due to either self-pollination or to biparental inbreeding (mating of relatives); in the case of Sitka spruce, it appears to be largely due to the latter as a function of the proximity of related individuals.

SGS can also provide information on the size and distribution of individual clones in plants that reproduce vegetatively, usually through horizontal above- or below-ground spreading of individual genotypes. Estimating population sizes is difficult if the extent of clonality is not known, and it is often not easy to phenotypically distinguish multiple stems of a clone from multiple individuals. Travis et al. (2004) used analysis of SGS of the salt marsh plant smooth cordgrass to determine the extent of clonal structure in populations of different ages. They found that populations of younger plants had less clonal diversity than older populations, and also had relatively high rates of self-pollination, but concluded that inbreeding depression was likely high as there were fewer inbred individuals in older cohorts.

9.3 Genetic divergence among populations and gene flow

The distribution of genetic variation among populations results from the interaction of the opposing effects of genetic drift, which acts to cause subpopulations to diverge, and the cohesive effects of gene flow, which acts to make subpopulations more similar to each other. All neutral loci in the genome are expected to show similar patterns of divergence (i.e., F_{ST}). Nevertheless, we will see that the effects of natural selection on individual loci can cause them to display very different patterns of divergence among subpopulations.

9.3.1 Complete isolation

Let us initially consider a large random mating population that is subdivided into many completely isolated demes. We will consider the effect of this subdivision on a single locus with two alleles. Assume all HW conditions are valid except for small population size within each individual isolated subpopulation. Genetic drift will occur in each of the isolated demes; eventually, each deme will become fixed for one allele or the other.

What is the effect of this subdivision on our two measures of the genetic characteristics of populations: allele frequencies and genotype frequencies? If the initial allele frequency of the A allele in the large, random mating population was p, the allele frequency in our large, subdivided population will still be p because we expect p of the isolates to become fixed for the A allele, and $(1 - p)$ of the isolates to become fixed for the a allele. Thus, subdivision (nonrandom mating) itself has no effect on overall allele frequencies.

We can see this effect in the guppy example from Table 6.3 where 16 subpopulations were founded by a single male and female from a large population. Genetic drift within each subpopulation acted to change allele frequencies at the two loci (*AAT-1* and *PGM-1*) for four generations. However, the average allele frequencies over the 16 subpopulations at both loci are very close to the frequencies in the large founding population. Therefore, allele frequencies in the populations as a whole were not affected by subdivision.

However, the subdivision into 16 separate subpopulations did affect the genotypic frequencies of this population. We can use the F-statistics approach developed in the previous section to describe this effect at the *AAT-1* locus. In this case, H_S is the mean expected heterozygosity averaged over the 16 subpopulations (0.344), and H_T is the HW heterozygosity (0.492) using the allele frequency averaged over all subpopulations (0.562). Therefore,

$$F_{ST} = 1 - (H_S/H_T) = 1 - (0.344/0.492) = 0.301$$

In words, the average heterozygosity of individual guppies in this population has been reduced by 30% because of the subdivision and subsequent genetic drift within the subpopulations.

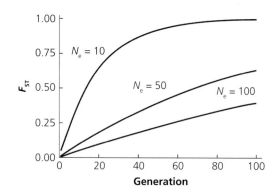

Figure 9.4 Expected increase in F_{ST} over time (generations) among completely isolated populations of different sizes using Equation 9.9.

We know from Section 6.3 that heterozygosity will be lost by genetic drift at a rate of $1/2N_e$ per generation. Therefore, we expect F_{ST} among completely isolated populations to increase as follows (modified from Equation 6.3):

$$F_{ST} = 1 - \left(1 - \frac{1}{2N_e}\right)^t \qquad (9.9)$$

where N_e is the effective population size of each subpopulation and t is the number of generations (Figure 9.4). The application of this expression to real populations is limited because it assumes a large number of equal size subpopulations and constant population size.

9.3.2 Gene flow

In most cases, there will be some genetic exchange (gene flow) among demes within a species. We must therefore consider the effects of such partial isolation on the genetic structure of species. Let us first consider the simple case of two demes (A and B) of equal size that are exchanging individuals in both directions at a rate m. Therefore, m is the proportion of individuals reproducing in one deme that were born in the other deme. In this case:

$$q'_A = (1 - m)q_A + mq_B$$

$$\text{and} \qquad (9.10)$$

$$q'_B = mq_A + (1 - m)q_B$$

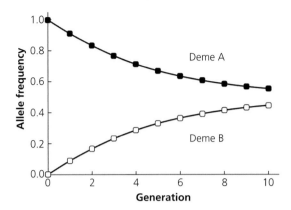

Figure 9.5 Expected changes in allele frequencies in two demes that are exchanging 10% of their individuals each generation ($m = 0.10$) using Equation 9.10.

For example, consider two previously isolated populations that begin to exchange migrants at a rate of $m = 0.10$ (10% exchange per generation). Assume that the allele frequency in population A is 1.0, and in population B it is 0.0. This model can be used to predict the effects of gene flow between these two populations as shown in Figure 9.5. Equilibrium will be reached when $q_A = q_B$, and q^* will be the average of the initial allele frequencies in the two demes; in this case; $q^* = 0.5$.

In general, there are two primary effects of gene flow:

1. Gene flow reduces genetic differences between populations.
2. Gene flow increases genetic variation within populations.

Gene flow among populations is the cohesive force that holds together geographically separated populations into a single evolutionary unit—the species. In the rest of this chapter we will consider the interaction between the homogenizing effects of gene flow and the action of genetic drift and natural selection that cause populations to diverge.

9.4 Gene flow and genetic drift

In the absence of other evolutionary forces, any gene flow between populations will bring about genetic homogeneity. With lower amounts of gene flow it will take longer, but eventually all populations will become genetically identical. However, we saw in Section 6.1 that genetic drift causes isolated subpopulations to diverge genetically. Thus, the actual amount of divergence among subpopulations is primarily a balance between the homogenizing effects of gene flow, making subpopulations more similar, and the disruptive effects of drift, causing divergence among subpopulations. We examine this using a series of models for different patterns of gene flow. All of these models will necessarily be much simpler than the actual patterns of gene flow in natural populations.

9.4.1 Island model

We will begin with the simplest model, which combines the effects of gene flow and genetic drift. Assume that a population is subdivided into a series of ideal populations, each of local effective size N, which exchange individuals at a rate of m. Specifically, each subpopulation contributes an equal number of genes to a migrant pool. The total proportion of individuals in a subpopulation from the migrant pool is m per generation, and the rest $(1 - m)$ are drawn from the local population. This model is called the **island model of migration** (Figure 9.6).

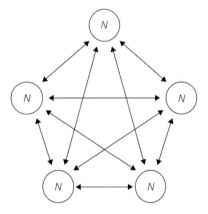

Figure 9.6 Pattern of exchange among five subpopulations under the island model of migration. Each subpopulation of local effective size N exchanges migrants with the other subpopulations with equal probability. More specifically, each subpopulation contributes an equal number of genes to a migrant pool. The total proportion of individuals in a subpopulation from the migrant pool is m per generation, and the rest $(1 - m)$ are drawn from the local population.

As before, we will measure divergence among subpopulations (demes) using F_{ST}. Genetic drift within each deme will act to increase divergence among demes; that is, increase F_{ST}. However, migration between demes will act to decrease F_{ST}. As long as $m > 0$, there will be some steady-state (equilibrium) value of F_{ST} at which the effects of drift and gene flow will be balanced.

Sewall Wright (1969) has shown that at equilibrium under the island model of migration with an infinite number of demes:

$$F_{ST} = \frac{(1 - m)^2}{[2N - (2N - 1)(1 - m)^2]} \qquad (9.11)$$

Fortunately, if m is small this approaches the much simpler:

$$F_{ST} \approx \frac{1}{(4mN + 1)} \qquad (9.12)$$

This approximation provides an accurate expectation of the amount of divergence under the island model (Figure 9.7). For example, the exact expected equilibrium value of F_{ST} with one migrant per generation ($mN = 1$) using Equation 9.11 is 0.199. The approximate value using Equation 9.12 is 0.200; the value resulting from the simulation shown in Figure 9.7 with 20 subpopulations ($F_{ST} = 0.215$) is very close to this expected value. One important result of this analysis is that very little gene flow is necessary for populations to be genetically connected (Section 19.4).

Equations 9.11 and 9.12 assume an infinite number of subpopulations. The expected value of F_{ST} at equilibrium can be corrected as shown to take into account a finite number (n) of subpopulations (Slatkin 1995):

$$F_{ST} \approx \frac{1}{(4mNa + 1)} \qquad (9.13)$$

where:

$$a = \left(\frac{n}{n - 1}\right)^2$$

This effect is small unless there are very few subpopulations. For example, with 20 subpopulations, the expected value of F_{ST} with Equation 9.13 is 0.184, rather than 0.200 with Equation 9.12.

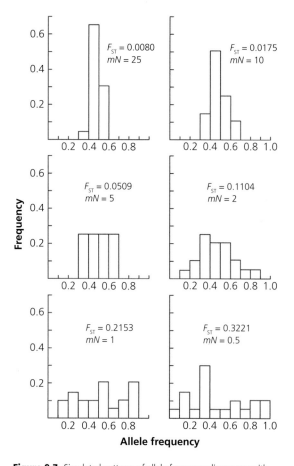

Figure 9.7 Simulated patterns of allele frequency divergence with the island model of migration showing the effect of different amounts of migration. Computer simulations were carried out with 20 subpopulations ($N = 200$) and different amounts of migration (mN). Redrawn from Allendorf & Phelps (1981).

Equation 9.12 also provides a surprisingly simple result: the amount of divergence among demes, as measured by F_{ST}, depends only on the number of migrant individuals (mN), and not the proportion of exchange among demes (m). The dependence of divergence only on the number of migrants irrespective of population size may seem counterintuitive. Remember, however, that the amount of divergence results from the opposing forces of drift and migration. The larger the demes are, the slower they are diverging through drift; thus, proportionally fewer migrants are needed to counteract the effects of drift. Small demes diverge rapidly through drift, and thus proportionally more migrants are needed to counteract drift.

This result has important implications for our interpretation of observed patterns of genetic divergence among populations (e.g., Figure 9.1). We expect to find approximately the same amount of divergence among demes of size 200 with $m = 0.025$ as we do with demes of size 50 with $m = 0.1$ ($0.025 \times 200 = 0.1 \times 50 = 5$ migrants per generation). One important outcome of this result is that species with larger local populations (N) are expected to have less genetic divergence among populations than species with the same amount of genetic exchange (m) with smaller local populations (Lowe & Allendorf 2010a). We will see in Section 19.4 that this effect has important implications for understanding the relationship between genetic divergence and demographic connectivity among populations.

9.4.2 Stepping-stone model

In natural populations, migration is often greater between subpopulations that are near each other

Figure 9.8 Pattern of exchange among subpopulations under the single-dimension stepping-stone model of migration. Each subpopulation of size N exchanges $m/2$ individuals with each adjacent subpopulation.

(Slatkin 1987). This violates the assumption of equal probability of exchange among all pairs of subpopulations with the island model of migration. The **stepping-stone model of migration** was introduced (Kimura & Weiss 1964) to take into account both short-range migration (which occurs only between adjacent subpopulations) and long-range migration (which occurs at random between subpopulations). Linear stepping-stone models (Figure 9.8) are useful for modeling populations with a one-dimensional linear structure, as occurs along a river, valley, or a mountain ridge, for example. Two-dimensional stepping-stone models are useful for modeling populations with a grid structure (or two-dimensional checker board pattern) across the landscape.

The mathematical treatment of the stepping-stone model is much more complex than the island model. In general, migration in the stepping-stone model is less effective at reducing differentiation caused by drift because subpopulations exchanging genes tend to be genetically similar to each other. Therefore, there will be greater differentiation (i.e., greater F_{ST}) among subpopulations with the stepping-stone than the island model for the same amount of genetic exchange (m). In addition, in the stepping-stone model, adjacent subpopulations should be more similar to each other than geographically distant populations (Figure 9.1). With the island model of migration, genetic divergence will be independent of geographic distance (Figure 9.9).

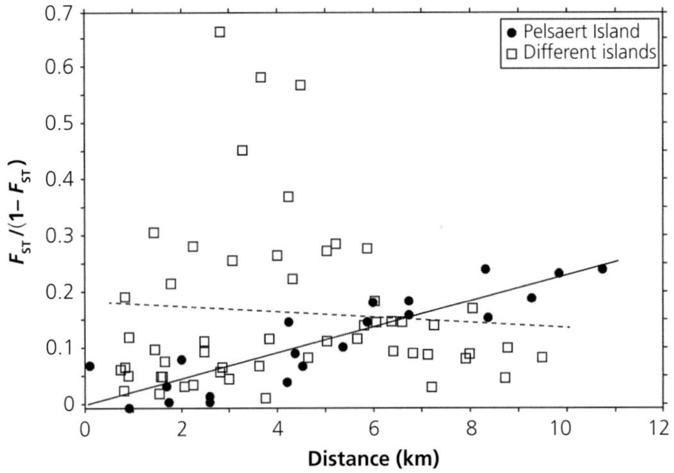

Figure 9.9 Observed relationship between genetic divergence $F_{ST}/(1 - F_{ST})$ and geographical distance in populations of the intertidal snail *Austrocochlea constricta* off the coast of Western Australia at 17 polymorphic allozyme loci. The open boxes are comparisons between pairs of populations on different islands. The dashed line is a regression line fitted through the open boxes (different islands). There is no relationship here between genetic divergence and geographical distance, as expected with the island model of migration. The closed circles show comparisons between populations found on a single large island (Pelsaert Island). These closed circles show the "isolation by distance" pattern (black solid line) expected with either the stepping-stone model of migration or isolation by distance in a continuously distributed population. From Johnson & Black (2006).

9.5 Continuously distributed populations

In some species, individuals are distributed continuously across large landscapes (e.g., coniferous tree species across boreal forests) and are not subdivided into discrete subpopulations by barriers to gene flow (Figure 9.10). Nonetheless, gene flow can be limited to relatively short distances leading to increasing genetic differentiation as the geographical distance between individuals becomes greater (Section 9.2.1). This effect was originally called isolation by distance by Wright (1943). Today, many papers use the term isolation by distance in a more general sense to refer to any pattern where genetic divergence increases with geographical distance (e.g., Figure 9.9).

The mathematics of the distribution of genetic variation in continuously distributed populations is complex (Felsenstein 1975; Epperson 2007). It is impossible to identify and sample discrete population units because no sharp boundaries exist. In this case, the **neighborhood** has been defined as the area from which individuals can be considered to be drawn at random from a random mating population (Wright 1943, 1946). This model assumes that dispersal distances are normally distributed about a mean of zero. In this case, Wright's neighborhood

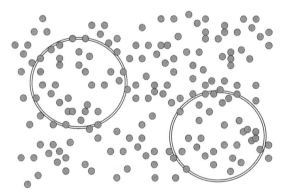

Figure 9.10 Continuous distribution of individuals where no sharp boundaries separate individuals (gray dots) into discrete groups. Nonetheless, genetic isolation arises over geographic distance because nearby individuals are more likely to mate with each other than with individuals that are farther away (isolation by distance). We can place a circle of the appropriate neighborhood size anywhere (see two circles) and individuals inside will represent a panmictic group in HW proportions (i.e., a genetic neighborhood).

size (NS = the effective number of parents in a neighborhood) is:

$$NS = 4\pi\sigma^2 D \qquad (9.14)$$

where D is density (number of individuals per unit area), π is pi, and σ^2 is the mean squared parent–offspring axial distance (i.e., along one axis). NS can be thought of as the number of reproducing individuals in a circle of radius 2σ, and a circle of this size would include about 87% of the parents of individuals at the center (Wright 1946).

Chambers (1995) provides a helpful overview of the conservation implications of isolation by distance in maintaining genetic variation in continuously distributed populations. It is difficult to come up with simple applications of this model to conservation. However, it does allow us to estimate the geographic distance at which individuals will become genetically differentiated due to limited gene flow (e.g., Figure 9.3). For example, if the mean gene flow distance is 1 km, then we would expect substantial genetic differentiation between individuals separated by, say, 5–10 km (Manel et al. 2003).

9.6 Cytoplasmic genes and sex-linked markers

Uniparentally inherited cytoplasmic genes and sex-linked markers generally show different amounts of differentiation among populations than autosomal loci for several reasons. First, they usually have a smaller effective population size than autosomal loci and therefore show greater divergence due to genetic drift. In addition, differences in migration rates between males and females can cause large differences between cytoplasmic genes and sex-linked markers compared with autosomal loci.

9.6.1 Cytoplasmic genes

We can estimate F_{ST} values to compare the amount of allelic differentiation for nuclear and mitochondrial genes. However, since **mitochondrial DNA (mtDNA)** is haploid, individuals are **hemizygous** rather than homozygous or heterozygous.

We generally expect more differentiation at mtDNA and **chloroplast DNA (cpDNA)** genes than for nuclear genes because of their smaller effective

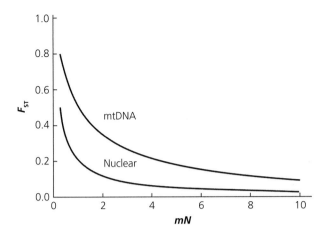

Figure 9.11 Expected values of F_{ST} with the island model of migration for a nuclear locus (Equation 9.12) and mtDNA (Equation 9.15) assuming equal migration rates in males and females.

size. That is, the greater genetic drift with smaller effective population size will bring about greater differentiation than populations that are connected by the same amount of gene flow. If migration rates are equal in males and females, then we expect the following differentiation for mtDNA with the island model of migration (Birky et al. 1983):

$$F_{ST} \approx \frac{1}{(mN + 1)} \tag{9.15}$$

This expression is sometimes written to consider only females:

$$F_{ST} \approx \frac{1}{(2mN_f + 1)} \tag{9.16}$$

where N_f is the number of females in the population. Equations 9.15 and 9.16 are identical if there are an equal number of males and females in the population ($N_f = N_m$) because $2N_f = N$. Thus, with equal migration rates in males and females, we expect approximately two to four times as much allele frequency differentiation at mitochondrial genes than at nuclear genes (Figure 9.11). We can see this effect in the study of sockeye salmon shown in Section 9.6.3; the F_{ST} at mtDNA was greater than the F_{ST} at all but one of the 20 nuclear loci examined.

This difference in F_{ST} for a nuclear locus and maternally inherited mtDNA is expected to be greater for species in which migration rates of males are greater than those of females (Hedrick et al. 2013b; see Example 9.3). Similarly, we often see much greater F_{ST} values for cpDNA than nuclear genes in plants because most of the gene flow is via pollen (Petit et al. 2005). Larsson et al. (2009) present a valuable summary of the statistical power for detecting genetic divergence using nuclear and cytoplasmic markers.

9.6.2 Sex-linked loci

Genes on the Y chromosome of mammals present a parallel situation to mitochondrial genes. The Y chromosome is haploid and is only transmitted through the father. Thus, the expectations that we just developed for cytoplasmic genes also apply to Y-linked genes except we must substitute the number of males for females. Comparison of the patterns of differentiation at autosomal, mitochondrial, and Y-linked genes can provide valuable insight into the evolutionary history of species and current patterns and drivers of gene flow (Example 9.4).

9.7 Gene flow, genetic drift, and natural selection

We will now consider the effects of natural selection on the amount of genetic divergence expected among subpopulations in an island model of migration (Allendorf 1983). We previously concluded that the amount of divergence, as measured by F_{ST}, is dependent only upon the product of migration rate and deme size (mN). Does this simple principle hold when we combine the effects of natural selection with the island model of migration? As we will see shortly, the answer is no.

Example 9.3 Sex-biased dispersal of great white sharks

The great white shark is globally distributed in temperate waters off continental shelves (Figure 9.12). Relatively little is known about the ecology and demography of this species because of its rarity, large size, and mobility. Pardini et al. (2001) examined both mtDNA and microsatellite genotypes in great white sharks collected off the coasts of South Africa, Australia, and New Zealand.

Figure 9.12 Great white shark near Guadalupe Island, Mexico. Photo courtesy of Elias Levy, CC BY 2.0, Flickr.

Comparison of the control region of mtDNA revealed two major **haplogroups** (A and B) that have ~4% sequence divergence. A haplogroup is a group of similar haplotypes that share a common ancestor. No differences in haplogroup frequencies were found between sharks from Australia or New Zealand. However, sharks from South Africa were extremely divergent from sharks in Australia and New Zealand ($F_{ST} = 0.85$):

Population	Type A	Type B
Australia & NZ	48	1
South Africa	0	39

In striking contrast to this result, no allele frequency differences were found at five microsatellite loci among these regions.

Pardini et al. (2001) concluded that female great white sharks are philopatric and that males undertake long transoceanic movements. However, study of transoceanic movement with electronic tags and photographic identification indicates that females, as well as males, make transoceanic movements between these areas (Bonfil et al. 2005). Therefore, the difference between divergence of mtDNA and nuclear markers is apparently not based on differences in transoceanic migrations of males and females, but results from whether or not these migrants become reproductively integrated into the recipient population.

Example 9.4 Y chromosome isolation in a shrew hybrid zone

A Y chromosome microsatellite locus and 10 autosomal microsatellite loci were genotyped across a hybrid zone between the Cordon and Valais races of the common shrew in western France (Balloux et al. 2000). There is a contact zone where the two races occur on either side of a stream. Gene flow is somewhat limited, but the two races show relatively little divergence at the autosomal microsatellite loci ($F_{ST} = 0.02$; Brünner & Hausser 1996; Balloux et al. 2000).

Almost all gene flow in these shrews appears female mediated, and male hybrids are generally unviable. No alleles were shared across the hybrid zone at the Y-linked locus (Figure 9.13). However, the F_{ST} value between races at the Y-linked microsatellite locus is just 0.19; this low value does not reflect the absence of alleles shared between races because of the high within-race heterozygosity at this locus (this effect is discussed in Section 9.7). R_{ST} is an analog of F_{ST} that takes the relative length of microsatellite alleles into consideration, assuming that under the **stepwise mutation model**, alleles that are more similar in length are more closely related genealogically. The strong divergence of populations on either side of the stream is reflected in the R_{ST} value of 0.98. It is important to incorporate allele length (for microsatellites) or other genealogical information about alleles (Section 9.8.1), when H_S is high, because mutations likely contribute to population differentiation when populations are long-isolated, as in this example.

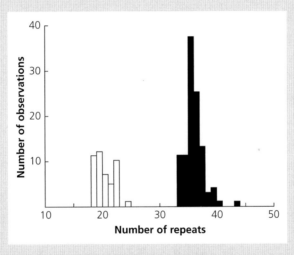

Figure 9.13 Allele frequency distribution of a Y-linked microsatellite locus (*L8Y*) in the European common shrew. Males of the Cordon race are represented by black bars, and males of the Valais race by white bars. From Balloux et al. (2000).

9.7.1 Heterozygous advantage

Assume the following fitnesses hold within each deme:

$$
\begin{array}{cccc}
 & AA & Aa & aa \\
\text{Fitness} & 1-s & 1.0 & 1-s
\end{array}
$$

Because of its complexity, one of the best ways to explore this system is using computer simulations. We will examine the results of simulations combining natural selection with the island model of migration in which there are 20 subpopulations. Natural

selection will act to maintain a stable equilibrium of $p^* = 0.5$ in each deme. Thus, this model of selection will reduce the amount of divergence among demes (Table 9.1). The greater the value of s, the greater will be the reduction in F_{ST}. Even relatively weak selection can have a marked effect on F_{ST}; for example, see $s = 0.01$ and $mN = 0.5$ in Table 9.1. It is also apparent that genetic divergence among demes is no longer only a function of mN. For a given value of mN, natural selection becomes more effective, and thus F_{ST} is reduced, as population size increases.

Table 9.1 Simulation results of steady-state F_{ST} values for 20 demes with selective neutrality ($s = 0.00$) or heterozygous advantage ($s > 0$) in which both homozygous phenotypes have a reduction in fitness of s. Each value is the mean of 20 independent simulations. Expected F_{ST} values from Equation 9.12. From Allendorf (1983).

			mN				
	0.5	1	2	5	10	25	N
Expected	0.333	0.200	0.111	0.048	0.024	0.010	
$s = 0.00$	0.307	0.204	0.124	0.042	0.020	—	25
	0.335	0.183	0.108	0.048	0.026	0.012	50
	0.322	0.188	0.106	0.044	0.025	0.010	100
$s = 0.01$	0.283	0.164	0.067	0.050	0.022	—	25
	0.243	0.153	0.082	0.041	0.023	0.012	50
	0.178	0.124	0.093	0.038	0.036	0.011	100
$s = 0.05$	0.193	0.126	0.071	0.044	0.024	—	25
	0.133	0.107	0.062	0.034	0.024	0.009	50
	0.083	0.107	0.043	0.024	0.018	0.011	100
$s = 0.10$	0.122	0.104	0.053	0.043	0.022	—	25
	0.094	0.076	0.050	0.031	0.021	0.009	50
	0.041	0.029	0.032	0.022	0.010	0.007	100

Table 9.2 Simulation results of steady-state F_{ST} values for 20 demes with differential directional selection. Each value is the mean of 20 independent simulations. One homozygous genotype has a reduction in fitness of t in 10 demes; the other homozygous genotype has the same reduction in fitness in the other 10 demes. Heterozygotes have a reduction in fitness of one-half t in all demes. Expected F_{ST} values from Equation 9.12. From Allendorf (1983).

			mN				
	0.5	1	2	5	10	25	N
Expected	0.333	0.200	0.111	0.048	0.024	0.010	
$t = 0.00$	0.307	0.204	0.124	0.042	0.020	—	25
	0.335	0.183	0.108	0.048	0.026	0.012	50
	0.322	0.188	0.106	0.044	0.025	0.009	100
$t = 0.01$	0.334	0.170	0.107	0.056	0.022	—	25
	0.298	0.119	0.100	0.038	0.026	0.010	50
	0.300	0.185	0.115	0.035	0.023	0.010	100
$t = 0.05$	0.356	0.186	0.120	0.050	0.022	—	25
	0.462	0.268	0.149	0.055	0.026	0.011	50
	0.595	0.423	0.198	0.063	0.021	0.012	100
$t = 0.10$	0.470	0.245	0.163	0.047	0.029	—	25
	0.624	0.365	0.261	0.077	0.036	0.013	50
	0.805	0.657	0.443	0.159	0.063	0.019	100

9.7.2 Divergent directional selection

Assume the following relative fitnesses in a population consisting of 20 demes:

	AA	Aa	aa
Demes 1 – 10	1	$1 - (t/2)$	$1 - t$
Demes 11 – 20	$1 - t$	$1 - (t/2)$	1

This pattern of divergent directional selection will act to maintain allele frequency differences among demes so that large differences can be maintained even with extensive genetic exchange. Again, selection is more effective with larger demes (Table 9.2).

9.7.3 Comparisons among loci

Gene flow and genetic drift are expected to affect all loci uniformly throughout the genome. However, the effects of natural selection will affect loci differently depending upon the intensity and pattern of selection and effective population size. As we have noted, even fairly weak natural selection can have a substantial effect on divergence among large subpopulations. Therefore, surveys of genetic differentiation at many loci throughout the genome can be used to detect **outlier loci** that are candidates for the effects of natural selection.

Detecting locus-specific effects is critical because only genome-wide effects inform us reliably about population demography and phylogenetic history, whereas locus-specific effects can help identify genes important for fitness and adaptation. An example of a locus-specific effect is differential directional selection whereby one allele is selected for in one environment but the allele is disadvantageous in a different environment. This selection would generate a large allele frequency difference (high F_{ST}) only at this locus relative to neutral loci throughout the genome (Example 9.5). For example, just a 10% selection coefficient favoring different alleles in two environments can generate large differences with this pattern of selection between the selected locus ($F_{ST} = 0.657$) and neutral loci ($F_{ST} = 0.188$), as shown in Table 9.2 with local population sizes of $N = 100$ and $mN = 1$.

Knowledge of the breeding structure of fish stocks is crucial for developing and implementing effective management strategies that are urgently needed to maintain sustainable fisheries (Chapter 15). However, population genetic studies of marine fishes often have failed to detect genetic differences, even between apparently geographically isolated subpopulations for which there is evidence of some

Example 9.5 Use of adaptive loci to detect genetic subdivision in marine fishes

Almost no genetic differentiation (F_{ST} = 0.003) was found at nine neutral microsatellite loci in Atlantic cod, but substantial differentiation (F_{ST} = 0.261) was found at the *Pan* I locus (Pampoulie et al. 2006), which previous studies have shown to be under natural selection (Pogson & Fevolden 2003). The utility of divergent markers such as this for stock analysis would be greatly reduced if such differences were not stable and changed over a few generations. Fortunately, comparison of current patterns of genetic differentiation using **otoliths** (inner ear organs) going back up to 70 years demonstrated that these allele frequency differences have been stable and therefore can be used as a reliable marker for stock identification (Nielsen et al. 2007).

Other loci have been found in Atlantic cod to be useful in describing genetic differentiation among populations. Nielsen et al. (2009) screened 98 gene-associated **single nucleotide polymorphisms (SNPs)** in Atlantic cod. Eight of these SNPs demonstrated exceptionally high F_{ST} values and were considered to be subject to directional selection in local demes, or closely linked to loci under selection. Even on a limited geographical scale between the nearby North Sea and Baltic Sea populations, four loci displayed evidence of adaptive divergence. Analysis of archived otoliths from one of these populations indicated that these allele frequencies were stable over 24 years.

A similar result has been found with European flounder. Little genetic differentiation was found among subpopulations at nine microsatellite loci (F_{ST} = 0.02). However, substantial differentiation (F_{ST} = 0.45) was present at a heat-shock locus (*Hsc70*) that was selected as a candidate gene because of its known function (Hemmer-Hansen et al. 2007). Population genomic approaches allow us to identify genes involved in adaptive traits without prior information about which traits are important in the species in question. These adaptive genes can then be employed to describe SGS for species in which neutral genetic markers are not informative.

reproductive isolation (Waples 1998). This failure results from the large population sizes and high gene flow among stocks of many marine fishes. Even very low exchange rates among stocks with large population sizes will be sufficient to eliminate genetic evidence of population differentiation at neutral loci. Detecting outlier loci under differential directional selection among populations can be an important tool for understanding the genetic population structure of species such as marine fishes that show little population differentiation for neutral markers (Example 9.5).

It is crucial to identify outlier loci not only because such loci might be under selection and help us to understand adaptive differentiation, but also because outlier loci can severely bias estimates of population parameters (e.g., F_{ST} or the number of migrants). Most estimates of population parameters assume that loci are neutral. For example, Allendorf & Seeb (2000) found with sockeye salmon that a single outlier locus with extremely high F_{ST} could bias estimates of the mean F_{ST} from 0.09 up to 0.20 (Figure 9.14). This bias more than doubles the F_{ST} estimate!

9.8 Limitations of F_{ST} and other measures of subdivision

F_{ST} was developed by Sewall Wright long before the use of molecular markers to describe genetic variation in natural populations, and he assumed that loci had just two alleles. Nei (1977) developed G_{ST} as an analog to F_{ST} when allozyme loci with more than two alleles were first used to describe population genetic structure. Here we consider some of the limitations of F_{ST} as a measure of genetic divergence among populations.

9.8.1 Genealogical information

One important limitation of F_{ST} values (and related measures like G_{ST}) is that they do not consider the identity of alleles (i.e., genealogical degree of relatedness). For example, in the common shrew in Example 9.4, the F_{ST} for a Y-linked microsatellite is only 0.19 across a hybrid zone between races even though the two races share no alleles at this locus. An examination of Figure 9.10 shows that all of the alleles on either side of the hybrid zone are more similar to each other than to any of the

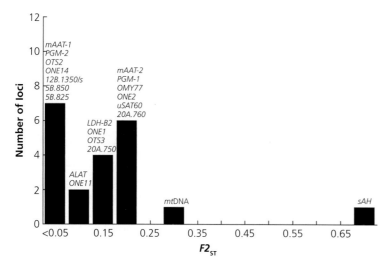

Figure 9.14 Genome-wide versus locus-specific effects, and the identification of outlier loci that are candidates for being under selection. Gene flow and genetic drift lead to similar genome-wide allele frequency differentiation (F_{ST}) among four populations for 19 nuclear loci with an F_{ST} less than 0.20. The $F2_{ST}$ values are based on two alleles at each locus so that loci with different heterozygosities can be compared (see Section 9.8.2). One nuclear locus (sAH) has a much greater $F2_{ST}$ and is a candidate for being under natural selection. From Allendorf & Seeb (2000).

alleles on the other side of the hybrid zone. A measure related to F_{ST}, called R_{ST}, uses information on the length of alleles at microsatellite loci (Slatkin 1995), and assumes that each mutation changes an allele's length by only one repeat unit (see stepwise mutation in Section 12.1.2 and Example 9.4). In this hybrid zone, R_{ST} is much higher than F_{ST} (R_{ST} = 0.98).

Another measure of differentiation that uses information on allele genealogical relationships is Φ_{ST} (Excoffier et al. 1992; Michalakis & Excoffier 1996). Measures using genealogical information (like Φ_{ST} and R_{ST}) use the degree of differentiation between alleles as a weighting factor that increases the metric (e.g., F_{ST}) proportionally to the number of mutational differences between alleles.

Gaggiotti & Foll (2010) have presented a method to estimate population-specific F_{ST} values (F_{ST}'s) rather than global or pairwise F_{ST} values. They define F_{ST}'s as the probability that two genes chosen at random from the population share a common ancestor within that population. This allows for differences in local population sizes and migration rates, unlike the island model. Their approach has the potential to be extremely valuable in interpreting patterns of genetic population structure from a conservation perspective. For example, Cosentino et al. (2012) have used F_{ST}'s values to interpret the metapopulation structure of tiger salamanders. They found that small, isolated wetland populations with low genetic diversity

tended to have greater F_{ST}'s values, showing F_{ST}'s can help identify small, isolated populations.

9.8.2 High heterozygosity within subpopulations

F_{ST} (and its analog G_{ST}) has limitations when using loci with high mutation rates and high heterozygosities, such as microsatellites. F_{ST} is biased downward when variation within subpopulations (H_S) is high. The source of this bias is obvious: when variation within populations is high, the proportion of the total variation distributed between populations can never be very high (Hedrick 1999; Meirmans & Hedrick 2011). For example, if H_S = 0.90, F_{ST} cannot be higher than 0.10 (1–0.90 = 0.10; see Equation 9.3).

Allendorf & Seeb (2000) used $F2_{ST}$ to compare F_{ST} between different marker types to test if they showed different patterns of genetic divergence among subpopulations because of natural selection (Figure 9.14). $F2_{ST}$ is estimated by using the frequency of the most common allele at each locus and **binning** (i.e., combining) all other alleles into a single allele frequency, as recommended by McDonald (1994). The advantage of $F2_{ST}$ is that it allows valid comparison of divergence at different loci because it is based on two alleles at all loci. The disadvantage of $F2_{ST}$ is that much information is lost because of the binning together of alleles. However, it is possible to estimate an $F2_{ST}$ for each allele by binning all other alleles (Chakraborty & Leimar 1987; Bowcock et al. 1991).

Hedrick (2005) introduced G'_{ST}, which is G_{ST} divided by its maximum possible values with the same overall allele frequencies. Thus, G'_{ST} has a range from 0 to 1, and was designed to be independent of H_S (although see Ryman & Leimar 2008). If H_S is high, then G'_{ST} can be much greater than G_{ST}. G'_{ST} is designed to be a standardized measure of G_{ST}, which accounts for different levels of total genetic variation at different loci (Meirmans & Hedrick 2011).

Jost (2008) was quite critical of the use of F_{ST} and G_{ST} as measures of differentiation, and he introduced D, which is a similar measure to G'_{ST}. D differs from G'_{ST} in that G_{ST} measures deviations from panmixia, while D measures deviations from complete differentiation (Whitlock 2011). D and G'_{ST} behave quite differently (Heller & Siegismund 2009; Ryman & Leimar 2009). D is zero when all populations are identical, and it monotonically increases with increasing divergence between populations and goes to one as different populations have no shared alleles.

Both G'_{ST} and Jost's D have been used increasingly in the literature since they were introduced. Nevertheless, they do not solve the problem of measuring divergence at loci with high within-subpopulation heterozygosity that they were designed to address. In addition, they have serious problems themselves (Ryman & Leimar 2008, 2009; Heller & Siegismund 2009; Whitlock 2011). They both are insensitive to genetic drift and gene flow when the mutation rate is high relative to the migration rate (Ryman & Leimar 2008, 2009; Whitlock 2011). Furthermore, D is specific to the locus being measured even with selective neutrality, and so little can be inferred about the population demography from estimating D. Moreover, neither G'_{ST} nor D estimates a quantity that can be interpreted in terms of population genetic theory. Jost (2009) showed that D is a useful measure of differentiation, but also warns that it should not be used for measuring migration. In contrast, F_{ST} measures a fundamental parameter in population genetics theory (Holsinger & Weir 2009). F_{ST} provides insights on the evolutionary processes (e.g., gene flow, drift, selection) that influence structuring

of genetic variation among populations, and also on the demographic history of populations (Wright 1951; Holsinger & Weir 2009). We may have imperfect statistical estimators of this quantity (such as G_{ST} for loci with high mutation rates), but the underlying quantity is inherently interesting, and these biases can be addressed with other techniques (e.g., R_{ST} in the case of microsatellite loci).

Verity & Nichols (2014) compared the usefulness of G_{ST}, G'_{ST}, and Jost's D for measuring population genetic differentiation. Results suggested G_{ST} is the most informative statistic for loci with low mutation rates like SNPs. However, for markers with high mutation rates and for inferring demographic history (e.g., distinguishing between **migration–drift equilibrium** with a high mutation rate, and nonequilibrium with a low mutation rate), G_{ST} should be used along with D or G'_{ST}. The authors also suggested using G_{ST} along with D or G'_{ST} when analyzing genomic datasets with many loci and a wide range of mutation rates to dissect the confounding signals from mutation, migration, and population size.

Wang (2015) showed that G_{ST} underestimates differentiation when mutation is more important than gene flow, and H_S is high. Importantly, Wang (2015) showed that markers with higher H_S have lower G_{ST} values, resulting in a negative correlation between G_{ST} and H_S across loci. For example, highly heterozygous microsatellite markers showed a strong negative relationship between locus-specific H_S and G_{ST} in Atlantic salmon, but SNPs showed no such relationship (Figure 9.15). Wang (2015) concluded that the correlation between G_{ST} and H_S across loci can be used to determine whether estimates of population subdivision are being underestimated due to mutational effects. Wang's (2015) paper is important because it provides a test (regression) and software for use on any empirical dataset to determine if G_{ST} estimates are likely biased by high H_S loci. This bias is more of an issue with microsatellite data than with SNP data due to the high mutation rates of the former.

Whitlock (2015) explained that when mutation is more important than gene flow, the process of mutation erases the historical gene flow information at a locus. That is, if mutation occurs at a higher rate than migration, a locus will retain little or no

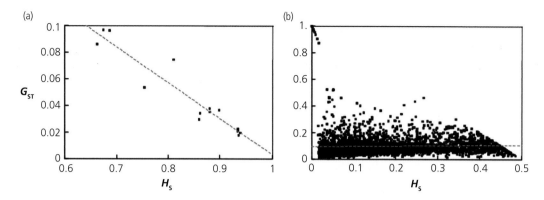

Figure 9.15 The relationship between single-locus estimates of G_{ST} and H_S in Atlantic salmon populations for (a) 15 microsatellite loci and (b) 3,129 SNPs. For highly heterozygous markers like microsatellites, H_S is negatively correlated with G_{ST} ($r^2 = 0.90$, $P < 0.001$), but for markers with lower heterozygosity like SNPs, there is no relationship. From Wang (2015).

signal from gene flow. He also pointed out that Wang's (2015) approach could be applied to published datasets to retrospectively ask if G_{ST} measurements from past studies are biased and potentially unreliable.

9.8.3 Other measures of divergence

Another widely used measure of population genetic differentiation is Nei's genetic distance (D; Nei 1972). This measure will increase linearly with time for completely isolated populations under the infinite allele model of mutation with selective neutrality. D is often used and appears to perform relatively well for populations connected by gene flow (Paetkau et al. 1997). Nei's (1978) unbiased D provides a correction for sample size. This correction is not so important for comparison between species, but can be for conservation in cases where intraspecific populations are being compared. Without this correction, poorly sampled populations will on average appear to be the most divergent. Another reliable and widely used measure of genetic distance is Cavalli-Sforza & Edwards' chord distance (Cavalli-Sforza & Edwards 1967). There are numerous other genetic distance measures (e.g., see Paetkau et al. 1997) that are less widely used and beyond the scope of this book.

9.8.4 Hierarchical structure

Populations are often structured at multiple hierarchical levels; for example, locally and regionally.

For example, several subpopulations (demes) might exist on each side of a barrier such as a river or mountain ridge. Here, two hierarchical levels are (1) the local deme level, and (2) the regional group of demes on either side of the river (Figure 9.16). It is useful to identify such hierarchical structures and to quantify the magnitude of differentiation at each level to help guide conservation management (e.g., identification of management units and evolutionarily significant units; see Chapter 20). For example, if regional populations are highly differentiated but local demes within regions are not, managers should often prioritize translocations between local demes and not between regional populations.

Hierarchical structure is often quantified using hierarchical F-statistics that partition the variation into local and regional components; that is, the proportion of the total differentiation due to differences between subpopulations within regions (F_{SR}), and the proportion of differentiation due to differences between regions (F_{RT}). Hierarchical structure is also often quantified using **AMOVA** (analysis of molecular variance; Excoffier et al. 1992), which is analogous to the standard statistical approach ANOVA (analysis of variance). Sherwin et al. (2017) has proposed using a hierarchical approach to describe genetic variation that is similar to Shannon's entropy-based diversity, which is the standard for ecological communities. Gaggiotti et al. (2018) proposed a diversity index to describe diversity at all hierarchical levels of spatial subdivision for genes, populations, and species. The

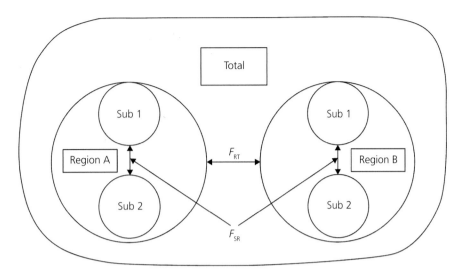

Figure 9.16 Organization of hierarchical population structure with two levels of subdivision: subpopulations within regions (F_{SR}) and regions within the total species (F_{RT}). Each region has two subpopulations. F_{SR} is the proportion of the total diversity due to differences between subpopulations within regions. F_{RT} is the proportion of the total diversity due to differences between regions.

index behaves consistently across different hierarchical levels to promote comparative biodiversity studies from genes to species.

9.9 Estimation of gene flow

Gene flow is important to measure in conservation biology because low or reduced gene flow can lead to local inbreeding and inbreeding depression, whereas high or increased gene flow can limit local adaptation and cause outbreeding depression. Measuring and monitoring gene flow can help to maintain viable populations (and metapopulations) in the face of changing environments and habitat fragmentation. Renewed gene flow (following isolation) can result in "**genetic rescue**", through heterosis in "hybrid" offspring (Bell et al. 2019). Finally, rates of gene flow in animals are correlated with rates of dispersal; thus, knowing rates of gene flow can help predict the likelihood of recolonization of vacant habitats following extirpation or over harvest (i.e., "demographic rescue"). However, over 90% of the gene flow in plant species is due to pollen movement and not seed dispersal (Petit et al. 2005).

Rates of gene flow can be estimated several ways using molecular markers. First, indirect estimates of average migration rates (mN) can be obtained from

(1) allele frequency differences (F_{ST}) among populations, (2) the proportion of **private alleles** in populations, or (3) a likelihood-based approach using information both on allele frequencies and private alleles. The migration rate estimate is an average over the past tens to hundreds of generations.

Second, direct estimates of current dispersal rates can be obtained using genetic tagging and a mark–recapture approach that directly identifies individual immigrants by identifying their "foreign" genotypes (i.e., genotypes unlikely to originate from the local gene pool). This approach can give estimates of migration rates in the current generation. We now discuss the indirect and direct (assignment test) approaches, in turn.

9.9.1 F_{ST} and indirect estimates of mN

We can estimate the average number of migrants per generation (mN) by using the island model of migration (Figure 9.4; Guest Box 9). For example, an F_{ST} of 0.20 yields an estimate of one migrant per generation ($mN = 1$), under the island model of migration (Figure 9.11, nuclear markers). Less differentiation ($F_{ST} = 0.10$) leads to a higher estimate of gene flow ($mN = 2$). Equation 9.12 can be rearranged to allow estimation of the average number of migrants (mN)

from F_{ST}, under the island model, as follows:

$$\widehat{mN} \approx \frac{(1 - F_{ST})}{4F_{ST}} \qquad (9.17)$$

The following assumptions are required for interpreting estimates of mN from the simple island model:

1. An infinite number of populations of equal size (Section 9.4.1).
2. That N and m are the same and constant for all populations (thus migration is symmetrical).
3. Selective neutrality and no mutation.
4. That populations are at migration–drift equilibrium (a dynamic balance between migration and drift).
5. Demographic equality of migrants and residents (i.e., natives and migrants have equal probability of survival reproduction).

The assumptions of this simple model are unlikely to hold in natural populations. This has led to criticism of the usefulness of mN estimates from the island model approach (Whitlock & McCauley 1999). Nonetheless, performance evaluations using both simulations and analytical theory suggest that the approach gives reasonable estimates of mN even when certain assumptions are violated (Slatkin & Barton 1989; Mills & Allendorf 1996).

A major limitation of estimating mN from F_{ST} is that F_{ST} must be moderate to large ($F_{ST} > 0.05$–0.10). This is because the variance in estimates of F_{ST} (and thus confidence intervals on mN estimates) is high at low F_{ST}. Confidence intervals on a single mN estimate could range, for example, from <10 up to 1,000 (depending on the number and variability of the loci used). This high variance is unfortunate because managers often need to know if, for example, mN is 5 versus 50, because 50 would be high enough to allow recolonization and demographic rescue on an ecological time scale, whereas 5 might not. The high variance of mN estimates at low F_{ST}, along with the model assumptions, means that we often cannot interpret mN estimates literally; instead, often we use mN to roughly assess the approximate magnitude of migration rates (e.g., "high" versus "low").

Another limitation of indirect approaches is that few natural populations are at migration–drift equilibrium, primarily because many generations are required to reach equilibrium. For example, if a population becomes fragmented, but N remains large, drift will be weak. In this case, many generations are required for F_{ST} to increase to the equilibrium level. The approximate number of generations required to approach equilibrium is given by the following expression: $1/[2m + 1/(2N)]$. If N is large and m is small, the time to equilibrium is large. Thus, F_{ST} will increase slowly in large recently isolated population fragments, and the effects of reduced gene flow will not be detectable by indirect methods until after many generations of isolation. In such a case, direct estimates of gene flow are preferred, to complement the indirect estimates. Nonetheless, for conservation genetic purposes, fragments with large N are relatively less crucial to detect because they are relatively less susceptible to rapid genetic change.

If N is small, then drift will be rapid and we might detect increased F_{ST} after only a few generations. Such a scenario of severe fragmentation is obviously the most important to detect for conservation biologists. It will also be the most likely to be detectable using an indirect (e.g., F_{ST}-based) genetic monitoring approach.

In summary, although mN estimates from F_{ST} must be interpreted with caution, they can provide useful information about gene flow and population differentiation. Nonetheless, the use of different and complementary methods (several indirect plus direct methods) is recommended (Neigel 2002).

9.9.2 Private alleles

Another indirect estimator of mN is the private allele method (Slatkin 1985). A private allele is one found in only one population. Slatkin showed that a linear relationship exists between mN and the average frequency of private alleles. This method works because if gene flow (mN) is low, populations will have numerous private alleles that arise through mutation, for example. The time during which a new allele remains private depends only on migration rates, such that the proportion of alleles that are private decreases as migration rate increases. If gene flow is high, private alleles will be uncommon.

This method could be less biased than the F_{ST}-island model method, when using highly polymorphic markers, because it apparently is less sensitive to problems of **homoplasy** created by back mutations than is the F_{ST} method (Allen et al. 1995). Homoplasy is most likely when using loci with high rates of mutation and back mutation, like some microsatellites (e.g., evolving under the stepwise mutation model).

For example, Allen et al. (1995) studied gray seals and obtained estimates of mN of 41 using the F_{ST} method, 14 using the R_{ST} method, and 5.6 from the private allele method. The lowest mN estimate might arise from the private allele method because this method could be less sensitive to homoplasy, which causes underestimation of F_{ST} or R_{ST}, and thus overestimation of mN (Allen et al. 1995). The values of mN from this study must be interpreted with caution as the assumptions of the island model are probably not met and mN values are fairly high and thus have a high variance. Furthermore, the reliability of the private allele method has not been thoroughly investigated for loci with potential homoplasy (e.g., microsatellites).

In another study, mN estimates from allozyme markers were highly correlated with dispersal capability among 10 species of ocean shore fish (Waples 1987). Three estimators of mN were compared: Nei & Chesser's F_{ST}-based method (F_{STn}), Weir & Cockerham's F_{ST}-based method (F_{STw}), and the private allele method. The two F_{ST}-based estimators gave highly correlated estimates of mN, whereas the private allele method gave less correlated estimates). This lower correlation could result from a low incidence of private alleles in some species. These species were studied with up to 19 polymorphic allozymes with heterozygosities ranging from 0.009 to 0.087 (mean 0.031). Low polymorphism markers might be of little use with the private allele method because very few private alleles might exist. More studies are needed comparing the performance of different mN estimators (e.g., likelihood-based methods) and different marker types (microsatellites versus allozymes or SNPs).

9.9.3 Maximum likelihood and the coalescent

A maximum likelihood estimator of mN was published by Beerli & Felsenstein (2001). This method is useful because, unlike classical methods, it does not assume symmetric migration rates or identical population sizes. Furthermore, likelihood-based methods use all the data in their raw form (Section A5), rather than a single summary statistic, such as F_{ST}. The statistic F_{ST} does not use information such as the proportion of alleles that are rare. Thus, the likelihood method should give less biased and more precise estimates of mN than classic moments-based methods (Beerli & Felesenstein 2001). Indeed, an empirical study on garter snakes (Bittner & King 2003) suggests that **coalescent** methods are likely to give more reliable estimates of mN than F_{ST}-based methods, because the F_{ST}-based methods are more biased by lack of migration–drift equilibrium and changing population size.

Beerli & Felsenstein (2001, p. 4568) state that "Maximum likelihood methods for estimating population parameters, as implemented in *MIGRATE* and *GENE-TREE*, will make the classical F_{ST}-based estimators obsolete . . ." While this is likely true for some scenarios, new methods and software should be used in conjunction with the classical methods until performance evaluations have thoroughly validated the new methods (e.g., Section A9). A problem with evaluating the performance of the many likelihood-based methods is they are computationally slow, especially when working with genomic datasets with many markers. Access to high-performance computing facilities is often needed. This makes the validation of methods difficult because validation requires hundreds of estimates for each of the numerous simulated scenarios (i.e., different migration rates and patterns, population sizes, mutation dynamics, and sample sizes).

The software program *MIGRATE* (Beerli 2006) provides likelihood-based estimates of mN and is widely used to estimate historical migration rates from molecular genetic data (see also *GENE-TREE* from Bahlo & Griffiths 2000; see also Hey 2010). *MIGRATE* assumes effective population sizes and migration rates between populations are constant over the coalescence period (about $4N_e$ generations). Thus, *MIGRATE* provides estimates of historical m.

The coalescent modeling approach (a "backward-looking" strategy of simulating genealogies) is usually used in likelihood-based analysis in population genetics (Section A10). The coalescent is useful because it provides a convenient and computationally efficient way to generate random genealogies for different gene flow patterns and rates. The efficiency of constructing coalescent trees is important because likelihood (Section A5) involves comparisons of enormous numbers of different genealogies in order to find those genealogies (and population models) that maximize the likelihood of the observed data. The coalescent also facilitates the extraction of genealogical information from data (e.g., divergence patterns between microsatellite alleles or DNA sequences), by easily incorporating both random drift and mutation into population models. Traditional estimators of gene flow sometimes do not use genealogical information, and are based on "forward-looking" models for which simulations are slow and probability computations are difficult.

Crandall et al. (2019) suggest that F-statistics can fail to reliably characterize population genetic structure in species with high gene flow, large population sizes, and low differentiation (e.g., $F_{ST} < 0.01$) because of the nonlinear relationship between gene flow and F-statistics. However, coalescent model-based approaches can reliably characterize population structure in high gene flow species with low F_{ST}. For example, coalescent approaches can distinguish demographically independent populations in marine species with high gene flow (noise) that overwhelms the differentiation signal using traditional F-statistics. With the availability of data from thousands of loci (e.g., microhaplotypes), coalescent approaches can help delineate demographically independent populations and the geographic scale appropriate for management in high gene flow species with low differentiation.

9.9.4 Assignment tests and direct estimates

Direct estimates of migration (mN) can be obtained by directly observing migrants moving between populations. Direct estimates of mN have been obtained traditionally by marking many individuals after birth and following them until they reproduce or by tracking pollen dispersal by looking for the spread of rare alleles or morphological mutants in seeds or seedlings. The number of dispersers that

breed in a new (nonnatal) population then becomes the estimated mN.

An advantage of direct estimates is that they detect migration patterns of the current generation without the assumption of population equilibrium (migration–mutation–drift equilibrium). This allows up-to-date monitoring of movement and more reliable detection of population fragmentation (reduced dispersal) without waiting for populations to approach equilibrium.

An important limitation of direct estimates is that they might not detect pulses of migrants that can occur only every 5–10 years, as in species where dispersal is driven by cyclical population demography or periodic weather conditions. Unlike direct estimates, indirect estimates of mN estimate the average gene flow over many generations and thus will incorporate effects of pulse migration. For example, 10 migrants once every 10 generations will have the same impact on indirect mN estimates, as will 1 migrant per generation each of 10 generations.

Another limitation of direct estimates is that they often cannot estimate rates of genetically successful gene flow. Direct estimates of mN only assume that an observed migrant will reproduce and pass on genes (with the same probability as a local resident individual). However, migrants might have a reduced mating success if they cannot obtain a local territory, for example. Alternatively, migrants might have exceptionally high mating success if there is a "rare male" or "foreign individual" advantage. Furthermore, immigrants could produce offspring more fit than local individuals if heterosis occurs following crossbreeding between immigrants and residents. Heterosis can lead to more gene flow than expected from Neutral Theory, for any given number of migrants (see "genetic rescue," Section 19.5). Direct observations generally only estimate dispersal and not gene flow (i.e., migration) unless we assume observed migrants reproduce because direct observation of migrants generally does not detect local mating success.

Unfortunately, direct estimates of mN are difficult to obtain using traditional field methods of capture–mark–recapture. Following individuals from their birth place until reproduction is extremely difficult or impossible for many species.

Assignment tests offer an attractive alternative to the traditional capture–mark–recapture approach to estimating mN directly. For example, we can genotype many individuals in a single population sample, and then determine the proportion of "immigrant" individuals; that is, individuals with a foreign genotype that are unlikely to have originated locally. For example, a study of the inanga (a widespread southern hemisphere fish) revealed that one individual sampled in New Zealand had an extremely divergent mtDNA haplotype, which was very similar to the haplotypes found in Tasmania (Figure 9.17). It is likely that the individual (or one of its maternal ancestors) originated in Tasmania and migrated to New Zealand. The inanga spawns in freshwater, but spends part of its life history in the ocean.

One problem with using only mtDNA is we cannot estimate male-mediated migration rates (because mtDNA is maternally inherited). Further, the actual migrant could have been the mother or grandmother of the individual sampled. We could test if the migrant or its mother was the actual immigrant by genotyping many autosomal markers

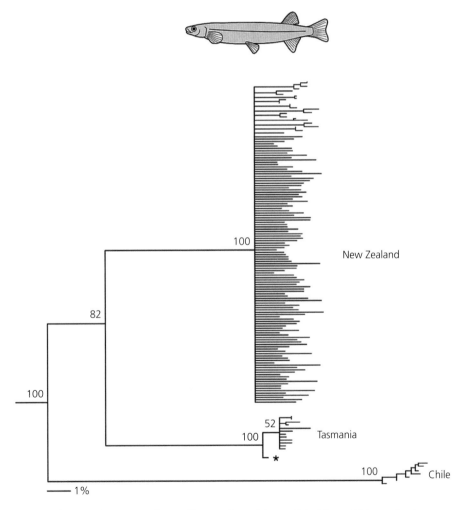

Figure 9.17 Detection of a migrant between populations of inanga using a phylogram derived from mtDNA control region sequences. One mtDNA type (marked with a star) sampled in New Zealand was very similar to the mtDNA types found in Tasmania. This suggests that a small amount of gene flow occurs between New Zealand and Tasmanian populations. From Waters et al. (2000).

(e.g., microsatellites). For example, if a parent was the migrant then only one-half of the individual's genome (alleles) would have originated from another population (and not the Y chromosome). We can estimate the proportion of an individual's genome arising from each of two parental populations via **admixture** analysis using available algorithms and software (e.g., Pritchard et al. 2000).

Assignment tests based on multiple autosomal makers are useful for identifying immigrants. For example, for a candidate immigrant, we first remove the individual from the dataset and then compute the expected frequency of its genotype (p^2) in each candidate population of origin by using the observed allele frequencies (p) from each population (Figure 9.18). If the likelihood for one population is significantly higher than the other, we "assign" the individual to the most-likely population. The likelihood can be computed as the frequency of the genotype in the population (expected under HW proportions). Computing the multilocus assignment likelihood requires multiplying single-locus probabilities (multiplication rule), and thus requires the assumption of independence among loci (e.g., no gametic disequilibrium).

The power of assignment tests increases with the amount of differentiation among subpopulations.

Therefore, outlier loci with high differentiation (Example 9.5) can be extremely valuable for individual assignment (Hansen et al. 2007; Ackerman et al. 2011) For example, Karlsson et al. (2011) were interested in developing genetic markers to detect potentially harmful introgression from farmed Atlantic salmon into wild populations in Norway. They found very low overall genetic divergence (F_{ST} = 0.016) at 4,514 SNP loci. However, they identified a set of 200 SNP loci having much higher F_{ST} values (0.094), apparently due to domestication selection in the farmed fish. They then developed a panel of 60 SNPs that collectively are diagnostic in identifying individual salmon as being farmed or wild, regardless of their populations of origin.

Parentage analysis can also be used to detect the movement of individuals between locations. Christie et al. (2010) genotyped adult and recently settled juveniles of the exploited fish species yellow tang off of the coast of the Island of Hawai'i at 20 microsatellite loci. Parentage analysis detected 4 juveniles that dispersed between 15 and 184 km from the location of their parents. Two of these juveniles dispersed from marine protected areas (MPAs) demonstrating the effectiveness of MPAs in this species to replenish unprotected areas.

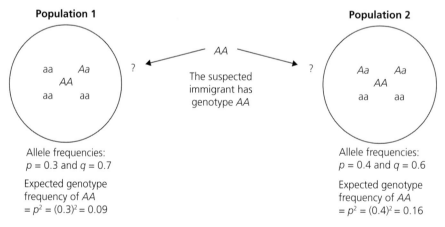

Figure 9.18 Simplified example of using an assignment test to identify an immigrant (*AA*). We first remove the individual in question from the dataset and then compute its expected genotype frequency (p^2) in each population using the observed allele frequencies for each population, assuming HW proportions. If the individual with the genotype *AA* was captured in Population 1 but its expected genotype frequency is far higher in Population 2, then we could conclude the individual is an immigrant. Assignment tests are relatively simple but powerful if many loci are used. Note that obtaining the multilocus likelihoods generally requires multiplication of single-locus probabilities (multiplication rule), and thus requires independent loci in **gametic equilibrium**.

9.9.5 Current versus historical gene flow

Differentiating between current and historical gene flow is crucial for conservation and informing management decisions. For example, if historical gene flow was high but contemporary gene flow is lower, populations might be suffering from fragmentation, isolation, and loss of genetic variation. Fortunately, computational methods exist to estimate both current and historical gene flow from genetic markers.

Chiucchi & Gibbs (2010) estimated contemporary and historical gene flow among 19 populations sampled across the range of the endangered massasauga rattlesnake. Both contemporary and historical migration rates among populations were low and similar in magnitude. A test of models of population history favored a model of long-term migration–drift equilibrium. This suggested this species has persisted in small, isolated populations and that recent habitat fragmentation has had little effect on the genetics of these snakes. These results also suggest rattlesnake populations might suffer relatively little from small population size compared with a species that recently became fragmented into small populations.

Stevens et al. (2018) investigated historical and contemporary gene flow in the grey-crowned babbler, a threatened Australian woodland bird, to test if habitat fragmentation affected gene flow.

Long-term and contemporary gene flow rates were estimated using *Migrate-n* (Beerli 2006) and *BayesAss* (Wilson & Rannala 2003). They detected equal amounts of long-term historical gene flow from east to west and vice versa in the state of Victoria (Figure 9.19). However, contemporary gene flow was largely unidirectional, with a much higher rate than historical levels from west to east, but nonexistent from east to west. This study suggested landscape connectivity and gene flow have been reduced and that connectivity is likely insufficient to maintain genetic variation and long-term population persistence. A similar study of the California gnatcatcher also reported recently reduced gene flow related to habitat fragmentation (Vandergast et al. 2019).

9.10 Population subdivision and conservation

Understanding the genetic population structure of species is essential for conservation and management. The techniques to study genetic variation and the genetic models that we have presented in this chapter allow us to rather quickly understand the genetic population structure of any species of interest. Understanding the amount of genetic differentiation among populations is crucial when developing a captive breeding program

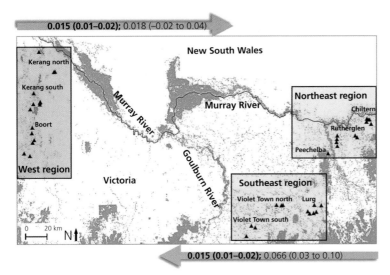

Figure 9.19 Directional long-term and contemporary gene flow rates per generation between the eastern (blue shading) and western (orange shading) subregions of the grey-crowned babbler in northern Victoria, Australia. Long-term gene flow rates are indicated in bold and contemporary gene flow is in plain text on the two arrows. Confidence interval values are given in parentheses. The map shows sample sites (solid black triangles); study regions (black rectangular outlines); subpopulations (labeled); and tree cover (gray shading; vegetation cover >2 m in height). From Stevens et al. (2018).

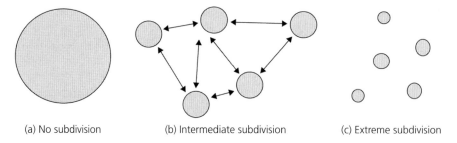

(a) No subdivision   (b) Intermediate subdivision   (c) Extreme subdivision

Figure 9.20 Range of possible population subdivision. Intermediate subdivision (b) provides the greatest possibility for local adaptation, yet with enough gene flow and large enough local effective sizes to prevent loss of genetic variation and inbreeding depression.

(Chapter 21) or selecting individuals to be moved among populations for either **demographic** or genetic rescue (Chapter 19), or **assisted migration** (Chapter 16).

An understanding of SGS within populations can provide important information for their conservation and management. Data on SGS can be used to indirectly estimate dispersal rates and neighborhood sizes (e.g., Fenster et al. 2003), which are laborious and time-consuming to estimate directly though tracking dispersal. For example, Solmsen et al. (2011) found strong SGS along 7 km of a dry riverbed in female African striped mice, but low SGS in males of the same species. From these data, they concluded that males disperse farther than females, and that males of lower fitness (lower body weight) disperse farther than larger males. Information about SGS can inform reserve design. If a population has strong SGS and if only a portion of a population is conserved, more genetic variation will be lost than for a population with no SGS in which variation is distributed randomly across the population.

However, the application of information on genetic population structure of a species is often not straightforward and is sometimes controversial. For example, how "distinct" does a population have to

be to be considered a **distinct population segment** in order to be listed under the US ESA? The application of genetic information to identify appropriate units for conservation and management is considered in detail in Chapter 20.

Population subdivision influences the evolutionary potential of a species; that is, the ability of a species to evolve and adapt to environmental change. To understand this, it is helpful to consider the extremes of subdivision (Figure 9.20). For example, a species with no subdivision would have such high gene flow that local adaptation would not be possible (Figure 9.20a). Thus, the total range of types of multilocus genotypes would be limited. On the other hand, if subdivision is extreme then new beneficial mutations that arise will not readily spread across the species (Figure 9.20c). Furthermore, subpopulations may be so small that genetic drift overwhelms natural selection. Thus local adaptation is limited and random change in allele frequencies dominates so that harmful alleles may drift to high frequency or go to fixation. An intermediate amount of population subdivision will result in substantial genetic variation both within and between local populations (Figure 9.20b); this population structure has the highest probability of long-term survival.

Guest Box 9 A decade of tiger conservation genetics in the Indian subcontinent
Uma Ramakrishnan

Like large carnivores worldwide, hunting records and shrinking ranges suggest that tiger populations have declined historically (Figure 9.21). Tigers now retain only 5% of their historical range (Walston et al. 2010). About 65% of the world's remaining 5,000 or so wild tigers inhabit the Indian subcontinent, making it an important region for conservation attention. Early studies on tiger phylogeography showed genetic differences between extant subspecies (Luo et al. 2004). However, studies like Luo et al. (2004) did not focus on population structure and genetic variation within subspecies. I started independent research in 2005, and as a young conservation geneticist, it felt important to understand population structure and genetic variation in Indian tigers.

Tigers are rare, elusive, and endangered, and they had not been studied in the wild because acquiring biological samples that yield genetic information through DNA is difficult. In initial work on Indian tigers, we sampled tiger nuclear and mitochondrial genetic variation across the Indian subcontinent **noninvasively** (Mondol et al. 2009). Our analyses (based on a very small set of samples and genetic markers) suggested that Indian tigers retain more than half of extant genetic diversity of the species. Coalescent analyses suggested that extant genetic diversity is retained despite a precipitous, likely human-induced population crash ~200 years ago in India (Mondol et al. 2009), and apparent population structure. Genetic data from historical tiger skins (primarily from the Museum of Natural History, London) allowed us to compare past population genetic variation (up to 125 years ago) with that of modern tigers (Mondol et al. 2013). Population genetic analyses suggested that existing tigers have lost substantial mtDNA variation, and that populations are more structured today than in historical times (Mondol et al. 2013).

It bothered me that inference of population structure was shaky at best. Looking at where tigers lived, there definitely seemed to be landscapes, sets of protected areas that were closer to others. These landscapes should be identifiable genetically as discrete populations. Our power to detect population structure could be increased by increasing sample size or the number of loci typed. Increasing sample size across India is difficult. The tiger genome was sequenced in 2013 (Cho et al. 2013), and NGS technologies revolutionized our ability to type thousands of markers. We were able to use 10,000 SNPs from across the genome of 38 individuals and identify between three and five genetic populations of tigers across India (Natesh et al. 2017). Importantly, we identified isolated and genetically impoverished populations (e.g., Ranthambore in Western India). We also identified greater genetic variation in populations from central India that include several tiger reserves. We inferred that small and isolated populations, like Ranthambore, could be affected by detrimental effects of inbreeding, while populations that have high variation could be made up of several subpopulations connected by gene flow.

Figure 9.21 Tigers in India survive cheek by jowl with over 1.3 billion people. A tigress in Ranthambore Tiger Reserve walks toward the reservoir. An erstwhile hunting lodge in typical pink stone architecture is visible in the background. Photo courtesy of Prasenjeet Yadav.

Guest Box 9 Continued

But is gene flow and connectivity facilitated in tiger populations in central India? How? We sampled 9 tiger reserves and collected 580 fecal samples that yielded 116 unique individuals to investigate population substructure. Our analyses suggested three subpopulations within this landscape. Landscape genetics approaches suggested correlations between genetic data and resistance to movement offered by landscape features (human settlements, forest cover, roads). Human impacts on the landscape (such as high-traffic roads, urban settlements) significantly impact movement of tigers here (Joshi et al. 2013; Thatte et al. 2018). We used simulations to explore the best strategies to sustain tiger connectivity in the future within the central Indian landscape (Thatte et al. 2018). These simulations identified which populations are more likely to go extinct, but also indicated how to minimize extinction probability in the coming century (Figure 9.22). Our results reveal that securing corridors with new protected sites within them is the best way to sustain gene flow and minimize extinction. Since our work, legal battles argued based on the need to maintain tiger connectivity have resulted in construction of animal passageways under a national highway, and recent research shows this is being used for movement.

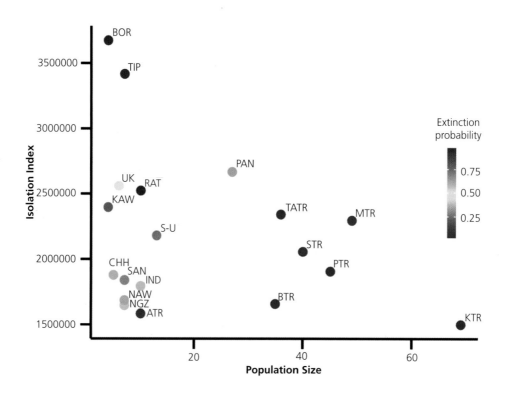

Figure 9.22 Plot showing extinction probability after 100 years as a function of population size (average size, allowing for increase in tiger numbers over the next century) and isolation index, calculated as the average cost distance between populations. Each point represents a population and the color represents its average extinction probability across scenarios. From Thatte et al. (2018).

Guest Box 9 Continued

We have just completed an intensive study on Ranthambore, and whole genome resequencing data from 57 Indian tigers reveal that tigers here are almost twice as inbred as those in other parts of India. Isolated tiger populations had lower loads of predicted deleterious alleles than populations in central India. This could possibly be inferred as purging in Ranthambore. On the other hand, the frequency of predicted deleterious alleles in Ranthambore is high indicating these shared inbred stretches with putative deleterious alleles could cause inbreeding depression in the near future.

In summary, our work has shown that studies of population structure go beyond descriptive population genetics to suggest actionable management interventions. Maintaining gene flow between central Indian tiger reserves will require management of linear intrusions like roads and where they are distributed. Management of the negative impacts of inbreeding will soon become a reality for some tiger populations. Even within India, where tigers are doing relatively well overall, conservation genetics gives us a glimpse of the varied futures of these populations.

CHAPTER 10

Beyond Individual Loci

Field cricket, Example 10.2

Population geneticists recently have devoted much attention to the topic of gametic disequilibrium. The analysis of multiple-locus genotypic distributions can provide a sensitive measure of selection, genetic drift, and other factors that influence the genetic structure of populations.

(David W. Foltz et al. 1982, p. 80)

We thus pass from a point-theory to a strand-theory.

(R.A. Fisher 1965, p. 95)

We have so far considered one locus at a time. Population genetic models become much more complicated when two or more loci are considered simultaneously (Slatkin 2008; Sved & Hill 2018). Many of our genetic concerns in conservation can be dealt with from the perspective of individual loci. Nevertheless, there are a variety of situations in which we must concern ourselves with the interactions between multiple loci, especially when using genomic data where many markers are no longer independent of each other. For example, genetic drift in small populations can generate nonrandom associations between genotypes at multiple loci. Therefore, the consideration of multilocus genotypes can provide powerful methods for detecting the effects of genetic drift in natural populations. It is more important than ever to understand the interpretation of multilocus genotypes to take advantage of the great power in genomic techniques that allow the description of genotypes at thousands of loci.

In addition, genotypes over many loci can be used to identify individuals genetically because the genotype of each individual (with the exception of identical twins or clones) is genetically unique if enough loci are considered. This genetic "fingerprinting" capability has many potential applications in understanding populations, estimating population size (Chapter 18), and applying genetics to problems in forensics (Chapter 22).

The nomenclature of multilocus genotypes is particularly messy and often inconsistent. It is difficult to find any two papers (even by the same author!) that use the same symbols and nomenclature for multilocus genotypes. Therefore, we have made a special effort to use the simplest possible nomenclature and symbols that are consistent as possible with previous usage in the literature.

The term **linkage disequilibrium** is commonly used to describe the nonrandom association between alleles at two loci (Box 10.1). However, this term is misleading because unlinked loci can be in so-called "linkage disequilibrium." Things are complicated enough without using misnomers that lead to additional confusion when considering multilocus models. The term **gametic disequilibrium** is a much more descriptive and appropriate term to use in this situation. We have chosen to use gametic disequilibrium in order to reduce confusion.

Conservation and the Genomics of Populations, Third Edition. Fred W. Allendorf, *et al.*, Oxford University Press.
© Fred W. Allendorf, W. Chris Funk, Sally N. Aitken, Margaret Byrne, and Gordon Luikart (2022). DOI: 10.1093/oso/9780198856566.003.0010

Box 10.1 Linkage or gametic disequilibrium?

The terminology to describe the nonrandom association (i.e., correlation) between genotypes or alleles at multiple loci within a population has changed over the years. Lewontin & Kojima (1960, p. 459) introduced the term "linkage disequilibrium." They used "disequilibrium" because, if there is any recombination at all ($r > 0$), genotypes at two loci will eventually be randomly associated "in the absence of any evolutionary pressure such as selection." The term "linkage" apparently was chosen because the early treatments of this problem assumed that the loci were linked (e.g., Geiringer 1944).

As the theory of two-locus systems developed, it quickly became clear that the term linkage disequilibrium was inappropriate, and confusing because unlinked loci can be nonrandomly associated in many situations (e.g., small population size, population subdivision, hybridization, etc.). The term "gametic disequilibrium" began to be used in the literature within a few years (e.g., Fraser 1967). The theory of two loci is based upon analysis of gametic frequencies (Section 10.1); thus, this term is much more appropriate. Crow & Kimura (1970) used the phrase "gametic phase imbalance," but this phrase is awkward and unwieldy.

Gametic disequilibrium became common in the literature through the 1980s and peaked about 1990 (Hedrick 1987; Lewontin 1988). However, since then, linkage disequilibrium has become overwhelmingly more common. This has contributed to increasing confusion in the literature. The authors of this book have reviewed papers in which the authors test for "linkage disequilibrium," and incorrectly state that they tested for linkage. Some authors have distinguished between linkage and gametic disequilibrium with the view that nonrandom associations between syntenic loci is linkage disequilibrium, and it is gametic disequilibrium if the loci are on different chromosomes. However, in many cases, it is not known if the loci are syntenic or not.

The authors of this book have had some entertaining discussions on which term we should use. One of us, SNA, started a lively Twitter thread on this topic, which has led to some interesting suggested alternative terms (e.g., **interlocus allelic love**). In conclusion, we have decided to use the term gametic disequilibrium because we believe that reducing misunderstanding is more important than following convention.

We will first examine general models describing associations between loci and their evolutionary dynamics from generation to generation. We will then explore the various evolutionary forces that cause nonrandom associations between loci to come about in natural populations (genetic drift, natural selection, population subdivision, and hybridization). Finally, we will compare various methods for estimating associations between loci in natural populations.

10.1 Gametic disequilibrium

We now focus our interest on the behavior of two autosomal loci considered simultaneously under all of our Hardy–Weinberg (HW) equilibrium assumptions. We know that each locus individually will reach a neutral equilibrium in one generation under HW conditions. Is this true for two loci considered jointly? We will see shortly that the answer is no.

Loci on different chromosomes will be unlinked ($r = 0.5$) so that heterozygotes at both loci ($AaBb$) will produce all four gametes (AB, Ab, aB, and ab) in equal frequencies (Box 4.1; Chapter 4). Two loci that are close together on the same chromosome are generally linked so that the frequency of the parental gamete types (AB and ab in Figure 10.1) will be greater than the frequency of the nonparental gametes ($r < 0.5$). Some loci on the same chromosome can be far enough apart so that there is enough recombination to produce equal frequencies of all four gametes so they are unlinked ($r = 0.5$). Two loci that are on the same chromosome are **syntenic**, whether they are linked ($r < 0.5$) or unlinked ($r = 0.5$).

Allele frequencies are insufficient to describe genetic variation at multiple loci. Fortunately, however, we do not have to keep track of all possible genotypes. Rather, we can use the gamete frequencies to describe nonrandom associations between alleles at different loci. For example, in the case of

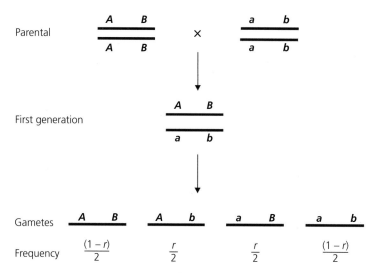

Parental

First generation

Gametes

Frequency

Figure 10.1 Outline of gamete formation in F_1 hybrids between two parents homozygous for different alleles at two loci. The gametes produced by the F_1 hybrids are affected by the rate of recombination (r). These four gametes will be equally frequent (25% each) for unlinked loci ($r = 0.5$). There will be an excess of parental gametes (AB and ab in this case) if the loci are linked ($r < 0.5$).

two loci that each has two alleles, there are just two allele frequencies, but there are nine different genotype frequencies. However, we can describe this system with just four gamete frequencies.

Let G_1, G_2, G_3, and G_4 be the frequencies of the four gametes AB, Ab, aB, and ab, respectively, as shown below. If the alleles at these loci are associated randomly then the expected frequency of any gamete type will be the product of the frequencies of its two alleles:

$$\begin{array}{ccl}
\text{Gamete} & \text{Frequency} & (10.1) \\
AB & G_1 = (p_1)\,(p_2) & \\
Ab & G_2 = (p_1)\,(q_2) & \\
aB & G_3 = (q_1)\,(p_2) & \\
ab & G_4 = (q_1)\,(q_2) &
\end{array}$$

where (p_1; q_1) and (p_2; q_2) are frequencies of the alleles (A; a) and (B; b), at locus 1 and 2, respectively. The expected frequencies of two-locus genotypes in a random mating population can then be found as shown in Table 10.1.

D is used as a measure of the deviation from random association between alleles at the two loci (Lewontin & Kojima 1960). D is known as the coefficient of gametic disequilibrium and is defined as:

$$D = (G_1\,G_4) - (G_2\,G_3) \quad (10.2)$$

Table 10.1 Genotypic array for two loci showing the expected genotypic frequencies in a random mating population.

	AA	*Aa*	*aa*
BB	$G_1{}^2$	$2G_1G_3$	$G_3{}^2$
Bb	$2G_1G_2$	$2G_1G_4 + 2G_2G_3$	$2G_3G_4$
bb	$G_2{}^2$	$2G_2G_4$	$G_4{}^2$

or:

$$D = G_1 - p_1\,p_2 \quad (10.3)$$

If alleles are associated at random in the gametes (as in Equation 10.1), then the population is in gametic equilibrium and $D = 0$. If D is not equal to zero, the alleles at the two loci are not associated at random with respect to each other, and the population is said to be in gametic disequilibrium (Example 10.1). For example, if a population consists only of a 50:50 mixture of the gametes AB and ab, then:

$$G_1 = 0.5$$

$$G_2 = 0.0$$

$$G_3 = 0.0$$

$$G_4 = 0.5$$

and:

$$D = (0.5)\,(0.5) - (0.0)\,(0.0) = +0.25$$

Example 10.1 Genotypic frequencies with and without gametic disequilibrium

Let us consider two loci at which allele frequencies are $p_1 = 0.4$ ($q_1 = 1 - p_1 = 0.6$) and $p_2 = 0.7$ ($q_2 = 1 - p_2 = 0.3$) in two populations. The two loci are randomly associated in one population, but show maximum nonrandom association in the other. The gametic frequency values below show the case of random association of alleles at the two loci (gametic equilibrium, $D = 0$) and the case of maximum positive disequilibrium ($D = +0.12$; see Section 10.1.1 for an explanation of the maximum value of D).

Gamete	$D = 0$	D(max)
A B	$(p_1)(p_2) = 0.28$	0.40
A b	$(p_1)(q_2) = 0.12$	0.00
a B	$(q_1)(p_2) = 0.42$	0.30
a b	$(q_1)(q_2) = 0.18$	0.30

In a random mating population, the following genotypic frequencies will result in each case as shown below. The expected genotypic frequencies with $D = 0$ are shown without brackets, and the expected genotypic frequencies with maximum positive gametic disequilibrium are shown in square brackets:

	AA	Aa	aa	Total
BB	0.08	0.24	0.18	0.49
	[0.16]	[0.24]	[0.09]	[0.49]
Bb	0.07	0.20	0.15	0.42
	[0]	[0.24]	[0.18]	[0.42]
bb	0.01	0.04	0.03	0.09
	[0]	[0]	[0.09]	[0.09]
Total	0.16	0.48	0.36	
	[0.16]	[0.48]	[0.36]	

Notice that each locus is in HW proportions in the populations either with or without gametic disequilibrium.

The amount of gametic disequilibrium (i.e., the value of D) will decay from generation to generation as a function of the rate of recombination (r, see

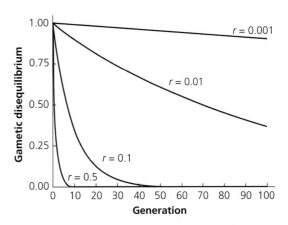

Figure 10.2 Expected decay of gametic disequilibrium (D_t/D_0) with time for various amounts of recombination (r) between the loci from Equation 10.3.

Box 4.1) between the two loci.

$$D' = D(1 - r) \qquad (10.4)$$

So that after t generations:

$$D_t = D_0(1 - r)^t \qquad (10.5)$$

If the two loci are not linked (i.e., $r = 0.5$), the value of D_t will be halved each generation until equilibrium is reached at $D = 0$. Linkage ($r < 0.5$) will delay the rate of decay of gametic disequilibrium. Nevertheless, D eventually will be equal to zero, as long as there is some recombination ($r > 0$) between the loci. However, if the two loci are tightly linked, it will take many generations for them to reach gametic equilibrium (Figure 10.2).

We therefore expect that nonrandom associations of genotypes between loci (i.e., gametic disequilibrium) would be much more frequent between tightly linked loci. For example, Zapata & Alvarez (1992) summarized observed estimates of gametic disequilibrium between five allozyme loci in several natural populations of *Drosophila melanogaster* on the second chromosome. The effective frequency of recombination is the mean of recombination rates in females and males. Only pairs of loci with less than 15% recombination showed consistent evidence of gametic disequilibrium. In contrast, recent studies of species of conservation interest have found much greater gametic disequilibrium, even between loci on different chromosomes (Section 10.8).

10.1.1 Other measures of gametic disequilibrium

D is a less than ideal measure of the relative amount of disequilibrium at different pairs of loci because the possible values of D are constrained by allele frequencies at both loci. The largest possible positive value of D is either p_1q_2 or p_2q_1, whichever value is smaller; and the largest negative value of D is the lesser value of p_1p_2 or q_1q_2. We can see that the largest positive value of D occurs when G_1 is maximum. p_1 is equal to G_1 plus G_2, and p_2 is equal to G_1 plus G_3. Therefore, the largest possible value of G_1 is the smaller of p_1 and p_2. We can see this in Example 10.1 in which the largest positive value of D occurs when G_1 is equal to p_1, which is less than p_2. Once the values of G_1, p_1, and p_2 are set, all of the other gamete frequencies must follow.

This allele frequency constraint of D reduces its value for comparing the amount of gametic equilibrium for the same loci in different populations or for different pairs of loci in the same population. For example, consider two pairs of loci in complete gametic disequilibrium. In case 1, both loci are at allele frequencies of 0.5, while in case 2, both loci are at allele frequencies of 0.9. The following gamete frequencies result:

Gamete	Frequencies	
	Case 1	Case 2
AB	0.5	0.9
Ab	0.0	0.0
aB	0.0	0.0
ab	0.5	0.1

The value of D in case 1 will be +0.25, while it will be +0.09 in case 2.

Several other measures of gametic disequilibrium have been proposed that are useful for various purposes (Hedrick 1987). A useful measure of gametic disequilibrium should have the same range regardless of allele frequencies. This will allow comparing the amount of disequilibrium among pairs of loci with different allele frequencies.

Lewontin (1964) suggested using the parameter D' to circumvent the problem of the range of values

being dependent upon the allele frequencies:

$$D' = \frac{D}{D_{\max}} \qquad (10.6)$$

Thus, D' ranges from zero to one for all allele frequencies. However, even D' is not independent of allele frequencies, and, therefore, is not an ideal measure of gametic disequilibrium (Lewontin 1988). Nevertheless, the D' coefficient is a useful tool for the estimation and comparison of the extent of overall disequilibrium among many pairs of multiallelic loci (Zapata 2000).

The correlation coefficient (R) between alleles at the two loci also has been used to measure gametic disequilibrium.

$$R = \frac{D}{\left(p_1q_1p_2q_2\right)^{1/2}} \qquad (10.7)$$

R has a range of values between -1.0 and $+1.0$. However, this range is reduced somewhat if the two loci have different allele frequencies. Both D' and R will decay from generation to generation by a rate of $(1 - r)$, as does D, because they are both functions of D.

10.1.2 Associations between cytoplasmic and nuclear genes

Just as with multiple nuclear genes, nonrandom associations between nuclear loci and **mitochondrial DNA (mtDNA)** genotypes may occur in populations, as shown in the following table where A and a are alleles at a nuclear locus and M and m are haplotypes at a mtDNA locus.

Gamete	Frequency
AM	G_1
Am	G_2
aM	G_3
am	G_4

Again, D is a measure of the amount of gametic equilibrium and is defined as in Equation 10.2. D between nuclear and cytoplasmic genes will decay at a rate of one-half per generation, just as for two

Example 10.2 Cytonuclear disequilibrium in a hybrid zone of field crickets

Hybrid zones occur where two genetically distinct taxa are sympatric and hybridize to form at least partially fertile progeny (Section 13.2.3). Observations of the distribution of multilocus genotypes within hybrid zones and the patterns of introgression across hybrid zones can provide insight into the patterns of mating and the fitnesses of hybrids that may contribute to barriers to gene exchange between taxa.

Harrison & Bogdanowicz (1997) describe gametic disequilibrium in a hybrid zone between two species of field crickets, *Gryllus pennsylvanicus* and *G. firmus*. These two species hybridize in a zone that extends from New England to Virginia in the USA. Analyses of four anonymous nuclear loci, allozymes, mtDNA, and morphology at three sites in Connecticut indicate that nonrandom associations between nuclear markers, between nuclear and mtDNA (Figure 10.3), and between genotypes and morphology persist primarily because of more frequent matings between parental types. That is, the crickets at these three sites in this hybrid zone appear to be primarily parental with a few F$_1$ individuals and even fewer later generation hybrids.

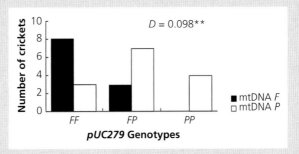

Figure 10.3 Gametic disequilibrium between mtDNA and a nuclear locus (*pUC279*) in a hybrid zone between two species of field crickets, *Gryllus pennsylvanicus* (P) and *Gryllus firmus* (F). The mtDNA from *G. firmus* (F) is significantly more frequent for homozygotes (*FF*) for the *G. firmus* nuclear allele. ** $P < 0.01$. Redrawn from Harrison & Bogdanowicz (1997).

These two species of field crickets are genetically similar. There are no fixed diagnostic differences at allozyme loci, and more than 50 anonymous nuclear loci had to be screened to find four that were diagnostic. These two taxa meet the criteria for species according to some **species concepts** but not others. Regardless, the long-term persistence of parental types throughout an extensive hybrid zone indicates that these species are clearly distinct biological units.

unlinked nuclear genes. That is,

$$D' = D(0.5) \qquad (10.8)$$

and, therefore,

$$D_t = D(0.5)^t \qquad (10.9)$$

For an empirical example of nonrandom association between nuclear and mtDNA loci, see Example 10.2.

10.2 Small population size

Nonrandom associations between loci will be generated by sampling effects in small populations.

We can see this readily in the extreme case of a bottleneck of a single individual capable of reproducing by selfing because a maximum of only two gamete types can occur within a single individual. Conceptually, we can imagine the four gamete frequencies to be analogous to four alleles at a single locus. Changes in gamete frequencies from generation to generation caused by drift will often result in nonrandom associations between alleles at different loci. The expected value of D due to drift is zero. Nevertheless, drift-generated gametic disequilibria may be great and are equally likely to be positive or negative in sign. For example, genome-wide investigations in humans have found that large blocks of gametic disequilibrium occur throughout the genome in human populations. These blocks of disequilibrium are thought to have arisen during an extreme population bottleneck that occurred some 25,000–50,000 years ago (Reich et al. 2001).

Gametic disequilibrium produced by a single generation of drift may take many generations to decay. Therefore, we would expect substantially more drift-generated gametic disequilibrium between closely linked loci. In fact, the expected amount of disequilibrium for closely linked loci is:

$$E(R^2) \approx \frac{1}{1 + 4Nr} \qquad (10.10)$$

where R^2 is the square of the correlation coefficient (R) between alleles at the two loci (Equation 10.7) (Hill & Robertson 1968; Ohta & Kimura 1969). For unlinked loci, the following value of R^2 is expected (Weir & Hill 1980):

$$E(R^2) \approx \frac{1}{3N} \qquad (10.11)$$

Guest Box 10 discusses the use of genotype frequencies at many loci to estimate effective population size in natural populations using Equation 10.11.

10.3 Natural selection

Let us examine the effects of natural selection with constant fitnesses at two loci each with two alleles. We will designate the fitness of a genotype to be w_{ij}, where i and j are the two gametes that join to form a particular genotype. There are two genotypes that are heterozygous at both loci (AB/ab and Ab/aB); we will assume that both double heterozygotes have the same fitness (i.e., $w_{23} = w_{14}$).

	AA	Aa	aa
BB	w_{11}	w_{13}	w_{33}
Bb	w_{12}	$w_{23} = w_{14}$	w_{34}
bb	w_{22}	w_{24}	w_{44}

The frequency of the AB gamete after one generation of selection will be:

$$G_{1'} = \frac{G_1 (G_1 w_{11} + G_2 w_{12} + G_3 w_{13} + G_4 w_{14}) - r w_{14} D}{\bar{w}} \qquad (10.12)$$

where $\bar{w}$ is the average fitness of the population. We can simplify this expression by defining $\bar{w}_i$ to be the average fitness of the ith gamete.

$$\bar{w}_i = \sum_{j=1}^{4} G_j w_{ij} \qquad (10.13)$$

and then:

$$\bar{w} = \sum_{i=1}^{4} G_i \bar{w}_i \qquad (10.14)$$

and:

$$G_{1'} = \frac{G_1 \bar{w}_1 - r w_{14} D}{\bar{w}} \qquad (10.15)$$

We can derive similar recursion equations for the other gamete frequencies.

$$G_{2'} = \frac{G_2 \bar{w}_2 + r w_{14} D}{\bar{w}}$$

$$G_{3'} = \frac{G_{3'} \bar{w}_3 + r w_{14} D}{\bar{w}} \qquad (10.16)$$

$$G_{4'} = \frac{G_4 \bar{w}_4 - r w_{14} D}{\bar{w}}$$

There are no general solutions for selection at two loci. That is, there is no simple formula for the equilibria and their stability. However, a number of specific models of selection have been analyzed. The simplest of these is the additive model where the fitness effects of the two loci are summed to yield the two-locus fitnesses. Another simple case is the multiplicative model where the two-locus fitnesses are determined by the product of the individual locus fitnesses. In both of these cases, heterozygous advantage at each locus is necessary and sufficient to ensure stable polymorphisms at both loci.

In some cases, the multilocus fitness cannot be predicted by either the additive or multiplicative combination of fitnesses at individual loci (Phillips 2008). Such interaction between loci is referred to as **epistasis** (i.e., the interaction of different loci such that the multiple locus phenotype is different than that predicted by simply combining the effects of each individual locus). The study of epistasis, or interactions between genes, is fundamentally important to understanding the structure and function of genetic pathways and the evolutionary dynamics of complex genetic systems.

A detailed examination of the effects of natural selection at two loci, including epistasis, is beyond the scope of our consideration. Interested readers are directed to appropriate population genetics sources (e.g., Hartl & Clark 1997; Phillips 2008; Hedrick 2011). We will consider two situations of selection at multiple loci that are particularly relevant for conservation.

10.3.1 Genetic hitchhiking

Natural selection at one locus can affect closely linked loci in many ways. Let us first consider the case where directional selection occurs at one locus (B) and the second locus is selectively neutral (A). The following fitness set results:

$$
\begin{array}{cccc}
 & AA & Aa & aa \\
BB & w_{11} & w_{11} & w_{11} \\
Bb & w_{12} & w_{12} & w_{12} \\
bb & w_{22} & w_{22} & w_{22}
\end{array}
$$

where $w_{11} < w_{12} < w_{22}$.

Imagine that the favored b allele is a new mutation at the B locus. In this case, the selective advantage of the b allele may carry along either the A or a allele, depending upon which allele is initially associated with the b mutation. This is known as **genetic hitchhiking** and will result in a so-called **selective sweep**. The magnitude of this effect depends on

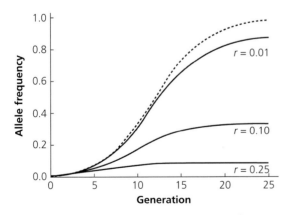

Figure 10.4 Effect of hitchhiking on a neutral locus that is initially in complete gametic disequilibrium with a linked locus that is undergoing directional selection ($w_{11} = 1.0$; $w_{12} = 0.75$; $w_{22} = 0.5$). r is the recombination rate between the two loci. The dashed line shows the expected change at the selected locus.

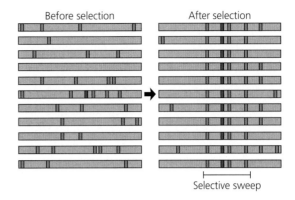

Figure 10.5 A selective sweep for a mutation (red) that quickly goes to fixation by natural selection. Neutral variants (blue) closely linked to the site under selection will also go to fixation through hitchhiking. This results in reduced genetic variation in the region around the selected site. As genetic distance increases, recombination will act to reduce this effect. From Sætre & Ravinet (2019).

the **selection differential**, the amount of recombination (r), and the initial gametic array (Figure 10.4). A selective sweep will reduce the amount of variation at loci that are tightly linked to the locus under selection (Figure 10.5).

For example, Kardos et al. (2015b) detected a selective sweep in bighorn sheep at *RXFP2*, a gene that strongly affects horn size in domestic sheep. The massive horns carried by bighorn rams appear to have evolved in part via strong positive selection at *RXFP2* in the past 2 million years since their divergence from domestic sheep.

10.3.2 Associative overdominance

Selection at one locus also can affect closely linked neutral loci when the genotypes at the selected locus are at an equilibrium allele frequency. Consider the case of heterozygous advantage where, using the previous fitness array, $w_{11} = w_{22} = 1.0$ and $w_{12} = (1 + s)$. The effective fitnesses at the A locus are affected by selection at the B locus (s) and D; the marginal fitnesses are the average fitness at the A locus considering the two-locus genotypes. These would be the estimated fitnesses at the A locus if only that locus were observed. If D is zero then all the genotypes at the A locus will have the same fitness. However, if there is gametic disequilibrium (i.e., D is not equal to zero), then heterozygotes at the A locus will experience a selective advantage because of selection at the B locus.

This effect has been called **associative overdominance** (Ohta 1971) or **pseudo-overdominance** (Carr & Dudash 2003). This pattern of selection has also been called marginal overdominance (Hastings 1981). However, **marginal overdominance** has more generally been used for the situation where genotypes experience multiple environments and different alleles are favored in different environments (Wallace 1968). This situation can lead to an overall greater fitness of heterozygotes even though they do not have a greater fitness in any single environment.

Heterozygous advantage is not necessary for linked loci to experience associative overdominance. Heterozygous individuals at a selectively neutral locus will have higher average fitnesses than homozygotes if the locus is in gametic disequilibrium with a locus having deleterious recessive alleles (Ohta 1971).

We can see this with the genotypic arrays in Example 10.1. Let us assume that the b allele is a recessive lethal (i.e., fitness of the bb genotype is zero). In the case of gametic equilibrium ($D = 0$), exactly q_2^2 ($0.3 \times 0.3 = 0.09$) of genotypes at the A locus have a fitness of zero. Thus, the mean or marginal fitness at the A locus is $1 - 0.09 = 0.91$. However, in the case of maximum positive disequilibrium, only the aa genotypes have reduced fitness because the AA and Aa genotypes do not occur in association with the bb genotype. Thus, the fitness of AA, Aa, and aa are 1, 1, and 0.75. There are many more aa than AA homozygotes in the population; therefore, Aa heterozygotes have greater fitness than the mean of the homozygotes.

10.3.3 Genetic draft

We saw in Section 10.3.1 that directional selection at one locus can reduce the amount of genetic variation at closely linked loci following a selective sweep. This is a special case of a more general effect in which selection at one locus will reduce the effective population size of linked loci. This has been termed the **Hill–Robertson effect** (Hey 2000) because it was first discussed in a paper that considered the effect of linkage between two loci under selection (Hill

& Robertson 1966). Observations with *Drosophila* have found that regions of the genome with less recombination tend to be less genetically variable as would be expected with the Hill–Robertson effect (Charlesworth 1996).

This effect has potential importance for conservation genetics. For example, we would expect a strong Hill–Robertson effect for mtDNA where there is no recombination. A selective sweep of a mutant with some fitness advantage could quickly fix a single haplotype and therefore greatly reduce genetic variation. Therefore, low variation at mtDNA may not be a good indicator of the effective population size experienced by the nuclear genome.

Gillespie (2001) has presented an interesting consideration of the effects of hitchhiking on regions near a selected locus. He has termed this effect **genetic draft** and has suggested that the stochastic effects of genetic draft may be more important than genetic drift in large populations. In general, it would reduce the central role thought to be played by effective population size in determining the amount of genetic variation in large populations. The potential effects of genetic draft seem to not be important for the effective population sizes usually of concern in conservation genetics.

10.4 Population subdivision

Population subdivision will generate nonrandom associations (gametic disequilibrium) between alleles at multiple loci if the allele frequencies differ among subpopulations at both loci. This is an extension to two loci of the Wahlund principle, the excess of homozygotes caused by population subdivision at a single locus, to two loci (Section 9.1) (Sinnock 1975). In general, for k equal-sized subpopulations:

$$D = \bar{D} + cov(p_1, p_2) \qquad (10.17)$$

where $\bar{D}$ is the average D value within the k subpopulations (Nei & Li 1973; Prout 1973).

This effect is important when two or more distinct subpopulations are collected in a single sample. For example, many populations of fish living in lakes consist of several genetically distinct subpopulations that reproduce in different tributary streams. Thus, a single random sample taken of the fish living in the lake will comprise several separate demes. Makela & Richardson (1977) have described the detection of multiple genetic subpopulations by an examination of gametic disequilibrium among many pairs of loci.

Cockerham & Weir (1977) introduced a composite measure of gametic disequilibrium that partitions gametic disequilibrium into two components: the usual measure of gametic disequilibrium, D, plus an added component that is due to the nonrandom union of gametes caused by population subdivision (D_B).

$$D_C = D + D_B \qquad (10.18)$$

In a random mating population, D and D_C will have the same value. We will see in the next section that the composite measure is of special value when estimating gametic disequilibrium from population samples. Campton (1987) has provided a helpful discussion of the derivation and use of the composite gametic disequilibrium measure.

10.5 Hybridization

Hybridization between populations, subspecies, or species will result in gametic disequilibrium. Figure 10.1 can be viewed as the resulting genotypes and gametes in the first two generations of hybridization. The F_1 hybrid will be heterozygous for all loci at which the two taxa differ. The gametes produced by the F_1 hybrid will depend on the linkage relationship of the two loci. If the two loci are unlinked, then all four gametes will be produced in equal frequencies because of recombination.

Table 10.2 shows the genotypes produced by hybridization between two taxa that are fixed for different alleles at two unlinked loci. This assumes that the two taxa are equally frequent and mate at random. We can see here that gametic disequilibrium (D) will be reduced by exactly one-half each generation. For unlinked loci, recombination will eliminate the association between loci in heterozygotes. However, only one-half of the population in a random mating population will be heterozygotes in the first generation. Recombination in the two homozygous genotypes will not have any effect. Therefore, gametic disequilibrium (D) will be reduced by exactly one-half each generation. A similar effect will occur in later generations even though more genotypes will be present. That is, recombination will only affect the frequency of gametes produced in individuals that are heterozygous at both loci ($AaBb$).

Gametic disequilibrium will decay at a rate slower than one-half per generation if the loci are linked. Tight linkage will greatly delay the rate of decay of D. For example, it will take an expected 69 generations for D to be reduced by one-half if there is 1% recombination between loci (Equation 10.5).

Gametic disequilibrium also will decay at a slower rate if the population does not mate at random and there is positive assortative mating of the parent types. This will reduce the frequency of double heterozygotes in which recombination can act to

Table 10.2 Expected genotype frequencies and coefficient of gametic disequilibrium (D) in a random mating hybrid swarm.

Genotypes	Parental	First generation	Second generation	Third generation	Equilibrium
			Genotype frequencies		
$AABB$	0.500	0.250	0.141	0.098	0.063
$AABb$			0.094	0.118	0.125
$AAbb$			0.016	0.035	0.063
$AaBB$			0.094	0.118	0.125
$AaBb$		0.500	0.312	0.267	0.250
$Aabb$			0.094	0.118	0.125
$aaBB$			0.016	0.035	0.063
$aaBb$			0.094	0.118	0.125
$aabb$	0.500	0.250	0.141	0.098	0.063
D	—	+0.250	+0.125	+0.063	0.000

reduce gametic disequilibrium. We can see this using Equation 10.18. In this case, the D component of the composite measure (D_C) will decline at the expected rate, but D_B will persist depending upon the amount of assortative mating. Random mating in a hybrid population can be detected by testing for HW proportions at individual loci.

These two alternative explanations of persisting gametic disequilibrium in a hybrid can be distinguished. Assortative mating will affect all pairs of loci (including cytoplasmic and nuclear associations) while the effect of linkage will differ between pairs depending upon their rate of recombination. Example 10.2 describes the multilocus genotypes in a natural hybrid zone between two species of crickets. In this case, most genotypes are similar to the parental taxa and gametic disequilibrium persists over all loci because of assortative mating. Forbes & Allendorf (1991) have described a hybrid swarm in which mating is at random (all loci are in HW proportions), but gametic disequilibrium persists at linked loci (Example 10.3).

We will examine hybridization and its genotypic effects again in Chapter 13 when we consider the effects of hybridization on conservation.

10.6 Estimation of gametic disequilibrium

There is no simple way to estimate gametic equilibrium values from population data (Kalinowski & Hedrick 2001; Barton 2011; Hui & Burt 2020). As described in the next section, even the simplest case of two alleles at a pair of loci is complicated. Estimation becomes more difficult for loci that have more than two alleles. There are a total of $n(n-1)/2$ pairwise combinations of loci if we examine n loci. So with 10 loci, each with just 2 alleles, there are a total of 45 combinations of two-locus gametic equilibrium values to estimate.

10.6.1 Two loci with two alleles each

Let us consider the simplest case of two alleles at a pair of loci (see genotypic array in Table 10.1). The gamete types (e.g., AB or Ab) cannot be observed directly but must be inferred from the diploid genotypes. For example, $AABB$ individuals can only

result from the union of two AB gametes, and $AABb$ individuals can only result from the union of an AB gamete and an Ab gamete. Similar inferences of gametic types can be made for all individuals that are homozygous at one or both loci. In contrast, gamete frequencies cannot be inferred from double heterozygotes ($AaBb$) because they may result from either union of AA and bb gametes or Ab and aB gametes. Consequently, gametic disequilibrium cannot be calculated directly from diploids.

Several methods are available to estimate gametic disequilibrium values in natural populations when the two gametic types of double heterozygotes cannot be distinguished. The simplest way is to ignore them, and simply estimate D from the remaining eight genotypic classes. The problem with this method is that double heterozygous individuals may represent a large proportion of the sample (Example 10.3), and their exclusion from the estimate will result in a substantial loss of information.

The best alternative is the **expectation maximization** (EM) algorithm, which provides a maximum likelihood estimate of gamete frequencies assuming random mating (Hill 1974). We previously used the EM approach in the case of a null allele where not all genotypes could be distinguished at a single locus (Section 5.4.2). This approach uses an iteration procedure along with the maximum likelihood estimate of the gamete frequencies:

$$\hat{G}_1 = [\frac{1}{2N}][2N_{11} + N_{12} + N_{21} \tag{10.19}$$

$$+ \frac{N_{22}\hat{G}_1(1 - \hat{p}_1 - \hat{p}_2 + \hat{G}_1)}{\hat{G}_1(1 - \hat{p}_1 - \hat{p}_2 - \hat{G}_1) + (\hat{p}_1 - \hat{G}_1)(\hat{p}_2 - \hat{G}_1)}]$$

where N is the sample size, N_{11} is the number of $AABB$ genotypes observed, N_{12} is the number of $AABb$ genotypes observed, and N_{21} is the number of $AaBB$ genotypes observed (Hedrick 2011, p. 585). This expression is not as opaque as it first appears. The first three sums in the right-hand parentheses are the observed numbers of the G_1 gametes in genotypes that are homozygous for at least one locus. The fourth value is the expected number of copies of the G_1 gamete in the double heterozygotes.

Example 10.3 Gametic disequilibrium in a hybrid swarm

Forbes & Allendorf (1991) studied gametic disequilibrium in a **hybrid swarm** of cutthroat trout (Figure 10.6). They observed the following genotypic distribution between two closely linked diagnostic allozyme loci. At both loci, the upper-case allele (*A* and *B*) designates the allele fixed in the Yellowstone cutthroat trout and the lower-case allele (*a* and *b*) is fixed in westslope cutthroat trout. The expected genotypes with **gametic equilibrium** ($D = 0$) are presented in parentheses. There is a large excess of both parental gamete types (*AABB*) and (*aabb*). The allele frequencies at the two loci are $p_1 = 0.589$ and $p_2 = 0.518$:

	LDH-A2			
ME-4	**AA**	**Aa**	**aa**	**Total**
BB	7	0	0	7
	(2.6)	(3.6)	(1.3)	
Bb	3	12	0	15
	(4.8)	(6.8)	(4.9)	
bb	0	1	5	6
	(1.2)	(2.2)	(1.0)	
Total	10	13	5	

Figure 10.6 Westslope cutthroat trout in Lake Rogers, Montana. Photo courtesy of John Ashley.

The estimated value of D in this case is 0.213 using the **expectation maximization** (EM) method described in Section 10.6.1, and $D' = 1.000$. The estimated gamete frequencies are presented below:

Gamete	$D = 0$	$D = 0.213$
A B	$(p_1)(p_2) = 0.305$	0.518
A b	$(p_1)(q_2) = 0.284$	0.071
a B	$(q_1)(p_2) = 0.213$	0.000
a b	$(q_1)(q_2) = 0.198$	0.411

Thus, we see that even though mating is at random in this hybrid swarm, gametic disequilibrium persists for many generations when loci are linked.

Bilton et al. (2018) have presented a method for estimating pairwise gametic disequilibrium in random mating populations with genomic data using the method of Hill (1974). Their method takes into account errors resulting from undercalled heterozygotes because of allelic dropout, as well as sequencing errors.

We need to make an initial estimate of gamete frequencies and then iterate using this expression. Our initial estimate can either be the estimate of gamete frequencies with $D = 0$, or we can use the procedure described in the previous paragraph to initially estimate D from the remaining eight genotypic classes. The other three gamete frequencies can be solved directly once we estimate G_1 and the single-locus allele frequencies. Iteration can sometimes converge on different gamete values depending upon the initial gamete frequencies (Excoffier & Slatkin 1995). Kalinowski & Hedrick (2001) present a detailed consideration of the implications of this problem when analyzing datasets with multiple loci.

It is crucial to remember that the EM algorithm assumes random mating and HW proportions. The greater the deviation from expected HW proportions, the greater the probability that this iteration will not converge on the maximum likelihood estimate. Stephens et al. (2001) have provided an algorithm to estimate gamete frequencies that assumes that the gametes in the double heterozygotes are likely to be similar to the other gametes in the samples. This method is likely to be less sensitive to nonrandom mating in the population being sampled.

10.6.2 More than two alleles per locus

The numbers of possible multilocus genotypes expand rapidly when we consider more than two alleles per locus. For example, there are six genotypes and three gametes types at a single locus with three alleles. Therefore, there are $6 \times 6 = 36$ diploid genotypes and $3 \times 3 = 9$ possible combinations of gametes at two loci each with three alleles. D values for each pair of alleles at two loci can be estimated and tested statistically (Kalinowksi & Hedrick 2001). The EM iteration procedure is more likely to converge to a value other than the maximum likelihood solution as the number of alleles per locus

increases. Therefore, it is important to initiate the iteration from many different starting points with highly polymorphic samples.

10.7 Strand theory: Junctions and chromosome segments

The ability to sequence large sections of chromosomes provides the opportunity to interpret multiple locus genetic data using entirely new conceptual approaches (Thompson 2018). It is now possible to use sequence data to identify chromosomal segments originating from different ancestral chromosomes. Junctions are the points at which the chromosome of origin changes because of historical recombination events during meiosis. For example, Figure 10.7 shows the inheritance and transmission of a single ancestral chromosome over two generations. One junction in each meiotic event has resulted in a chromosome with segments originating from three different ancestral chromosomes.

Chromosomal strand theory was developed by R.A. Fisher (1949) in the context of understanding

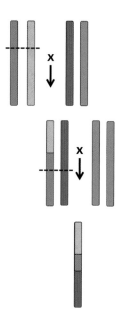

Figure 10.7 The inheritance and transmission of a chromosome over two generations. One junction (dashed lines) in each meiotic event has resulted in a final chromosome with segments originating from three different ancestral chromosomes. The gray shaded chromosomes do not contribute to the resulting chromosome.

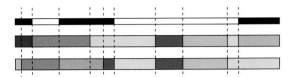

Figure 10.8 Two homologous chromosomes in a single individual sampled from a population a few generations after founding. Different colors represent different ancestral chromosomes. The dotted vertical lines indicate the location of junctions. The two white sections of the bar above the chromosomes indicate regions originating from the same ancestral chromosomes. These are regions of the genome where this individual is IBD (see Section 17.2) because of inbreeding. Redrawn from Chapman & Thompson (2002).

the effects of close inbreeding in lines of laboratory mice (Thompson 2018). For example, Figure 10.8 shows two hypothetical chromosomes sampled from a random mating population a few generations after founding (Chapman & Thompson 2002). It is astounding that this theory was envisioned and developed over 50 years before techniques were available to provide empirical data that could be used to apply this theory to real populations.

The number and distributions of lengths of these segments provide the opportunity to estimate a variety of fundamental population genetic parameters: inbreeding coefficients, migration rates, and effective population size. We will use this conceptual approach in understanding gene flow and hybridization (Chapter 13) and inbreeding (Chapter 17).

10.7.1 Microhaplotypes

The conceptual framework of strands rather than points can also be applied to the interpretation of multiple **single nucleotide polymorphisms (SNPs)**. Most SNPs are bi-allelic and thus individually have somewhat limited power in comparison with highly polymorphic microsatellite loci that often have many alleles. However, multiple SNPs that occur within the same small region can be genotyped jointly from high-throughput short-read DNA sequences to derive multi-allelic microhaplotype markers (Kidd et al. 2014). Microhaplotypes have been defined as a locus with two or more SNPs that occur within a short segment of DNA (e.g., 200 base pairs (bp)) that can be covered by a single

sequence read and collectively define a multiallelic locus (Kidd & Speed 2015).

Figure 10.9 shows a hypothetical example of the interpretation of three bi-allelic SNPs within a 40-bp region. There are eight possible microhaplotypes in a region containing three bi-allelic SNPs ($2 \times 2 \times 2 = 8$). However, only four of these possible eight microhaplotypes are present in this sample of four individuals. If these SNPs were in gametic equilibrium, then all eight microhaplotypes would be equally frequent. However, strong gametic disequilibrium is often present between closely linked SNPs.

Microhaplotypes are especially useful in situations where multiple allele loci are more powerful. For example, Box 12.1 considers the use of loss of alleles to detect population bottlenecks. Baetscher et al. (2018) have shown that the use of microhaplotypes in the kelp rockfish, a nearshore marine fish, provides large increases in power to identify kin relationships from the same amount of DNA sequence data. Kidd et al. (2015) demonstrate that microhaplotypes are extremely valuable for a number of forensic applications.

10.8 Multiple loci and conservation

Understanding multiple locus genotypes is especially important in conservation because small population size will generate nonrandom relationships between loci. Substantial gametic disequilibrium has been found even between unlinked pairs of loci on different chromosomes in many species (Example 10.4; Bensch 2006; Slate & Pemberton 2007). This is perhaps not unexpected. We saw in Section 10.2 that small population size in itself can produce substantial amounts of gametic disequilibrium. In addition, the rate of hybridization between subpopulations in many species has also increased because of human activities (Slate & Pemberton 2007; see Chapter 13). Thus, many of these populations that are of conservation interest might have substantial gametic disequilibrium because of hybridization, population subdivision, or small population size.

The interpretation of multilocus genotypes is becoming increasingly important for conservation because of the ability to screen many loci. The more loci examined, the more pairs of loci we are likely to

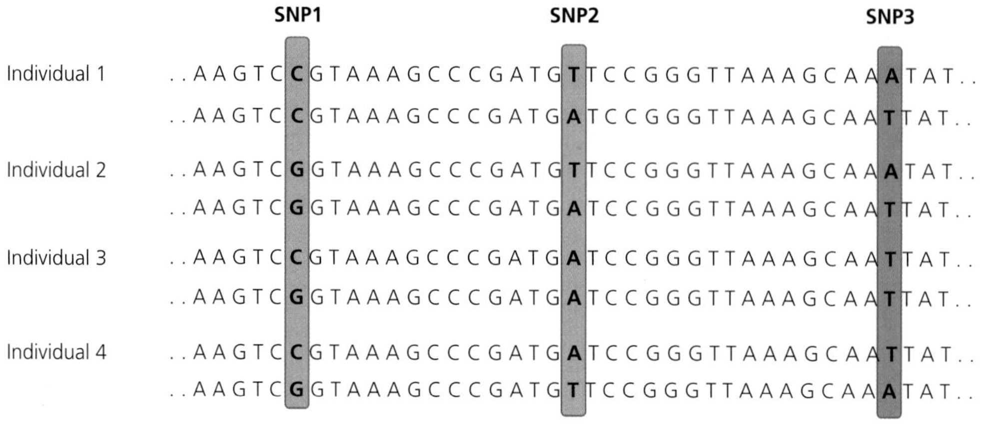

Microhaplotype 1	**C T A**	0.125
Microhaplotype 2	**C T T**	--
Microhaplotype 3	**C A A**	--
Microhaplotype 4	**C A T**	0.375
Microhaplotype 5	**G T A**	0.250
Microhaplotype 6	**G T T**	--
Microhaplotype 7	**G A A**	--
Microhaplotype 8	**G A T**	0.250

Figure 10.9 Hypothetical example of the interpretation of multiple closely linked SNPs as microhaplotypes. Above is shown the sequences of 40 bp in four individuals with three bi-allelic SNPs. The genotypes of the four individuals are CTA/CAT, GTA/GAT, CAT/GAT, and CAT/GTA at the individuals SNPs. Below are shown all eight possible microhaplotypes for the three SNPs. Only four of these possible haplotypes are actually present. The frequencies of these four microhaplotypes in this sample of four individuals are in the far right column.

Example 10.4 Extensive gametic disequilibrium at microsatellite loci in the Siberian jay

Li & Merilä (2010) estimated gametic disequilibrium between 103 microsatellite loci in a semi-isolated population of Siberian jay from western Finland. This subpopulation has been the subject of a long-term field study for over 35 years.

A linkage map for this population was constructed from pedigrees through direct field observations in combination with verification of parentage using microsatellite genotypes (Jaari et al. 2009). Recombination rates were estimated by the examination of 311 progeny fathered by 85 males and mothered by 95 females. A total of 107 microsatellite loci were assigned to one Z chromosome-specific and nine autosomal linkage groups. Ten loci could not be assigned to any linkage group. Six of the loci were found to be sex linked; three of these were in a pseudoautosomal region found on both the Z and W chromosomes, and three were Z chromosome-specific. As has been found in many species (Otto & Payseur 2019), there was less recombination in males than in females. On average, there was 28% greater recombination in females than in males. Figure 10.10 shows the comparative linkage map for one of the autosomal linkage groups.

A total of 97 autosomal and the 6 sex-linked loci were genotyped to estimate gametic disequilibrium in the wild population (Li & Merilä 2010). As expected, the amount of gametic disequilibrium between pairs of linked loci declined as the rate of recombination increased (Figure 10.11). Unlike the data from *Drosophila* described in Section 10.1, substantial gametic disequilibrium was found between pairs of loci separated by much more than 15% recombination. Significant ($P < 0.05$) gametic disequilibrium was even found in 83% of unlinked marker pairs on different chromosomes. As expected, the amount of gametic disequilibrium between pairs of loci on different chromosomes and unlinked pairs of loci on the same chromosome was quite similar: $D' = 0.356$ versus $D' = 0.354$.

Continued

Example 10.4 *Continued*

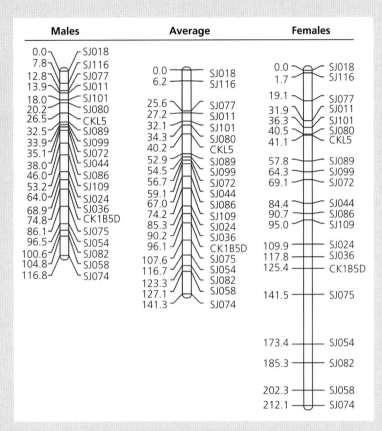

Figure 10.10 Linkage group 2 of the Siberian jay for males, females, and the average recombination rates of males and females. The names of the loci are on the right, and the total map distances on the left in centimorgans (cM). There is greater recombination for this linkage group in females than in males, as indicated by the greater distances between loci in the female. From Jaari et al. (2009).

The overall amount of gametic disequilibrium in the population is surprisingly high. This gametic disequilibrium probably results at least partially from the small effective population size of this population (N_e = 170, Fabritius 2010). In addition, pedigree analysis over many generations revealed five different extended family groups in this population. Such subdivision is expected to increase gametic disequilibrium, as we saw in Section 10.4.

The substantial gametic disequilibrium in this population has important implications, regardless of its cause. These observations also emphasize again how misleading it is to use the term "linkage disequilibrium" to refer to nonrandom associations between loci, as we discussed in Box 10.1. In this case, most pairs of loci found to be in "linkage" disequilibrium are actually not linked.

Continued

Example 10.4 *Continued*

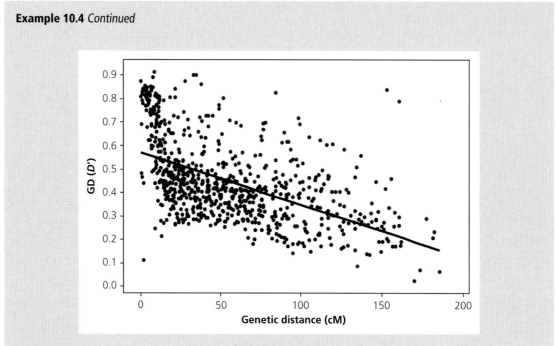

Figure 10.11 Gametic disequilibrium, as measured by D' (Equation 10.6), for syntenic pairs of loci separated by different amounts of recombination in the Siberian jay. Data from Li & Merilä (2010).

sample that are on the same chromosome and are in gametic disequilibrium.

In addition, advances in data analysis have revealed the presence in many species of chromosomal inversions that suppress recombination (Section 3.1.6). Such inversions have been found to be associated with local adaptation and life history variation in many species (Wellenreuther & Bernatchez 2018; Box 15.2). For example, Petrou et al. (2021) genotyped 6,718 SNP loci in over 1,000 Pacific herring from spawning aggregations

along the Pacific Coast of North America. Overall genome-wide differentiation was low ($F_{ST} = 0.014$), but 116 outlier loci were detected (mean $F_{ST} = 0.11$) that were strongly correlated with time of spawning. Plots of gametic disequilibrium versus physical distance indicated the presence of major inversions on four (7, 8, 12, and 15) of the 26 pairs of chromosomes (Figure 10.12). Many of the outlier loci responsible for local adaptation were found within these inversions.

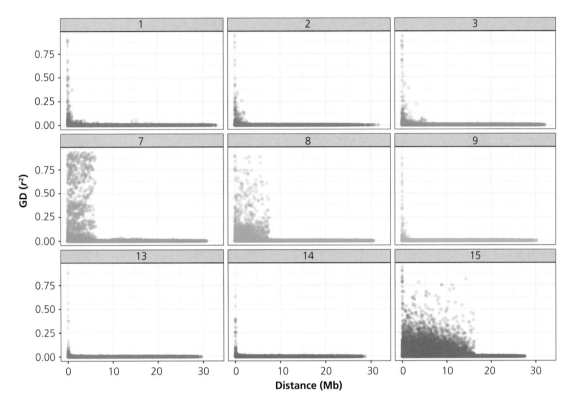

Figure 10.12 GD (r^2) versus the physical distance (Mb) between pairs of SNP loci in Pacific herring (Petrou et al. 2021). Nine of the 26 pairs of chromosomes in this species are shown. The average genome-wide recombination rate is 2.54 cM/Mb in the closely related Atlantic herring (Peterrsson et al. 2019). Thus, 10 Mb corresponds approximately to 25 cM. High gametic disequilibrium over long physical distances is present on chromosomes 7, 8, and 15 because of chromosomal inversions. F_{ST} outlier loci associated with spawning time among 23 spawning aggregations tended to map within these inversions. Figure courtesy of Eleni Petrou.

Guest Box 10 Estimation of effective population size using gametic disequilibrium with genomic data
Robin S. Waples

For many decades, use of genetic data to estimate N_e focused on the temporal method, which uses changes in allele frequency in temporally spaced samples (Waples 1989). Because gametic disequilibrium arises by drift in all finite populations at a rate inversely proportional to N_e (Equation 10.11), genetic estimates of effective size also can be obtained from individual samples. Initially, the gametic disequilibrium method was thought to have little practical relevance because of the high variance associated with gametic disequilibrium at individual pairs of loci (Hill 1981). However, the bias-corrected *LDNE* method developed by Waples and Do (2008) showed that robust estimates can be obtained by combining data for multiple pairs of loci, and within a few years Palstra and Fraser (2012) reported that most genetic estimates of N_e used single-sample methods, either *LDNE* or the sibship method of Wang (2009).

Extensive computer simulations have demonstrated the following regarding the gametic disequilibrium method:

- Precise estimates can be obtained with even modest amounts of data (10–20 microsatellite loci or ~100 SNPs) for relatively small populations ($N_e \leq$ a few hundred; Waples and Do 2010).
- Estimating N_e is challenging in large populations because the genetic signal (proportional to $1/N_e$) is weak (Waples 2016; Marandel et al. 2019), but this can be overcome to some extent using large numbers of individuals and incorporating life history information, which can place an upper bound on $\hat{N}_e$ (Waples et al. 2018).
- Missing data reduces precision but does not cause bias, provided that missingness is independent of genotype (Peel et al. 2013).

- Rare alleles can upwardly bias $\hat{N}_e$ but this can be controlled by setting an allele frequency cutoff (Waples & Do 2010).
- Effects of age structure can be accounted for based on the species' life history traits (Waples et al. 2014).
- In metapopulations (see Chapter 18), the gametic disequilibrium method estimates local N_e unless migration rate is relatively high (Waples & England 2011; Gilbert & Whitlock 2015). For continuously distributed populations, samples taken from within the breeding window estimate Wright's neighborhood size (Neel et al. 2013).
- The gametic disequilibrium method can rapidly detect small N_e associated with a bottleneck (England et al. 2010).

Genomics-scale datasets create new opportunities, as well as future challenges for the gametic disequilibrium method. For example, linkage downwardly biases estimates of contemporary N_e, unless it is accounted for based on genome size (Waples et al. 2016). On the other hand, as detailed linkage information becomes more readily available for nonmodel species, it is increasingly feasible to estimate N_e back in time, based on analogs to Equation 10.11 that include a term for recombination rate (Hollenbeck et al. 2016; Lehnert et al. 2019; Santiago et al. 2020). For large numbers of SNPs (10^3–10^6), the theoretical precision of the gametic disequilibrium method becomes arbitrarily high; however, these pairwise comparisons are not independent, and exactly how much this reduces precision is difficult to quantify (Waples et al. 2016). Preliminary results suggest that in most cases use of more than a few thousand SNPs does not help much to reduce variance of $\hat{N}_e$.

CHAPTER 11

Quantitative Genetics

Axolotl, Section 11.3

Quantitative genetic approaches offer evolutionary predictions that would be of great value, both in deepening scientific understanding of evolutionary process and in informing measures to address issues of pressing societal concern.

(Ruth G. Shaw 2019, p. 2)

Most of the major genetic concerns in conservation biology, including inbreeding depression, loss of evolutionary potential, genetic adaptation to captivity, and outbreeding depression, involve quantitative genetics.

(Richard Frankham 1999, p. 237)

Most phenotypic differences among individuals within natural populations are quantitative rather than qualitative. Some individuals are larger, stronger, or can run faster than others. Such qualitative phenotypic differences cannot be classified based on single characteristics, such as wrinkled or smooth peas, and are not determined by a single change in DNA sequence. The inheritance of quantitative traits is usually complex, and many genes are involved (i.e., they are **polygenic**). In addition to genetics, the environment to which individuals are exposed will also affect their phenotype. These traits often affect overall fitness and reflect adaptation to environments. Understanding genetic variation in quantitative traits is important for conservation, as natural selection acts directly on phenotypes, not on genotypes. The single-locus genetic models that we have been using until this point are inadequate for understanding this variation. Instead of considering only the effects of one or two genes at a time, we will expand our examination to inheritance of polygenic traits, and partition the genetic basis of such phenotypic variation into various sources using statistical procedures. We will also explore

how newer approaches are revealing the genomic basis of quantitative phenotypes.

Farmers no doubt understood for millennia that many traits are heritable, and that selecting more productive plants or animals as breeding stock produced better yields. The formal study of quantitative genetics began shortly following the rediscovery of Mendel's principles to resolve the controversy of whether discrete Mendelian factors (genes) could explain the genetic basis of continuously varying characters (Lynch & Walsh 1998). The theoretical basis of quantitative genetics was developed primarily by R.A. Fisher (1918) and Sewall Wright (1921). Empirical aspects of quantitative genetics were generally developed from applications to improve domesticated animals and agricultural crops (Lush 1937; Falconer & Mackay 1996).

Models of quantitative genetics have been used for understanding genetic variation in natural populations only in the past few decades, and the past decade has seen the emergence of new genomic and computational tools for this purpose. The abundance of genomic markers now available makes it

Conservation and the Genomics of Populations, Third Edition. Fred W. Allendorf, *et al.*, Oxford University Press.
© Fred W. Allendorf, W. Chris Funk, Sally N. Aitken, Margaret Byrne, and Gordon Luikart (2022). DOI: 10.1093/oso/9780198856566.003.0011

possible to identify **quantitative trait loci** (QTLs)—specific chromosomal regions that influence variation in quantitative traits (Barton & Keightley 2002), genes within those regions, and even **single nucleotide polymorphisms (SNPs)** within those genes that contribute to quantitative trait variation and adaptation (Stinchcombe & Hoekstra 2008; Stapley et al. 2010). Understanding the evolutionary effects of QTLs will allow us to improve our understanding of how genes influence phenotypic variation and improve our understanding of the **genetic architecture** of phenotypic variation (i.e., the number of genes involved and the distribution of variation in those genes among and within populations).

The principles of quantitative genetics can also be applied to a variety of problems in conservation (reviews by Kruuk 2004; Kramer & Havens 2009; Shaw 2019). Pink salmon on the west coast of North America have become smaller at sexual maturity over a period of 25 years. This apparently resulted from the effects of a size-selective fishery in which larger individuals had a higher probability of being caught, but this effect will only reduce body size across generations if size is at least partially genetically determined (Ricker 1981). Understanding the quantitative genetic basis of traits is essential for predicting genetic and genomic changes that are likely to occur in captive propagation programs as populations become adapted to captivity, or to determine whether adequate adaptive variation exists for a population to adapt to new environmental conditions or threats (Sgrò et al. 2011). Quantitative genetic studies of population differentiation can also be used to select well-adapted source populations for ecological restoration or reintroductions.

This chapter provides a conceptual overview of the application of quantitative genetics to problems in conservation. Until recently, we had to draw on quantitative genetic experiments with model species in laboratory studies and with crop species, but the number of recent phenotypic and genomic studies of quantitative genetic variation in natural populations is growing rapidly. Detailed consideration of quantitative genetic principles can be found in Falconer & Mackay (1996), Lynch & Walsh (1998), and Walsh & Lynch (2018).

11.1 Heritability

Quantitative traits include those phenotypes with continuous distributions (e.g., weight, height, or even gene expression levels); those that are **meristic**, that is, with values restricted to integers (e.g., number of vertebrae or seeds per fruit); and **threshold characters** that fall into a few discrete states (e.g., alive, diseased, or dead). Many quantitative traits are distributed approximately normally within populations. It takes relatively few loci, combined with a small amount of environmental variance, to generate a normal distribution (Figure 11.1). So while many quantitative traits are polygenic, the distributions of phenotypes do not tell us how many genes underlie a trait.

As we discussed in Chapter 2, phenotypes are the joint products of genotypes and environments. The total amount of phenotypic variation for a quantitative trait within a population can be thought of as arising from two major sources: environmental differences among individuals and genetic differences among individuals. Writing this statement in the form of a simple mathematical model, we have:

$$V_P = V_E + V_G \qquad (11.1)$$

where V is variance, a statistical measure of variation equal to the standard deviation squared, and V_P, V_E, and V_G are the phenotypic, environmental, and genetic variances of a trait, respectively. The genetic differences among individuals (V_G) can be attributed to three different sources of variance:

V_A = additive effects

(effects of allele substitution at each locus)

V_D = dominance effects

(effects of interactions between alleles at each locus)

V_I = epistatic effects

(effects of interactions between loci)

Therefore,

$$V_P = V_E + V_G \qquad (11.2)$$
$$= V_E + V_A + V_D + V_I$$

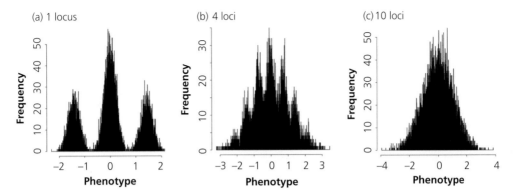

Figure 11.1 Simulated distributions of phenotypes resulting from one, four, and 10 loci, all with $V_A = 1$ and $V_E = 0.05$ showing how relatively few loci and a small amount of environmental variance can generate an approximately normal distribution of phenotypes within a population. From Coop (2019).

11.1.1 Broad-sense heritability

Heritability is a measure of the extent to which phenotypic differences among individuals are due to genetic differences. To determine the heritability of a trait, you need to evaluate the similarity of phenotypes among related individuals compared with unrelated individuals. You do not, however, need to know anything about the genes that underlie phenotypic variation. **Broad-sense heritability** (H_B) is the proportion of the phenotypic variability that results from genetic differences among individuals. That is:

$$H_B = \frac{V_G}{V_P} = \frac{V_A + V_D + V_I}{V_P} \qquad (11.3)$$

For example, Sewall Wright removed virtually all of the genetic differences between guinea pigs within lines by continued sister–brother matings for many generations. The total phenotypic variance (V_P) in the population as a whole (consisting of many separate inbred lines) for the amount of white spotting was 573. The average variance within the inbred lines was 340; this must be equal to V_E because genetic differences among individuals within the lines were removed through inbreeding.

$$V_P = V_E + V_G$$
$$573 = 340 + V_G$$
$$V_G = 573 - 340 = 233$$

and

$$H_B = 233/573 = 0.409$$

11.1.2 Narrow-sense heritability

Conservation biologists are often interested in predicting how a population will respond to selection when individuals differ in survival and reproductive success (Stockwell & Ashley 2004). Similarly, animal and plant breeders are often interested in improving the performance of agricultural species for specific traits of interest (e.g., growth rate or egg production). However, broad-sense heritability may not provide a good prediction of the response if genetic variation includes nonadditive allelic effects on phenotypes. A trait may not respond to selective differences, even though variation for the trait is largely based upon genetic differences between individuals, if that variation is due to dominance or epistasis (Section 11.2.1).

Narrow-sense heritability (H_N) is more commonly used than broad-sense heritability because the former provides a measure of the genetic resemblance between parents and offspring or between other related individuals, and therefore predicts the response of a trait to selection (Figure 11.2). H_N is the proportion of the total phenotypic variation that is due only to **additive genetic variation** (V_A) among individuals.

$$H_N = \frac{V_A}{V_P} \qquad (11.4)$$

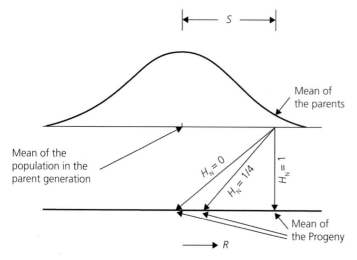

Figure 11.2 Illustration of the meaning of narrow-sense heritability based upon a selection experiment and Equation 11.8. The x-axis represents phenotype, and the y-axis represents frequency of individuals within a population. If there is no response to selection, then $H_N = 0$; if the mean of the progeny from selected parents is equal to the mean of the selected parents, then $H_N = 1$. Modified from Crow (1986).

If all genetic variation for a trait is additive, then H_B and H_N will be equal. If not specified, heritability usually (but not always) refers to narrow-sense heritability, both in this chapter and in the published literature.

Heritability is another area where the nomenclature and the symbols used in publications can cause confusion. Narrow-sense heritability is often represented by h^2 and broad-sense heritability sometimes by H^2. The square in these symbols is in recognition of Wright's (1921) original description of the resemblance between parents and offspring using his method of path analysis in which, under the additive model of gene action, an individual's phenotype is determined by $h^2 + e^2$, where e^2 represents environmental effects and h^2 is the proportion of the phenotypic variance due to the genotypic value (see appendix 2 in Lynch & Walsh 1998). We have chosen not to use these symbols in the hope of reducing possible confusion.

11.1.3 Estimating heritability

Heritability can be estimated in natural or captive populations by several different methods that all depend upon comparing the relative phenotypic similarity of individuals with known pedigrees based on long-term observations or estimated genetic relationships based on genomic markers. For a detailed simulation comparing these methods see de Villemereuil et al. (2013).

11.1.3.1 Parent–offspring regression

One of the most direct ways to estimate heritability is by regressing the progeny phenotypic values on the parental phenotypic values for a trait. The narrow-sense heritability can be estimated by the slope of the regression of the offspring phenotypic value on the mean of the two parental values (called the mid-parent value).

For example, Alatalo & Lundberg (1986) estimated the heritability of tarsus length in a natural population of the pied flycatcher using data from 338 nest boxes in Sweden. The narrow-sense heritability was estimated by twice the slope of the regression line of the progeny value on the maternal value. The slope is doubled in this case because only the influence of the maternal parent was considered. The male parents were not included in this analysis because previous results had shown that nearly 25% of progeny were the result of extra-pair copulations rather than mating with the social father. Heritability was estimated to be 0.496 (2 × 0.248) in an examination of nests in which nestlings were reared by their maternal mother (Figure 11.3a). In addition, 54 clutches were exchanged as eggs between parents to separate genetic and post-hatching environmental effects. There was no detectable resemblance at all between foster mothers and their progeny, suggesting **maternal effects** during incubation and juvenile growth are small (Figure 11.3b). If this study were repeated today, birds might be genotyped to confirm or exclude

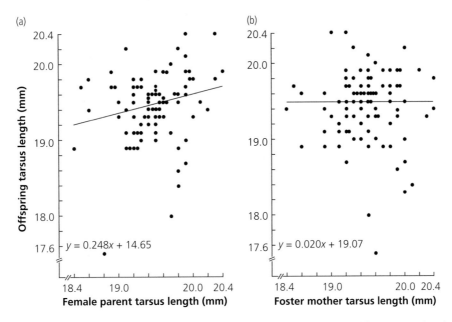

Figure 11.3 Mother–offspring regression estimation of heritability of tarsus length in the pied flycatcher: (a) regression with mother so the narrow-sense heritability is twice the slope (H_N = 0.496), and (b) regression with foster mother. Each point represents the mean tarsus length of progeny from one nest. From Alatalo & Lundberg (1986).

social fathers, and the animal model analytical framework (described in Section 11.1.3.3) would likely be used to better estimate quantitative genetic parameters.

The relatively high heritability in this example is somewhat typical of estimates for morphological traits in natural populations (Roff 1997). Our reanalysis in Chapter 2 of Punnett's (1904) data on vertebrae number in velvet belly sharks results in a heritability estimate of 0.63 ($P < 0.01$; Figure 2.5). Similarly, high heritabilities for a variety of meristic traits (vertebrae number, fin rays, etc.) have been reported in many fish species (reviewed in Kirpichnikov 1981). Many morphological characters in bird species also have high narrow-sense heritabilities (e.g., Table 11.1).

11.1.3.2 Progeny testing

Another approach for estimating heritability is to use phenotypic similarity among half- or full-sibling progeny to evaluate quantitative genetic variation and estimate the **breeding value** of parents in an approach called "progeny testing." In animal breeding, such an approach is traditionally

called the "sire model," where the breeding value of a male parent is determined from the performance of his offspring. Animal breeding often focuses on fathers since they can usually produce more offspring for testing than mothers; for example, in cattle. With plants, it is usually easier to collect seed from a known mother plant.

Table 11.1 Heritability estimates from parent–offspring regression for morphological traits for three species of Darwin's finches in the wild (Grant 1986). Heritability ranges between zero and one. However, estimates of heritability can be greater than 1.0; for example, the slopes of the regression of progeny on mid-parent values for weight and bill length were greater than one in *Geospiza conirostris*.

Character	*Geospiza fortis*	*Geospiza scandens*	*Geospiza conirostris*
Weight	0.91***	0.58	1.09***
Wing cord length	0.84***	0.26	0.69*
Tarsus length	0.71***	0.92***	0.78**
Bill length	0.65***	0.58*	1.08***
Bill depth	0.79***	0.80*	0.69***
Bill width	0.90**	0.56*	0.77**

(* $P < 0.05$; ** $P < 0.01$; *** $P < 0.001$).

To separate genetic and environmental variances for a phenotypic trait, particularly when genetic variances are smaller than environmental variances in wild populations, individuals are ideally raised and phenotyped in a common environment with replication. **Common garden** experiments, a term used even for animals, have designs that facilitate the statistical separation of genetic and environmental effects (Section 2.5). They are common for plant species, both wild and domesticated, as well as for domesticated and model animal species. However, these experiments are difficult or impossible to conduct for most wild animal species, or for highly endangered plants, and may produce results that are quite different from those in the wild.

Progeny testing methods to estimate heritability from common gardens use the extent to which related individuals carry genes that are identical by descent. For example, if you collect seed from individual plants of an outcrossing species, plants grown from those seeds will all have the same mother but can have different fathers. If you assume that the seeds are half-siblings, then on average, they will share one-quarter of their genes with each other from their mother by descent, so the variance among half-sib families (V_F) reflects one-quarter of the additive genetic variation (V_A):

$$V_A = 4V_F \qquad (11.5)$$

and:

$$H_N = \frac{4V_F}{V_P} \qquad (11.6)$$

This approach can bias heritability estimates upward if maternal effects exist, for example, nongenetic differences among mother plants in seed weight that influence phenotypic traits, or if some individuals from the same mother are full- rather than half-siblings. If known full-sibling families are used, then additive and nonadditive genetic variation cannot be separated.

11.1.3.3 Animal model

A big step forward in the estimation of genetic variance components and heritabilities for populations in the wild has been the development of a more analytically complex method that uses a maximum

likelihood mixed model approach called the "**animal model**" (Kruuk 2004). The name refers to the estimation of additive genetic value of individual animals rather than groups of individuals with the same relatedness. It is of great use in natural populations as one can include data for individuals with a wide range of relationships to each other, rather than, for example, just half- or full siblings. It is now widely used in both plant and animal breeding programs as it is more flexible and produces more accurate estimates that use more information than the parent–offspring or progeny approach (de Villemereuil et al. 2013). This model uses multigenerational pedigree or relatedness information to partition phenotypic variation into additive genetic and environmental components, using **restricted maximum likelihood (REML)** or a **Markov chain Monte Carlo (MCMC)** approach. Best linear unbiased prediction models are used to estimate variance components (see Kruuk 2004 for details). With the animal model, environmental variation in the wild can be further partitioned into various sources, including the effects of common maternal environments.

The animal model can be applied to populations in the wild with known pedigrees, typically populations of mammals or birds that are the focus of long-term studies that tag individual animals and record their parentage. Pedigrees can also be inferred from genetic markers, especially if individuals are genotyped for a large number of loci (Wang 2002; Jones & Wang 2010). For example, Pigeon et al. (2016) combined these approaches, using a pedigree based on observed mother–offspring relationships and microsatellite marker-inferred father–offspring relationships to study selection responses due to hunting in a bighorn sheep population. They found that horn size had declined as a result of intense hunting, then plateaued but did not fully recover after hunting pressure was reduced.

Additive genetic variation and heritability of skeletal size-related traits were estimated for polar bears in the western Hudson Bay region using the animal model. As polar ice decreases, polar bears are subjected to longer fasting periods, and Malenfant et al. (2018) wanted to examine the potential for evolution of body size in response to changing selection pressures. They combined data for 4,449 bears measured over a 45-year period. A subset of 859 bears was genotyped for 5,433 SNPs. The pedigree used in this analysis combined field observations and genetic markers. They estimated heritabilities of 0.38–0.48 for skull size and body length traits. They also analyzed bear girth, which had a lower heritability, likely due to trait plasticity (e.g., correlated with timing since last meal) as well as greater measurement error than the other traits.

The complexity of genetic and environmental effects on phenotypic variance and response to selection has been well illustrated for a well-studied island population of Soay sheep in Scotland using the animal model. Complete records of animal births, deaths, and phenotypes have been kept since 1985. Wilson et al. (2006) used the animal model to study natural selection in this population under varying environmental conditions. They found that when environmental conditions are harsh, selection for increased birthweight is strong but the response is constrained by low genetic variance. When conditions were favorable, genetic variance was higher; however, selection was weak. Fluctuating selection pressures and differences among environments in phenotypic expression of genetic variation may limit rates of evolution but maintain genetic variation.

While relatedness estimates can be based either on social or observed pedigrees or on genome-wide relatedness, the latter is advantageous as: (1) observed pedigrees often have errors; and (2) genome-wide relatedness can account for the actual similarity of genomes rather than the average similarity expected for a given relationship. For example, a pair of full siblings would be expected, on average, to have 50% of their DNA for autosomal chromosomes in common, assuming each of the siblings has an equal chance of inheriting either copy of a gene from each parent. However, some pairs of full siblings will share more than 50%, and others by chance will share less (Figure 17.7).

The animal model can be informed by this variation in genome-wide relatedness estimates for estimating additive genetic variation and heritability. For example, Perrier et al. (2018) estimated heritabilities for four morphological traits in blue tits, using either the social pedigree or the genome-wide relatedness. Heritability estimates were slightly higher

for the genome-wide relatedness approach due to absence of erroneous relationships, greater precision, and the greater information provided about genomic similarity. Heritability estimates rose with the number of SNPs used to estimate the genomic relationship matrix (GRM) until plateauing between 10,000 and 15,000 SNPs. However, Bérénos et al. (2014) found for Soay sheep that genome-wide relatedness estimates based on over 37,000 SNPs did not improve heritability estimates over using the observed maternal relationships and microsatellite-derived paternal relationships for this well-studied population.

11.1.4 Genotype-by-environment interactions

We saw in Figure 2.8 that the environment can have a profound effect on the phenotypes of yarrow plants resulting from a particular genotype. We also saw evidence for local adaptation with reciprocal transplants in *Gilia capitata* (Figure 8.1) in which plants had greater fitness in their native habitat. In a statistical sense, these are examples of interactions between genotypes and environments. We can expand our basic model to include these important interactions as follows:

$$V_P = V_E + V_G + V_{G \times E} \qquad (11.7)$$

Genotype-by-environment (G × E) interactions are of major concern in conservation biology when translocating individuals to alleviate inbreeding depression (genetic rescue, Chapter 19), when source populations are being chosen for ecological restoration, and when reintroducing captive populations into the wild (Chapter 21).

While common garden experiments in controlled environments typically yield higher heritability estimates, they may not produce phenotypes that are typical of natural environments. Genotype–environment interactions may be particularly strong between artificial and natural environments. For example, flowering time is a trait with high heritability for many plant species in common garden experiments. Anderson et al. (2014) studied flowering time in Drummond's rock-cress **recombinant inbred lines** in six laboratory and two field environments (Figure 11.4). While genotype-by-environment interactions were small

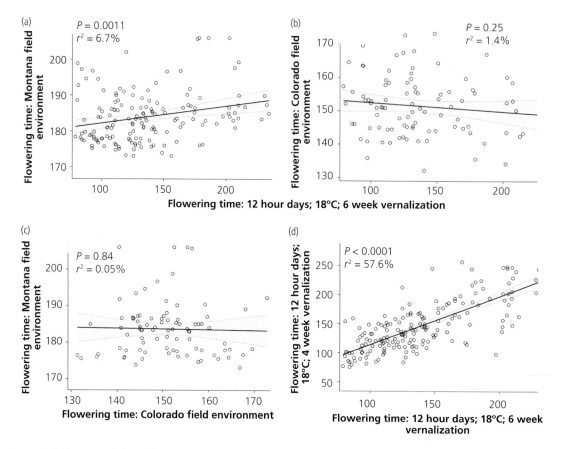

Figure 11.4 Genetic correlations of flowering times in Drummond's rockcress recombinant inbred lines grown in six laboratory and two field experiments show strong genotype-by-environment interactions. Each point shows the flowering time of one genotype in two environments. While genetic correlations between lab environments are relatively strong (d), meaning that genotypes rank similarly in different controlled environments, they are weak (a) or nonsignificant (b) between the laboratory and the field environments, and nonsignificant between field environments (c). From Anderson et al. (2014).

and **genetic correlations** high among the highly controlled laboratory environments, none of the artificial environments predicted flowering time in the field well, and correlations among field environments were weak to nonexistent.

Species that show strong genotype-by-environment interactions will likely need more populations conserved to capture the adaptive genetic diversity present in the species as a whole, and will need more local populations used as sources of individuals for restoration to avoid maladaptation. Reciprocal transplant experiments are needed to quantify the extent of G × E on phenotypes. Genomic markers cannot characterize genotype-by-environment interactions, although

genotype–environment associations can provide some signals of the drivers of selection (e.g., climate variables) and the loci involved (Hoban et al. 2016). We will see in Chapter 21 that many traits that are advantageous in captivity may greatly reduce the fitness of individuals in the wild.

11.2 Selection on quantitative traits

Evolutionary change by directional natural selection can be thought of as a two-step process. First, there must be phenotypic variation for the trait that results in differential survival or reproductive success (i.e., fitness). Second, there must be additive genetic variation for the trait ($H_N > 0$; Fisher 1930).

Heritability in the narrow sense can also be estimated by the response to selection. This is usually called the "realized" heritability (Figure 11.5). If a trait does not respond at all to directional selection, then there may be no additive genetic variation and $H_N = 0$ (or there are other genetic constraints such as tradeoffs among traits). If the mean of the selected progeny is equal to the mean of the selected parents, then $H_N = 1$. Generally, the mean of the selected progeny will be somewhere in between these two extremes ($0 < H_N < 1.0$). Francis Galton, a cousin of Charles Darwin (Provine 2001), coined the expression "regression" to describe the general tendency for progeny of selected parents to "regress" toward the mean of the unselected population (Box 1.1).

In this case, heritability is the response to selection divided by the total selection differential:

$$H_N = \frac{R}{S} \tag{11.8}$$

where S is the selection differential, defined as the difference in the means between the selected parents and the whole population, and R is the response to selection, which is the difference between the mean of the progeny of the selected parents and mean of the whole population in the previous generation. Equation 11.8 is often written in the form of the breeder's equation to allow prediction of the expected genetic gain (equivalent to response to selection R) from artificial selection:

$$R = H_N S \tag{11.9}$$

Figure 11.5 illustrates two generations of artificial selection for a trait with a heritability of 0.33. The breeder selects by truncating the population and uses only individuals above a certain threshold as breeders. The mean of the progeny from these selected parents will regress two-thirds (1–0.33) of the way toward the original population mean.

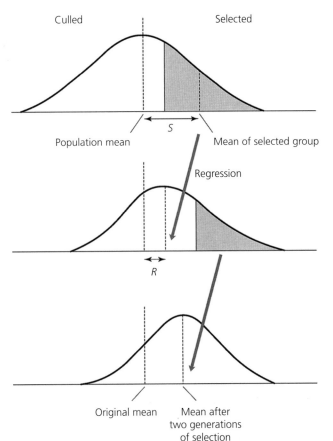

Culled   Selected

S

Population mean   Mean of selected group

Regression

R

Original mean   Mean after
two generations
of selection

Figure 11.5 Two generations of selection for a trait with a realized heritability of 0.33. The progeny mean moves one-third (0.33) of the distance from the population mean toward the mean of the selected parents each generation. From Crow (1986).

11.2.1 Heritabilities and allele frequencies

Heritability for a particular trait is not constant. It will vary among populations that are genetically divergent. It will also vary within a single population under different environmental conditions or over time. In addition, heritability within a population will not be constant even within the same environment over many generations, because it is influenced by allele frequencies that will change with selection. However, if many genes are involved with predominantly additive effects, heritability will stay more similar over time than if few genes and nonadditive effects are involved.

Let us examine this effect with a simple single-locus model in which all phenotypic variation has a genetic basis, with notation for fitness (w) following Chapter 8, and w_{A1} and w_{A2} indicating the additive effects of alleles A_1 and A_2 on fitness. The frequency of A_1 and A_2 are p and q, respectively. In a single-locus case such as this, heritability can be calculated directly by calculating the appropriate variances:

$$V_G = p^2(w_{11} - \bar{w})^2 + 2pq(w_{12} - \bar{w})^2 + q^2(w_{22} - \bar{w})^2 \tag{11.10}$$

$$V_A = 2[p(w_{A1} - \bar{w})^2 + q(w_{A2} - \bar{w})^2] \tag{11.11}$$

where

$$w_{A1} = \frac{p^2(w_{11}) + pq(w_{12})}{p} \tag{11.12}$$

and:

$$w_{A2} = \frac{pq(w_{12}) + q^2(w_{22})}{q} \tag{11.13}$$

V_A is very closely related to average heterozygosity for the gene determining this trait, and it will also be maximum at $p = q = 0.5$. Since we have assumed that V_E is zero, the narrow-sense heritability is the additive genetic variance divided by the total genetic variance.

$$H_N = \frac{V_A}{V_G} \tag{11.14}$$

We can use this approach to estimate heritability in the example of height in a plant determined by a single locus with two alleles, A_1 and A_2. We can use Equations 11.10–11.14 to estimate heritability over all possible allele frequencies (Figure 11.6). Here we consider three cases: (A) purely additive alleles, where the heterozygotes are intermediate to the homozygotes; (B) complete dominance of allele A_2,

where the heterozygotes have the same phenotype as the taller homozygotes (A_2A_2); and (C) overdominance, where the heterozygotes are taller than either homozygote.

Genotype	Additive (A)	Dominant (B)	Overdominant (C)
A_1A_1	20 cm	20 cm	20 cm
A_1A_2	22 cm	24 cm	24 cm
A_2A_2	24 cm	24 cm	20 cm
H_N	Always 1.0	High when q is low	0 when $p = 0.5$ >0 otherwise

There are several important features of the relationship between allele frequency and heritability in this simple model (Figure 11.6). If all of the genetic variance is additive (case A), heritability is always 1.0. In the case of dominance (case B), heritability is high when the dominant allele (A_2) is rare but low when this allele is common. Selection for a high-frequency dominant allele is not effective since most individuals will be homozygotes or heterozygotes with the same phenotype, so heritability will be low. Finally, in the case of overdominance (case C), heritability is high when either of the alleles is rare because increasing the frequency of a rare allele will increase the frequency of heterozygotes and thus the population will be taller in the next generation. That is, the population will respond to selection when either allele is rare, and therefore heritability will be high. However, the frequency of heterozygotes is greatest when the two alleles are equally frequent, and heritability will be zero.

While these single-locus examples are useful for illustrating the principles of quantitative genetics, heritability, and the effects of additive and nonadditive genetic variation, the vast majority of phenotypic traits with a continuous distribution are determined by several to many genes as well as affected by environmental variation. Since alleles at different loci combine to generate a phenotype, different genotypes may result in the same phenotype. A classic study of DDT resistance in *Drosophila* provides an example of this (Crow 1957). A DDT-resistant strain was produced by raising flies

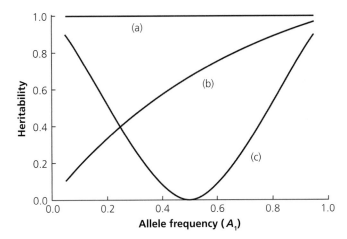

Figure 11.6 Hypothetical example of heritability (V_A/V_G) varying with allele frequency and degree of dominance for a trait determined by a single locus with two alleles: (a) the additive case where heterozygotes are intermediate to both homozygotes, (b) complete dominance so that the heterozygotes have the same phenotype as the taller homozygotes (A_2A_2), and (c) the case of overdominance where the heterozygotes are taller than either of the homozygotes. It is assumed that all of the variability in height is genetically determined (i.e., $V_E = 0$).

in a large experimental cage with the inner walls painted with DDT. The concentration of DDT was increased over successive generations as the flies became more and more resistant, until over 60% of the flies survived doses of DDT that initially killed over 99% of all flies. Flies from the resistant strain were then mated to flies from laboratory strains that had not been selected for DDT resistance. The F_1 flies were mated to produce an F_2 generation that had all possible combinations of the three major chromosomal pairs, as identified by marker loci on each chromosome (Figure 11.7).

The results of this elegant experiment demonstrate that genes affecting DDT resistance occur on all three chromosomes. The addition of any one of the three chromosomes (Figure 11.7) increases DDT resistance. Furthermore, the effects of different chromosomes are cumulative, contributing to additive genetic variation; that is, the more copies of chromosomes from the resistant strain, the more resistant the F_2 flies were to DDT. For another case of polygenic control of a quantitative trait as well as the genomic architecture of that trait, see Example 11.1 on the phenotypic and genomic effects of an experimental size-selective fishery on Atlantic silverside fish.

11.2.2 Genetic correlations

Genetic correlations are a measure of the genetic associations of different traits. Such correlations can result from two underlying causes. First, many genes affect more than a single trait and so they can cause simultaneous effects on different aspects

of the phenotype; this is known as **pleiotropy**. For example, a gene that increases growth rate is likely to affect both stature and weight, and this pleiotropy will cause genetic correlations between the two traits. **Gametic disequilibrium**, the nonrandom association of alleles at different loci (Chapter 10), can also result in genetic correlations between traits. It can result from close physical linkage of genes on a chromosome, from population structure, or from the effects of differential selection in different populations. Selection (either natural or artificial) for a particular trait will often result in a correlated response in the mean value of another trait because of genetic correlations.

Figure 11.10 demonstrates a genetic correlation between two meristic traits in an experimental population of pink salmon. The parental values for one trait, pelvic rays, were a fairly good predictor of the progeny phenotypes for another trait, pectoral rays ($P < 0.05$). The reciprocal regression (progeny pelvic rays on mid-parent pectoral rays) is also significant ($P < 0.01$). The actual point estimate of the genetic correlation between these traits (0.64) takes into account both of these parental–progeny relationships (Funk et al. 2005):

$$r_A = \frac{\text{cov}_{XY}}{\sqrt{\text{cov}_{XX}\text{cov}_{YY}}} \qquad (11.15)$$

where cov_{XY} is the covariance that is obtained from the product of the value of trait X in parents and the value of trait Y in progeny, and cov_{XX} and cov_{YY} are the progeny–parent covariance for traits X and Y separately.

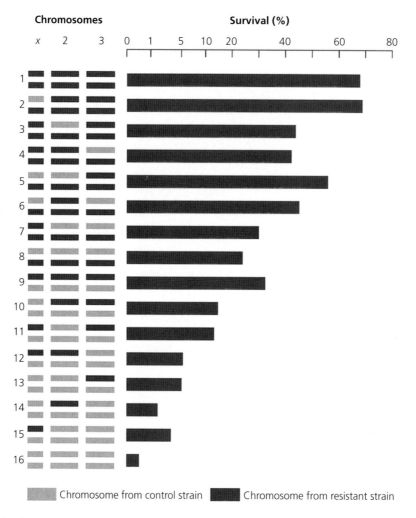

Figure 11.7 Results of an experiment demonstrating that numerous genes with largely additive effects determine DDT resistance in *Drosophila* (Crow 1957). The addition of any one of the chromosomes from the strain selected for DDT resistance increases DDT resistance. Redrawn from Crow (1986).

We can see from Figure 11.10 that if we select for parents with many pelvic rays, the number of pectoral rays in the progeny will increase. Genes that decrease developmental rate tend to increase counts for a suite of meristic traits in the closely related rainbow trout (Leary et al. 1984). Additive genetic correlations can also be calculated from progeny experiments using the additive genetic covariance cov_A between traits X and Y, and V_A for trait X and trait Y in the denominator (Falconer & Mackay 1996). Genetic correlations can also be used to quantify genotype-by-environment interaction; that

is, the extent to which the same trait is expressed by related individuals in different environments (e.g., Figure 11.4).

When two traits are genetically correlated in a direction that means selection increasing fitness for one trait will also increase fitness for the other, the correlation is reinforcing and will enhance evolutionary response (Etterson & Shaw 2001; Figure 11.11). In contrast, if two traits are negatively correlated such that an increase in fitness in one results in a decrease in fitness in the other, the correlation is antagonistic, and evolutionary

Example 11.1 An experiment with Atlantic silverside fish illustrates the highly polygenic nature and complex genomic changes that underlie responses to size-selective harvests

Therkildsen et al. (2019) subjected fish from two replicate populations from the center of the species' latitudinal range to five generations of selection for larger and smaller body size, allowing only the largest or smallest 10% of fish to reproduce each generation (Figure 11.8). After five generations, the lines selected for greater size weighed nearly twice as much as those selected for smaller size. Through low-coverage whole genome sequencing, fish from the original populations as well as the up- and down-selected lines were genotyped for 2.36 million SNPs.

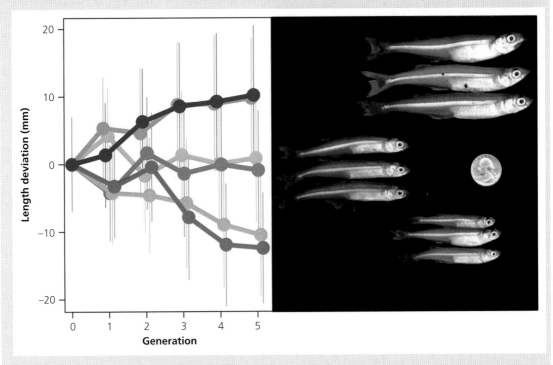

Figure 11.8 Observed changes in the size of Atlantic silversides from five generations of selection for greater length (blue populations), shorter length (orange), and control (green). From Therkildsen et al. (2019).

They identified over 10,000 SNPs with an allele frequency difference between the up- and down-selected lines of over 0.3, more than twice as many as expected based on simulations of neutral evolution. The length change in fish in selected lines was associated with the number of fast- or slow-growing alleles per genotype (Figure 11.9a). Many of these SNPs appear to also be involved in local adaptation as they had high population differentiation among unselected natural populations (1,596 with $F_{ST} > 0.25$ and 357 with $F_{ST} > 0.5$). However, the overlap in SNPs with large shifts in allele frequency as a result of selection differed considerably between the two replicate populations. The two down-selected populations shared only 0.9% of the same SNPs with large responses to selection, while the up-selected populations shared just 2.3% of these SNPs. In one of the down-selected populations, there was a block of 9,348 SNPs in strong linkage disequilibrium on one chromosome that changed in frequency from 0.05 to 0.6 in five generations.

The selected populations in this experiment also lost genetic diversity faster than the control populations. The selected populations with 100 fish available to spawn each generation lost 23–27% of their polymorphic sites and 7–9% of nucleotide diversity, while the control populations lost 17–20% of their polymorphic sites and 5% of nucleotide diversity due to genetic

Continued

Example 11.1 *Continued*

drift. This experiment demonstrates how size-selective harvesting, a common practice in some fisheries as well as in hunting, can greatly impact the genomic make-up of a species, altering both phenotypes and genotypes profoundly, and reducing genetic diversity over relatively few generations.

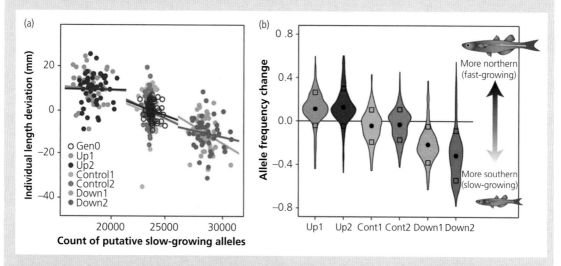

Figure 11.9 Parallel genomic shifts with selection for length in Atlantic silversides. (a) The number of slow-growing alleles per genotype was correlated with the average length of fish. (b) The fast-growing alleles were more common in northern populations and the slow-growing alleles were more common in southern populations. From Therkildsen et al. (2019).

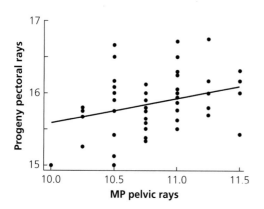

Figure 11.10 Genetic correlation between meristic traits in pink salmon. Regression of number of rays in the pectoral fins of progeny on the mid-parent (MP) values of number of rays in the pelvic fins (slope = 0.342; $P < 0.023$). Data from Funk et al. (2005).

responses will be inhibited. Strong antagonistic correlations often reflect biological tradeoffs among traits, and will slow or prevent responses to

selection despite the presence of additive genetic variation for each trait involved.

11.3 Finding genes underlying quantitative traits

The field of quantitative genetics developed to allow the genetic analysis of traits affected by multiple loci for which it was impossible to identify individual genes having major phenotypic effects. Formal genetic analysis could only be performed for traits in which discrete (qualitative) phenotypes (round or wrinkled seeds, pigmented or albino coloring of animals, etc.) could be identified to test their mode of inheritance. Classic quantitative genetics treated the genome as a black box and employed estimates of a variety of statistical parameters (e.g., heritabilities, genetic correlations, and the response to selection) to describe the genetic basis of continuous (quantitative) traits. However, the actual genes affecting these traits could not

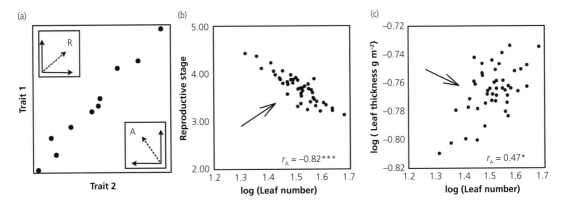

Figure 11.11 Genetic correlations among traits can enhance or inhibit response to selection. (a) Hypothetical positive genetic correlation (r_A) between two traits. Each point represents the breeding value of one genotype for each of the two traits. There are two selection scenarios. Reinforcing (R) selection acts in the same direction on the traits, enhancing the evolutionary response. Selection on one trait will increase fitness in the other. Antagonistic (A) selection acts in the opposite direction for the two traits, inhibiting the response. (b) and (c) show traits with antagonistic genetic correlations for the plant *Chamaecrista fasciculata*. The arrow indicates the vector of joint selection. From Etterson & Shaw (2001).

be identified with this approach. Alan Robertson (1967) described this phenomenon as the "fog of quantitative variation."

The fog has now lifted. High-throughput sequencing has made it possible to identify the chromosomal regions, genes, and in some cases, SNPs underlying quantitative traits, even for genes with a relatively small effect on genotypes. Understanding the effects of these potentially adaptive loci allows us to determine how specific genes influence phenotypic variation, directly assess adaptive genetic variation, and determine how many genes affect phenotypic traits of particular interest. High-throughput sequencing has facilitated this in four ways, by: (1) reducing the cost and time of genotyping many individuals for many markers; (2) allowing for the precise mapping of those genetic markers within the genome; (3) making feasible the testing of associations between particular markers and phenotypes within pedigreed populations that allow for mapping markers within the genome based on recombination rates; and (4) allowing for the precise estimation of pedigrees and relatedness in natural populations (Stapley et al. 2010).

11.3.1 QTL mapping

QTL mapping involves searching for polymorphic genetic markers that segregate with phenotypic variation for a trait. Traditionally, a high-density

genetic linkage map was developed through genotyping large numbers of F_1 or F_2 progeny of a controlled cross between phenotypically different parents for a large number of genetic markers that are distributed across the genome. Those markers or pairs of adjacent markers that segregate with variation for the phenotypic trait of interest must be located close to one or more causal genes affecting that trait. QTL mapping is not usually seeking the causal gene or sequence variation within the gene that results in phenotypic variation, but is rather seeking a marker linked to that gene.

Moving from a QTL to finding a causal gene within the region involved proved to be quite difficult using a mapping approach (Mackay 2001; Stinchcombe & Hoekstra 2008). New genome-wide association approaches with whole genome or reduced representation library approaches (exome sequencing, **restriction-site associated DNA sequencing (RADseq)**, etc.) that do not require a pedigreed set of samples have largely replaced QTL mapping for natural populations. However, genetic mapping remains a useful way to co-locate markers and genes when reference genomes are nonexistent or poorly assembled (e.g., the large and highly repetitive genomes of conifers; De La Torre et al. 2014).

Detecting QTLs in a sample from a population requires the existence of gametic disequilibrium between the causal and marker loci. Therefore, the

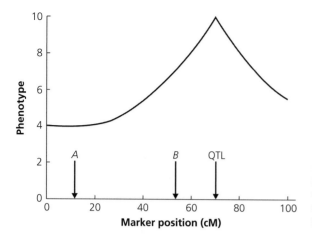

Figure 11.12 Hypothetical effect of a QTL on the mean phenotype of two syntenic marker loci on a chromosome containing a QTL at map position 70. Locus *A* is unlinked and is in gametic equilibrium with the QTL. Locus *B* is linked and is in gametic disequilibrium with the QTL.

tighter the linkage, the more likely a QTL will be detected (Chapter 10). Figure 11.12 illustrates QTL mapping for a chromosome that contains a QTL with two alleles, where the Q^+ allele causes a greater phenotypic value than the Q^- allele. If the marker locus is close to the QTL, then genotypes at the marker locus (*B* in this case) are likely to have different mean phenotypes because of gametic disequilibrium. Loci farther away on the same chromosome (*A* in this case) are less likely to be in gametic disequilibrium because of independent segregation, and therefore all three genotypes at the marker will have the same phenotypic mean. QTLs are often population-specific, and even pedigree-specific, because of the reliance on gametic disequilibrium that may exist in one population but not another, and because markers that segregate in one population may be fixed in another (Mackay 2001).

QTL mapping was used to determine the genetic basis of the derived lifecycle mode of paedomorphosis in the Mexican axolotl (Voss 1995; Voss & Shaffer 1997). Mexican axolotls do not undergo metamorphosis and have a completely aquatic lifecycle. In contrast, closely related tiger salamanders undergo metamorphosis in which their external gills are absorbed and other changes occur before they become terrestrial. Voss (1995) crossed these two species and found that their F_1 hybrids underwent metamorphosis. He then backcrossed the F_1 hybrids to a tiger salamander and found evidence for a single major gene or chromosomal region controlling most of the variation

in this life history difference, with additional loci having smaller effects. A dominant allele for the gene *MET* was found to be associated with the metamorphic phenotype in the tiger salamander, and the recessive *met* allele from the Mexican axolotl was associated with paedomorphosis (Smith, J.J. et al. 2005). This gene also contributes to continuous variation for the timing of metamorphosis in the tiger salamander (Voss & Smith 2005).

The increased availability of SNP markers and decreased cost of genotyping have facilitated QTL mapping studies in model and domesticated species, but the need for pedigreed families limits this approach in natural populations. With the advent of high-throughput sequencing, the search for the genetic basis of adaptive-trait variation shifted from QTL mapping to candidate gene and then to whole genome scan approaches. If a city is used as a metaphor for a genome, QTL mapping searches for the neighborhood or street containing a gene or genes affecting a phenotypic trait, while candidate gene approaches or whole genome scans are seeking the specific house addresses of the genes involved.

11.3.2 Candidate gene approaches

Candidate genes are genes of known function that are suspected to have a substantial influence on a phenotypic trait of interest. While the QTL approach is top-down, in that it begins with the phenotype in order to search for responsible genes, the candidate gene approach begins with the genes implicated in phenotypic trait variation in other species and searches for phenotypic effects of polymorphisms in those genes in target species.

The DNA sequences, functions, and effects on phenotypes of many genes are now known for fully sequenced model organisms. To search for the genetic basis of quantitative trait variation in natural populations of nonmodel species, candidate genes are first sequenced in a small number of individuals to identify SNPs. SNPs are then genotyped or candidate genes are resequenced in many individuals. The SNPs can then be tested for significant associations with phenotypic traits, or for significant associations with environmental factors, after controlling for relatedness and population structure in an approach called **association mapping**.

Candidate gene approaches have been particularly successful in identifying polymorphisms associated with phenotypic variation in coat color in wild mammal and bird populations. For example, the coloration of beach mice is locally adapted to the color of soil in the area they are found. Hoekstra

et al. (2006) identified a single amino acid change in the candidate gene *melanocortin 1 receptor* (*MC1R*) that explained 10–36% of the phenotypic variation in coat color (Figure 11.13). See Stinchcombe & Hoekstra (2008) and Stapley et al. (2010) for more information on applications of candidate genes and genome scans to understanding adaptation in wild populations.

11.3.3 Genome-wide association mapping

The third, and most common, approach to identify genetic polymorphisms underlying phenotypic variation is whole genome scans by genotyping many SNPs across the genome, and to test for associations with phenotypes in a method called **genome-wide association study** (GWAS). This approach requires no *a priori* knowledge about individual gene function, but requires considerable genomic resources and computational power. Markers can be selected at random, for example, through RADseq approaches, or can be based on exome capture and sequencing or

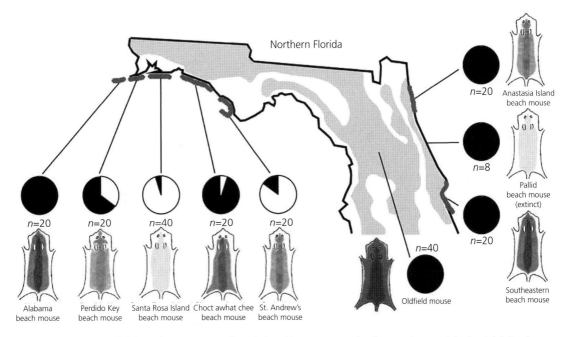

Figure 11.13 Frequency of alleles at the *MC1R* gene and association with average coat color phenotype in one mainland and eight beach mouse subspecies in northern Florida. The *MC1R* locus explains 10–36% of the phenotypic variation in coat color in beach mice. The light gray area represents the distribution of the mainland subspecies, and the gray areas indicate the distributions of the beach mouse subspecies. Circles indicate the relative allele frequencies of the light (white) and dark (black) alleles identified. From Hoekstra et al. (2006).

expressed sequence tags (ESTs) from transcriptome sequencing, increasing the chance of markers being linked to loci under selection. The sequence they are found within can be compared to genes of known function using the **Basic Local Alignment Search Tool** *(BLAST)*.

Some studies with model organisms now use whole genome resequencing, and this is feasible for species with small genomes (e.g., birds, insects, and some fish), but remains beyond available resources for most species with larger genomes. GWAS requires large numbers of markers to saturate the genome, as well as large sample sizes. Ideally, a large set of noncandidate loci are used to control for neutral population structure as well as relatedness (Santure & Garant 2018). Until recently, it was not feasible to genotype individuals with a sufficiently high marker density for GWAS to be effective, and obtaining sufficient sample sizes is still a constraint. However, genotyping and sequencing costs have decreased rapidly, and GWAS is now within reach for many studies of natural populations (see Santure & Garant 2018, and papers in the same special issue).

Most GWAS experiments test the strength of association of individual SNPs while controlling for population structure and relatedness. A common way to visualize the results of a GWAS, if that species or a closely related species has a well-assembled genome, is to plot the $-\log_{10}$(probability of association) against the chromosome and SNP position in a **Manhattan plot** (Section 4.2). For example, in a study of the model nitrogen-fixing leguminous plant species *Medicago trunculata*, many SNPs were identified that were associated with flowering date or with N-fixing *Rhizobium* nodules on the lower roots (Figure 11.14).

The significantly associated SNPs identified in GWAS studies collectively often do not explain all of the additive genetic variance for phenotypic traits. This is called the **missing heritability problem**. This has particularly been an issue in medicine. It arises due to a combination of: (1) strict statistical cutoffs needed to account for multiple comparisons of many markers resulting in loci with small effects not being identified; (2) interactions among loci that are not considered; and (3) low-frequency

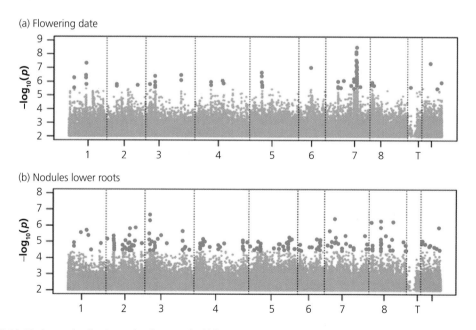

Figure 11.14 Manhattan plot showing results of a GWAS for (a) flowering date and (b) root nodules containing nitrogen-fixing *Rhizobium* symbionts in the leguminous plant *Medicago trunculata*. The $-\log10$(probability) of association between individual SNPs and phenotypes is plotted by chromosome number. SNPs significantly associated with phenotypes are shown in green. T represents all genome assembly contigs that could not be assembled into chromosomes of the reference genome. From Stanton-Geddes et al. (2013).

alleles such as deleterious alleles have effects that are hard to detect and quantify, but collectively may explain considerable variation. When all SNPs are considered without strict statistical cutoffs, the genetic variance explained is much higher; for example, using a genome-wide complex trait analysis (GCTA) that evaluates the genetic similarity of unrelated individuals.

The average effect of allele substitution at an associated SNP or marker is estimated as the **effect size** of that locus. Polygenic scores for genotypes can be estimated by summing effects of alleles across all associated SNPs for individual genotypes or populations. A simpler approach is to simply count the number of alleles in a genotype with a positive effect on a phenotypic trait rather than using effect sizes (e.g., Therkildsen et al. 2019; Example 11.1; Figure 11.9b). Single-locus tests have some limitations, such as the inability to test for epistasis or genotype-by-environment interactions, and some studies use machine learning algorithms, such as *random forest*, to test multiple SNPs and their interactions (Brieuc et al. 2018).

While most phenotypic traits important for adaptive phenotypic variation are highly polygenic, GWAS approaches have found some loci of large effect that could be of importance for conservation. For example, the timing of migration of anadromous fish to the spawning grounds has a large effect on population structure, and runs of salmon at different times of year need to be managed as genetically distinct populations. Large effect loci for run timing have been found for steelhead salmon (Waples & Lindley 2018), Atlantic salmon (Cauwelier et al. 2018), Pacific salmon (Prince et al. 2017), and Chinook salmon (Thompson et al. 2020). In the case of Atlantic salmon, using a SNP panel of 52,731 markers, a single QTL was identified that explained 24% of the variation in run timing. Figure 11.15 is a Manhattan plot showing just one region on chromosome 9 with a high probability of association with run timing. Genetic markers within this region could be of great use in identifying the population of fish sampled in the ocean from unknown runs in monitoring or research.

Even when a major effect QTL is detected, the remainder of variation in that trait may be substantial and polygenic, and that major effect gene may modify the effects of the other loci involved. In barn owls, there is a white allele that is associated with paler color than the rufous allele for *MC1R* (San-Jose et al. 2017). Birds that are homozygous for the rufous allele show less phenotypic and additive genetic variance for several other color-related traits than birds that are heterozygous or homozygous for the white allele.

Repeatability of GWAS results has only been moderate (Schielzeth et al. 2018). In general, major effect QTLs (including chromosomal inversions) are detected in multiple independent studies. However, loci of small effect are less consistently detected, and the number of small-effect loci detected will vary considerably with sample size and marker number. GWAS results vary among populations, in part because associations between markers and causal loci may vary. In addition, associations

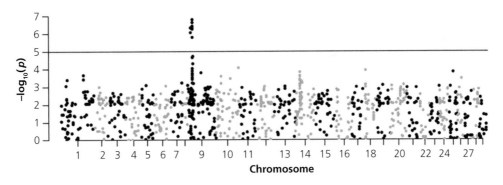

Figure 11.15 Manhattan plot of GWAS results for run timing in Atlantic salmon showing a major effect region on chromosome 9. The horizontal line represents the $P < 0.05$ significance threshold after a correction for the false discovery rate. From Cauwelier et al. (2018).

may vary with environments due to genotype-by-environment interaction. Results from a GWAS approach within a population can differ markedly from those using QTL mapping within a small pedigreed population as the former will include greater genetic variation, larger sample sizes, often more genetic markers, and have more statistical power. For example, Hansson et al. (2018) conducted a GWAS on wing length in great reed warblers. A previous QTL study had mapped a large effect locus on chromosome 2, while the GWAS found no major outliers on chromosome 2, and instead found associated markers distributed across the genome.

Some genes and markers may be detected through GWAS in some environments but not others due to **conditional neutrality**, where they are associated with phenotypic traits in some environments but have no effect on phenotypes in others. Fournier-Level et al. (2011) used a GWAS of ~213,000 SNPs to identify loci associated with local adaptation to climate in *Arabidopsis* plants originating from over 900 locations across Europe and grown in four common gardens in different countries. They found that SNPs conferring higher fitness in one environment often had no effects on fitness in other environments.

Another approach to GWAS is to pool phenotypes at either end of the phenotypic distribution, then sequence those phenotypic pools and look for SNPs with substantial differences in allele frequency between the two pools. This approach is analogous to the case-control approach used in medicine to identify genes involved in disease resistance or susceptibility in humans, comparing groups of diseased versus healthy individuals. By pooling individuals prior to sequencing, costs are reduced, but information for individual genotypes is lost. Micheletti & Narum (2018) demonstrated this approach with Chinook salmon, identifying genes associated with age of maturity as well as with phenotypic differences between sexes.

11.4 Loss of quantitative genetic variation

The loss of genetic variation (heterozygosity and allelic diversity) via genetic drift will affect quantitative as well as neutral genetic variation. As a result, the ability of small populations to adapt and evolve is expected to be lower than that of large populations, as genetic drift will erode the amount of additive genetic variation available for selection, and new beneficial mutations will accumulate more slowly (Willi et al. 2006). However, the rate at which genetic variation will be lost will depend upon a large number of factors, including the number of loci affecting a trait, the amount of dominance or epistasis, and the strength and type of selection.

11.4.1 Effects of genetic drift and bottlenecks

We saw in Chapter 6 that heterozygosity at neutral loci will be lost at a rate of $1/(2N_e)$ per generation. We can relate the effects of genetic drift on allele frequencies at a locus with two alleles to additive quantitative variation as follows (Falconer & Mackay 1996):

$$V_A = \sum 2pq[a + d(q - p)]^2 \qquad (11.16)$$

where p and q are allele frequencies so that $2pq$ is the expected frequency of heterozygotes, a is half the phenotypic difference between the two homozygotes, and d is the dominance deviation. This model results in the simple prediction that V_A, and therefore H_N, will be lost at the same rate as neutral heterozygosity in the case where all of the variation is additive ($d = 0$).

Can heterozygosity for neutral markers predict within-population quantitative variation and evolvability in small populations? In a comprehensive review of limits to adaptive potential in small populations, Willi et al. (2006) summarized laboratory studies that maintained either high levels of inbreeding or small effective population sizes over 10 generations, and estimated H_N (Figure 11.16). At low levels of inbreeding or larger effective population sizes, heritabilities were often higher than in outbred control lines, suggesting that variation due to dominance or epistasis was being expressed, and combined with drift, was inflating V_A and H_N.

This increased phenotypic variation and heritability is unlikely to enhance adaptive responses to selection as it is based largely on the expression of recessive deleterious alleles (Willi et al. 2006). At higher levels of inbreeding or smaller effective population sizes, heritability estimates decreased below

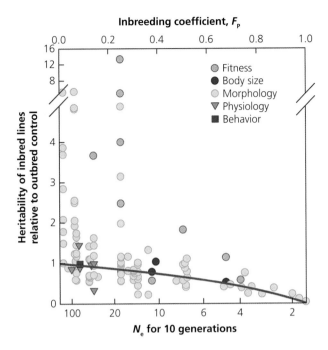

Figure 11.16 The effects of inbreeding coefficient (upper axis) or the effective population size that would result in the same inbreeding coefficient (lower axis) on heritability estimates (y-axis) in laboratory studies of experimental populations. Study organisms include *Drosophila* (10 studies), house flies (five studies), *Tribolium* beetles (two studies), two plant species, laboratory mice, and a butterfly. Heritability estimates are expressed as a ratio relative to outbred control populations. The solid line indicates the expected decline for purely additive genetic variation. From Willi et al. (2006).

that of outbred controls and tracked expected values based on an additive genetic model with no selection, paralleling predictions for heterozygosity. This suggests that the effects of population bottlenecks on additive genetic variation in the short term are unpredictable and trait-specific, with heritability often increasing. However, in the longer term, the genetic variation available for adaptation is expected to decline, reducing the capacity for populations to adapt to new conditions.

Field studies of the relationship between population size and population V_A have produced mixed results, with many not meeting theoretical explanations of reduced additive genetic variation in smaller populations (Wood et al. 2016; Hoffmann et al. 2017). While some studies have found that smaller populations have lower V_A and H_N than larger populations, others have found no relationship. A recent analysis of over 80 species by Wood et al. (2016) found no reduction in H_N with decreasing population size. Possible reasons why population size and V_A are not correlated in some field studies include: (1) the current population size may have decreased recently and the genetic effect of that decline has not yet manifested; (2) V_A is often imprecisely estimated; (3) if stabilizing rather than

directional selection is acting, then V_A will not decrease until population size is very low; (4) spatial and temporal environmental variability will maintain variation, and habitats may become more variable with fragmentation; (5) gene flow among populations may be increasing effective population size; and (6) the smallest, most vulnerable populations may have already been extirpated prior to being sampled (Willi et al. 2006; Wood et al. 2016; Hoffmann et al. 2017). Based on these results, it should not be assumed that all small populations lack additive genetic variation and therefore cannot adapt to new conditions, although they will be facing other genetic challenges. We discuss the importance of population size in maintaining additive genetic variance further, and the debate about how large populations need to be, in Chapter 18.

11.4.2 Effects of selection

In general, directional selection will reduce additive genetic variation for a trait, so alleles that increase fitness should increase in frequency until they reach fixation, while alleles associated with reduced fitness will decline in frequency and eventually be lost from the population. Therefore, we expect that traits that are strongly associated with fitness should have lower heritabilities than traits that are under weak or no natural selection. A vast body of empirical information in many species supports this expectation (Mousseau & Roff 1987; Roff & Mousseau 1987). For example, Kruuk et al. (2000) found a strong negative association between the heritability of traits and their association with fitness in a wild population of red deer. However, Houle (1992) analyzed *Drosophila* heritability estimates from many studies, and concluded that life history traits have lower H_N, not due to lower V_A, but due to higher environmental variance, V_E, resulting in greater phenotypic variance, V_P. He suggested that the additive genetic coefficient of variation is a better measure of adaptive variation than H_N (Section 11.1.3).

We do not expect strong selection for a trait to remove all additive genetic variation for highly polygenic traits. A famous long-term experiment in maize in Illinois has been selecting separate lines for high and low oil and protein content for over 100 years (Dudley 2007). Oil and protein content continue to increase, on average, in lines selected for increasing content (Figure 11.17). The average oil content of high-oil lines is now more than four-fold higher than the initial population, and protein has increased three-fold! Less surprising is that the

lines selected for low oil and low protein eventually plateaued, although they still contained genetic variation as evidenced by the responses to selection in the reverse low-selection lines (see RLO line in Figure 11.17). QTL mapping studies have shown that many loci are involved in determining both oil and protein content. Explanations for the long-term continued response to selection include: (1) low initial frequencies of alleles conferring higher oil or protein content at many of these loci; (2) release of **epistatic variation** over time; (3) mutations in QTLs underlying these traits adding phenotypic variation; and (4) a change in environments over time, for example, increased availability of nitrogen from fertilization may have allowed alleles for higher protein to be expressed and selected (Dudley 2007).

11.5 Divergence among populations

Quantitative genetics can also be used to understand genetic differentiation among local populations within species (Merilä & Crnokrak 2001). This approach has been especially interesting when used to understand the patterns of local adaptation caused by differential natural selection acting on heritable traits. We also expect local populations of species to differ for quantitative traits because of the effects of genetic drift. Understanding the relative

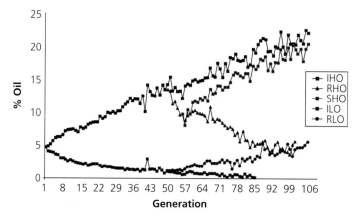

Figure 11.17 The effects of over a century of divergent selection for oil content in corn showing the persistence of additive genetic variation. All selection lines started with the same open-pollinated corn cultivar in 1896. Each line indicates selection on a different strain of corn. IHO = Illinois High Oil, was selected continually for increased oil content for 106 generations. RHO = Reverse High Oil was selected first for high oil for 48 generations, and then for low oil for the remaining generations. SHO = Switchback High Oil was selected for high oil for 48 generations, low oil for seven generations, and then high oil for the remaining generations. ILO = Illinois Low Oil, was selected for decreased oil content for 106 generations. RLO = Reverse Low Oil was selected for low oil content for 48 generations and then for high oil content for the remaining generations. From Dudley (2007).

importance of genetic drift and natural selection as determinants of population differentiation is an important goal when studying quantitative traits.

We saw in Chapter 9 that the amount of genetic differentiation among populations is often estimated by the fixation index F_{ST}, which is the proportion of the total genetic variation that is due to genetic differentiation among local populations. An analogous measure of population differentiation for quantitative traits has been termed Q_{ST} (Spitze 1993):

$$Q_{ST} = \frac{V_{AB}}{2V_{AW} + V_{AB}} \qquad (11.17)$$

where V_{AB} is the additive genetic variation due to differences among populations, and V_{AW} is the mean additive genetic variation within populations. Q_{ST} is expected to have the same value as F_{ST} if it is estimated from the allele frequencies at the loci affecting the quantitative trait under investigation. This assumes that the local populations are in **Hardy–Weinberg (HW) proportions** ($F_{IS} = 0$) and the loci are in gametic equilibrium.

Q_{ST} is more difficult to estimate than F_{ST} because it requires common garden experiments containing multiple populations to distinguish genetic differences among populations from environmental influences on the trait. It also requires known genetic relationships among individuals within those populations to allow for partitioning within-population variation into genetic and environmental components similar to estimating heritability. These requirements make it difficult to estimate Q_{ST} except in those plant and animal species that can be readily raised in experimental conditions.

For species where Q_{ST} is difficult to estimate, some researchers have used the surrogate P_{ST}, an estimate of the proportion of total phenotypic variation due to additive genetic differences among populations (Raeymaekers et al. 2007; Leinonen et al. 2008):

$$P_{ST} = \frac{cV_B}{cV_B + 2\,H_N V_W} \qquad (11.18)$$

The scalar variable c is the proportion of total phenotypic variation that is estimated to be due to additive genetic effects. Environmental effects can obscure the true amount of quantitative trait divergence if it is estimated from wild phenotypes.

Population divergence can be overestimated in cases where phenotypic variation is primarily a plastic response to the environment, or underestimated in cases where environmental effects result in reduced phenotypic variation, even if genetic divergence is high. A meta-analysis by Leinonen et al. (2008) concluded that studies using P_{ST} do not tend to yield higher estimates than common garden Q_{ST} studies that use estimated additive genetic variation within and among populations, but Brommer (2011) suggests P_{ST} and F_{ST} comparisons should be interpreted very conservatively.

The comparison of F_{ST} and Q_{ST} can provide valuable insight into the effects of natural selection on quantitative traits. F_{ST} is relatively easy to measure for a wide variety of genetic markers that are generally assumed not to be strongly affected by natural selection. Thus, the value of F_{ST} for selectively neutral markers (e.g., randomly selected intergenic SNPs, microsatellites) depends only on local effective population sizes (genetic drift) and dispersal (gene flow) among local populations. Consequently, differences between F_{ST} and Q_{ST} can be attributed to the effects of natural selection on the quantitative trait.

There are three possible relationships between F_{ST} and Q_{ST}. First, if $Q_{ST} > F_{ST}$, then the degree of differentiation in the quantitative trait exceeds that expected by genetic drift alone, and consequently, directional natural selection favoring different phenotypes in different populations must have been involved to achieve this much differentiation. Second, if Q_{ST} and F_{ST} estimates are roughly equal, the observed degree of differentiation for the quantitative trait is the same as expected with genetic drift alone. Finally, if $Q_{ST} < F_{ST}$, the observed differentiation is less than expected on the basis of genetic drift alone, and natural selection must be favoring the same mean phenotype in different populations. While these comparisons are simple qualitatively, making statistically rigorous comparisons is challenging. A single trait provides a single Q_{ST} estimate; yet this must be compared with the distribution of F_{ST} estimates across many loci rather than only to the mean to determine whether the difference between the two is significant (Whitlock 2008; Whitlock & Guillaume 2009).

Laboratory experiments with house mice have supported the validity of the Q_{ST} approach to understand the effects of natural selection on quantitative traits (Morgan et al. 2005). Comparisons of Q_{ST} and F_{ST} in laboratory lines with known evolutionary history generally produced the correct evolutionary inference in the interpretation of comparisons within and between lines. In addition, Q_{ST} was relatively greater than F_{ST} for those traits for which strong directional selection was applied between lines.

11.6 Quantitative genetics and conservation

How should we incorporate quantitative genetic and genomic information into conservation and management programs? Quantitative genetics has not played a major role in conservation genetics to date. Some have argued that quantitative genetic approaches may be more valuable than molecular genetics in conservation, since quantitative genetics allows us to study traits associated with fitness rather than just markers that are neutral or nearly neutral with respect to natural selection (Storfer 1996; Shaw 2019). This might be particularly important in light of anthropogenic climate change (Kramer & Havens 2009; Sgrò et al. 2011; see Chapter 16). Others suggest that quantitative genetic parameters are too difficult to estimate and have errors too large to be useful, and that genome-wide variation can provide a reasonable estimate of evolutionary potential in the absence of knowledge of specific adaptive traits or large-effect loci (Harrisson et al. 2014).

Ideally, we would like to be able to predict the capacity of a population to evolve for quantitative phenotypes and fitness in the longer term. The heritability of a trait and the breeder's equation can predict the short-term response to selection of the same population under the same conditions over one or a few generations, but it varies with the environment as well as with the population under consideration (Bay et al. 2017; Kardos & Luikart 2021).

A few different quantitative genetic metrics have been proposed for predicting the long-term adaptive potential of a population. Quantitative geneticists have long debated the best parameter to use for this purpose. Houle (1992) and Hansen & Houle (2008) proposed that either the **additive coefficient of variation** (CV_A) or its square, called **evolvability**, could be used for this purpose. The additive coefficient of variation is simply the square root of the additive genetic variance scaled by the trait mean $\bar{x}$:

$$CV_A = \frac{\sqrt{V_A}}{\bar{x}} \qquad (11.19)$$

Evolvability (e_μ, sometimes abbreviated I_A) is the additive genetic variance scaled by the phenotypic mean squared.

$$e_\mu = \frac{V_A}{\bar{x}^2} \qquad (11.20)$$

Evolvability can be interpreted as the proportional change expected in a trait mean under a unit strength of selection (Hansen et al. 2011). It can be compared among not just traits, but also among organisms and studies. However, single-trait estimates of evolvability ignore the constraints resulting from genetic correlations with other traits. Hansen et al. (2019) developed an approach to estimating conditional evolvabilities that take into account such constraints.

Evolvability is often expressed as a percentage. Evolvability can show quite different patterns among traits than heritability. Houle (1992) showed that while life history traits tend to have lower heritabilities than morphological trait, this pattern is reversed for evolvability as life history traits have both higher additive genetic variance and higher total phenotypic variance than morphological traits. As a result, heritabilities and evolvabilities are almost uncorrelated.

Shaw (2019) proposed that the best prediction of the capacity for adaptive evolution in a quantitative trait, based on Fisher's Fundamental Theorem of Natural Selection, is:

$$\frac{V_A(W)}{\bar{W}} \qquad (11.21)$$

where W is lifetime fitness as measured by reproduction. Fitness, of course, is a function of many other phenotypic traits. Kulbaba et al. (2019) demonstrated the evolution of this parameter by estimating its change over three consecutive years in three populations of the annual plant *Chamaecrista fasciculata*. The additive genetic variance for fitness varied greatly from year to year. They predicted changes in fitness each generation using

$V_A(W)/\bar{W}$, and showed evidence of potential for evolutionary rescue for two declining populations with mean fitness less than one. They argue that this experimental approach is useful for characterizing adaptive potential in the wild for species where lifetime fitness can be estimated.

Quantitative genetics also provides an invaluable conceptual basis for understanding other factors affecting evolutionary potential in populations. For example, quantitative genetics is essential for understanding inbreeding depression (Chapter 17). If inbreeding depression is caused by a few loci with major effects, then the alleles responsible for inbreeding depression may be "purged" from a population. However, inbreeding depression caused by many loci with small effects will be extremely difficult to purge (see discussion in Section 17.5). For example, in populations of Engelmann, white, and hybrid spruce, 13% of all alleles observed in exome resequencing had a bioinformatically inferred deleterious amino acid substitution (Conte et al. 2017). Most of these alleles were at very low frequency, as would be expected for deleterious alleles. It would not be feasible to purge all of these deleterious alleles from a population.

Similar considerations come into play with such important issues as the loss of evolutionary potential in small populations, and the minimum effective population size required to maintain adequate genetic variation to increase the probability of long-term persistence of threatened populations and species (Chapter 18). Conservation breeding programs are only an option for altering traits and increasing fitness if those traits have nonzero heritability. For example, the American chestnut has negligible genetic variation in resistance to the introduced chestnut blight, and so selection among American chestnuts can't produce resistant phenotypes (Steiner et al. 2017). Instead, avenues being pursued include hybridization with the Chinese chestnut and genetic modification.

Small populations experiencing prolonged bottlenecks will likely eventually lose additive genetic variation. While genetic variation can increase in the short term following a bottleneck, this effect is not expected to increase the adaptive potential of populations in the longer term (Willi et al. 2006). Individuals in small, threatened populations often

have reduced mean fitness due to factors including environmental stress and inbreeding, and the loss of adaptive variation will further contribute to their risk of population extirpation and species extinction.

The application of quantitative genetics to conservation has differed between plants and animals. With plant species, there is a history of using common garden experiments and reciprocal transplant studies to guide conservation activities. Quantitative genetics is used to identify suitable seed source populations for species and ecosystem restoration, quantify how far seed can be moved from a collection population to restoration sites or for genetic rescue without resulting in maladaptation, and evaluate the potential for breeding for resistance to introduced insects and diseases (Hufford & Mazer 2003). This is particularly the case for tree species, but increasingly for other types of plants used in restoration or of conservation concern as well. Quantitative genetic approaches and common garden experiments are being used to predict the capacity for adaptation and impacts of climate change on local populations (Aitken et al. 2008; Kramer & Havens 2009; Guest Box 11). Information on additive and nonadditive genetic variation, heritability and extent of resistance to an introduced pathogen has also been critical for the development of conservation strategies in some cases. For example, whitebark pine and limber pine populations are being inoculated with the introduced fungal pathogen causing white pine blister rust in common gardens to determine the quantitative genetic basis and frequency of resistance (Schoettle & Sniezko 2007). This information has also been used to model the demographic viability of populations under various scenarios to inform conservation strategies (Schoettle et al. 2011).

For most animal species, common garden experiments are not feasible, and obtaining estimates of heritability or Q_{ST} in the wild can be problematic and may not be informative. However, genomic techniques now allow estimation of quantitative genetic parameters using marker-based approaches. Studies of model organisms are generating candidate genes for investigation in natural populations of related species. GWAS approaches are now feasible in the wild (Santure & Garant

2018). Heritability estimates in the wild are improving due to marker-based pedigree reconstruction and use of the animal model. The large standard errors associated with heritability estimates (Storfer 1996) and the statistical challenges of Q_{ST} analysis still make the use of neutral molecular estimates for population diversity and divergence more straightforward for many species. Nevertheless in some cases, estimates of quantitative genetic parameters for specific adaptive traits provide useful information about the capacity of populations to adapt to new abiotic conditions and new biotic challenges.

In captive breeding programs, there is a considerable risk that populations will become adapted to captivity, and as a result, become less adapted to their natural environments (Section 21.5). Captive populations may be subjected to directed artificial selection (e.g., for docility of animals), unconscious selection (e.g., if only some individuals reproduce well in captivity), or selection resulting from the captive environment itself (including lack of predators and ample resources). Breeding strategies can be developed to reduce or avoid these effects of captive breeding on phenotypic traits, including equalizing family sizes or minimizing kinship (Williams & Hoffman 2009).

11.6.1 Response to selection in the wild

Quantitative genetics provides a framework for understanding and predicting the possible effects of selection acting on wild populations (Guest Box 8). For example, many commercial fisheries target particular age or size classes within a population (Box 15.2). In general, larger individuals are more likely to be caught than smaller individuals. The expected genetic effect (i.e., response, R) of such selectivity on a particular trait depends upon the selection differential (S), and the narrow-sense heritability (H_N), as we saw in Equation 11.9. For example, the mean size of pink salmon caught off the coast of North America declined between 1950 and 1974 (12 generations), apparently because of size-selective harvest; ~80% of the returning adult pink salmon were harvested in this period (Ricker 1981). The mean reduction in body weight in 97 populations over these years was ~28%. A size-selective fishery experiment with Atlantic silverside fish reveals the profound changes across the genome

that can result from such a size-selective fishery in Example 11.1 (Therkildsen et al. 2019).

Realized responses to artificial selection in laboratory experiments or in animal or crop breeding programs have generally been close to those predicted by this simple approach (Shaw 2019). However, such predicted responses are often not realized in wild populations (Pujol et al. 2018). This could be due to: (1) oversimplifying assumptions about the genetic control of a trait ignoring nonadditive sources of genetic variation (i.e., dominance and epistasis); (2) not adequately accounting for environmental variation and complexity or possible correlations between genotypes and environments; (3) not adequately accounting for population demography; (4) fluctuating selection pressures over time; or (5) genetic correlations with other traits under selection. In experiments or breeding populations, the experimenter or breeder chooses the traits under selection, whereas in nature, selection may be acting on traits other than those humans choose to study (Shaw 2019).

As we have seen, additive genetic variation is necessary for a response to natural selection. Empirical studies have found that virtually every trait that has been studied has some additive genetic variance (i.e., $H_N > 0$; Roff 2003). Thus, evolutionary change may be slowed by a loss of genetic variation for a trait, but it is not expected to be prevented in most cases. An exception to this has been found in a rainforest species of *Drosophila* in Australia (Hoffmann et al. 2003). These authors found no response to selection for resistance to desiccation after 30 generations of selection! A parent–offspring regression analysis estimated a narrow-sense heritability of zero with the upper 95% confidence value of 0.19. This result is especially puzzling since there are clinal differences between populations for this trait. This suggests there has been a history of selection and response for this trait. Similarly, Baer & Travis (2000) suggested that the lack of genetic variation was responsible for a lack of response to artificial selection for acute thermal stress tolerance in a live-bearing fish.

Genetic correlations among traits can also constrain the response to selection in the wild when there is substantial additive genetic variation for a trait (Section 11.2.2). For example, Etterson & Shaw (2001) studied the evolutionary potential of three populations of a native annual legume in tallgrass prairie fragments in North America to respond to the warmer and more arid climates predicted by global climate models. Despite substantial heritabilities for the traits under selection, between-trait genetic correlations antagonistic to the direction of selection will limit the adaptive evolution of these populations in response to warming. The predicted rates of evolutionary response, taking genetic correlations into account, were much slower than the predicted rate of change with heritabilities alone.

11.6.2 Can molecular genetic variation within populations estimate quantitative variation?

As we saw in Section 11.4, we expect an equivalent loss of molecular heterozygosity and additive genetic variation during a bottleneck. Therefore, heterozygosity may provide a good estimate of the loss of quantitative variation. For example, de Villemereuil et al. (2019) found the hihi, an endangered New Zealand passerine bird, had a lack of genomic diversity in RADseq data, low trait heritability, and negligible additive genetic variation for fitness. Some studies from natural populations have also found that molecular genetic variation supports this result (Example 11.2). However, a meta-analysis by Reed & Frankham (2001) found no relationship between heterozygosity and heritability across 19 studies.

The relationship between molecular and quantitative variation is not so simple. Bottlenecks have been found to increase the heritability for some traits, but will reduce additive genetic variation for others and are always expected to reduce molecular genetic variation. In addition, different types of genetic variation will recover at different rates from a bottleneck because of different mutation rates (Section 12.5). Quantitative traits have a high effective mutation rate because they are polygenic (i.e., a mutation at any one of the causal loci can result in variation for that trait). This means that quantitative genetic variation may recover from a bottleneck more quickly than most molecular variation (Lande 1996; Lynch 1996; Example 11.2).

In addition, even populations with fairly low effective population sizes can maintain enough additive genetic variation for substantial adaptive evolution (Lande 1996). We also expect a weak correlation between molecular and quantitative genetic variation because of statistical sampling. Substantial variation in additive genetic variation among small populations is expected (Lynch 1996). There may also be large differences among quantitative traits in both among- and within-population variation because they will be under different selection pressures, with some under divergent selection resulting in greater differentiation among populations, and some under stabilizing selection with the same phenotype favored in all populations. Therefore, the amount of molecular genetic variation within a population should be used with caution to make inferences about quantitative genetic variation (Example 11.2).

11.6.3 Does population divergence for molecular markers estimate divergence for quantitative traits?

Is F_{ST} a good predictor of adaptive divergence, Q_{ST}? Leinonen et al. (2008) conducted a meta-analysis of published F_{ST} and Q_{ST} estimates. The correlation between the two was positive but weak. Adaptive divergence of populations was better predicted by F_{ST} for species with greater population differentiation; this relationship broke down for low-F_{ST} species, suggesting that neutral markers predict adaptive divergence poorly for widespread species with large populations and high levels of gene flow (e.g., wind-pollinated tree species and marine fish). Differentiation in quantitative traits (Q_{ST}) typically exceeds differentiation at molecular markers (F_{ST}). This suggests a prominent role for natural selection in determining patterns of differentiation at QTLs.

A comparison of quantitative and molecular markers is a useful approach for understanding the role of natural selection and drift in determining patterns of differentiation in natural populations. Nevertheless, caution is needed in making these comparisons and in generalizing their results. Rigorous statistical comparisons are challenging, as previously discussed (Whitlock & Guillaume 2009). It is also somewhat surprising that Q_{ST} almost always exceeds F_{ST} (Leinonen et al. 2008) given that so many quantitative traits seem to be under stabilizing selection. This result suggests that different local populations almost always

Example 11.2 A tale of two pines: lack of molecular variation corresponds to a lack of quantitative variation in red pine but not stone pine

Conifers are generally genetically variable species, with high variation for molecular markers and quantitative traits due to large populations, low F_{ST} values due to high levels of gene flow via pollen, and relatively high Q_{ST} values for some quantitative traits indicating strong adaptation to local climates (Savolainen et al. 2007; see also Example 2.3 for Sitka spruce). Two pines, red pine and stone pine, do not fit this norm. The red pine is a common species with a broad range across eastern North America. The stone pine is a common circum-Mediterranean species well known for its edible seeds, which are a common source of "pine nuts".

Molecular genetic variation is extremely low in red pine. Allozyme studies have found little or no variation (Allendorf et al. 1982; Simon et al. 1986; Mosseler et al. 1991). The estimated expected heterozygosity at the species level was 0.001 at 27 allozyme loci (Allendorf et al. 1982). Some genetic variation has been found in chloroplast microsatellites from red pine (Echt et al. 1998). For example, Walter & Epperson (2001) found genetic variation in 10 chloroplast microsatellite loci in individuals collected throughout the range of red pine. Only six chloroplast haplotypes were found, and 78% of all trees had the same haplotype. Studies have also detected very little genetic variation for quantitative characters in this species (Fowler & Lester 1970).

Stone pine is nearly as genetically depauperate as red pine for nuclear genetic markers. It has an expected heterozygosity for allozymes at the species level of just 0.015 (Fallour et al. 1997). It has even less chloroplast diversity than red pine, with just four haplotypes found using 12 cpDNA microsatellite markers across the species range (Vendramin et al. 2008). Of 34 populations sampled, 29 were fixed for a single haplotype. Furthermore, only 21 SNPs were found in more than 55,000 bp of sequence in total from 77 loci (Jaramillo-Correa et al. 2020). This species has also accumulated a substantial genetic load, with higher frequencies of alleles bioinformatically predicted to be deleterious than the parapatric maritime pine. However, quantitative variation has been found among populations for a number of growth traits in a common garden experiment (Court-Picon et al. 2004), and broad-sense heritability (H_B) has been estimated as 0.17 and 0.20 for cone weight and seed production, respectively, in a grafted clone bank (Vendramin et al. 2008).

Both of these pines likely underwent a severe and prolonged bottleneck associated with glaciation within the past 20,000 years, but the decline of stone pine is estimated to have started much earlier than that, around 1 million years ago (Jaramillo-Correa et al. 2020). There has not been enough time for these species to recover genetic variation for molecular genetic markers. However, stone pine appears to have maintained some quantitative genetic variation through this decline or has recovered some variation through mutations at loci underlying those traits. This illustrates the complementary nature of molecular genetic markers and quantitative traits, and some of the uncertainties around predicting one from the other.

have different optimum phenotypic values. Another interpretation is that the optimum mean value is similar in different populations but that environmental differences result in different combinations of genotypes producing a similar phenotype (e.g., see countergradient selection in Section 2.7.1). This may also result from a bias in the subset of traits selected for study in a given organism. Researchers are more likely to phenotype populations for a trait they suspect is related to local adaptation.

Once again, in the absence of quantitative genetic information, the amount of molecular genetic variation between populations should be used carefully to make inferences about quantitative genetic variation. Substantial molecular genetic divergence between populations suggests some isolation between these populations, and therefore provides strong evidence for the opportunity for adaptive divergence. And it is fair to say that some adaptive differences are likely to occur between populations that have been isolated long enough to accumulate substantial molecular genetic divergence. However, the reverse is not true. Lack of molecular genetic divergence should not be taken to suggest that adaptive differences do not exist. As we saw in Section 9.7, even fairly weak natural selection can have a profound effect on the amount of genetic divergence among populations.

Guest Box 11 How genome-enhanced breeding values can assist conservation of tree populations facing climate warming
Victoria L. Sork

Oaks are the most speciose and abundant tree genus of northern hemisphere forest ecosystems (Cavender-Bares 2016; Kremer & Hipp 2020). In California, the iconic valley oak (Figure 11.18), which occupies oak savannah and riparian forests, has already experienced significant losses of landcover since the arrival of the Europeans. For the surviving populations, their ability to survive future climate change will depend on the extent to which they are locally adapted to the current climate environment. In 2012, United States Forest Service geneticist Jessica Wright and I led the effort to establish two common gardens starting with 11,000 acorns collected from 94 locations all over California. Our research teams have compared phenotypes and growth of trees collected from different locations to assess the extent of genetic differences that might reflect local adaptation. We were particularly concerned about the ability of oaks to grow in warmer climates over the next 50 years (Sork et al. 2010).

We discovered a surprising result through an analysis of relative growth rates of 8-year-old trees in our common gardens: valley oak populations are poorly adapted to the current climate (Browne et al. 2019). In fact, they would grow better if their local climates were cooler than what they now experience. This observed lag in climate adaptation could be due to factors that delay the response to climate warming over the past 20,000 years, such as the long life span of oaks, or to competing selective pressures from biotic factors such as pathogens. This lag presents a serious challenge to future oak populations because a reduction in growth rate can seriously affect the fitness of trees by reducing seed production, as acorn reproduction is tied to tree size. Furthermore, reduction in growth rates could also reduce the carbon sequestration services of our forests.

In California, restoration of oak habitat damaged by fires or prior landscape modification often requires the replanting of trees. The question is where to source seed for those plantings. We used genomic sequencing data to model the genomic basis of high growth rate phenotypes through genomic-estimated breeding values (GEBVs: the sum of the effects of genome-wide markers capturing variation in a target trait). We first used a GWAS framework to estimate

Figure 11.18 Valley oak tree growing in an oak–savannah ecosystem located at Sedgwick Reserve, California, USA. Valley oak is the largest oak in California, with some individuals growing to over 30 m tall. Valley oaks are foundation trees and provide habitat and food for a variety of animals, including squirrels, birds, deer, and insects. Valley oaks provide services to humans too, filtering water and providing shady places to escape the heat. The reference genome based on DNA from this tree has allowed the identification of genetic markers used for quantitative genetic and conservation genomic studies of this species and other related oaks. Photo courtesy of Andy Lentz.

genotype-by-environment interactions of relative growth rates with maximum temperature (T_{max}) difference across 12,357 SNPs. We calculated GEBVs based on the maternal genotype of each planted tree. We then identified genetic variants that are associated with positive relative growth rates at warmer temperatures. Under realistic emissions scenarios, our models predicted that, by 2070, average temperatures in the state are projected to be up to 4.8°C degrees warmer than they were a century ago. We found that the predicted changes in growth rates of local populations with existing GEBVs would mostly be negative throughout the species range except at higher elevations and in northern regions (Figure 11.19). This study indicates that climate change can negatively impact trees, whether or not it causes mortality, by creating further maladaptation to temperature through a reduction in growth rates.

The task is then to identify appropriate seed sources that could be "pre-adapted" to future climate conditions. If seed sources are selected based on GEBVs, trees in a much larger portion of the species range should show increased growth (Figure 11.19). These findings demonstrate that assisted gene flow (Aitken & Whitlock 2013; Chapter 16) based on a breeding value approach could result in tree populations that are more likely to thrive as climate change continues to warm California. This study, which is unique in the integration of common garden and genomic data, needs to be continued as the trees age but its early results demonstrate the potential of genome-assisted management of tree populations. Eventually, the goal is to replace common garden experiments, which can take 10–20 years, with more easily available genomic data so that effective conservation plans can be developed over a shorter time frame.

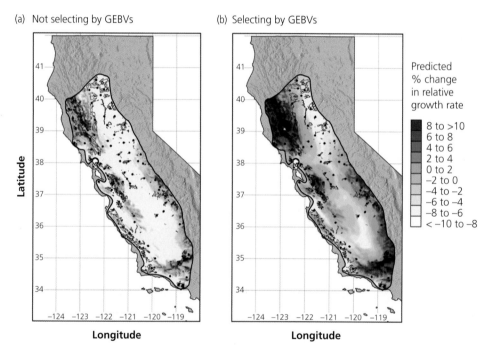

Figure 11.19 Landscape distribution of genomic-estimated breeding values (GEBVs) and predicted changes in relative growth rate for valley oak across the species range (black outline of colored region indicates contemporary valley oak range). (a) Predicted changes in relative growth rates by 2070–2099 under a business-as-usual emissions scenario (RCP 8.5) based on current distribution of GEBVs across valley oak populations. GEBVs were estimated to capture polygenic variation in relative growth rates across temperature differences estimated from a common garden experiment. Black circles indicate sampled localities. (b) Predicted changes in relative growth rates by 2070–2099 for a scenario where individuals with the highest GEBV within 25 km of each planting site are used as a seed source. From Brown et al. (2019).

CHAPTER 12

Mutation

Minke whale, Example 12.2

Mutation is the ultimate source of all the genetic variation necessary for evolution by natural selection; without mutation evolution would soon cease.

(Michael C. Whitlock & Sarah P. Otto 1999, p. 295)

Mutations can critically affect the viability of small populations by causing inbreeding depression, by maintaining potentially adaptive genetic variation in quantitative characters, and through the erosion of fitness by accumulation of mildly deleterious mutations.

(Russell Lande 1995, p. 782)

Mutations are errors in the transmission of genetic information from parents to progeny. The process of mutation is the ultimate source of all genetic variation in natural populations. Nevertheless, this variation comes at a cost because most mutations that have phenotypic effects are harmful (deleterious). Mutations occur both at the chromosomal level and the molecular level. As we will see, mutations may or may not have a detectable effect on the phenotype of individuals.

An understanding of the process of mutation is important for conservation for several reasons. The amount of **standing genetic variation** within populations is largely a balance between the gain of genetic variation from mutations and the loss of genetic variation from genetic drift. Thus, an understanding of mutation is needed to interpret patterns of genetic variation observed in natural populations.

Moreover, the increased homozygosity of deleterious mutations is the primary source of **inbreeding depression** (Chapter 17). The frequency of deleterious mutations results from a balance between mutation and natural selection (Section 12.3). We have seen that natural selection is less effective in small populations (Section 8.6). Therefore, deleterious mutations will tend to accumulate more rapidly in small populations; this effect can further threaten the persistence of small populations (Section 18.6). On the other hand, the rate of adaptive response to environmental change is proportional to the amount of standing genetic variation for fitness within populations. Thus, long-term persistence of populations may require large population sizes in order to maintain important adaptive genetic variation.

Unfortunately, there is little empirical data available about the process of mutation because mutations are rare. The data that are available generally come from model organisms (e.g., mice, *Drosophila*, or *Arabidopsis*) that are selected because of their short generation times and suitability for raising a large number of individuals in the laboratory. However, we must be careful in generalizing results from such model species; the very characteristics that make these organisms suitable for these experiments, including short generation lengths, may make them less suitable for generalizing to other species.

How common are mutations? On a per-locus or per-nucleotide level they are rare. For example, the rate of mutation for a single nucleotide in plants and animals ranges from 10^{-7} to 10^{-10} per year, and 10^{-7} to 10^{-9} per generation (Lynch et al. 2016). This

Conservation and the Genomics of Populations, Third Edition. Fred W. Allendorf, *et al.*, Oxford University Press.
© Fred W. Allendorf, W. Chris Funk, Sally N. Aitken, Margaret Byrne, and Gordon Luikart (2022). DOI: 10.1093/oso/9780198856566.003.0012

means that only one in a billion or so gametes, on average, will have a mutation at a specific base pair. Mutation rates vary considerably among species (Figure 12.1; Hanlon et al. 2019). Species with longer generation lengths on average have lower mutation rates per year, but also have higher rates per generation. However, from a genome-wide perspective, mutations are actually very common. The genome of most species consists of billions of base pairs. Therefore, it has been estimated (Lynch et al. 1999) that each individual may possess hundreds of new mutations! Fortunately, almost all of these mutations are in nonessential regions of the genome and have no phenotypic effect.

In this chapter, we consider the processes resulting in mutations and examine the expected relationships between mutation rates and the amount of genetic variation within populations. We examine evidence for both harmful and advantageous mutations in populations. Finally, we examine the effects of mutation rates on the rate of recovery of genetic variation following a population bottleneck.

12.1 Process of mutation

Chromosomes and DNA sequences are normally copied exactly during the process of replication and are transmitted to progeny. However, sometimes errors occur that produce new chromosomes or new DNA sequences. Empirical information on the rates of mutation is hard to come by because mutations are so rare. It used to be necessary to study thousands of progeny to detect mutational events. Thus, estimating the rates of mutations or describing the types of changes brought about by mutation was generally very difficult. Most of our direct information about the process of mutation until recently came from model organisms (Example 12.1). However, next-generation sequencing has allowed estimation of mutation rates in pedigrees

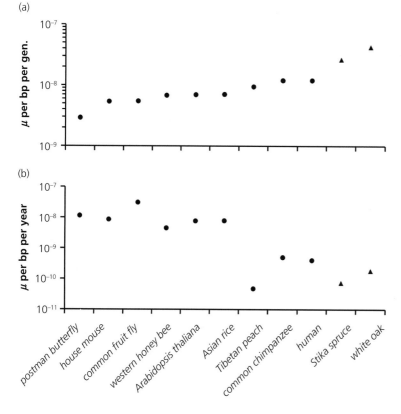

Figure 12.1 Mutation rates (a) per generation and (b) per year for multicellular species. Per-year rates were calculated using a reported age, an estimate of generation time, or an average of two estimates of generation time. For the trees, the somatic mutation rate was estimated in exceptionally long-lived specimens (triangles). Modified from Hanlon et al. (2019).

Example 12.1 Coat color mutation rate in mice

Schlager & Dickie (1971) presented the results of a direct and massive experiment to estimate the rate of mutation to five recessive coat color alleles in mice: albino, brown, nonagouti, dilute, and leaden. They examined more than 7 million mice in 28 inbred strains. Overall, they detected 25 mutations in over 2 million gene transmissions for an average mutation rate of 1.1×10^{-5} per gene transmission. As expected, the **reverse mutation rate** (from the recessive to the dominant allele) was much lower, $\sim 2.5 \times 10^{-6}$ per gene transmission. The reverse mutation rate is expected to be lower because there are more ways to eliminate the function of a gene than to reverse a defect. This assumes that the recessive coat color mutations are caused by mutations that result in a loss of function.

lineages of vegetative cells; for example, flowers produced at the top of the plant or ends of branches. In long-lived or large organisms such as trees, somatic mutations are a substantive source of variation. Ally et al. (2010) found evidence that somatic mutations in long-lived aspen clones reduce fertility and may eventually lead to senescence.

Plomion et al. (2018) compared whole genome sequences of three different branch tips from a single mature white oak tree, and found 46 somatic mutations. They were also able to demonstrate the transmission of nine of these mutations to progeny by sequencing acorns. Hanlon et al. (2019) used targeted sequence capture to sequence about 10 megabases of the exome from the bottom and the top of 20 exceptionally tall, old-growth Sitka spruce trees (see back cover). While they found few mutations, the resulting estimated rate of somatic mutations (ignoring mutations during meiosis) translated into one of the highest mutation rates found for a eukaryote (Figure 12.1). One old-growth Sitka spruce could have on the order of 100,000 different somatic mutations scattered throughout its crown, although only a very small subset could be passed on through pollen or seed. Despite the high number of somatic mutations detected, the rate per year and per meter of height growth is lower than expected given their size, suggesting that the stem cells in tree apical meristems may have evolved mechanisms to reduce mutations, or do

from nonmodel species such as wolves and flycatchers, both of which reported mutation rates of 4.6×10^{-9} mutations per site per generation (Smeds et al. 2016; Koch et al. 2019).

Mutations can occur in germline cells, and these gametic mutations occur primarily during meiosis. In animals, only gametic mutations result in genetic variation passed on to progeny. However, in plants, mutations during mitosis in somatic cells can result in variation that can be passed on because reproductive structures are produced by

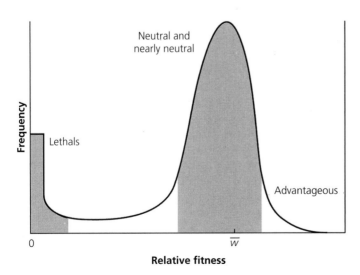

Figure 12.2 Hypothetical frequency of the fitness of new mutations relative to the mean fitness of the population ($\overline{w}$). Redrawn from Hedrick (2011).

not themselves divide as frequently as previously thought (Orr et al. 2020).

Most mutations with phenotypic effects tend to reduce fitness (Figure 12.2). Thus, as we will see in Chapter 18, the accumulation of mutations can decrease the probability of survival of small populations. Nevertheless, rare beneficial mutations are important for adaptive evolutionary change (Elena et al. 1996). In addition, work with the plant *Arabidopsis thaliana* has suggested that nearly half of new spontaneous mutations increase fitness (Shaw et al. 2002); however, this result has been questioned in view of data from other experiments (Bataillon 2003; Keightley & Lynch 2003). Work in *Escherichia coli* shows that intermediate amounts of mutation were beneficial for adaptation to novel environments, but this effect was not present at high mutation rates (Sprouffske et al. 2018).

Mutations occur randomly, but there is evidence that some aspects of the process of mutation may be an adaptive response to environmental conditions. There has been an ongoing controversy that mutations in prokaryotes may be directed toward particular environmental conditions (Lenski & Sniegowski 1995). There is some evidence that the rate of mutations in eukaryotes may increase under stressful conditions and thus create new genetic variability that may be important in adaptation to changing environmental conditions (Capy et al. 2000). However, the relationship between stress and mutation is complex, and stress does not necessarily increase mutation rate (Ferenci 2019). In addition, there is little evidence for an influence of the environment on the effects of new mutations (Halligan & Keightley 2009).

12.1.1 Chromosomal mutations

We saw in Chapter 3 that rates of chromosomal evolution vary tremendously among different taxonomic groups. There are two primary factors that may be responsible for differences in the rate of chromosomal change: (1) the rate of chromosomal mutation, and (2) the rate of incorporation of such mutations into populations (Rieseberg 2001). Differences among taxa in rates of chromosomal

change may result from differences in either of these two effects.

White (1978) estimated a general mutation rate for chromosomal rearrangements of the order of one per 1,000 gametes in a wide variety of species from lilies to grasshoppers to humans. Lande (1979) considered different forms of chromosomal rearrangements in animals and produced a range of estimates between 10^{-4} and 10^{-3} per gamete per generation. There is evidence in some groups that chromosomal mutation rates may be substantially higher than this. For example, Porter & Sites (1987) detected spontaneous chromosomal mutations in five of 31 male lizards that were examined.

The apparent tremendous variation in chromosomal mutation rates suggests that some of the differences among taxa could result from differences in mutation rates. In addition, there is some evidence that chromosomal polymorphisms may contribute to increased chromosomal mutation rates. That is, chromosomal mutation rates may be greater in chromosomal heterozygotes than homozygotes (King 1993). We will see in Section 12.1.4 that genomes with more **transposable elements** may also have higher chromosomal mutation rates.

12.1.2 Molecular mutations

There are several types of **molecular mutations** in DNA sequences: (1) substitutions, the replacement of one nucleotide with another; (2) recombinations, the exchange of a sequence from one homologous chromosome to the other; (3) deletions, the loss of one or more nucleotides; (4) insertions, the addition of one or more nucleotides; and (5) inversions, the rotation by $180°$ of a double-stranded DNA segment of two or more base pairs (Graur & Li 2000).

Mutation rates are sometimes estimated indirectly by an examination of rates of substitutions over evolutionary time in regions of the genome that are not affected by natural selection. The expected rate of substitution per generation will be equal to the mutation rate for selectively neutral mutations (Kimura 1983). Using this approach, the average rate of mutation in mammalian nuclear DNA has been estimated to be $3-5 \times 10^{-9}$ nucleotide substitutions per base pair per year (Graur & Li 2000), which is not far off estimates from short-term experiments (Figure 12.1). However, the substitution rate varies enormously among different regions of the nuclear genome.

The rate of mutation also differs between genomes. In mammalian mtDNA, mutation rate has been estimated to be at least 10 times higher than the average nuclear rate (Brown et al. 1982), while the rate of mutation in the plant mitochondrial genome is estimated to be 40–100 times slower in plants than in animals (Palmer 1992). In angiosperm plants, the mutation rate of the chloroplast genome is estimated to be five times slower than the nuclear rate and the mitochondrial rate is 16 times slower than that in the nuclear genome (Drouin et al 2008). The chloroplast genome has a slower mutation rate than the mitochondrial genome in protists (Smith 2015). However, these are general overall rates and different regions have different rates. For example, genes and spacer regions in the chloroplast genome show quite variable rates of variation making them useful for estimation of relationships at different taxonomic levels (Shaw et al. 2014).

Whole genome sequencing has made estimating mutation rate much more straightforward. Rates can be estimated through mutation accumulation experiments with model organisms (Lynch et al. 2016). In these experiments, isogenic (genetically identical) individuals are used to start multiple lineages over multiple generations with extreme bottlenecks produced by either clonal reproduction of one individual or mating of two individuals within lines. Mutations accumulate in each lineage. At the end of the experiment, individuals from each lineage are whole genome sequenced, and mutations identified. From this, the mutation rate can be calculated. **Single nucleotide polymorphisms (SNPs)** are generally bi-allelic and follow the infinite-sites mutation model (Morin et al. 2004). The per base pair per generation rate of all organisms is $<10^{-7}$, but it is as low as 10^{-9} or 10^{-10} in some unicellular eukaryotes and bacteria. Lynch et al. (2016) hypothesize that lower mutation rates have been able to evolve in these unicellular organisms due to their very large effective population sizes.

Table 12.1 Mutations at the *OGO1c* microsatellite locus in pink salmon (Steinberg et al. 2002). Approximately 1,300 parent–progeny transmissions were observed in 50 experimental matings. Mutations were found only in the four matings shown. The mutant allele is indicated by bold-face type and the most likely progenitor of the mutant allele is underlined. All of the putative mutations differ by one repeat unit from their most likely progenitor. The overall mutation rate estimated from these data is 3.9×10^{-3} (5/1,300).

Dam	Sire	Progeny genotypes				
a/b	c/d	a/c	a/d	b/c	b/d	Mutant genotypes
342/350	408/4<u>74</u>	1	1	3	3	342/**478**
295/366	303/<u>362</u>	1	2	4	2	295/**366**
269/420	346/<u>450</u>	8	16	10	8	420/**446** (2)
348/348	309/<u>448</u>	5	4	0	0	348/**444**

The rate of mutation at microsatellite loci is much greater than other regions of the genome because of the presence of simple sequence repeats (Li et al. 2002). Two mechanisms are thought to be responsible for mutations at microsatellite loci: (1) mispairing of DNA strands during replication, and (2) recombination. Estimates of mutation rates at microsatellite loci have generally been approximately 10^{-3} mutants per generation (Ellegren 2000a; Table 12.1). Microsatellite mutations appear largely to follow the stepwise mutation model (SMM) where single repeat units are added or deleted with near equal frequency (Valdes et al. 1993; Figure 12.3). However, the actual mechanisms of microsatellite mutation are more complicated than this simple model (Estoup & Angers 1998; Li et al. 2002; Anmarkrud et al. 2008).

The mutation rate for protein coding loci (e.g., allozymes) is relatively low. In addition, not all DNA mutations will result in a change in the amino acid sequence because of the inherent redundancy of the genetic code. Nei (1987, p. 30) reviewed the literature on direct and indirect estimates of mutation rates for allozyme loci. Most direct estimates of mutation rates in allozymes have failed to detect any mutant alleles; for example, Kahler et al. (1984) examined a total of 841,260 gene transmissions from parents to progeny at five loci and failed to detect any mutant alleles. General estimates of mutation rates for allozyme loci are on the order of 10^{-6} to 10^{-7} mutants per gene transmission (Nei 1987).

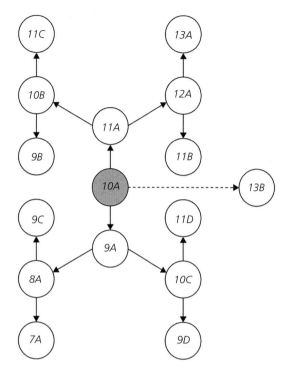

Figure 12.3 Pattern of mutation for microsatellites beginning with a single ancestral allele (shaded circle in the middle) with 10 repeats. Most mutations are a gain or loss of a single repeat (stepwise mutation model, SMM). The dashed arrow shows a multiple-step mutation from 10 to 13 repeats. Alleles are designated by the number of repeats and a letter that distinguishes homoplastic alleles, which are alike in state (number of repeats) but differ in origin.

New mutations sometimes occur in clusters and have been associated with hotspots for crossing over during meiosis (Chan & Gordenin 2015). Woodruff & Thompson (1992) found as many as 20% of new mutations in *Drosophila* represented clusters of identical mutant alleles sharing a common pre-meiotic origin. Cluster mutations at microsatellite loci have been found in several other species (Jones et al. 1999; Steinberg et al. 2002). The occurrence of clustered mutations results in nonuniform distributions of novel alleles in a population that could influence interpretations of mutation rates and patterns as well as estimates of genetic population structure. For example, Woodruff et al. (1996) have shown that mutant alleles that are part of clusters are more likely to persist and be fixed in a population than mutant alleles entering the population independently.

12.1.3 Quantitative characters

As we saw in Chapter 11, the amount of genetic variation in quantitative characters for morphology, physiology, and behavior that can respond to natural selection is measured by the additive genetic variance (V_A). The rate of loss of additive genetic variance due to genetic drift in the absence of selection is the same for the loss of heterozygosity (i.e., $1/2N_e$). The effective mutation rate for quantitative traits is much higher than the rate for single gene traits because a mutation at any one of the many loci underlying a polygenic trait can affect a quantitative trait. The input of additive genetic variance per generation by mutation is V_m. The expected genetic variance at equilibrium between these two factors is $V_A = 2N_e V_m$ (Lande 1995, 1996).

Estimates of mutation rates for quantitative characters are rare and imprecise. It is thought that V_m is roughly on the order of 10^{-3} V_A (Lande 1995). However, some experiments suggest that the great majority of these mutations are highly detrimental and therefore are not likely to contribute to the amount of standing genetic variation within a population. Thus, the effective V_m responsible for much of the standing variation in quantitative traits in natural populations may be an order of magnitude lower, 10^{-4} V_A (Lande 1996; Barton & Keightley 2002; Mackay 2010).

12.1.4 Transposable elements, stress, and mutation rates

Much of the genome of eukaryotes consists of sequences associated with transposable elements that possess an intrinsic capability to make multiple copies and insert themselves throughout the genome (Graur & Li 2000). Transposable elements were first discovered by Nobel Laureate Barbara McClintock, who studied these "jumping genes" in maize (McClintock 1950; Ravindran 2012). Approximately half of the human genome consists of DNA sequences associated with transposable elements (Lynch 2001). This activity is analogous to the "cut and paste" mechanism of a word processor. Transposable elements are potent agents of **mutagenesis** (Kidwell 2002). For example, Clegg & Durbin (2000) found that nine of 10 mutations affecting flower color in the morning glory were the result of the insertion of transposable elements into genes. A consideration of the molecular basis of transposable elements is beyond our consideration (see Chapter 7 of Graur & Li 2000). Nevertheless, the mutagenic activity of these elements is of potential significance for conservation.

Transposable elements can cause a wide variety of mutations. They can induce chromosomal rearrangements such as deletions, duplications, inversions, and reciprocal translocations. Kidwell (2002, p. 2219) has suggested that "transposable elements are undoubtedly responsible for a significant proportion of the observed karyotypic variation among many groups." In addition, transposable elements are responsible for a wide variety of substitutions in DNA sequences, ranging from insertion of the transposable element sequence to substitutions, deletions, and insertions of a single nucleotide (Kidwell 2002).

Stress has been defined as any environmental change that drastically reduces the fitness of an organism (Hoffmann & Parson 1997). McClintock (1984) first suggested that transposable element activity could be induced by stress. A number of transposable elements in plants have been shown to be activated by stress (Grandbastien 1998; Capy et al. 2000). Some transposable elements in *Drosophila* have been shown to be activated by heat stress, but other studies have not found an effect of heat shock (Capy et al. 2000). In addition, hybridization has also been found to activate transposable elements and cause mutations (Kidwell & Lisch 1998).

12.2 Selectively neutral mutations

Many mutations in DNA sequence have no phenotypic effect so that they are neutral with regard to natural selection (e.g., mutations in noncoding regions). In this case, the amount of genetic variation within a population will be a balance between the gain of variation by mutation and the loss by genetic drift. The distribution of neutral genetic variation among populations is primarily a balance between these two forces. Gene flow among subpopulations retards the process of differentiation until eventually a steady state may be reached

between the opposing effects of gene flow and genetic drift. However, the process of mutation may also contribute to allele frequency divergence among populations in cases where the mutation rate approaches the gene flow rate.

12.2.1 Genetic variation within populations

The amount of genetic variation within a population at equilibrium will be a balance between the gain of variation as a function of the neutral mutation rate (μ) and the loss of genetic variation by genetic drift as a function of effective population size (N_e).

We will first consider the so-called infinite allele model (IAM) in which we assume that every mutation creates a new allele that has never been present in the population. This model is appropriate if we consider variation in DNA sequences. Even for a relatively short DNA sequence, there are a very large number of possible allelic states, as each nucleotide site can be occupied by one of four bases (A, T, C, or G). For example, there are over 1 million possible alleles if we just consider 10 base pairs (bp) ($4^{10} = 1,048,576$).

The average expected heterozygosity (H) at a locus (or over many loci with the same mutation rate) is:

$$H = \frac{4N_e\mu}{(4N_e\mu + 1)} = \frac{\theta}{\theta + 1} \qquad (12.1)$$

where μ is the neutral mutation rate and $\theta = 4N_e\mu$ (Kimura & Crow 1964). The multiple parameter $4N_e\mu$ is an important expression in population genetics theory and is called the **population-scaled mutation rate** (θ). This relationship can be used to estimate effective population size if we know the mutation rate (Example 12.2).

The much greater variation at microsatellite loci compared with SNPs and allozymes results from the differences in mutation rates that we discussed in Section 12.1.2. Figure 12.4 shows the equilibrium heterozygosity for microsatellites, allozymes, and 100 bp of DNA using mutation rates of 10^{-4} and 10^{-6}, respectively, and Equation 12.1. Thus, we expect a heterozygosity of 0.038 at allozyme loci (or for a 100-bp sequence) and 0.80 at microsatellite

---

Example 12.2 How many whales are there in the ocean?

Equation 12.1 can also be used to estimate the effective population size of natural populations if we know the mutation rate (μ). For example, Roman & Palumbi (2003) estimated the historical (pre-whaling) number of humpback, fin, and minke whales in the North Atlantic Ocean by estimating θ for the control region of **mitochondrial DNA (mtDNA)**. In the case of mtDNA, $\theta = 2N_{e(f)}\mu$ because of maternal inheritance and haploidy. Roman & Palumbi (2003) used a range of mutation rates based on observed rates of divergence between mtDNA of different whale species.

Their genetic estimates of historical population sizes for humpback, fin, and minke whales were far greater than those previously calculated, and are 6–20 times higher than the current population estimates for these species. This discrepancy is crucial for conservation because the International Whaling Commission management plan uses the estimated historical population sizes as guidelines for setting allowable harvest rates. We should be careful using estimates of N_e with this approach because there are a host of pitfalls (e.g., how reliable are our estimates of mutation rate?). Roman & Palumbi (2003) provide a useful discussion of the limitations of this method for estimating N_e. Palsbøll et al. (2013) have provided an insightful critique of using this method to estimate historical population sizes.

---

loci with an effective population size of 10,000. However, we also expect a substantial amount of variation in heterozygosity among loci, especially loci with lower mutation rates (Figure 12.5).

The heterozygosity values for microsatellite loci in Figures 12.4 and 12.5 are likely to be overestimates because of several important assumptions in this expectation. Microsatellite mutations tend to occur in steps of the number of repeat units. Therefore, each mutation will not be unique, but rather will be to an allelic state (say 11 copies of a repeat) that already occurs in the population. This is called

homoplasy in which two alleles that are identical in state have different origins (e.g., alleles *11C* and *11D* in Figure 12.3). Therefore, the actual expected heterozygosity is less than predicted by Equation 12.1. Allozymes also tend to follow a stepwise model of mutation (Ohta & Kimura 1973), but homoplasy will be less common because of the smaller number of alleles present in a population due to the lower mutation rate.

The type of point mutation that occurs when a single nucleotide is replaced by another depends on whether a purine (A or G) is replaced by a

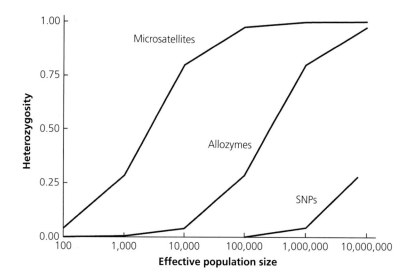

Figure 12.4 Expected heterozygosity in populations of different size using Equation 12.1 for microsatellites ($\mu = 10^{-4}$), allozymes ($\mu = 10^{-6}$), and SNPS ($\mu = 10^{-8}$).

(a)

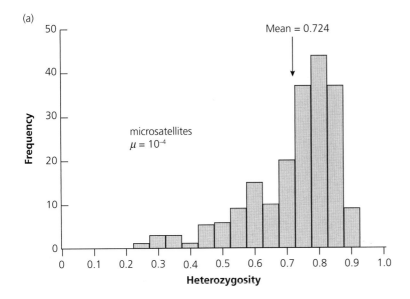

(b)

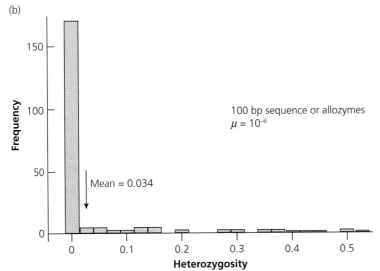

Figure 12.5 Simulated heterozygosities at 200 loci in a population with N_e = 10,000 and the IAM of mutation produced with the program *EASYPOP* (Balloux 2001). (a) Microsatellite loci with $\mu = 10^{-4}$; (b) Allozyme loci with $\mu = 10^{-6}$. A 100-bp DNA sequence would be expected to have a similar distribution to allozymes. The expected heterozygosities are (a) 0.800 and (b) 0.038 (Equation 12.1).

purine or by a pyrimidine (G or C), and vice versa. Mutations replacing a purine with a purine or a pyrimidine with a pyrimidine are called transition mutations. Mutations replacing a purine with a pyrimidine, or vice versa, are called transversion mutations. Although there are twice as many possible transversions as transitions, transitions are more common because of the similarity of the chemical structures substituted.

It is also important to remember that the mutation rates used here (μ) are the neutral mutation rates. Mutations in DNA sequence within some regions of the genome are likely not to be selectively neutral. Therefore, different regions of the genome will have different effective neutral mutation rates, even though the actual rate of molecular mutations may be the same. For example, mutations in protein coding regions may affect the amino acid sequence of an essential protein and thereby reduce fitness. Transversion mutations within coding regions are more likely to result in amino acid substitutions than transitions. Such mutations will not be neutral and therefore will not contribute to the amount of variation maintained by our model of drift–mutation equilibrium considered here. In these regions, so-called **purging selection** stops these mutations reaching high frequencies in a population. In contrast, mutations in DNA sequence in regions of the genome that are not functional are much more likely to be neutral. This expectation is supported by empirical results; exons, which are the coding regions of protein loci, are much less variable than the introns, which do not encode amino acids (Graur & Li 2000).

12.2.2 Population subdivision

The process of mutation may also contribute to allele frequency divergence among populations (Ryman & Leimar 2008). The relative importance of mutation on divergence (e.g., F_{ST}) depends primarily upon the relative magnitude of the rates of migration and of mutation. Under the IAM of mutation with the island model of migration (Crow & Aoki 1984), the expected value of F_{ST} at equilibrium is approximately:

$$F_{ST} = \frac{1}{(1 + 4Nm + 4N\mu)} \qquad (12.2)$$

However, greater mutation rates will increase F_{ST} when new mutations are not dispersed at sufficient rates to attain equilibrium between genetic drift and gene flow. Under these conditions, new mutations may drift to substantial frequencies in the population in which they occur before they are distributed among other populations via gene flow (Neigel & Avise 1993). Divergence for markers like SNPs and allozymes with relatively low mutation rates is unlikely to be affected by differences in mutation rates unless the subpopulations are completely isolated.

In general, mutations will have an important effect on population divergence only when the migration rates are very low (say 10^{-3} or less) and the mutation rates are unusually high, as for microsatellite markers (10^{-3} or greater; Nichols & Freeman 2004; Epperson 2005; Wang 2015). However, as we saw in Section 9.8, F_{ST} will underestimate genetic divergence at loci with very high within-deme heterozygosities (H_S; Hedrick 1999). Large differences in H_S caused by differences in mutation rates among loci (e.g., Steinberg et al. 2002) can result in discordant estimates of F_{ST} and G_{ST} among microsatellite loci. This may result in an underestimation of both the degree of genetic divergence among populations if all loci are pooled for analysis and the estimation of F_{ST} (Olsen et al. 2004b). See Section 9.8.2 and Wang (2015) for a solution to this problem.

We saw in the previous section that long-term N_e can be estimated using the amount of heterozygosity in a population if we know the mutation rate. However, we also know from Chapter 9 that the amount of gene flow affects the amount of genetic variation in a population. Therefore, estimates of N_e using Equation 12.1 may be overestimates because they reflect the total N_e of a series of populations connected by gene flow rather than the N_e of the local population (Guest Box 7). Consider two extremes. In the first, a population on an island is completely isolated from the rest of the members of its species ($mN = 0$). In this case, estimates of N_e using Equation 12.1 will reflect the local N_e. In the other extreme, a species consists of a number of local populations that are connected by substantial gene flow (say $mN = 100$); in this case the estimates of N_e using Equation 12.1 will reflect the combined N_e of all populations (Table 12.2).

12.3 Harmful mutations

Most mutations that affect fitness have a detrimental effect (Figure 12.2). Natural selection acts to keep these mutations from increasing in frequency. Consider the joint effects of mutation and selection at a single locus with a normal allele (A_1) and a mutant allele (A_2) that reduces fitness as shown below:

A_1A_1	A_1A_2	A_2A_2
1	$1 - hs$	$1 - s$

Table 12.2 Estimates of effective population size (N_e) with computer simulations using Equation 12.1 in a series of 20 subpopulations (local N_e = 200) that are connected by different amounts of gene flow with an island model of migration (EASYPOP, Balloux 2001). A mutation rate of 10^{-4} was used to simulate the expected heterozygosities at 100 microsatellite loci. The simulations began with no genetic variation in the first generation and ran for 10,000 generations. $F_{ST}*$ is the expected F_{ST} with this amount of gene flow corrected for a finite number of populations (Mills and Allendorf 1996). $\widehat{N}_e$ is the estimated effective population size based upon the mean expected local heterozygosity (H_S) using Equation 12.1.

mN	H_T	H_S	F_{ST}	$F_{ST}*$	$\widehat{N}_e$
0	0.814	0.076	0.907	1.000	205
0.5	0.665	0.477	0.283	0.311	2,274
1.0	0.635	0.516	0.187	0.184	2,667
2.0	0.621	0.558	0.100	0.101	3,156
5.0	0.618	0.592	0.041	0.043	3,630
10.0	0.606	0.594	0.020	0.022	3,665

where s is the reduction in fitness of the homozygous mutant genotype and h is the degree of dominance of the A_2 allele. A_2 is recessive when $h = 0$, dominant when $h = 1$, and partially dominant when h is between 0 and 1.

If the mutation is recessive ($h = 0$), then at equilibrium:

$$q^* = \sqrt{\frac{\mu}{s}} \qquad (12.3)$$

When A_2 is partially dominant, q will generally be very small and the following approximation holds:

$$q^* \approx \frac{\mu}{hs} \qquad (12.4)$$

See Lynch et al. (1999) for a consideration of the importance of mildly deleterious mutations in evolution and conservation.

12.4 Advantageous mutations

Genetic drift plays a major role in the survival of advantageous mutations even in extremely large populations. That is, most advantageous mutations will be lost during the first few generations because new mutations will always be rare. The initial frequency of a mutation will be one over the total number of gene copies at a locus (i.e., $q = 1/2N$). Even a greatly advantageous allele that is recessive will have the same probability of initial persistence in a population because the advantageous homozygotes will not occur until the allele happens to drift to a relatively high frequency. For example, a new mutation will have to drift to a frequency over 0.20 before even 5% of the population will be homozygotes with the selective advantage. Therefore, the great majority of advantageous mutations that are recessive will be lost.

Dominant advantageous mutations have a much greater chance of surviving the initial period because their fitness advantage will immediately be effective in heterozygotes that carry the new mutation. However, even most dominant advantageous mutations will be lost within the first few generations because of genetic drift. For example, over 80% of dominant advantageous mutations with a selective advantage of 10% will be lost within the first 20 generations (Crow & Kimura 1970, p. 423). This effect can be seen in a simple example. Consider a new mutation that arises that increases the fitness of the individual that carries it by 50%. Even if the individual that carries this mutation contributes three progeny to the new generation, there is a 0.125 probability that none of the progeny will carry the mutation because of the vagaries of Mendelian segregation ($0.5 \times 0.5 \times 0.5 = 0.125$).

Gene flow and spread of globally advantageous mutations may be an important cohesive force in evolution (Rieseberg & Burke 2001). Ehrlich & Raven (1969) argued in a classic paper that the amounts of gene flow in many species are too low to prevent substantial differentiation among subpopulations by genetic drift or local adaptation so that local populations are essentially independently evolving units in many species. We saw in Chapter 9 that even one migrant per generation among subpopulations can cause all alleles to be present in all subpopulations. However, even much lower amounts of gene flow can be sufficient to cause the spread of an advantageous allele (say $s >$ 0.05) throughout the range of a species (Rieseberg & Burke 2001). The rapid spread of such advantageous alleles may play an important role in maintaining

Box 12.1 Detection of bottlenecks with the heterozygosity excess test

There are a variety of tests available to detect past population bottlenecks (see Peery et al. 2012). For example, we saw in Example 6.2 with a graphical qualitative approach that the absence of rare alleles in a population sample indicates the effects of a recent bottleneck. The heterozygosity excess test is a quantitative test of this same effect based upon Equation 12.1. As we have seen, heterozygosity at selectively neutral loci results from an equilibrium between mutation and genetic drift (Equation 12.1). The expected heterozygosity also can be calculated from the observed number of alleles and the sample size of individuals, assuming neutrality and mutation–drift equilibrium (Cornuet & Luikart 1996). In nonbottlenecked populations that are near mutation–drift equilibrium, the expected heterozygosity equals the measured Hardy–Weinberg (HW) heterozygosity. But if a population has experienced a recent bottleneck, the mutation–drift equilibrium is transiently disrupted and the heterozygosity measured at a locus will exceed the heterozygosity computed from the number of alleles sampled (Maruyama & Fuerst 1985). That is, bottlenecks generate a "heterozygosity excess" because alleles are generally lost faster than heterozygosity during a bottleneck (Section 6.4).

For example, Ramstad et al. (2013) tested for bottlenecks using the heterozygosity excess test with 15 microsatellite loci in four human-founded island populations of little spotted kiwi, including the Long Island population in Example 6.4. Little spotted kiwi were once found throughout the North and South Islands of New Zealand, but they declined rapidly in the 1800s, and were extinct on the North Island by 1900 owing to introduced predators and an enormous trade in their skins for export to Europe. They remained common on the South Island until the early 1900s, but they then declined rapidly owing to predation from stoats, cats, and dogs, and were virtually extinct by the 1980s. They were saved from extinction by translocation of five birds from the South Island of New Zealand to Kapiti Island in 1912. The Kapiti Island population now numbers over 1,000 birds and has provided founders for several new populations. All four of these populations displayed significant heterozygosity excess ($P < 0.025$; Figure 12.6).

The difference between an excess of heterozygotes and heterozygosity excess has sometimes been confused in the literature. The former compares the number of observed heterozygotes with that expected under HW proportions given the allele frequencies of the population. The latter compares mean expected heterozygosity observed in a population relative to that expected of a nonbottlenecked population with the same number of alleles under mutation–drift equilibrium.

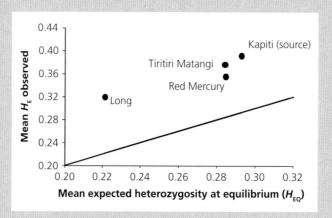

Figure 12.6 Relationship between mean expected heterozygosity observed (H_E) and expected at mutation–drift equilibrium (H_{EQ}) at 15 microsatellite loci for four island populations of little spotted kiwi. The line represents equality between H_E and H_{EQ}. All four populations displayed significant heterozygosity excess ($P < 0.025$), with the most extreme signal in the sample from Long Island. From Ramstad et al. (2013).

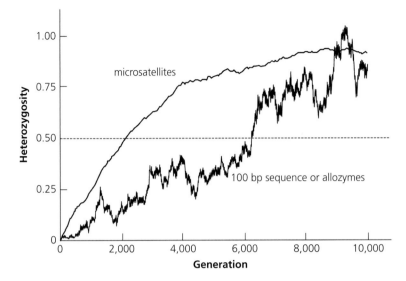

Figure 12.7 Simulated recovery of heterozygosity at 100 loci in a population of 5,000 individuals following an extreme bottleneck using *EASYPOP* (Balloux 2001). The initial heterozygosity was zero. The mutation rates are 10^{-4} for microsatellites and 10^{-6} for a 100-bp DNA sequence or for allozymes. Heterozygosity is standardized as the mean heterozygosity over all 100 loci divided by the expected equilibrium heterozygosity using Equation 12.1 (0.670 and 0.020, respectively).

the genetic integration of subpopulations connected by small amounts of genetic exchange.

12.5 Recovery from a bottleneck

We have seen that population bottlenecks will have a greater effect on allelic diversity than on heterozygosity. This effect can be used to detect the presence of previous bottlenecks (Box 12.1). The rate of recovery of genetic variation from the effects of a bottleneck will depend primarily on the mutation rate (Lynch 1996). The equilibrium amount of neutral heterozygosity in natural populations (Equation 12.1) will be approached at a time scale equal to the shorter of $2N_e$ or $1/(2\mu)$ generations (Kimura & Crow 1964).

Mutation rates affect how quickly genetic diversity will accumulate following a strong bottleneck. Microsatellites have a mutation rate of ~10^{-3} per generation, while a 100-bp DNA sequence or an allozyme will have mutation rates of ~10^{-6} per generation. Simulations of 100 typical loci show the expected heterozygosity at microsatellite loci returned to 50% of that expected at equilibrium after 2,000 generations in populations of 5,000 individuals (Figure 12.7). It took approximately three times as long at the loci with mutation rates of

10^{-6} per generation. In this case, $1/(2\mu)$ is 5,000 generations for microsatellites and 100 times that for the other markers. However, $2N_e$ is 10,000 for all types of markers.

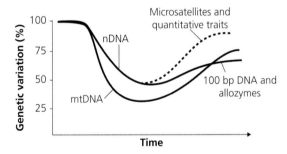

Figure 12.8 Diagram of relative expected effects of a severe population bottleneck on different types of genetic variation. The smaller N_e for mtDNA causes more genetic variation to be lost during a bottleneck. The rate of recovery following a bottleneck is largely determined by the mutation rate (Section 12.5).

As we saw in Section 12.1.3, the estimated mutation rate (V_m) for phenotypic characters affected by many loci (quantitative characters) is much higher than expected for a single gene. Therefore, we would expect quantitative genetic variance for phenotypic characters to be restored at rates comparable with those of microsatellites (Lande 1996).

Thus, recovery of microsatellite variation following a severe bottleneck may be a good measure of the recovery of polygenic variation for fitness traits.

Figure 12.8 provides a simplistic representation of the effects of a severe population bottleneck on different sources of genetic variation. Microsatellites, allozymes, and quantitative traits are all expected to lose genetic variation at approximately the same rates. However, mtDNA will lose genetic variation more rapidly because of its smaller N_e. The rates of recovery of variation will depend upon the mutation rates for these different sources of genetic variation.

Guest Box 12 Mutation, inbreeding depression, and adaptation
Philip W. Hedrick

Mutation is the original source of genetic variation, whether it be advantageous variants, neutral variants, or detrimental variants. Most mutations have virtually no effect on fitness or a slightly detrimental one and, as a result, the majority of mutations are lost due to the large chance effects of genetic drift for alleles at low frequency. However, there is an important group of mutations that have a higher deleterious impact on fitness and another group which have an advantageous impact on fitness. Both these latter categories of mutations are particularly important in conservation because deleterious variants are the cause of inbreeding depression (Hedrick & Garcia-Dorado 2016) and advantageous variants are the basis of future adaptation.

Let us first examine an example of deleterious variation and the genetic basis of inbreeding depression from a genomic survey of 28 offspring produced by selfing from a single eucalyptus tree in a species (*Eucalyptus grandis*) that ordinarily does not self-fertilize (Hedrick et al. 2016). In this parental tree, 9,560 SNPs were heterozygous and were examined in the progeny. If there were no selection, 50% would be expected to be homozygous for one of the two parental alleles, or identical by descent, at each of these loci, and 50% of the progeny would be expected to

be heterozygous. However, in these 28 progeny, only 34% of these markers were homozygous and 66% were heterozygous, a difference from expectations that was present on all 11 chromosomes (Myburg et al. 2014).

As an example of these results, Figure 12.9 gives the observed proportion of the three genotypes for each of the 1,019 SNPs along chromosome 1. Except for a short region on the far right end of this chromosome, the proportion of the two homozygotes combined is much less than 0.5 and averages around 30%, while the proportion of heterozygotes is much greater than 0.5 and averages around 70%. These effects appear to be the result of strong selection at many genes along the chromosome that cause high mortality when made homozygous by one generation of self-fertilization. The greatest reduction in homozygosity is expected for markers that are closely linked to genetic mutants with high detrimental effects. It is likely that mutations at more than 100 genes over the genome, many with a substantial effect on viability, are contributing to inbreeding depression in this example.

Mutation importantly is also the source of new genetic variation for future adaptation. An example is in the flightless Galápagos cormorant, only first discovered in 1898, and

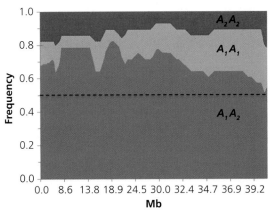

Figure 12.9 The observed proportions of heterozygotes (blue) and the two homozygotes (red and green) for 1,019 SNPs along chromosome 1 for 28 progeny produced by self-fertilization from a single parent in *Eucalyptus grandis*. The broken line gives the expected proportion of heterozygotes in the absence of selection. From Hedrick et al. (2016).

present only on the two youngest and most western of the Galápagos Islands (Hedrick 2019). The Galápagos cormorant is the only flightless cormorant of the ~40 recognized species of cormorants. Figure 12.10 shows the body and wing structure of the flightless Galápagos cormorant on the right and the closely related flighted double-crested cormorant on the left. Note that as well as rudimentary wings, about one-third of the size expected for a cormorant of that size, the flightless cormorant has a much larger body than the double-crested cormorant.

The western shores of the two most western Galápagos Islands are quite nutrient rich and appear to be the only suitable Galápagos environments that provide abundant and reliable food for the Galápagos cormorant. It is assumed that flightless cormorants descended from a flighted cormorant, colonized this new niche, and subsequently adapted

to it by losing the ability to fly and becoming highly adapted to foraging in this environment where there are no native mammalian predators.

In an effort to determine the genetic mutants resulting in flightlessness, Burga et al. (2017) identified function-altering variants at multiple genes, none of which were found in related flighted cormorants. These variants are at genes that influence limb development and bone growth, cause genetic deformities in humans, and are all probably new mutations. All of the variants they identified at 11 different genes were fixed for the potentially function-altering variants in the flightless cormorants. In other words, mutations at multiple genes in the ancestor of the flightless cormorant, which were probably detrimental for a cormorant with flight, became adaptive and were fixed in this new environment.

Figure 12.10 An illustration of a double-crested cormorant on the left and the flightless cormorant from the Galápagos Islands on the right showing the rudimentary wings and larger body size of the flightless cormorant. From Hedrick (2019); drawing by Katie Bertsche.

PART III

Evolutionary Response to Anthropogenic Changes

CHAPTER 13

Hybridization

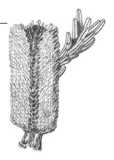

Banksia, Section 13.3

Hybridization, with or without introgression, frequently threatens populations in a wide variety of plant and animal taxa because of various human activities.

(Judith M. Rhymer & Daniel Simberloff 1996, p. 103)

Genomic data provide a special opportunity to characterize the history of hybridization and the genetic basis of speciation.

(Bret A. Payseur & Loren H. Rieseberg 2016, p. 2337)

Rates of **hybridization** and **introgression** have increased dramatically worldwide because of widespread intentional and incidental translocations of organisms and habitat modifications by humans. Hybridization has contributed to the extinction of many species through direct and indirect means (Levin et al. 1996; Allendorf et al. 2001). The severity of this problem has been underestimated by conservation biologists (Rhymer & Simberloff 1996). The increasing pace of the three interacting human activities that contribute most to increased rates of hybridization (introductions of plants and animals, fragmentation, and habitat modification) suggests that this problem will become even more serious in the future (Kelly et al. 2010). For example, increased turbidity in Lake Victoria, Africa, has reduced color perception of cichlid fishes and has interfered with the mate choice that produced reproductive isolation among species (Seehausen et al. 1997). Similarly, increased turbidity because of land development and forest harvesting has led to increased hybridization among stickleback species in British Columbia, Canada (Wood 2003).

On the other hand, hybridization is a natural part of the evolutionary process. Hybridization has long been recognized as playing an important role in the evolution of plants (Rieseberg 2011; Figure 13.1). In addition, recent studies have found that hybridization has also played an important role in the evolution of animals (Taylor & Larson 2019). Several reviews have emphasized the creative role that hybridization may play in adaptive evolution and speciation (e.g., Grant & Grant 1998; Seehausen 2004) and many examples show that hybridization can facilitate rapid evolutionary change (Arnold et al. 2012). Many early conservation policies generally did not allow protection of hybrids. However, increased appreciation of the important role of hybridization as an evolutionary process has caused a re-evaluation of these policies. Determining whether hybridization is natural or anthropogenic is crucial for conservation, but is often difficult (Allendorf et al. 2001).

Hybridization provides an exceptionally tough set of problems for conservation biologists (Ellstrand et al. 2010). The issues are complex

Conservation and the Genomics of Populations, Third Edition. Fred W. Allendorf, *et al.*, Oxford University Press.
© Fred W. Allendorf, W. Chris Funk, Sally N. Aitken, Margaret Byrne, and Gordon Luikart (2022). DOI: 10.1093/oso/9780198856566.003.0013

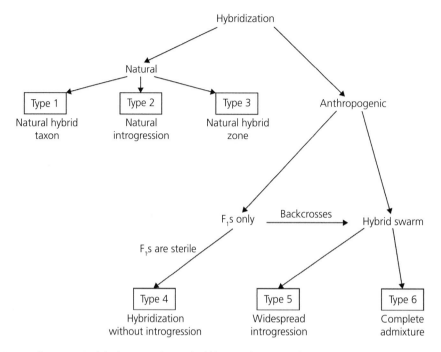

Figure 13.1 Framework to categorize hybridization. Each type should be viewed as a general descriptive classification used to facilitate discussion rather than a series of strict, all-encompassing divisions. Types 1–3 represent hybridization events that are a natural part of the evolutionary legacy of taxa; these taxa should be eligible for protection. Types 4–6 divide anthropogenic hybridization into three categories that have different consequences from a conservation perspective. From Allendorf et al. (2001).

and controversial, beginning with the seemingly simple task of even defining hybridization (Harrison 1993). Hybridization is often used to refer to interbreeding between species (e.g., Grant & Grant 1992b). However, we consider that this taxonomically restrictive use of hybridization can be problematic (especially since it is sometimes difficult to agree on what is a species). We have adopted the more general definition of Harrison (1990) that includes matings between "individuals from two populations, or groups of populations, which are distinguishable on the basis of one or more heritable characters."

The term "hybrid" itself sometimes has a negative connotation in conservation. In the USA, a proposed policy to treat hybrids and hybridization under the US Endangered Species Act (ESA) used the term "intercross" (suggested by John Avise) and "intercross progeny" rather than hybrid to avoid the connotations of the term hybrid (USFWS and NOAA 1996a).

In this chapter, we first present and discuss genetic methods for detecting and evaluating hybridization. We next consider the role that natural hybridization has played in the process of evolution. We then consider the possible harmful effects of anthropogenic hybridization. Finally, we consider several key aspects of hybridization with regard to conservation.

13.1 Detecting and describing hybridization

Early studies on the detection of hybrid individuals relied upon morphological characteristics until the mid-1960s. However, not all morphological variation has a genetic basis, and the amount of morphological variation within and among populations is often greater than recognized (Campton 1987; Payseur & Rieseberg 2016). The detection of hybrids using morphological characters generally assumes that hybrid individuals will be phenotypically intermediate to parental individuals (Smith 1992). This is often not the case because hybrids can be highly variable in phenotype across hybrid classes as

Example 13.1 Hybridization between the threatened Canada lynx and bobcat

The Canada lynx is a wide-ranging felid that occurs in the boreal forests of Canada and Alaska (Schwartz et al. 2004). The southern distribution of native lynx extends into the northern contiguous USA from Maine to Washington State. The Canada lynx is listed as Threatened under the US ESA. Canada lynx are elusive animals and their presence routinely has been detected by genetic analysis of **mitochondrial DNA (mtDNA)** from hair and fecal samples (Mills et al. 2001). In 2001, a trapper was prosecuted for trapping a lynx. The trapper thought it was a bobcat, while the biologist registering the pelt and the enforcement officer processing the case thought it was a lynx. Initial analysis showed the sample had the mitochondrial DNA of a lynx. However, recognizing that mitochondrial DNA could only determine the female parent of the cat, Schwartz et al. (2004) designed an assay that could detect hybridization between bobcats and lynx. Hybridization between these species had never been confirmed in the wild.

The controversial sample was a hybrid. In addition, one of the other samples from a carcass and one hair sample collected on a putative lynx backcross were also identified as a hybrid using microsatellite analysis. The hybrids were identified as having one lynx-diagnostic allele and one bobcat-diagnostic allele (Figure 13.2). A heterozygote with one allele from each parental species is expected in an F_1 hybrid (although some F_2 hybrids will also be heterozygous for species-diagnostic alleles at some loci). The species-diagnostic alleles were identified by analyzing microsatellites in 108 lynx and 79 bobcats across North America, far away from potential hybridization zones between the two species. In addition, mtDNA analysis revealed that all hybrids had lynx mothers (i.e., lynx mtDNA). Further analysis of samples from Maine and New Brunswick showed four additional hybrids. However, screening of hundreds of lynx samples from the Rocky Mountains (another area where the two species co-occur) revealed no hybrids there.

These data have important conservation implications. First, bobcat trapping is legal, while it is illegal to trap lynx anywhere in the contiguous USA. The trapping of bobcats in areas where Canada lynx are present could be problematic because both lynx and lynx–bobcat hybrids can be incidentally taken from extant populations. On the other hand, any factors that may favor bobcats in lynx habitat may lead to the production of hybrids and thus be potentially harmful to lynx recovery. Efforts need to be undertaken to describe the extent, rate, and nature of hybridization between these species, and to understand the ecological context in which hybridization occurs.

well as having variable fitness (Payseur & Rieseberg 2016). Furthermore, individuals from hybrid swarms (populations in which all individuals are hybrids by varying numbers of generations of backcrossing with parental types and mating among hybrids) that contain a large proportion of their genes from one of the parental taxa may be morphologically indistinguishable from that parental taxon (Leary et al. 1996; Brisbin & Peterson 2007). Morphological characteristics often do not allow determination of whether an individual is a first-generation hybrid (F_1), a backcross, or a later-generation hybrid. These distinctions are crucial in conservation situations to enable effective management of hybrid situations involving rare species. For example, if a population has not become a hybrid swarm and still contains a reasonable number of parental individuals, it could potentially be recovered by removal of hybrids or by a captive breeding program.

Detection of hybridization is becoming much easier through the application of various molecular techniques, particularly with the advent of genomics. While improved molecular data can be collected with relative ease, interpreting the evolutionary significance of hybridization and determining the role of hybrid populations in developing conservation plans remain difficult. According to one review: "It is an understatement to say that hybridization is a complex business!" (Stone 2000, p. 354).

13.1.1 Diagnostic loci

The use of molecular genetic markers greatly improves the identification and characterization of hybridized populations. Genetic analysis of hybrids and hybridization has generally been based on loci at which the parental taxa have different allele frequencies. **Diagnostic loci** that are

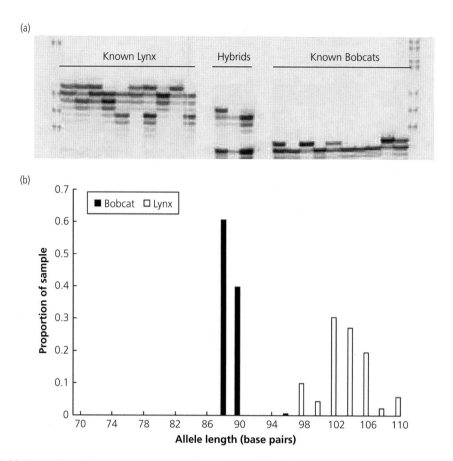

Figure 13.2 (a) Microsatellite gel image showing genotype profiles (locus *Lc106*) for 10 lynx, 10 bobcat, and three putative hybrids. Dark bands represent alleles; lighter bands are "stutter" bands. Outer lanes show size standards. (b) Allele frequencies for locus *Lc106* in bobcat and lynx. This locus is diagnostic because the allele size ranges do not overlap between species. From Schwartz et al. (2004).

fixed or nearly fixed for different alleles in two hybridizing populations have been most useful, although hybridization can also be detected using multiple loci at which the parental types differ in allele frequency (Cornuet et al. 1999). Example 13.1 demonstrates the use of diagnostic loci to identify F_1 hybrids between Canada lynx and bobcats.

Trout species occur throughout the rivers and streams of North America, and interspecific hybridization is common. The examination of genotypes at many diagnostic loci can be used to understand the consequences of hybridization between species (Figure 13.3). First-generation hybrids (F_1s) will be heterozygous for alleles from the parental taxa at all diagnostic loci (interspecific heterozygosity). Later-generation hybrids may result either from matings between hybrids

or backcrosses between hybrids and one of the parental taxa. The absence of genotypes expected in later-generation hybrids suggests that the F_1 hybrids are sterile or have reduced fertility. The following two cases of hybridization with different pairs of trout species demonstrate the contrasting results depending upon whether or not the F_1 hybrids are fertile or sterile.

The loss of native cutthroat trout by hybridization with introduced rainbow trout has been recognized as a major threat for over 100 years in the western USA (Allendorf & Leary 1988). The westslope cutthroat trout is one of eight major subspecies of cutthroat trout (Allendorf & Leary 1988). The geographical range of westslope cutthroat trout is the largest of all cutthroat trout subspecies and Westslope cutthroat trout are genetically highly

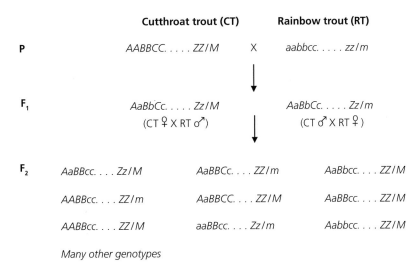

Figure 13.3 Outline of genotypic analysis of hybridization. Alleles present in cutthroat trout at diagnostic nuclear loci are designated by capital letters and the alleles in the rainbow trout by lower-case letters. The parental (P) mtDNA haplotypes are designated by M and m, respectively. Two types of F$_1$ individuals are produced, one with the cutthroat trout mtDNA haplotype and one with the rainbow trout mtDNA haplotype.

divergent at both nuclear and mitochondrial genes from the other subspecies.

Hybridization of westslope cutthroat and rainbow trout generally results in the formation of random mating populations in which all individuals are hybrids by varying numbers of generations of backcrossing with parental types and mating among hybrids (i.e., hybrid swarms). In addition, the Yellowstone cutthroat trout has been widely introduced outside of its native range and has hybridized with other cutthroat trout subspecies, including westslope cutthroat trout. Table 13.1 shows genotypes at eight diagnostic nuclear loci between native westslope cutthroat trout and Yellowstone cutthroat trout introduced into Forest Lake, Montana, in a representative sample of 15 individuals. All but one of these 15 fish are homozygous for both westslope and Yellowstone alleles at different loci. Each individual in this sample appears to be a later-generation hybrid. Thus, the fish in this lake are a hybrid swarm.

In contrast, hybridization between endangered bull trout and introduced brook trout does not produce hybrid swarms, and mostly only F$_1$ individuals are found. Bull trout are legally protected as threatened in the USA under the US

ESA. Hybridization with introduced brook trout is potentially one of the major threats to the persistence of bull trout. Table 13.2 shows genotypes at eight diagnostic nuclear loci between native bull trout and brook trout introduced into Mission Creek, Montana, in a sample of 15 individuals that were selected to be genetically analyzed because they appeared to be hybrids. Eleven of the 15 fish in this sample contained alleles from both species, indicating that they are indeed hybrids. However, in striking contrast to the hybrid swarm shown in Table 13.1, 10 of the 11 hybrids were heterozygous at all eight loci, suggesting that they are F$_1$ hybrids. It is extremely unlikely that a later-generation hybrid would be heterozygous at all loci. For example, there is a 0.50 probability that an F$_2$ hybrid will be heterozygous at a diagnostic locus. Thus, there is a probability of $(0.50)^8 = 0.004$ that an F$_2$ individual will be heterozygous at all eight loci.

The F$_1$ hybrids in this sample have both bull and brook trout mtDNA, indicating that both reciprocal crosses produced hybrids (Figure 13.3). This general pattern has been seen throughout the range of bull trout. Almost all hybrids appear to be first-generation (F$_1$) hybrids with very little evidence of F$_2$ or backcross individuals (Kanda et al. 2002).

Table 13.1 Genotypes at mtDNA and eight diagnostic nuclear allozyme loci in a sample of native westslope cutthroat trout, Yellowstone cutthroat trout, and their hybrids from Forest Lake, Montana (Allendorf & Leary 1988). Heterozygotes are *WY*, while individuals homozygous for the westslope cutthroat trout allele are indicated as *W*, and individuals homozygous for the Yellowstone cutthroat trout allele are indicated as *Y*. All individuals in this sample are later-generation hybrids; thus, the fish in this lake are a hybrid swarm.

No.	mtDNA	Nuclear encoded loci							
		Aat1	*Gpi3*	*Idh1*	*Lgg*	*Me1*	*Me3*	*Me4*	*Sdh*
1	Y	W	W	WY	W	W	W	W	Y
2	Y	W	WY	WY	WY	Y	W	WY	Y
3	W	WY	Y	Y	W	Y	WY	Y	WY
4	W	Y	W	WY	WY	W	Y	W	WY
5	Y	Y	Y	Y	WY	WY	WY	Y	Y
6	Y	WY	Y	W	WY	W	W	W	Y
7	W	WY	WY	Y	W	WY	W	W	W
8	W	WY	Y	WY	WY	Y	W	Y	Y
9	W	Y	Y	WY	WY	W	WY	WY	W
10	W	WY	Y	WY	WY	WY	Y	W	Y
11	Y	Y	W	W	WY	W	Y	W	Y
12	W	W	WY	Y	WY	W	WY	WY	Y
13	Y	W	Y	W	Y	W	WY	W	W
14	Y	Y	Y	WY	WY	WY	WY	WY	W
15	W	WY	Y	WY	Y	W	Y	WY	W

Table 13.2 Genotypes at mtDNA and eight diagnostic nuclear loci in a sample of native bull trout, brook trout, and their hybrids from Mission Creek, Montana (data from Kanda et al. 2002). Heterozygotes are *LR*, while individuals homozygous for the bull trout allele are indicated as *L*, and individuals homozygous for the brook trout allele are indicated as *R*. Individuals are identified in the Status column as bull trout (BL), brook trout (BR), or hybrids on the basis of their genotype.

No.	mtDNA	Nuclear encoded loci								Status
		Aat1	*Ck-A1*	*IDDH*	*sIDHP-2*	*LDH-A1*	*LDH-B2*	*MDH-A2*	*sSod-1*	
1	L	LR	LR	LR	LR	LR	LR	LR	LR	F$_1$
2	L	LR	LR	LR	LR	LR	LR	LR	LR	F$_1$
3	L	R	R	LR	LR	LR	LR	LR	R	F$_1$ × BR
4	L	L	L	L	L	L	L	L	L	BL
5	L	LR	LR	LR	LR	LR	LR	LR	LR	F$_1$
6	R	LR	LR	LR	LR	LR	LR	LR	LR	F$_1$
7	L	LR	LR	LR	LR	LR	LR	LR	LR	F$_1$
8	R	LR	LR	LR	LR	LR	LR	LR	LR	F$_1$
9	R	LR	LR	LR	LR	LR	LR	LR	LR	F$_1$
10	L	LR	LR	LR	LR	LR	LR	LR	LR	F$_1$
11	R	LR	LR	LR	LR	LR	LR	LR	LR	F$_1$
12	R	LR	LR	LR	LR	LR	LR	LR	LR	F$_1$
13	R	R	R	R	R	R	R	R	R	BR
14	R	R	R	R	R	R	R	R	R	BR
15	R	R	R	R	R	R	R	R	R	BR

The near absence of progeny from hybrids of bull and brook trout might result from the sterility of the hybrids, their lack of mating success, the poor survival of their progeny, or combinations of these factors. Over 90% of the F$_1$ hybrids are male, suggesting some genetic incompatibility between these two genomes.

13.1.2 Using many single nucleotide polymorphism loci to detect hybridization

Population assignment based upon large numbers of **single nucleotide polymorphism (SNP)** loci provides a powerful approach to identify admixture and ancestry of putative hybrid individuals,

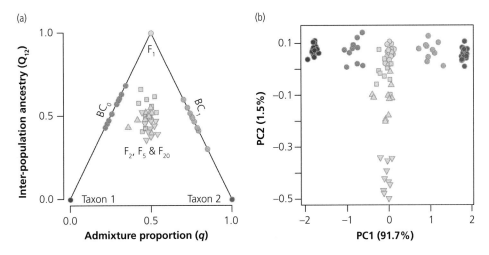

Figure 13.4 Representation of simulated identification of parental and hybrid generations based on level of admixture from parental taxa. (a) Plot of admixture proportion (hybrid index) versus interpopulational ancestry (interpopulational heterozygosity) showing separation of parental individuals of taxa 1 and 2 at the base, F_1 hybrids at the apex, backcross hybrids along the axes, and F_2, F_5, and F_{20} hybrids in the center. (b) Principal components analysis of genetic variation among individuals showing separation of parental, F_1, backcross, and F_2 hybrid individuals on the PC1 axis and separation of advanced generation hybrids (F_2, F_5, and F_{20}) on the PC2 axis. From Gompert & Buerkle (2016).

particularly beyond the F_1 stage. These techniques are useful even when the loci examined are not diagnostic and the putative parental populations are not available to provide baseline information.

The identification of F_1, F_2, backcross, and advanced hybrid individuals enables characterization of the patterns of hybridization and interpretation of hybridization dynamics (Figure 13.4). Detection of hybrid individuals often focuses on use of hybrid indices or admixture analyses that identify proportions of parental ancestry in individuals and can be powerful when combined with simulation approaches (Gompert & Buerkle 2016). A hybrid index, or admixture proportion, based on proportion of alleles inherited from each parental species is often used to identify likely hybrid classes of individuals across a hybrid zone, based on a continuum of hybridity. In a hybrid index, one parent has an index value of 0 and the other 1, F_1s have an index of 0.5, and backcrosses to the parents have indices of around 0.25 and 0.75. Information on hybrid admixture can be further refined by considering the hybrid index in conjunction with levels of interspecific heterozygosity where the parents have values of 0, F_1s have a value

of 1, and later-generation hybrids have intermediate values.

There are now many other analytical approaches that can be used to characterize the often complex dynamics of hybridization across many scenarios. Payseur & Rieseberg (2016) summarize these approaches in relation to analyses of rate and timing of gene flow and ancestry of loci. Understanding the extent and timing of gene flow is critical for differentiating between hybridization following secondary contact and speciation with gene flow (Payseur & Rieseberg 2016).

13.1.3 Gametic disequilibrium

The distribution of gametic disequilibrium (D) between pairs of loci is helpful to describe the distribution of hybrid genotypes and to estimate the "age" of hybridized populations. Recently hybridized populations will have high D because they will contain parental types and many F_1 hybrids. By contrast, hybrid swarms that have existed for many generations will have genotypes that are randomly associated among loci. This will occur rather quickly for unlinked loci because D

will decay by one-half each generation (Equation 10.4). However, nonrandom association of alleles at different loci will persist for many generations at pairs of loci that are closely linked. Barton (2000) has provided single measures of gametic disequilibrium that can provide a meaningful measure to compare the amount of gametic disequilibrium at a number of unlinked loci in hybrid swarms.

13.2 Natural hybridization

Consideration of the role of hybridization in systematics and evolution goes back to Linnaeus and Darwin (see discussion in Arnold 1997, p. 6). Botanical and zoological workers have tended to focus on the two opposing aspects of hybridization. Botanists have generally accepted hybridization as a pervasive and important aspect of evolution (e.g., Stebbins 1959; Grant 1963). They demonstrated that many plant taxa have hybrid origins and that hybridization is an important mechanism for the production of new species and novel adaptations (Mallet 2007). In contrast, early evolutionary biologists working with animals were very interested in the evolution of reproductive isolation leading to speciation (Mayr 1942; Dobzhansky 1951). They emphasized that hybrid offspring were often relatively unfit, and that this led to the development of reproductive isolation and eventually speciation.

Analyses of whole genomes are now showing that hybridization has played a much greater role in evolution than had been recognized previously (Taylor & Larson 2019). For example, genomic analyses indicate some 2–3% of the genomes of modern European and Asian humans originated from Neandertals (Petr et al. 2019). Melanesians and aboriginal Australians trace some 3–4% of their DNA to introgression from Denisovans (Reich et al. 2010). These findings suggest that the ancestors of modern humans hybridized with Neandertals and Denisovans some tens of thousands of years ago (Green et al. 2010).

Recent genomic analysis has shown that interspecific hybridization has played a widespread role in the evolution of canids (Gopalakrishnan et al. 2018). These authors used whole genome resequencing of 48 individuals representing almost all extant canid species. They found the lowest genome-wide heterozygosity in the Ethiopian wolf, African hunting dog, and dhole; these species all have small population sizes. Most interesting, however, they found evidence of several hybridization events between species (Figure 13.5). Specifically, they found genomic regions that indicated gene flow between the ancestors of the dhole and African hunting dog and the gray wolf, coyote, golden jackal, and African golden wolf.

13.2.1 Intraspecific hybridization

Intraspecific hybridization in the form of gene flow among populations has several important effects. It has traditionally been seen as the cohesive force that holds species together as units of evolution (Mayr 1963). This view was challenged by Ehrlich & Raven (1969), who argued that the amount of gene flow observed in many species is too low to prevent differentiation through genetic drift or local adaptation.

The resolution to this conflict is the recognition that even very small amounts of gene flow can have a major cohesive effect. We saw in Chapter 9 that an average of one migrant individual per generation with the island model of migration is sufficient to make it likely that all alleles will be found in all populations. That is, populations may diverge quantitatively in allele frequencies, but qualitatively the same alleles will still be present. We saw in Chapters 9 and 12 that just one migrant per generation can greatly increase the local effective population size.

Rieseberg & Burke (2001) have presented a model of species integration that considers the effects of the spread of selectively advantageous alleles. They have shown that new mutations that have a selective advantage will spread across the range of a species much faster than selectively neutral mutations with low amounts of gene flow. They have proposed that it is the relatively rapid spread of highly advantageous alleles that holds a species together as an integrated unit of evolution.

In contrast, high gene flow can reduce fitness and restrict the ability of populations to adapt to local conditions. Genetic swamping occurs when

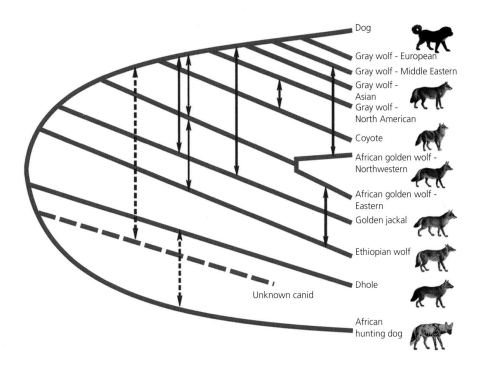

Figure 13.5 Phylogeny of seven canid species based upon complete genome sequences (the dog and gray wolf are the same species). Hybridization events are indicated with red arrows, and dotted red arrows show possible gene flow events detected but not previously reported. From Gopalakrishnan et al. (2018).

gene flow causes the loss of locally adapted alleles or genotypes (Lenormand 2002). This effect may be greatest in populations in low density areas where there is high gene flow from more densely populated areas (García-Ramos & Kirkpatrick 1997). In such cases, the continued immigration of locally unfit genotypes reduces the mean fitness of a population and potentially could lead to what has been called a **hybrid sink** effect. This is a self-reinforcing process in which immigration produces hybrids that are unfit, which then reduces local density and increases the immigration rate (Lenormand 2002).

Hybridization sometimes produces phenotypes that are extreme or outside the range of either parental type. This has been called **transgressive segregation**. Rieseberg et al. (2003) have shown that sunflower species that are found in extreme habitats tend to be ancient interspecific hybrids. They argue that new genotypic combinations resulting from hybridization have led to the ecological divergence

and success of these species. A review of many hybrid species concluded that transgressive phenotypes are generally common in plant populations of hybrid origin (Rieseberg et al. 1999).

13.2.2 Interspecific hybridization

Hybridization and introgression between species may occur more often than usually recognized. Whitney et al. (2010) reviewed the literature and concluded that interspecific hybridization is widespread among plants; on average, nearly 10% of all plant species have been documented as producing hybrids. Interestingly, although it has generally been considered that there is less hybridization in animals, Mallet (2005) concluded that 6–10% of animal species hybridize, and analysis by Grant & Grant (1992b) indicates nearly 10% of birds hybridize. Bush (1994) defined speciation as a process of divergence of lineages that are sufficiently distinct from one another to follow independent evolutionary paths. Many independent lineages are capable of hybridizing and exchanging genes (introgression) for quite long times without losing their phenotypic identities.

Interspecific hybridization can be an important source of genetic variation for some species. Grant & Grant (1998) have studied two species of Galápagos finches on the volcanic island of Daphne Major beginning in 1973. They have found that hybridization between the two species (medium ground finch and cactus finch) has been an important source of genetic variation for the rarer cactus finch species. And they have suggested that their results may apply to many species.

Such introgression is especially important for island populations in which the effective population size is restricted because of isolation and the amount of available habitat. Two species of land snails (*Partula*) occur sympatrically on the island of Moorea in French Polynesia. In spite of being markedly different both phenotypically and ecologically, estimates of genetic distance between some sympatric populations of these species based on molecular markers are lower than is typical for conspecific comparisons for these taxa on different islands. Clarke et al. (1998) concluded that this apparent paradox was best explained by "molecular leakage, the convergence of neutral and mutually advantageous genes in two species through occasional hybridization."

Adaptive introgression occurs when genes that increase fitness are transferred from one taxon to another by introgressive hybridization (Arnold 2004). Adaptive introgression has become recognized as an important factor in the evolution of many species (Suarez-Gonzalez et al. 2018; Marques et al. 2019). For example, Song et al. (2011) found that German house mice carry a segment of DNA from Algerian mice that contains an allele that produces a blood-clotting protein that provides resistance to warfarin, an anticoagulant used as a rodent poison. This allele is found at high frequency in German house mice where the two species overlap, apparently due to molecular leakage accompanied by strong selection for warfarin resistance. Interestingly, Abi-Rached et al. (2011)

recently suggested that modern humans acquired a **major histocompatibility complex (MHC)** allele important for disease resistance through introgression with Denisovans.

Organelle DNA seems particularly prone to introgression and molecular leakage (Ballard & Whitlock 2004). There are many examples of cases where the mtDNA molecule of one species has completely replaced the mtDNA of another species in some populations without any evidence of nuclear introgression (mtDNA capture; Good et al. 2008). For example, the mtDNA in a population of brook trout in Lake Alain in Québec is identical to the Québec Arctic char genotype, yet these brook trout are morphologically indistinguishable from normal brook trout and have diagnostic brook trout alleles at nuclear loci (Bernatchez et al. 1995). Roca et al. (2005) have found that mtDNA genotypes can be misleading in describing the relationship between populations and species of elephants because of one-way introgression. Recurrent backcrossing of female hybrids between savannah and forest elephants to savannah elephant males results in populations that have the nuclear genome of savannah elephants with forest elephant mtDNA.

Introgression of organelle DNA can lead to confusion in taxonomic relationships. Such introgression is often identified through incongruence in phylogenetic studies. This introgression usually represents historical hybridization, often related to changes in distribution during glacial cycles. The alpine whitefish complex in Europe is proposed to represent adaptive radiations following expansion from glacial refugia where introgressive hybridization occurred (Hudson et al. 2011). Five radiations occur across lake systems with up to six species present within individual lakes, but species share two mitochondrial haplotypes distributed across the radiations. A similar pattern of geographic haplotype sharing occurs for **chloroplast DNA (cpDNA)** in some congeneric plant species. For example, in Tasmania, Australia, 14 of 17 different species of eucalypts studied were shown to share identical cpDNA haplotypes in the same geographic area (McKinnon et al. 2001).

13.2.3 Hybrid zones

An interspecific hybrid zone is a region where two species are sympatric and hybridize to form at least partially fertile progeny (Example 10.2). Hybrid zones usually result from secondary contact between species that have diverged in **allopatry**. Barton & Hewitt (1985) reviewed 170 reported hybrid zones and concluded that hybrids were selected against in most hybrid zones that have been studied. Nevertheless, some hybrid zones appear to be stable and persist over long periods of time through a balance between dispersal of parental types and selection against hybrids (Harrison 1993). Hybrid zones may act as selective filters that allow introgression of only selectively advantageous alleles between species (Martinsen et al. 2001). Such hybrid zones usually comprise mixtures of individuals, including pure individuals and those with a range of admixture from F_1 to later-generation hybrids.

Arnold (1997) has proposed three models to explain the existence of a stable hybrid zone without **genetic swamping** of one or both of the parental species. In the Tension Zone Model, first- and second-generation hybrids are less fit than the parental types, but a balance between dispersal into the hybrid zone and selection against hybrids produces an equilibrium with a persistent, narrow hybrid zone containing F_1 individuals but few or no F_2 or beyond hybrids. This model does not depend upon the ecological differences between habitats of the two parental types. In the Bounded Hybrid Superiority Model, hybrids are fitter than either parental species in environments that are intermediate to the parental habitats, but are less fit than the parental species in their respective native habitats (Example 13.2). The Mosaic Model is similar to the Bounded Hybrid Superiority Model, but the parental habitats are patchy rather than there being an environmental gradient between two spatially separated parental habitats. Under both models, theory predicts that hybridization and backcrossing would occur for many generations, creating admixed populations containing individuals varying in their proportions of genetic material from the parental species.

Hybridization is usually thought of as occurring between two species or divergent lineages. However, hybridization also can occur across multiple species. Analysis of a hybrid zone between three species of montane lizards in eastern Australia showed hybridization primarily occurred between one species (*Pseudemoia cryodroma*) and each of another two (*P. entrecasteauxii* and *P. pagenstecheri*). There was evidence these hybridizations formed a bridge for hybridization between the other two species (*P. entrecasteauxii* and *P. pagenstecheri*) that do not hybridize when the third species is absent (Haines et al. 2016). Three-way hybridization has also been identified among Antarctic, Subantarctic, and New Zealand fur seals colonizing Macquarie Island following cessation of harvesting that led to extinction of the native Antarctic fur seal population (Lancaster et al. 2006).

13.2.4 Hybrid taxa

Hybridization is a natural part of the evolutionary process. For example, many extant species are derived from hybrid ancestors (Mallet 2007; Schumer et al. 2018). Approximately one-half of all plant species have been derived from polyploid ancestors, and many of these polyploid events involved hybridization between species or between populations within the same species (Stebbins 1950; Spoelhof et al. 2017). Evidence suggests that all vertebrates went through an ancient polyploid event that might have involved hybridization (Lynch & Conery 2000). Other major vertebrate taxa have gone through additional polyploid events. For example, all salmonid fishes (trout, salmon, char, whitefish, and grayling) went through an ancestral polyploid event some 25–50 million years ago (Allendorf & Waples 1996).

Hybrid speciation can also occur without change in ploidy level and requires high fitness in the hybrids and development of mechanisms providing reproductive isolation to disrupt ongoing gene flow with the parental species (Renault et al. 2014). Reproductive isolation can occur through disruptions to mating systems such as changes in flower color pattern, flowering times and pollinators in

Example 13.2 Genetic analysis of a hybrid zone between Sitka and white spruce

Sitka spruce is a major component of forest ecosystems along coastal areas of northwestern North America with cool wet environments. White spruce is common across inland areas of Canada in environments characterized by cold wet winters and warm dry summers. A hybrid zone occurs in northwestern British Columbia in transitional areas between these ecosystems along river valleys with moist cool climates. Genomic analysis of the hybrid zone using several hundred SNPs indicates that the stable, narrow hybrid zone is most likely maintained by hybrid superiority limited to environments that are intermediate to the ecological niches of the parental species (Bounded Hybrid Superiority Model).

The hybrid index values show a range from zero to one, indicating extensive hybridization, and the interspecific heterozygosity values are intermediate, indicating individuals are mainly backcrosses to each parent and later-generation hybrids rather than F_1s (in which case the interspecific heterozygosity would have a value of one; Figure 13.6). The presence of later-generation hybrids (beyond the F_1 and F_2 generations) is consistent with hybridization following putative expansion since the last glacial

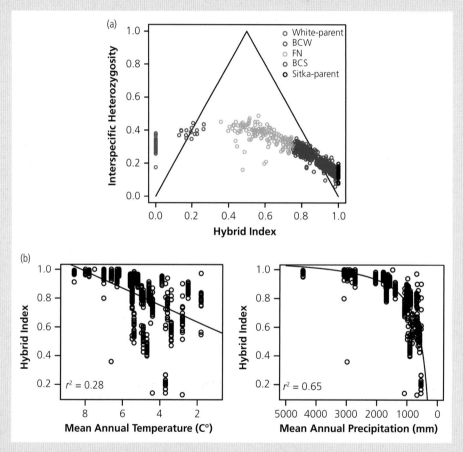

Figure 13.6 Analysis of hybridization between white and Sitka spruce. (a) Interspecific heterozygosity (*y*-axis) versus hybrid index (*x*-axis) for individuals from the hybrid zone (hybrid index 0 = white spruce and hybrid index 1 = Sitka spruce, and interspecific heterozygosity of 0 = parents and interspecific heterozygosity of 1 = F_1). BCW denotes backcross to white parent, BCS denotes backcross to Sitka parent, FN denotes advanced generation hybrid. (b) Relationship between hybrid index (*y*-axis) and geographic and climatic variables (*x*-axis) for 721 individuals from the hybrid zone. From Hamilton et al. (2013).

Continued

Example 13.2 *Continued*

maximum. Backcrosses to Sitka spruce were common, with relatively few backcrosses to white spruce. The distribution of backcrosses and advanced generation hybrids showed a geographical cline across the hybrid zone that was concordant with the climatic gradient from the maritime to continental climate of the two parental species. A patchy distribution of hybrid index estimates corresponding to environmental variation as expected with the Mosaic Model was not observed.

The hybrid index was highly correlated with climatic variables, particularly precipitation, and distance along drainage features of the system (Figure 13.6). The hybrid superiority in transitional habitats could result from a combination of higher cold and drought hardiness of the white spruce and the higher growth potential of the Sitka spruce. Hybrids also show some evidence for transgressive segregation (Section 13.2.1) of cold hardiness phenotypes, with hybrid individuals being more cold hardy than either parental species under testing at $-8°C$ (Hamilton et al. 2013). Given the relatively steep cline observed in hybrid index across the maritime/continental climate **ecotone**, local adaptation should be managed in this hybrid zone through limiting longitudinal seed transfer for reforestation. The high level of genetic diversity in hybrid populations in this zone may facilitate rapid adaptation to climate change (Aitken et al. 2008).

plants, or fertility times, body size, color patterning, and behavior in animals. Other factors can lead to geographic isolation, such as occupying different habitat.

Some hybrid taxa of vertebrates are unisexual. For example, unisexual hybrids between the northern redbelly dace and the finescale dace occur across the northern USA (Angers & Schlosser 2007). Reproduction of such unisexual species is generally asexual or semisexual, and they are often regarded as evolutionary dead ends. However, it appears that some tetraploid bisexual taxa had their origins in a unisexual hybrid (e.g., all salmonid fish).

Asexual and hybrid taxa provide some interesting challenges to conservation. For example, some species of corals appear to be long-lived first-generation hybrids that primarily reproduce asexually (Vollmer & Palumbi 2002). However, even rare backcrossing and introgression between hybrid corals and their parental species can blur species boundaries (Miller & van Oppen 2003; Vollmer & Palumbi 2007). The morphology of intraspecific corals can be amazingly variable, and **polyphyly** of morphologically defined coral species appears to be common (Forsman et al. 2009; Pinzón & LaJeunesse 2011). Understanding reproductive boundaries is essential for setting conservation priorities for corals.

13.3 Anthropogenic hybridization

The increasing pace of introductions of plants and animals and habitat modifications has caused increased rates of hybridization among plant and animal species. The introduction of plants and animals outside of their native range clearly provides the opportunity for hybridization among taxa that were reproductively isolated. Habitat modification can also increase opportunities for hybridization, and it is sometimes not appreciated just how much habitat modification has increased rates of hybridization. Human involvement has been attributed in 72% of cases in which hybridization is identified as a risk for extinction (Todesco et al. 2016).

In many cases, it is difficult to identify whether hybridization is "natural" or the direct or indirect result of human activities. In some cases, authors have referred to hybridization events resulting from habitat modifications as natural since they do not involve the introduction of species outside of their native range. Decline in abundance itself because of anthropogenic changes also promotes hybridization among species because of the greater difficulty in finding mates. In both of these cases, we consider that hybridization is the indirect result of human activities.

Wiegand (1935) was perhaps the first to suggest that introgressive hybridization is observed most frequently in habitats disturbed by humans. The creation of extensive areas of new habitats around the world has the effect of breaking down mechanisms of isolation between species (Rhymer & Simberloff 1996). For example, two native *Banksia* species in Western Australia hybridize only in disturbed habitats where more vigorous growth has extended the flowering seasons of both species and removed asynchronous flowering as a major barrier to hybridization (Lamont et al. 2003). In addition, taxa that can adapt quickly to new habitats may undergo adaptive genetic change very quickly. It now appears that many of the most problematic invasive plant species have resulted from hybridization events (Ellstrand & Schierenbeck 2000; Gaskin & Schaal 2002; Blair & Hufbauer 2010). This topic is considered in more detail in Chapter 14.

Increased turbidity in aquatic systems because of deforestation, agricultural practices, and other habitat modifications has increased hybridization among aquatic species that use visual cues to reinforce reproductive isolation (Wood 2003). This has threatened sympatric species on the western coast of Canada (Kraak et al. 2001) and cichlid fish species in Lake Victoria (Seehausen et al. 1997). It is estimated that nearly half of the hundreds of species in Lake Victoria have gone extinct in the past 50 years primarily because of the introduction of the Nile perch in the 1950s (Goldman 2003). The waters of this lake have grown steadily murkier, in part due to algal blooms resulting from the decline of cichlids. Mating between species now appears to be widespread and this classic example of adaptive radiation is now threatened (Goldman 2003).

Many other forms of habitat modification can lead to hybridization (Rhymer & Simberloff 1996; Seehausen et al. 2008). For example, the modification of patterns of water flow may bring species into contact that have been previously geographically isolated. It is likely that hybridization will continue to present a problem for conservation. Global environmental change may further increase the rate of hybridization between species in cases where it allows geographic range expansion. Kelly et al. (2010) have suggested that the melting of polar ice could cause increased frequency of hybridization among polar species. Seehausen (2006) suggested that loss of environmental heterogeneity causes a loss of biodiversity through increased hybridization (i.e., **reverse speciation**).

Hybridization can contribute to the decline and eventual extinction of species in two general ways. In the case of sterile or partially sterile hybrids, hybridization results in loss of reproductive potential and may reduce the population growth rate below that needed for replacement (**demographic swamping**). In the case of fertile hybrids, genetically distinct populations may be lost through genetic mixing (genetic swamping).

The effect of hybridization on extinction risk depends on a range of confounding factors. In a literature survey to identify and rank factors affecting extinction risk through hybridization, Todesco et al. (2016) found that human activities have a strong influence on the risk of extinction through hybridization arising from activities such as introduction of nonnative taxa, including intentional release of captive-bred individuals, and habitat disturbance. As expected, these risks were reduced when there were strong reproductive barriers between taxa with potential to hybridize, and increased when there was demographic inequity leading to genetic and demographic swamping. They noted that genetic swamping was more frequently reported than demographic swamping. Figure 13.7 provides a diagrammatic representation of the effects of demographic and genetic swamping in two color morphs of a hypothetical plant. Demographic swamping leads to extinction of the rare species in a population, and genetic swamping leads to extinction of the pure parental genotypes.

13.3.1 Hybridization without introgression

Many interspecific hybrids are sterile so that introgression does not occur. For example, matings between horses and donkeys produce mules, which are sterile because of chromosomal pairing problems during meiosis. Sterile hybrids are evolutionary deadends. Nevertheless, the production of these hybrids reduces the reproductive potential of populations and can contribute to the extinction of species. This is commonly referred to as demographic swamping (Wolf et al. 2001).

Hybridization represents a demographic threat to native species even without the occurrence of genetic admixture due to introgression (type 4, Figure 13.1). In this case, hybridization is not a threat through genetic mixing, but wasted reproductive effort that could pose a demographic risk. For example, females of the European mink hybridize

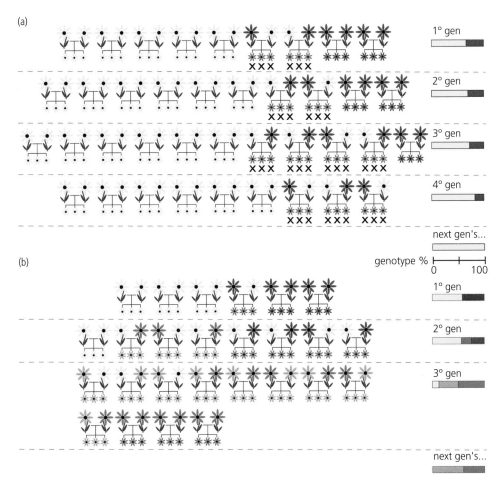

(a)

1° gen

2° gen

3° gen

4° gen

next gen's...

(b)

genotype %   0   100

1° gen

2° gen

3° gen

next gen's...

Figure 13.7 Representation of outcomes of hybridization between common plants represented by yellow flowers and rare plants represented by red flowers. (a) Demographic swamping where unfit hybrids (dark orange-flowered F_1s) do not survive in the population. (b) Genetic swamping where F_1 (dark orange-flowered) and backcross hybrids (light orange-flowered) are fertile and replace pure parental genotypes. Proportion of common, rare, and hybrid genotypes in the population over generations are represented in the colored bars to the right. From Todesco et al. (2016).

with males of the introduced North American mink. Embryos are aborted so that hybrid individuals are not detected, but wastage of eggs through hybridization has accelerated the decline of the European species (Rozhnov 1993).

The presence of primarily F_1 hybrids should not jeopardize protection of populations affected by hybridization. However, actions may be required to determine conditions that favor the native species to protect and improve its status and reduce the wasted reproductive effort of hybridization. In extreme (and expensive) cases, it may be possible to selectively remove all of the hybrids, allowing

the native species to recover and maintain a viable population.

For example, hybridization between the endangered bull trout and introduced brook trout does not lead to introgression, as only F_1s are produced (Table 13.2). Bull trout in Crater Lake National Park, Oregon, occur in a single stream, Sun Creek (Buktenica 1997). Introduced brook trout far outnumbered bull trout in this stream in the early 1990s and threatened to completely replace the native bull trout. The US National Park Service undertook a plan to recover this population by removing the brook trout. An effort was made to capture and identify every fish in this stream. Brook trout were identified by visual observation and removed. Putative bull trout and hybrids were sampled by fin clips for genetic analysis and held until genetic testing revealed the identity of each individual (Spruell et al. 2001). Sun Creek was then chemically treated to remove all fish from the stream. After treatment, the bull trout were placed back into Sun Creek. This population is currently increasing in abundance and distribution.

13.3.2 Hybridization with introgression

In many cases, hybrids are fertile and may displace one or both parental taxa through the production of hybrid swarms. This phenomenon has generally been referred to as genetic swamping, but also as genetic assimilation, and is considered to lead to genetic extinction or genomic extinction. Waddington (1961) used the term "genetic assimilation" to mean a process in which phenotypically plastic characters that were originally "acquired" become converted into inherited characters by natural selection (Pigliucci & Murren 2003). Thus, we do not use the term "genetic assimilation" to mean "genetic swamping" to avoid confusion.

Genomic extinction is a more appropriate term than the phrase genetic extinction. It is not genes or single-locus genotypes that are lost by hybridization; it is combinations of genotypes over the entire genome that are lost. Genomic extinction results in the loss of the legacy of an evolutionary lineage. That is, the genome-wide combination of alleles and genotypes that have evolved over evolutionary time will be lost by genetic swamping through introgression with another lineage. In a review of studies on hybridization and extinction, Todesco et al. (2016) found that where hybridization was considered to be a contributor to extinction threat, genetic swamping was the most commonly identified cause.

Perhaps surprisingly, introgression and admixture can spread even if hybridized individuals have reduced fitness. Population models (Epifanio & Philipp 2001) indicate that introgression may spread even when hybrids have severely reduced fitness (e.g., just 10% that of the parental taxa). This occurs because the production of hybrids is unidirectional, a sort of **genomic ratchet** (the Epifanio–Philipp

effect). That is, all of the progeny of a hybrid will be hybrids. Thus, the frequency of hybrids within a local population may increase even when up to 90% of the hybrid progeny do not survive. The increase in the proportion of hybridized individuals in the population can occur even when the proportion of admixture in the population (i.e., the proportion of alleles in a hybrid swarm that come from each of the hybridizing taxa) is constant.

Hybridization can also spread rapidly when hybrids have reduced reproductive success if hybrid individuals are more likely to disperse than nonhybrids. For example, Kovach et al. (2015) found that hybrids between native cutthroat trout and introduced rainbow trout were more likely to disperse than native cutthroat trout. Shine et al. (2011) have referred to this process as "**spatial sorting**." Lowe et al. (2015) have considered the implications of spatial sorting for the spread of hybridization.

13.3.3 Hybridization between wild species and their domesticated relatives

Hybridization between wild species and their domesticated relatives can be especially problematic because it can be difficult to detect the hybrids (Ellstrand 2003). Ellstrand et al. (1999) found that 12 of the world's 13 most valuable food crops hybridize with their wild relatives somewhere in the world. Such hybridization can be harmful because of reduced fitness of the wild population, as well as genetic swamping. Hybridization with domesticated species has increased risk of extinction in several wild species, including wild relatives of two of the world's 13 most important crops (rice and cottonseed). Such hybridization is also a concern because introgression between a crop and a related weedy taxon can produce a more aggressive weed (Ellstrand et al. 2010). Gene flow from a crop to a wild relative has been implicated in enhanced weediness in wild relatives of seven of the world's 13 most important crops (Ellstrand et al. 1999).

Hybridization between wild animal species and their domesticated relatives is also a conservation concern (e.g., Brisbin & Peterson 2007). Randi (2008) has reviewed the effects of such introgression in European populations of wolves, wildcats, rock partridges, and red-legged partridges. He has concluded introgressive hybridization is sometimes very common. He has recommended a number of steps to preserve the integrity of the gene pools of wild populations, including assessing the extent of hybridization, placing high conservation priority on nonhybridized populations, and enforcing strict controls on the genetics of game species raised in captivity and released for restocking. Introgression of alleles from domesticated relatives may increase adaptive potential. For example, Anderson et al. (2009) found that haplotypes for darker pelage in North American gray wolves confer an adaptive advantage in forested areas and might have been acquired through introgression from domestic dogs.

Hybridization between wild and hatchery populations of fish is a major conservation problem for many species (Araki & Schmid 2010). For example, up to 2 million Atlantic salmon are estimated to escape from salmon farms each year. Hybrids (F_1, F_2, and backcrosses) between farm and wild fish all show reduced survival compared with wild salmon (McGinnity et al. 2003). Nevertheless, farm and hybrid salmon show faster growth rates as juveniles and therefore may displace juvenile wild salmon. The repeated escapes of farmed salmon present a substantial threat to remaining wild populations of Atlantic wild salmon through introgression. This issue is considered in more detail in Chapter 21.

13.3.4 Hybridization and climate change

Changing climates are leading to changing species distributions on a macro and micro climatic scale (Chunco 2014). These changes in climate are leading to changes in interactions among species at spatial and temporal scales. Changing interactions among species usually occur at the edges of their distribution, with new opportunities for hybridization where the leading or trailing edges of species distributions change differentially. Studies suggest that species distributions change faster at leading edges compared with trailing edges, providing overlaps in distribution that may increase hybridization (Chunco 2014). Changes in temporal factors also mean that reproductive opportunities among previously isolated species will increase chances of hybridization, and changes in behavior

will also lead to increased chances of hybridization. Of course, changes in these factors may also decrease current cases of hybridization. As we saw in Example 13.2, hybridization may also provide new opportunities for adaptation and colonization of new environments that may increase species capacity to respond to changing climates. Hybridization may also be used to facilitate evolutionary change in conservation management for climate change (Chapter 16).

Hybrid zones that are influenced by climatic factors provide an opportunity to study how populations respond to changes in these climatic variables and track change in hybrid boundaries (Taylor et al. 2015). Analysis of the hybrid zone in tiger swallowtail butterflies in North America shows northward introgression of alleles for larval detoxification of host plants and diapause (Scriber 2011). Genetic analysis of the threatened westslope cutthroat trout populations in western North America using historical and contemporary samples from 1978 to 2008 shows an increase in hybridization with the introduced rainbow trout. This expansion of hybridization across the catchment is associated with increased temperature and reduced spring flushing that are more favorable to rainbow trout (Muhlfeld et al. 2014).

13.4 Fitness consequences of hybridization

Hybridization may have a wide variety of effects on fitness (Arnold & Martin 2010). In the case of **heterosis**, or **hybrid vigor**, hybrids have enhanced performance or fitness relative to either parental taxon. In the case of outbreeding depression, the hybrid progeny have lower performance or fitness than either parent (Lynch & Walsh 1998). Both heterosis and outbreeding depression have many possible causes, and the overall fitness of hybrids results from an interaction among these different effects. To further complicate matters, much of the heterosis that is often detected in F_1 hybrids is lost in subsequent generations so that a particular cross may result in heterosis in the first generation and outbreeding depression in subsequent generations.

There are two primary mechanisms that may reduce the fitness of hybrids. The first mechanism is genetic incompatibilities between the hybridizing taxa; this has been referred to as both **intrinsic outbreeding depression** and endogenous selection. Outbreeding depression may also result from reduced adaptation to environmental conditions by hybrids; this has been referred to as **extrinsic outbreeding depression** and also as exogenous selection. With endogenous selection, fitness effects are independent of environments, while with exogenous selection, hybrids may have lower fitness than parental types in some environments, but higher fitness than parental types in other environments.

13.4.1 Hybrid superiority

Heterosis occurs when hybrid progeny have higher fitness than either of the parental types. In many regards, heterosis is the opposite of inbreeding depression. The primary cause of heterosis is the sheltering of deleterious recessive alleles in hybrids. In addition, increased heterozygosity will increase the fitness of hybrid individuals for loci where the heterozygotes have a selective advantage over homozygous genotypes. Heterosis is greatest in the F_1 hybrids. Heterosis will be diminished in subsequent generations when heterozygosity is reduced and homozygosity is increased because of Mendelian segregation.

Subdivision of natural populations (Chapter 9) can provide the appropriate conditions for heterosis. Different deleterious recessive alleles will drift to relatively high frequencies in different populations. Therefore, progeny produced by matings between immigrant individuals are expected to have greater fitness than resident individuals. This effect is expected to result in a higher effective migration rate because immigrant alleles will be present at much higher frequencies than predicted by neutral expectations (Ingvarsson & Whitlock 2000; Whitlock et al. 2000; Morgan 2002).

Experiments involving immigration into inbred, laboratory populations of African satyrine butterflies have revealed surprisingly strong heterosis (Saccheri & Brakefield 2002). Heterosis led to immigrants being, on average, over 20 times more successful in contributing descendants to the fourth generation than were inbred nonimmigrants. The disproportionately large impact of some immigrants suggests that rare immigration events may be very important in evolution, and that heterosis may drive their fitness contribution.

Hybrids can have a more lasting fitness advantage because they possess advantageous traits from both parental populations (Example 13.2). Lewontin & Birch (1966) suggested many years ago that hybridization can provide new variation that allows adaptation to new environments. This may be an important mechanism for adaptation to rapidly changing environments; for example, anthropogenic climate change. As we saw in Section 13.2.1, Rieseberg et al. (2003) found that hybridization between sunflower species produced progeny that are adapted to environments very different from those occupied by the parental species. This was associated with the hybrids possessing new combinations of traits. Choler et al. (2004) found hybrids between two subspecies of an alpine sedge in the Alps that occurred only in marginal habitats for the two parental subspecies.

Similarly, studies on Darwin's finches have shown that hybrids can have a fitness advantage under changed environmental conditions. A very strong El Niño event in 1982–1983 resulted in a significant change in plant species and size of seed available on the island of Daphne Major, giving increased advantage to hybrids between medium ground finches and common cactus finches and between medium ground finches and small ground finches that had previously had limited survival (Grant & Grant 1993).

13.4.2 Intrinsic outbreeding depression

Intrinsic outbreeding depression results from genetic incompatibilities between hybridizing taxa, which may be chromosomal or genic. Reduced fitness of hybrids can result from heterozygosity for chromosomal differences between populations or species (Chapter 3). Differences in chromosomal number or structure may result in the production of **aneuploid** gametes that result in reduced survival of progeny. We saw in Table 3.3 that hybrids between races of house mice with different chromosomal arrangements produce smaller litters in captivity. Hybrids between chromosomal races of the threatened owl monkey from South America show reduced fertility in captivity (De Boer 1982).

Reduced fitness of hybrids can also result from genetic interactions between genes originating in different taxa (Whitlock et al. 1995). Dobzhansky (1948) first used the word "coadaptation" to describe reduced fitness in hybrids between different geographic populations of the fruit fly *Drosophila pseudoobscura*. This term became controversial (and somewhat meaningless) following Mayr's (1963) argument that most genes in a species are coadapted because of the integrated functioning of an individual.

Reduced fitness of hybrids can potentially occur because of the effects of genotypes at individual loci. Perhaps the best example is that of the direction of shell coiling in snails (Johnson 1982). Shells of some species of snails coil either to the left (sinistral) or to the right (dextral). Variation in shell coiling direction occurs within populations of snails of the genus *Partula* on islands in the Pacific Ocean. Many species in this genus are now threatened with extinction because of the introduction of other snails (Mace et al. 1998). The variation in shell coiling in many snail species is caused by two alleles at a single locus (Sturtevant 1923; Johnson 1982). Snails that coil in different directions find mating difficult or impossible. Thus, the most common phenotype (sinistral or dextral) in a population will generally be favored leading to the fixation of one type or the other. Hybrids between sinistral and dextral coiling populations may have reduced fitness because of the difficulty of mating with snails of the other type (Johnson et al. 1990).

Analysis of populations in hybrid zones between the native California tiger salamander and the introduced barred tiger salamander in California showed distorted ratios at a locus associated with embryonic mortality where the heterozygote appears to be selected against in the F_2 generation (Fitzpatrick et al. 2009). It is interesting to note that this embryonic mortality has not been sufficient to prevent maintenance of localized hybrid zones in this species.

Outbreeding depression may result from genic interactions between alleles at multiple loci (epistasis; Whitlock et al. 1995). That is, alleles that enhance fitness within their parental genetic background may reduce fitness in the novel genetic background produced by hybridization. Such interactions between alleles were first known as **Dobzhansky–Muller incompatibilities** because they were first described by these two famous *Drosophila* geneticists (Johnson 2000), but are now referred to as Bateson–Dobzhansky–Muller incompatibilities. Dobzhansky referred to these interactions as coadapted gene complexes. Such interactions are thought to be responsible for the evolution of reproductive isolation and eventually speciation.

There are few empirical examples of specific genes that show such Bateson–Dobzhansky–Muller incompatibilities. Rawson & Burton (2002) have presented an elegant example of functional interactions between loci that code for proteins involved in the electron transport system of mitochondria in an intertidal copepod *Tigriopus californicus*. A nuclear gene encodes the enzyme cytochrome *c* (CYC) while two mtDNA genes encode subunits of cytochrome oxidase (COX). CYC proteins isolated from different geographic populations each had significantly higher activity in combination with the COX proteins from their own source population. These results demonstrate that proteins in the electron transport system form coadapted combinations of alleles and that disruption of these coadapted gene complexes leads to functional incompatibilities that may lower the fitness of hybrids.

Self-fertilization in plants has long been recognized as potentially facilitating the evolution of adaptive combinations of alleles at many loci. Many populations of primarily self-fertilizing plants are dominated by a few genetically divergent genotypes that differ at multiple loci. Parker (1992) has shown that hybrid progeny between genotypes of the highly self-fertilizing hog peanut have reduced fitness (Figure 13.8). These genotypes naturally co-occur in the same habitats and the hybrid progeny have reduced fitness in a common garden. Thus, the reduced fitness of hybrids apparently results from Bateson–Dobzhansky–Muller incompatibilities between genotypes.

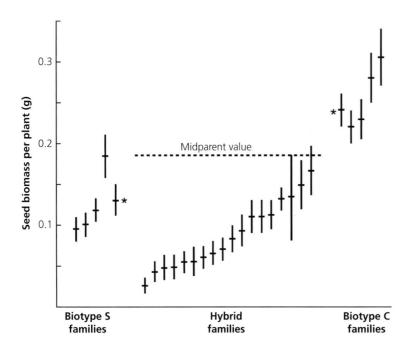

Figure 13.8 Fitness as measured by lifetime seed biomass of parental genotypes (biotypes S and C) and their hybrids in the highly self-fertilizing hog peanut. The two parental families marked with asterisks are the parents of the hybrids. From Parker (1992).

13.4.3 Extrinsic outbreeding depression

Extrinsic outbreeding depression results from the reduced fitness of hybrids because of loss of adaptation by ecologically mediated selection. The escape gaits of hybrids between the closely related white-tailed and mule deer provide an interesting example of outbreeding depression (Lingle 1992). White-tailed deer gallop to escape rapidly from predation. In contrast, mule deer "stott" by using long high bounds to escape predators. F_1, and other generation hybrids, between these species have an intermediate gait that is not as effective as either of the gaits of the parental species in making a quick escape.

Increased susceptibility to diseases and parasites is an important potential source of outbreeding depression because of the importance of disease in conservation and the complexity of immune systems and their associated gene complexes. Sage et al. (1986) studied a hybrid zone between two species of mice in Europe (*Mus musculus* and *M. domesticus*). Hybrids had significantly greater loads of pinworm (nematodes) and tapeworm (cestodes) parasites than either of the parental taxa. A total of 93 mice were examined within the hybrid zone.

Fifteen of these mice had exceptionally high numbers of nematodes (>500) while 78 mice had "normal" numbers of nematodes (<250). Fourteen of the 15 mice with high nematode loads were hybrids, while 37 of the 78 mice with normal loads were hybrids ($P < 0.005$). Cestode infections showed a similar pattern in hybrid and parental mice.

Currens et al. (1997) found that hybridization with introduced hatchery rainbow trout native to a different geographic region increased the susceptibility of wild native rainbow trout to myxosporean parasites. Similarly, Goldberg et al. (2005) found that hybrid largemouth bass from two genetically distinct subpopulations were more susceptible to largemouth bass virus. Parris (2004) found that hybrid frogs show increased susceptibility to emergent pathogens compared with the parental species.

Muhlfeld et al. (2009) used parentage analysis to estimate the reproductive success (fitness) of hybrids between nonnative rainbow and native westslope cutthroat trout. They found that small amounts of admixture markedly reduced fitness, causing reproductive success to decline by ~50% with only 20% individual admixture for both females and males (Figure 13.9). The

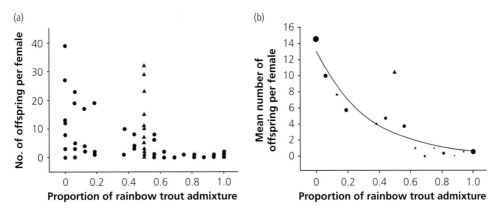

Figure 13.9 Effect of hybridization on fitness of native cutthroat trout. (a) Number of offspring per female versus the proportion of nonnative rainbow trout admixture. The plot includes 61 mothers and 397 juvenile assignments. Circles represent westslope cutthroat trout, rainbow trout, and later-generation hybrids, and triangles represent first-generation (F_1) hybrids. (b) Bubble plot of the mean number of offspring per female plotted against the proportion of rainbow trout admixture. The size of the circles is proportional to sample size. The mean value for first-generation hybrids is shown as a triangle, but these points were not included in the regression curve fitted through the points. From Muhlfeld et al. (2009).

exception to this observation was first-generation (F_1) hybrids, which showed much greater fitness than predicted on the basis of their proportion of admixture (Figure 13.9). This suggests that the sheltering of deleterious recessive alleles overcame the outbreeding depression in this population (Section 13.4.1).

13.4.4 Long-term fitness effects of hybridization

There is evidence that hybridization can increase fitness in the long term (tens of generations) even in cases where there is substantial outbreeding depression (Templeton 1986; Carney et al. 2000). For example, Hwang et al. (2011) found substantial reduction in fitness for many replicates of interpopulation hybrids of the intertidal copepod *T. californicus* in the first several generations. However, two of four long-term replicates showed equal fitness to parental types and two showed greater fitness than the parental types. Thus, in some cases, the increase in genetic variation from hybridization can lead to greater fitness in the long term even when there is substantial outbreeding depression. In rapidly changing environments, the increased genetic variation from hybridization might facilitate long-term adaptation. However, in some situations, the reduction in fitness can persist for long periods of time (Johnson et al. 2010a).

13.5 Hybridization and conservation

There are several key aspects of hybridization in relation to conservation. Protection of hybrids and reducing hybridization are an ongoing topic of debate and are key points for policy and legislation. Hybridization is a natural part of evolution and taxa that have arisen through natural hybridization may be eligible for protection. Nevertheless, increased anthropogenic hybridization is causing the extinction of many taxa (species, subspecies, and locally adapted populations) by both replacement and genetic mixing, and conservation policies often aim to reduce anthropogenic hybridization. These conflicting approaches mean that developing policies to deal with the complex issues associated with hybridization has been challenging (Allendorf et al. 2001; Ellstrand et al. 2010). Another key aspect of hybridization relating to conservation is the intentional use of hybridization to achieve conservation outcomes for rare or declining species and in ecological restoration (Section 19.5).

13.5.1 Protection of hybrids

Hybrids present many difficult problems for conservation. Increased rates of hybridization caused by human activities are one of the greatest threats to biodiversity. On the other hand, natural

Example 13.3 A rare plant species identified as a hybrid

Rare species are often known from a limited number of specimens, which can hinder taxonomic understanding, particularly in remote areas that may not have been well surveyed, such as northwestern Australia. *Seringia* is a genus of purple-flowered long-lived shrubs common in northern Australia. *Seringia exastia* and *S. katatona* are two poorly known species from the remote Kimberley region of northern Western Australia. Recent surveys discovered additional populations of both species and also revealed extensive morphological variation between the two species and a third, more widespread species, *S. nephrosperma*. This raised questions about possible hybridization among the species, and whether they are distinct species.

Genomic analysis of more than 5,000 SNP loci derived from a reduced representation approach was used to investigate species boundaries and hybridization within this group (Binks et al. 2020). Initial analysis suggested that individuals previously recognized as *S. katatona* were hybrids between *S. exastia* and *S. nephrosperma*, even though the other species do not tend to occur at the same sites. The identification of hybrids is consistent with the morphology of *S. katatona*, which is intermediate in diagnostic characters between *S. exastia* and *S. nephrosperma*. Further analysis of recent collections that were morphologically intermediate identified these, and the original samples, as F_1, F_2, and backcross individuals (Figure 13.10).

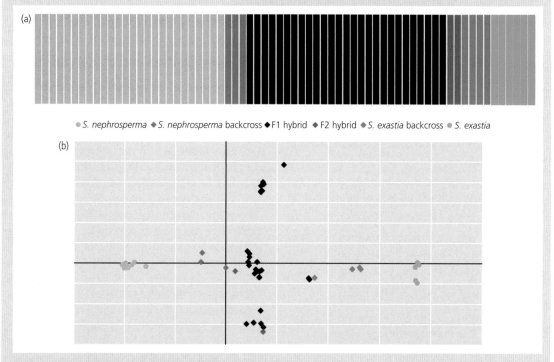

Figure 13.10 Identification of individuals to hybrid classes in *Seringia* hybrid populations based on genomic data. (a) Assignment of individuals (vertical bars) to parental and hybrid classes (colored according to class) using *NewHybrids* analysis of 189 SNP loci with fixed differences among parents. (b) Principal coordinates analysis of parental and hybrid individuals based on 5,289 SNP loci. From Binks et al. (2020).

The analysis also found extensive clonality within populations of *S. katatona*, enabling the populations to persist through asexual reproduction despite low fertility of F_1 hybrids. Analysis also showed that the rare *S. exastia* was not differentiated from the more widespread *S. elliptica*, and these taxa were synonymized under the older valid name of *S. exastia*. These taxonomic revisions present new information that will enable reconsideration of the current conservation status of these taxa and inform their management.

hybridization has played an important role in the evolution of many species. Distinguishing between anthropogenic and natural hybrids can sometimes be very difficult. Genomic analysis provides a powerful tool to study recent hybridization that was not previously detectable.

In Western Australia, the high plant diversity has led to instances where hybrid individuals have been recognized as separate taxa and provided conservation protection. Genetic analysis is now enabling recognition of these as early-generation hybrids (Walker et al. 2009; Binks et al. 2020; Bradbury et al. 2021), particularly in eucalypts where clonality through an extensive underground lignotuber facilitates persistence of F_1 hybrids for thousands of years (Robins et al. 2020). Hybrids may be considered for conservation listing under the Biodiversity Conservation Act 2016 (and previously under policy) if they satisfy three criteria: (1) they must be a distinct entity, that is, the progeny are consistent within the agreed taxonomic limits for that species group; (2) they must be capable of being self-perpetuating, that is, not reliant on parent stock for replacement; and (3) they are the product of a natural event, that is, both parents are naturally occurring and cross-fertilization was by natural means. Thus, hybrid-derived taxa would be provided protection but hybrid zones maintained between species would not be protected (Example 13.3).

Protection of hybrids in the USA was considered relatively early in conservation, yet has had a controversial history (Haig & Allendorf 2006). In May 1977, the US Department of the Interior's Office of the Solicitor issued a statement that "because it defines 'fish or wildlife' to include any offspring without limitation, the Act's plain meaning dictates coverage of hybrids of listed animal species. The legislative history buttresses this conclusion for animals and also makes clear its applicability to plants." However, response from the US Fish and Wildlife Service (July 1977) indicated ". . . since the Act was clearly passed to benefit endangered species, . . . it must have meant the offspring of two listed species and was not meant to protect a hybrid where that protection would in fact cause jeopardy to the continued existence of a species." The Solicitor responded (August 1977 and reaffirmed in 1983),

stating that "hybrids of listed species are not protected under the ESA" because he had learned there was the potential for a listed species to be harmed by hybridization. Overall, the US Fish and Wildlife Service's early position was to "discourage conservation efforts for hybrids between taxonomic species or subspecies and their progeny because they do not help and could hinder recovery of endangered taxa."

This series of correspondences and decisions that denied ESA protection for organisms with hybrid ancestry became known as the "Hybrid Policy" (O'Brien & Mayr 1991). O'Brien and Mayr pointed out that we would lose invaluable biological diversity if the ESA did not protect some subspecies or populations that interbreed (e.g., the Florida panther), or taxa derived from hybridization (e.g., the red wolf). Further, Grant & Grant (1992b) pointed out that few species would be protected by eliminating protection for any species interbreeding since so many plant and animal species interbreed to some extent. Discussions such as these and the need to rescue the endangered Florida panther contributed to the US Fish and Wildlife Service suspending the Hybrid Policy in December 1990.

A proposed policy on hybrids was published in 1996 (USFWS & NOAA 1996a), but this "Intercross Policy" has not been approved. Thus, no official policy provides guidelines for dealing with hybrids under the ESA (Allendorf et al. 2001). The absence of a policy probably results from the difficulty in writing a hybrid policy that would be flexible enough to apply to all situations, but which would still provide helpful recommendations.

While there is no formal policy on hybrids, other policy in the USA provides for use of scientific evidence and discretion among wildlife service officers. A survey by Lind-Riehl et al. (2016) found field officers tended to base their decisions on scientific evidence and consensus, although hybridization was generally considered a threat to rare species.

A review of conservation policy instruments in North America by Jackiw et al. (2015) found that many do not explicitly account for hybrids and those that do generally consider conservation of hybrids when they provide a benefit to at-risk species. Jackiw et al. (2015) proposed a framework for consideration of conservation of hybrids based on ecological and ethical considerations. The framework leads to conservation of hybrids when they do not threaten other species, provide benefit to at-risk species, or have been generated for conservation purposes.

Other authors have considered the ecological context of hybrids. Wayne & Shaffer (2016) have provided a framework for consideration of protection of hybrids as a starting point for evaluation of the complex issues involved in conservation of hybrids. This framework focused on consideration of three factors associated with natural or anthropogenic hybridization, ecosystem function of hybrids in relation to endangered species, and restoration of habitat to favor natural selection of the native entity.

Another aspect to consider is the dynamics of co-dependent communities associated with hybrids. Invertebrate communities occurring on plant hybrids are known to differ compared with those on their parental species, and in some cases species may have diverged sufficiently to be considered separate taxa. For example, analysis of mites of the *Aceria parapopuli* species complex occurring on Fremont cottonwood, narrowleaf cottonwood, and their hybrids showed differentiation of form on the hybrid (Evans et al. 2008).

13.5.2 Ancient hybrids versus recent hybridization

Protection of hybrid-derived taxa often depends on whether they are taxa that are derived from natural hybridization, or whether they represent recent anthropogenic hybridization. As we have seen, it can be challenging to identify hybridization definitively, particularly where it involves more long-term hybridization, and it can be difficult to distinguish from lineage sorting and coancestry.

This is readily evident in exploration of the role of hybridization among canids in North America (Figure 13.5). There has been much debate over whether there are two primary species (gray wolves and western coyotes) with other lineages derived as hybrids between these species, or whether eastern wolves represent a separate species derived from a third ancestral species (see vonHoldt et al. 2016 versus Hohenlohe et al. 2017). Genomic analysis has

been undertaken to shed further light on this, but can still be challenging to interpret in the context of complex evolutionary relationships. An analysis of SNP data and associated modeling was interpreted to provide support for a third ancestral species giving rise to eastern wolves (Rutledge et al. 2015). In contrast, analysis of sequenced genomes across the broader suite of canid species supported two primary species with eastern wolves and other coyote and wolf populations as hybrid-derived lineages (Sinding et al. 2018).

13.5.3 How much admixture is acceptable?

Consideration of the formal protection of hybrids also requires agreement on the level of hybridization that is acceptable in protected species, particularly in the context of intentional hybridization for genetic or evolutionary rescue. This is a difficult issue that has both a philosophical and practical aspect to it. Should an individual with both bison and cattle genes be considered a bison? If so, what proportion of cattle genes is acceptable? Some have argued that only "pure" bison with no admixture should be protected as bison (Marris 2009). Moreover, most bison herds have been found to be admixed with cattle so that potentially valuable genetic variation could be lost if these herds were no longer protected as bison (Halbert & Derr 2007). Others have argued that a "small" proportion of admixture should not disqualify populations from being protected. Admixture might also result in individuals with lower fitness. For example, Derr et al. (2012) found that individuals with cattle mtDNA were smaller than individuals with bison DNA in two bison herds that had substantial introgression from cattle.

The amount of admixture that precludes protection is situation specific. Setting some arbitrary limit of admixture below which a population will be considered "pure" is problematic. First, estimating the proportion of admixture precisely is difficult because of a limited number of diagnostic markers. In addition, it is often hard to distinguish between a small proportion of admixture (e.g., <5%) and natural polymorphisms that might exist in some populations. Finally, setting an arbitrary threshold could give way to further erosion of the genetic integrity of the parental taxon by constantly lowering the definition of "pure." If 5% is acceptable, why not 6% or 10%?

Several factors need to be considered when assessing the potential value of a population hybridized because of human activities. One factor is how many pure populations of the taxon remain. The smaller the number of pure populations, the greater the conservation and restoration value of any hybridized populations. Another factor to consider is whether the continued existence of hybridized populations poses a threat to remaining pure populations. The greater the perceived threat, the lower the value of the hybridized population. Finally, if harmful ecosystem-level effects result from hybridization, the value of hybrids is clearly much lower (Schweitzer et al. 2011).

The intentional hybridization of the Norfolk Island owl with its related subspecies has enabled persistence of the genes of this taxon within a small population, despite the taxon itself being listed as extinct. Garnett et al. (2011) argue that this species should not be considered extinct because approximately one-half of the nuclear genome and all of the mitochondrial genome is preserved in hybrids.

Dingoes are an iconic animal in Australia, but have experienced significant hybridization with wild dogs such that it can be difficult to tell a dingo from a dog hybrid (Guest Box 13). There has been much debate over whether dingoes can continue to be recognized as a separate species and be eligible for conservation protection (Jackson et al. 2019; Smith et al. 2019).

The creation of hybrid swarms between native cutthroat and introduced rainbow trout is widespread in the western USA. For example, most local populations of native westslope cutthroat trout are now hybrid swarms with rainbow trout. What proportion of admixture must be present before a population should no longer be considered westslope cutthroat trout? Some have argued that only nonhybridized populations should be included as westslope cutthroat trout in the unit to be considered for listing under the US ESA (Allendorf et al. 2004). Westslope cutthroat trout are a monophyletic lineage that has been evolutionarily isolated from other taxa for 1–2 million years (Allendorf & Leary 1988). This time of isolation and the amount of genetic divergence corresponds to that usually seen between congeneric species of fish. Only nonhybridized populations that still contain the westslope cutthroat trout genome that has evolved in isolation are likely to possess the local adaptations important for long-term persistence. This assessment has been supported by recent work showing that even small amounts of admixture reduce fitness (Figure 13.9). Unfortunately, the USFWS (2003) finding on westslope cutthroat trout was that populations with less than 20% admixture from rainbow trout would be considered westslope cutthroat trout for listing under the US ESA.

13.5.4 Predicting outbreeding depression

Some populations of conservation concern are small or have gone through a recent bottleneck, and therefore contain little genetic variation. In some cases, it might be helpful to increase genetic variation in these populations through intentional intraspecific hybridization. Use of intentional hybridization and management of genetic risk requires consideration of outbreeding depression, yet this is much less predictable than inbreeding depression (Frankham et al. 2011; Section 19.5). The intensity of inbreeding depression differs, but all populations have some deleterious recessive alleles that will reduce fitness when they become homozygous because of inbreeding. However, the effects of matings between genetically distinct populations depend upon a combination of factors. Fitness can increase because of the sheltering of deleterious recessive alleles (heterosis). On the other hand, fitness can decrease because of genetic incompatibilities (intrinsic outbreeding depression) or loss of local adaptations (extrinsic outbreeding depression).

Guest Box 13 Hybridization in Australian dingoes
Danielle Stephens, Peter J.S. Fleming, and Oliver F. Berry

Dingoes are iconic apex predators in Australia and readily interbreed with modern domestic dogs, which arrived in 1788. Genomics has revealed the extent of hybridization between dingoes and modern dogs and its drivers with remarkable clarity. Yet, the dingo case study illustrates how human values, along with science, can be central to how hybridization is viewed and managed.

What is a dingo? Dingoes are an early domesticated form of the wolf (Pollinger et al. 2010; Zhang et al. 2020). They were brought to Australia ~4,500 years ago through Southeast Asia and quickly spread across the continent. For many Australians, the dingo is a valued national icon, and there has been long-standing concern about their hybridization with other dogs (Daniels & Corbett 2003). Reflecting this, in some parts of Australia dingoes are protected by law.

Yet, how far has hybridization progressed, and how can dingoes be protected when their physical appearance is sometimes a poor indicator of ancestry (Figure 13.11; Elledge et al. 2009)? A genomic approach offered solutions to these questions, but still faced challenges since after ~200 years of potential hybridization the provenance of "pure" dingoes could not be guaranteed, and because their expansive range across Australia likely included genetic structure (Wilton et al. 1999; Cairns et al. 2018).

Stephens et al. (2015) used a learning samples method (Pritchard et al. 2000) with *a priori* dog and dingo reference samples to iteratively reveal distinct dog and dingo genetic groups. The test was validated through simulation and applied to 3,637 individuals collected across Australia. Globally, this was the largest hybridization mapping program undertaken for any species.

Figure 13.11 Wild canids photographed in remote regions where dingo purity levels are very high (clockwise from top left: Kimberley; northern Strzelecki Desert; central Strzelecki Desert; Kakadu). Which ones do you think are dingoes? See Section 21.3 for the answer. Photographs courtesy of Rob Davis/Kimberley Land Council, Paul Meek, Ben Allen, and Peter Fleming.

Guest Box 13 Continued

It revealed that hybridization follows the density and duration of post-colonial human settlement, with high levels in eastern Australia and low levels in the sparsely populated interior (Figure 13.12). Although 46% of individuals sampled were pure dingoes and most regions contained pure dingoes, all regions contained hybrids (i.e., type 5, Figure 13.1; Allendorf et al. 2001). Interestingly, virtually all free-living dogs had some dingo ancestry.

Valuable as these results were, the problem of how to manage dingoes remains complicated. First, the cultural and legal significance of dingoes varies with context (Fleming et al. 2014). Dingoes are considered iconic by many Australians and are totemic for some First Nations people, while some Australians revile them because of their predation on livestock. These attitudes focus on distinct cultural and functional aspects of the dingo, and consequently, people differ in how they perceive hybridization.

From one cultural heritage perspective, dingoes represent Australia's unique pre-colonial biota, which merits

protection, so dingo purity is important, and hybridization is a threat. Yet for some, purity of phenotype is more important than genotype. In addition, for some indigenous peoples, both dogs generally and dingoes are culturally significant.

Functional perspectives on dingoes also vary depending whether the focus is on livestock and faunal predation or a valued ecosystem function. In neither case is there compelling evidence that hybrid status significantly alters function (Crowther et al. 2020), so functional attitudes are agnostic to the threat posed by hybridization. Indeed, hybrids persist in abundance in regions where they are killed to protect livestock and where environmental conditions can be extreme (e.g., desert or alpine). Thus, hybridization does not obviously reduce the fitness of dingoes, which remain adaptable generalists—a trait that has made wild canids one of the most widespread taxa on Earth.

What do you think we should conserve: genetically "pure" dingoes, dingo-like hybrids, or wild canids (of whatever breeding) for ecological functions?

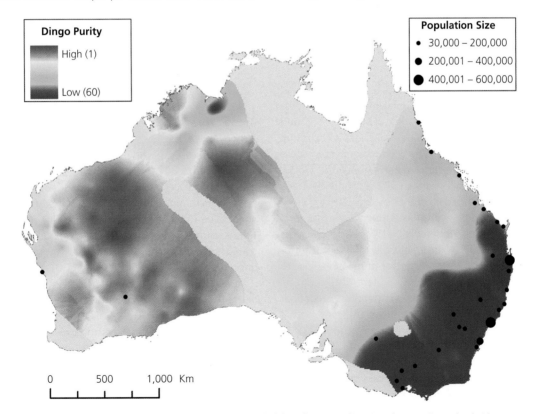

Figure 13.12 Spatial representation of the purity of dingoes sampled throughout Australia. Point values were interpolated with an ordinary kriging method. Unsampled areas are shown in gray. Percentage of dingo ancestry was determined by model-based clustering. Major towns (>25,000 residents, excluding capital cities) and their population sizes are shown for association with hybridization. From Stephens et al. (2015).

CHAPTER 14

Invasive Species

King brown snake eating a cane toad, Guest Box 14

Biological invaders are now widely recognized as one of our most pressing conservation threats.

(Ingrid M. Parker et al. 2003, p. 60)

The synergism arising from combining ecological, genetic, and evolutionary perspectives on invasive species may be essential for developing practical solutions to the economic and environmental losses resulting from these species.

(Ann K. Sakai et al. 2001, p. 323)

Invasion by **nonindigenous** (**alien**) species is recognized as second only to loss of habitat and landscape fragmentation as a major cause of loss of global biodiversity (Walker & Steffen 1997), and the spread of nonindigenous species is accelerating (Seebans et al. 2017). The economic effect of these species is a major concern throughout the world. For example, an estimated 50,000 nonindigenous species established in the USA cause major environmental damage and economic losses that total more than an estimated US$125 billion per year (Pimentel 2000). Conservative estimates of annual damage and costs of control of invasive species in Europe are €12 billion (Scalera 2010). Effects on native species, such as competition, predation, disease, and hybridization are significant factors for many invasive species (Scalera et al. 2012). In Australia, feral cats have been implicated in the extinction of 20 mammal species and the predation of over 388 species of birds, reptiles, mammals, and frogs per year (Doherty et al. 2015; Woinarski et al. 2019). Management and control of nonindigenous

species is perhaps the biggest challenge that conservation biologists will face in the next few decades.

A chapter on **invasive species** might at first seem out of place in a book on conservation genetics. However, we have chosen to include this chapter for several reasons. An understanding of the ecological genetics of invasive species provides insights into a range of factors such as the source and number of introduced populations, and their dynamics and spread (Le Roux & Wieczorek 2009). This information is important for developing methods of eradication or control that are critical for conservation efforts.

In addition, the study of species introductions offers exceptional opportunities to answer fundamental questions in population genetics that are important for the conservation of species. For example, how crucial is the amount of genetic variation present in introduced populations for their establishment and spread? Comparisons of genetic diversity and adaptive capacity in invasive species and endangered species offer insights into

Conservation and the Genomics of Populations, Third Edition. Fred W. Allendorf, *et al.*, Oxford University Press.
© Fred W. Allendorf, W. Chris Funk, Sally N. Aitken, Margaret Byrne, and Gordon Luikart (2022). DOI: 10.1093/oso/9780198856566.003.0014

their evolutionary trajectories. A recent review found higher potential for adaptive evolution in invasive species due to rapid population growth following bottlenecks, lower genetic load, greater variability in environmental conditions, and reduced selection from natural enemies and competition (Colautti et al. 2017).

Molecular genetic analysis of introduced species (including diseases and parasites) can provide valuable information for their control (Walker et al. 2003; Criscione et al. 2005). Understanding the "epidemiology of invasions" (Mack et al. 2000) is crucial to control current and prevent future invasions. Understanding the source(s) of the introduced population, the frequency with which a species is introduced into an area, the size of each introduction, the subsequent pattern of spread, and the role of adaptation in the success of invasive species are important in order to develop effective mechanisms of control. However, observing such events is challenging and assessment of the relative frequency of introductions or pattern of spread is extremely difficult. Molecular markers are a tool with which to answer these questions, and genomics has provided much greater power than was previously available (Chown et al. 2015; Fitak et al. 2019; Roe et al. 2019).

There is evidence that native species evolve and adapt to the presence of invasive species. For example, red-bellied black snakes exposed to invasive cane toads in northern Australia have evolved taste aversion, greater physiological tolerance of the toxin, and smaller heads (Phillips & Shine 2004, 2006). Native species sometimes change their response to predators, use different habitats, and exhibit other adaptations that allow them to persist in invaded areas (Strauss et al. 2006). The ability of a population to respond to selection from invaders depends upon the effects of the invader, the presence of appropriate genetic variation, and the history of previous invasions. Adaptive change in native populations can diminish the effects of invaders and potentially promote coexistence between invaders and natives. Genomics can help identify native populations that are likely resistant to invasive species and populations that are most susceptible and needing management (Fitak et al. 2019).

An understanding of genetics may also help predict which species are most likely to become invasive. There are two primary stages in the development of an invasive species (Figure 14.1). The first stage is the introduction, colonization, and establishment of a nonindigenous species in a new

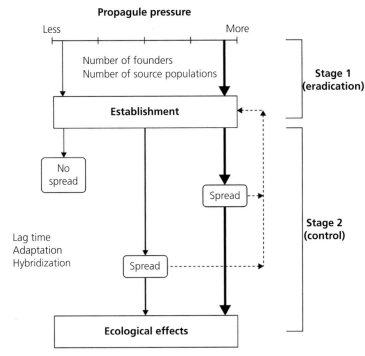

Figure 14.1 The two stages of invasion that generally coincide with different management responses. Propagule pressure is a continuum with greater pressure leading to increased chance of establishment and spread with shorter lag times. If spread involves small groups of dispersing individuals, each group must be able to establish in a different area. Establishment or subsequent spread may be inhibited where groups reach the limits of particular environmental conditions. From Allendorf & Lundquist (2003).

area. In other words, the introduced species must arrive, survive, and establish. The second stage is the spread and replacement of native species by the introduced species. The genetic principles that may help us predict whether or not a non-indigenous species will pass through these two stages to become invasive are the same principles that apply to the conservation of species and populations threatened with extinction: (1) genetic drift and the effects of small populations; (2) gene flow, hybridization, and genetic rescue; and (3) natural selection and adaptation. Mutation and recombination can also play a role; for example, in producing novel genotypes and adaptation.

In this chapter, we consider the possible importance of genetic change in the establishment and spread of invasive species. We also examine the significant role that hybridization may play in the development of invasive species. We consider ways in which genetic understanding may be applied to help predict which species are likely to be successful invaders and to help control invasive species. We discuss the use of genetic techniques to understand emerging diseases. Finally, we discuss the opportunities that may arise from new genetic technologies.

14.1 Why are invasive species so successful?

Not all introduced species become invasive (Sakai et al. 2001). A general observation is that only one out of every 10 introduced species becomes established, and only one out of every 10 newly established species becomes invasive. Therefore, roughly only one out of every 100 introduced species becomes a pest. The next few sections consider what factors may influence whether a species becomes established and invasive.

Invasive species provide an exceptional opportunity for basic research into the population biology and short-term evolution of species. Many of the best examples of rapid evolutionary change come from the study of introduced populations (Lee 2002; Prentis et al. 2008). For example, the fruit fly *Drosophila subobscura* evolved a north–south cline in wing length just 20 years after introduction into North America that paralleled the pattern present in their native Europe of increased wing length with

latitude (Huey et al. 2000). Similarly, two species of goldenrods evolved a cline in flowering time that resembled a cline in their native North America after being introduced into Europe (Weber & Schmid 1998). A study of invasive red lionfish in the Gulf of Mexico employed 50,000 **single nucleotide polymorphisms (SNPs)** to reveal multiple outlier loci in genomic regions that are putatively under selection with genes facilitating the invasion (Burford Reiskind et al. 2019).

Many unresolved central issues in the application of genetics to conservation—such as the inbreeding effects of small populations and the importance of local adaptation—can be experimentally addressed with introduced species. Two apparent paradoxes emerge from comparison of conclusions regarding the effects of small population size and local adaptation in species of conservation concern with successful invasions by introduced species.

14.1.1 Why are invasive species that have gone through a founding bottleneck so successful?

Much of the concern in conservation genetics relates to the potential harmful effects of small population sizes. The loss of genetic variation through genetic drift and the inbreeding effect of small populations contribute to the increased extinction rate of these populations (e.g., Frankham & Ralls 1998). However, colonization of introduced species often involves a population bottleneck since the number of initial colonists is often small. Thus, a newly established population is likely to be much less genetically diverse than the population from which it is derived (Barrett & Kohn 1991).

The reduced genetic diversity can have two harmful consequences. First, inbreeding depression may limit population growth, and lower the probability that the population will persist. Second, reduced genetic diversity will limit the ability of introduced populations to evolve in their new environments. Thus, we face a paradox: if population bottlenecks are harmful, then why are invasive species that have gone through a founding bottleneck so successful?

One answer to this paradox is that introduced species often have greater genetic variation than native species because they are a mixture of several source populations (Example 14.1; Section 14.2.3.2).

Example 14.1 Lizard invasions associated with increased genetic variation

The brown anole is a small lizard that is native to the Caribbean (Kolbe et al. 2004). It has been introduced widely throughout the world (Hawaii, Taiwan, and the mainland USA). Introduced populations often reach high population densities, show exponential range expansion, and are often competitively superior to as well as a predator of native lizards. The brown anole first appeared in the Florida Keys in the late 19th century. Its range did not expand appreciably for 50 years, but widespread expansion throughout Florida began in the 1940s and increased in the 1970s.

Genetic analyses of **mitochondrial DNA (mtDNA)** suggests that at least eight introductions have occurred from across this lizard's native range into Florida (Kolbe et al. 2004), and microsatellite analysis shows interbreeding between the introduced lineages (Kolbe et al. 2008). This has resulted in admixture from different geographic source populations and has produced populations that are substantially more genetically variable than native populations. Studies on physiological variation have also shown high variability in traits associated with thermal minimum, rate of water loss, and metabolic rate across the invaded distribution likely due to multiple invasion sources (Kolbe et al. 2014). Moreover, recently introduced brown anole populations around the world originated from Florida, and some have maintained these elevated levels of genetic variation.

There is a strong observed effect of propagule pressure on the invasiveness of species. That is, the clear association between the greater number of introduced individuals and the number of release events and the probability of an introduced species becoming invasive suggests that many invasive species are not as genetically depauperate as expected.

Plant species can avoid the reduction in genetic variation associated with colonization by asexual reproduction (Barrett & Husband 1990). Many invasive plant species reproduce through vegetative reproduction strategies, including layering, rhizomes, root suckering, and fragmentation (Baker 1995; Calzada et al. 1996), or by **apomixis**, which is the formation of seed without meiosis or fertilization. In all cases, the effects of inbreeding depression are avoided because the progeny are genetically identical to the parental plants. In addition, many invasive plant species are polyploids and can reproduce by self-pollination. Allopolyploids can maintain genetic variation in the form of fixed heterozygosity because of genetic divergence between their ancestral genomes (Brown & Marshall 1981).

Many invasive species can reproduce rapidly, thus generating large numbers of individuals and overcoming demographic impacts of small populations. Invertebrates, amphibians, and fish can lay large numbers of eggs and many invasive plants produce large numbers of seeds either sexually or through apomixis.

14.1.2 Why are introduced species that are not locally adapted so successful at replacing native species?

Local adaptation is often an important concern in the conservation of threatened species (McKay & Latta 2002). That is, adaptive differences between local populations are expected to evolve in response to selective pressures associated with different environmental conditions. The presence of such local adaptations in geographically isolated populations often plays an important role in the management of threatened species (Crandall et al. 2000).

When a species invades a new locality it will almost certainly face a novel environment. However, many introduced species often outcompete and replace native species. For example, introduced brook trout are a serious problem in the western USA where they often outcompete and replace ecologically similar native trout species. However, the situation is reversed in the eastern USA, where brook trout are native. They are in serious jeopardy because of competition and replacement from introduced rainbow trout that are native to the western USA. Thus, we face a second paradox: if local adaptation is common and important, then why are introduced species so successful at replacing native species?

A variety of explanations have been proposed to explain why introduced species often outperform indigenous species. First, some species may be intrinsically better competitors because they evolved in a more competitive environment. Second, the absence of enemies (e.g., herbivores in the case of plants) allows nonindigenous species to have more resources available for growth and reproduction and thereby outcompete native species. Siemann & Rogers (2001) found that an invasive tree species, the Chinese tallow tree, had evolved increased competitive ability in its introduced range. Invasive genotypes were larger than native genotypes and produced more seeds; however, they had lower quality leaves and invested fewer resources in defending them. Thus, there are a number of reasons why introduced species may fare well even though native species may be locally adapted.

In addition, local adaptation of native populations might only be essential during periodic episodes of extreme environmental conditions (e.g., winter storms, drought, or fire). Wiens (1977) has called these episodes "ecological crunches" and has suggested they are commonplace and limit the value of short-term studies of competition and fitness. For example, Rieman & Clayton (1997) have suggested that the complex life histories of some fish species (mixed migratory behaviors, etc.) are adaptations to periodic disturbances such as fire and flooding. Thus, introduced species may be able to outperform native species in the short term (a few generations) because the performance of native species in the short term is constrained by long-term adaptations that come into play every 50 or 100 years. Changing climate can also reduce local adaptation and might favor nonnative (e.g., warm-adapted) species.

14.2 Genetic analysis of introduced species

Molecular genetic analysis of introduced species can provide valuable information about their origin. In addition, a study of the amount and distribution of genetic variation in introduced species can provide valuable insight into the mechanisms of establishment and spread. In some cases, even identifying the species of invasive organisms may be difficult without genetic analysis. In other cases, distinct populations of a native species may become invasive when introduced into a new ecosystem. Such populations would technically not be considered alien since conspecific populations were already present. Nevertheless, such populations may become invasive when introduced outside of their natural area (Genner et al. 2004). Current regulations dealing with invasive organisms are based upon species classification. However, recognizing biological differences between populations within the same species is also important for control of invasive species.

14.2.1 Molecular identification of invasive species

In some cases, genetic identification may be necessary to identify the species of introduced organisms. In recent years molecular identification of species has increased greatly (Chapter 22) and this is highly applicable to identification of invasive species. For example, the thrip *Scirtothrips dorsalis* has a rapidly expanding global distribution and is a recent introduction to the USA. This species feeds on a large number of agricultural crops, and causes extensive damage to them. The species as currently recognized is considered to be a cryptic species complex with at least three morphologically indistinguishable taxa (Dickey et al. 2015). Application of the *COI* barcoding locus and development of next-generation sequencing loci identified 11 cryptic entities within the complex with three being invasive.

Genetics also allows the detection of an invasive lineage that co-occurs with a native species. For example, Genner et al. (2004) used mtDNA to detect a nonnative morph from Asia of the gastropod *Melanoides tuberculata* in Lake Malawi, Africa, which now co-occurs with native forms of the same species. This nonnative morph was not present in historical collections and appears to be spreading rapidly and replacing the indigenous form. More recent genetic analysis in a number of lakes and rivers in Africa has revealed the extent of the presence of cryptic invasive lineages of *M. tuberculata* with shell morphology that overlaps with that of native lineages (Van Bocxlaer et al. 2015).

14.2.2 Molecular identification of origins of invasive species

Genomics has provided greater power to identify the origin of invasive species and to determine whether there have been multiple introductions. This is particularly important for managing introduction pathways, identifying biological control agents, and determining whether invasive species are undergoing adaptation. *Passiflora foetida* is a pantropical invasive species now common in Southeast Asia and the Pacific, and it has become a highly significant transformer weed in northern Australia (Yockteng et al. 2011). It is a climbing vine and can rapidly become completely dominant in ecosystems, overtopping tall trees, although it appears to have greater invasiveness in the west of the continent than in the east. It is native to Central and South America, and genomic analysis shows the diversity is consistent with three independent origins of the species into Australia from Ecuador/Peru, Brazil, and the Caribbean, with only one of these lineages becoming widespread. This information is now guiding the search for biological control agents.

14.2.3 Distribution of genetic variation in invasive species

Examination of published descriptions of the amount and patterns of genetic variation in introduced species reveals two contrasting patterns. In the first, introduction and establishment are often associated with a population bottleneck, or bottlenecks, so that introduced populations have less genetic variation than populations in the native range of the species. Under some circumstances, population bottlenecks are associated with invasion (Example 14.2). In the second pattern, introduction and establishment are associated with admixture of more than one local population in the native range, or multiple introductions over time, so that populations in the introduced range have greater genetic variation. The bottleneck and admixture situations result in very different patterns of genetic variation in introduced species. Genetic analysis can assist in identifying the two patterns and understanding their invasion dynamics.

Example 14.2 Loss of genetic variation in an introduced ant species promotes a successful invasion

Ants are among the most successful, widespread, and harmful invasive taxa. Highly invasive ants sometimes form unicolonial supercolonies in which workers and queens mix freely among physically separate nests. By reducing costs associated with territoriality, such unicolonial species can attain high worker densities, allowing them to achieve interspecific dominance.

Tsutsui et al. (2000) examined the behavior and population genetics of the invasive Argentine ant in its native and introduced ranges. They demonstrated with microsatellites that population bottlenecks have reduced the genetic diversity of introduced populations. The number of alleles in the invasive range was half that of the native range, reduced from 59 to 30 alleles, and the average expected heterozygosity decreased by 68%, from 0.639 to 0.204. This loss is associated with reduced intraspecific aggression among spatially separate nests, and leads to the formation of interspecifically dominant supercolonies. In contrast, native populations are more genetically variable and exhibit pronounced intraspecific aggression. These findings provide an example of how a genetic bottleneck associated with introduction can lead to widespread ecological success.

14.2.3.1 Bottleneck model

Introduced species often have few founders so that genetic variation is reduced by the **founder effect**. The rapa whelk, a predatory marine gastropod, presents an extreme example of this (Chandler et al. 2008). Sequences for two mtDNA genes (a total of 1,292 base pairs (bp)) revealed 110 haplotypes in 178 individuals from eight populations in its native range of China, Korea, and Japan. However, all 106 individuals sampled from 12 introduced populations throughout Europe and North America had a single haplotype that was observed in four of the 178 native rapa whelks sequenced. Unfortunately, there are no nuclear gene data for this species, and mtDNA might not be representative.

Genomic analysis of mosquitofish using ddRAD (a type of RAD sequencing; Section 4.5.1) shows

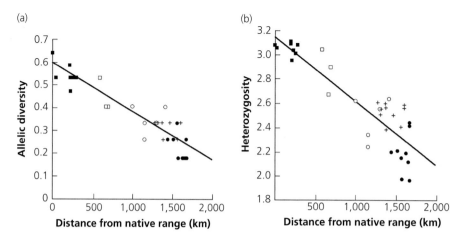

Figure 14.2 Relationship in an invasive gallwasp between distance from its native Hungary and (a) allelic diversity and (b) expected heterozygosity. Populations further from the native range show reduced genetic variation and patterns of allele frequency differentiation that suggest a stepping-stone invasion process rather than multiple introductions from its native range. From Stone & Sunnucks (1993).

highly reduced heterozygosity in introduced populations in Europe (H_e = 0.047) compared with native populations in America (H_e = 0.149), highlighting the effects of bottlenecks and drift on diversity (Vera et al. 2016). Serial bottlenecks as invasive populations expand from a local source in a stepping-stone process would lead to a strong pattern of isolation by distance among populations, and a serial reduction of diversity, as seen at 13 allozyme loci in the invasive populations of gallwasp in Europe (Stone & Sunnucks 1993; Figure 14.2).

However, small numbers of founders may still provide sufficient standing variation for adaptation. For example, the green anole was introduced to the Ogasawara Islands of Japan from Florida in the 1960s. Genomic resequencing has led to estimates that the effective size of the founding population was less than 50, yet the invasive populations show evolution of longer hindlimb traits associated with changes in microhabitat such as tree branches and distance from the ground (Tamate et al. 2017). Genomic analysis showed selection on five candidate genes with functions associated with muscle development and contraction and metabolic processes suggesting selection on variation within the founding population.

Green anoles have also expanded from subtropical Florida into temperate North America, and

Bourgeois & Boissinot (2019) tested for signatures of selection associated with this northward expansion. They analyzed 29 whole genome sequences and identified genes involved in behavior and response to cold that had frequent signals of selection. They also detected signatures of balancing selection at immune genes, and genes involved in neuronal and anatomical development. This study suggests that successful colonization of northern environments was linked to behavioral shifts and physiological adaptation.

Similarly, in the invasive plant *Hypericum canariense*, which is native to the Canary Islands, invasive populations in California and Hawai'i showed reduced genetic diversity from bottlenecks, yet exhibited evolution of phenotypic variation in growth and flowering time in the invasive populations (Dlugosch & Parker 2008). Analysis of genic SNPs in the plant Pyrenean rocket showed evidence for selection on flowering time genes, suggesting rapid genetic adaptation during establishment and spread in the invasive range (Vandepitte et al. 2014).

14.2.3.2 Admixture model

In contrast to the bottleneck model, many introduced species actually have greater variation in comparison with populations from the native range because their founders come from different local populations within the native range (Roman & Darling 2007). Admixing individuals from genetically divergent populations will increase genetic variation by converting genetic differences between populations to genetic variation between individuals within populations (Example 14.1). As we saw in Chapter 9, the total heterozygosity (H_T) within a species can be partitioned into genetic variation within and among subpopulations, so admixture of populations that are genetically differentiated will increase genetic variation in populations of an invasive species compared with that in its native range.

This effect can be seen in many introduced populations (Roman & Darling 2007). For example, approximately 400 chaffinches were imported from England into New Zealand between 1862 and 1877. Overwintering birds from several populations on the European continent were included in the birds collected for introduction. Baker (1992) reported that chaffinches from eight populations in New Zealand have an average heterozygosity that is 38% greater (0.066 versus 0.048) than 10 native European populations at 42 allozyme loci. As expected, chaffinches in New Zealand have greatly reduced differentiation among subpopulations (F_{ST} = 0.04) compared with chaffinches in their native Europe (F_{ST} = 0.22).

Genomic analysis of two salmonid species, Chinook salmon and brook trout, introduced into South America from North America shows similar levels of diversity and less differentiation in the introduced populations compared with the native populations, demonstrating the capacity of multiple introductions to maintain genetic diversity (Narum et al. 2017). In addition, some evidence for divergent selection was observed in the introduced populations.

14.2.4 Mechanisms of reproduction

Molecular genetic analysis also can be used to determine if an introduced plant species is reproducing sexually or asexually, and the ploidy level of introduced plants. Further examination can determine how many different clonal lineages are present if an invader is reproducing asexually. This information, along with an understanding of genetic population structure, is essential for the development of effective control measures for invasive weed species.

Cogongrass is a highly invasive species on a global scale that reproduces both sexually and asexually. Genomic analysis of invasive population in the southeast USA has shown the presence of five clonal lineages among the southeast US populations, indicating multiple introductions of this species and a clonal lineage of a congeneric grass that is also invasive (Example 14.3).

> **Example 14.3** Clonal reproduction and multiple introductions of invasive cogongrass
>
> Cogongrass, also known as speargrass, is a diploid C4 grass that is present on all continents, although the original native range is unknown. It has become highly invasive in many countries. Like many invasive grasses, as well as having an outcrossed mating system, it has vigorous vegetative growth and produces a dense impenetrable matrix of rhizomatous growth. It is a transformer weed, changing the ecosystem, particularly the fire ecology, and has shown to be resistant to herbicides. It was introduced into the southeastern USA as a forage grass and for use as a soil stabilizer, and now occurs from North Carolina to Texas.
>
> Burrell et al. (2015) undertook genomic analysis using 2,320 SNP and **insertion–deletion (indel)** loci generated by genotyping-by-sequencing of 449 samples from invasive populations from 13 locations in the southeastern USA, and identified five clonal lineages among these populations (Figure 14.3). The analysis also identified one of these as a clonal lineage of the congener Brazilian satintail. There was no evidence of hybridization between clonal lineages even though there is overlap in their geographical distribution.
>
> The analysis also included samples putatively from Japan and the Philippines, and showed that two of the lineages from the USA were present in the samples from Japan and none were present in the samples from the Philippines. The lineage of satintail was the same as that sampled from Brazil. The lineages are clonal in the putative native range, giving capacity to spread rapidly in a new environment even though they have low genetic diversity. The presence of multiple lineages indicates multiple introductions to the southeastern USA.

(a)

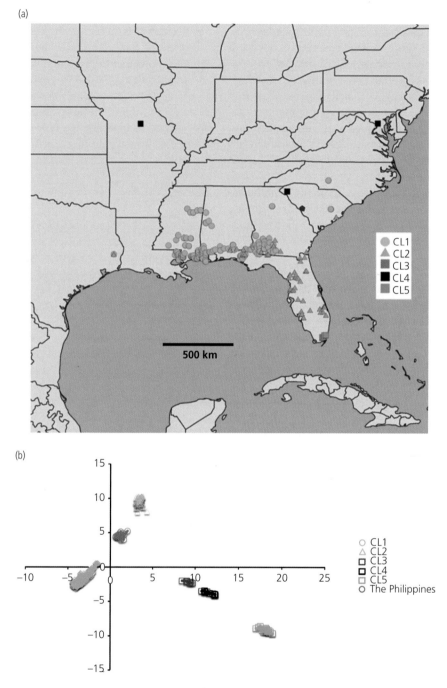

(b)

Figure 14.3 Genomic analysis of cogongrass revealed five clonal lineages. (a) Distribution of clonal lineages. (b) Principal component analysis (PCA) based on analysis of 2,320 SNPs and indels showing the relationships among clonal lineages. From Burrell et al. (2015).

The Bermuda buttercup is native to South Africa, but invasive in other areas with Mediterranean climates. It is polyploid, with tetraploid and pentaploid forms, and can reproduce clonally as well as sexually. The native populations are mainly tetraploid. Genetic analysis shows multiple introductions in the invasive range. Invasive populations in the western Mediterranean have a mixture of pentaploid and tetraploid forms, with a high frequency of the clonal pentaploid form compared with the rarity of this form in the native range (Ferrero et al. 2015).

Parthenogenesis (Greek for "virgin birth") is a form of asexual reproduction that occurs in animals without fertilization from a male. More than 80 taxa of fish, amphibians, and reptiles are now known to reproduce by parthenogenesis (Neaves & Baumann 2011). There are a variety of genetic mechanisms underlying parthenogenesis. Some result in completely homozygous progeny (automictic parthenogenesis). For example, automictic parthenogenesis has been reported in Komodo dragons (Watts et al. 2006), hammerhead sharks (Chapman et al. 2007), and zebra finches (Schut et al. 2008). This form of parthenogenesis is not likely to facilitate successful invasion because of the homozygosity of the progeny. However, apomictic parthenogenesis results in progeny that are genetically identical to their mother and may be heterozygous, and therefore has the potential to facilitate successful invasion (Schwander & Crespi 2009).

North American crayfish have been successful invaders throughout the world, especially in Europe where they have had major harmful effects on native crayfish. Buřič et al. (2011) discovered that females of the spiny-cheek crayfish (a successful invader of Europe) held in isolation are as reproductively successful as females held with males. All of the progeny from isolated females were genetically identical to the mothers at seven microsatellite loci; progeny from females held with males had genotypes indicating sexual reproduction. These authors have suggested that this reproductive plasticity might contribute to the great invasive success of these species. Similarly, in apomixis in plants, the formation of seed without meiosis or fertilization produces progeny that are genetically identical to the mother, providing rapid reproduction of the successful genotype.

14.2.5 Quantitative genetic variation

Although much information can be gained from molecular markers, characterization of the genetic variation controlling those life history traits most directly related to establishment and spread is also crucial. These traits are likely to be under polygenic control with strong interactions between the genotype and the environment; they cannot be analyzed directly with molecular markers, although mapping quantitative trait loci (QTLs) affecting fitness, colonizing ability, or other traits affecting invasiveness may be possible (Barrett 2000). For example, variation in the number of rhizomes producing above-ground shoots, a major factor in the spread of the noxious weed Johnsongrass, is associated with three QTLs (Paterson et al. 1995). This knowledge may provide opportunities for predicting the location of corresponding genes in other species and for growth regulation of major weeds.

Application of the methods of quantitative genetics could be useful for those species in which information can be obtained from progeny tests, parent–offspring comparisons, or use of the animal model (Chapter 11). For example, one could compare the additive genetic variance/covariance structure of a set of life history traits of different populations to evaluate the role of genetic constraints in the evolution of invasiveness. Comparisons of the heritability of a trait could be made among different, newly established populations or between invasive populations and the putative source population. Consideration of both the genetic and ecological context of these traits is critical, given the potentially strong interaction of genetic and environmental effects (Barrett 2000).

Quantitative genetics can also be useful in detecting adaptation and its importance in successful invasion, and is being combined with genomics in integrated approaches. The plant family Asteraceae has many globally significant invasive species that show evidence of changes in traits associated with invasiveness. Comparative analysis of transcriptomes among invasive and native genotypes of species of Asteraceae showed evidence

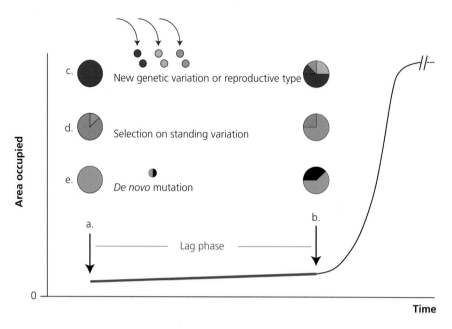

Figure 14.4 Invasive species frequently show a lag time prior to accelerated expansion of the area occupied. Genetic changes during the lag time can contribute to the expansion. Potential genetic changes at a single locus between (a) initial establishment and (b) spread are shown from (c) increase in genetic variation due to immigration of new genotypes, (d) selection on standing variation, and (e) *de novo* adaptive mutation. From Bock et al. (2015).

for selection among loci in the invasive genotypes, although there was no evidence for parallel changes in **orthologous** genes among species. This suggests no common genomic responses in invasive species, but rather that there are multiple pathways for adaptation in these species (Hodgins et al. 2015).

14.3 Establishment and spread of invasive species

One common feature of invasions is a lag from the time of initial colonization to the onset of rapid population growth and range expansion (Sakai et al. 2001; Figure 14.1). This lag time is often interpreted as an ecological phenomenon (the lag phase in an exponential population growth curve). Lag times are also expected if evolutionary change is an important part of the colonization process. This process could include additional introductions of new variation, selection acting on standing variation, and/or *de novo* adaptive mutation (Figure 14.4). It appears likely that in many cases, there are genetic constraints on the probability

of a successful invasion, and the lag times of successful invasions could be a result of the time required for adaptive evolution to overcome these constraints (Ellstrand & Schierenbeck 2000; Mack et al. 2000).

14.3.1 Propagule pressure

Propagule pressure has emerged as the most important factor for predicting whether or not a nonindigenous species will become established (Kolar & Lodge 2001). Propagule pressure includes both the number of individuals introduced and the number of release events or invasions. Propagule pressure is expected to be an important factor in the establishment of introduced species on the basis of demography alone. Propagule pressure has played an important role in causing introgression of genes from invasive species into native species (e.g., Muhlfeld et al. 2017).

There are two primary ways in which the genetics of an introduced species may be affected by propagule pressure. First, a greater number of founding individuals would be expected to reduce the effect of any population bottleneck so that the newly established population would have greater genetic variation. Multiple introductions over time, even from the same locations, will also increase genetic diversity. Second, and perhaps most importantly, different releases may originate from different source populations. Therefore, hybridization between individuals from genetically divergent native populations may result in introduced populations having more genetic variation than native populations of the same species (Section 14.4).

14.3.2 Spread

Many recently established species often persist at low, and sometimes undetectable, numbers and then "explode" to become invasive years or decades later (Sakai et al. 2001). Adaptive evolutionary genetic changes may explain the commonly observed lag time that is seen in many species that become invasive (García-Ramos & Rodríguez 2002). Many of the best examples of rapid evolutionary change come from the study of recently introduced populations (Lee 2002; Prentis et al. 2008; Feng et al. 2019). Changes in environmental conditions in the invaded location, such as changing climate or new disturbance regimes, may also provide opportunities for expansion and increased spread of invasive species that have previously persisted at low levels. Some introduced species persist at low abundance until another interacting species is introduced, such as species that are a food source (e.g., Ellis et al. 2011).

14.4 Hybridization as a stimulus for invasiveness

Hybridization can play an important role in introduced species becoming invasive. A meta-analysis by Hovick & Whitney (2014) found some evidence for association of hybridization with invasiveness, weediness, or range expansion, although experimental tests are lacking for many species. As we have seen in the previous sections, many species become invasive only (1) after an unusually long lag time after initial arrival, and (2) after multiple introductions. Ellstrand & Schierenbeck (2000) have proposed that hybridization between species,

or genetically divergent source populations, may serve as a stimulus for the evolution of invasiveness on the basis of these observations. Ellstrand & Schierenbeck (2000) proposed four genetic mechanisms to explain how hybridization can stimulate invasiveness:

1. Evolutionary novelty. Hybridization can result in the production of novel genotypes and phenotypes that do not occur in either of the parental taxa. Evolutionary novelty can result either from the combination of different traits from both parents or from traits in the hybrids that transgress the phenotypes of both parents (transgressive segregation; Section 13.2.1).
2. Genetic variation. An increase in the amount of genetic variation may in itself be responsible for the evolutionary success of hybrids. That is, the greater genetic variation (heterozygosity and allelic diversity) in hybrid populations may provide more opportunity for natural selection to bring about adaptive evolutionary change.
3. Heterosis. Many invasive plant species have genetic or reproductive mechanisms that stabilize first-generation hybridity and thus may fix genotypes at individual or multiple loci that demonstrate heterosis. These mechanisms include **allopolyploidy**, **permanent translocation heterozygosity**, **agamospermy**, and clonal reproduction. The increased fitness resulting from fixed heterozygosity may contribute to the invasiveness of many plant species.
4. Reduction in genetic load. As we have seen, small isolated populations will accumulate deleterious recessive mutations so that mildly deleterious alleles become fixed, and this leads to a slow erosion of average fitness (Chapter 17). Hybridization between populations would reduce this mutational genetic load (Whitlock et al. 2000; Morgan 2002). Ellstrand & Schierenbeck (2000) have suggested that the increase in fitness of this effect may under some circumstances be sufficient to account for invasiveness.

Hybridization is involved in evolution of the common cordgrass that has been identified as one of the world's worst invasive species by the International Union for the Conservation of Nature (Lowe et al. 2000). It is a perennial salt marsh grass that has been planted widely to stabilize tidal mudflats. Its invasion and spread caused the exclusion of native plant species and the reduction of suitable feeding habitat for wildfowl and waders. This species originated by chromosome doubling of the sterile hybrid between the eastern hemisphere *Spartina maritima* and the western hemisphere *S. alterniflora*. Genetic analysis has found an almost total lack of genetic differences among individuals. However, the allopolyploid origin of this species has resulted in fixed heterozygosity at all loci for which these two parental species differed.

14.5 Eradication, management, and control

Understanding the biology of invasive species is not necessary before taking action (Simberloff 2001). Simberloff has described this as a policy of "Shoot first, ask questions later" (see also Ruesink et al. 1995). This recommendation is in agreement with basic population biology. The best way to reduce the probability that an introduced species becomes invasive is to eliminate it before it becomes abundant, widespread, and has had sufficient time to evolve any adaptations that may allow it to outcompete or otherwise impact native species. Nevertheless, understanding population biology, genetics, and evolution is important for development of control and management of invasive species in situations where eradication is not feasible, and may be helpful to predict the potential for invasive species to evolve responses to management practices. It may also be critical in the case of cryptic invasive species that cannot be morphologically differentiated from native species.

Genetics can also contribute to policy in management and control of invasive species. Regulations generally have not taken into account that some genotypes may be more invasive than others of the same species. The standards set by the International Plant Protection Convention that importation cannot be restricted for entities that are already widespread and not the object of an "official" control program have generally been applied at the species level (Baskin 2002), although there is growing recognition that this policy may need to be applied at the subspecies or genotype level.

Given observed variation in invasiveness in many species, constraints on future introductions of additional strains of established alien species would be prudent, both to avoid introduction of strains with greater invasiveness and to prevent opportunities for increased invasiveness through hybridization or improved genetic diversity.

14.5.1 Units of eradication

Eradication of introduced species is a potentially valuable ecological restoration strategy (Myers et al. 2000). The eradication of rats, mice, and other introduced species is becoming increasingly common on oceanic islands and isolated portions of continental land masses. The New Zealand Department of Conservation applied 120 metric tons (120,000 kg) of poison bait onto Campbell Island, a large Subantarctic island (11,300 ha) located ~700 km south of the New Zealand mainland. A survey of the island in 2003 found no trace of brown rats on the island, and an incredible recovery of bird and insect life. Eradication of feral cats and goats on Dirk Hartog Island off the coast of Western Australia is allowing restoration of the diverse mammal fauna that was previously abundant on the island prior to white settlement (Algar et al. 2020).

Successful eradication requires a low risk of recolonization. Isolated island populations or populations limited to isolated "habitat islands" have low risk of recolonization. However, other islands or regions that display no distinct geographical structure or barriers are more problematic. The eradication of a portion of a population, or a sink population within a **metapopulation**, would result in rapid recolonization and a waste of resources (see examples in Myers et al. 2000). Fewster et al. (2011) have used the amount of genetic divergence between island and mainland populations to predict the likelihood of successful reinvasion of ship rats in New Zealand.

Genetic analysis can provide important information about the source of reinvasion as well. For example, several species of mammalian pests were removed from Rangitoto Island in the Hauraki Gulf of New Zealand in 2009 (Veale et al. 2012). A single stoat was found on the island in 2010, and it was not known if this individual was a remnant from

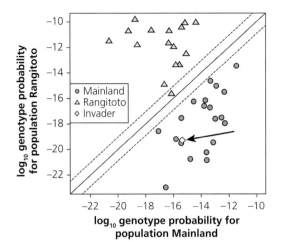

Figure 14.5 Molecular genetic analysis of a stoat found on Rangitoto Island 1 year after eradication. Log posterior probability plot for stoats from the pre-eradication population (*y*-axis) and the mainland population (*x*-axis). Triangles indicate pre-eradication stoats, circles indicate mainland stoats, and stoat found on the island 1 year after eradication is indicated by a diamond and an arrow. Points below the solid diagonal line have greater posterior probability of belonging to the mainland than to Rangitoto. Points outside the dashed diagonal lines have over nine times greater posterior probability of belonging to one population than the other. The post-eradication stoat is ~10,000 times as likely to have originated from the mainland as from the pre-eradication population. From Veale et al. (2012).

the eradicated populations or a migrant that swam at least 3 km from the mainland. Genetic analysis clearly determined that this individual was a new migrant, rather than a remnant from the eradicated population (Figure 14.5).

Genetics can be used to identify isolated reproductive units that are appropriate groups for eradication or control on the basis of patterns of genetic divergence (Calmet et al. 2001; Robertson & Gemmell 2004; Adams et al. 2014). Little genetic differentiation between spatially isolated populations is indicative of significant gene flow, while significant differentiation between adjacent populations indicates limited dispersal. Examination of the patterns of genetic variation can allow for the identification of distinct population units with negligible immigration. With appropriate care, these population units could be eradicated with little chance of recolonization. The identification of "units of eradication" is an interesting analog to the identification of units of conservation as seen in

Chapter 20 (Robertson & Gemmell 2004). Genetic analysis would also allow distinction between an eradication failure (i.e., recovery by a few surviving individuals) and recolonization.

Brown rats were unintentionally introduced to South Georgia Island in the Antarctic when commercial sealing started there in the late 1700s and have devastated the island's avifauna. Robertson & Gemmell (2004) examined 18 microsatellite loci in two populations of brown rats separated by a glacier on South Georgia Island in the Antarctic. The eradication of rats from the entire island was a daunting task because of its great size (400,000 ha). However, appropriate rat habitat is limited to coastal regions that are often separated by glaciers, permanent snow and ice, and icy waters. If such barriers preclude dispersal, then each discrete population could be considered as a discrete eradication unit and eradicated sequentially. These two populations showed substantial genetic differentiation that allowed individuals to be readily assigned to the correct population of origin. Robertson and Gemmell concluded that there was little gene flow between these populations and that these populations comprised distinct eradication units that could be eradicated sequentially with low risk of cross-recolonization. These two populations have since been eradicated sequentially so that brown rats no longer occur on South Georgia Island (Bruce Robertson, personal communication).

14.5.2 Genetics and biological control

Invasive species can undergo rapid adaptive evolution during the process of range expansion. Here, such evolutionary change during invasions has important implications for biological control programs (Wilson 1965). The degree to which such evolutionary processes might affect biological control efficacy remains largely unknown (Müller-Scharer et al. 2004). The rate of change in response to natural selection is proportional to the amount of genetic variation present (Fisher 1930). Therefore, the amount of heterozygosity or allelic diversity at molecular markers that are likely to be neutral with respect to natural selection may provide an indication of the amount of genetic variation at loci that potentially could be involved in response to a control agent. For example, greatly reduced molecular variation indicating a small effective population size during the founding event would give an expectation of reduced variation at adaptive loci. In contrast, greater molecular variation indicating introductions from multiple source populations would lead to expectations of substantial adaptive genetic variation to escape the effects of a control agent.

The first applications of genetics in invasive species control have been in association with the sterile insect technique. One approach has been to introduce genotypes that could subsequently facilitate control, or render the pest innocuous (Foster et al. 1972). Another approach was to release genotypes with chromosomal aberrations whose subsequent segregation would result in reduced fertility and damage the population (Foster et al. 1972). Later efforts using the sterile insect technique have used transgenic insects homozygous for repressible female-specific lethal effects (Thomas et al. 2000; Dyck et al. 2005). Release of males with this system may be an effective mechanism of control of some insect pests. Similar approaches to production of sterile males and females have also been developed for control of invasive fish (Thresher et al. 2014) as well as sex biasing systems where the female progeny are lethal (Bajer et al. 2019).

Hodgins et al. (2009) have modeled the use of a selfish genetic element found in plants called **cytoplasmic male sterility (CMS)** for controlling invasive weeds. CMS is caused by mutations in the mitochondrial genome that sterilize male reproductive organs. They found that the introduction of a CMS allele can cause rapid population extinction, but only under a restricted set of conditions. They conclude that this approach would work only with species where pollen limitation is negligible, inbreeding depression is high, and the fertility advantage of females over hermaphrodites is substantial.

14.5.3 Pesticides and herbicides

Invasive species can often evolve quickly in response to human control efforts (Example 14.4). Therefore, application of genetic principles is important for developing effective controls for invasive species. The evolution of resistance to insecticides and herbicides has increased rapidly in many species over the past 50 years (Denholm et al. 2002). Reducing the evolution of resistance to chemical control measures in invasive species will require an understanding of the origin, selection, and spread of resistant genes. The amount of genetic diversity can indicate the potential for evolution of resistance to pesticides and herbicides. Comparison of the genomes of insect species such as *Drosophila* and malaria vector mosquitoes should aid the development of new classes of insecticides, and should also allow the lifespan of current pesticides to be increased (Hemingway et al. 2002).

14.5.4 Gene editing and gene drive

Gene editing and **gene drive** are new genetic technologies that have potential to provide significant outcomes in conservation, particularly for the control of invasive pests and the reduction of toxicity or resistance to disease. Gene drive is a process where one allele of a gene is inherited at a higher frequency than the other, rather than alleles being inherited in a Mendelian manner. This allows the favored allele to rapidly spread through a population (Esvelt et al. 2014; Gantz et al. 2015). Gene drives can be engineered into organisms through the insertion of a selfish genetic element using genetic modification techniques or through gene editing with **CRISPR-Cas9** (clustered regularly interspaced short palindromic repeats and CRISPR-associated protein 9). CRISPR-Cas9 evolved as an immune system in prokaryotes that recognizes and cuts DNA from pathogens. This system of sequence recognition and cutting has been harnessed in biotechnology to insert, delete, or alter DNA sequences in specific locations within genomes—a far more precise system than traditional genetic modification through gene insertion. Jennifer Doudna and Emmanuelle Charpentier were awarded the Nobel Prize in Chemistry in 2020 for the discovery of CRISPR-Cas9 gene editing. This technique can be used in combination with gene drive to introduce alleles that disrupt or modify existing genes; for example,

Example 14.4 Evolution of herbicide resistance in the invasive aquatic plant hydrilla

In the early 1950s, a female form of a dioecious strain of hydrilla was released into the surface water of Florida in Tampa Bay and spread rapidly throughout the state (Michel et al. 2004). Today hydrilla is one of the most serious aquatic weed problems in North America. This invasive plant can rapidly cover thousands of contiguous hectares, displacing native plant communities and causing significant damage to ecosystems.

Hydrilla has been controlled by the sustained use of the herbicide fluridone in lake water for several weeks. An apparent decrease in the effectiveness of fluridone to control hydrilla has been observed in a number of lakes. Evolution of herbicide resistance was considered unlikely in the absence of sexual reproduction. Nevertheless, a major effort was undertaken in 2001 and 2002 to test for herbicide-resistant hydrilla in 200 lakes throughout Florida. No within-site variation in fluridone resistance was detected. Approximately 90% of the lakes contained fluridone-sensitive hydrilla. However, three phenotypes of fluridone-resistant hydrilla were discovered in 20 water bodies of central Florida. A hydrilla phenotype with low resistance was found in eight lakes, the phenotype with intermediate resistance was found in seven lakes, and the most resistant phenotype was found in five lakes.

Sequencing of the phytoene desaturase locus (*pds*) indicated that the three fluridone-resistant types had different amino acid substitutions making them two to five times less sensitive to fluridone. It appears that fluridone resistance has arisen through somatic mutations that caused a single biotype to quickly become the dominant type within a lake. The establishment of herbicide-resistant biotypes as the dominant forms in these lakes was not anticipated. Asexually reproducing plants are under strong uniparental constraints that limit their ability to respond to environmental changes (Holsinger 2000).

The future expansion of resistant biotypes poses significant environmental challenges in the future. Weed management in large water bodies relies heavily on fluridone, the only approved synthetic herbicide available for systemic treatments of lakes in the USA. Current plans include regular monitoring to detect resistance and prevent the spread of these herbicide-resistant biotypes.

resulting in sterility of pests or pathogens and halting their spread. Gene drive is being investigated as a means to control malaria-carrying mosquitoes, providing a significant health benefit (Hammond et al. 2016).

Gene drive is also being considered for applications in agriculture and in wildlife conservation (Rode et al. 2019). An application that has gained considerable attention is the management of mammal pests through disruption of the sex-determining pathway by inheritance of the male-determining *sry* gene or through shredding of the X chromosome (Esvelt et al. 2014; Harvey-Samuel et al. 2017). This would lead to skewing of the sex ratio in a population until all animals are male and the population is extirpated. Such an application is appealing in wildlife management, as it is species-specific and avoids issues of nontarget impacts; however, the technology still requires significant research to overcome technical hurdles associated with development of resistance (Prowse et al. 2017).

While the potential benefits to wildlife management from effective pest animal control are significant, there are also significant concerns over use of gene drives due to the potential for their spread outside the intended areas. This could lead to the extinction of species; for example, from pest possums in New Zealand to native possums in Australia, or from pest cane toads in Australia to native populations in South America. There is significant research going into technical solutions to control gene drives (Dhole et al. 2018; Sudweeks et al. 2019). There is recognition from all sectors, including scientists, ethicists, governments, and community organizations that development of gene editing requires evaluation in a risk management framework (Dearden et al. 2018). There are numerous international frameworks being developed for regulation of this technology. These environmental risk assessments will require a great deal of knowledge of the genetics and ecology of the target organisms (Moro et al. 2018). CRISPR-Cas9 might also facilitate amplification of **environmental DNA**

(eDNA) from water or soil samples to facilitate early detection of invasive species (e.g., Phelps 2019).

14.6 Emerging diseases and parasites

An emerging disease or parasite is one that has appeared in a population for the first time, or that existed previously but is rapidly increasing in **prevalence** or geographic range. Emerging parasites, including bacteria, viruses, protozoans, fungi, and helminths, are increasingly problematic for many wildlife and plant populations of conservation concern. An example is the bacterium *Yersinia pestis* that causes the plague in small mammals and humans worldwide. Black-footed ferrets can directly contract *Y. pestis* and this disease has extirpated numerous prairie dog populations, consequently threatening black-footed ferrets, which depend on prairie dogs as prey. Loss of prairie dogs and their influence modifying landscapes also could reduce biodiversity of entire grassland ecosystems. Genetic analysis of *Y. pestis* **isolates** indicated that some human *Y. pestis* isolates were genetically similar to *Y. pestis* originating from Colorado chipmunks and from the "chipmunk flea," suggesting that chipmunks are potential sources of human *Y. pestis* infection in Colorado (Figure 14.6).

DNA markers offer enormous potential to understand the causes and consequences of parasite infection, including the emergence, spread, persistence, and evolution of infectious disease (Archie et al. 2009). Genetic markers are increasingly used to identify the source population of origin of outbreaks, trace the pathway of spread, predict future transmission corridors and barriers following climate change, and identify genes underlying phenotypes or fitness of successful invaders (Section 19.2.1; Roe et al. 2019).

14.6.1 Detection and quantification of disease vectors

DNA markers are increasingly used to detect insect, fungal, and microbial parasites in samples from the environment and from tissues or noninvasive

samples from individual organisms (e.g., Beja-Pereira et al. 2009a). Detection is a crucial first step in determining **prevalence** of disease vectors and the effects of infection on individual fitness and population persistence. For example, Lindner et al. (2011) developed a **quantitative polymerase chain reaction** (qPCR) diagnostic test to identify and track white-nose syndrome (WNS), an infectious disease that has caused major reductions in bat populations in western North America. Over 1,000,000 bats from six species in eastern North America have died from WNS since 2006, several species of bats may become endangered or extinct, and the disease is spreading rapidly (Foley et al. 2011). The qPCR test detects low-level amounts of the disease-causing fungal pathogen *Pseudogymnoascus destructans* on bats and in soil samples collected from bat hibernacula. The qPCR test and a subsequent more specific assay that also detects close relatives of the fungus (Shuey et al. 2014) are crucial to allow assessment of effects of the fungus on individual viability, characterization of the fungal lifecycle, and tracking of the spread across North America. Similarly, qPCR analyses have been developed to detect presence and abundance of two viruses causing disease and mortality in red squirrels, with gray squirrels being carriers of the virus (Dale et al. 2016). qPCR assays also help detect the world's most common bacterial **zoonoses**, *Brucella* spp., which cause brucellosis, a significant disease leading to abortion, inflammation of gonads, and reduced reproductive success in wild and domestic animals.

Genomic approaches such as metagenomics and metabarcoding are increasingly used for parasite and pathogen detection. For example, a single metabarcoding test allows detection of multiple parasite species, discovery of new invasive parasite taxa, and simultaneous detection of insect vector species spreading microbial pathogens (Batovska et al. 2018; Piper et al. 2019).

14.6.2 Tracking origins of infectious disease outbreaks

DNA markers can help determine the species or population of origin of disease outbreaks among

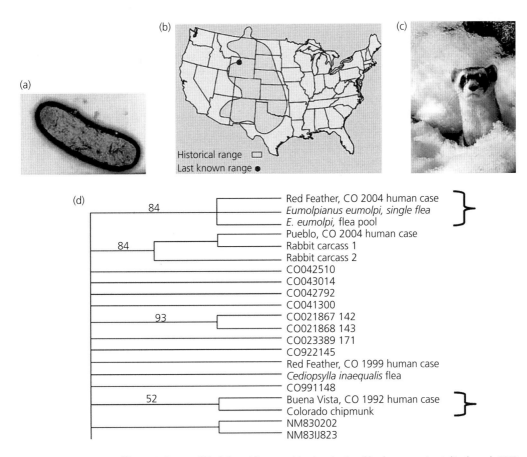

Figure 14.6 (a) *Yersinia pestis*. (b) Historical range of black-footed ferrets and last location in wild to become extirpated in the early 1980s. (c) Black-footed ferret. (d) Consensus maximum parsimony tree based on 17 microsatellite loci of *Y. pestis* plague isolates from humans, Colorado chipmunks, chipmunk fleas, and rabbits from Colorado or New Mexico (CO- or NM-labeled branch tips). The two numbers following the state abbreviation indicate the year collected. Numbers on the branches of the tree are jackknife support values. Brackets indicate closely related isolates found in humans and chipmunks. From Lowell et al. (2009).

wild and domestic animals. Identifying the source of infectious disease outbreaks is difficult, especially for pathogens that infect multiple wildlife species, such as brucellosis. Genome sequencing of *Brucella abortus* isolates from bison, elk, and cattle in the Greater Yellowstone Ecosystem show multiple transmissions between elk and bison, and identified elk as a key reservoir of the pathogen and the origin of recent outbreaks of brucellosis in cattle in the Greater Yellowstone Area (Kamath et al. 2016). This illustrates the power of genetics to assess the origin of disease outbreaks, which are increasing worldwide following habitat fragmentation, climate change, and expansion of human and livestock populations.

Sudden oak death is a disease caused by the oomycete *Phytophthora ramorum*. Since the mid-1990s, this pathogen has caused extensive mortality in several species of oaks, particularly coastal live oak and tanoaks in coastal California and southern Oregon, and widespread infection in larch and other conifers in Europe (Grünwald et al. 2019). It is not known where this pathogen originated, but it is thought to have been introduced to North America and Europe from different source populations, possibly in Asia. Genetic and genomic studies have shed considerable light on the population structure and the role of human activity in spreading the pathogen and causing this epidemic (Example 14.5).

Example 14.5 Origins, spread, and quarantine strategies of sudden oak death

Sudden oak death affects several species of oak as well as rhododendron, larch, and conifers, and is caused by the oomycete pathogen *Phytophthora ramorum*. It was first detected in the 1990s and has caused extensive mortality in forests of the USA and Europe. It has a high economic impact on trade and cost of control efforts, and significant effects on ecosystems and food webs. Genetic and genomic studies have provided information on the origins and spread of this pathogen (Grünwald et al. 2019).

The pathogen has two mating types (A1 and A2), although these have only been found in Asia and this may represent the center of origin of the pathogen. Sexual reproduction has not been found in the outbreaks in North America and Europe, where four clonal lineages have been identified. These lineages represent three highly diverged **clades** based on a range of genetic markers (Figure 14.7; Grünwald et al. 2009), each likely derived from the introduction of a single clone, and all are currently reproducing clonally. One lineage is widespread in both forests and plant nurseries in North America (NA1), another has only been found in a few nurseries in California and Washington State (NA2), and the third is found primarily in Europe (EU1), but occurs sporadically in North American plant nurseries. NA1 and NA2 are both mating type A2, and so are not compatible with each other. When EU1 samples of the A1 mating type were detected in North American nurseries, there was fear that this would lead to sexual reproduction and recombination with NA1 or NA2, and great efforts were made to sanitize nurseries to prevent this from happening. It now appears that the EU1 lineage may be sexually incompatible with the North American lineages, from which it has long been separated.

Population genetic analysis (Goss et al. 2009; Prospero et al. 2009) has clearly shown that this pathogen has spread in North America through the long-distance transport of horticultural plants between nurseries and subsequent infection of native forests. Studies of mating type in lineages have also made it possible to establish quarantine zones and possibly prevent much broader dissemination of the pathogen.

14.6.3 Assessing transmission routes

Molecular genetic analysis to infer pathogen transmission patterns (**molecular epidemiology**) is increasingly used to assess, predict, and manage risks of disease spread. Transmission routes of parasites are now widely investigated using population genetics and phylogenomic data (e.g., Kozakiewicz et al. 2018; Dupuis et al. 2019; Fitak et al. 2019; Hamelin & Roe 2019).

Biek et al. (2007) used phylogenetic analysis of raccoon rabies virus (RRV) to distinguish seven genetic lineages of the virus. Each of the seven lineages exhibited a general direction of spread relative to the first reported cases in West Virginia, USA. The spread routes toward the southeast, southwest, and east were each associated with one particular viral lineage, whereas spread in the northwest and northeast direction were each marked by two different groups. Counties sampled 5–25 years after their first raccoon rabies cases consistently yielded viruses of the same genetic lineages that had colonized that area initially. This phylogeographic study illustrates the potential usefulness of sampling and sequencing years after an outbreak to help infer the routes of spread, which could help conservation geneticists understand, manage, and predict future disease transmission.

A more recent study of RRV used whole genome sequencing and fine spatial scale phylogenetic cluster analysis to explore the role of geography and alternative carnivore hosts in the dynamics of RRV spread. Sequencing of 160 RRV samples provided evidence of skunk-to-skunk and skunk-to-raccoon transmission but no evidence for skunks as alternative reservoir hosts (Nadin-Davis et al. 2018). These examples highlight the tremendous potential of genetic and genomic approaches to inform the biology and control of emerging diseases and parasites, and of invasive species more generally.

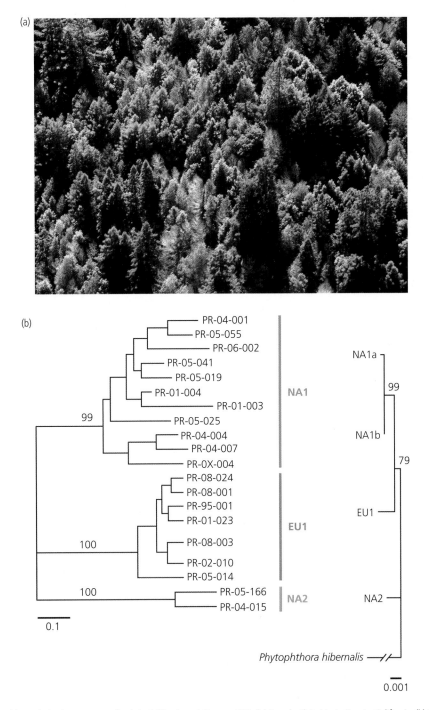

Figure 14.7 Sudden oak death causes tree death in California and Oregon, USA. (a) Tree death in Marin County, California. (b) Genetic analysis of isolates from the invasive range has identified three divergent lineages of the pathogen, named NA1, EU1, and NA2. Left shows a neighbor-joining tree based on six microsatellite loci; right shows a maximum likelihood tree based on approximately 5 kb of sequence from eight mitochondrial regions. From Grünwald et al. (2019).

Guest Box 14 Rapid evolution of introduced cane toads
Richard Shine and Lee Ann Rollins

The cane toad is one of the largest anurans, with adults sometimes weighing more than 1 kg (Shine 2010; Figure 14.8). Native to South and Central America, these voracious predators were brought to northeastern Australia in 1935 in a misguided attempt at biological control of insect pests. The toads rapidly spread across tropical and subtropical Australia, killing many native predators. The lack of native toads in Australia meant that most species of native predators had no evolutionary history of exposure to the distinctive toxins of the invader. Many lizards, snakes, crocodiles, and marsupials have been killed when they tried to eat toads, and populations of some species have crashed dramatically (e.g., yellow-spotted monitor, front cover; Shine 2010).

Rapid evolutionary changes have already occurred both in the toads and in their Australian predators. In the toads, the biggest changes have been in traits that increase rate of invasion. The cane toad invasion has been like a giant footrace across thousands of kilometers, and in every generation the fastest-dispersing toads inevitably will be the ones closest to the invasion front. These fast dispersers breed with each other, producing offspring that in some cases are even quicker than either parent—and so, the rate at which cane toads have invaded has increased steadily over the years. In the 1950s and 1960s, the toad front moved at about 10 km

per year, but it has been accelerating ever since and now averages around 50–60 km per year (Phillips et al. 2006; Shine et al. 2011).

Radio-tracked toads at the invasion front often move more than 1 km per night in a highly directional way, whereas toads in established populations move much less, without any consistent dispersal direction (Alford et al. 2009). By raising offspring of toads from different populations under identical conditions, Phillips et al. (2010) showed that this dispersal shift is genetically based: offspring inherit the dispersal rates of their parents.

To clarify the mechanisms underlying this rapid shift in heritable traits, we have developed genomic resources (e.g., multi-tissue transcriptome, Richardson et al. 2018; a reference genome, Edwards et al. 2018) that, paired with the ecological research, make toads an excellent model to study evolution at the molecular level. The molecular studies have confirmed that environmental (climatic) factors, as well as the invasion process *per se,* have shaped evolutionary trajectories in cane toads (Selechnik et al. 2019). Although neutral diversity decreases across the invasion trajectory, diversity at loci putatively under selection increases in some invaded sites, illustrating that standing genetic variation is not reduced by sequential introductions. In turn, that result

Figure 14.8 A cane toad on a gibber plain at Longreach, Queensland, Australia. Photo courtesy of Rick Shine.

Guest Box 14 Continued

shows the importance of studying coding loci rather than neutral loci to determine the genetic health of populations.

A genomic approach also can identify targets for future control of invasives, with positive impacts on vulnerable native species. For example, metatranscriptomic analysis revealed novel viruses within the genome of the toad, showing ancestral viral infection with subsequent genomic integration (Russo et al. 2018). Such viruses may suggest novel agents for biocontrol. More generally, we are investigating the degree to which epigenetic modifications of gene expression (rather than shifts in allele frequencies) may underpin the dramatically rapid rate of phenotypic evolution in cane toads. For example, we have experimentally manipulated methylation levels to investigate the proximate bases of phenotypic plasticity in toads from across their Australian range (Sharma et al. 2013).

Managers need to consider the possibility of rapid evolutionary changes, both in invasive species and in the native fauna and flora that are affected by invaders (Ashley et al. 2003). For example, understanding the evolved acceleration of invasion fronts can help managers predict how quickly the invading species will reach areas of specific conservation concern, and hence how urgently they need to manage those areas to combat invader impact. If we understand the evolutionary processes that rapidly reshape an invader as it colonizes new areas, we may be able to design new and more effective approaches to control that invader.

CHAPTER 15

Exploited Populations

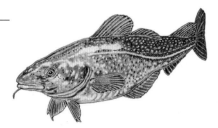

Atlantic cod, Box 15.2

I suggest that minimizing the impact of sport hunting on the evolution of hunted species should be a major preoccupation of wildlife managers. **(Marco Festa-Bianchet 2003, p. 191)**

Human actions cause rapid evolutionary change in many species, but the underlying genomic basis remains poorly understood. **(Nina O. Therkildsen et al. 2019, p. 487)**

Humans have **harvested** animals and plants from the wild since the beginning of our species. Darwin discussed the lack of wildness of birds on islands where they have not been hunted by humans (1859, p. 231). He was especially impressed by the many bird species that he saw on the Galápagos that did not respond to the presence of humans (Darwin 1839, p. 475).

There is mounting evidence that **overexploitation** has led to the direct demographic extinction of many populations and species (Burney & Flannery 2005). Genetic changes brought about by **exploitation** pose a less visible threat than direct extinction. Nevertheless, such genetic changes can greatly increase the complexity of managing populations in order to maintain sustainability (Palkovacs 2011; Dunlop et al. 2015; Hutchings & Kuparinen 2020).

Many natural resource managers have been reluctant to accept the potential for harvest to cause genetic change, and many are doubtful that any such changes are harmful (Harris et al. 2002; Conover 2007; Hutchings & Fraser 2008). However, intense and prolonged mortality caused by exploitation will inevitably result in genetic change. Harvest need not be selective to cause genetic change;

uniformly increasing mortality independent of phenotype will select for earlier maturation (Law 2007). Genetic changes caused by exploitation can increase extinction risks and reduce recovery rates of overharvested populations (Olsen et al. 2004a; Walsh et al. 2006; Hutchings & Kuparinen 2020).

Most of the concern in the literature about genetic changes caused by exploitation has focused on marine and freshwater fish populations and hunted ungulate populations. However, an incredible variety of wild animal and plant populations are exploited by humans: terrestrial game birds, waterfowl, whales, snakes, turtles, land snails, a wide range of marine invertebrates (anemones, sea urchins, sponges, sea cucumbers, jellyfish), marine birds, kangaroos, forest primates, trees, cycads, plants with attractive flowers, butterflies and other colorful insects, and so on (Examples 15.1 and 15.2). The same concerns of genetic change elicited by harvest apply to all of these species. For example, the size at sexual maturity in rock lobsters off the west coast of Australia declined substantially over 35 years (Figure 15.1). This change apparently is partially an evolutionary response to extremely high annual exploitation rates of adults (~75%),

Conservation and the Genomics of Populations, Third Edition. Fred W. Allendorf, *et al.*, Oxford University Press.
© Fred W. Allendorf, W. Chris Funk, Sally N. Aitken, Margaret Byrne, and Gordon Luikart (2022). DOI: 10.1093/oso/9780198856566.003.0015

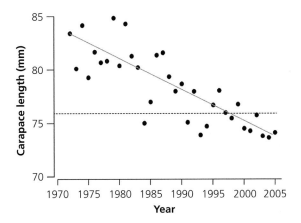

Figure 15.1 Observed decline in mean carapace length of rock lobsters captured in the fishery near Jurien off the coast of Western Australia from 1972 to 2005 (Melville-Smith & de Lestang 2006). Only animals with a carapace length of greater than 76 mm (dotted line) can be legally harvested. This decline apparently is partially an evolutionary response to extremely high annual exploitation rates of adults (~75%), combined with the required minimum carapace length. From Allendorf et al. (2008).

combined with a required minimum carapace length of 76 mm in harvested animals (Melville-Smith & de Lestang 2006).

Understanding the genetic changes and evolutionary responses of exploited populations is crucial for the planning of sustainable management of natural biological resources. In this chapter, we consider how harvest can affect the genetics of wild populations by reducing effective population size, selecting for certain phenotypes, changing spatial population structure, and releasing individuals. Finally, we consider the management and recovery of exploited populations from such genetic changes.

15.1 Loss of genetic variation

Harvesting can cause the loss of genetic variation in several ways: (1) direct reduction of population size (N_C), (2) reduction of the N_e/N_C ratio by increasing variance in individual reproductive success, and (3) reduction of the number of migrants into a local population (Section 15.3). Kuparinen et al. (2016) have provided a very useful review of the effects of harvesting on N_e and the loss of genetic variation. Genetic variation can be reduced by these effects even in very large populations (Box 15.1).

The magnitude of local N_e is determined by demographic factors including N_C (census population size), sex ratio, and the mean and variance of lifetime number of progeny produced by males and females (Chapter 7). Harvest often targets specific sex or age classes and thereby can reduce the effective population size, and increase the rate of loss of genetic variation. This effect is often exacerbated by ongoing habitat loss resulting in decreased population size and increased isolation. Many papers report reduced genetic variation in a wide variety of exploited species (Table 15.1). For example, harvesting to produce rodent poison and making ropes has accelerated the rate of loss of genetic variation in isolated populations of a long-lived cycad tree that is endemic to Cuba (Pinares et al. 2009).

Bishop et al. (2009) reported that uncontrolled hunting of Nile crocodiles in the Okavango Delta, Botswana, in the mid- to late 20th century substantially reduced the census and effective population size. They estimated that the current census size is less than 10% of the historical population size, and inferred a recent loss of genetic variation using the bottleneck test of Piry et al. (1999) with seven microsatellite loci. They also estimated that the contemporary N_e/N_C ratio is 0.05. Simulations indicated that ongoing removal of adults will cause continued loss of allelic diversity and heterozygosity in this population.

Management operates in calendar time (e.g., years), whereas knowing N_e allows the prediction of the loss of heterozygosity per generation. When considering loss of variation over calendar time, a small N_e can be compensated by a large generation interval (G; Section 7.9.2). Therefore, consideration of the effects of management on loss of genetic variation over time should not be restricted, as they often are, to N_e alone because effects of G are equally important (Sæther et al. 2009). For example, Ryman et al. (1981) found that different harvest regimes for moose can have strong effects on both effective population size and generation interval (Figure 7.9). Populations with smaller N_e tended to lose heterozygosity at a slower rate because those effects of hunting that reduced N_e (e.g., harvesting young animals) also tended to increase the generation interval. That is, hunted populations with relatively smaller N_e and a longer generation interval would

Example 15.1 Rapid evolution in a hunted snake

The Japanese mamushi snake has been heavily hunted for personal and commercial use because of its medicinal and nutritional value, and it has been killed indiscriminately in large numbers in more recent years (Figure 15.2). Many populations of this snake are declining or have been extirpated. Sasaki et al. (2009) compared several phenotypes in four local populations that have been hunted regularly with six populations that have not been hunted.

Figure 15.2 Adult female mamushi snake from Japan. Photo courtesy of Kiyoshi Sasaki.

Mamushi in hunted populations were smaller and had increased reproductive effort. In addition, snakes from hunted populations fled at greater distances from an approaching human (Figure 15.3). Heritability estimates for body size and avoidance of humans were all substantially greater than zero ($P < 0.05$). Sasaki et al. (2009) concluded that predation by humans selected for traits that would normally be disadvantageous under natural conditions in the absence of humans.

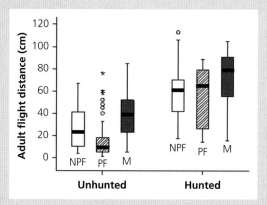

Figure 15.3 Escape (flight) distance of adult mamushi snakes in hunted ($n = 89$) and unhunted ($n = 198$) populations (NPF, nonpregnant female; PF, pregnant female; M, male). Bars show median and 25th and 75th percentiles (interquartile range) and whiskers extending to the highest and lowest values, excluding outliers (open circle, values >1.5 times the interquartile range) and extreme values (asterisk, values >3 times the interquartile range). From Sasaki et al. (2009).

Box 15.1 Loss of genetic variation in overharvested marine fishes

Figure 15.4 Approximately 400 tons of Inca scad caught by a purse seiner off the coast of Peru in 1997. This is one of the overfished populations included in Pinsky & Palumbi (2014). Photograph courtesy of C.O. Rojas, National Oceanic and Atmospheric Administration, US Department of Commerce.

Pinsky & Palumbi (2014) performed a meta-analysis on genetic variation in 140 species of highly abundant marine fish species (Figure 15.4). They found that overfished populations had ~2% reduced heterozygosity and 12% reduced allelic richness compared with populations that are not overfished. They also performed simulations which suggested that their estimates likely underestimate the actual loss of rare alleles by a factor of three or four. This important paper shows that the harvesting of marine fish can have genetic effects that threaten the long-term sustainability of this valuable resource. This evidence suggests that overharvest has reduced genetic variation in a wide range of marine fishes.

Are such small losses of genetic variation likely to be important? An average 2% reduction of heterozygosity does not sound like much. Nevertheless, this is the amount of heterozygosity expected to be lost in a single generation if N_e is 25 ($-1/2N_e$ per generation). Alternatively, this amount of loss of heterozygosity is expected with an N_e of ~250, if

the loss is spread out over 10 generations. Such bottlenecks could have a major effect on a population's viability, but intense bottlenecks are more likely to have harmful genetic effects than the same amount of loss of heterozygosity spread out over many generations (England et al. 2003). Thus, it is important to know the time frame of loss in the harvested populations surveyed by Pinsky & Palumbi (2014) to understand their likely effects.

The observed 12% loss of allelic diversity could have important effects, especially for loci having many alleles. Loci associated with adaptation often have more alleles than neutral loci because of balancing selection (e.g., the major histocompatibility complex). As expected, greater loss of allelic diversity has been observed at "adaptive" loci that have more alleles than neutral loci (Sutton et al. 2011). Such loss might be more significant because of the need for future adaptation in view of climate change affecting ocean conditions (e.g., temperature, acidity, etc.).

lose genetic variation over time (not generations) more slowly than some populations with large N_e and shorter generation interval. Similar interactions between generation interval and N_e were found for harvest of young bison in Yellowstone National Park (Pérez-Figueroa et al. 2012).

Harvest regulations can reduce the N_e/N_C ratio and thereby increase the rate of loss of heterozygosity per generation without having any detectable effect on census population size. Male-biased or male-only harvest is practiced in many species of ungulates and some marine crustaceans (e.g.,

Table 15.1 Examples of loss of genetic variation in exploited populations.

Species	Observation
African elephant	Intense hunting in the early 1900s combined with slow post-bottleneck recovery and lack of gene flow into Addo Elephant National Park (South Africa) is associated with reduced microsatellite heterozygosity and allelic diversity. In contrast, the Krueger National Park population recovered faster due to immigration after a similar hunting-induced bottleneck and has nearly double the heterozygosity and allelic diversity (Nystrom et al. 2006).
Antarctic fur seal	The Antarctic fur seal was hunted to the brink of extinction by the end of the 19th century. Variation at microsatellite loci suggested that at least four relict populations survived commercial harvesting. Coalescent analysis suggested that all of these populations experienced a severe bottleneck. This conclusion was supported by evidence of severe bottlenecks based upon heterozygosity excess tests (Box 12.1) (Paijmans et al. 2020).
Arctic fox	The Arctic fox population in Scandinavia probably numbered 10,000 historically, but heavy hunting pressure associated with a profitable fur trade in the early 20th century rapidly reduced the population to a few hundred individuals. Analysis of ancient DNA revealed that this population has lost ~25% of the microsatellite alleles and four out of seven mtDNA haplotypes (Larson et al. 2002).
New Zealand snapper	Microsatellite heterozygosity and alleles per locus declined between 1950 and 1988 after commencement of a fishery on this population, in spite of an estimated standing population of well over 3 million fish (Hauser et al. 2002).
Northern elephant seal	The northern elephant seal was heavily hunted and declared extinct in the 19th century. However, the species has since repopulated from a single colony on Guadalupe Island, Mexico. Variation at microsatellite loci demonstrated a severe loss of genetic variation caused by the demographic bottleneck. Much of the genetic variation in the contemporary population has arisen by mutations since the bottleneck some 13 generations in the past (Abadia-Cardoso et al. 2017).
Red deer	Deer in both open and fenced hunted Spanish populations have lower levels of microsatellite heterozygosity than deer from protected areas (Martinez et al. 2002).
Sea otter	Analysis of ancient DNA reveals that all the current populations examined exhibit considerably lower heterozygosities at microsatellite loci than samples predating the population size bottleneck caused by extensive fur trading in the 18th and 19th centuries (Larson et al. 2002).
Sika deer	Three out of seven mitochondrial DNA haplotypes in Hokkaido, Japan, were lost during a 200-year bottleneck caused by heavy hunting reinforced by heavy snow in two winters (Nabata et al. 2004).
Tule elk	The Tule elk of the Central Valley of California, USA, dwindled in 50 years from about half a million down to fewer than 30 animals in 1895 through habitat loss, hunting, and poaching set about by the gold rush. Approximately 60% of heterozygosity was lost, and the present population exhibits little genetic variation (McCullough et al. 1996).
White seabream	Mediterranean populations in areas protected from fishing have significantly greater microsatellite allelic richness than those from nonprotected areas (Perez-Rusafa et al. 2006).

lobsters), and a skewed sex ratio among breeders might severely reduce effective population size. Sex ratios of less than one adult male for 10 adult females are not uncommon in hunted populations of deer and elk (Scribner et al. 1991; Noyes et al. 1996), and are probably common in other species where males are selectively hunted. For example, males comprised less than 1% of all adult elk in the Elkhorn Mountains of Montana in 1985 (Lamb 2010). In addition, harvest regulations can also increase the variance in reproductive success. For example, female brown bear, moose, and wild boar are protected by regulations in Sweden when accompanied by subadults. These policies will also result in the individuals surviving the hunting season being more closely related than expected by chance, thereby further decreasing N_e (e.g., Ryman et al. 1981).

Marine fish and invertebrates generally have much larger census and effective population sizes than terrestrial vertebrates. However, heterozygosity can be lost even in populations with large census population sizes because N_e is often much smaller than the census size in many marine species (Section 7.10; Hauser & Carvalho 2008; Waples 2016).

In addition, loss of allelic diversity can have harmful effects in large, exploited marine populations where the loss of heterozygosity due to harvest is minimal (Ryman et al. 1995b). Allelic diversity is more sensitive than heterozygosity to dramatic reductions in population size, and N_e is a poor predictor of the rate of loss of allelic diversity. That is, populations with the same N_e can lose allelic diversity at very different rates. The reason for this effect is that allelic diversity is affected not only by N_e, but also by N_C (Crow & Kimura 1970, p. 455). Thus, reducing N_C from, for example, millions down to thousands, might have no effect on heterozygosity, but could result in a decline in allelic diversity. In contrast, greater harvest of males through hunting in ungulates could have a limited effect on allelic diversity while reducing heterozygosity because N_C can remain large even when N_e is reduced due to increasingly skewed sex ratios favoring females (Section 7.2).

Tropical tree species often occur at low population densities in species-diverse forests, and are subjected to size-limited selective harvesting, where the largest trees are logged periodically. Degen et al. (2006) simulated the genetic and demographic effects of more- and less-intensive harvesting over four centuries on genetic diversity in four tropical tree species in French Guiana. The higher-intensity harvesting scenario reduced population sizes in three of four species, but the effects on genetic diversity were small due to overlapping generations and high seed and pollen dispersal. However, they did not evaluate whether this diameter-limit logging would result in selection for slower growth.

The loss of genetic variation will also be influenced by gene flow among subpopulations that comprise a metapopulation. Estimating the effective population size of a metapopulation is extremely complex (Section 19.1.2). In addition, harvesting might have unexpected effects on the overall N_e of a metapopulation. For example, Hindar et al. (2004) found that small subpopulations within a metapopulation of Atlantic salmon contribute more per spawner to the overall effective population size than large subpopulations, and harvesting of the subpopulations jointly in mixed-stock fisheries has

a relatively larger demographic effect on small than large populations. Consequently, the mixed-fishery harvest could reduce metapopulation N_e far more than expected.

15.2 Unnatural selection

Fisheries, wildlife, and forest management is at risk of reducing productivity in wild populations because exploitation often removes phenotypes that are favored by natural and sexual selection. Darimont et al. (2009) have concluded that harvested populations show some of the most abrupt trait changes ever detected in wild populations. These changes, which include average declines of almost 20% in size and shifts in life history traits of nearly 25%, have important conservation and economic implications. Thus, accounting for selection that acts counter to natural adaptive processes is therefore an important component of a comprehensive and effective sustainable management strategy. For example, Hanlon (1998) has concluded that commercial harvest of squid on the spawning grounds imposes sexual selection that could reduce recruitment and affect the long-term sustainability of these fisheries.

The reduction in the frequency of the silver morph of the red fox between 1834 and 1933 in eastern Canada was apparently the first documented change over time resulting from selective harvest (Elton 1942). J.B.S. Haldane used these data to provide one of the first estimates of the strength of selection in a wild population using his then recently developed mathematical models of the effects of selection at a single locus (Haldane 1942). The fur of the homozygous silver morph (*RR*) was worth approximately three times as much as the red fur of the heterozygous cross (*Rr*) or homozygous (*rr*) fox to the furrier, and, therefore, was more likely to be pursued by hunters.

The frequency of the desirable silver morph declined from 16% in 1830 to 5% in 1930 (Figure 15.5). Haldane concluded that this trend could be explained by a slightly greater harvest rate of the silver than the red and cross phenotypes. The lines in Figure 15.5 show the expected change in phenotypic frequencies, assuming that the relative fitness of the silver phenotype was 3% less than both the homozygous red and cross phenotypes,

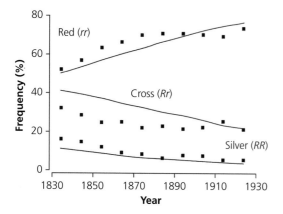

Figure 15.5 Reduction in frequency of the silver morph of the red fox in eastern Canada resulting from the preferential harvest by hunters of the more valuable silver morph. The points represent data presented by Elton (1942). The lines represent the expected change in frequencies of the three phenotypes via selection at a single locus assuming that the silver fox morph has a 3% survival disadvantage per generation relative to the red and cross morphs as modeled by Haldane (1942). The initial frequency of the R allele was 0.3 and the mean generation interval was 2 years. Modified from Allendorf & Hard (2009).

and the generation interval was 2 years. This example demonstrates the importance of understanding the underlying genetic architecture of a trait in order to predict the effects of harvest. The frequency of the heterozygous cross phenotype declined even though it had higher fitness than the silver phenotype. The favored cross phenotype is expected to decline in frequency because of the reduction in frequency of the R allele resulting from the lower fitness of the silver type.

Plant populations can also be subject to genetic changes as a result of phenotypic selection by humans during harvest. For example, harvesting of the bulb of the lily species *Frittilaria delavayi* has resulted in increased frequency of camouflaged flower colors as more visible flower colors make it easier for harvesters to locate the plants (Example 15.2).

Selective genetic changes resulting from exploitation are inevitable because increasing mortality will result in selection for earlier maturation, even if harvest is independent of phenotype (Figure 15.7). Moreover, harvesting of wild populations is inevitably phenotypically nonrandom (Law

2007). That is, individuals of certain phenotypes (e.g., sizes, behaviors, or colors) are more likely than others to be removed from a wild population by harvesting. Such selective harvest will bring about genetic changes in harvested populations if the favored phenotype has at least a partial genetic basis (i.e., $H_N > 0$). In addition, such changes are likely to both reduce the frequency of desirable phenotypes and can reduce productivity (Box 15.2). It is now recognized that fisheries-induced evolution can cause genetic changes in life history, behavior, and morphology, as well as affect the productivity of harvested populations (Hutchings & Kuparinen 2020).

The rate of genetic change by exploitative selection depends upon the amount of additive genetic variation for the trait (heritability). As discussed in Chapter 11, the rate of change in a quantitative trait is described by the breeder's equation (Equation 11.9):

$$R = H_N S$$

where H_N is the narrow-sense heritability, S is the selection differential (the difference in the phenotypic means between the selected parents and the whole population), and R is the response (the difference in the phenotypic means between the progeny generation and the whole population in the previous generation). In the case of exploitative selection, S will be affected both by the intensity of harvest (the proportion of the population harvested) and the phenotypic selectivity (e.g., the removal of only the largest individuals) of the harvest.

Attempts to predict the expected effects of exploitative selection have generally assumed that life history traits are affected by many loci with small effects (Kuparinen & Hutchings 2017). However, the **genetic architecture** underlying a trait can dramatically affect how a population responds to exploitative selection.

Genomic analysis now makes it possible to understand the genetic architecture underlying the inheritance of traits. For example, age at sexual maturity in Atlantic salmon was found to be controlled largely by a single locus (*vgll3*) with two alleles (Barson et al. 2015). This locus explained 39% of the variation in the age at sexual maturity. Homozygotes for the early maturity allele tend to mature at the age

Example 15.2 Genetic response to commercial harvest in an alpine plant

Natural populations of plants are often harvested for a variety of uses. However, genetic effects of such harvest have received less attention than the harvest of animals. The bulb of *F. delavayi* has been harvested by humans for use in Chinese traditional medicinal purposes for over 2,000 years in the Hengduan Mountains of China and Nepal (Niu et al. 2021). The price of these bulbs has increased in recent years, and was about US$480 per kilogram in 2020. Approximately 3,500 individuals are required to harvest just 1 kg of bulbs.

This plant is a perennial herb living in alpine scree slopes. It has leaves only at a young age and produces a single flower each year after the fifth year. Leaf and flower color varies among populations from gray to brown to green (Figure 15.6). Gray or brown types appear well camouflaged, while green individuals are much more conspicuous.

Figure 15.6 Green (left) and brown (right) flower phenotypes of the fritillary lily *Fritillaria delavayi* in China. In populations where bulbs are heavily harvested, the frequency of the more visible green morph has declined (Niu et al. 2021).

Niu et al. (2021) tested the relationship between background matching of the leaves and the intensity of harvest pressure. They found a significant relationship between color matching and collection intensity, indicating that plants in populations that experience heavier harvesting are more cryptic to humans. Flower color also varied but was not included in their analysis because many plants were collected without flowers. Nonetheless, flower color also matched the background very well in the camouflaged populations.

These authors also tested the prediction that improved match to the background results in longer detection times by using an online citizen science experiment. Subjects were asked to locate a fritillary target as quickly as possible in a series of photos to simulate the collection process by collectors. As expected, targets with better camouflage required longer times to be located. Try it yourself at www.plant.sensoryecology.com.

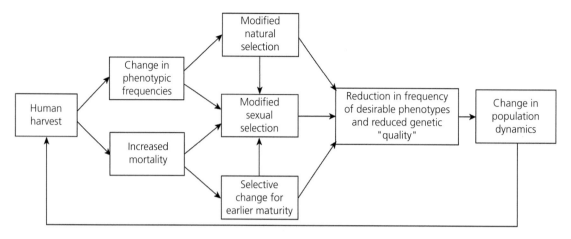

Figure 15.7 Human harvest can have a variety of direct and indirect genetic effects on populations, and it has the potential to affect the future yield and viability of exploited populations. From Allendorf & Hard (2009).

of one sea winter, whereas homozygotes for the late maturity allele often mature only after three sea winters. There are also differences between males and females in the effect of this locus. Female heterozygotes tended to postpone maturity, whereas male heterozygotes matured earlier.

The models of Kuparinen & Hutchings (2017) found that a trait architecture of many loci with small effect results in evolution of earlier maturation. In contrast, single-locus control causes divergent and disruptive evolution of age at maturity without a clear phenotypic trend but a wide range of alternative evolutionary trajectories and greater trait variability within trajectories. These results demonstrate that the range of evolutionary responses to selective fishing can be wider than previously thought and that a lack of a phenotypic trend need not imply that evolution has not occurred. These findings suggest that understanding and predicting population responses to exploitative selection require understanding the underlying genetic architecture of selected traits.

There are many examples in the literature of phenotypic changes in exploited populations that might be the result of exploitative selection (Table 15.2; Example 15.3). However, it has been difficult to determine if observed phenotypic changes over time indicate genetic change or are caused by other factors such as relaxing density-dependent effects on individual growth due to reductions in

population density or abiotic factors such as temperature affecting growth and development (Fenberg & Roy 2008). A review critically evaluated the observed evidence for a genetic basis of such phenotypic change and concluded that establishment of practices for routinely monitoring and sampling harvested fish stocks is vital for the detection and management of fisheries-induced evolution (Kuparinen & Merilä 2007).

Many harvest regimes of trees, fish, and wildlife selectively remove larger individuals. Life history theory predicts that this should select for maturation at a younger age and smaller size (Marshall & Bowman 2007). This prediction is concordant with the long-term trend toward earlier maturation that has been observed for many commercially exploited fish stocks. However, such trends might also be explained by phenotypic plasticity as a direct response to decreased population size, or long-term environmental changes.

The selective harvesting of large, straight trees from native forests has occurred for centuries or millennia. The genetic effects of this historical logging have been suggested as the reason for the crooked stems and long branches of the Cedar of Lebanon, and for the poor form of Scots pine along waterways in Sweden and Finland, but these hypotheses have not been scientifically tested (Savolainen & Kärkkäinen 1992). Cornelius et al. (2005) evaluated the extent to which contemporary logging of natural

Box 15.2 Fisheries-induced evolution

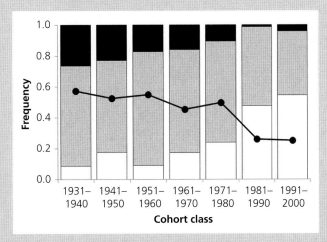

Figure 15.8 Long-term changes in frequencies of the *Pan* I genotypes of Icelandic stock of Atlantic cod (white, *AA*; gray, *AB*; black, *BB*) for different cohort classes (of 10 years each) collected in 1948–2002. The black line shows the frequency of the *B* allele. Adapted from Jakobsdóttir et al. (2011)

The harvesting of fish populations is certain to cause genetic change over time because harvest is almost always non-random. That is, individuals of certain size, morphology, or behavior are more likely than others to be removed from the population by harvesting. Such selective removal will bring about genetic change in harvested populations if the selected phenotype has at least a partial genetic basis (i.e., H_N > 0). In addition, harvest need not be selective to cause genetic change; uniformly increasing mortality independent of phenotype will select for earlier maturation (Law 2007). Experiments (Example 11.1) have demonstrated how size-selective harvesting can change the genomic makeup of a species, altering both phenotypes and genotypes profoundly, as well as reducing genetic diversity over relatively few generations. From a management perspective, the key question is how such inevitable genetic changes will affect the sustainable productivity of the harvested population.

Cloudsley Rutter (1902) was perhaps the first biologist to consider and write about the genetic effects of harvest on a fishery in his classic treatise on the Chinook salmon of the Sacramento River, California, USA. He suggested that the value of this fishery would diminish if the medium size and larger fishes were harvested, leaving only the smaller fish to reproduce. He therefore recommended that mesh size of capture-nets should be designed to allow the larger fish to escape and reach the spawning grounds. Ricker (1995) provided a comprehensive review of the possible effects of fisheries-induced evolution on the five species of Pacific salmon.

Nelson & Soulé (1987) and Policansky (1993) provided valuable early warnings about the possible harmful effects of harvest on fish populations. Heino et al. (2015) have provided an excellent review of fisheries-induced evolution. **Probabilistic maturation reaction norms** (PMRNs) have been used to help disentangle genetic from **phenotypic plasticity** effects on maturation (Heino et al. 2002; Dieckmann & Heino 2007). Reconstructing PMRNs from historical data in exploited populations has provided evidence for fishery-induced selection. However, some have argued that because PMRNs do not fully account for physiological aspects of maturation, the observed shifts might reflect directional environmental effects on maturation rather than genetic changes (Marshall & McAdam 2007). It is impossible in most circumstances to completely disentangle genetic and plastic effects. Nevertheless, the use of PMRNs provides a useful method to determine if genetic effects are at least partially responsible for an observed phenotypic change over time.

Changes in gene frequencies at the *Pantophysin* locus (*Pan* I) in the Icelandic stock of Atlantic cod are a clear example of genetic change caused by fishing (Jakobsdóttir et al. 2011). Icelandic Atlantic cod are distributed over the continental shelf all around Iceland. Genetic analysis of archived otoliths from the period of intense fishing (1948–2002) revealed dramatic changes in the frequencies of genotypes at *Pan* I, whereas no changes were observed at six microsatellite loci. The changes in *Pan* I genotype frequencies were quite dramatic, with the frequency of *BB* declining

---

Box 15.2 Continued

from 26% in the 1930s to 5% in the 1990s, and the frequency of *AA* increasing to above 50% during the same time period (Figure 15.8). The *BB* genotype was more common among the older spawning cod, while the *AA* genotype was most frequently observed among the younger age classes

(Figure 15.9). Hemmer-Hansen et al. (2013) reported that the *Pan* I locus in Atlantic cod occurs in a 20-cM inversion that distinguished between migratory and stationary ecotypes.

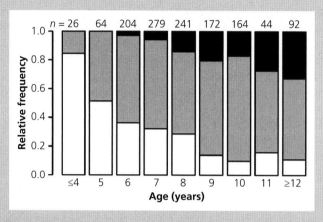

Figure 15.9 Age-specific distributions of *Pan* I genotype frequencies (white, *AA*; gray, *AB*; black, *BB*) in samples of Icelandic cod collected from the fishery between 1948 and 2002. The corresponding sample size is listed above each column. From Jakobsdóttir et al. (2011).

---

mahogany populations might result in selection for growth or form. They concluded that given the low heritability of size and form phenotypic traits in natural forests ($H_N \leq 0.1$) that the phenotypic response to logging is likely to be less than 5% following a single logging event. While these effects may accumulate with multiple logging events and over generations, Cornelius et al. (2005) suggested that erosion of genetic diversity from reduced population sizes is a much greater concern in this tropical tree.

An interesting historical note is that Patrick Matthew published a book in 1831 that contained a forerunner of Darwin's concept of natural selection (Wells 1973). Matthew was aware of the great variation present within a single species of tree. He also pointed out that different species and varieties were well adapted for growing in certain habitats. He argued that foresters should use seed from the

"largest, most healthy, and luxuriant growing trees" (Matthew 1831, p. 126).

15.3 Spatial structure

Virtually all harvested species have local breeding groups (subpopulations) that are somewhat reproductively isolated. In some cases, even different species have been managed as a single "**stock**" (Guest Box 15). Harvesting a group of individuals that is a mixture of several subpopulations can result in the overharvest and extirpation of one or more subpopulations. This will not be recognized unless the subpopulations are identified separately and individuals from population mixtures are assigned to subpopulations. Recent genomic analysis has found that management units of harvested fish populations often do not correspond to the genetic population structure (Mullins et al. 2018;

Table 15.2 Examples of phenotypic changes that could have resulted from exploitative selection.

Species	Trait(s)	Observation
African elephant	Tusk length	Tusk length in African elephants from southern Kenya declined by 22% in males and 37% in females between 1966–1968 and 2005–2013. Severe ivory harvesting began during the 1970s and 1980s. Pattern of tusk size among related individuals suggested that tusk size is heritable ($H_N > 0$). The authors concluded that selection of large tuskers by poachers resulted in the decline in tusk size (Chiyo et al. 2015).
American plaice	Body size Age at maturation	Three fish stocks with historically different levels of exploitation showed the same long-term shift toward maturation at younger ages and smaller sizes. This situation warrants further investigation to determine if these stocks are truly demographically and genetically independent (Barot et al. 2005).
Atlantic cod	Body size Age at maturation Growth rate	Survey data prior to the collapse of the fisheries in 1992 showed a significant genetic shift toward earlier maturation and at smaller sizes. Probabilistic maturation reaction norms (PMRNs) were used to account for any confounding effects of phenotypic plasticity. In the 1970s, fishing selection targeted slow-growing individuals. Later in the 1980s, the net mesh size was increased, resulting in a bigger catch of large, faster-growing individuals. Application of a quantitative genetic model showed a reduction of length-at-age between cohorts of offspring and parents as a result of exploitative selection (Swain et al. 2007).
Atlantic salmon	Time of spawning Body size	Earlier running fish experienced greater harvest by anglers. Allozyme and mitochondrial DNA data from four populations in Spain showed that the late-running individuals that escaped harvest were genetically distinct and significantly smaller. Catch records in Ireland extending back many decades and recent electronic counter data show a reduction in the abundance of early migrants and a decline in size of late migrants (Consuegra et al 2005).
European grayling	Age at maturation	Gillnet fishing is suggested to have caused a constant reduction in the age and length at maturity in separate populations founded from the same common ancestors (Haugen et al. 2001).
Japanese mamushi snake	Body size Reproductive effort Escape behavior	Hunted local populations showed a variety of changes in morphology, reproduction, and behavior compared with populations that were not hunted. Hunted populations had fewer vertebrae, produced more and smaller offspring, had increased reproductive effort, and in nature fled at greater distances from an approaching human (Sasaki et al. 2009).
Moose	Delayed birth date Body size	Pedigree analysis over a 20-year period detected delayed birth dates in males and reduced body mass as calf in females as a selective response caused by hunting (Kvalnes et al. 2016).
North Sea plaice	Age at maturation Body size	The reaction norms for age and length at maturation showed a significant trend toward younger age and shorter body length (Grift et al. 2003).
Northern pike	Body size	Over a period of four decades, selective harvesting targeted large individuals and directional natural selection favors large body size. The result of these two opposing forces is stabilizing selection, but a reduction in overall fitness (Edeline et al. 2007).
Red kangaroo	Body size	Hunters target the larger individuals in a group and there is evidence that average size has declined (Croft 1999).

Example 15.3 Phenotypically selective harvest within and among local subpopulations

Phenotypically selective harvest of mixed populations comprising individuals from many contributing subpopulations can result in both exploitative selection within subpopulations and differential intensity of harvest on those subpopulations. For example, hundreds of reproductively isolated local subpopulations of sockeye salmon contribute to the Bristol Bay fishery in Alaska (Hilborn et al. 2003). There is a gillnet fishery in Bristol Bay that harvests these subpopulations before the salmon return to their home spawning grounds in freshwater. This mixed-stock fishery has the potential to harvest selectively depending upon run timing, body size, body shape, and life history (primarily age at sexual maturity).

Quinn et al. (2007) examined daily records in two fishing districts in Bristol Bay for evidence of temporally selective harvest over a 35-year period. They found that earlier migrants experienced lower capture rates in the fishery than later migrants. The timing of the run has become earlier over this period as expected in response to selection favoring individuals (and subpopulations) that arrive earlier. This observed phenotypic change apparently results from genetic changes both within subpopulations, resulting in earlier run timing of individuals, and among subpopulations with differential harvest intensity favoring earlier arriving subpopulations.

The gillnet fishery also results in selection within subpopulations and differential intensity of harvest on subpopulations because the effectiveness of gillnets in capturing migrating fish is dependent upon body size and shape (Hamon et al. 2000). Fish that are too small are able to escape by swimming through the mesh, and fish above the target size-class are too large to be wedged in the mesh. Therefore, capture by gillnets is likely to be selective on age at sexual maturity, size at age of sexual maturity, and body depth. Subpopulations that spawn in different habitat types show consistent differences for all of these characteristics (Quinn et al. 2007). For example, males from lake-spawning subpopulations generally have much deeper body depth than males from stream-spawning subpopulations.

Hamon et al. (2000) compared harvested fish with fish that escaped the fishery and returned to the spawning grounds to determine the relationship between morphology and fitness caused by selective capture in gillnets (Figure 15.10). They found that the effects of gillnet selectivity within subpopulations were strongly influenced by variability in age at reproduction. Subpopulations with mixed-aged fish at maturity experienced disruptive selection, with smaller and larger fish having the greatest fitness. In contrast, subpopulations predominantly of a single age-class experienced directional selection favoring smaller fish. In addition, differences in morphology among subpopulations resulted in large differences in harvest intensity. Some subpopulations experienced virtually no fishing mortality, while others sustained high mortality due to harvest (>70%).

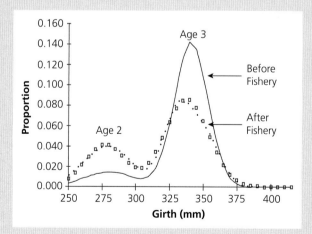

Figure 15.10 Distribution of girths before and after the fishery for sockeye salmon in Little Togiak Lake, Alaska. Solid lines represent the distributions of girths before the fishery. Squares indicate the distributions of girths after the fishery. Modified from Hamon et al. (2000).

Continued

Example 15.3 *Continued*

Their analysis indicated that mixed age classes are subject to disruptive selection, but that single age-class populations experienced directional selection. The effect of this selection depends on cumulative selection pressures, which probably include natural and sexual selection on this trait. They concluded that selection by gillnets is a strong selective force affecting body size and shape within populations.

Leone et al. 2019). Genomic analysis also has detected adaptive loci that show greater differentiation among populations that are useful for identifying populations which show little genetic differentiation at neutral loci (Bekkevold et al. 2015; Section 20.6.1).

To manage populations sustainably, we need to know what constitutes the harvested population and how it is genetically delineated (Palsbøll et al. 2007; Example 9.5; Section 22.4.2). Extirpation of some subpopulations is likely to directly reduce overall population productivity. In addition, Schindler et al. (2010) have shown that productivity of subpopulations of sockeye salmon can change dramatically over time as environmental conditions change (Example 19.3). Therefore, ensuring long-term productivity of the overall metapopulation depends on conserving all subpopulations, including the currently less productive ones. In addition, reduction in the size or density of subpopulations might decrease the number of migrants among subpopulations and cause increased genetic drift and loss of genetic variation. Harvest can also increase the rate of gene flow into certain subpopulations and cause genetic swamping and loss of local adaptations. An understanding of this population genetic substructure at different points of a species' life history is necessary to predict the potential effects of harvest on genetic subdivision.

Harvest of mixed populations is common in migratory waterfowl, marine mammals, ungulates, and many other species. For example, Pacific salmon are generally harvested in the ocean in mixed stocks that comprise many reproductively isolated subpopulations that spawn in freshwater (Schindler et al. 2010). Understanding which subpopulations are contributing to the harvest is essential to avoid overharvest and extirpation of some local subpopulations while others experience very little harvest whatsoever (Example 15.4).

Extirpation of subpopulations through overharvesting has been observed in both marine and freshwater fish (Nelson & Soulé 1987; Dulvy et al. 2003). For example, the number of streams contributing substantially to production of four salmon species in southern British Columbia suffered a severe decline between 1950 and 1980 (Walters & Cahoon 1985). In general, the subpopulations that are less productive and the least resilient to exploitation have been the first to disappear (Loftus 1976). Moreover, stocks with the most desirable characteristics often experience the greatest exploitation (Example 15.4).

Example 15.4 Genetic analysis of century-old notebooks reveals subpopulation-specific patterns of decline

The Skeena River watershed in Canada is composed of 13 separate population complexes of sockeye salmon (Price et al. 2019). A commercial fishery began at the mouth of the Skeena River in 1877, and a scale-collection program was commenced in 1912 by Robert Gibson (Ogden 2019). Gibson took basic data on a sample of fish and affixed a scale from each fish into his notebooks. His notebooks were discovered in 1997 and have provided DNA samples that have allowed the reconstruction of the stock structure of the population complex over the past 100 years (Price et al. 2019).

Individual fish were assigned to one of the 13 population units on the basis of their genotypes at 12 microsatellite loci. The total number of wild adult sockeye salmon returning to the Skeena River is ~25% of that during historical times. Individual populations in the Skeena watershed have declined in abundance by between 56 and 99% since 1913. The populations that showed the most dramatic declines tended to have larger body size. This suggests that fisheries selectivity may have contributed to variation in declines among populations.

Some plant species used by humans have been overexploited and are threatened in the wild, yet are widely cultivated agriculturally. The orchid genus *Vanilla* provides an example of this. Aromatic species in this genus are widely used for the flavor of their pods. Approximately 95% of the world's commercial vanilla production uses *V. planifolia*, a species native to Mexico, Guatamala, and Belize, and historically used by the Aztecs (Bory et al. 2008). This species is now cultivated and propagated around the world in tropical areas. However, remaining wild populations are extremely small and at very low population density, with one individual per 2–10 km^2 in two areas surveyed in Mexico. One expert spent 20 years searching in southern Mexico and only found 30 plants! These small populations are seriously threatened by deforestation and overharvesting yet represent a valuable genetic resource to the vanilla industry from which to develop new varieties.

Exploitation can also increase gene flow or hybridization among subpopulations and potentially swamp local adaptations. Overexploitation could reduce the density of local subpopulations and allow for more immigration from nearby subpopulations less affected by exploitation. This could bring about the genetic swamping of the remnants of exploited subpopulations and thereby reduce fitness. Studies of red deer report that a change of fine-scale genetic structure appears to be associated with changes in harvest management (Nussey et al. 2005; Frantz et al. 2008).

Examination of genetic samples collected over time (i.e., genetic monitoring, Chapter 23) is the most powerful way to detect genetic changes caused by harvest. For example, the Flamborough Head population of North Sea Atlantic cod apparently went through a decline in genetic variation followed by genetic swamping between 1954 and 1998, based upon genetic variation at three microsatellite loci using otolith samples archived over this period (Hutchinson et al. 2003). Genetic diversity declined between 1954 and 1970, indicating reduced effective population size, apparently resulting from harvest. Genetic variation increased after this period because of increased immigration during a period of exceptionally high exploitation. Thus, the original genetic

characteristics of the Flamborough Head population likely have been lost.

15.4 Effects of releases

Large-scale exploitation of wild animals and plants through fishing, hunting, and logging often depends on augmentation through releases of individuals raised in captivity (Laikre et al. 2010b). Such releases are performed worldwide in vast numbers. Augmentation can be demographically and economically beneficial, but can also cause four types of adverse genetic change to wild populations (Figure 15.11):

1. Loss of genetic variation.
2. Loss of adaptations.
3. Change of population composition.
4. Change of population structure.

While adverse genetic impacts are recognized and documented in fisheries, little effort is devoted to actually monitoring them. In forestry and wildlife management, genetic risks associated with releases are largely neglected. We outline key features of programs to effectively monitor consequences of such releases on natural populations.

15.4.1 Genetic effects of releases

Even releases that do not result in gene flow can have genetic consequences if they reduce local population size (e.g., through competition or disease transmission, or through wasted reproductive effort by native individuals that mate with captive-bred individuals but do not produce viable offspring). The main concern in these cases is that changes to naturally existing genetic diversity within and among populations can reduce viability and productivity of exploited populations. This could cause a problem both in the short term by reducing individual fitness and in the long term by reducing the capacity for populations to evolve and adapt to future conditions.

Introgression from introduced to native populations has been documented in a number of species subject to large-scale releases (Laikre et al. 2010b). Although risks to native gene pools have

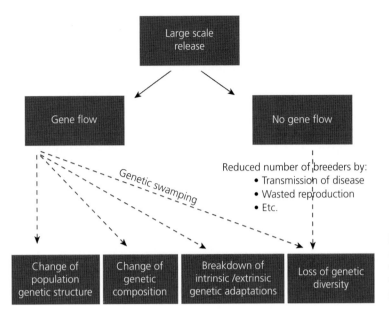

Figure 15.11 Primary pathways by which large-scale releases can change genetic characteristics between natural populations (population genetic structure) and within populations. From Laikre et al. (2010b).

been recognized for a variety of species, the most intense attention has focused on Pacific salmon, Atlantic salmon, and brown trout (Waples & Drake 2004). In Denmark, these concerns have resulted in a ban of all releases of salmonids originating from anything but the local population (Nielsen & Hansen 2008). However, this is the exception even for salmonids. Surprisingly little monitoring of the effects of mass releases has occurred in forestry and wildlife management, but there are a few examples such as eucalypt populations used in Australian forestry (Barbour et al. 2008) and exotic game birds in southern Europe (Barilani et al. 2005).

15.4.1.1 Loss of genetic variation

Wild populations might also lose genetic variation if their effective population size is reduced due to increased mortality caused by parasites or diseases transmitted by released individuals. A classic example refers to the effects of the unintended introduction of the parasite *Gyrodactylus salaris* (an ectoparasite living on the skin of Atlantic salmon) into Norway with juvenile salmon imported from Sweden in the early 1970s (Peeler et al. 2006). Substantial mortality has been observed in wild populations following spread of parasites through alien

populations in salmon in Japan, red-legged partridge in Spain, tortoises in the USA, red deer in Italy, and rabbits in France and southern Europe (Daszak et al. 2001; Laikre et al. 2010b). Spread of parasites and deadly disease is often associated with release of individuals to supplement harvested wild populations.

Individuals released for supportive breeding (Section 21.7) can cause loss of genetic diversity by reducing N_e. Typically, relatively few parents are brought into captivity for reproduction; these parents often contribute disproportionately large numbers of genes to the next generation in the wild, potentially resulting in increased rates of inbreeding and genetic drift in the total population. Reduced genetic variation has been observed in populations of salmonid fishes subject to supportive breeding (Hansen et al. 2000).

15.4.1.2 Breakdown of adaptations

Releases can reduce adaptation by causing loss of extrinsic or intrinsic adaptation. Loss of fitness can occur when alleles that confer local adaptation are replaced by ones that are locally nonadaptive. Gene flow from a nonlocal source population can cause breakup of **coadapted gene complexes**; that is, alleles at multiple loci that work synergistically to increase fitness (intrinsic adaptation). Because this breakup is caused by recombination, loss of adaptation generally occurs only in the F_2 generation and beyond and can be much more difficult to detect than loss of extrinsic adaptation. Empirical examples from wild populations show that both types of adaptation can be lost by gene flow from genetically divergent populations (Laikre et al. 2010b). Fitness effects can be insidious: in some studies, increased F_1 fitness due to heterosis has been followed by decreased fitness in F_2 or later generations as co-adapted gene complexes are eroded (e.g., Muhlfeld et al. 2009, but see Hwang et al. 2011 and Frankham et al. 2011).

A 37-year study of Atlantic salmon in Ireland found that naturally spawning farmed fish depress wild recruitment and disrupt the capacity of natural populations to adapt to higher water temperatures associated with climate change (McGinnity et al. 2009). Similar results have been found with Atlantic salmon on the east coast of Canada (Sylvester et al. 2019). Hansen et al. (2009) examined Danish populations of brown trout subject to hatchery supplementation for 60 years and found evidence for selection in the wild against alleles associated with nonnative hatchery fish. Muhlfeld et al. (2009) showed that nonnative rainbow trout that hybridized with native cutthroat trout had high F_1 reproductive success. However, in subsequent generations fitness declined by nearly 50% (compared with fitness of native trout) following 20% introgression of nonnative genes (Section 13.4.3).

15.4.1.3 Change of population genetic structure

Most natural animal and plant species are structured into genetically distinct populations because of restricted gene flow, genetic drift, and local adaptation. Since large-scale releases can affect these microevolutionary processes, they can alter genetic structuring of natural populations, but such effects have rarely been monitored (Laikre et al. 2010b). Complete replacement of native gene pools of Mediterranean brown trout with introduced populations of Atlantic origin has occurred over large areas (Araguas et al. 2009). Slovenian populations of Adriatic grayling have been stocked with fish originating from the Danube River for over four decades,

and levels of introgression are so high (40–50%) that few native individuals can be found (Sušnik et al. 2004).

15.4.2 Effects on species and ecosystem diversity

Genetic changes to native populations can have consequences that extend beyond the affected species. Evidence is accumulating from **community genetics** studies to show that genetic changes to one species can affect other species as well as entire communities and ecosystems (Whitham et al. 2008; Crutsinger 2016). For instance, genetic characteristics of individual plant populations can affect the composition of arthropods (Crutsinger et al. 2006), and foraging behavior of beavers is affected by genetic makeup of the *Populus* species on which they feed (Bailey et al. 2004).

There is evidence that high levels of genetic diversity increase resilience of species and ecosystems, and that genotypic diversity can complement the role of species diversity in a species-poor coastal ecosystem, and thus help buffer ecosystems against extreme climatic events. Genetic variation was positively correlated with recovery of seagrass ecosystems following overgrazing and climatic extremes (Worm et al. 2006). Reusch et al. (2005) conducted manipulative field experiments and found that increasing genotypic diversity of the cosmopolitan seagrass *Zostera marina* enhanced biomass production, plant density, and invertebrate faunal abundance, despite near-lethal water temperatures.

15.4.3 Monitoring large-scale releases

Genetic monitoring of releases should aim to provide answers to the following key questions:

1. What are the genetic characteristics of the natural population(s) prior to the release?
2. Do releases alter these characteristics?
3. What are the biological consequences?

Such monitoring should be included as a basic part of any program for commercial or other releases.

Assessments of risk–benefit tradeoffs are most effective if conducted prior to release activities.

Waples & Drake (2004) outline a framework for elements of comprehensive risk–benefit analysis that should be conducted prior to fish stock enhancement programs. Similarly, Barbour et al. (2008) discuss strategies for assessing risks of pollen-mediated gene flow from translocated species and hybrids *Eucalyptus globulus* plantations into native populations. These studies show that different risk–benefit assessment protocols are needed for different taxa and should be refined to fit particular species.

For releases that have already been carried out, an idealized monitoring design often cannot be followed. However, sometimes archived material such as fish scales, animal skins, herbarium samples, or DNA can help address questions regarding genetic composition prior to release. Within forestry, **provenance trials** have been used since the 19th century to identify populations with economically important characteristics. Such traditional tree breeding programs are aimed at examining performance of trees from different geographic localities (provenances) to find the best sources of seed for selective breeding and planting. They have usually resulted in the use of relatively local populations for reforestation of native species due to local adaptation. Geographic source materials for provenance trials are thus known, and existing trial stands can be used to study long-term effects of plantations, such as gene flow into neighboring, native populations (König et al. 2002).

15.5 Management and recovery of exploited populations

The most difficult political and economic decision in harvest management is to reduce the current take in order to increase the likelihood of long-term sustainability. This decision is especially difficult when taking actions to halt or reverse historical declines will come at the cost of economic hardship for dependent communities (Walters & Martell 2004). Management measures to reduce harmful long-term genetic effects are most likely to be adopted by managers if they also help to meet short-term management objectives. For example, maintaining large, old individuals within populations provides both short- and long-term benefits (Birkeland & Dayton 2005).

The emphasis on disentangling genetic and plastic mechanisms of phenotypic change is crucial from a basic scientific perspective, but is less important from a strictly management perspective. It is not necessary to prove that an observed phenotypic shift in a wild population is an evolutionary response to harvest in order to apply evolutionary principles to management. Moreover, complete disentanglement of genetic and plastic responses is difficult, except in laboratory experiments, which have limited applicability to management of harvested wild populations (Hilborn 2006).

We recommend assuming that some genetic change due to harvest is inevitable and to apply basic genetic principles combined with molecular genetic monitoring to develop management plans for harvested species. This approach can be especially powerful if archived samples that have been collected over time are available for analysis. Such archived samples are available for many species of fish (scales and otoliths), mammals (bones, teeth, and skin), birds (feathers and skin), and plants (leaves, stems, and reproductive parts).

The molecular genetic analysis of samples collected over a period of time has tremendous untapped potential to inform and guide management of exploited populations. Genetic monitoring can provide a window into the past as the examples of genetic swamping of the Flamborough Head population of North Sea cod (Hutchinson et al. 2003) and the loss of genetic variation in New Zealand snapper illustrate (Hauser et al. 2002). Analysis of contemporary samples alone would not have uncovered these important consequences of past exploitation.

15.5.1 Loss of genetic variation

Small populations are most likely to be affected by the loss of genetic variation due to excessive harvest because of their smaller effective population size. Management actions that reduce effective population size below threshold values where loss of genetic variation might have harmful effects should be avoided. As we have seen, substantial loss of genetic variation can occur even when census population sizes are very large because the genetically effective population size is often much smaller than the census size in many harvested species of marine fishes and invertebrates. The only way to detect such "cryptic" loss of genetic variation of exploited populations is empirical observation of genetic variation over time (Luikart et al. 1998). Genetic monitoring programs can provide a powerful means to detect loss of genetic variation if enough marker loci are used (Chapter 23).

15.5.2 Unnatural selection

Lowering rates of exploitation is the most direct way to reduce the effects of exploitative selection. Kuparinen & Festa-Bianchet (2017) have reviewed the evidence for harvest-induced evolution in aquatic and terrestrial systems. They conclude that there is strong evidence for evolutionary effects in both terrestrial and aquatic systems, and that these changes are generally harmful. Nevertheless, the effects of such evolutionary changes can be managed effectively by simply reducing harvest intensity. Hutchings & Kuparinen (2020) have concluded that, in general, the effects of fisheries-induced evolution are less important than the demographic effects of overfishing itself.

Consideration should also be given to management approaches that spread the harvesting across the distribution of age- and size-classes, or target the intermediate sized individuals by establishing an upper size limit on individuals (especially for long-lived species). These actions will both reduce the long-term effects of exploitative selection and increase the number of older females that produce more and higher-quality offspring in the short term (Birkeland & Dayton 2005). However, upper size limits might reduce N_e because individuals surviving to the size where they are "safe" will contribute a disproportionately large number of progeny, and this is expected to increase the variance of family size (Ryman et al. 1981). This effect on N_e might be substantial in some cases depending on the age distributions before and after introducing the limit. However, the expected effect on heterozygosity over calendar time would be more complicated because this harvest strategy could also lead to longer generation intervals, as mentioned previously.

The effects of selection can sometimes be reduced by harvesting after reproduction by changing either the time or location of harvesting. For example, the northeast Arctic stock of Atlantic cod uses the Barents Sea for feeding but spawns further south off the northwest coast of Norway (Heino 1998). Harvesting on spawning grounds rather than feeding grounds would select for delayed maturation, and results in increased sustainable yield. In contrast, harvesting on the feeding grounds would select for early maturation because late-spawning fish might be harvested before they mature. However, there are challenges in making this biological solution socially palatable due to its potential economic impact on the fishing fleet and markets through increased seasonality of harvest and supply.

Recovery following relaxation, or even reversal, of exploitative selection will often be much slower than the initial accumulation of harmful genetic changes (Heino 1998; Loder 2005; de Roos et al. 2006). This is because harvesting often creates strong selection differentials while relaxation of this selective pressure will generally result in only mild selection in the reverse direction (however, see Conover et al. 2009). de Roos and colleagues (2006) used an age-structured fishery model to show that exploitation-induced evolutionary regime shifts can be irreversible under likely fisheries management strategies such as belated or partial fishery closure. This effect has been termed "Darwinian debt" (Loder 2005), and has been suggested to have general applicability (Tenhumberg et al. 2004). That is, time scales of evolutionary recovery are likely to be much slower than those on which undesirable evolutionary changes occur. However, gene flow has the potential to accelerate the rate of recovery by restoring alleles or multiple-locus genotypes associated with the trait. For example, trophy hunting might reduce or eliminate alleles for large horn size, but gene flow from national parks with no hunting might eventually restore alleles associated with large horn size (Example 15.5).

15.5.3 Subdivision

The importance of individually managing reproductively isolated populations is obvious and has long been recognized in fisheries (Rich 1939). Nevertheless, application of this understanding is often

Example 15.5 Selective harvest reduces horn size in bighorn sheep

Bighorn sheep hunting regulations sometimes have required that rams must reach a minimum horn size before they can be harvested. At Ram Mountain, Alberta, Canada, 57 rams were harvested under such a regime over 30 years corresponding to an average harvest of ~40% of the legal-sized rams per year (Coltman et al. 2002). A legal-sized ram for harvest was defined as having a minimum degree (length) of horn curl of 4/5-curl. Harvested rams were an average of 6 years old, and some were only 4 years old. However, their peak reproductive age was 8 years old so that there was selection against large horn size. In addition, horn size is highly heritable (H_N = 0.69; Coltman et al. 2003; Pigeon et al. 2016). Rams with the highest **breeding value** (calculated as twice the expected deviation of each individual's offspring from the population mean) were harvested at a younger age than rams with lower breeding value. Consequently, average horn length declined by near one-third over 20 years (Figure 15.12).

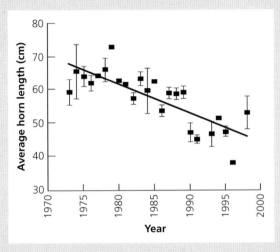

Figure 15.12 Decline in mean horn length (± standard error (SE)) of 4-year-old rams (n = 119) over 30 years in the Ram Mountain population of bighorn sheep. From Coltman (2013).

This study suggests that removing rams with genes for large horn size before peak reproductive age contributed to a decline in horn size. This "unnatural" selection can be avoided in part by harvesting only larger rams (e.g., "full curl"), and fewer rams per year to reduce artificial selection pressure. This research has led to changes in harvest regulations (e.g., fewer and only full-curl rams) to reduce selection pressure (Pigeon et al. 2016). This study has also led to debate regarding how often hunting causes an evolutionary response (e.g., Festa-Bianchet & Mysterud 2018; Heffelfinger 2018; Kardos et al. 2018a).

complex and has proven difficult. For example, the concept of setting a maximum sustained yield in fisheries was developed to ensure long-term sustainability. However, if applied to a mixed-stock fishery, this policy is likely to result in a ratchet-like loss of the less productive local reproductive subpopulations (Larkin 1977).

There are two main approaches to this problem:

1. Harvest subpopulations individually.

2. Use genetic monitoring to determine the contribution of each subpopulation to a mixed harvest.

Genetic analysis of such mixed harvests can provide rapid and accurate estimates of the contribution of different subpopulations (Smith, C.T et al. 2005). For example, samples of the Bristol Bay sockeye salmon fishery are analyzed shortly after capture and the results of a mixed-stock analysis are radioed to the fleet every other day so that the locations of harvest can be adjusted (Elfstrom et al. 2006; Dann et al. 2013).

15.5.4 Protected areas

No-take protected areas have great potential for reducing the effects both of loss of genetic variation and harmful exploitative selection (Sala & Giakoumi 2018). Models of reserves in both terrestrial (Baskett et al. 2005) and marine (Kritzer & Sale 2004) systems support this approach for a wide variety of conditions. However, the effectiveness of such reserves on exploited populations outside of the protected area depends upon the amount of interchange between protected and nonprotected areas and upon understanding the pattern of genetic subdivision (Palumbi 2003; Dawson et al. 2006). It has been suggested that as exploitation pressure intensifies outside protected areas, local protection could select for decreased dispersal distance, thereby increasing isolation and fragmentation, and potentially reduce the genetic capacity of organisms to respond to future environmental changes (Coltman 2008). However, on balance, having more protected areas should mitigate more harvest-associated genetic problems than they create.

Guest Box 15 Baltic Sea flounder: cryptic species, undetected stock structure, and the decline of a local fishery
Paolo Momigliano and Juha Merilä

Flounders (*Platichthys* spp.), one of the main fishery targets within the Baltic Sea, were once very abundant in the Gulf of Finland (Figure 15.13 top left). However, commercial catch data (ICES 2017) and fishery-independent surveys (Jokinen et al. 2015) suggest that present-day abundance is about 10% of what it used to be. It was recently discovered (Momigliano et al. 2017, 2019) that the Baltic Sea flounder fishery targets two morphologically indistinguishable flounder species: the European flounder (*P. flesus*), which spawns pelagic eggs (in upper layers of open sea) in deep offshore

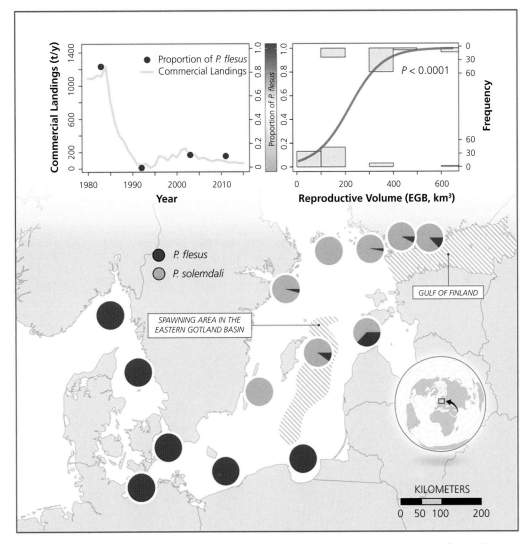

Figure 15.13 The map shows the contemporary proportion of the European flounder (*Platichthys flesus*) and the Baltic flounder (*P. solemdali*) in different regions of the Baltic Sea (modified from Jokinen et al. 2019). The top-left insert shows the proportion of the European flounder in the Gulf of Finland and total commercial landings as a function of time (modified from Momigliano et al. 2019). The top-right insert shows the results of a logistic regression between the proportion of European flounder in the Gulf of Finland and the reproductive volume in the Eastern Gotland Basin, with a clear "threshold" in reproductive volume after which most of the individuals caught in the Gulf of Finland are likely to be European flounder (modified from Momigliano et al. 2019). Graphics courtesy of Bohao Fang.

Guest Box 15 Continued

basins, and the newly described Baltic flounder (*P. solemdali*; Momigliano et al. 2019), which spawns demersal eggs (at or near the ocean bottom) in coastal areas. Demographic modeling based on genomic data suggests that the two species originated via two independent invasions of the Baltic Sea (Momigliano et al. 2017), and population genetic analyses revealed strong reproductive isolation in syntopy as well as high differentiation in genomic regions associated with reproductive isolation (Momigliano et al. 2017; Jokinen et al. 2019).

A reconstruction of the genetic makeup of flounder species in the Gulf of Finland through time based on DNA from archived inner ear bones (otoliths) leads to an unexpected discovery: when flounder catches were high, European flounder dominated local assemblages, while it nearly completely disappeared by the 1990s as fishery landings dropped dramatically (Figure 15.13, top left). This undetected change in species composition and the drop in landings closely match the worsening environmental conditions in European flounder's northernmost spawning ground (Jokinen et al. 2019; Momigliano et al. 2019) within the Eastern Gotland Basin (Figure 15.13, top right), likely attributable to anthropogenic eutrophication of the Baltic Sea. The suitability of a spawning site for reproduction is encapsulated by its reproductive volume—that is, the volume of water in which environmental conditions (in this case oxygen concentration and salinity) are suitable for reproduction. The proportion of European flounder in the Gulf of Finland flounder assemblage, as well as commercial landings, are correlated with the reproductive volume in the Eastern Gotland Basin during the preceding years (Figure 15.13), suggesting that the Gulf of Finland population of European flounder is a sink population relying on larval subsidies from southern spawning grounds. Therefore, the likely cause of the disappearance of European flounder from the Gulf of Finland was a cessation of larval supply: the drivers of the local declines of the Gulf of Finland lie hundreds of kilometers away from it.

This story teaches us about the importance of uncovering unobserved stock structure and studying spatiotemporal changes in the relative contribution of different stock components, as well as the underlying environmental causes, to manage marine resources in the age of rapid anthropogenic change. The decline of the Gulf of Finland flounders was caused by the near disappearance of one species in a multispecies fishery, which went completely unnoticed for decades (Momigliano et al. 2019) and was likely unrelated to local fishery and environmental pressures.

CHAPTER 16

Climate Change

Whitebark pine, Section 16.7

> Climate change is one of the largest threats to biodiversity of our times. Only when we, as a planet, adopt a socio-economic strategy that will allow organisms to adapt in pace with the changes in their environment can we prevent severe loss of species due to global climate change.
>
> **(Marcel E. Visser 2008, p. 657)**

> Although the geographical range limits of species are dynamic and fluctuate over time, climate change is impelling a universal redistribution of life on Earth. For marine, freshwater, and terrestrial species alike, the first response to changing climate is often a shift in location, to stay within preferred environmental conditions.   **(Gretta T. Pecl et al. 2017, p. 1)**

The global distribution of most plant and animal species is largely determined by climate. Genetic variation among populations within species for phenotypic traits involved in local adaptation is also frequently associated with climate. It is abundantly clear that climate is changing rapidly as a result of increasing greenhouse gases in the atmosphere (IPCC 2014). For those species whose ranges are determined primarily by climate, this means that suitable conditions are moving away from where they are currently found to new geographic areas (Ackerly et al. 2010). Novel climates comprising new combinations of temperature and precipitation may develop, and some climatic combinations may cease to exist (Mahony et al. 2017). Climate change is presenting a large challenge to conserving populations and species at risk, and will test the abilities of populations to withstand rapid changes through phenotypic plasticity, adaptation to new conditions, or shift through range migration to favorable conditions elsewhere.

Some species may be quite resilient in the face of climate change and established individuals may persist in locations outside of the climatic niche of that species. For example, some perennial plants such as trees that live for hundreds or even thousands of years can persist as adults through periods that might be climatically unsuitable for establishment or reproduction. Species that spread and reproduce clonally may be unable to sexually reproduce outside of their climatic niche, but may be able to persist clonally. The alpine sedge *Carex curvula* was estimated to have clones as old as 2,000 years in the Alps (Steinger et al. 1996). One trembling aspen clone in Utah named Pando that covers 43 hectares has been estimated to be up to 10,000 years old, with root systems thought to have persisted through a wide range of past climates (although the health and extent of this clone is declining; Rogers & McAvoy 2018). However, if conditions are too unfavorable for sexual reproduction in such environments, long-lived individuals and the populations

Conservation and the Genomics of Populations, Third Edition. Fred W. Allendorf, *et al.*, Oxford University Press.
© Fred W. Allendorf, W. Chris Funk, Sally N. Aitken, Margaret Byrne, and Gordon Luikart (2022). DOI: 10.1093/oso/9780198856566.003.0016

they comprise may become functionally extirpated, the "living dead," unable to adapt and evolve.

Species will differ in their capacity to tolerate, adapt, or migrate in the face of climate change. Interacting species may also migrate in different geographic directions as a result of differences in climatic niches determined by different climatic variables (Davis & Shaw 2001; Ackerly et al. 2010). This is likely to cause major shifts in interspecific interactions including competition, parasitism, predator–prey, and mutualism (Traill et al. 2010b; Van der Putten et al. 2010). Species may be limited in their migration capacity by the rate of migration of a symbiont (e.g., plants limited by the absence of mycorrhizae (Johnson et al. 2013) or pollinators (Potts et al. 2016)). Hybrid zones may move geographically, and reproductively compatible species may come into contact or lose connectivity (Buggs 2007). Hybridization can threaten a species' genetic integrity and persistence, but may also provide the genetic variation required to adapt to new climatic conditions (Hamilton & Miller 2016; Kovach et al. 2016; Chapter 13).

A population whose climatic niche is moving rapidly as a result of climate change may respond in the following ways:

1. **Phenotypic plasticity**: Individuals in the population might tolerate new conditions through changing their morphology, physiology, or behavior, and persist.
2. **Adaptation**: The population might adapt to the new climatic conditions through natural selection.
3. **Migration**: Individuals in the population might move and track the climatic niche geographically.
4. **Extirpation**: The population may become inviable and become extirpated.

In general, two or more of these processes are likely to occur simultaneously (Davis & Shaw 2001; Merilä & Hendry 2014). Only some combination of phenotypic plasticity and adaptation will likely prevent local extirpation of species facing rapid environmental change (Gienapp et al. 2008).

In this chapter, we explore the alternative outcomes of climate change for populations, with a particular emphasis on phenotypic plasticity and adaptive responses. We review how phenotypic, genomic, and environmental data can be used to evaluate the capacity for adaptation or plasticity, and predict the extent of maladaptation under projected future conditions. We also consider management options in the face of climate change, and discuss the implications of these actions for the conservation of populations.

16.1 Predictions and uncertainties of future climates

Conservation strategies are usually focused on a species or ecosystem of concern within a fixed geographic context (Hansen et al. 2010). Populations targeted for conservation usually occur within defined geographic areas or regions that are often geopolitically bounded (i.e., within states, provinces, or countries). *In situ* conservation within parks, protected areas, and ecological reserves often forms the background of such strategies, with fixed land or aquatic areas maintained in a relatively natural state and managed to conserve plant and animal populations, and their biological communities. If climate change results in protected areas no longer containing suitable habitat, population extirpations and, ultimately, species extinctions could result. For example, Ackerly et al. (2010) modeled the future climates of over 500 protected areas in the San Francisco Bay Area of California. By 2100, only eight of these conservation areas were predicted to have temperatures within the current observed range. At a global scale, Loarie et al. (2009) predicted that current climatic conditions would persist within the boundaries of current protected areas for the next century in only 8% of protected areas globally. Urban (2015) estimated as many as one in six species globally would be at risk of extinction with climate change under status quo policies and emissions. It is clear there is a need to consider climate change and the capacity for species to adapt or migrate as part of conservation planning.

Both the paleoecological record and the recent history of contemporary populations provide evidence of past species responses to climatic changes. While geologists and paleoecologists have documented large global and regional climatic shifts over geological time, on a conservation timescale, climatic changes were not considered relevant to conservation until relatively recently (Hansen et al. 2010). Biological responses to climate change have been detected in many natural populations (Parmesan 2006; Merilä & Hendry 2014; Otto 2018), and these responses will increase given that the extent of measured climatic changes over recent decades pales in comparison to modeled projections for the remainder of this century (IPCC 2014).

In order to interpret the implications of climate change for conservation, it is useful to understand the scientific sources of, and the uncertainty associated with, the predictions. Two types of models are combined to map the geographic distributions of suitable climatic conditions for species or populations: **global climate models (GCMs)** and **bioclimatic envelope models (BEMs)**. GCMs are complex mathematical models of global thermodynamics and atmospheric or oceanic circulation (IPCC 2007). These models are used to predict future climatic conditions for different levels of emissions of greenhouse gases. There is considerable uncertainty around predictions of future climates for two reasons: (1) GCMs differ in their scientific assumptions and methodologies, and thus produce different predictions; and (2) it is highly uncertain what the trajectory of greenhouse gas emissions will be over time, depending on how quickly humankind is able to curb carbon dioxide and other greenhouse gas emissions. The projections from different GCMs are often combined in **ensemble models** that take into account model uncertainty (Taylor et al. 2012).

The trajectory of greenhouse gas increases in the atmosphere and their subsequent radiative forcing of climate used in GCMs and ensemble models is now standardized using the International Panel on Climate Change's (IPCC) **Representative Concentration Pathways (RCPs)** that are based on different future emissions scenarios: RCP 2.6 is a very stringent scenario with global emissions declining starting in 2020 and projected warming limited to below 2°C; RCP 4.5 is an intermediate warming scenario; and RCP 8.5 is based on emissions continuing to rise throughout this century, with projected global warming of 4.3°C (IPCC 2014). While all these emission scenarios predict profound biological impacts, the consequences for biodiversity under RCP 8.5 are catastrophic (Urban 2015). Unfortunately, every recent year that has passed without substantial emission reductions has made it more likely that this is the trajectory the planet is on.

16.2 Phenotypic plasticity

Phenotypic plasticity occurs when the same genotype produces different phenotypes in different environments. Recall from Chapters 2 and 11 that a phenotype, particularly for a quantitative trait, is the product of both a genotype and the environment it is in. If the environment changes, an individual's phenotype may change with no genetic change. Environmental conditions can affect phenotypes for phenology (Example 16.1), growth, morphology, or reproductive traits. Phenotypic plasticity can be observed within an individual if the phenotype changes in response to environmental changes within its lifespan, or between generations if genetic makeup does not change, on average, yet the environment does. Individuals and populations can vary genetically in their capacity for phenotypic plasticity.

Phenotypic plasticity can increase the environmental tolerances of genotypes, but whether it also shields genotypes from selection and slows adaptation to new conditions, or generates new phenotypes on which selection can act, has been a source of some debate (Ghalambor et al. 2007; Reed et al. 2010; Catullo et al. 2019). Plasticity can be adaptive, moving phenotypes toward a new optimum in new environments, or can be maladaptive, causing phenotypes to shift toward lower fitness. In some environments, cryptic genetic variation may be expressed under new conditions, allowing selection to act. Plasticity can also play an important role in stabilizing populations demographically and allowing them to persist while **evolutionary rescue** takes place (Catullo et al. 2019; Figure 16.1).

Evolutionary rescue is the process whereby a population recovers demographically following an environmental change as a result of natural selection. However, plasticity can also slow adaptation by reducing the effects of selection. Phenotypic plasticity is an effective mechanism for individuals to adjust to short-term environmental shifts, but is likely to be insufficient in the long term to allow populations to cope with predicted climate change, and will not produce as great a shift in phenotype

---

Example 16.1 Phenological responses to climate change are already widespread

One of the most commonly observed types of phenotypic plasticity related to climate is variation in **phenology**, the timing of biological events. In seasonally variable climates, the timing of transitions, for example, between growth and dormancy, or between summer and winter coloring, is often adapted to local climatic conditions. The phenology of many species has already changed in response to shifts in climate over recent decades (Parmesan 2006). Many phenological traits, such as timing of bud break or flowering date in plants, are based on the accumulation of heat sum in spring, and thus have inherent plasticity in responding to rising temperatures. Species with these traits may compensate for warmer conditions without genetic changes. Many animal behaviors and life history events are also temperature dependent. Long-term records of migratory bird arrival dates and egg-laying dates have been pivotal in understanding relationships between climate and life history events. For example, Gienapp et al. (2008) summarized data on egg-laying date from 15 studies and five bird species using records up to 70 years long. In all cases, birds had laid eggs earlier in recent years by an average of 9.5 days. The primary cause is likely phenotypic plasticity, with birds responding to warmer conditions by laying eggs earlier.

While phenological traits often show plasticity with temperature, they can also vary genetically and be locally adapted to climate. Snowshoe hares molt to change coat color from brown (summer) to white (winter) for camouflage. The timing of molt in fall and spring shows both phenotypic plasticity from year to year within locations, as well as geographic variation, with populations in colder locations molting earlier in fall and later in spring than populations in warmer locations (Zimova et al. 2020). Climate change has resulted in winters becoming milder and reduced the duration of snow cover. Hares that molt too early in fall or too late in spring are mismatched with their environments and suffer 3–7% higher weekly mortality due to predation than hares that are matched with ground cover (Zimova et al. 2016). Some snowshoe hares have adapted to warmer winters by staying brown all winter.

In some cases, polymorphisms within genes affecting variation in phenology have also been identified. The winter brown trait in snowshoe hares is the result of a cis-regulated polymorphism at the *Agouti* gene that changes gene expression at the time of the autumn molt (Jones et al. 2020). This polymorphism originated from introgression from black-tailed jackrabbits, and is the common phenotype of snowshoe hares in Oregon, Washington, and British Columbia. Other examples of genes affecting phenology include polymorphism in the flowering time gene *PtFT2*, which explains 65% of the variation in bud-set timing of European aspen trees (Wang et al. 2018), and variation in the *Clock* gene that predicts migration phenology in birds, salmon, and other species (Saino et al. 2015; Merlin & Liedvgal 2019).

---

as a microevolutionary response in the longer term (Gienapp et al. 2008; Visser 2008).

Not all phenotypic effects of climate warming are harmful for plant and animal species. In low temperature-limited environments, conditions are more favorable for growth for some species. Yellow-bellied marmots are large montane rodents that eat alpine vegetation. Between 1976 and 2008, the pre-hibernation weight of yellow-bellied marmots increased by 4–6% in a Colorado population, and the probability of survival increased with weight for both juvenile and mature animals (Ozgul et al. 2010). The authors attributed the fatter pre-hibernation condition of the marmots primarily to two phenotypically plastic phenological shifts: earlier emergence from hibernation leading to a longer growing season, and earlier weaning of pups. A

similar phenomenon has been recorded in long-term forestry provenance trials, where trees from northern locations planted in warmer climates have greater growth capacity than they do in their home provenances (e.g., Wang et al. 2010).

Phenotypic plasticity in gene expression can be a source of thermal resilience in a warming climate. While substantial genetic shifts can take generations to impact average fitness in a population, the up- or down-regulation of genes can occur in minutes, hours, or days. Plasticity of gene expression can also be a genetic target of selection. Now that RNA sequencing using next-generation and third-generation sequencing platforms has become an efficient method of studying gene expression genome-wide, researchers are starting to understand the importance of gene expression

(a) Population responses with low plasticity and moderate heritability

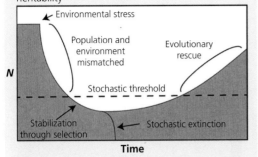

(b) Population response with low heritability and low plasticity

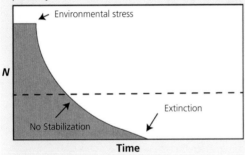

(c) Population response with high plasticity and moderate heritability

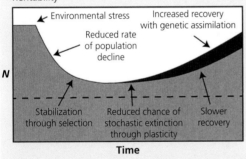

Figure 16.1 Possible responses of a closed population over time in response to an abrupt change in environment. (a) Evolutionary rescue without plasticity: a population with little phenotypic plasticity for a key trait under selection, but moderate heritability for that trait, may initially decline in size, and then stabilize and recover as a result of adaptation to the new environment. (b) Extirpation: a population without phenotypic plasticity or heritable variation for a key trait may become extirpated if it falls below a stochastic threshold for population size. (c) Evolutionary rescue with plasticity: populations with both phenotypic plasticity and genetic variation for a key trait may be buffered by plasticity while they adapt, slowing the decline but also slowing population recovery. From Catullo et al. (2019).

plasticity for organisms to tolerate climate variability. For example, Sandoval-Castillo et al. (2020) studied gene expression in ecotypes of rainbowfish and found variation among these ecotypes in their transcriptional response to projected climate warming and their tolerance of thermal stress (Example 16.2).

16.3 Epigenetic effects

Phenotypic plasticity often results from **epigenetic** effects, which are modifications to DNA structure, but not DNA sequences that result in changes in gene expression (Section 4.10). Epigenetic marks on DNA, including methylation and histone modification, change levels of gene expression and can alter phenotypes as a result. These effects can be transient; can be inherited mitotically, persisting within an individual through cell divisions; or, for some species, can be inherited meiotically and persist across generations. Epigenetic variation in phenotypes can arise from interactions between epigenetic processes and the environment; from interactions between epigenetic processes and DNA sequence; or from epimutations that arise randomly (Angers et al. 2020). These processes can play an important role in allowing organisms to survive environmental fluctuations. However, the extent to which epigenetic effects might be inherited over generations, and whether these might increase population persistence in a changing climate, is unclear. Transgenerational inheritance is more likely to occur in plants than in mammals as epigenetic marks are generally reset during gamete formation in the latter (Eirin-Lopez & Putnam 2019). While there has been considerable speculation about the importance of transgenerational epigenetic effects for adaptation in a changing climate, there has been relatively little conclusive research (Perrone & Martinelli 2020).

Epigenetic effects on phenotypes could complement or supplant adaptation in rapidly changing climates for some populations or species through phenotypic plasticity. However, Visser (2008) pointed out that in a rapidly changing environment, environments of offspring will differ from those of parents, particularly for long-lived organisms, and **maternal effects**, including epigenetics, may be insufficient for phenotypes to keep up with climate.

Example 16.2 Phenotypic plasticity in gene expression is correlated with thermal tolerance in Australian rainbowfish

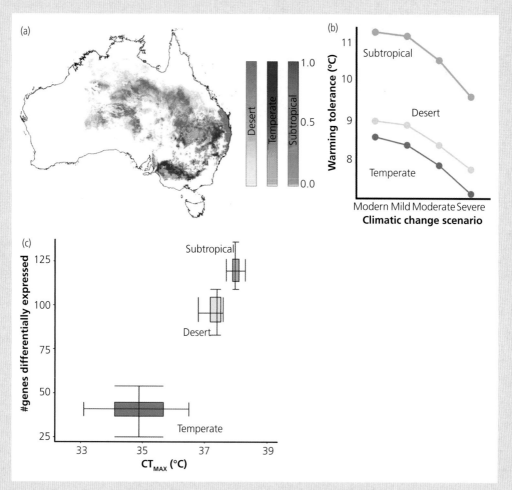

Figure 16.2 Plasticity in gene expression in three ecotypes of Australian rainbowfish. (a) Distribution model of desert, subtropical, and temperate rainbowfish ecotypes, based on presence records and nine bioclimatic variables. (b) Projected average physiological sensitivity of the three ecotypes to different climate change scenarios based on temperature tolerance. (c) Relationship between maximum critical temperature (CT_{MAX}) of ecotypes and the number of genes differentially expressed between moderate and high temperatures. From Sandoval-Castillo et al. (2020).

Plastic changes in gene expression in response to changing thermal conditions are relatively common. Local adaptation of populations to temperature regime is also relatively common. Sandoval-Castillo et al. (2020) quantified gene expression in temperate, subtropical, and desert ecotypes of Australian rainbowfish under control (21°C) and warming (33°C) treatments (Figure 16.2a). After 14 days at these temperatures, RNA was extracted from liver samples from fish and sequenced. The short-term maximum thermal tolerance (CT_{MAX}) of fish from each ecotype was also determined in separate experiments.

Continued

Example 16.2 *Continued*

From the RNA sequencing, 34,815 **unigenes** were identified, of which 28,483 were present in all three ecotypes. A unigene is a bioinformatically determined gene identified from RNA transcript sequencing. Gene expression profiles were relatively consistent among individuals from the same ecotype, and varied considerably among ecotypes. In total, 236 unigenes were expressed differentially in the control and warming treatment, showing plasticity in expression. However, the ecotypes varied greatly in the number of genes showing differential expression: the temperate ecotype had 27 differentially expressed genes, the desert ecotype had 84, and the subtropical ecotype had 109 (Figure 16.2b). The number of differentially expressed genes increased with the maximum critical temperature CT_{MAX} that each ecotype could withstand: 34.9°C for the temperate ecotype, 37.2°C for the desert ecotype, and 38.0°C for the subtropical ecotype (Figure 16.2c).

Eirin-Lopez & Putnam (2019) identified questions that remain about the importance of epigenetics for conservation and restoration of populations in a changing climate, including: How are environmental signals translated into epigenetic responses, and do these differ with predictability of environments? Is the **epigenome**, the complete set of epigenetic marks across a genome, linked to the microbiome? Are epigenetic effects inherited transgenerationally? Does phenotypic plasticity resulting from epigenetic effects enhance or retard rates of evolution?

Parental environments can also affect offspring in nonepigenetic ways. Characteristics of the environment during reproduction or development can affect reproductive traits or maternal care in ways that affect the phenotype of offspring. These effects are termed maternal effects. For example, in plants, the quality of the maternal environment often affects seed size, with plants producing larger seeds when conditions are more favorable for growth, and larger seeds in turn producing plants with more rapid juvenile growth due to the availability of additional nutritional resources in the seeds.

16.4 Adaptation to climate change

The capacity for populations to adapt to new environmental conditions is substantial, as evidenced by the ability of many plant and animal species to adapt to local conditions during or since postglacial recolonization. As long as genetic variation exists for genes affecting phenotypic traits involved in adaptation to climate and populations are demographically sustainable, then **genetic assimilation** can occur through selection imposed by new conditions. However, the rates of anthropogenic climate change predicted for this century far outpace those seen during the Pleistocene. Here, we first summarize the theoretical basis of evolutionary rescue, the rapid adaptation of populations to new environmental conditions (Figure 16.1), and then review examples of adaptive responses to global warming.

16.4.1 Theoretical predictions of capacity for adaptation

Quantitative geneticists have developed theoretical predictions of the maximum rate of environmental change a population can withstand and recover from through evolutionary rescue (Figure 16.1). This rate is determined by the amount of genetic variation in the population, the fecundity and effective size of the population, environmental stochasticity, and the strength of selection. Beyond a threshold rate of change, the population will be unable to adapt quickly enough to keep up with the necessary rate of change in optimum phenotype to maintain fitness. Over time, the population will lag increasingly behind its climate fitness optimum. When the **adaptational lag** between the average population phenotype and the optimum phenotype with the highest fitness becomes too great, extirpation is likely.

Lynch & Lande (1993) developed an equation to predict this threshold rate of change (k_c, in **haldanes**, which are units of phenotypic standard deviations, σ_p) based on the strength of stabilizing selection within populations (σ^2_w) and the maximum rate of population increase (r_{max}):

$$\frac{k_c}{\sigma_p} = \sigma_p \sqrt{2 \, r_{max}/\sigma_w^2} \qquad (16.1)$$

This equation predicts that species with low fecundity ($r_{max} = 0.5$) can only tolerate a rate of environmental change per generation corresponding to about 0.1 haldanes or less. This threshold may vary considerably among species. However, small populations experiencing demographic stochasticity might only be able to adapt to changes corresponding to 0.01 haldanes per generation (Bürger & Lynch 1995). Populations for which this is true have very little chance of keeping pace with climate change through adaptation even if they contain genetic variation for relevant genes and traits.

On the other hand, species with high fecundity, such as a pioneer tree species that can produce 10,000 seeds/tree, could have an r_{max} of 9.2 and could theoretically tolerate a change in optimum phenotype up to ~0.42 haldanes per generation (Aitken et al. 2008). Kopp & Matuszewski (2013) concluded that in most cases, the capacity for genetic change per generation that is sustainable is less than 0.1 haldanes. Since the maximum rate of climate change that a population can tolerate is on a per-generation basis, long-lived organisms are experiencing a much higher rate of climate change per generation than those with short generations, and will need to adapt more per generation to persist.

16.4.2 Phenotypic approaches for detecting adaptation to climate change

While it is easy to speculate that phenotypic changes in behavior, morphology, physiology, or phenology observed in a warming climate might reflect genetic changes produced by natural selection, in reality, it is not a simple matter to test this hypothesis. Merilä & Hendry (2014) describe four methods based on phenotypic data for inferring if such changes are genetic.

Animal model approaches use quantitative genetic analyses of populations with known or inferred relatedness to partition changes in phenotypic variation into genetic and plastic components (Section 11.1.3.3). Genomic analyses or pedigree information (where available) can be used to estimate relatedness among individuals. Common

garden experiments allow for the separation of genetic, environmental, and genetic × environmental plasticity. If they are repeated over time, then climate-associated shifts in phenotypes can be quantified. For example, Bonnet et al. (2019) used an animal model analysis to partition the advancement of parturition (birthing) dates in a long-studied population of red deer on the Isle of Rum into genetic and plastic components using an animal model approach (Example 16.3).

Resurrection studies compare phenotypes for old versus contemporary populations in common gardens, for example, using stored seeds from plants or dormant eggs from invertebrates, to evaluate the extent of genetic change over time. In such studies, it is important to control for maternal effects. This is usually accomplished by rearing a generation of individuals from each source (old and recent) in a common environment before conducting the experimental common garden. For example, Franks et al. (2007) compared phenotypes of field mustard plants grown from seed collected before and after a five-year drought in California. They found that flowering dates had advanced substantially during that period as an adaptation to drought in a population found in a normally wet location, but a population native to a dry location did not experience a genetic change in date of flowering over the same period.

Experimental evolution studies expose experimental populations to different environmental conditions associated with climate change, including increased carbon dioxide, warmer temperatures, or ocean acidification, and then compare populations exposed to different treatments for genetically based phenotypic changes. These types of studies have been useful for testing theoretical expectations of adaptation to climate change and evolutionary rescue with model organisms; however, they reduce climate change to one or a few environmental factors in artificial environments, and so are less useful for understanding capacity for adaptation to changing climates in species of conservation concern in natural environments.

Space-for-time substitutions assess phenotypic differences among populations originating from different climates within common gardens, and extrapolate from those differences to estimate the

Example 16.3 Separating the effects of plasticity versus adaptation in advancing red deer parturition date

Phenotypic changes in response to climate change are often the result of both phenotypic plasticity and genetic effects. In a population of red deer on the Isle of Rum in northwest Scotland, parturition (birth) date has advanced by an average of 4.2 days per decade since 1980 (Bonnet et al. 2019; Figure 16.3). Bonnet et al. (2019) used quantitative genetic animal models to separate the genetic and plastic contributions to this advancement of parturition phenology, and explore alternative hypotheses for causes of genetic changes, including selection, genetic drift, gene flow, and inbreeding. Parturition date was heritable, with additive genetic variance responsible for 17% of phenotypic variance. Females that gave birth earlier had higher lifetime breeding success, on average. A best linear unbiased prediction (BLUP) estimated that genetic changes accounted for 2.1–2.4 days of advancement in parturition over the total study period. There was little evidence that genetic drift contributed to this genetic effect (Figure 16.4). Both inbreeding and gene flow were estimated to have very small genetic effects that, on average, would delay rather than advance birth date over the time period studied. The authors also predicted the response using the secondary theorem of natural selection (STS in Figure 16.4), based on the additive genetic covariance between birth date and fitness. This simple approach estimated that the 4.9 days of advancement over the study period was due to evolution.

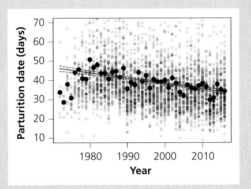

Figure 16.3 Temporal trend in the birth date (days after 1 May) of red deer on the Isle of Rum, Scotland. Large dots are annual means and small gray dots are individual dates of birth, with darker colors representing more calves born on a single day. The red lines indicate the slope and 95% confidence interval. From Bonnet et al. (2019).

Continued

Example 16.3 *Continued*

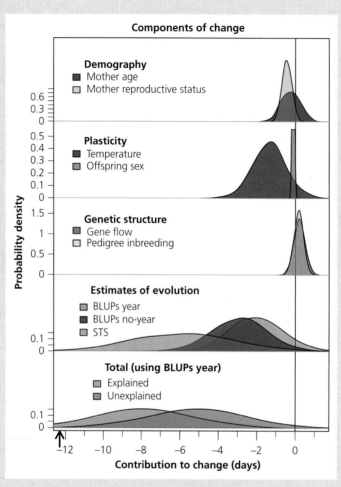

Figure 16.4 Components of change in birth date of red deer over time. The posterior distributions from a BLUP are shown. Both phenotypic plasticity to temperature (red) and evolution (pink and purple) contributed to the advancement in birth date shown in Figure 16.3. STS is an analysis based on the secondary theorem of natural selection and the additive genetic covariance between birth date and fitness. The BLUP analysis including all factors above explained the majority of variation in birth date (green). From Bonnet et al. (2019).

necessary extent of phenotypic change needed to adapt to new climates. Ideally, these studies use a reciprocal transplant approach with common gardens on multiple sites where all populations are phenotyped at all sites, both "home" and "away." In this way, phenotypic plasticity can also be quantified. While this approach can estimate the difference in phenotypes between current and future climates, it has some limitations. First, the population currently inhabiting an analog of the anticipated future climate is assumed to be locally adapted to that climate. Secondly, the capacity for adapting to that future climate is not measured directly, but is inferred, for example, through using the breeder's equation to estimate response to selection. This approach has been used widely in understanding local adaptation to climate and adaptational lags in tree populations in field common garden experiments called **provenance trials** that include individuals from many populations planted on multiple sites (e.g., Example 16.4).

Detecting phenotypic changes using any of these methods and separating adaptation from phenotypic plasticity is challenging; proving that changes are adaptive and that climate change is causal is also difficult. Merilä & Hendry (2014) summarize adaptive and plastic phenotypic responses to climate change to date for 25 studies of plants, fish, birds, mammals, and insects that used temporal rather than space-for-time approaches. This survey found significant genetic changes in 16 of these studies, and phenotypic plasticity in 11. Fourteen of these studies were able to show that the genetic or plastic responses were adaptive; that is, they increased fitness in new environments.

If adaptation is occurring in response to climate change, is it happening rapidly enough? Franks et al. (2014) reviewed studies of phenotypic responses of plants in response to climate change. Of the 38 studies included in this assessment, all found either genetic or plastic responses, and 26 of these studies found both. Just 12 of the studies assessed whether these responses were sufficient to keep pace with climate change. A sobering two-thirds of these studies concluded that the responses are inadequate.

There are several reasons why adaptation may not keep pace with climate change:

Countergradient variation: If environmental and genetic effects on phenotypes are in opposition, that is, if genetic and environmental clines have opposing signs along a climatic gradient, then the effects of selection may be invisible as phenotypes will not change.

Lack of genetic variation: If a population lacks genetic variation for a phenotypic trait under selection, then no adaptive response can occur. This may be a major problem in small populations.

Antagonistic genetic correlations: If two traits under selection have an antagonistic genetic correlation (i.e., if selection increases fitness in one trait but decreases fitness in another due to pleiotropy or gametic disequilibrium), then evolutionary responses will be slow (Etterson & Shaw 2001; Hellmann & Pineda-Krch 2007).

Inbreeding: Small and declining populations are likely to have substantial levels of inbreeding, and this may also hamper adaptation.

Gene flow: If individuals or gametes (e.g., pollen) move between populations inhabiting and adapted to different climatic conditions, the extent to which gene flow facilitates or slows adaptation will depend on the source and recipient population climates as well as the size of the populations; that is, the relative strength of gene flow from one population to another (Davis & Shaw 2001). Gene flow may slow adaptive responses unless the source populations of immigrants or gametes inhabit climates similar to the new climates faced by a population.

16.4.3 Genomic approaches for predicting adaptation to climate change

Genomic approaches for understanding local adaptation to climate and projecting maladaptation as a result of climate change are developing rapidly (Capblancq et al. 2020). The first step in these approaches is to identify a focal subset of **single nucleotide polymorphisms (SNPs)** that have evidence for associations with climate or climate-relevant phenotypic traits, or are population differentiation outliers. Then projections are made of the degree to which climate change will generate a genetic offset or mismatch between current allele frequencies for those SNPs and future local climates.

Example 16.4 Local maladaptation along an elevational gradient due to climate change

Reciprocal transplant experiments designed to inform projections about climate change responses substitute spatial variation in current environments for projected temporal changes in climate (space-for-time substitution). Anderson & Wadgymer (2020) conducted a series of common garden experiments across an elevational gradient with the perennial plant Drummond's rockcress to understand local adaptation of populations to climate and responses to climate change. These experiments included provenance trials, where common gardens contain individuals sampled from many locations and planted on multiple sites; and **reciprocal transplant experiments**, where local populations from all experimental sites are planted on all sites. In some experiments, environments were also manipulated by removing or adding snow, as snow persistence is an important driver of species niche and population adaptation in montane environments. Over 100,000 plants were included across all experiments!

Low-elevation populations established, survived, and reproduced more than local populations, especially in snow removal treatments (Figure 16.5). This indicates local populations, and especially low-elevation populations, are becoming maladapted under climate change. In treatments where snow was added, local populations outperformed nonlocal populations, demonstrating recovery of local adaptation to pre-anthropogenic climate change conditions. The authors conclude that upslope migration could facilitate population persistence and expansion in a changing climate.

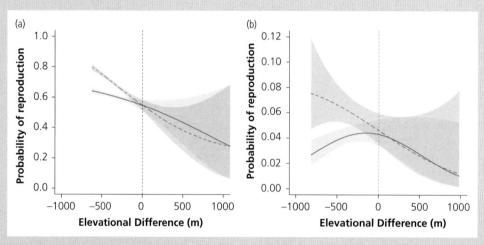

Figure 16.5 The probability of reproduction of Drummond's rockcress plants in common gardens across an elevational gradient with and without snow removal in (a) 2013 and (b) 2014, expressed in terms of the elevational difference between the population elevation of origin and the common garden. Plants moved up in elevation (negative elevational difference) had greater probability of reproduction than those moved down, especially with snow removal. The blue line indicates the control treatment of natural snowpack, while pink indicates snow removal. The variation in results between the 2 years reflects differences in weather. From Anderson & Wadgymar (2020).

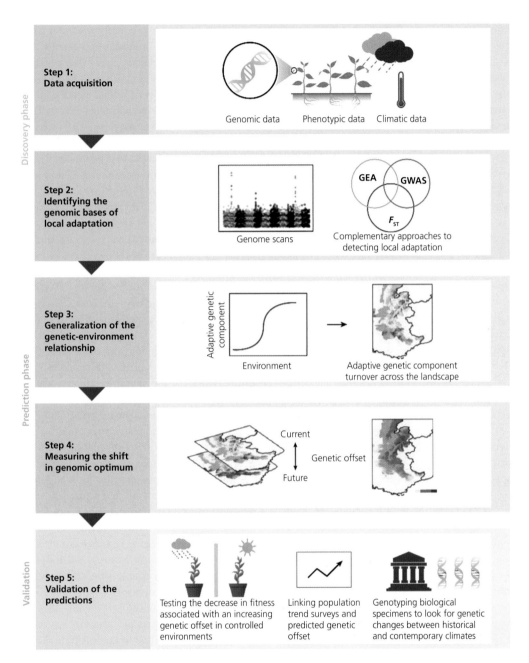

Figure 16.6 The different steps of a study performing genome-informed predictions of climate (mal)adaptation. Steps 1 and 2 correspond to the discovery phase; steps 3 and 4 describe the prediction phase; and step 5 gives ideas for validating the predictions. From Capblancq et al. (2020).

Finally, predictions are validated with an independent dataset (Capblancq et al. 2020; Figure 16.6).

The most common approach for SNP discovery of a set of climate-associated loci is to identify **genotype–environment associations (GEAs)** between SNPs and climate variables (Flanagan et al. 2018; Capblancq et al. 2020). First, individuals

from different populations are genotyped for a large number of SNPs genome-wide, using genome resequencing or a reduced representation approach such as exome capture or **restriction site-associated DNA sequencing (RADseq)**. Next, GEAs are identified while controlling for population structure. The environmental variables involved in these associations provide information on what climate variables are the primary drivers of local adaptation to climate. In some cases, the annotations of genes associated with SNPs can provide further evidence of their role in climate adaptation (e.g., genes in plants involved in photoperiod responses, or behavioral genes involved in migration timing in animals). However, in many cases the phenotypes associated with GEA SNPs are unknown as many genes in nonmodel organisms lack annotations, and associations between polymorphisms and SNPs may be due to linkage rather than causal.

Outlier tests, such as genome scans for F_{ST} outliers, can also be used to identify loci with elevated divergence of allele frequencies among populations. This type of analysis has the disadvantage of not being informed by variation in climate or climate-associated phenotypes, and has some other limitations (Hoban et al. 2016). However, in some studies candidate loci for climate adaptation have been identified as outliers and subsequently validated (Capblancq et al. 2020).

Genome-wide association study (GWAS) approaches (Section 11.3.3) can also be used to identify climate-relevant SNPs associated with key phenotypic traits such as phenology or thermal tolerance (Capblancq et al. 2020). Approaches developed for plant and animal breeding can be used to estimate genomic-estimated breeding values (GEBVs) for phenotypic traits that are sensitive to climate or involved in local adaptation, and to predict phenotypes under different climate scenarios (Guest Box 11). An advantage of the GWAS approach is that the effects of SNPs on relevant phenotypes can be quantified; however, a disadvantage is that usually only a small number of the many possible physiological, behavioral, or morphological traits that may relate to climate adaptation are phenotyped.

Once a set of candidate SNPs for climate adaptation has been identified using one of these methods, allele frequencies for these SNPs can be used to generalize the relationships between genotypes and environments, and to predict maladaptation in the context of future climates (Figure 16.6). Projections of future climatic conditions from GCMs can be used to evaluate the genetic offset between current allele frequencies or GEBVs in populations and those frequencies currently associated with future climatic conditions (Capblancq et al. 2020).

One approach for predicting maladaptation estimates the **risk of nonadaptedness** (RONA) for individual loci (Rellstab et al. 2016). This analysis can be applied individually to many loci to estimate the number of alleles needed in a population to adapt to new climates (Exposito-Alonso et al. 2018). Another approach characterizes the frequency turnover of many alleles simultaneously along climate gradients, and estimates the multilocus **genetic offset** between current genomic composition of populations and future climatic conditions along that gradient. To do this, a method called "gradient forests" that was developed to model species turnover in community ecology was repurposed to assess the turnover of alleles along climatic gradients and predict the genetic offset between current populations and future climates (Fitzpatrick & Keller 2015). Redundancy analysis (RDA) can also be used to estimate genetic offsets (Capblancq et al. 2020).

The RONA and genetic offset methods described above predict the extent of maladaptation to future climates based on relatively simple correlations between candidate SNPs and climate, or between SNPs and phenotypes. They assume: (1) that loci are correctly identified as climate-associated; and (2) that populations are optimally adapted to current or recent climates. Ideally, these predictions should be validated with an independent dataset and approach. This can be done using common gardens, population simulation studies, or studies of historical records or museum specimens (Capblancq et al. 2020). For example, Mahony et al. (2020) determined that those climate variables identified as the most important drivers of local adaptation in a GEA study of over a million SNPs were also the best explanatory predictors of variation in growth rate among populations in a lodgepole pine provenance trial.

Bay et al. (2018) used the *gradient forests* program to test for local adaptation to climate and predict **genomic vulnerability** in the yellow warbler, a migratory North American songbird with a wide geographic distribution. They found good concordance between their projections of maladaptation and population declines found in the North American Breeding Bird Survey, validating their results (Example 16.5). Guest Box 16 provides an example estimating capacity for adaptation in populations of coral.

Example 16.5 Genomic estimates of maladaptation in yellow warblers correlate with declines in observed breeding populations

It is challenging or impossible to implement the phenotypic approaches described in Section 16.4.2 for many animals, including migratory songbirds. Genomic approaches offer another avenue for characterizing local adaptation to climate and for predicting capacity for adaptation to new conditions. Yellow warblers are a migratory songbird and breed across much of the USA and southern Canada. Bay et al. (2018) genotyped birds from 21 breeding locations for over 100,000 SNPs using a RADseq approach. They then tested these SNPs for GEAs with climate variables for locations sampled. They found that the climate variables with the strongest associations with SNP frequencies were related to precipitation at breeding sites. A total of 187 SNPs were associated with all top three climate variables. One of the strongest associations between genotype and climate was a SNP on chromosome 5 that was associated with the top three climate variables. This SNP is upstream of two genes with known associations with migration in birds, *DRD4* and *DEAF1*. *DRD4* is a dopamine receptor, and polymorphisms in this gene are associated with novelty-seeking behavior in birds and some other animals.

  Bay et al. (2018) then predicted genomic vulnerability of populations by estimating the genetic offset between current allele frequencies and genomic composition associated with projected climates for 2050 under RCP2.6 (Figure 16.7a). They then compared the estimated genomic vulnerability of populations with population trend estimates based on data from the North American Breeding Bird Survey, and found a negative relationship between their estimate of genomic vulnerability and modeled population trend, validating their predictions (Figure 16.7b, c).

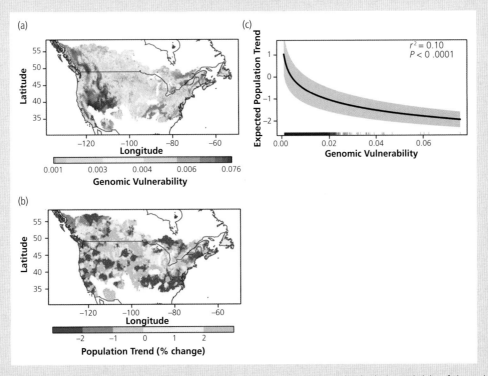

Figure 16.7 Genomic vulnerability of yellow warblers to projected climate change is associated with a higher probability of observed population decline. (a) Estimated genomic vulnerability under 2050 RCP2.6 projections. (b) Population trend estimates (percentage change per year) based on the North American Breeding Bird Survey. (c) Generalized additive model showing the relationship between genomic vulnerability and population trend. From Bay et al. (2018).

16.5 Species range shifts

16.5.1 Modeling species distribution

Species distribution models (SDMs) are models based on the statistical relationship between species occurrence and local climate. They are useful for generating hypotheses about the potential effects of climate change on natural species distributions, and for evaluating management options for conserving species in a changing climate. SDMs have also been used to predict biodiversity hotspots (Koch et al. 2017) and occurrences of endangered species (McCune 2016). Most SDMs first determine the multivariate climatic tolerances of a species by analyzing geographic records of species presence, or both presences and absences, in conjunction with the distribution of temperature and moisture variables. The resulting models can then be used to project where those climatically favorable conditions will occur under future climate scenarios. These models can range from simple to complex, and model selection is challenging (Merow et al. 2014).

A subset of SDMs uses information on physiological climatic tolerances or phenotypic traits of species or populations rather than using existing distributions to predict future distributions (Pearson & Dawson 2003). These mechanistic or physiological SDMs have the advantage of predictions being independent of current species distributions; however, they do not reflect the effects of interspecific competition and other types of species interactions on occurrence.

One criticism of most SDMs has been the lack of inclusion of population-level genetic or demographic data. Including population variation can change the results of SDMs considerably. There are several approaches that have been used that incorporate phenotypic, genomic, or demographic data in SDMs. For example, Benito Garzón et al. (2019) developed an approach that uses phenotypic data from common garden experiments on multiple sites to characterize local adaptation and phenotypic plasticity, and incorporated that information in an SDM. They found this approach projected less impact of climate change on tree species than SDMs based on species occurrence alone. Razgour et al.

(2019) used RADseq data to classify populations of each of two European bat species into those adapted to cold-wet conditions and those adapted to hot-dry conditions (Example 16.6). They also found that ignoring differences in population climatic niches resulted in more dire projections of loss of species range than models taking local adaptation into account.

The predictions from SDMs are often presented as maps illustrating current observed species distributions and future projected distributions of climate-based habitat. Such maps are valuable visual tools for education and conservation. However, they should be interpreted with caution. They often depict areas where climatic conditions are predicted to be favorable without considering the biological capacity for populations to adapt to new conditions *in situ* or migrate to areas that are climatically favorable. The nonclimatic qualities of new potential habitats are also generally ignored.

The rate at which climates are predicted to shift geographically with climate change varies geographically and topographically. As a result, the speed with which populations or species need to geographically shift to track their climatic niche will vary. This ecological movement of species or populations is called migration, but should not be confused with genetic migration (i.e., gene flow). Loarie et al. (2009) estimated "velocities of climate change," defined as the horizontal rate of migration in kilometers per year required to track temperature changes, for biomes globally. Their estimates averaged 0.42 km/year, and ranged from 0.08 km/year for tropical and subtropical coniferous forest biomes to 1.28 km/year for flooded grassland, mangrove, and desert biomes. Predicted velocities were also lowest in mountainous terrain due to the rapid changes in temperature over short elevational distances. Population persistence and species survival will depend on their capacity to keep pace with these moving climates.

Migration is often considered to be the expected response for species to stay within current climate envelopes and SDMs assume that the species' current climatic envelope defines where they can persist. Yet phylogeographic studies of plants in nonglaciated environments, such as California and

Example 16.6 Incorporating local adaptation in SDMs

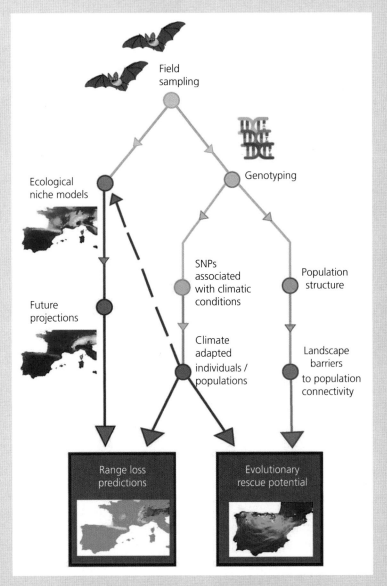

Figure 16.8 Framework for incorporating within-species climate adaptation into species distribution modeling to project future range losses and potential for evolutionary rescue. From Razgour et al. (2019).

SDMs often treat populations across a species range as homogeneous, without considering intraspecific variation in climate adaptation and population climatic niches. Razgour et al. (2019) used genomic data to incorporate intraspecific variation in SDMs for two cryptic Mediterranean bat species, *Myotis escalerai* and *M. crypticus*. First, they used a RADseq approach to genotype 220 *M. escalerai* individuals from 67 locations, and 58 *M. crypticus* from 48 locations for ~18,000 SNPs (Figure 16.8). A GEA approach was used to detect a genomic signature of adaptation to maximum temperatures of the warmest month and

Continued

Example 16.6 *Continued*

precipitation of the warmest quarter. Populations within each species were then classified into one of two ecotypes: hot-dry or cold-wet. SDMs were run on these ecotypes separately, as well as jointly.

For a worst-case climate change scenario using ensemble GCM projections for RCP 8.5, *M. escalerai* was projected to lose 47% of its current range. When the SDM was run on the two ecotypes, the projected range loss was just 19% (Figure 16.9). Similarly, *M. crypticus* was projected to lose 87% of its range based on a species wide SDM, and 58% when the two ecotypes were modeled separately.

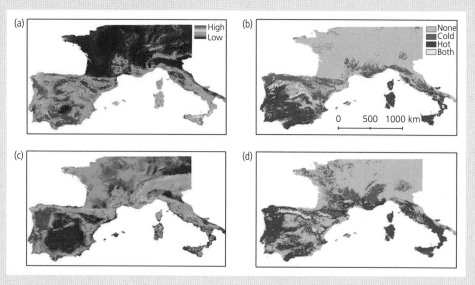

Figure 16.9 Species distribution models (SDMs) for the bat *M. escalerai* illustrating the effect of including intraspecific climate adaptation into projections. Panels (a) and (c) are based on models of the full species range showing climatic suitability. Panels (b) and (d) show areas favorable for the cold-wet ecotype (blue); the hot-dry ecotype (red); or both ecotypes (yellow). Panels (a) and (b) are for current climates, while (c) and (d) are projections based on RCP 8.5 in 2070. From Razgour et al. (2019).

southwestern Australia, have shown evidence of localized persistence from at least the mid-Pleistocene in some cases, and lack of major historical migration to track their climate niches (e.g., Sork 2016; Sampson et al. 2018). This may suggest that plant species that have localized persistence through historical times may tolerate broader climatic environments than shown by their current climate envelopes.

16.5.2 Observed species range shifts

Species ranges shift toward cooler locations through population expansion and migration at the leading edge of the range, and population extirpation at the rear edge. Migration of populations is a function of population growth and dispersal coupled with the distribution, structure, and occupancy by other species of available habitat (Higgins & Richardson 1999). The ability of populations to move to new habitat depends greatly on the life history characteristics of the species, including age at reproductive maturity, fecundity, and mechanism of offspring dispersal, as well as the degree of fragmentation of suitable habitat. There are several causes of lags in range shifts (Alexander et al. 2018). First, there can be **dispersal lags** as a result of reproduction and dispersal limitations; that is, a species can't disperse quickly into newly available climatically suitable habitat. Next, there may be **establishment lags** due to resistance of the new resident community, missing mutualists, or disturbance. Finally, there may be **extinction lags** at the trailing edge of species ranges as species persist for some time before becoming extirpated, preventing community turnover. Natural barriers to dispersal such as mountain ranges, large water bodies, and deserts, as well as human-modified landscapes such as agricultural and urban areas, can slow or prevent range shifts.

Species range shifts in response to climate change are already widespread (Parmesan 2006; Chen et al. 2011). For example, evidence of range shifts is abundant for Antarctic bird species. Species dependent on sea ice, such as Adelie and emperor penguins, have shown marked shifts to higher latitudes. Open-water feeding penguins in the meantime have invaded more southern areas as ice shelves have contracted or collapsed. Many Arctic species have also shown range shifts, some resulting in major ecosystem transitions. For example, shrub species have been invading Arctic tundra in both North America and parts of Russia to the extent that fundamental changes in the albedo, carbon balance, and hydrology of those regions is predicted (Tape et al. 2006). Chen et al. (2011) estimated that species distributions had already shifted, on average, upwards in elevation by 11 m, and toward higher latitudes by 16.9 km, and that in general, species experiencing the greatest warming had the largest range shifts.

For many species, dispersal and establishment lags will slow the advance of the leading edge of species migration, but extirpations will result in the contraction of the trailing species margin, resulting in shrinking species ranges. This type of range contraction was observed by Kerr et al. (2015) in an analysis of 67 bumblebee species in North America and in Europe.

As trees and other plants are often foundational to ecosystems, their migration rates may directly affect the migration potential of other species. Pollen records since the last glacial maximum have provided important evidence about the past migration of plant populations. Cores are extracted from bogs that contain a chronosequence of pollen in sediments, providing records of past species or genus presence. However, in a study of tree species in

eastern North America, genetic data provided evidence that fossil pollen records did not capture the existence of small populations in more northern glacial refugia, and thus overestimated past migration rates (McLachlan & Clark 2004). The genetic data indicated that tree migration rates during postglacial recolonization in North America average only about 100 m per year, which is an order of magnitude slower than what would be needed to track climates spatially if climates warm to the extent projected in the next century (McKenney et al. 2007).

The capacity for migration is likely to have a genetic component and interactions with adaptation (Davis et al. 2005). For example, poorly adapted phenotypes are likely to have lower fecundity and fewer dispersers. Also, selection can act on phenotypic traits that affect dispersal ability, for example, seed wing size or fruit characteristics for plants, or wing size for insects. As a result, individuals that are less fit are less likely to migrate, and individuals with higher dispersal capabilities are more likely to migrate.

Invasive species often have short generation times and are well adapted for long-distance dispersal (Chapter 14). Because of this, they may have a greater capacity than noninvasive species for migrating as the climate changes, which could lead to invasive species dominating more ecosystems in a rapidly changing climate. Many invasive species are early successional, and may become more problematic with climate warming, competing with and reducing the capacity of later-successional species to persist.

16.6 Extirpation and extinction

If populations cannot tolerate rapid climate change through plasticity, or adapt to new conditions, or migrate to track their climatic niche spatially, they will become extirpated. Species extinction may follow. To understand the relationship between climatic change and extinction, it is useful again to look back at the later Pleistocene. Between 50,000 and 3,000 years before the present, 65% of large (>44 kg) mammal genera went extinct (Nogués-Bravo et al. 2010). There is some evidence that this high extinction rate resulted from climate changes in the late Quaternary, although the roles of climatic change versus anthropogenic factors such as hunting are hotly debated. Nogués-Bravo et al. (2010) analyzed paleorecords of climate and large mammal distributions over this period. They estimated "climatic footprints" for this period based on the magnitude of climatic change. Their analysis revealed that continents other than South America that experienced higher rates of climate change, and had larger climatic footprints, during this period had higher extinction rates.

Arctic and alpine populations are at high risk of species extirpation due to anthropogenic climate change. This is in part due to the greater rates of change predicted and observed, to date, for high-latitude areas, and in part due to a lack of adjacent habitat that remains climatically favorable. A low-vagility species inhabiting a mountaintop has nowhere to go in a warming environment. Nine out of 25 re-censused populations of American pika, a small alpine mammal in the Great Basin of the western USA, were extirpated between the 1930s and 2007 (Beever et al. 2010). The populations that were extirpated occupied significantly lower-elevation habitats and experienced higher levels of heat stress than those that persisted, suggesting climate may have played a major role in these extirpations.

The greater the magnitude of climate change, the higher the number of climate-related extinctions will be. Urban (2015) projected the proportion of species that will be lost under different climate change scenarios based on 131 studies, most of which used SDMs, but also included a small number of mechanistic models, species–area relationships, and expert opinion. If warming is limited to 2°C (i.e., the target of the Paris Agreement), 5.2% of species were at risk of extinction. However, if we continue on the emissions pathway described by RCP 8.5, with 4.3°C warming this century, 15.7% of the species are projected to be lost. However, this simple approach does not take into account intraspecific variation, capacity for adaptation to new conditions, dispersal barriers, or other complicating factors.

Predicting the risk of species extinction due to climate change is difficult. In general, species with larger geographic ranges have lower extinction risks. This is called extinction resistance. However, if a species contracts to only a few populations,

then what will matter is threat tolerance of those remaining populations, not extinction resistance. A species that has a recently contracted range may not have any higher threat tolerance than a species that has always had a small range and relatively few populations (Waldron 2010).

Populations in different parts of a species range will face different degrees of threat from climate change, depending on the position of the local environment within the species niche, population history, and contemporary patterns of gene flow across species ranges (Davis & Shaw 2001; Figure 16.10). **Rear edge populations** (those at the warm or dry margin of a migrating species range) are at the greatest risk of extirpation during broad-scale shifts in species range. These populations may already inhabit extreme environments at the margin of the species niche. They also may be older lineages with relatively high among-population divergence and high regional diversity as they may be located closer to glacial refugia (Hampe & Petit 2005). In contrast, theory suggests that **leading edge populations**, those at the cool margin of a species climatic niche that are likely to expand with warming, should experience increased fitness with climate change. Gibson et al. (2009) have made compelling arguments on conserving leading edge populations as they will play a key role in migration and species range shifts, potentially expanding into newly habitable areas. Hampe & Petit (2005) advocate for a conservation focus on rear edge populations as they are likely to be reservoirs of within-population genetic diversity and to show greater among-population divergence than younger populations.

If a species has a greater abundance in the center of the species range, population genetic theory has shown this is likely to result in greater gene flow from central to peripheral populations than the reverse (Kirkpatrick & Barton 1997; Davis & Shaw 2001). Populations at the leading edge of migration will receive gene flow from more central populations that carry alleles pre-adapted to the warmer conditions in the center of the species range. In a static environment, this means that leading edge populations occupy environments colder than optimal for their average genotypes. Warming will move the climate closer to optimal for the average

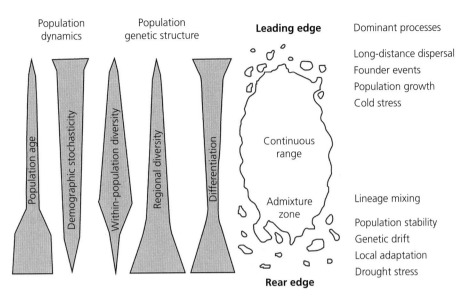

Figure 16.10 Population features and relevant processes at the leading and rear edge of species ranges. The width of gray bars shown on the left indicates the quantity of features at the corresponding position within the range. From Hampe & Petit (2005).

phenotypes and genotypes of those leading edge populations. In contrast, rear edge populations were likely to have already inhabited climates warmer than optimal under pre-global warming conditions due to gene flow from more central populations in cooler environments. As those environments warm, the average genotypes and phenotypes of rear edge populations will become less fit.

16.7 Management in the face of climate change

To maintain viable populations, conservation strategies should maintain or enhance the ability of populations to adapt to new conditions or migrate to new locations where conditions have become favorable. Hansen et al. (2010) proposed four basic principles for adapting *in situ* **conservation** strategies to a changing climate:

1. Protect adequate and appropriate space.
2. Reduce nonclimatic stresses.
3. Use adaptive management to implement and test climate change adaptation strategies.
4. Reduce the rate and extent of climate change to reduce overall risk.

Of these, (1) and (2) relate to genetic factors. Protected area reserve design will become increasingly important. Large protected areas that include a broad range of habitats and climates are more likely to contain suitable future habitat for extant species under new conditions than small or homogeneous reserves (Ackerly et al. 2010). Environmentally heterogeneous protected areas offer greater opportunities for migration over shorter distances into favorable microsites than more homogeneous areas. Large and heterogeneous protected areas are also more likely to contain numerous populations with correspondingly higher genetic diversity on which selection can act.

Climate change may exacerbate other genetic risks to small and declining populations. Climate warming may reduce reproductive rates or increase mortality. If maladaptation due to climate change results in a decline in population size, then other genetic factors such as inbreeding and genetic drift will increasingly come into play, and further decrease the genetic variation present for an adaptive response to new climatic conditions. Monitoring threatened populations will become an even more important aspect of conservation.

Maintaining genetic connectivity through gene flow has long been considered important for conserving genetic diversity, but habitat for migration between reserves has become increasingly important for both adaptation and migration of populations with climate change. While gene flow can impede local adaptation under stable environmental conditions, in a rapidly changing environment, gene flow may be a source of genetic variation containing alleles pre-adapted to new conditions. Fragmented landscapes will pose a major barrier to climate-driven population migration.

16.7.1 Assisted migration

It is clear that many species and populations will not be able to migrate sufficiently rapidly to track their current climates spatially, while other more vagile species have already migrated into newly climatically favorable geographic areas outside of their natural ranges, and should not be treated as invasive species in those areas (Urban 2020). This has opened a hot debate among conservation biologists on the merits and risks of **assisted migration** for species that cannot keep up with climate change.

The language around translocating individuals and species as a conservation tool for changing climates is rather muddied. Here we define assisted migration as the intentional translocation of individuals within or outside of the natural species range. Assisted migration includes **assisted gene flow**, the intentional translocation of individuals *within* a species' range to facilitate adaptation to anticipated local conditions (Aitken & Whitlock 2013); and **assisted colonization**, the movement of individuals *outside* of the natural species' range (Hoegh-Guldberg et al. 2008).

Assisted colonization introduces species into ecosystems where they have not previously occurred, at least in recent history; thus the risks are primarily ecological rather than genetic. These ecological effects are complex, unpredictable, and potentially negative (Ricciardi & Simberloff 2009; Webber et al. 2011; Van der Putten 2012). The risks and benefits of assisted colonization depend on the scale of movement and the extent to which species in recipient ecosystems have previously coexisted with the species being introduced on

evolutionary timescales (McLachlan et al. 2007; Mueller & Hellmann 2008). Proponents of assisted colonization insist that it is a needed tool for preventing some species extinctions. For example, Swarts & Dixon (2009) proposed assisted colonization as a necessary conservation tool for many terrestrial orchids. Opponents argue that assisted colonization will lead to invasive species problems and the destabilization of ecosystems through the introduction of new species, and suggest the precautionary principle is in order (Ricciardi & Simberloff 2009). It is clear that decisions around assisted colonization need to evaluate both the risks of introducing species to new ecosystems and the risks of doing nothing. Unintentional assisted colonization is already common for garden plants (e.g., in Europe, Van der Veken et al. 2008). Well-meaning individuals are also moving endangered species, perhaps without thorough scientific consideration. For example, a volunteer conservation group called the Torreya Guardians has already moved the endangered species Torrey pine 600 km without a scientific assessment of risks and benefits (Schwartz et al. 2009b).

In other cases, assisted colonization is being tested experimentally as a conservation strategy for tree species in field experiments that will be closely monitored and assessed over time. An assisted migration experiment was established for the endangered species whitebark pine to determine if the species distribution model accurately predicted areas of climatically suitable habitat outside of the species range, and to determine if whitebark pine could be established in these climatic refugia (McLane & Aitken 2012). The Mexican sacred fir provides high-elevation overwintering habitat for monarch butterflies in the Trans-Mexican Volcanic Belt. Two field tests have been established at elevations above the natural distribution of sacred fir to test whether monarch butterfly habitat can be established for the future given climate warming (Carbajal-Navarro et al. 2019).

Population genomics has a role to play in informing assisted colonization in a conservation framework. Knowledge of patterns of local adaptation in relation to climate will inform conservationists about what source populations are more likely to provide more successful migrants for a recipient

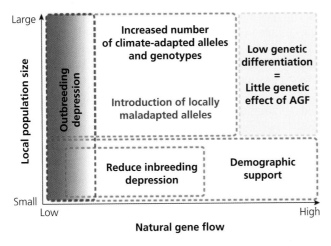

Figure 16.11 Positive and negative effects of assisted gene flow vary with historical levels of gene flow between source and recipient populations and with the size of the recipient population. Each box indicates a parameter combination for which certain genetic or demographic forces are important. The red box indicates that the risk of negative effects of outbreeding depression likely outweighs the potential positive effects of increased adaptive capacity, and the white boxes indicate that the positive effects likely outweigh the negative ones. The yellow box indicates lower-priority situations for assisted gene flow, where high natural gene flow likely provides sufficient genetic variation for evolutionary rescue. From Aitken & Whitlock (2013).

environment, how much genetic diversity will be available for adaptation, whether **outbreeding depression** is a likely outcome of population translocation, and whether translocated species can hybridize with congeneric species following **assisted range expansion**. Genetic risk frameworks can be used to identify and manage risks associated with outbreeding depression and hybridization (Byrne et al. 2011).

Assisted gene flow involves moving pre-adapted individuals from their native origins to locations with favorable climates within the current species range (Aitken & Whitlock 2013). The ecological effects of assisted gene flow should be less risky and somewhat more predictable than those of assisted colonization since the focal species is already present in the recipient ecosystem. As a result, it is less controversial than assisted colonization, as it does not involve species introductions. Assisted gene flow has the greatest potential to facilitate or accelerate evolutionary rescue in a changing climate in moderate to large populations that are locally adapted to recent historic climates and have not been separated for long enough for outbreeding depression to be a substantive risk (e.g., not subspecies or varieties; Figure 16.11). In small populations of conservation concern, assisted gene flow may be useful for reducing inbreeding depression or providing demographic support (i.e., conditions under which assisted gene flow becomes equivalent to **genetic rescue**; Section 19.5). However, small populations are less likely to be locally

adapted than large populations (Leimu & Fisher 2008).

There are some genetic risks associated with assisted gene flow. First, if the source and recipient populations have been long separated in evolutionary terms, there is the potential for outbreeding depression, a reduction in fitness resulting from individuals from different populations mating due to genetic incompatibilities that may arise following hybridization between lineages (Chapter 13). Second, translocated individuals may be locally maladapted to nonclimatic factors, such as soil types for plants, mutualistic or parasitic relations, or available food sources for animals. If populations are not differentiated or locally adapted, then assisted gene flow may provide some demographic support if populations are small, but if populations are sufficiently large, assisted gene flow will have little effect (yellow region of Figure 16.11).

Consideration of future climates is also becoming common in ecosystem restoration activities. Provenance strategies for sourcing of seeds have generally moved toward maximizing genetic diversity through collections from multiple sites, and are now frequently considering sourcing seeds from sites with warmer, drier climates than currently occurring at the restoration site. In restoration, this application of the assisted gene flow concept is generally known as **climate-adjusted provenancing** (Prober et al. 2015).

Assisted gene flow only makes sense if populations are locally adapted to climate. How common is local adaptation in general, and local adaptation to climate in particular? Hereford (2009) analyzed results from 74 common garden experiments including both plants and animals, and found local populations had greater or comparable fitness to nonlocal populations in 71% of comparisons. Leimu & Fischer (2008) also found some evidence for local adaptation in 71% of plants alone. Many field-based studies of local adaptation do not identify the specific environmental factors that differ among sites and drive local adaptation. Of the studies Hereford analyzed, 27 discussed climatic factors that differ among sites, 35 did not indicate factors that differed, and 12 focused on nonclimatic environmental factors. GEA approaches with genomic data can identify climatic and other environmental drivers of local adaptation, but do not directly test for the relative fitness of individuals from local versus nonlocal populations.

16.7.2 *Ex situ* conservation

Traditional *in situ* approaches to conservation in parks and protected areas may be inadequate for conserving populations of some species in a rapidly changing climate. Rear edge populations will be at greatest risk of population extirpation. *In situ* approaches to conservation are unlikely to be successful in these areas if populations are collapsing due to abiotic conditions or shifts in biotic interactions resulting from climatic changes. An alternative strategy for conserving the genetic diversity of these populations is *ex situ* conservation. For example, the vast Svalbard Global Seed Vault, sometimes nicknamed the Doomsday Vault, is dug into a mountain in Norway's Arctic and stores seeds for plant species (Charles 2006). Embryos or sperm of animals can similarly be cryopreserved.

While climate change is an existential threat to many species, and to biodiversity broadly, it is not yet a major consideration for conservation management decisions for species at risk. An analysis of recovery plans for 146 species listed under Canada's Species at Risk Act found only 27% listed climate change as a threat, and it was the first listed threat in only 3% of plans (McCune et al. 2013). A parallel analysis in the USA analyzed conservation plans for 2,733 rare plant species and found less than 5% addressed climate change as a threat (Hernández-Yáñez et al. 2016). This is likely because while climate change projections show impacts will be high this century, most species currently face greater immediate threats due to habitat degradation, land use change, invasive species, and other impacts. In Australia, close to 60% of recovery plans for 100 plants and animals listed climate change as a current or potential threat, and 22 identified conservation actions to help mitigate this threat (Hoeppner & Hughes 2019). The authors of this last study considered that all recovery plans should be climate-informed, and that bolder actions will be needed for conserving some species as climate change accelerates.

Some target species may benefit from assisted gene flow, assisted colonization, *ex situ* conservation, and other management actions. Nevertheless, it will be impossible to translocate entire ecosystems with the full suite of species that they contain. High-fecundity, high-vagility species are rarely the focus of conservation efforts, and these species are more likely to move and persist in a rapidly changing climate. Population genomic studies can be used to model risks of maladaptation under climate change, and to inform conservation management decisions and allocation of resources. Of course, the best conservation strategy for all species is to curb the production of greenhouse gases and slow climate change to a rate that species can tolerate through phenotypic plasticity in the shorter term, and adapt or move in response to climate change in the longer term.

Guest Box 16 Genomic prediction of coral adaptation to warming
Rachael A. Bay

Coral reefs are perhaps one of the most striking examples of the broad-scale effects of climate change on ecosystems. Increases in ocean temperature of just a few degrees can lead to coral bleaching, the disassociation of the coral and its algal symbiont, which can ultimately result in mortality. However, across many bleaching events researchers have observed striking variation in susceptibility to bleaching; two corals right next to one another can express extremely different bleaching phenotypes (Figure 16.12). Such observations raise the question of whether standing genetic variation within populations might allow adaptation to future ocean warming.

Bay et al. (2017) used computer simulations to estimate the potential for adaptation in a single population of the coral *Acropora hyacinthus*. In a previous study, they identified 114 SNPs associated with the ability to survive increased temperatures across a reef in American Samoa (Bay et al. 2014). Next, they sampled additional individuals from the same species in Rarotonga, Cook Islands, and estimated genotypes at the same SNPs. They found that the alleles associated with thermal tolerance in American Samoa were also present in Rarotonga, though generally at lower frequencies. This suggests that the Rarotonga population harbors standing genetic variation that could be beneficial for adaptation to future climate change conditions. This is

especially important as Rarotonga is a much more southern, colder population, suggesting populations inhabiting colder reefs already harbor genetic variation important for persistence under climate change.

Starting with the frequencies of thermal tolerance alleles in the Rarotonga population, Bay et al. (2017) simulated selection based on the amount of warming predicted by the IPCC. The IPCC provides four different climate scenarios ranging from strongly mitigated emissions (RCP 2.6) to "business as usual" (RCP 8.5). Thermal tolerance in these simulations was determined by an additive polygenic effect—individuals with more thermal tolerance alleles across the 114 SNPs were given higher thermal tolerance scores. Under simulated severe warming, the populations of *A. hyacinthus* were projected to go extinct before the year 2100 (Figure 16.12). However, under mitigated climate scenarios the populations persisted to the end of the century. Further, the frequency of thermal tolerance alleles increased over time, showing adaptation.

Using this framework, the authors tested the effects of assisted migration on adaptation and persistence. With no intervention, populations simulated under the climate scenario RCP 6.0 went extinct by the end of the century. However, when individuals with high thermal tolerance scores were added to the gene pool, the population did not go extinct as

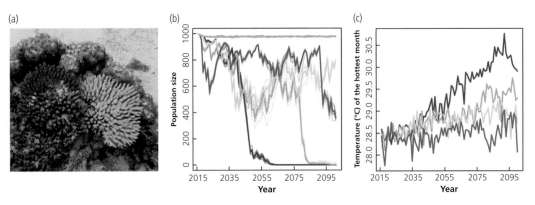

Figure 16.12 Genetic simulations projecting climate change response in *Acropora hyacinthus*. (a) Two corals of the species *A. hyacinthus* with different bleaching outcomes during a warming event. (b) Simulated population response to IPCC warming scenarios based on additive genetic variation. (c) Simulated effect of different levels of assisted migration, all under RCP 6.0. Photo courtesy of Rachel Bay. From Bay et al. (2017).

Guest Box 16 Continued

the increase in beneficial standing genetic variation aided in adaptation (Figure 16.12). Although the simulated rates of assisted migration were high (at least 1% of population size), these results demonstrate the possibility for assisted gene flow to mitigate responses to ocean warming, at least at local scales.

This work demonstrates the importance of the rate of warming in adaptation and persistence of populations under future climate conditions. Additionally, the framework shows how computer simulations can be used alongside genomic data to project the consequences of future climate change.

PART IV

Conservation and Management

CHAPTER 17

Inbreeding Depression

Monkeyflower, Example 17.1

Probably the oldest observation about population genetics is that individuals produced by matings between close relatives are often less healthy than those produced by mating between more distant relatives.

(Anthony R. Ives & Michael C. Whitlock 2002, p. 454)

Inbreeding depression, the harmful effects of inbreeding on the fitness of individuals, is widespread among plants and animals, with recent genomic studies revealing an even greater impact on individual fitness than previously thought.

(Claudio Bozzuto et al. 2019, p. 1359)

In 1965, the eminent population geneticist Richard C. Lewontin reviewed a book entitled *The Theory of Inbreeding* by Sir Ronald A. Fisher. In his review, Lewontin emphasized the central importance of inbreeding for understanding population genetics, as well as the difficulty of understanding "inbreeding" (Lewontin 1965, p. 1800):

Notions of inbreeding lie at the very heart of genetics of sexual organisms, and every discovery in classical and population genetics has depended on some sort of inbreeding experiment. But a full understanding of the theory and ramifications of inbreeding always seems to evade us, just.

Inbreeding is one of those topics that appears relatively straight-forward, but becomes more complex the deeper we examine it.

The term "inbreeding" is used to mean many different things in population genetics. Jacquard (1975) described five different effects of nonrandom mating that are measured by **inbreeding coefficients**. The multiple uses of "inbreeding" can sometimes lead to confusion, so it is important to be precise when using this term.

Conservation and the Genomics of Populations, Third Edition. Fred W. Allendorf, *et al.*, Oxford University Press.
© Fred W. Allendorf, W. Chris Funk, Sally N. Aitken, Margaret Byrne, and Gordon Luikart (2022). DOI: 10.1093/oso/9780198856566.003.0017

Templeton & Read (1994) have described three different phenomena of special importance for conservation that are all measured by inbreeding coefficients:

1. genetic drift (F_{ST}, Section 9.1);
2. nonrandom mating within local populations (F_{IS}; Section 9.1);
3. the increase in genome-wide homozygosity (F) resulting from matings between related individuals (e.g., father–daughter mating in ungulates, or matings between cousins in birds).

Keller & Waller (2002) provide an exceptionally clear presentation of these three different uses of inbreeding (see their box 1). We will focus on the third meaning in this chapter.

Inbreeding (i.e., mating between related individuals) will occur in both large and small populations. In large populations, inbreeding may occur by self-fertilization or by nonrandom mating because of a tendency for related individuals to mate with each other. For example, in many tree species, nearby individuals are more likely to be related than trees farther apart, and have a higher probability of mating with each other because of geographic proximity (Section 9.2; Hall et al. 1994). However, substantial inbreeding will occur in small populations even with random mating because all or most individuals within a small population will be related. In an extreme example of a population of size two, after one generation, only brother–sister matings are possible. In a slightly larger population with 10 breeders, even the most distantly related individuals will be cousins after only a few generations. This has been called the "**inbreeding effect of small populations**" (Crow & Kimura 1970, p. 101).

Inbred individuals generally have reduced fitness in comparison to noninbred individuals from the same population because of their increased homozygosity. Inbreeding depression is the reduction in fitness (or phenotype value) of progeny from matings between related individuals relative to the fitness of progeny between unrelated individuals (Example 17.1). Inbreeding depression in natural populations contributes to the extinction of populations under some circumstances (Chapter 18; Allendorf & Ryman 2002; Keller & Waller 2002; Bozzuto et al. 2019).

The importance of inbreeding depression has been debated since the time of Darwin. Even today, the importance of inbreeding depression is sometimes questioned by some conservationists and policy-makers (Nonaka et al. 2019). Nevertheless, there is now overwhelming evidence that inbreeding depression is an important consideration in the persistence of populations (Frankham 2010; Spurgin & Gage 2019; Chapter 18).

Genomic analysis has revolutionized our understanding of the effects of inbreeding depression in wild populations (Hedrick & García-Dorado 2016). It is now possible to estimate the proportion of the genome that is **identical by descent** (IBD) in individuals using strand theory, which we first considered in Section 10.7, and other methods. In general, much more severe inbreeding depression has been uncovered than was found using previous approaches. In this chapter, we review the evidence for inbreeding depression and consider how it can be detected and measured to predict its effects in natural populations.

17.1 Inbreeding

An individual is "inbred" if its mother and father share a common ancestor. This definition must be put into perspective because any two individuals in a population are related if we trace their ancestries back far enough. We must, therefore, define inbreeding relative to some "base" population in which we assume all individuals are unrelated to one another. We usually define the base population operationally as those individuals in a pedigree beyond which no further information is available (Ballou 1983).

Inbred individuals will have increased homozygosity and decreased heterozygosity over their entire genome because they might receive two copies of the same allele that was present in a common ancestor of their parents. Such an individual is IBD at that locus (i.e., **autozygous**). The inbreeding coefficient (F) of an individual is the proportion of the genome that is IBD.

An autozygous individual will be homozygous unless a mutation has occurred in one of the two copies descended from a parental common ancestor. The alternative to being autozygous is **allozygous**. Allozygous individuals possess two

Example 17.1 Inbreeding depression in the monkeyflower

The monkeyflower is a self-compatible wildflower that occurs throughout western North America, from Alaska to Mexico. Willis (1993) studied two annual populations of this species on adjacent mountains about 2 km apart in the Cascade Mountains of Oregon. Seeds were collected from both populations and germinated in a greenhouse. Hand pollinations produced self-pollinations and pollinations from another randomly chosen plant from the same population. Seeds resulting from these pollinations were germinated in the greenhouse, and randomly chosen seedlings were transplanted back into their original population. The transplanted seedlings were marked and followed throughout the course of their life. Cumulative inbreeding depression through several life history stages was estimated by the proportional reduction in fitness in selfed versus out-crossed progeny ($1 - w_s/w_o$). Substantial inbreeding depression was detected in both populations (Figure 17.1). While there were marked population differences in inbreeding depression at early life history stages, cumulative effects across the length of the experiment were similar. A similar set of seedlings was maintained in the greenhouse. The amount of inbreeding depression in the greenhouse was similar to that found in the wild.

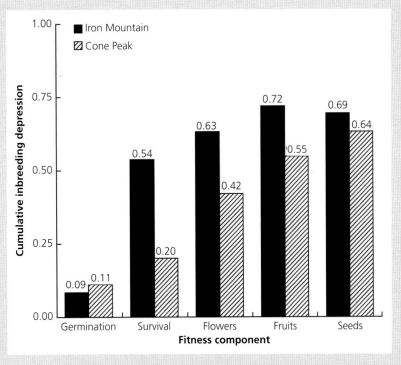

Figure 17.1 Cumulative inbreeding depression in two wild populations of the monkeyflower. Inbreeding depression was measured as the proportional reduction of fitness in progeny produced by selfing versus outcrossing ($1 - w_s/w_o$). From Willis (1993).

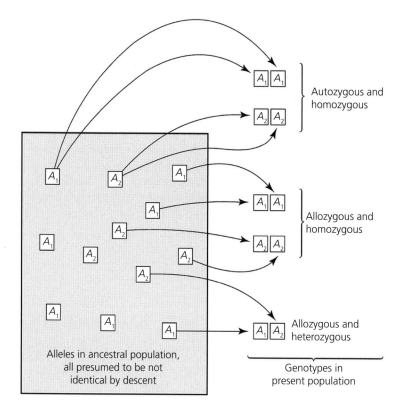

Figure 17.2 Patterns and definitions of genotypic relationships with pedigree inbreeding. Autozygous individuals in the present population contain two alleles that are IBD from a single gene in the ancestral population. In contrast, allozygous individuals contain two alleles derived from different genes in the ancestral population. Redrawn from Hartl & Clark (1997).

alleles descended from different ancestral alleles in the base population. Figure 17.2 illustrates the relationship of the concepts of autozygosity, allozygosity, homozygosity, and heterozygosity.

Is it really necessary to introduce these two new terms? Yes. Autozygosity and allozygosity are related to homozygosity and heterozygosity, but they refer to the descent of alleles through Mendelian inheritance rather than the molecular state of the allele in question. This distinction is important when considering the effects of inbreeding on individuals. We assume that the founding individuals in a pedigree are allozygous for two unique alleles. However, these allozygotes may either be homozygous or heterozygous at a particular locus, depending upon whether the two alleles are identical in state or not. For example, an allozygote would be homozygous if it had two alleles that are identical in DNA sequence.

We can see this using Figure 12.3. An individual with one copy of the *10A* allele and one copy of the *10C* allele (i.e., *10A/10C*) would be homozygous in state for 10 repeats, but would be allozygous. In contrast, an individual with two copies of the *10B* allele (*10B/10B*) would be homozygous and autozygous.

17.1.1 The pedigree inbreeding coefficient

The pedigree inbreeding coefficient (F_P) is the expected increase in homozygosity for inbred individuals relative to the base population (i.e., the founders of the pedigree). F_P is also the expected decrease in heterozygosity throughout the genome. F_P ranges from zero (for noninbred individuals) to one (for totally inbred individuals).

Several methods are available for calculating the pedigree inbreeding coefficient. We will use the method of path analysis developed by Sewall Wright (1922). Figure 17.3 shows the pedigree of an inbred individual, X. By convention, females are represented by circles, and males are represented by squares in pedigrees. Diamonds are used either to represent individuals whose sex is unknown or to represent individuals whose sex is not of concern.

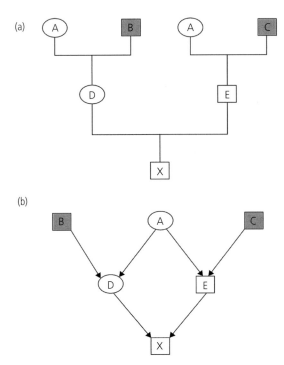

Figure 17.3 Calculation of the pedigree inbreeding coefficient (F_P) for individual X using path analysis. (a) Conventional representation of pedigree for an individual whose mother and father had the same mother (individual A). (b) Path diagram to represent this pedigree to calculate the inbreeding coefficient. Shaded individuals in (a) need not be included in (b) because they are not part of the path through the common ancestor (A) and therefore do not contribute to the inbreeding of individual X.

What is the pedigree inbreeding coefficient of individual X in Figure 17.3a? The first step is to draw the pedigree as shown in Figure 17.3b so that each individual appears only once. Next, we examine the pedigree for individuals who are ancestors of both the mother and father of X. If there are no such common ancestors, then X is not inbred, and $F_P = 0$. In this case, there is one common ancestor, individual A. Next, we trace all of the paths that lead from one of X's parents, through the common ancestor, and then back again to the other parent of X. There is only one such path in Figure 17.3 (D<u>A</u>E); it is helpful to keep track of the common ancestor by underlining.

The pedigree inbreeding coefficient of an individual can be calculated by determining N, the number of individuals in the loop (not including

the individual of concern) containing the common ancestor of the parents of an inbred individual. If there is a single loop then:

$$F_P = (1/2)^N (1 + F_{P(CA)}) \qquad (17.1)$$

where $F_{P(CA)}$ is the pedigree inbreeding coefficient of the common ancestor. The term $(1 + F_{P(CA)})$ is included because the probability of a common ancestor passing on the same allele to two offspring is increased if the common ancestor is inbred. For example, if the inbreeding coefficient of an individual is 1.0, then it will always pass on the same allele to two progeny. If there is more than one loop, then the inbreeding coefficient is the sum of the F_P values from the separate loops.

$$F_P = \sum [(1/2)^N (1 + F_{P(CA)})] \qquad (17.2)$$

In the present case (Figure 17.3), there is only one loop with $N = 3$ and the common ancestor (A) is not inbred, therefore:

$$F_{P(X)} = (1/2)^3 (1 + 0) = 0.125$$

This means that individual X is expected to be IBD at 12.5% of his loci. Or, stated another way, the expected heterozygosity of individual X is expected to be reduced by 12.5%, compared with individuals in the base population. See Example 17.2 for calculating F when the common ancestor is inbred ($F_{P(CA)} > 0$).

Figure 17.5 shows a complicated pedigree obtained from a long-term population study of the great tit in the Netherlands (van Noordwijk & Scharloo 1981). They have shown that the hatching of eggs is reduced by ~7.5% for every 10% increase in F_P. Ten different loops contribute to the inbreeding of the individual under investigation (Table 17.1). The total pedigree inbreeding coefficient of this individual is 0.1445.

On an interesting historical note, both Sewall Wright and Charles Darwin displayed inbreeding within their immediate families. Sewall Wright's parents were first cousins (Provine 1986, p. 1); Wright calculated his own F_P value as 0.0625. Charles Darwin and his wife Emma Wedgwood were first cousins (Berra et al. 2010). In addition, Charles and Emma also shared a set of distant grandparents going back several generations (Figure 17.6). An analysis of the Darwin/Wedgwood family tree has concluded that

Example 17.2 Calculating pedigree F (F_P)

Figure 17.4 shows a pedigree in which a common ancestor of an inbred individual is inbred. What is the pedigree inbreeding coefficient of individual K in this figure?

There is one loop that contains a common ancestor of both parents of K (I<u>G</u>J). Therefore, using Equation 17.2, $F_{P(K)} = (1/2)^3(1 + F_{P(G)})$. The common ancestor in this loop, G, is also inbred; there is one loop with three individuals through a common ancestor for individual G (D<u>B</u>E). Therefore, $F_{P(G)} = (1/2)^3(1 + F_{P(B)}) = 0.125$. Individual B is not inbred ($F_{P(B)} = 0$) since she is a founder in this pedigree. Therefore, $F_{P(G)} = 0.125$, and $F_{P(K)} = (1/2)^3(1.125) = 0.141$.

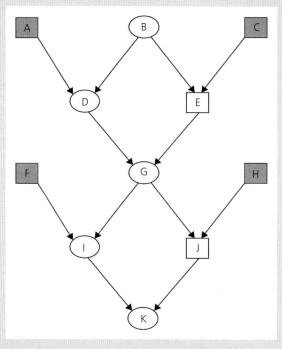

Figure 17.4 Hypothetical pedigree in which the common ancestor (G) of an inbred individual's (K) parents is also inbred. See Example 17.2. Shaded individuals are not part of the path through either of the common ancestors (individuals B and G) and therefore do not contribute to the inbreeding of individuals G and K.

inbred children had significantly reduced probability of surviving through childhood compared to noninbred children (Berra et al. 2010).

17.1.2 Expected versus realized proportion of the genome IBD

For many years, F_P was considered to be the most useful measure of individual inbreeding (Pemberton 2008). However, F_P has two major shortcomings. First, it does not account for inbreeding caused by more distant ancestors not included in the reference pedigree. In addition, F_P

is an imprecise predictor of the actual proportion of the genome identical by descent (F) in a given individual because the proportion of the genome IBD will vary substantially among individuals with the same pedigree by chance alone:

They have all the same "Coefficient of Inbreeding", but differ in the extent of homozygosity.

(R.A. Fisher 1965, p. 97)

We will apply Fisher's strand theory that we considered in Section 10.7 to understand the variability in F of individuals with the same pedigree.

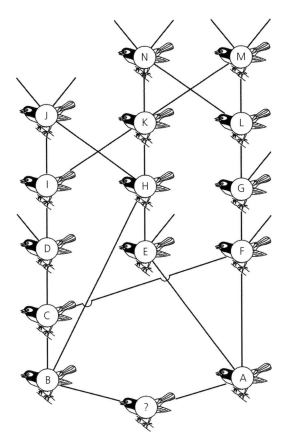

Figure 17.5 Complex pedigree from a wild population of great tits in the Netherlands. The pedigree inbreeding coefficient of the bottom individual (?) is 0.1445 (see Table 17.1). From van Noordwijk & Scharloo (1981).

Table 17.1 Calculation of the pedigree inbreeding coefficient of individual (?) at the bottom of the pedigree shown in Figure 17.5. The common ancestor in each path is bolded and underlined.

Path	Length (N)	Common Ancestor	$F_{P(CA)}$	$(1/2)^N(1+F_{P(CA)})$
BC**F**A	4	F	0	0.0625
B**H**EA	4	H	0	0.0625
BCDI**K**HEA	8	K	0	0.0039
BCDI**J**HEA	8	J	0	0.0039
BHK**M**LGFA	8	M	0	0.0039
BHK**N**LGFA	8	N	0	0.0039
BCDIK**N**LGFA	10	N	0	0.0010
BCDIK**M**LGFA	10	M	0	0.0010
BCFGLM**K**HEA	10	M	0	0.0010
BCFGL**N**KHEA	10	N	0	0.0010
Total	—	—	—	0.1445

common grandfather of both parents (Figure 17.7a) to understand the source of the variability in F of individuals with the same pedigree. These sibs all have an F_P of 0.125. That is, they each have a 0.125 probability of being IBD at a locus because their parents shared a common father. Figure 17.7b shows F_{ROH} in eight individuals produced by this mating based on simulations (Kardos et al. 2016b). Ten pairs of homologs are shown in each individual. Each chromosome is 50 cM long, and the lengths range from 10 Mb to 100 Mb in steps of 10 Mb. F_{ROH} ranges between 0.08 and 0.25 in these eight individuals.

There is large variability in F among these sibs because of the tendency for large tracts of chromosomes to be transmitted together during meiosis. In the illustration of meiosis in Figure 3.6, two of the four gametes contain a chromosome that was transmitted without recombination. That is, the entire chromosome was transmitted intact. This pattern of inheritance is common. For example, Limborg et al. (2015) captured all four products of single meiotic events by **gynogenesis** genome-wide in sockeye salmon. They found that nearly one-half of all chromosomes were transmitted without a recombination event because of strong **crossover interference** (i.e., a crossover reduces the probability of a neighboring crossover; see their figure 2). A recent review has found that such strong crossover interference is common in a wide variety of species and that at least one, and not many more than one, crossover occurs

Figure 10.8 shows two homologous chromosomes in a single individual sampled from a population a few generations after founding. The colors of these "strands" represent different ancestral chromosomes originating from the founders. We can see that there are five regions (i.e., **chromosomal tracts**) of these chromosomes where this individual is IBD that represent ~57% of the total length of these chromosomes. This individual is autozygous in these regions and will be homozygous unless a mutation occurs. These so-called **runs of homozygosity** (ROH) can be summed to estimate the proportion of this chromosome that is IBD ($F_{ROH} = 0.57$).

Let us consider the example of a family produced by matings between half-sibs that share a

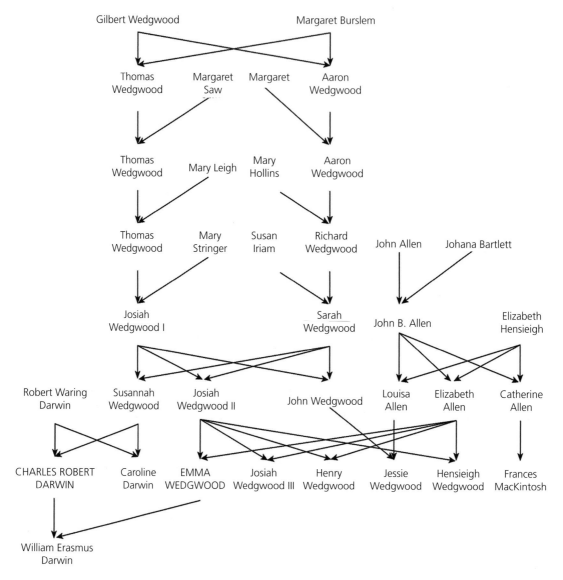

Figure 17.6 Pedigree of the Darwin/Wedgwood dynasty showing just one (William Erasmus Darwin) of Charles Darwin and Emma Wedgewood's 10 children. Modified from Berra et al. (2010).

per pair of chromosomes per meiosis in most taxa (Otto & Payseur 2019).

We can see the result of this pattern of transmission in Figure 17.7. The top chromosome in pair 1 of the top-left individual consists of two large tracts of the two chromosomes carried by the common grandfather. The bottom chromosome in this pair does not contain any tracts from the grandfather. Thus, the F value for this chromosome pair is zero.

Contrast this with the third chromosome pair in the same individual. The top chromosome in this pair was inherited from the grandfather without recombination, and the bottom chromosome consists of two tracts from the common grandfather. Thus, the F value for this chromosome pair is ~65%. The proportion of both of these chromosomes expected to be IBD is 0.125. However, the actual proportion of these two chromosomes IBD is zero and 0.65.

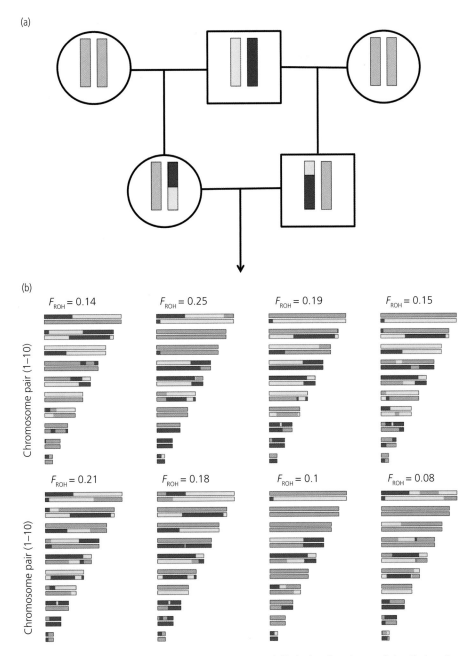

Figure 17.7 (a) Pedigree of eight offspring produced by a mating between two half-sibs that share the same father. The homologous chromosomes in the common parent are color-coded yellow and purple. (b) Results of simulations showing the proportion of the genome IBD in individuals produced by the mating above. Ten pairs of homologs are shown for each individual. Each chromosome is 50 cM long, and the physical lengths range from 10 Mb to 100 Mb in steps of 10 Mb. The yellow and purple segments represent the two homologous chromosomes in the common father. The gray represents chromosomal segments derived from the two parents that are not shared by the half-sibs.

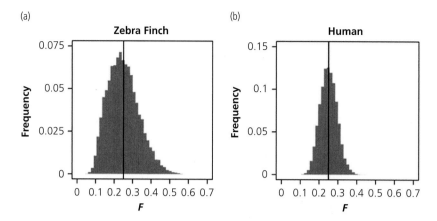

Figure 17.8 Simulated variation in realized proportion of the genome IBD (i.e., F) for offspring of matings between full sibs. Panel (a) show the results for zebra finches and panel (b) shows the results for humans. The black lines indicate the pedigree inbreeding coefficients F_P. From Knief et al. (2017).

In general, the fewer sections of the genome that are transmitted independently, the greater the variation of the actual IBD with the same F_P (i.e., greater Mendelian noise). We can see this by considering the two extremes. Imagine a genome that consists of a single pair of chromosomes with no recombination. In this case, an inbred individual will either be completely IBD or will not be IBD at all. In the other extreme, consider a genome in which each locus is inherited independently. In this case, F_P will do an excellent job of predicting the actual proportion of the genome IBD (law of large numbers). Thus, the variability of F among individuals with the same pedigree is highest in species with fewer chromosomes and shorter genetic maps (Franklin 1977; Hill & Weir 2011).

Knief et al. (2017) compared the realized proportion of the genome IBD in two species with high (zebra finch) and low (humans) amounts of Mendelian noise. Half of the autosomal genome of the zebra finch consists of only six pairs of chromosomes that have extremely low recombination rates in their centers. In contrast, recombination in humans is distributed quite uniformly along the 22 pairs of autosomes. Knief et al. (2017) simulated the distribution of IBD in these two species for matings between full sibs ($F_P = 0.25$; Figure 17.8). As expected, there was much greater variability for F in the zebra finch than for humans. These authors also show that the same number of molecular markers provides more precise estimates of F in zebra finches

than in humans because there will be more markers per chromosome in finches. They conclude that in species with high Mendelian noise, relatively few markers will often be more informative about inbreeding and fitness than large pedigrees.

17.2 Estimation of F with molecular markers

We must know the realized proportion of the genome that is IBD (F) in individuals because of inbreeding to understand the effect of inbreeding on fitness in natural populations. Pedigree-based analyses, Section 17.1.1, have traditionally been the cornerstone of studies on understanding the effects of inbreeding (Pemberton 2008). The use of genomic methods has dramatically improved our ability to measure individual inbreeding and detect inbreeding depression in the wild. As we saw in Section 17.1.2, the realized proportion of the genome IBD in individuals is poorly predicted by F_P because of limited pedigree depth, linkage, and random Mendelian segregation. Additionally, pedigrees are difficult to obtain for natural populations because they require reliable parentage information across several generations (Pemberton 2008).

Fortunately, it is now possible to genotype thousands of loci or sequence the genomes of many individuals in any natural population (Guest Box 17). This huge amount of molecular genetic

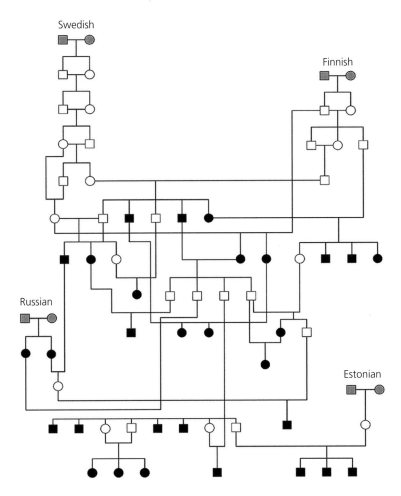

Figure 17.9 Pedigree of a captive gray wolf population. The black symbols are those animals included in the study evaluating the use of 29 microsatellite loci to estimate inbreeding coefficients. The gray individuals are the four founder pairs (assumed $F_P = 0$) from four countries. From Hedrick et al. (2001).

data can be used to precisely measure individual inbreeding via analysis of genetic variation across individual genomes (Pemberton et al. 2012; Kardos et al. 2016b), and to study inbreeding depression without the need to conduct parentage analysis over many generations. Molecular marker-based measures of F range from simple estimates of individual heterozygosity (Section 17.2.1) to more advanced methods that use mapped loci to estimate F via identification of IBD chromosome segments as stretches of homozygous genotypes at mapped **single nucleotide polymorphisms (SNPs)** (Section 17.2.2).

17.2.1 Using unmapped loci to estimate *F*

Estimates of F based on individual heterozygosity rely on the concept that individuals whose parents are more closely related will have lower heterozygosity on average across the genome due to the presence of IBD chromosome segments. The pedigree inbreeding coefficient, F_P, is the expected increase in homozygosity due to identity by descent. As we saw in the previous example, the offspring of parents that share a common grandfather have an F_P of 0.125, and thus they are expected to have only 87.5% of the heterozygosity of their parents. Therefore, we expect individual heterozygosity (H) at many loci to be reduced by a value of F.

For example, 29 microsatellite loci were examined in a captive population of gray wolves (Ellegren 1999) for which the complete pedigree is known (Figure 17.9). The distribution of individual heterozygosity ranged from about 0.20 to 0.80. The pedigree inbreeding coefficient was significantly correlated with heterozygosity (Figure 17.10; $r^2 = 0.52$, $P < 0.001$).

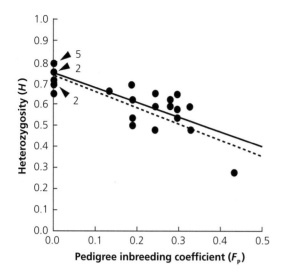

Figure 17.10 Relationship between individual heterozygosity (*H*) at 29 microsatellite loci and pedigree inbreeding coefficient (*F*ₚ) in a captive wolf population. The solid line represents the regression of *H* on *F*ₚ, the dashed line is the expected relationship between *H* and *F*ₚ, assuming an *H* of 0.75 in noninbred individuals. From Ellegren (1999).

The availability of thousands of genetic markers or more now makes it possible to use this approach in many taxa. F_H measures the excess in the observed number of homozygous genotypes within an individual relative to the mean number of homozygous genotypes expected under random mating (Keller et al. 2011):

$$F_H = \frac{O(Hom) - E(Hom)}{L - E(Hom)}$$

Where $O(Hom)$ and $E(Hom)$ are the observed and expected Hardy–Weinberg (HW) numbers of homozygous loci in an individual and L is the number of loci examined. The average $O(Hom)$ is expected to equal $E(Hom)$ in random mating populations. F_H can thus be negative when the parents are less related than expected with random mating. The distribution of F_H will be centered near zero in a sample of noninbred individuals.

17.2.2 Using mapped loci to estimate *F*

We saw in Figure 17.7 that *F* can be estimated by F_{ROH}, which is the sum of all regions where a pair of homologous chromosomes in an individual are IBD.

This can be applied to wild populations by genotyping individuals at many SNPs that are mapped. Figure 17.11 shows heterozygosity at mapped SNP loci on five chromosomes in four wolves from Scandinavia (Guest Box 17).

17.3 Causes of inbreeding depression

Inbreeding depression can result from either increased homozygosity or reduced heterozygosity (Crow 1948). This may sound like double talk, but there is an important distinction to be made here. Increased homozygosity leads to expression of a greater number of deleterious recessive alleles in inbred individuals, thereby lowering their fitness. Reduced heterozygosity reduces fitness of inbred individuals at loci where the heterozygotes have a selective advantage over homozygotes (heterozygous advantage or overdominance; Section 8.2.2). As we discuss in the following text, there is more evidence that the primary problem is increased homozygosity rather than reduced heterozygosity. In Box 17.1 we discuss the various mechanisms contributing to overall **genetic load**, and their relationship to inbreeding load.

The primary cause of inbreeding depression is an increase of homozygosity for deleterious recessive alleles (Charlesworth & Willis 2009). The probability of being homozygous for rare alleles increases surprisingly rapidly with inbreeding. Consider the effect of an average pedigree inbreeding coefficient of $F_P = 0.10$ on expression of a recessive lethal allele at a frequency of $q = 0.10$. The proportion of heterozygotes will be reduced by 10% compared with $F_P = 0$, and each of the homozygotes will be increased by half of that amount (Section 9.1):

	AA	Aa	aa
Expected	$p^2 + pqF_P$	$2pq - 2pqF_P$	$q^2 + pqF_P$
$F_P = 0$	0.810	0.180	0.010
$F_P = 0.10$	0.819	0.162	0.019

Thus, the expected proportion of individuals to be affected by this deleterious allele (*a*) will nearly double (from 0.010 to 0.019) with just a 10% increase in inbreeding. The increase in the number of affected

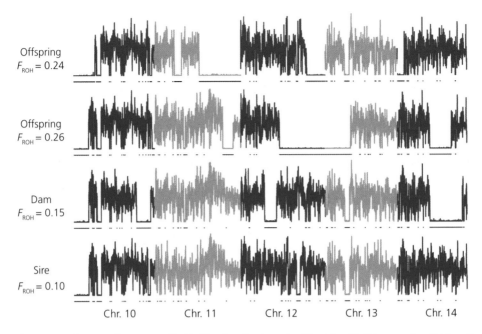

Figure 17.11 Heterozygosity across chromosomes 10–14 of four Scandinavian wolves (Guest Box 17). Heterozygosity, which is on the *y*-axis, ranges from zero to one for each individual. The black lines at zero heterozygosity show ROH that are greater than 100 kb. The two top individuals are the offspring of the two lower individuals. From Kardos et al. (2018b).

individuals is even greater for less frequent alleles (Crow & Kimura 1970, p. 74). For example, we expect over 10 times the number of homozygous individuals for a deleterious recessive allele at a frequency of $q = 0.01$ when $F_P = 0.10$ compared with $F_P = 0$.

It is crucial to know the mechanisms causing inbreeding depression because it affects the ability of a population to "adapt" to inbreeding. A population could adapt to inbreeding if inbreeding depression is caused by deleterious recessive alleles that potentially could be removed (purged by selection). However, inbreeding depression caused by heterozygous advantage cannot be purged because overdominant loci will always suffer reduced fitness as homozygosity increases due to increased inbreeding.

Both increased homozygosity and decreased heterozygosity are likely to contribute to inbreeding depression, but it is thought that increased expression of deleterious recessive alleles is the more important mechanism (Carr & Dudash 2003; Charlesworth & Willis 2009). For example, Remington & O'Malley (2000) performed a genome-wide evaluation of inbreeding depression caused by selfing during embryonic viability in loblolly pines. Nineteen loci were found that contributed to inbreeding depression. Sixteen loci showed predominantly recessive action. Evidence for heterozygous advantage was found at three loci.

17.4 Detection and measurement of inbreeding depression

Some inbreeding depression is expected in all species (Hedrick & Kalinowski 2000). Deleterious recessive alleles are present in the genome of all species because they are continually introduced by mutation, and natural selection is inefficient at removing them because most copies are "hidden" phenotypically in heterozygotes that do not have reduced fitness (Sections 8.2.1 and 12.3). We, therefore, expect all species to show some inbreeding depression due to the increase in homozygosity of recessive deleterious alleles. For example, Figure 17.12 shows inbreeding depression for infant survival in a captive population of callimico monkeys.

Box 17.1 What is genetic load?

Genetic load is the relative difference in fitness between the theoretically fittest genotype within a population and the average genotype in that population (Crow 1970; Wallace 1991). There are many underlying mechanisms for genetic load in populations. The term "load" was first used by Muller (1950) in his consideration of the possible effects of increased mutation rates because of nuclear weapons. In this case, the **mutation load** is the reduction in fitness caused by the presence of deleterious alleles introduced by mutation. We saw in Section 8.2 that natural selection is not very effective at removing deleterious recessive alleles introduced by mutation because most of the deleterious alleles are hidden in heterozygotes.

There are four primary sources of genetic load that we need to consider in conservation:

Mutation load: The decrease in fitness caused by the accumulation of deleterious mutations (Section 12.3).

Segregation load: The decrease in fitness caused by heterozygous advantage.

When the fittest genotype is heterozygous, homozygotes with lower fitness will be produced by Mendelian segregation (Section 8.2).

Drift load: The decrease in fitness caused by the increase in frequency of deleterious alleles resulting from genetic drift. Drift load is caused by the increase in frequency of deleterious alleles that are maintained in a population at equilibrium between mutation and selection (Section 12.3). In the extreme, alleles that contribute to drift load can become fixed in small populations. This fixed genetic load will cause the reduction in fitness of all individuals in small populations.

Migration load: The reduction in fitness caused by the migration into a population of individuals that are less adapted to the local environment than native individuals (Section 9.7.2).

The **inbreeding load** is the reduction in fitness of inbred individuals. As we considered above, this reduction is caused by both increase in homozygosity of deleterious recessive alleles and the reduction in heterozygosity at loci with heterozygous advantage. Thus, the inbreeding load results from a combination of mutation and segregation load (Charlesworth & Willis 2009). See Wallace (1991) for a clear discussion of several other types of genetic load.

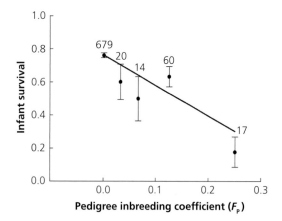

Figure 17.12 Relationship between inbreeding (F_P) and infant survival in captive callimico monkeys. Callimico show a 33% reduction in survival resulting from each 10% increase in inbreeding ($P < 0.001$). Data are from 790 captive-born callimico, 111 of which are inbred. The numbers above the bars are the number of individuals studied at each inbreeding level. From Lacy et al. (1993).

17.4.1 Lethal equivalents

The effects of inbreeding depression on survival are often measured by the mean number of "**lethal equivalents**" (LEs) per diploid genome. A lethal equivalent is a set of deleterious alleles that would cause death if homozygous. Thus, one lethal equivalent may be a single allele that is lethal when homozygous, two alleles each with a probability of 0.5 of causing death when homozygous, or 10 alleles each with a probability of 0.10 of causing death when homozygous.

We can see the effect of one LE in the example of a mating between full sibs that will produce a progeny with an F_P of 0.25 (Figure 17.13). Individuals A and B each carry one lethal allele (a and b, respectively). The probability of individual E being homozygous for the a allele is $(1/2)^4 = 1/16$; similarly, there is a 1/16 probability that individual E will be homozygous bb. Thus, the probability of E

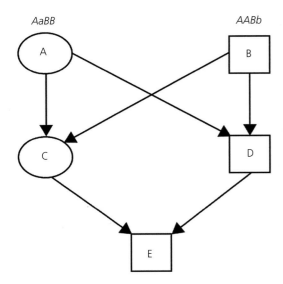

AaBB   *AABb*

Figure 17.13 Effect of a single lethal equivalent on survival of inbred progeny produced by a mating between full sibs ($F_P = 0.25$). Individuals A and B each carry one lethal allele (*a* and *b*, respectively). The probability of E being homozygous for the *a* allele is $(1/2)^4 = 1/16$; similarly, there is a 1/16 probability that individual E will be homozygous *bb*. Therefore, the probability of E *not* being homozygous for a recessive allele at either of these two loci is $(15/16)(15/16) = 0.879$. This demonstrates that 1 LE per diploid genome will result in approximately a 12% reduction $(1 - 0.879)$ in survival of individuals with an F_P of 0.25.

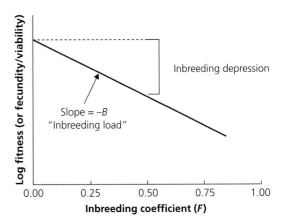

Figure 17.14 Relationship between inbreeding coefficient (*F*) and reduction in fitness. Inbreeding depression is the reduction in fitness of inbred individuals and measured by the number of LEs per gamete (*B*). The bracket shows the reduction in fitness (inbreeding depression) predicted for individuals with an *F* of 0.5. Redrawn from Keller & Waller (2002).

not being homozygous for a recessive allele at either of these two loci is $(15/16)(15/16) = 0.879$. Thus, one LE per diploid genome will result in approximately a 12% reduction (1–0.879) in survival of individuals with an F_P of 0.25.

The number of LEs present in a species or population is generally estimated by regressing survival on the inbreeding coefficient (Figure 17.14). The effects of inbreeding on the probability of survival, *S*, can be expressed as a function of F_P (Morton et al. 1956):

$$S = e^{-(A + BF_P)}$$

$$\ln S = -A - BF_P \tag{17.3}$$

where e^{-A} is survival in an outbred population, and *B* is the inbreeding load, the rate at which fitness declines with inbreeding (Hedrick & Miller 1992). *B* is the reduction in survival expected in a completely homozygous individual. Therefore, *B* also estimates the average number of LEs per gamete, and 2*B* is the number of LEs per diploid individual. *B* is estimated

by the slope of the weighted regression of the natural log of survival on F_P. The callimico monkeys shown in Figure 17.12 have an estimated 7.90 LEs per individual ($B = 3.95$; Lacy et al. 1993).

17.4.2 Estimates of inbreeding depression

Estimation of inbreeding depression is extremely complex. There are a variety of methods in use for measuring the proportion of the genome IBD, and a variety of models of the relationship between fitness and *F* or F_P. Many of these models do not provide unbiased estimates of the number of LEs. Therefore, estimates in the literature should be compared with caution (Nietlisbach et al. 2019).

As we discussed in Section 6.7.3, Kathy Ralls and her colleagues at the US National Zoo provided the first estimates of inbreeding depression in a conservation context (Ralls et al. 1979; Ralls & Ballou 1983). The range of LEs per individual estimated for mammal populations ranges from about zero to 30 in captivity (Ralls et al. 1988). The median number of LEs per diploid individual for captive mammals was estimated to be 3.14. This corresponds to about a 33% reduction of juvenile survival, on average, for offspring with an inbreeding coefficient of 0.25. However, this is an underestimate of the magnitude of inbreeding depression in populations because it considers the reduction of fitness for only one life history stage (juvenile survival), and ignores all others (e.g., adult survival, embryonic survival, fertility, etc.). As we saw in Example 17.1, inbreeding depression becomes greater as we consider more life history stages.

In addition, captive environments are less stressful than natural environments, and stress typically increases inbreeding depression (see below). A meta-analysis estimated an overall average of 12 diploid LEs in wild mammal and bird populations over the life history of species (O'Grady et al. 2006). As we will see in Chapter 18, this amount of inbreeding depression can have a substantial effect on the viability of populations.

There are fewer estimates of the number of LEs using pedigree analysis in plant species. Most studies of inbreeding depression in plants compare selfed and outcrossed progeny from the same plants (Example 17.1). In this situation, inbreeding depression is usually measured as the proportional reduction in fitness in selfed versus outcrossed progeny ($\delta = 1 - w_s/w_o$). These can be converted by:

$$\delta = 1 - \frac{w_s}{w_o} = 1 - e^{-B/2} \qquad (17.4)$$

and:

$$B = -2\ln(1 - \delta) \qquad (17.5)$$

Some plant species show a tremendous amount of inbreeding depression. For example, Figure 17.15 shows estimates of inbreeding depression for embryonic survival in 35 individual Douglas-fir trees based upon comparison of the production of viable seed by selfing and crossing with pollen from unrelated trees. On average, each tree contained ~10 LEs. This is equivalent to over a 90% reduction in embryonic survival of progeny produced by selfing! However, perhaps more interesting is the wide range of LEs in different trees (Figure 17.15). Conifers in general seem to have high inbreeding depression, perhaps because they are typically outcrossing species with large effective population sizes, but inbreeding depression is especially great in Douglas-fir (Sorensen 1999). Myburg et al. (2014) reported a similar number of LEs (12) in the selfed progeny of the tree *Eucalyptus grandis*.

Most studies of inbreeding depression have been made in captivity or under controlled conditions, but experiments with both plants (e.g., Dudash 1990) and animals (e.g., Jiménez et al. 1994) have found that inbreeding depression is more severe in natural environments. For example, Jiménez et al. (1994) estimated 0.5 LEs in the white-footed mouse in captivity, while they estimated 12.6 LEs in the wild. Estimates of inbreeding depression in captivity are thus likely to severely underestimate the true effect of inbreeding in the wild. Example 17.3 presents an interesting experiment in which inbreeding depression was measured in the same population under captive and wild conditions in different stages of the life cycle.

Crnokrak & Roff (1999) reviewed the empirical literature on inbreeding depression for wild species and compared it with inbreeding depression in captive populations (Ralls et al. 1988, p. 262). They found that the cost of inbreeding in terms of survival was much higher in wild than in captive mammals. They concluded that in general, "the cost of inbreeding under natural conditions is much higher than under captive conditions." This is supported by recent studies of the impacts of inbreeding in the wild on the alpine ibex (Bozzuto et al. 2019; Guest Box 18) and house sparrows (Niskanen et al. 2020).

In addition, there is evidence that inbreeding depression is more severe under environmental stress and extreme events (e.g., extreme weather, pollution, or disease; Armbruster & Reed 2005). Bijlsma et al. (1997) found a synergistic interaction between stress and inbreeding with laboratory *Drosophila* so that the effect of environmental stress is greatly enhanced with greater inbreeding. Schou et al. (2015) found similar results with *Drosophila* across a continuum of nutritional stress. Plough (2012) found that inbreeding depression in the Pacific oyster was greater with nutrient-poor diets. He concluded that stress-associated increases in selection against individual deleterious alleles cause greater inbreeding depression with stress.

17.4.3 Estimates of inbreeding depression with marker-based estimates of *F*

In general, greater inbreeding depression has been detected using marker-based estimates of F than with F_P. As we have seen, there is a great deal

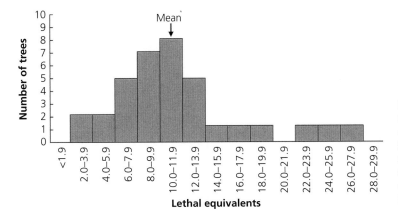

Figure 17.15 Inbreeding depression measured by the observed number of LEs for embryonic survival of seeds from 35 individual Douglas-fir parental trees based upon comparison of the number of viable seeds resulting from self-pollination compared with pollination by unrelated trees. Redrawn from Sorensen (1969).

of variability in the proportion of the genome IBD with the same F_P. Complete genome sequencing based on high-quality genome assemblies makes it possible to measure F with essentially no error because an individual can in principle be scored as heterozygous or homozygous at virtually every position in the genome.

For example, Huisman et al. (2016) compared inbreeding depression estimated with F_P to that detected with over 30,000 SNPs to estimate F in red deer. They estimated F genomically with F_{GRM}, which uses the correlation between uniting gametes across all markers and is highly correlated with average marker homozygosity. When F was estimated using genomic data (F_{GRM}), they found inbreeding depression for breeding success in both sexes, birth weight, and juvenile survival. They also found a strong relationship between maternal F and offspring survival from birth to age 2 years.

However, they detected inbreeding depression for only juvenile survival when using a deep and comparatively complete pedigree to estimate F_P. The genomic estimates of inbreeding depression for lifetime reproductive success in both sexes were startling high (Figure 17.17). Lifetime reproductive success was reduced by 83.5% for individuals with $F = 0.125$. This is equivalent to approximately 38 LEs per individual!

17.4.4 Founder-specific inbreeding effects

The intensity of inbreeding depression can differ greatly depending on which individual happens

to be the founder—that is, a common ancestor of the parents of inbred individuals (Lacy et al. 1996; Lacy & Ballou 1998; Casellas et al. 2008). This suggests that the genetic load is unevenly spread among founder genomes and supports the notion that inbreeding depression sometimes results from major effects at a few loci. We can see this clearly in Figure 17.15, which shows that the number of LEs for embryonic survival in individual Douglas-firs varied from fewer than four to more than 25. The effects of inbreeding depression will be much greater if the common ancestor of both parents carried 25 in comparison with 4 LEs.

This effect also has been detected by using the founder-specific partial F coefficient (Lacy et al. 1996; Gulisija et al. 2006). This allows the increase in homozygosity due to inbreeding to be attributed to a particular founder. For example, the overall pedigree inbreeding coefficient of the individual at the bottom of the pedigree shown in Figure 17.5 is 0.1445. This can be partitioned into the following founder-specific partial F_P coefficients for the six common ancestors: $F_{P:F} = 0.0625$, $F_{P:H} = 0.0625$, $F_{P:K} = 0.0039$, $F_{P:J} = 0.0039$, $F_{P:M} = 0.0059$, and $F_{P:N} = 0.0059$.

A study with Ripollesa domestic sheep found that most of the inbreeding depression resulted from individuals being IBD for genes from just two of nine founders (Casellas et al. 2009). Managing founder-specific inbreeding depression using partial inbreeding coefficients could be extremely effective in cases in which inbreeding depression results primarily from a few loci with major effects; such

Example 17.3 Inbreeding depression for marine survival in anadromous rainbow trout (steelhead)

Thrower & Hard (2009) performed an elegant experiment to estimate the amount of inbreeding depression for a salmonid fish under captive and wild conditions. They were motivated by the absence of studies on inbreeding depression in salmonid fish except under hatchery conditions. Twenty wild anadromous steelhead (15 females and five males) were captured in 1996 from Sashin Creek in southeast Alaska, and mated in a hatchery. The progeny of these fish were then used in an experiment to measure inbreeding depression.

The F_1 progeny of the initial wild steelhead were used to produce an F_2 generation over five different years. Inbreeding depression was estimated by comparing the performance of F_2 progeny that were outbred ($F = 0$) and inbred fish produced by full-sib mating ($F_P = 0.25$). These F_2 progeny were raised in the hatchery during their freshwater life history phase and then released into the ocean. Surviving progeny were captured when they returned to spawn after spending 2 or 3 years in the ocean.

In captivity, no consistent difference was found in the growth or survival of outbred versus inbred progeny over the 5-year course of the experiment. In great contrast, consistent differences were found in the marine survival of noninbred and inbred progeny (Figure 17.16). The survival of noninbred progeny was significantly greater ($P < 0.001$) in all 5 years. On average, the marine survival of inbred progeny was reduced by 71%. This is equivalent to 10.8 diploid LE per individual just for this single phase of the life history of these fish.

These results emphasize that measuring inbreeding depression in captivity can be a poor indicator of inbreeding depression in the wild environment. Specifically, these results indicate that inbreeding depression can be a major hazard to the persistence of small wild populations of salmonid fishes.

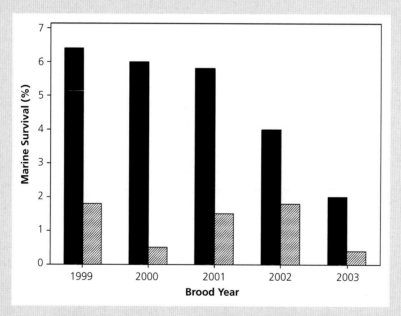

Figure 17.16 Marine survival rates for anadromous rainbow trout (steelhead) released from a hatchery in Alaska. Noninbred fish (black bars) had significantly greater marine survival than inbred fish (hatched bars) in all 5 years of the experiment. Inbred fish ($F_P = 0.25$) were produced by mating full sibs. The overall relative marine survival of inbred fish was ~29% of the noninbred fish. Redrawn from Thrower & Hard (2009).

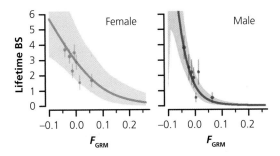

Figure 17.17 Inbreeding depression for lifetime breeding success in red deer. F_{GRM} is a marker-based estimate of F. Points show observations, grouped into seven bins using seven approximately equal subsets. Lines show the fitted models, and shaded areas show the 95% confidence intervals. From Huisman et al. (2016).

partial inbreeding coefficients could be useful when selecting potential matings in a captive population (e.g., Sheikhlou et al. 2020).

17.4.5 Are there species without inbreeding depression?

Some have suggested that some species or populations are unaffected by inbreeding (e.g., Shields 1993). However, deleterious recessive alleles will be present in every population because of the process of mutation (Section 12.3). Therefore, every population will have some inbreeding depression. In addition, lack of statistical evidence for inbreeding depression does not demonstrate the absence of inbreeding depression (Box 20.3). This is especially true because of the low power to detect even a substantial effect of inbreeding in many studies because of small sample sizes and confounding factors (Kalinowski & Hedrick 1999). In a comprehensive review of the evidence for inbreeding depression in mammals, Lacy (1997, p. 320) was unable to find "statistically defensible evidence showing that any mammal species is unaffected by inbreeding." Inbreeding depression for disease resistance has even been found in the close inbreeding naked mole-rat, which for many years was thought to be impervious to inbreeding depression (Example 17.4).

17.5 Genetic load and purging

Small populations can be **purged** of deleterious recessive alleles by natural selection (Templeton

& Read 1984; Crnokak & Barrett 2002; Hedrick & García-Dorado 2016). Deleterious recessive alleles may reach substantial frequencies in large random mating populations because most copies are present in heterozygotes and are therefore not affected by natural selection. For example, over 5% of the individuals in a population in HW proportions will be heterozygous for an allele that is homozygous in only 1 out of 1,000 individuals. Such alleles will be more exposed to natural selection in inbred or small populations and thereby be reduced in frequency or eliminated. Thus, populations with a history of inbreeding because of nonrandom mating (e.g., self-pollinating plants) or small N_e (e.g., a population bottleneck) may be less affected by inbreeding depression because of the reduction in frequency of deleterious recessive alleles (i.e., purging).

17.5.1 Effectiveness of purging

The efficacy of purging depends on the **genetic architecture** of inbreeding depression and on the amount and intensity of inbreeding (García-Dorado 2012). We expect purging to be most effective if inbreeding depression is caused by alleles with a major effect at a few loci. However, purging will be less likely to be effective if inbreeding depression is caused by alleles with a slightly deleterious effect at many loci throughout the genome, as we saw in Example 17.3. Experimental results support this expectation that purging will not be effective against nonlethal deleterious alleles in smaller populations (e.g., Bersabé & García-Dorado 2013). A comparison of complete genome sequences in six ibex species found evidence for purging of highly deleterious recessive alleles, but not mildly deleterious alleles (Grossen et al. 2020).

Reduced differences in fitness between inbred and noninbred individuals within a population after it has gone through a bottleneck is not evidence for purging (Figure 17.19). Many nonlethal deleterious alleles might become fixed in such populations by genetic drift. The fixation of these alleles will cause a reduction in fitness of all individuals following the bottleneck relative to the individuals in the population before the bottleneck (**drift load**, Box 17.1). However, inbreeding depression will appear to be reduced following fixation of deleterious

Example 17.4 Inbreeding depression revealed by a disease challenge in the habitually inbreeding naked mole-rat

The naked mole-rat is a burrowing rodent native in east Africa that lives in eusocial colonies of 75–80 individuals in complex systems of burrows in arid deserts (Maree & Faulkes 2008). The tunnel systems built by naked mole-rats can stretch up to 5 km in cumulative length.

The eusocial naked mole-rat has been considered a classic example of a habitual inbreeder that is not affected by inbreeding depression because of purging: "The only mammal species that has been shown to undergo continuous close inbreeding with no obvious effects of inbreeding depression" (Bromham & Harvey 1996, p. 1083). A microsatellite study found that over 80% of all mating occurs between first-degree relatives in the wild (Reeve et al. 1990). No evidence of inbreeding depression was found in 25 years of captive breeding (Ross-Gillespie et al. 2007).

A virulent enteric coronavirus swept unchecked through a captive naked mole-rat study population, causing acute diarrhea, dehydration, and severe enteric hemorrhaging (Ross-Gillespie et al. 2007). No attempts were made to medicate infected animals. Mortality was monitored daily and dead animals were removed immediately for identification and confirmation of the presence of disease symptoms.

The severe symptoms associated with the coronavirus killed 161 of 365 animals (44%) in just 8 weeks. Survival was significantly lower among more inbred animals (Figure 17.18). Offspring produced by half-sibling ($F_P = 0.125$) and full-sibling ($F_P = 0.250$) parent pairs were two to three times more likely to die than the offspring of unrelated parents ($F_P = 0$). Inbreeding depression in survival corresponded to 2.3 diploid LEs per individual. This study demonstrates how loss of genetic heterozygosity through inbreeding may render populations vulnerable to local extinction from emerging infectious diseases even when other indications of inbreeding depression are absent.

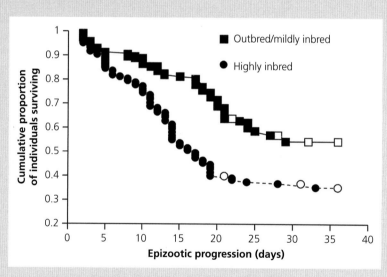

Figure 17.18 Proportional survival of highly inbred ($F_P \geq 0.25$) versus outbred and mildly inbred naked mole-rats ($0 \leq F_P \leq 0.125$) through the course of the coronavirus outbreak. Open markers denote values for which the likelihood function was adjusted to include still-living individuals. From Ross-Gillespie et al. (2007).

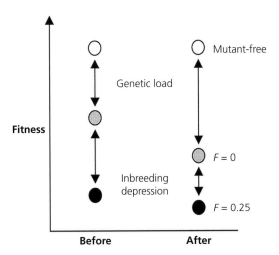

Figure 17.19 Diagram showing decreased inbreeding depression in a population before and after a bottleneck in which all of the inbreeding depression is due to increased homozygosity for deleterious recessive mutant alleles. The open circles show the average fitness of hypothetical "mutant-free" individuals that have no deleterious alleles. The shaded circles show the average fitness of individuals produced by random mating. The dark circles show the average fitness of individuals with an F of 0.25. This illustrates that reduced inbreeding depression following a bottleneck can be caused by an increase in the fixed genetic load (lowering the fitness of individuals with $F = 0$) rather than by purging.

alleles because of depressed fitness of outbred individuals ($F = 0$) rather than by increased inbred fitness (**fitness rebound**; Byers & Waller 1999). To test for purging, the fitness of individuals in the post-bottleneck population must be compared with the fitness of individuals in the pre-bottleneck population. Alternatively, we might test for purging by comparing the fitness of offspring from resident (relatively inbred) individuals versus outbred offspring of crosses between residents and individuals from the pre-bottleneck population.

Reviews have found little evidence for purging in plant and animal populations (Gulisija & Crow 2007; Leberg & Firmin 2008). Byers & Waller (1999) found that only 38% of 52 studies in plant populations showed evidence of purging. And when purging was found, it removed only a small proportion of the total inbreeding depression (roughly 10%). These authors concluded that "purging appears neither consistent nor effective enough to reliably reduce inbreeding depression in small and inbred populations" (Byers & Waller 1999, p. 505).

Reviews of evidence for purging in animals have come to similar conclusions. Ballou (1997) found evidence for a slight decline in inbreeding depression in neonatal survival among descendants of inbred animals in a comparison of 17 captive mammal species. However, he found no indication for purging in weaning survival or litter size in these species. Boakes et al. (2007) reviewed the effects of inbreeding in 119 captive populations of mammals, birds, reptiles, and amphibians. Inbreeding depression for neonatal survival was significant across the 119 populations, although the severity of inbreeding depression varied greatly among taxa. Purging was found to be significant in 14 of the populations, and purging had a significant effect when the entire dataset was analyzed. However, the change in inbreeding depression due to purging averaged across the 119 populations was just 1%. Both Ballou (1997) and Boakes et al. (2007) have concluded that purging is not likely to be strong enough to be of practical use in eliminating inbreeding depression in populations of conservation interest.

Speke's gazelle has been cited as an example of the effectiveness of purging in reducing inbreeding depression in captivity (Templeton & Read 1984). However, Willis & Wiese (1997) have concluded that this apparent purging may have been due to the data analysis rather than purging itself; this interpretation has been disputed by Templeton & Read (1998). Ballou (1997), in his reanalysis of the Speke's gazelle data, found that purging effects were minimal and nonsignificant. Perhaps most importantly, he found that the inbreeding effects in Speke's gazelle were the greatest in any of the 17 mammal species he examined. Kalinowski et al. (2000) have concluded that the apparent purging in Speke's gazelle is the result of a temporal change in fitness and not a reduction in inbreeding depression (see response by Templeton 2002).

We would expect purging to be very effective in haplodiploid species in which males are haploid so that deleterious recessive alleles are exposed to natural selection in males every generation and therefore should be purged relatively efficiently. Nevertheless, inbreeding depression is substantial in a variety of haplodiploid taxa: the insect order Hymenoptera (Antolin 1999), mites (Saito et al. 2000), and rotifers (Tortajada et al. 2009).

Some papers have presented evidence for a reduction in the intensity of inbreeding depression in the first few generations of inbreeding (e.g., Fox et al. 2008; Larsen et al. 2011). However, closer examination of these results indicates that the decrease in inbreeding depression is not due to "purging" as defined above, the selective removal of deleterious alleles contributing to inbreeding depression. Rather, the reduction in inbreeding depression observed in these experiments resulted primarily from the extinction of some experimental lines with greater inbreeding depression. For example, Lynch & Walsh (1998, pp. 274–276) summarized two sets of mouse inbreeding experiments in which full-sib mating resulted in a rapid reduction in litter size averaged over all inbred lines. However, only 10% of the original lines survived over this period and the mean litter size of all lines returned to pre-inbreeding levels after some 10 generations.

The empirical evidence is now clear: purging is not an effective mechanism to reduce inbreeding depression in the conservation of plants and animals (Leberg & Fermin 2008). Nevertheless, purging of alleles can become effective with long-term continuous inbreeding (Guest Box 6; Gulisija & Crow 2007). For example, Latter et al. (1995) found that *Drosophila* strains that experienced slow inbreeding over 200 generations had considerably less inbreeding depression. This might explain why populations of species that have had small effective population sizes have persisted over long periods of time.

17.5.2 Why is purging not more effective?

Failure of purging to decrease inbreeding depression can be explained by several mechanisms. First, purging is expected to be most effective in the case of lethal or semi-lethal recessive alleles (Lande & Schemske 1985; Hedrick 1994); however, when inbreeding depression is very high (more than 10 LEs), even lethals may not be purged except under very close inbreeding (Lande 1994). Second, the lack of evidence for purging is consistent with the hypothesis that a substantial proportion of inbreeding depression is caused by many recessive alleles with minor deleterious effects. Alleles with minor effects are unlikely to be purged by selection, because selection cannot efficiently target harmful

alleles when they are spread across many different loci and different individuals. Third, it is not possible to purge the genetic load (segregation load) at overdominant loci, as mentioned in Section 17.3. On the contrary, the loss of alleles at overdominant loci generally reduces heterozygosity and thus fitness.

Husband & Schemske (1996) found that inbreeding depression for survival after early development, reproduction, and growth was similar in selfing and nonselfing plant species (also see Ritland 1996). They suggested that this inbreeding depression is due primarily to mildly deleterious mutations that are not purged, even over long periods of time. Willis (1999) found that most inbreeding depression in the monkeyflower (Example 17.1) is due to alleles with small effect, and not to lethal or sterile alleles. Bijlsma et al. (1999) have found that purging in experimental populations of *Drosophila* is effective only in the environment in which the purging occurred because additional deleterious alleles were expressed when environmental conditions changed.

Ballou (1997) suggested that **associative overdominance** may also be instrumental in maintaining inbreeding depression. Associative overdominance occurs when heterozygous advantage or deleterious recessive alleles at a selected locus results in apparent heterozygous advantage at linked loci (Section 10.3.2; Pamilo & Pálsson 1998). Kärkkäinen et al. (1999) have provided evidence that most of the inbreeding depression in the self-incompatible perennial herb *Arabis petraea* is due to overdominance or associative overdominance.

Inbreeding depression due to heterozygous advantage cannot be purged. However, it is unlikely that heterozygous advantage is a major mechanism for inbreeding depression in general (Charlesworth & Charlesworth 1987). Nevertheless, there is strong evidence for heterozygous advantage at the immune system-related **major histocompatibility complex (MHC)** in humans (Thoss et al. 2011). Black & Hedrick (1997) found evidence for strong heterozygous advantage (nearly 50%) at both *HLA-A* and *HLA-B* in Indigenous South Americans. Carrington et al. (1999) found that heterozygosity at *HLA-A*, *-B*, and *-C* loci was associated with extended survival of patients infected with the human immunodeficiency virus

(HIV). Strong evidence for the selective maintenance of MHC diversity in vertebrate species comes from other approaches as well (Carrington et al. 1999; Califf et al. 2013).

17.5.3 Evidence for selection against homozygosity in inbred individuals

We saw in Guest Box 12.1 that the loss of heterozygosity in inbred individuals can be less than that expected because of selection against homozygosity. In this example, selfed progeny had average heterozygosity that was more than 30% greater than that expected for neutral loci. Similar results have been reported in other species. For example, Hemmings et al. (2012) found that inbreeding causes early death in the zebra finch. They found that within groups of individuals having the same pedigree, more homozygous individuals at 384 SNPs were more likely to die early, and thus were less likely to survive to sexual maturity.

Bensch et al. (2006) reported analogous results for breeding success and heterozygosity at 31 microsatellite loci in Scandinavian wolves. They found that for a given group of individuals with the same F_P, more heterozygous individuals at the 31 microsatellite loci were more likely to become a breeder. The loci contributing to the relationship between heterozygosity and breeding success were located throughout the genome on many chromosomes. This effect appears to have slowed the loss of heterozygosity in this population, based on comparing the increase in F_P with the decline in heterozygosity over 10 years.

17.6 Inbreeding depression and conservation

The effects of inbreeding depression resulting from bottlenecks are difficult to predict in any given population for a variety of reasons. Even if a population's history of inbreeding is known, it can be difficult to predict the cost of inbreeding on fitness. And, as discussed in Section 17.4.4, the reduction in fitness caused by inbreeding can vary greatly depending upon which individual is the common ancestor of an inbred individual's parents.

The inability to accurately predict the magnitude of inbreeding depression makes it difficult to incorporate inbreeding depression into models of population viability, even when a population's history and biology are well known. Nevertheless, managers must recognize that substantial inbreeding depression is likely to occur in any small population, especially under changing environments or stressful conditions (e.g., Coltman et al. 1999).

The application of genomics has transformed our understanding of inbreeding depression in wild populations. Nevertheless, it is still somewhat unclear how this understanding will assist efforts to minimize the harmful effects of inbreeding depression in conservation. Knowing if inbreeding depression arises from large effects of a few loci versus many loci with smaller effects (or some combination of the two) will help us understand how efficient purging is likely to be, but evidence to date suggests we can't count on it to rescue inbred populations at risk of extinction. It might also be helpful in selecting individuals for breeding or introductions based on their genotypes at particular loci.

Using the distribution of ROH to understand the demographic history of populations will be useful to determine if populations have reduced heterozygosity because of ancient or recent bottlenecks. This information will be very useful to differentiate inbreeding load versus drift load in the population (Box 17.1), and whether or not genetic rescue is likely to be helpful (Example 17.5).

Some skepticism about the importance of inbreeding depression was perhaps justified in the early days of conservation biology. For example, Graeme Caughley questioned a central role for genetics in conservation biology in his 1994 review of conservation biology. He argued that there was no evidence for "genetic malfunction" leading to the extinction of a population or species. Many used Caughley's review as a basis to dismiss genetics as of minor relevance to population viability compared with ecological and demographic factors. In the next chapter, we will consider the evidence that has accumulated over the more than 25 years since the review of Caughley that unambiguously demonstrates the importance of inbreeding in the viability of populations.

Example 17.5 ROH in killer whales provide a global perspective of demographic histories and inbreeding

Foote et al. (2021) sequenced 26 whole genomes from killer whales sampled throughout the world. They found greater heterozygosity in whales sampled nearer to the equator. Killer whales sampled at higher latitudes had hundreds of short ROH that apparently reflect greater background relatedness due to coalescence of haplotypes during bottlenecks associated with founder events during post-glacial range expansions. Individual inbreeding coefficients (F_{ROH}) provided evidence for low amounts of recent inbreeding in 21 of the 26 killer whales.

A female killer whale found in 2016 stranded on a beach on the west coast of Scotland was exceptional in that it was homozygous over 41.6% of its autosomes (Figure 17.20). The distribution of ROH length in this individual indicated many generations of inbreeding in this population. Photo-identification of individuals in this population over 19 years identified just 10 animals remaining in 2011. There was no detected recruitment through births during the study, and the population appeared to be in a slight decline. At the time of writing, only two killer whales remained in this population (Andrew Foote, personal communication).

Figure 17.20 One of the last killer whales in the Scottish population. This female (Lulu) was found stranded on a beach in 2016. Her genome was sequenced and is discussed in this example. Photo courtesy of Andrew Foote, © Hebridean Whale and Dolphin Trust.

Guest Box 17 The genomics of inbreeding depression in Scandinavian wolves
Marty Kardos

Several fundamental questions about inbreeding depression remain unanswered. Is inbreeding depression in the wild usually caused by many loci with small effects, or are loci with large effects often involved? How often does inbreeding substantially affect population viability? Clear answers to these questions remain elusive in part because it was, until recently, very difficult to reliably measure individual inbreeding in wild populations.

Much of our knowledge of inbreeding depression comes from studies that used controlled crosses or pedigrees to estimate individual inbreeding as the expected IBD proportion of the genome (F_P). However, there are several limitations to pedigree analysis. First, the requirement of multiple generations of parentage information has limited pedigree-based analysis in the wild almost exclusively to long-term studies of populations on islands or other isolated habitats (Pemberton 2008). Additionally, F_P can underestimate realized individual inbreeding because parents usually share recent ancestors not included in the pedigree. Lastly, stochasticity in Mendelian segregation and recombination can cause a high variance in realized inbreeding among individuals with the same F_P (e.g., Figure 17.7). Fortunately, the increasing

ability to sequence genomes provides the opportunity to precisely measure realized individual inbreeding (Hedrick & García-Dorado 2016). This development inspired us to use genome sequencing to study the genomic consequences of inbreeding in wild Scandinavian wolves (Kardos et al. 2018b).

The Scandinavian Peninsula was recolonized by wolves in the early 1980s after extirpation by hunting earlier in the 20th century (Åkesson et al. 2016). The population has been intensively monitored since recolonization, including the construction of a pedigree. We sequenced the genomes of 97 individuals with F_P ranging from 0, for immigrants, to >0.5 for some Scandinavian-born individuals. We quantified realized inbreeding as the proportion of the genome in ROH (F_{ROH}), where individuals inherited IBD chromosome segments from their parents (Figure 17.21). We tested how precisely F_P and heterozygosity estimated from small subsets of SNPs measured inbreeding calculated from the whole genome. Additionally, we searched for genomic regions showing molecular signatures of large-effect deleterious recessive alleles.

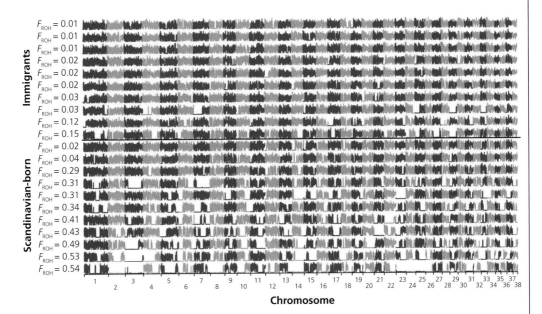

Figure 17.21 Heterozygosity plotted across the genomes of 21 wolves. A subset of Scandinavian-born individuals are shown below the horizontal black line, and a subset of immigrants are shown above the line. Alternating colors within a row represent different chromosomes. From Kardos et al. (2018b).

Guest Box 17 Continued

Realized inbreeding was high in many Scandinavian-born individuals (Figure 17.21). Remarkably, some individuals were homozygous across entire chromosomes (e.g., chromosomes 23–25 for the individual with $F_{ROH} = 0.54$). Some of the immigrants, which are assumed to be noninbred for the pedigree analysis, had $F_{ROH} > 0.1$ (Figure 17.21), substantially higher than expected for the offspring of first cousins. While F_P underestimated F_{ROH}, particularly for immigrants, F_P was strongly correlated with F_{ROH} ($r^2 = 0.86$; Figure 17.22). High precision of F_P is expected in canids due to the large number of chromosomes (38 autosomes), which is expected to result in a low variance in F_{ROH} among individuals with the same F_P (Franklin 1977; Stam 1980). However, inbreeding was more precisely measured with heterozygosity at as few as 500 SNPs than with the pedigree. This is consistent with the theoretical prediction that inbreeding and its fitness effects are best measured with molecular genomic data rather than with pedigrees (Kardos et al. 2015a; Knief et al. 2017). This is good news as it suggests that a few hundred or thousand loci, and little expense, are sufficient to precisely measure inbreeding and its fitness effects, at least in populations with high variance in realized inbreeding as in our study population.

Several genomic regions had fewer ROH than expected by chance. These regions are good candidates for carrying large-effect deleterious recessive alleles, which are expected to rarely be found homozygous in living individuals. While the fitness effects of these regions need to be confirmed with individual fitness data, this result suggests that deleterious recessive alleles with large effects were present in particular genomic regions in the population.

This and other studies (Hoffman et al. 2014; Palkopoulou et al. 2015; Stoffel et al. 2021) demonstrate the power of genomics to understand inbreeding depression in natural populations. Genomics is leading to a more complete understanding of the strength and genetic basis of inbreeding depression in the wild. My hope is that fitness and inbreeding can be measured in enough populations, and in diverse enough environments, to broadly characterize the genetic basis of inbreeding depression and its conservation implications. The challenge for the future is to determine how to best translate the additional information gained from genome-scale analyses of inbreeding depression into practical conservation action.

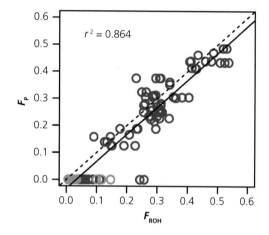

Figure 17.22 Relationship between the pedigree inbreeding coefficient (F_P) and realized genomic inbreeding (F_{ROH}) in 97 Scandinavian wolves. F_{ROH} is the estimated proportion of the genome in ROH that likely arose from ancestors in the last 10 generations. Orange circles represent immigrants, and Scandinavian-born individuals are shown in blue. The dashed-line shows where F_P and F_{ROH} are equal; the solid-line is the actual estimated linear regression. Points below the dashed-line represent individuals where F_P underestimated F_{ROH}. From Kardos et al. (2018).

CHAPTER 18

Demography and Extinction

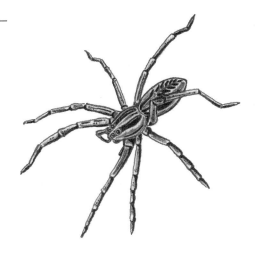

Wolf spider, Section 18.2

As some of our British parks are ancient, it occurred to me that there must have been long-continued close interbreeding with the fallow-deer (*Cervus dama*) kept in them; but on inquiry I find that it is a common practice to infuse new blood by procuring bucks from other parks.

(Charles Darwin 1896, p. 99)

What are the minimum conditions for the long-term persistence and adaptation of a species or a population in a given place? This is one of the most difficult and challenging intellectual problems in conservation biology. Arguably, it is the quintessential issue in population biology, because it requires a prediction based on a synthesis of all the biotic and abiotic factors in the spatial-temporal continuum.

(Michael E. Soulé 1987a, p.1)

The quote from Darwin above shows that both evolutionary biologists and wildlife managers have recognized for well over 120 years that the persistence of small isolated populations may be threatened by inbreeding. Nevertheless, the harmful effects of inbreeding and the importance of genetics in the persistence of populations have been controversial and remain so to this day (Teixeira & Huber 2021).

There are a variety of reasons for this controversy. Some have suggested that inbreeding is unlikely to have significant harmful effects on individual fitness in wild populations. Others have suggested that inbreeding may affect individual fitness, but is not likely to affect population viability (Caro & Laurenson 1994). Still others have argued that

genetic concerns can be ignored when estimating the viability of small populations because they are in much greater danger of extinction due to stochastic demographic effects (Lande 1988; Pimm et al. 1988). Finally, some have suggested that it may be best not to incorporate genetics into demographic models because genetic and demographic "currencies" are difficult to combine, and we have insufficient information about the effects of inbreeding in most wild populations (Beissinger & Westphal 1998).

The disagreement over whether or not genetics should be considered in demographic predictions of population persistence has been unfortunate and misleading. It is extremely difficult to separate genetic and environmental factors when

Conservation and the Genomics of Populations, Third Edition. Fred W. Allendorf, *et al.*, Oxford University Press.
© Fred W. Allendorf, W. Chris Funk, Sally N. Aitken, Margaret Byrne, and Gordon Luikart (2022). DOI: 10.1093/oso/9780198856566.003.0018

assessing the causes of population extinction. This is because inbreeding depression initially usually causes subtle reductions in birth and death rates that interact with other factors to increase extinction probability (Mills & Allendorf 1996). Obvious indications of inbreeding depression (severe congenital birth defects, monstrous abnormalities, or otherwise easily visible fitness deficiencies) are not likely to be detectable until after severe inbreeding depression has accumulated in a population.

Extinction is a demographic process that will be influenced by genetic effects under some circumstances. The key issue is to determine under what conditions genetic concerns are likely to influence population persistence (Nunney & Campbell 1993). There have been important recent advances in our understanding of the interaction between demography and genetics in order to improve the effectiveness of our attempts to conserve endangered species (e.g., Szűcs et al. 2017; Benson et al. 2019; Bozzuto et al. 2019).

Perhaps most importantly, we need to recognize when management recommendations based upon strict demographics or genetics may actually be in conflict with each other. For example, Ryman & Laikre (1991) have considered supportive breeding in which a portion of wild parents are brought into captivity for reproduction and their offspring are released back into the natural habitat where they mix with wild individuals. Programs like this are carried out in a number of species to increase population size and thereby temper stochastic demographic fluctuations. However, under some circumstances, supportive breeding may reduce effective population size and cause a reduction in heterozygosity that may have harmful effects on the population (Ryman & Laikre 1991). This demonstrates a conflict in that supplemental breeding can provide demographic benefits, yet be genetically detrimental.

The primary causes of species extinctions today are deterministic and result from human-caused habitat loss, habitat modification, overexploitation, invasive species, pollution, and now climate change (Caughley 1994; Lande 1999; Ripple et al. 2017). Reduced genetic diversity in plants and animals is generally a symptom of endangerment, rather than its cause (Holsinger et al. 1999). Nevertheless,

genetic effects of small populations have an important role to play in the management of many threatened species. For example, Ellstrand & Elam (1993) examined the population sizes of 743 sensitive plant taxa in California. Over 50% of the occurrences contained less than 100 individuals. In general, those populations that are the object of management schemes are often small and therefore are likely to be susceptible to the genetic effects of small populations. Many parks and nature reserves around the world are small and becoming so isolated that they are more like megazoos than healthy functioning ecosystems. Consequently, many populations will require management (including genetic management) to ensure their persistence (Ballou et al. 1994). Despite this, a small percentage of threatened species recovery plans consider genetic factors (Pierson et al. 2016). Increased extinction rates caused by loss of genetic variation in fragmented populations have led Ralls et al. (2018) to call for a paradigm shift in genetic management in which the default is to restore gene flow rather than keep populations isolated (Section 19.5.2).

In this chapter, we will consider the effects of inbreeding depression, loss of **evolutionary potential** (the ability of a population to adapt to environmental change), and several other genetic factors on the probability of persistence of small populations. It is important to be aware that genetic problems associated with small populations go beyond inbreeding depression and the loss of heterozygosity. We also synthesize the current debate about the population size needed to increase the probability of long-term persistence of populations. We finish the chapter by discussing the use of **population viability analysis** (PVA) to quantify the effects of multiple factors, including genetics, on population persistence. Throughout, we explain how genomics is improving our ability to incorporate genetic factors into predictions of extinction risk.

18.1 Estimation of population size

The number of individuals in a population is perhaps its most fundamental demographic characteristic. Accurate estimates of abundance or **census population size** (N_C) are essential for effective conservation and management (Sutherland 1996).

Moreover, the rate of loss of genetic variation in an isolated population will be primarily affected by the number of breeding individuals in the population. It seems that it should be relatively easy to estimate population size compared to the obvious difficulties of estimating other demographic characteristics, such as the sex-specific age distribution of individuals in a population. However, estimating the number of individuals in a population is usually difficult, even under what may appear to be straightforward situations (Luikart et al. 2010). For example, estimating the number of grizzly bears in the Yellowstone Ecosystem has been an especially contentious issue (Eberhardt & Knight 1996; Kamath et al. 2015). This is perhaps surprising for a large mammal that is fairly easy to observe and that occurs within a relatively small geographic area.

Genetic analyses can provide help in estimating the number of individuals in a population (Luikart et al. 2010). A variety of creative methods have been applied to this problem (Schwartz et al. 1998). For example, we saw in Section 12.2.1 that the amount of variation within a population can be used to estimate effective population sizes, which can be modified to estimate historical census population sizes (Roman & Palumbi 2003). Here, we consider two primary genetic methods for estimating population census size. Bellemain et al. (2005) provide an excellent comparison of these methods.

18.1.1 One-sample

The simplest method for estimating the minimum size of a population is from the number of unique genotypes observed. Kohn et al. (1999) used feces from coyotes to genotype three hypervariable microsatellite loci in a 15-km^2 area in California near the Santa Monica Mountains. They detected 30 unique multilocus genotypes in 115 feces samples. Thus, their estimate of the minimum N_C was 30.

The actual N_C of a population can be much greater than the number of genotypes detected depending on what proportion of the population was sampled. For example, it is likely that not all coyotes in this population were sampled in the collection of 115 feces. However, the estimate of total population size can be modified to take into account the probability of not sampling individuals. The cumulative number of unique multilocus genotypes (y) can be expressed as a function of the number of feces sampled (x), and the asymptote of this curve (a) can be estimated with iterative nonlinear regression with a computer to provide an estimate of local population size:

$$y = \frac{(ax)}{b + x} \qquad (18.1)$$

where b is the rate of decline in value of the slope (Kohn et al. 1999). In this case, the estimate was 38 individuals with a 95% confidence interval of 36–40 coyotes (Figure 18.1). Eggert et al. (2003) have provided an alternative estimator that behaves similarly to Equation 18.1 (Bellemain et al. 2005).

18.1.2 Two–sample: Capture–mark–recapture

A capture–mark–recapture approach can also be used with genetic data to estimate population size (Bellemain et al. 2005). The multilocus genotypes of individuals can be considered as unique "tags" that exist in all individuals and are permanent.

The simplest mark–recapture method to estimate population size is the Lincoln–Peterson index (Lincoln 1930):

$$N_C = \frac{(N1)(N2)}{R} \qquad (18.2)$$

where $N1$ is the number of individuals in the first sample, $N2$ is the number of individuals in the second sample, and R is the number of individuals recaptured in the second sample. For example, suppose that 10 ($N1$) animals were captured in the first sample, and one-half of 10 ($N2$) animals captured in the second sample were marked ($R = 5$). This would suggest that the 10 animals in the first sample represented one-half of the population so that the total population size would be 20.

$$N_C = \frac{(10)(10)}{5} = 20$$

Genetic capture–mark–recapture is potentially a very powerful method for estimating population size over large areas (Bellemain et al. 2005; Example 18.1). For example, in the largest noninvasive study ever, Kendall et al. (2008) used hair sampling of grizzly bears in Glacier National Park, USA, to estimate abundance, which was far higher than previously thought with impressively narrow

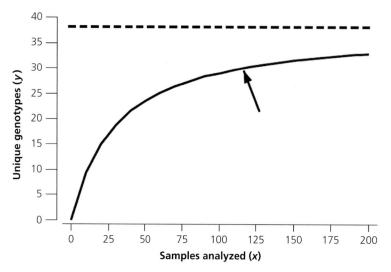

Figure 18.1 Use of rarefaction to estimate the number of coyotes from feces sampled in an area near Los Angeles. Plot of the average number of unique genotypes (y) discovered as a function of the number of samples (x) using the equation $y = (ax)/(b + x)$, where a is the population size asymptote, and b is a constant, which is the rate of decline in the value of the slope. Kohn et al. (1999) found 30 unique genotypes in 115 samples analyzed (see arrow), resulting in an estimate of 38 individuals in the population. Drawn using information in Kohn et al. (1999).

confidence intervals (CIs) (N_C = 240.6; 95% CI = 205–304). It also can be noninvasive (i.e., does not require handling or manipulating animals). However, there are a variety of potential pitfalls in noninvasive studies (Taberlet et al. 1999; Luikart et al. 2010). Perhaps the greatest problem is getting random samples of the population. That is, heterogeneous capture probabilities because of geography or behavior can cause serious estimation problems (Boulanger et al. 2008). Another potential problem is the failure to distinguish individuals due to using too few or insufficiently variable loci. For example, a new capture might be erroneously recorded as a recapture if the genotype is not unique (due to low marker polymorphism). This has been termed the **shadow effect** (Mills et al. 2000). The shadow effect can result in an underestimation of population size and will also affect CIs (Waits & Leberg 2000). Using thousands of genomic markers should reduce the shadow effect by decreasing the probability that multiple individuals will have the same genotype at all loci. In particular, sequencing regions with multiple **single nucleotide polymorphisms (SNPs)** to infer multiple, distinct **microhaplotype** alleles will greatly improve power to distinguish individuals and infer relationships among individuals (Section 10.7.1; Baetscher et al. 2018). Finally, genotyping errors can generate false unique genotypes and thereby cause an overestimate of population size (Waits & Leberg 2000).

Example 18.1 Genetic tagging of humpback whales

Palsbøll et al. (1997) used a genetic capture–mark–recapture approach to estimate the number of humpback whales in the North Atlantic Ocean. Six microsatellite loci were analyzed in samples collected on the breeding grounds by skin biopsy or from sloughed skin in 1992 and 1993. A total of 52 whales sampled in 1992 were recaptured in 1993 as shown below:

	Females	Males	Total
1992	231	382	613
1993	265	408	673
Recaptures	21	31	52

Substitution into Equation 18.2 provides estimates of 2,915 female and 5,028 male humpback whales in this population. Palsbøll et al. (1997) used a more complex estimator that has better statistical properties and estimated the North Atlantic humpback whale population to be 2,804 females (95% confidence interval of 1,776–4,463) and 4,894 males (95% confidence interval of 3,374–7,123). The total of 7,698 whales was in the upper range of previous estimates based on photographic identification.

18.1.3 Other methods for estimating census population size

The census size of a population can be estimated by identifying the number of parent–offspring pairs in samples of adults and juveniles (Skaug 2001; Nielsen et al. 2001; Bravington et al. 2016b). This method, termed **close-kin mark–recapture** (CKMR), is analogous to standard capture–mark–recapture methods, but it relies upon identifying the number of juveniles for which parents can be identified. In mark–recapture terms, each juvenile "marks" two adults, which might subsequently be recaptured, allowing us to estimate the number of adults. This technique has been used with minke whales (Skaug 2001), male humpback whales (Nielsen et al. 2001), bluefin tuna (Bravington et al. 2016a), and great white sharks (Hillary et al. 2018). Ruzzante et al. (2019) validated CKMR using data from brook trout, finding that it provides similar estimates of abundance to traditional mark–recapture methods.

We can also estimate the population census size and density of individuals at different spatial scales thanks to noninvasive genetic sampling coupled with spatially explicit capture–recapture (SECR) models. Unlike nonspatial capture–recapture methods, SECR does not require sampling over areas far larger than the average home range of the study species. Roffler et al. (2019) used SECR to monitor multi-year wolf abundance in densely forested areas in southeast Alaska. Between 2013 and 2015, population size estimates ranged from 221 to 89 wolves (95% CI = 130–378 to CI = 49.8–159.4). These CIs are far narrower than CIs from typical nongenetic methods, which greatly facilitates conservation management decision-making. This and other modeling approaches are being used to address an increasing number of questions in diverse species and landscapes (Qiao et al. 2019; Shyvers et al. 2019; Hooker et al. 2020).

18.2 Inbreeding depression and extinction

We saw in Chapter 17 that inbreeding depression is a universal phenomenon. In this section, we will examine when inbreeding depression is likely to affect population viability. Three conditions must hold for inbreeding depression to reduce the viability of populations:

1. Inbreeding must occur.
2. Inbreeding depression must occur.
3. The traits affected by inbreeding depression must reduce population viability.

Conditions 1 and 2 will hold to some extent in all small populations. As discussed earlier and below, matings between relatives must occur in small populations, and some deleterious recessive alleles will be present in all populations. However, condition 3 is the crux of the controversy. There is little empirical evidence that tells us when inbreeding depression will affect population viability and how important that effect will be.

For inbreeding depression to affect population viability, it must affect traits that influence population viability. For example, Leberg (1990) found that eastern mosquitofish populations founded by two siblings had a slower growth rate than populations founded by two unrelated founders. However, it has been difficult to isolate genetic effects in the web of interactions that affect viability in wild populations (Soulé & Mills 1998; Figure 18.2). Laikre (1999) noted that many factors interact when a population is driven to extinction, and it is generally impossible to single out "the" cause.

Some early authors asserted that there is no evidence for genetics affecting population viability (Caro & Laurenson 1994, p. 485):

Although inbreeding results in demonstrable costs in captive and wild situations, it has yet to be shown that inbreeding depression has caused any wild population to decline. Similarly, although loss of heterozygosity has detrimental impact on individual fitness, no population has gone extinct as a result.

This observation prompted several papers, reviewed in Section 18.2.1, that tested for evidence of the importance of genetics in population declines and extinction. Now there is clear consensus that inbreeding can reduce the viability of populations in the wild (Frankham 2010; Pierson et al. 2016).

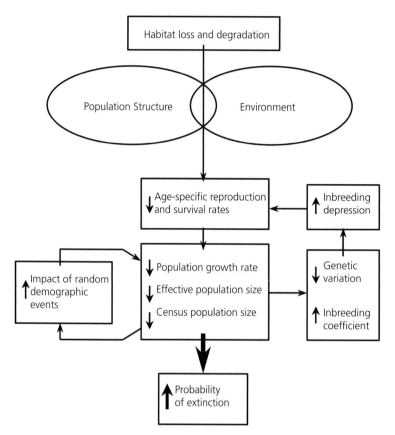

Figure 18.2 Simplified extinction *vortex* showing interactions between demographic and genetic effects of habitat loss and isolation that can cause increased probability of extinction. Redrawn from Soulé & Mills (1998).

18.2.1 Evidence that inbreeding depression affects population dynamics

Newman & Pilson (1997) founded a number of small populations of the annual plant deerhorn clarkia by planting individuals in a natural environment. All populations were founded by the same number of individuals (12); however, in some populations the founders were unrelated (high N_e treatment) and in some they were related (low N_e treatment). All populations were demographically equivalent (i.e., the same N_C) but differed in the effective population size (N_e) of the founding population. A significantly greater proportion of the populations founded by unrelated individuals persisted throughout the course of the experiment (Figure 18.3).

Saccheri et al. (1998) found that extinction risk of local populations of the Glanville fritillary butterfly increased significantly with decreasing heterozygosity at seven allozyme loci and one

microsatellite locus after accounting for the effects of environmental factors. Larval survival, adult longevity, and hatching rates of eggs were all reduced by inbreeding, and were thought to be the fitness components responsible for the relationship between heterozygosity and extinction. Nonaka et al. (2019) extended the study of Glanville fritillary butterflies to include entire metapopulations. The authors concluded that the negative fitness consequences of individual inbreeding depression can propagate through the hierarchy of spatial scales and reduce the persistence of metapopulations.

Westemeier et al. (1998) monitored greater prairie chickens for 35 years and found that egg fertility and hatching rates of eggs declined in Illinois populations after these birds became isolated from adjacent populations during the 1970s. These same characteristics did not decline in populations that remained large and widespread. These results

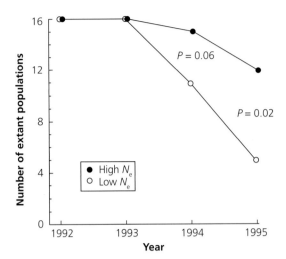

Figure 18.3 Population persistence curves for populations of deerhorn clarkia founded by related (low N_e) and unrelated founders (high N_e). All populations were founded with 12 individuals; however, in some populations the founders were unrelated (high N_e treatment) and in some they were related (low N_e treatment). From Newman & Pilson (1997).

suggested that the decline of birds in Illinois was at least partially due to inbreeding depression. This conclusion was supported by the observation that fertility and hatching success recovered following translocations of birds from the large adjacent populations.

Madsen et al. (1999) studied an isolated population of adders in Sweden that declined dramatically some 35 years ago and has since suffered from severe inbreeding depression. The introduction of 20 males from a large and genetically variable population of adders resulted in a dramatic demographic recovery of this population. This recovery was brought about by increased survival rates, even though the number of litters produced by females per year actually declined during the initial phase of recovery.

Bozzuto et al. (2019) studied 26 reintroduced alpine ibex populations in Switzerland and found that inbreeding reduced per capita population growth rates. Population growth rates were reduced by 71% in populations with high average inbreeding ($F \approx 0.2$) compared with no inbreeding. The study included data spanning 100 years and showed that inbreeding can have long-term demographic

consequences even when environmental variation is large (Guest Box 18).

The previous examples are from populations that were highly inbred and contained little heterozygosity. In an important more general result, Reed et al. (2007) found that reduction in fitness caused by inbreeding depression affected the population dynamics in seven wild populations from two species of wolf spiders. Differences in population growth rates were especially pronounced during stressful environmental conditions and in smaller populations (<500 individuals).

The relationships between genetic variation and rates of population decline and extinction are especially important when populations are faced with environmental stress (e.g., climate change, see Chapter 16). Negative inbreeding effects on population growth rates of reintroduced alpine ibex in Switzerland were more pronounced in climatically harsher environments (Bozzuto et al. 2019; Guest Box 18). Agashe (2009) also found that laboratory populations of flour beetles were more likely to go extinct when challenged with a new food resource if the founding population contained less genetic variation.

18.2.2 Are small populations doomed?

The concepts and results presented here should not be taken to mean that populations that have lost substantial genetic variation because of a bottleneck are somehow doomed or are not capable of recovery (Lesica & Allendorf 1992). An increase in frequency of some deleterious alleles and loss of genome-wide heterozygosity is inevitable following a bottleneck. However, the magnitude of these effects on fitness-related traits (survival, fertility, etc.) might not be large enough to constrain recovery.

For example, the tule elk of the Central Valley of California has gone through a series of bottlenecks since the 1849 gold rush (McCullough et al. 1996). Simulation analysis was used to estimate that tule elk have lost ~60% of their original heterozygosity (McCullough et al. 1996). Analyses of allozymes (Kucera 1991) and microsatellites (Williams et al. 2004) have confirmed relatively low genetic variation in tule elk. Nevertheless, the tule elk has shown a remarkable capacity for population growth, and today there are 22 herds totaling over 3,000 animals (McCullough et al. 1996). Tule elk still may be affected by the genetic effects of the bottleneck in the future if they face some sort of stress (Example 17.5). Similarly, the northern elephant seal that occurs in the northeast Pacific went through an extreme bottleneck due to hunting to near-extinction by the 1880s (Abadía-Cardoso et al. 2017). The bottleneck resulted in low genetic diversity, yet this has not prevented a dramatic demographic recovery in this species that has expanded to cover most of its original geographic range (Stoffel et al. 2018). Red pine provides another example of a widespread species that has very little genetic variation (Example 11.2).

Some have argued that the existence of species and populations that have survived bottlenecks is evidence that inbreeding is not necessarily harmful (Simberloff 1988; Caro & Laurenson 1994; Robinson et al. 2016). However, we need to know how many similar populations went extinct following such bottlenecks to interpret the significance of such observations. For example, the creation of inbred lines of mice usually results in the loss of many of the lines (Lynch & Walsh 1998, pp. 274–276).

This argument is similar to using the existence of 80-year-old smokers as evidence that cigarette smoking is not harmful. Only populations that have survived a bottleneck can be observed after the fact. Soulé (1987b) referred to this as the "fallacy of the accident." A grizzly bear biologist once argued to one of this book's authors that the Kodiak Island bear population has very low genetic variation, yet has persisted 100s or 1,000s of years; therefore, this biologist argued, bear populations must not suffer from inbreeding. This argument suffers from the "fallacy of the accident."

18.3 Loss of phenotypic variation

Inbreeding depression is not necessary for the loss of genetic variation to affect population viability. Reduction in variability itself, even without a reduction in individual fitness because of inbreeding, can reduce population viability (Conner & White 1999; Fox 2005). For example, we saw in Section 6.4 that there can be extensive loss of allelic diversity caused by bottlenecks that are not severe enough to affect the loss of heterozygosity and inbreeding. Loci

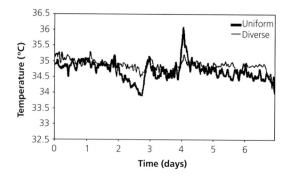

Figure 18.4 Hourly temperature variation in a genetically diverse and uniform honey bee colony. This graph shows the average hourly temperature for a representative pair of experimental colonies that differed only in the number of males with which the queen mated. The uniform colony queen mated with a single male; the diverse colony queen mated with multiple males. From Jones, J. C. et al. (2004).

associated with disease resistance (e.g., **major histocompatibility complex (MHC)**) often have many alleles (Section 6.7.2). Thus, loss of allelic diversity at loci associated with disease resistance is expected to increase the vulnerability of populations to extinction.

Honey bees present a fascinating example of the potential importance of genetic variation itself (Jones, J.C. et al. 2004). Honey bee colonies have different amounts of genetic variation depending on how many males the queen mates with. Brood nest temperatures tend to be more stable in colonies in which the queen has mated with multiple males (Figure 18.4). Honey bee workers regulate temperature by their behavior: they fan out hot air when the temperature is perceived as being too hot and cluster together and generate metabolic heat when the temperature is perceived as being too low. Increased genetic variation for response thresholds produces a more graded response to temperature and results in greater temperature regulation within the hive.

18.3.1 Life history variation

Individual differences in life history (age at first sexual maturity, clutch size, migration propensity, etc.) that have at least a partial genetic basis occur in virtually all populations of plants and animals. Many of these differences may have little effect on individual fitness because of a balance

> **Example 18.2** Effects of life history variation on population persistence in Pacific salmon
>
> Consider a hypothetical comparison of two streams for purposes of illustration. The first stream has separate odd- and even-year populations, as is typical for pink salmon. In the second stream, there is phenotypic (and genetic) variation for the time of sexual maturity so that ~25% of the fish become sexually mature at age one and 25% of the fish become sexually mature at age three; the remaining 50% of the population becomes mature at age two.
>
> All else being equal, we would expect the population with variability in age of return to persist longer than the two reproductively isolated populations. The effective population size (N_e) of the odd- and even-populations would be one-half the N_e of the single reproductive population with life history variability (Waples 1990). Thus, inbreeding depression would accumulate twice as rapidly in the two reproductively isolated populations than in the single variable population. The two smaller populations also would each be more susceptible to extinction from demographic, environmental, and catastrophic stochasticity. For example, a catastrophe that resulted in complete reproductive failure for one year would cause the extinction of one of the populations without variability.
>
> Greene et al. (2010) have provided empirical evidence for this effect. They compared population growth rates (as measured by recruits per spawner) and life history variation (length of freshwater and ocean residence) in nine populations of sockeye salmon from Bristol Bay, Alaska. There was an increasingly positive correlation between population growth rate and life history variation over time. The correlation was negative in the short term (less than 5 years), but increasingly positive from 5–20 years. These results suggest that in the short term, certain life history types are favored by natural selection each year, but the types that are favored change among years. Thus, populations with greater life history diversity are more stable over long periods of time. The authors suggested this **portfolio effect** of diversity is analogous to the financial stability expected from a diversified investment strategy.

or tradeoff between advantages and disadvantages. Nevertheless, the loss of this life history variability among individuals may reduce the likelihood of persistence of a population (Conner & White 1999; Fox 2005). For example, González-Suárez & Revilla (2013) found that mammal species with more variable body masses, litter sizes, ages at sexual maturity, and population densities were less vulnerable to extinction.

Pacific salmon return to freshwater from the ocean to spawn and then die. In most species, there are individual differences in age at reproduction that often have a substantial genetic basis (Hankin et al. 1993). For example, Chinook salmon usually become sexually mature at age 3, 4, or 5 years. The greater fecundity of older females (because of their greater body size) is balanced by their lower probability of survival to maturity. These different life history types have similar fitnesses. Pink salmon are exceptional in that all individuals become sexually mature and return from the ocean to spawn in freshwater at 2 years of age (Heard 1991). Therefore, pink salmon within a particular stream comprise separate odd- and even-year populations that are reproductively isolated (Aspinwall 1974). This life history variation in age at sexual maturity has important implications for persistence of Pacific salmon (Example 18.2).

18.3.2 Mating types and sex determination

The occurrence of separate genders or mating types is another case where the loss of phenotypic variation can cause a reduction in population viability without a reduction in the fitness of inbred individuals. Approximately 50% of flowering plant species have genetic incompatibility mechanisms (de Nettancourt 1977). In one of these self-incompatibility systems, an individual's mating type is determined by its genotype at the self-incompatibility (*S*) locus (Richards 1986). Pollen grains can only fertilize plants that do not have the same *S*-allele as carried by the pollen. Homozygotes cannot be produced at this locus, and the minimum number of alleles at this locus in a sexually reproducing population is three. Smaller populations are expected to maintain many fewer *S*-alleles than larger populations at equilibrium (Wright 1960).

Les et al. (1991) considered the demographic importance of maintaining a large number of S-alleles in plant populations. A reduction in the number of S-alleles because of a population bottleneck will reduce the frequency of compatible matings and may result in reduced levels of seed set. Demauro (1993) reported that the last Illinois population of the lakeside daisy was effectively extinct even though it consisted of ~30 individuals because all plants apparently belonged to the same mating type. Reinartz & Les (1994) concluded that some one-third of the remaining 14 natural populations of *Aster furactus* in Wisconsin had reduced seed sets because of a diminished number of S-alleles.

A similar effect can occur in the nearly 15% of animal species that are **haplodiploid** in which sex is determined by genotypes at one or more hypervariable loci (ants, bees, wasps, thrips, whitefly, certain beetles, etc.; Crozier 1971). Heterozygotes at the sex-determining locus or loci are female, and the **hemizygous** haploids or homozygous diploid individuals are male (Packer & Owen 2001). Diploid males have been detected in over 30 species of Hymenoptera, and evidence suggests that single-locus sex determination is common. Most natural populations have been found to have 10–20 alleles at this locus. Therefore, loss of allelic variation caused by a population bottleneck will increase the number of diploid males produced by increasing homozygosity at the sex-determining locus or loci.

Diploid males are often inviable, infertile, or give rise to triploid female offspring (Packer & Owen 2001). Thus, diploid males are effectively sterile, and will reduce a population's long-term probability of persisting both demographically and genetically (Zayed & Packer 2005; Hedrick et al. 2006). The decreased numbers of females will reduce the foraging productivity of the nest in social species or reduce the population size of other species. In addition, the skewed sex ratio will reduce effective population size and lead to further loss of genetic variation throughout the genome because of genetic drift.

A much weaker gender effect may occur in animal species in which sex is determined by three or more genetic factors. Leberg (1998) found that species with multiple-factor sex determination can experience large decreases in viability relative to

species with simple sex determination systems in the case of very small bottlenecks. This effect results from increased demographic stochasticity because of greater deviations from a 1:1 sex ratio, not because of any reduction in fitness. Multiple-factor sex determination is rare, but it has been described in fish, insects, and rodents. For example, a genomics study of zebrafish found regions on two chromosomes that had a major effect on sex determination (Bradley et al. 2011). There are no sex chromosomes in this species.

18.3.3 Phenotypic plasticity

Phenotypic plasticity has the potential to affect population viability when the environment is changing stochastically. Temporal variation in the climate can provide a challenge to the persistence of populations. Reed et al. (2010) developed an individual-based model in which phenotypes could respond to a temporally fluctuating environment and fitness depended on the match between the phenotype and a randomly fluctuating trait optimum. They found that when cue and optimum were tightly correlated, plasticity buffered absolute fitness from environmental variability, and population size remained high and relatively invariant. In contrast, when this correlation weakened and environmental variability was high, strong plasticity reduced population size, and populations with excessively strong plasticity had substantially greater extinction probability. They suggest that population viability analyses should include more explicit consideration of how phenotypic plasticity influences population responses to environmental change.

Limited phenotypic plasticity in timing of seasonal color molts in snowshoe hares means that they may become increasingly mismatched to the background as snow cover decreases due to climate change (Figure 18.5). Snowshoe hares undergo seasonal color molts to match their background for camouflage from predators. In the spring, they molt from white to brown, and in the fall, they molt from brown to white. Zimova et al. (2014) found minimal plasticity in response to mismatch between coat color and background color, which has high fitness costs from predation (Zimova et al.

Figure 18.5 Snowshoe hare with white coat color against mismatched snow-free background. Photo courtesy of L. Scott Mills.

2016). Hares also did not modify their behavior in response to color mismatch. The authors conclude that snowshoe hares will become increasingly vulnerable to predation due to limited plasticity in molt phenology and behavior unless coat color evolves in response to changing snow cover (Mills et al. 2013). Interestingly, a single gene, *Agouti*, controls winter coat color, which was discovered using whole genome sequencing (Jones et al. 2018). See Chapter 16 for more about phenotypic plasticity and climate change.

18.4 Loss of evolutionary potential

The loss in genetic variation caused by a population bottleneck can cause a reduction in a population's ability to respond adaptively to future environmental changes through natural selection. Bürger & Lynch (1995) predicted, on the basis of theoretical considerations, that small populations (N_e < 1,000) are more likely to go extinct due to environmental change because they are less able to adapt than are large populations.

The ability of a population to evolve is affected both by heterozygosity and the number of alleles present. Heterozygosity is relatively insensitive to bottlenecks in comparison with allelic diversity (Allendorf 1986). Heterozygosity is proportional to the amount of genetic variance at loci affecting quantitative variation (James 1971). Thus, heterozygosity is a good predictor of the potential of a

population to evolve immediately following a bottleneck. Nevertheless, the long-term response of a population to selection is determined by the allelic diversity either remaining following the bottleneck or introduced by new mutations (Robertson 1960; James 1971).

The effect of small population size on allelic diversity is especially important at loci associated with disease resistance (Guest Box 2). Small populations are vulnerable to extinction by epidemics, and loci associated with disease resistance often have an exceptionally large number of alleles. For example, Gibbs et al. (1991) described 37 alleles at the MHC in a sample of 77 adult blackbirds. Allelic variability at MHC is thought to be especially important for disease resistance (Edwards & Potts 1996; Black & Hedrick 1997). For example, Paterson et al. (1998) found that some microsatellite alleles within the MHC of Soay sheep are associated with parasite resistance and greater survival.

The effect of loss of variation due to inbreeding on response to natural selection has been demonstrated in laboratory populations of *Drosophila* by Frankham et al. (1999). They subjected several different lines of *Drosophila* to increasing environmental stress by increasing the salt (NaCl) content of the rearing medium until the line went extinct. Outbred lines performed the best; they did not go extinct until the NaCl concentration reached an average of 5.5% (Figure 18.6). Highly inbred lines went extinct at a NaCl concentration of 3.5%. Lines that experienced an expected 50–75% loss of heterozygosity due to inbreeding went extinct at a mean of ~5% NaCl. Thus, loss of genetic variation due to inbreeding made these lines less able to adapt to continuing environmental change.

In a more recent *Drosophila* experiment, Ørsted et al. (2019) tested the effects of inbreeding level and genome-wide variation on extinction risk and evolution of three traits (productivity, dry body mass, and egg-to-adult viability) in response to a stressful medium reduced in nutrition and increased in acidity. They found that evolutionary responses after 10 generations were better predicted by genomic variation than expected inbreeding levels. Lines with lower genetic diversity were also at greater risk of extinction. Both of these experiments provide strong empirical evidence supporting theory that

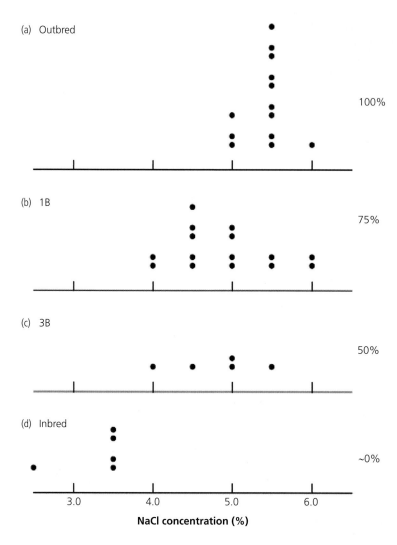

Figure 18.6 Results of an experiment demonstrating that loss of genetic variation can reduce a population's ability to respond by natural selection to environmental change. Lines of *Drosophila* (outbred, inbred, one generation of single pair mating (1B), and three generations of single pair mating (3B)) with different relative amounts of expected heterozygosity (as indicated on the right of the figure) were exposed to increasing NaCl concentrations. Each dot represents the NaCl concentration at which lines went extinct. Modified from Frankham et al. (1999).

predicts greater evolutionary potential and lower extinction risk in populations with higher genetic variation.

Recent genomic studies have demonstrated a loss of evolutionary potential in species of conservation concern. Thompson et al. (2019) found that anthropogenic habitat alteration affecting river conditions such as dams, logging, mining, and water diversions have selected against spring-run Chinook salmon in the Rogue River, Oregon, USA. Chinook salmon in the southern end of their range in North America have two distinct migratory strategies. Spring-run Chinook salmon enter rivers from the ocean in a sexually immature state during the spring, migrate high into watersheds where they hold over the summer, and then spawn in the fall. Fall-run Chinook salmon, in contrast, enter rivers in a sexually mature state in the fall and migrate directly to their spawning grounds where they spawn immediately. Surprisingly, migration phenotype is determined by a single locus (Prince et al. 2017; Thompson et al. 2019). Since dam construction in the Rogue River, the frequency of the allele that causes spring-run migration has declined precipitously. Models suggest that if selection against the spring-run phenotype continues, the spring-run allele could disappear entirely from this river basin, eroding evolutionary potential in this species.

New research combining genomics with phenotypic and fitness data has also demonstrated low evolutionary potential in an endangered bird from New Zealand, the hihi (Chen 2019; de Villemereuil et al. 2019). They found a lack of genome-wide molecular genetic diversity, low heritability of traits under selection, and low additive genetic variance in fitness. Together, these results indicate low evolutionary potential in the hihi, suggesting that these populations will not be able to adapt to environmental change.

18.5 Mitochondrial DNA

Research has suggested that mutations in mitochondrial DNA (mtDNA) might decrease the viability of small populations (Gemmell et al. 2004; Connallon et al. 2018). Mitochondria are generally transmitted maternally so that deleterious mutations that affect only males will not be subject to natural selection (Dowling et al. 2008), and empirical evidence has supported this expectation (Innocenti et al. 2011; Milot et al. 2017). Sperm are powered by a group of mitochondria at the base of the flagellum, and even a modest reduction in power output may reduce male fertility, yet have little effect on females. A study of human fertility has found that mtDNA haplogroups are associated with sperm function and male fertility (Ruiz-Pesini et al. 2000). The mitochondrial genome has also been found to be responsible for **cytoplasmic male sterility**, which is widespread in plants (Schnable & Wise 1998).

The viability of small populations may be reduced by an increase in the frequency of mtDNA genotypes that lower the fitness of males. Since females and males are haploid for mtDNA, it has not previously been recognized that mtDNA may contribute to the increased genetic load of small populations. The effective population size of the mitochondrial genome is generally only one-quarter that of the nuclear genome so that mtDNA mutations are much more sensitive to genetic drift and population bottlenecks than nuclear loci (Section 7.7).

Whether or not an increase in mtDNA haplotypes that reduce male fertility will affect population viability will depend on the mating system and reproductive biology of the particular population.

However, it seems likely that reduced male fertility may decrease the number of progeny produced under a wide array of circumstances. At a minimum, the presence of mtDNA genotypes that reduce the fertility of some males would increase the variability in male reproductive success and thereby decrease effective population size. This would increase the rate of loss of heterozygosity and other effects of inbreeding depression that can reduce population viability.

18.6 Mutational meltdown

Wright (1931, p. 157) first suggested that small populations would continue to decline in vigor slowly over time owing to the accumulation of deleterious mutations that natural selection would not be effective in removing because of the overpowering effects of genetic drift (Section 8.6). Several papers have considered the expected rate and importance of this effect for population persistence (Gabriel & Bürger 1994; Lande 1995; Coron et al. 2013). As deleterious mutations accumulate, population size may decrease further and thereby accelerate the rate of accumulation of deleterious mutations. This feedback process has been termed **mutational meltdown** (Lynch & Gabriel 1990).

Lande (1994) concluded that the risk of extinction through this process "may be comparable in importance to environmental stochasticity and could substantially decrease the long-term viability of populations with effective sizes as large as a few thousand." The expected timeframe of this process is hundreds or thousands of generations. Experiments designed to detect empirical evidence for this effect have had mixed results (e.g., Woodruff 2013).

18.7 Long-term persistence

When considering longer time periods, avoiding inbreeding depression is not enough for persistence. Environmental conditions are likely to change over time, and a viable population must be large enough to maintain sufficient genetic variation for adaptation to such changes. Evolutionary response to natural selection is generally thought to involve a gradual change of quantitative characters through allele frequency changes at the underlying loci, and discussions on the population sizes necessary to uphold evolutionary potential have focused on retention of additive genetic variation of such traits.

There is some disagreement among geneticists regarding how large a population must be to maintain "normal" amounts of additive genetic variation for quantitative traits (Franklin & Frankham 1998; Lynch & Lande 1998). The suggestions for the effective sizes needed to retain evolutionary potential range from 500 to 5,000. The logic underlying these contrasting recommendations is somewhat arcane and confusing. We, therefore, review some of the mathematical arguments used to support the conflicting views.

Franklin (1980) was the first to provide a direct estimate of the effective size necessary for retention of additive genetic variation (V_A) of a quantitative character. He argued that for evolutionary potential to be maintained in a small population, the loss of V_A per generation must be balanced by new variation due to mutations (V_m). V_A will be lost at the same rate as heterozygosity $(1/2N_e)$ at selectively neutral loci, so the expected loss of additive genetic variation per generation is $V_A/2N_e$. Therefore,

$$\Delta V_A = V_m - \frac{V_A}{2N_e} \qquad (18.3)$$

(see also Lande & Barrowclough 1987; Franklin & Frankham 1998). At equilibrium between loss and gain ΔV_A is zero, and:

$$N_e = \frac{V_A}{2V_m}. \qquad (18.4)$$

Using abdominal bristle number in *Drosophila* as an example, Franklin (1980) also noted that $V_m \approx 10^{-3}V_E$, where V_E is the environmental variance (i.e., the variation in bristle number contributed from environmental factors; Chapter 12). Furthermore, assuming that V_A and V_E are the only major sources of variation, the heritability (H_N, the proportion of the total phenotypic variation that is due to additive genetic effects; Chapter 11) of this trait is $H_N = V_A/(V_A + V_E)$, and $V_E/(V_A + V_E) = 1 - H_N$. Thus, Equation 18.4 becomes (Franklin & Frankham 1998):

$$N_e = \frac{V_A}{2 \times 10^{-3} V_E} = 500\frac{V_A}{V_E} = 500\frac{H_N}{1 - H_N} \qquad (18.5)$$

The heritability of abdominal bristle number in *Drosophila* is about 0.5. Therefore, the approximate effective size at which loss and gain of V_A are balanced (i.e., where evolutionary potential is retained) would be 500.

Lande (1995) reviewed the literature on spontaneous mutation and its role in population viability. He concluded that the approximate relation between mutational input and environmental variance observed for bristle count in *Drosophila* ($V_m \approx 10^{-3}V_E$) appears to hold for a variety of quantitative traits in several animal and plant species. However, he also noted that a large portion of new mutations seem to be detrimental, and that only about 10% are likely to be selectively neutral (or nearly neutral), contributing to the potentially adaptive additive variation of quantitative traits. Consequently, he suggested that a more appropriate value of V_m is $V_m \approx 10^{-4}V_E$, and that Franklin's (1980) estimated minimum N_e of 500 necessary for retention of evolutionary potential should be raised to 5,000.

In response, Franklin & Frankham (1998) suggest that Lande (1995) overemphasized the effects of deleterious mutations and that the original estimate of $V_m \approx 10^{-3}V_E$ is more appropriate. They argued that empirical estimates of V_m typically have been obtained from long-term experiments where a large fraction of the harmful mutations have had the opportunity of being eliminated such that a sizable portion of those mutations have already been accounted for. They also pointed out that in most organisms, heritabilities of quantitative traits are typically smaller than 0.5, and that this is particularly true for fitness-related characters. As a result, the quotient $H_N/(1 - H_N)$ in Equation 18.5 is typically expected to be considerably smaller than unity, which reduces the necessary effective size. Franklin & Frankham (1998) concluded that an N_e value of the order 500–1,000 would be generally appropriate.

Lynch & Lande (1998) criticized the conclusions of Franklin & Frankham (1998) and argued that much larger effective sizes are justified for the maintenance of long-term genetic security. They maintain that the problems with harmful mutations must be taken seriously. An important point is that a considerable fraction of new mutations is expected to be only mildly deleterious with a selective disadvantage of less than 1%. Such mildly deleterious mutations behave largely as selectively neutral ones and are not expected to be purged from the population by selective forces even at effective sizes of several hundred individuals. In the long run, the continued fixation of mildly deleterious alleles may reduce population fitness to the extent that it enters an **extinction vortex** (i.e., mutational meltdown; Lynch et al. 1995).

According to Lynch & Lande (1998) there are several reasons why the minimum N_e for long-term conservation should be at least 1,000. At this size, at least the expected (average) amount of additive genetic variation of quantitative traits is of the same magnitude as for an infinitely large population, although genetic drift may result in considerably lower levels over extended periods of time. Furthermore, Lynch & Lande (1998) considered populations with $N_e > 1,000$ highly unlikely to succumb to the accumulation of unconditionally deleterious alleles (i.e., alleles that are harmful under all environmental conditions) except on extremely long time scales. However, they also stressed that many single-locus traits, such as disease resistance, require much larger populations for the maintenance of adequate allele frequencies (Lande & Barrowclough 1987), and suggest that effective target sizes for conservation should be on the order of 1,000–5,000.

18.8 The 50/500 rule

The 50/500 rule was introduced by Franklin (1980). He suggested that as a general rule of thumb, in the short term, the effective population size should not be less than 50, and in the long term the effective population size should not be less than 500. The short-term rule was based upon the experience of animal breeders, who have observed that natural selection for performance and fertility can balance inbreeding depression if ΔF is less than 1%; this corresponds to an effective population size using $\Delta F = 1/2N_e$ (Equation 6.2). The basis of the long-term rule was discussed in detail in the previous section.

There are many problems with the use of simple rules such as this in a complicated world (Jamieson & Allendorf 2012). There are no real thresholds (e.g., 50 or 500) in this process; the loss of genetic variation is a continuous process. The theoretical and empirical basis for this rule is not strong and has been questioned and argued about repeatedly in the literature (Frankham et al. 2013; Jamieson & Allendorf 2013; Frankham et al. 2014a, b; Franklin et al. 2014). In addition, such simple rules can and have been misapplied. We once heard a biologist for a management agency use this rule to argue that genetics need not be considered in developing a habitat management plan that affected many species. After all, if N_e is less than 50, then the population is doomed so that we don't need to be concerned with genetics, and if N_e is greater than 50, then the population is safe so we don't need to be concerned with genetics.

Nevertheless, we believe that the 50/500 rule is a useful guideline for the management of populations (Jamieson & Allendorf 2012). Its function is analogous to a warning light on the dashboard of a car. If the N_e of an isolated population is less than 50, we should be concerned about possible increased probability of extinction because of genetic effects. However, there is experimental evidence with house flies that suggests that the N_e might have to be greater than 50 to escape extinction even in the short term (Reed & Bryant 2000). Thus, these numbers should not be used as targets but as indicators of potential risk. It is also important to remember that 50/500 is based only on genetic considerations. Some populations may face substantial risk of extinction because of demographic stochasticity before they are likely to be threatened by genetic concerns (Lande 1988; Pimm et al. 1988).

18.9 Population viability analysis

Predictive demographic models are essential for determining whether or not populations are likely to persist in the future. Such risk assessment is essential for identifying species of concern, setting priorities for conservation action, and developing effective recovery plans. For example, one of the criteria for being included on the International Union for Conservation of Nature (IUCN) Red List (IUCN 2001) is the probability of extinction within a specified period of time. Some quantitative analysis is needed to estimate the extinction probability of a taxon based on known life history, habitat requirements, threats, and any specified management options. This approach has come to play an important role in developing conservation policy (Shaffer et al. 2002).

PVA is the general term for modeling that takes into account a number of processes affecting population persistence to simulate the demography of populations in order to calculate the risk of extinction or some other measure of population viability (Ralls et al. 2002). The first use of this approach was by Craighead et al. (1973), who used a computer model of grizzly bears in Yellowstone National Park. They demonstrated that closing of the park dumps to bears and the park's approach to problem bears were driving the population to extinction. McCullough (1978) developed an alternative model that came to different conclusions about the Yellowstone grizzly bear population. Both of these models were **deterministic models** in which the same outcome will always result from the same initial conditions and parameter values (e.g., stage-specific survival rates).

Mark Shaffer (1981) developed the first PVA model that incorporated chance events (stochasticity) into population persistence while a graduate student at Duke University. Shaffer described four sources of uncertainty: **demographic stochasticity**, **environmental stochasticity**, **natural catastrophes**, and **genetic stochasticity** (also see Shaffer 1987).

Incorporation of stochasticity into PVA was a crucial step in attempts to understand and predict the population dynamics of populations. Many aspects of population dynamics are processes of sampling rather than completely deterministic (e.g., stage-specific survival, sex determination, and transmission of alleles in heterozygotes, etc.). The predictability of an outcome decreases in a sampling process as the sample size is reduced. For example, in a large population the sex ratio will be near 50:50. However, this might not be true in a small population in which a large excess of males may significantly reduce the population growth rate (Leberg 1998).

In addition, there will be synergistic interactions between demographic processes and genetic effects (Figure 18.2). Fluctuations in population size may result in genetic bottlenecks during which inbreeding may occur and substantial genetic variation may

be lost. Even if the population grows and recovers from the bottleneck, it will carry the legacy of this event in its genes. The loss of genetic variation during a bottleneck may have a variety of effects on demographic parameters (survival, reproductive rate, etc.). This may lead to large fluctuations in population size, increasing the probability of extinction. These interactions, shown in Figure 18.2, have been called extinction vortices (Gilpin & Soulé 1986) and consideration of these interactions is a central part of PVA (Lacy 2000b).

18.9.1 Incorporation of inbreeding depression into PVA

The primary way in which genetics has been incorporated into PVA is to include the effects of inbreeding depression. PVA requires information on birth and survival rates, reproductive rates, carrying capacity, severity of inbreeding depression, and many other factors. Due to the complexity of PVA models, a number of PVA packages are available (Brook et al. 2000). We have chosen to present results using *VORTEX* because of its flexibility, user friendliness, and widespread use (Miller & Lacy 2005; Lacy 2019). However, authors recently cautioned against using user-friendly PVA packages (or any PVA approach) for conservation decision-making unless the quality of the analysis has been thoroughly assessed and rigorous PVA guidelines are followed (Chaudhary & Oli 2020).

It is important to understand the basic structure of the model being used in order to interpret the results. Figure 18.7 shows the relationships among the primary life history, environmental, and habitat components used by *VORTEX*.

As we saw in Section 6.5, a population growing exponentially increases according to the equation

$$N_t = N_0 e^{rt} \qquad (18.6)$$

where N_0 is the initial population size ($t = 0$), N_t is the number of individuals in the population after t units of time (years in *VORTEX*), r is the exponential growth rate, and the constant e is the base of the natural logarithm (approximately 2.72). A population is growing if $r > 0$ and is declining if $r < 0$. Population size is stable if $r = 0$. Lambda (λ) is the factor

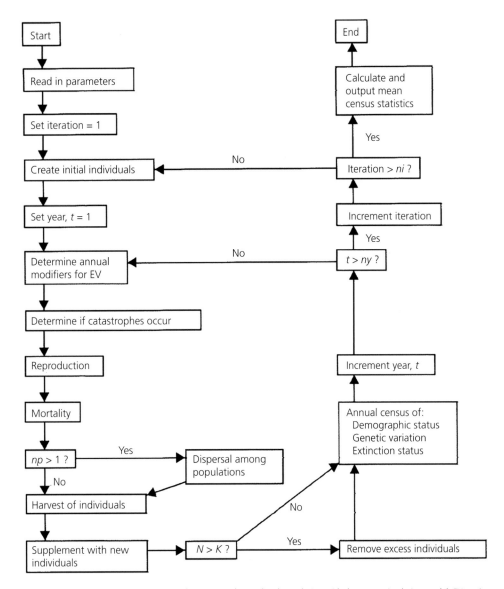

Figure 18.7 Flow chart of the primary components that occur within each subpopulation with the *VORTEX* simulation model. EV, environmental variation; *K*, carrying capacity; *N*, subpopulation size; *ni*, number of iterations; *np*, number of subpopulations; *ny*, years simulated; *t*, year. From Lacy (2000b).

by which the population increases during each time unit. That is,

$$N_{t+1} = N_t \lambda \qquad (18.7)$$

Let us use *VORTEX* to consider PVA of grizzly bears from the Rocky Mountains of the USA. Figure 18.8 shows a summary of the *VORTEX* input values used. The actual values are taken from Harris

& Allendorf (1989), but have been modified for use here.

These life history values result in a deterministic intrinsic growth rate (*r*) of 0.005 (λ = 1.005). Therefore, our simulated grizzly bear population is expected to increase by a factor of 1.005 each year (Figure 18.9). That is, if there are 1,000 bears in year *t* = 0 there will be 1,005 bears in year *t* = 1

```
VORTEX 9.42 -- simulation of population dynamics

1 population(s) simulated for 200 years, 1 iterations
Extinction is defined as no animals of one or both sexes.
No inbreeding depression
EV in reproduction and mortality will be concordant.

First age of reproduction for females: 5    for males: 5
Maximum breeding age (senescence): 30
Sex ratio at birth (percent males): 50
```

```
Population 1: Population 1

  Polygynous mating;
    % of adult males in the breeding pool = 100

  % adult females breeding = 33
  EV in % adult females breeding: SD = 9

  Of those females producing progeny, ...
    28.00 percent of females produce 1 progeny in an average year
    44.00 percent of females produce 2 progeny in an average year
    28.00 percent of females produce 3 progeny in an average year

  % mortality of females between ages 0 and 1 = 20
    EV in % mortality: SD = 4
  % mortality of females between ages 1 and 2 = 18
    EV in % mortality: SD = 4
  % mortality of females between ages 2 and 3 = 15
    EV in % mortality: SD = 4
  % mortality of females between ages 3 and 4 = 15
    EV in % mortality: SD = 4
  % mortality of females between ages 4 and 5 = 15
    EV in % mortality: SD = 4
  % mortality of adult females (5<=age<=30) = 12
    EV in % mortality: SD = 4

  (Same mortality values for males)

  Initial size of Population 1: 100
    (set to reflect stable age distribution)
  Carrying capacity = 1000
    EV in Carrying capacity = 0

Animals harvested from Population 1, year 1 to year 1 at 1 year intervals: 0

Animals added to Population 1, year 1 through year 1 at 1 year intervals: 0
```

Figure 18.8 *VORTEX* input summary for PVA of grizzly bears. This output has been slightly modified from that produced by the program. EV is the environmental variation for the parameter. Values used are modified from Harris & Allendorf (1989).

(1,000 × 1.005) and 1,010 bears in year $t = 2$, etc. The generation interval for grizzly bears is ~10 years. Therefore, this growth rate will result in just over a 5% increase in population size after one generation. It is important to look at the deterministic projections of population growth in any analysis with *VORTEX* or other PVAs. If r is negative, then λ will be less than one, and the population is in deterministic decline (the number of deaths outpace the

number of births) and will become extinct even in the absence of any stochastic fluctuations.

We can use *VORTEX* to examine how much stochastic variability in population growth we may expect (Figure 18.9). On average, Equation 18.7 does a good job of predicting growth rate. However, there is a wide range of results from each independent simulation run even though the same input values were used. The differences among runs

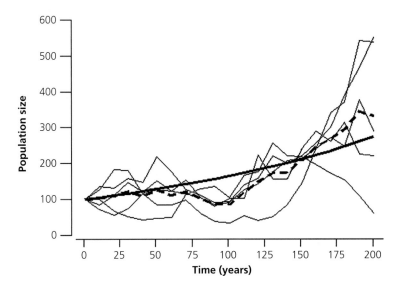

Figure 18.9 Stochastic variability in the growth of a grizzly bear population in five *VORTEX* simulation runs using input from Figure 18.8. The dark solid line is the expected growth rate with $r = 0.005$ ($\lambda = 1.005$) and an initial population size (N_0) of 100, using Equation 18.7. The dark dashed line is the mean of the five simulated populations.

result from *VORTEX* using random numbers to mimic the life history of each individual.

What will happen if we incorporate inbreeding depression into this model? Genetic effects due to inbreeding and loss of variation will come into play when the population size becomes small. For example, one of the runs reached an N of 34 after 100 years. This population then proceeded to grow very quickly and exceed 500 bears 90 years later. However, what would have happened if we had kept track of pedigrees within this population and then reduced juvenile survival as a function of the inbreeding coefficient (F)? Remember that the effective population size of grizzly bears is approximately one-quarter of the population size (Example 7.1). Therefore, the N_e was much smaller than 34 during this period. The increased juvenile mortality of progeny produced by the mating of related individuals would have hindered this population's recovery.

We can incorporate inbreeding depression with *VORTEX* by assigning a number of lethal equivalents (LEs) affecting survival during the first year of life. Figure 18.10 shows the average effects of inbreeding depression with these life history values on population persistence in 1,000 simulation runs for 0, 3, and 6 LEs per diploid genome. In the absence of any inbreeding depression (zero LEs), the persistence probability is similar in the first and second 100 years of the simulations. However, even moderate inbreeding depression (3 LEs) reduces the probability of population persistence by ~25% in the second century.

Simulations by Liao & Reed (2009) found that extinction times decreased some 23% when interactions between inbreeding and stress were included. Fox & Reed (2011) have suggested that Liao & Reed (2009) significantly underestimated the effect of inbreeding–stress interactions because they held the interaction constant rather than increasing it as stress increased. Inclusion of the inbreeding–stress interaction in viability modeling is crucial. This is especially important when populations are of intermediate size and are considered relatively safe from environmental and genetic stresses acting independently (Liao & Reed 2009).

Genomics will improve our ability to incorporate inbreeding depression into PVA by improving our ability to estimate inbreeding coefficients and inbreeding depression (Section 17.2). As we saw in Section 17.1.2, genomics has uncovered tremendous variation in realized inbreeding coefficients, even in individuals with the same pedigrees and pedigree inbreeding coefficients. A recent review of 160 PVAs reported that the majority of studies incorporated environmental (83.8%) and demographic (84.4%) stochasticity, but few incorporated genetic stochasticity (25.0%; Chaudhary & Oli 2020). It is

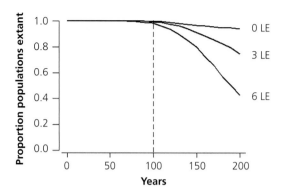

Figure 18.10 Effects of inbreeding depression on the persistence of a grizzly bear population based on *VORTEX* simulations using the values in Figure 18.8 except for the carrying capacity and initial population size. Each point represents the proportion of 1,000 simulated populations that did not go extinct during the specified time period. Simulated populations began with 200 bears and had a carrying capacity of 200. Inbreeding depression was incorporated as a different number of lethal equivalents (LEs) (zero, three, and six) that increased mortality in the first year of life.

crucial for PVAs and related vulnerability assessments to include all components or factors that influence population persistence, including genetic factors (Wade et al. 2017; Lacy 2019; Section 18.9.6). An important avenue for improving the realism of PVA will be to incorporate this variation in inbreeding coefficients into models.

18.9.2 Incorporation of evolutionary potential into PVA

Rapid climate change and other environmental change have led to an increasing recognition of the importance of incorporating evolutionary potential into PVA (Reed et al. 2011a; Pierson et al. 2015). Reed et al. (2011b) included adaptation in an individual-based model designed to predict the probability of extinction in Fraser River sockeye salmon from British Columbia, Canada (Figure 18.11). Adult sockeye salmon are highly sensitive to temperature when they migrate upstream to spawn. Thus, projected temperature increases under climate change could negatively impact migration, spawning, and population growth rates. Reed et al. (2011b) modeled the potential effect of evolution in migration timing to reduce these negative impacts. When they assumed heritability of migration timing was $H_N = 0.5$, their model predicted that migration timing from the ocean to the river would advance by ~10 days by 2100. The

evolution of earlier migration timing resulted in a much lower probability of **quasi-extinction** than when heritability of migration timing was set to zero, such that this trait could not evolve. The risk of quasi-extinction with evolution in migration timing was only 17% as high as the risk of quasi-extinction without evolution in this trait. Incorporation of evolution in their PVA thus has a dramatic effect on projected effects of climate change on population declines.

Incorporation of adaptation into PVA models is an exciting advance, but also challenging to do well for multiple reasons. First, the environmental changes that will impact a given species or population the most are usually difficult to predict. Researchers and managers cannot necessarily predict whether climate change, invasive species, habitat modification, pollution, overexploitation, or disease will be the most significant threat to a given population in the future. Second, many traits influence the fitness effects of environmental change and could adapt in response to this change. For example, a population of salmon could adapt to warming temperatures behaviorally (e.g., advanced migration timing, as we saw above) or physiologically (e.g., increased heat tolerance). Lastly, estimating the heritability and genetic basis of these traits is challenging for nonmodel organisms. Models suggest that the genetic basis of traits can strongly influence adaptive potential

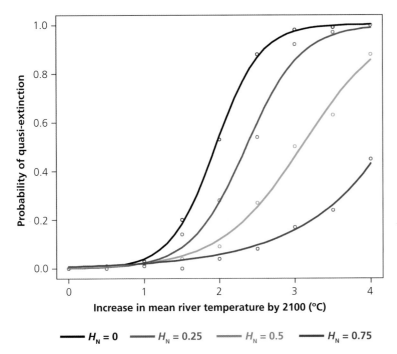

Figure 18.11 Predicted effects of evolution of migration timing on quasi-extinction risk (defined as the proportion of 100 replicate populations where <50 individuals remained by the year 2100) as a function of increases in river temperature by 2100 and different assumed heritability (H_N) for migration timing in Fraser River sockeye salmon. Each data point shows the means of 100 replicate population simulations. From Reed et al. (2011b).

and population viability. For example, polygenic architectures confer increased population viability compared with genetic architectures including large-effect loci (Kardos & Luikart 2021).

Genomics will help improve our ability to estimate heritability and characterize the genetic basis of traits (Chapter 11), but addressing the first two issues above will require a strong understanding of the ecology of and emerging threats to the species in question. Incorporating adaptation and evolution into population models is an important area for further research to improve the realism and accuracy of PVA models.

18.9.3 What is a viable population?

An evaluation of the viability of a population requires identifying the time horizon of concern and the required probability of persistence of remaining above some minimum population size. There is no generally accepted time horizon or level of risk with regard to species extinctions (Shaffer et al. 2002). The IUCN has offered standard criteria for placing taxa into categories of risk (Table 18.1). These

criteria include predictions of the probability of extinction, as well as a variety of other alternative criteria (e.g., reduction in population size, current geographic range, or current population size). For example, a species may be considered to be facing an extremely high risk of extinction in the wild (i.e., Critically Endangered) if its current population size is less than 50 mature individuals without performing a PVA. Authors recently suggested the IUCN and other organizations have not sufficiently incorporated genetic factors when placing taxa in risk categories (Garner et al. 2020; Laikre et al. 2020).

Early applications of PVA often set out to determine the minimum population size at which a population was likely to persist over some timeframe. The **minimum viable population (MVP)** concept was used to identify a goal or target for recovery actions. For example, one of the early grizzly bear recovery plans used the results of Shaffer's early work to set recovery targets for four of the six populations between 70 and 90 bears (Allendorf & Servheen 1986). This recommendation was based largely on simulation results of Shaffer

Table 18.1 Examples of demographic criteria for evaluating the results of population viability analyses.

Source	Status	Probability extinction	Timeframe
Shaffer (1978)	Minimum viable population (MVP)	<5%	100 years
Shaffer (1981)	MVP	<1%	1,000 years
Thompson (1991)	Endangered	>5%	100 years
Rieman et al. (1993)	Low threat	<5%	100–200 years
	High threat	>50%	100–200 years
EPBC Act*	Vulnerable	>10%	Medium-term future
	Endangered	>20%	Near future
	Critically endangered	>50%	Immediate future
IUCN	Vulnerable	>10%	100 years
	Endangered	>20%	20 years or 5 generations
	Critically endangered	>50%	10 years or 3 generations

* Australian Environment Protection and Biodiversity Conservation Act 1999

& Samson (1985), who reported that only 2% of grizzly bear populations beginning with 50 adults became extinct after 100 years. However, 56% of these populations became extinct after 115 years!

The term MVP has fallen out of favor for a variety of reasons. Many feel that the goal of conservation should not be to set a minimum number of individuals or minimal distribution of a species. Nonetheless, the MVP concept continues to be used to guide conservation decisions. Traill et al. (2010a) have argued on the basis of both demography and genetics that at least 5,000 individuals are required for populations to have an acceptable probability of long-term persistence. There is nothing new in this number itself, as discussed in Section 18.7, if we consider that the N_e is much smaller than census size (Section 7.10). However, these authors also suggested that conservation funding should be prioritized on the basis of a linear function of how far a species falls below the 5,000 threshold.

Clements et al. (2011) elaborated on this proposal and used the 5,000 threshold as the basis for their proposed SAFE index (Species Ability to Forestall Extinction). Flather et al. (2011) have disagreed and argued that the use of a single "magic" number of 5,000 individuals is overly simplistic and not useful. They also conclude that the proposed 5,000 threshold is not supported by either theory or empirical data. For example, the genetic justification for the 5,000 threshold is based on the need for N_e to be at least 500 assuming an N_e/N_C ratio

of 0.10 (Traill et al. 2010a). As we saw in Chapter 7, there is a great deal of variability among species in the N_e/N_C ratio, such that using 5,000 as a general guideline is not justified.

Discussion of guidelines for population sizes adequate for long-term persistence of populations from a genetics perspective will continue. Regardless of the precise value of this figure, there is agreement that the long-term goal for actual population size to ensure viability should be thousands of individuals, rather than hundreds. Nevertheless, the rigid application of these guidelines to specific cases is problematic. For example, one of the authors (C.J.A. Bradshaw) of the SAFE index paper (Clements et al. 2011) has argued in the public press that the kākāpō (Guest Box 6) is doomed to extinction because of its small population size (~100 individuals) no matter how many resources we invest (Jamieson & Allendorf 2012). Conservation priorities need to also consider what will be lost; the kākāpō has value as the world's only flightless, lek-building parrot and the sole representative of a monotypic genus and family (Section 1.3).

These examples highlight the problems with setting a minimum number of individuals to guide conservation decisions. However, the concept of MVP can be useful if we build in an appropriate margin of safety. Nevertheless, the term is not necessary as long as we define the timeframe and probability of persistence that we are willing to accept.

18.9.3.1 Demographic criteria

Table 18.1 lists a variety of demographic criteria that have been used or suggested in the literature. There is no correct set of universal criteria to be used. Setting the timeframe and minimum probability of persistence are policy decisions that need to be specific for the situation at hand. Nevertheless, biological considerations should be used to set these criteria. Shorter periods have been recommended because errors are propagated each time step in longer time periods (Beissinger & Westphal 1998). However, we should also be concerned with more than just the immediate future with which we can provide reliable predictions of persistence. The analogy of the distance we can see into the "future" using headlights while driving at night is appropriate here. We can only see as far as our headlights reach, but we need to be concerned about what lies beyond them (Shaffer et al. 2002). Population viability should be predicted on both short (say 10 generations) and long (more than 20 generations) timeframes.

There is a wide range of values presented in Table 18.1. The most stringent is the 99% probability of persistence for 1,000 years used by Shaffer in 1981. The IUCN values provide generally accepted standards and are fundamentally sound. They incorporate the concept of both short-term urgency (10 or 20 years) and long-term concerns. They also take into account that the appropriate timeframe will differ depending on the generation interval of the species under concern. For example, tuatara (Example 1.2) do not become sexually mature until after 20 years, and their generation interval is ~50 years. Therefore, 10 years is just one-fifth of a tuatara generation, but it would represent 10 generations for an annual plant.

18.9.3.2 Genetic criteria

Persistence over a defined time period is not enough. We are also concerned that the loss of genetic variation over the time period does not threaten the long-term persistence of the population or species under consideration. A variety of authors have suggested genetic criteria to be used in evaluating the viability of populations. Soulé et al. (1986) suggested that the goal of captive breeding programs should be to retain 90% of the heterozygosity in a population for 200 years. By necessity, these kinds of guidelines are somewhat arbitrary. Nevertheless, the genetic goal of retaining at least 90–95% of heterozygosity over 100–200 years seems reasonable for a PVA (Allendorf & Ryman 2002). A loss of heterozygosity of 10% is equivalent to a mean inbreeding coefficient of 0.10 in the population.

A genetic goal to retain 95% of allelic diversity in 200 years has been proposed for bison in Yellowstone National Park (Pérez-Figueroa et al. 2012). Computer simulations suggested that 95% of allelic diversity would be retained if population census size remained above 3,000, even with the high variance in male reproductive success in bison (Pérez-Figueroa et al. 2012). However, this goal would be achieved for loci with 5 alleles per locus but not for loci with 20 alleles, which would retain only ~83% of allelic diversity over 200 years. This exemplifies a challenge of using allelic diversity for conservation genetic goals. Similarly, a genetic goal to retain rare alleles (frequency ≤0.05) for 10 generations was established for the North Island brown kiwi, North Island robins, and red deer in New Zealand. Based on simulations, 11–30 immigrants per generation would be required to meet this genetic goal (Weiser et al. 2013).

18.9.4 Are plants different?

PVA has been used primarily with vertebrates. The *VORTEX* model itself was designed to model the life history of mammals and birds (Lacy 2000b). In his review of plant PVAs, Menges (2000) pointed out that previous reviews of PVAs included less than 1% plant species. Plants provide special challenges for PVA (e.g., seed banks, clonal growth, and periodic recruitment). Nevertheless, plant PVAs have proven useful in guiding conservation and management (Menges 2000).

In addition, most of the guidelines used to determine extinction risk and conservation status have been based on vertebrates, in which characteristics such as body size, fecundity, and geographic range have been found to be important (Knapp 2011). Davies et al. (2011) tested the effectiveness of using these largely vertebrate-based methods to estimate extinction risk with plants and argued that they do

Example 18.3 PVA of the Sonoran pronghorn

The pronghorn is endemic to western North America, and it has received high conservation priority because it is the only species in the family Antilocapridae. The Sonoran pronghorn is one of five subspecies and was listed as endangered under the US Endangered Species Act (ESA) in 1967 (Hosack et al. 2002). The pronghorn resembles an antelope in superficial physical characteristics, but it has a variety of unusual morphological, physiological, and behavioral traits (Byers 1997).

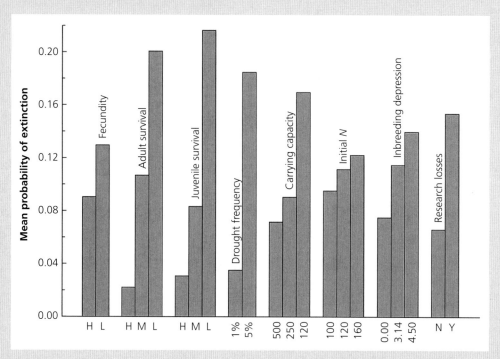

Figure 18.12 Results of PVA on the Sonoran pronghorn. The bars indicate the probability of extinction within 100 years for various values of eight parameters that were varied during sensitivity testing (H, high; M, medium; L, low; N, no; Y, yes). Extinction probabilities were fairly insensitive to some parameters (e.g., initial population size), but were greatly affected by others (e.g., adult and juvenile survival rates). From Hosack et al. (2002).

The Sonoran subspecies is restricted to approximately 44,000 hectares in southwestern Arizona. There were ~200 individuals in this population based on census estimates in the 1990s. A group of 22 biologists from a variety of federal, state, tribal, university, and environmental organizations convened a PVA workshop in September 1996. Nine primary questions and issues were identified as key to pronghorn recovery. All of these questions were explored with PVA simulation modeling during the workshop. The final three of those questions are presented below:

7. Can we identify a population size below which the population is vulnerable, but above which it could be considered for downlisting to a less threatened category?
8. Which factors have the greatest influence on the projected population performance?
9. How would the population respond (in numbers and in probability of persistence) to the following possible management actions: increase in available habitat; cessation of any research that subjects animals to the dangers of handling; exchange of some pronghorn with populations in Mexico; and supplementation of the wild population from a captive population?

Continued

Example 18.3 *Continued*

Estimates of the life history parameters used by *VORTEX* were provided by participants of the workshop. Some of the values were available from field data, but there were no quantitative data available for many parameters. For these parameters, the field biologists provided "best guesses." The participants performed a sensitivity analysis to evaluate the response of the simulated populations to uncertainty caused by varying eight parameters: inbreeding depression, fecundity, fawn survival, adult survival, effects of catastrophes, harvest for research purposes, carrying capacity, and size and sex/age structure of the initial population. Results indicated that the Sonoran pronghorn population had a 23% probability of extinction within 100 years using the best-parameter estimates. This probability increased markedly if the population fell below some 100 individuals. Sensitivity analysis indicated that fawn survival rates had the greatest effects on population persistence (Figure 18.12). Sensitivity analysis also indicated that short-term emergency provisioning of water and food during droughts would substantially increase the probability of population persistence. The workshop concluded that this population is at serious risk of extinction, but that a few key management actions could greatly increase the probability of the population persisting for 100 years.

not provide good estimates of extinction risk for the flora of the Cape of South Africa. They concluded that extinction risk is greater for young and fast-evolving lineages and cannot be predicted by comparison of life history traits.

Zeigler et al. (2013) reviewed 223 publications that describe plant PVAs and determined whether authors followed previous recommendations for improvement of PVAs. Among the publications, the typical model was a matrix population model parameterized with ≤5 years of demographic data. Few publications considered stochasticity, genetics, density dependence, seed banks, vegetative reproduction, dormancy, threats, or management strategies, suggesting a need to improve PVA applications in plants.

18.9.5 Beyond viability

Population viability analyses have great value beyond simply predicting the probability of extinction. Perhaps more importantly, PVA can be used to identify threats facing populations and identify management actions to increase the probability of persistence. This can be done by **sensitivity testing** in which a range of possible values for uncertain parameters are tested to determine what effects those uncertainties might have on the results. In addition, such sensitivity testing reveals which components of the data, model, and interpretation have the largest effect on population projections. This will indicate which aspects of the biology of the

population and its situation contribute most to its vulnerability and, therefore, which aspects might be most effectively targeted for management. In addition, uncertain parameters that have a strong impact on results are those which might be the focus for future research efforts, to better specify dynamics. Close monitoring of such parameters might also be important for testing the assumptions behind the selected management options and for assessing the success of conservation efforts (Example 18.3).

18.9.6 Complex models: Multiple species and environmental interactions

Population models have been criticized for their lack of complexity and ignoring risk factors that influence population persistence (e.g., Lacy et al. 2013; Prowse et al. 2013; Urban et al. 2016). Urban et al. (2016) suggested models should consider including six key factors or biological mechanisms to improve predictions of responses to environmental change. These include the following: species interactions, environmental change (e.g., predicted land use change), physiology (e.g., tolerance to warming), connectivity (dispersal, gene flow), evolution (adaptive potential), and demography (e.g., vital rates). Few PVA models consider more than two of these factors.

Lacy et al. (2013) introduced a "metamodel" approach to allow consideration of multiple diverse threats that act at different spatio-temporal scales, interact in complex ways, and include data from

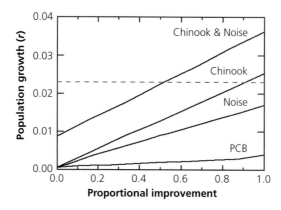

Figure 18.13 Predicted population growth for southern resident killer whales achieved by reduction of different threats. Threat reductions on the *x*-axis are scaled from no reduction in the given threat to the maximum reduction tested in PVAs: Chinook salmon abundance increased up to 1.3× the long-term mean; noise disturbance during feeding reduced from 85% to 0%; and polychlorinated biphenyls (PCBs) reduced from accumulation rates of 2 ppm/year to 0 ppm/year. The top line shows the predicted increase in population growth from a combination of increasing Chinook salmon abundance and reducing noise. The horizontal dashed line represents the conservation objective of 2.3% population growth. From Lacy et al. (2017).

different disciplines—from geography to disease epidemiology and landscape ecology. The meta-modeling approach links different individual models that each depicts a part of a complex system. Interactions among models are revealed as emergent properties of the system.

For example, Lacy et al. (2017) modeled the endangered killer whale population of the northeastern Pacific Ocean and considered the main threats, including limitation of preferred prey (Chinook salmon), anthropogenic noise and disturbance (which reduces foraging efficiency), and contaminants stored in the body, including polychlorinated biphenyls (PCBs). The modeling suggested that reducing acoustic disturbance by 50% combined with increasing Chinook by 15%

would allow the population to reach 2.3% growth (Figure 18.13). Inbreeding depression was included in the modeling and found to influence long-term population viability. The model also yielded estimates of effective population size that were 37% of the total census size, similar to the N_e estimated previously from genetic data.

Ultimately, the degree of complexity to incorporate into a PVA should be dictated by the question. No more complexity should be added than necessary to adequately capture the most important factors affecting population dynamics. The statistician George Box famously quipped that "all models are wrong, but some are useful." To be useful, models, including PVAs, should strive to strike a balance between realism and simplicity.

Guest Box 18 Inbreeding depression reduces population growth rates in reintroduced alpine ibex
Lukas F. Keller and Iris Biebach

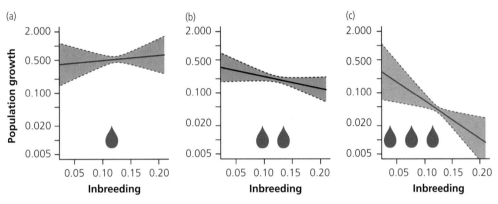

Figure 18.14 The effects of inbreeding on the growth of 26 reintroduced alpine ibex populations as a function of the amount of summer precipitation in their habitat. The figure provides a graphical illustration of the best-fitting statistical model relating population growth to the average degree of inbreeding in interaction with summer precipitation in the area of the European Alps where these ibex populations are found. Inbreeding did not impact population growth at minimum summer precipitation (a), but reduced population growth increasingly at mean (b) and maximum (c) summer precipitation. Lines represent estimates from the model and shading the 95% credible intervals. Population growth was estimated as the exponential population growth rate r and drawn on a logarithmic scale. Inbreeding was measured from microsatellite markers using population-specific F_{ST}. From Bozzuto et al. (2019).

Inbreeding depression, the detrimental effects of inbreeding on individuals, has been documented in a wide variety of species (Chapter 17). The demographic consequences of these effects at the population level, however, are far less well documented and understood. Why do we know so much less about the consequences of inbreeding on population growth rates, the parameter most critical for understanding the demographic consequences of inbreeding (Mills & Smouse 1994)?

First, empirical work requires a lot of data from several populations and over long time periods. Consequently, only some laboratory studies (e.g., Leberg & Firmin 2008; Szűcs et al. 2014, 2017) and very few field studies (Hanski & Saccheri 2006; Chapter 18) have been able to estimate inbreeding effects on population growth rates.

Second, theory predicts that the demographic consequences of inbreeding depend on genetic and ecological conditions (Keller et al. 2007), including the degree of intra- and interspecific competition, the type of trait, and the life history stages affected by inbreeding (e.g., Agrawal & Whitlock 2010, 2012). Under certain ecological conditions,

inbred populations may produce enough offspring to maintain population growth even if individuals experience substantial inbreeding depression (Wallace 1975; Saccheri & Hanski 2006). Thus, reduced population growth rates are not an automatic consequence of inbreeding depression among individuals.

Populations of a mountain ungulate, the alpine ibex, reintroduced to the European Alps over the past ~110 years provided us with an opportunity to quantify the effects of inbreeding on population growth rates. We chose 26 populations that had been monitored by the wildlife authorities since their reintroduction, and measured levels of inbreeding in these populations using genetic markers. In collaboration with colleagues, we then quantified the effects of inbreeding on the exponential population growth rate r using a modified version of Equation 18.6 (Bozzuto et al. 2019).

The negative effects of inbreeding were substantial (Figure 18.14). The most inbred population (with an average inbreeding coefficient of 0.21) exhibited a 71% reduction in population growth compared with a hypothetical, noninbred

Guest Box 18 Continued

population (Bozzuto et al. 2019). As is often the case with inbreeding depression at the individual level (Chapter 17), inbreeding interacted with environmental conditions: inbreeding reduced population growth the most in regions with high summer precipitation (Figure 18.14). Limitations in our study design prevent us from understanding the exact causes underlying this interaction. But the interaction parallels theoretical predictions that ecological processes modify the extent to which inbreeding affects populations (Agrawal & Whitlock 2012).

Alpine ibex have experienced several bottlenecks in the course of their near-extinction and reintroduction. While some of the highly deleterious alleles were purged during these bottlenecks, more mildly deleterious mutations accumulated, leading to a higher overall mutational burden than in related species (Grossen et al. 2020). This and the generally weak intraspecific competition (Sæther et al. 2007) are likely some of the reasons for the strong inbreeding effects on population growth in alpine ibex (Bozzuto et al. 2019).

Many endangered species will share some of these features and thus inbreeding will likely affect their population demography negatively. Therefore, conservation programs are well advised to minimize inbreeding effects in endangered populations.

CHAPTER 19

Population Connectivity

Mountain lion, Guest Box 19

An important case arises where local populations are liable to frequent extinction, with restoration from the progeny of a few stray immigrants. In such regions the line of continuity of large populations may have passed repeatedly through extremely small numbers even though the species has at all times included countless millions of individuals in its range as a whole.

(Sewall Wright 1940, p. 243)

The current reluctance to consider genetic rescue in conservation settings is not scientifically justified.

(Kathy Ralls et al. 2018, p.5)

The models of genetic population structure that we have examined to this point have assumed a connected series of equal-sized populations in which size is constant (e.g., Chapter 9). However, the real world is much more complicated than this. Local populations differ in size, and local populations of some species may go through local extinction events and then be recolonized by migrants from other populations. These events will have complex, and sometimes surprising, effects on the genetic population structure and evolution of species. Another complication is that individuals may not be structured as discrete populations, but may instead be continuously distributed. An important advance in population genetics in the last couple of decades is the development of a **landscape genetics** framework, which integrates population genetics, landscape ecology, and spatial statistics to characterize genetic variation in complex, real-world landscapes.

Understanding the genetic effects of habitat fragmentation is becoming increasingly important because of ongoing loss of habitat. Many species that historically were nearly continuously distributed across broad geographic areas are now restricted to increasingly smaller and more isolated patches of habitat. In some cases, supplementation is necessary to restore **functional connectivity** to small, isolated populations to infuse new genetic variation, reduce inbreeding depression, and restore evolutionary potential, a management strategy known as **genetic rescue**.

In this chapter, we will combine genetic and demographic models to understand the distribution

Conservation and the Genomics of Populations, Third Edition. Fred W. Allendorf, *et al.*, Oxford University Press.
© Fred W. Allendorf, W. Chris Funk, Sally N. Aitken, Margaret Byrne, and Gordon Luikart (2022). DOI: 10.1093/oso/9780198856566.003.0019

of genetic variation in species. We will also introduce how a landscape genetics perspective can be used to understand the effects of environmental heterogeneity and landscape features on genetic variation, gene flow, and other microevolutionary processes. Then, we will discuss genetic rescue as a management strategy to reestablish connectivity in small, fragmented populations. Finally, we will consider how **metapopulation** structure affects the viability of populations.

19.1 Metapopulations

Sewall Wright (1940) was the first to consider the effects of extinction of local populations on the genetics of species. Wright was interested in the effect that such local extinctions would have on genetic structure and evolution. He considered the case where local populations are liable to frequent extirpation and are restored with the "progeny of a few stray immigrants" (Figure 19.1).

Wright pointed out that such local extinctions and recolonization events would act as bottlenecks that would make the effective population size of a group of local subpopulations much smaller than expected based on the number of individuals present within the subpopulations. Therefore, many of the subpopulations would be derived from a few local subpopulations that persist for long time periods. In modern terms, the genes in many of the

subpopulations would **coalesce** to a single gene that was present in a "source" subpopulation in the relatively recent past.

The term metapopulation was introduced by Richard Levins (1970) to describe a "population of populations." In Levins' model, a metapopulation is a group of small populations that occupy a series of similar habitat patches isolated by unsuitable habitat. The small local populations have some probability of extinction (e) during a particular time interval. Empty habitat patches are subject to recolonization with probability (c) by individuals from other patches that are occupied. Metapopulation dynamics are a balance between extinction and recolonization so that at any particular time some proportion of patches are occupied (p) and some are extinct ($1 - p$). At equilibrium,

$$p^* = \frac{c}{c + e} \qquad (19.1)$$

The concept of metapopulations has become a valuable framework for understanding populations and species (Hanski & Gilpin 1997; Dobson 2003; Wade 2016). It is ironic that Levins (1969) originally developed this model in order to determine better strategies for controlling agricultural insect pests.

The general definition of a metapopulation is a group of local populations that are connected by dispersing individuals (Hanski & Gilpin 1991). More realistic models have incorporated differences in local population size and differential rates of

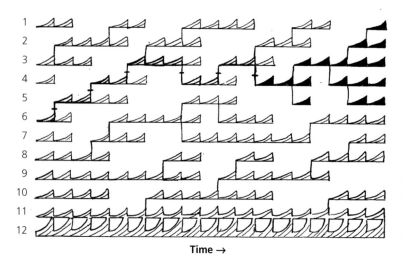

Figure 19.1 Original diagram from Wright (1940) of a species in which local populations are liable to frequent extinction and recolonization. Time (generations) proceeds from left to right. Twelve different local patches are represented by a horizontal row (numbered 1 through 12). Note that the bottom two local populations never go extinct whereas all others go extinct every two to nine time steps. For example, the subpopulation in patch 7 went extinct at the end of time steps 2, 6, and 15. The darkly shaded subpopulations in the upper right have passed through small groups of migrants six times.

Time →

exchange among populations as well as differential rates of extinction and colonization. In general, larger patches are less likely to go extinct because they will support larger populations. Patches that are near other occupied patches are more likely to be recolonized. In addition, immigration into a patch that decreases the extinction rate for either demographic or genetic reasons has been called the **rescue effect** (Brown & Kodric-Brown 1977; Ingvarsson 2001).

19.1.1 Genetic variation in metapopulations

It is important to consider both spatial and temporal scales when looking at the effective size of metapopulations. Slatkin (1977) described the first metapopulation genetic models. Hanski & Gilpin (1991) have described three spatial scales for consideration (Figure 19.2):

1. The **local scale** is the scale at which individuals move and interact with one another in their course of routine feeding and breeding activities.
2. The **metapopulation scale** is the scale at which individuals, or their gametes, infrequently disperse from one local population to another,

typically across habitat that is unsuitable for their establishment and breeding activities.
3. The **species scale** is the entire geographic range of a species; individuals typically have no possibility of moving to most parts of the range. Metapopulations on opposite ends of the range of a species do not exchange individuals, but they remain part of the same genetic species because of movement among intermediate metapopulations.

The effect of metapopulation structure on the pattern of genetic variation within a species depends upon the spatial and temporal scale under consideration. We can use our models of genetic subdivision introduced in Section 9.1 to see this relationship (Waples 2002). Effective population size is a measure of the rate of loss of heterozygosity over time. The **short-term effective population size** is related to the decline of the expected average heterozygosity within subpopulations (H_S). The **long-term effective population size** is related to the decline in expected heterozygosity if the entire metapopulation were **panmictic** (H_T).

Consider a metapopulation consisting of six subpopulations of 25 individuals each that are "ideal" as defined in Section 7.1 so that $N_e = N_c = 25$. The total population size of this metapopulation is $6 \times 25 = 150 = N_T$. The subpopulations are connected by migration under the island model of population structure so that each subpopulation contributes a proportion m of its individuals to a global migrant pool every generation, and each subpopulation receives the same proportion of migrants drawn randomly from this migrant pool (Section 9.4). The rate of decline of both H_S and H_T will depend upon the amount of migration among subpopulations (Figure 19.3).

In the case of complete isolation, the local effective population size is $N = 25$, and heterozygosity within each subpopulation declines at a rate of $1/2N = 1/50 = 2\%$ per generation. However, different alleles will be fixed by chance in different subpopulations. Therefore, expected heterozygosity in the global metapopulation (H_T) will become "frozen" and will not decline. This can be seen in Figure 19.4. Five of the six isolated subpopulations went to fixation within the first 100 generations.

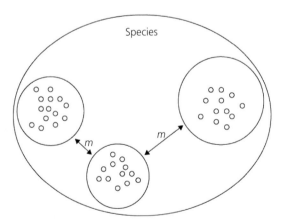

Figure 19.2 Hierarchical spatial organization of a species consisting of three metapopulations each consisting of a cluster of local populations that each exchange individuals. A small amount of gene flow between the three metapopulations (m) maintains the genetic integrity of the entire species. The two metapopulations on opposite ends of the range do not exchange individuals, but they remain part of the same genetic species because of movement via the intermediate metapopulation.

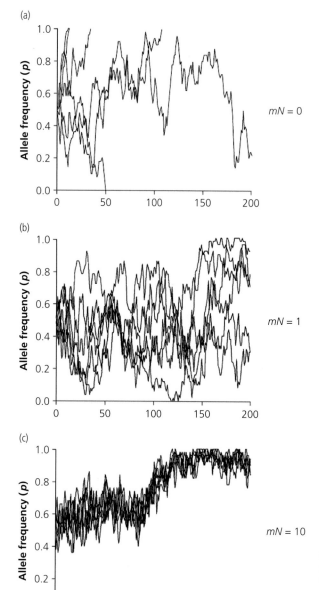

Figure 19.3 Changes in allele frequency in six subpopulations each of $N = 25$ connected by varying amounts of migration under the island model. Panel (a) shows the case of complete isolation ($m = 0$). Panel (b) shows the case of one migrant per generation ($mN = 1$; $m = 0.04$). Panel (c) shows the case of near-panmixia among subpopulations ($mN = 10$; $m = 0.4$). The graphs were drawn using the *Populus* simulation program (Alstad 2001).

At the other extreme of near-panmixia among the subpopulations, the local effective population size will be N_T so that heterozygosity within each subpopulation declines at a rate of $1/2 N_T = 1/300 = 0.3\%$ per generation. In this case, the heterozygosity in the global metapopulation (H_T) will decline at

the same rate as the local subpopulations. Eventually, all subpopulations will go to fixation for the same allele so that H_T will become zero (Figure 19.4).

Thus, complete isolation will result in a small short-term effective population size, but greater long-term effective population size. The case of

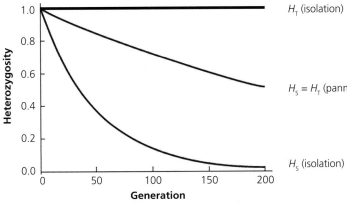

Figure 19.4 Expected decline in local (H_S) and total (H_T) heterozygosity in a population with six subpopulations of $N = 25$ each. In the case of effective panmixia ($mN = 10$) the decline in both local and global heterozygosities are equivalent and are equal to ($1/2N_T = 1/300$ per generation). In the case of complete isolation, local heterozygosity declines at a rate of $1/2N = 0.02$ per generation, but global heterozygosity is constant because random drift within local subpopulations causes the fixation of different alleles.

effective panmixia has the extreme opposite effect, which will result in greater short-term effective population size, but a smaller long-term effective population size (Figure 19.4).

Intermediate levels of gene flow have the best of both worlds (Figure 19.3). The introduction of new genes by migration will maintain greater heterozygosities within local populations than the case of complete isolation. However, a small amount of migration will not be enough to restrain the subpopulations from drifting to near fixation of different alleles. Therefore, a small amount of gene flow will maintain a similar amount of heterozygosity within local subpopulations as the case of effective panmixia, and will also maintain long-term heterozygosity at a similar rate as the case of complete isolation (Figure 19.4).

19.1.2 Effective size of a metapopulation

The effective population size of a metapopulation is extremely complex. Wright (1943) has shown that in the simplest case, when a metapopulation of size N_T is divided into many identical partially isolated islands that each contributes equally to migrant pool migration, then:

$$N_{eT} \approx \frac{N_T}{1 - F_{ST}} \qquad (19.2)$$

where N_{eT} is the long-term effective population size of the metapopulation (Nunney 2000; Waples 2002). Thus, increasing population subdivision (as measured by F_{ST}) will increase the long-term effective population size of the metapopulation.

Equation 19.2 also indicates that the effective size of the metapopulation will be greater than the sum of the N_e values of the subpopulations when there is divergence among the subpopulations. In fact, N_{eT} approaches infinity as m approaches zero (Figure 19.3a).

The validity of Equation 19.2 and our conclusions for natural populations depend upon the validity of our assumptions of no local extinction ($e = 0$) and N within subpopulations being constant and equal. However, in the classic metapopulation of Levins (1970), extinction and recolonization of patches (subpopulations) are common. Wright (1940) pointed out that in the case of frequent local extinctions, the long-term N_e can be much smaller than the short-term N_e because of the effects of bottlenecks associated with recolonization:

$$N_e \text{ (long)} \ll N_e \text{ (short)}$$

For example, the entire ancestry of the darkly shaded group of related populations in Figure 19.1 has "passed through small groups of migrants six times in the period shown" (Wright 1940, p. 244). Thus, these populations are expected to have low amounts of genetic variation, even though their current size may be very large.

This effect can be seen in Figure 19.5, which shows a metapopulation consisting of three habitat patches (Hedrick & Gilpin 1997). The local populations in all three patches initially have high heterozygosity. The population in patch 1 goes extinct and is recolonized by a few individuals from patch 2 in generation 20, resulting in low heterozygosity. The

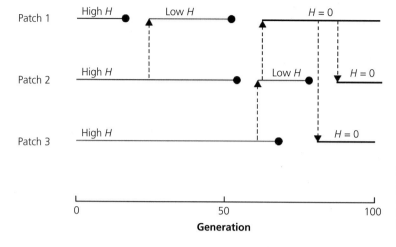

Figure 19.5 Effect of local extinction and recolonization by a few founders on the heterozygosity (*H*) in a metapopulation consisting of three habitat patches. Redrawn from Hedrick & Gilpin (1997).

population in patch 1 goes extinct and is recolonized again from patch 2. However, the few colonists from patch 2 have low heterozygosity because of an earlier extinction and recolonization in patch 2; this results in near-zero heterozygosity in patch 1. Patches 2 and 3 are later recolonized by migrants from patch 1 so heterozygosity is zero in the entire metapopulation.

Hedrick & Gilpin (1997) explored a variety of conditions with computer simulations to estimate long-term N_{eT} as a function of decline in H_T. They found that the rate of patch extinction (*e*) and the characteristics of the founders were particularly important. Slatkin (1977) described two extreme possibilities regarding founders. In the **propagule** pool model, all founders come from the same founding local population. In the migrant pool model, founders are chosen at random from the entire metapopulation. As expected, high rates of patch extinction greatly reduce N_{eT}. In addition, if vacant patches were colonized by a few founders, or if the founders came from the same subpopulation, rather than the entire metapopulation, H_T and N_{eT} are greatly reduced.

Relaxing the assumption that all subpopulations contribute an equal number of migrants also affects long-term N_{eT} as a function of decline in H_T. Nunney (1999) considered the case where differential productivity of the subpopulations brings about differential contributions to the migrant pool due to the accumulation of random differences among individuals in reproductive success. In this case, the effective size of a metapopulation (N_{eT}) is reduced

due to increasing F_{ST} by what Nunney (1999) has called **interdemic genetic drift**.

It is clear that the effect of metapopulation structure on the effective population size of natural populations is complex. The long-term effective population size may either be greater or less than the sum of the local N_e values depending on a variety of circumstances: rates of extinction and recolonization, patterns of migration, and the variability in size and productivity of subpopulations. It is especially important to distinguish between local and **global effective population** size because these two parameters often respond very differently to the same conditions. All of these factors should be considered in evaluating conservation programs for endangered species (Waples 2002).

19.2 Landscape genetics

Landscape genetics is an interdisciplinary field that aims to assess the influence of landscape features and environmental variables on dispersal, gene flow, other microevolutionary processes, and genetic variation (Manel et al. 2003; Holderegger & Wagner 2008; Balkenhol et al. 2016). Landscape genetics combines approaches from population genetics, landscape ecology, and spatial statistics. The approaches can involve novel individual-based statistical assessments that use genetic patterns such as spatial discontinuities to identify population boundaries or to group individuals into populations (Guillot et al. 2005).

Individual-based assessments help prevent incorrect delineation of populations, which can occur when populations are identified using the geographic location or physical characters of individuals sampled. *A priori* population identification using geographical or physical features is often subjective (Pritchard et al. 2000).

Individual-based landscape genetic approaches can provide finer-scale assessments of genetic structure than traditional population genetic approaches (Section 20.4.2). Such approaches are crucial for precise geographic localization of genetic discontinuities caused by **landscape resistance barriers** or **secondary contact zones**. Nonetheless, population-based approaches are also useful in landscape genetics (e.g., Epps et al. 2007; Robertson et al. 2018). Figure 19.6 presents three possible landscape genetic approaches to detect barriers to gene flow.

Continuously distributed populations (Section 9.5) are better investigated with landscape genetic models than metapopulation genetic models as the latter use an *a priori* grouping of individuals into local populations. This is important because most species are not distributed in discrete demes. Analyzing continuously distributed individuals as discrete groups can lead to erroneous inferences about genetic structure and connectivity (Section 20.4.2).

19.2.1 Landscape connectivity and complex models

Functional landscape connectivity (i.e., dispersal, gene flow, or disease transmission across landscapes) can be inferred using landscape genetic approaches (Kozakiewicz et al. 2018; Storfer et al. 2020). For example, Blanchong et al. (2008) used landscape genetic modeling to test for landscape features influencing the distribution of chronic wasting disease (CWD) in white-tailed deer in Wisconsin, USA. The features tested were the Wisconsin River running east–west through the northern third of the region and US Highway 18/151 running east–west through the southern third of the region. Genetic differentiation between deer populations was greatest, and CWD prevalence lowest, in areas separated by a river, indicating that rivers reduced spread of disease from the geographic area of the

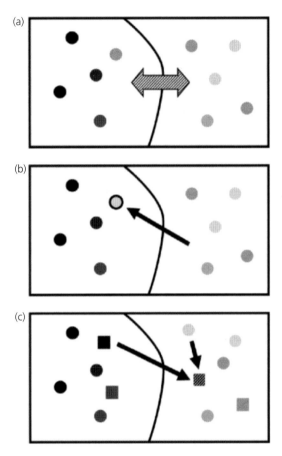

Figure 19.6 Three landscape genetic approaches to infer barriers to gene flow. The black line refers to a landscape feature, such as a river; the filled circles refer to the adult individuals; and different gray shadings of these circles refer to genetically similar genotypes. (a) Approach based on genetic distances between pairs of individuals. Here, the genetic distances between all pairs of individuals are determined and correlated with landscape structure. In the present case, the largest genetic distances occur across the landscape feature; that is, the hypothesized barrier (hatched double-headed arrow). (b) Recent gene flow assessed by assignment tests (Section 9.9.4). The individuals are grouped *a priori* into two populations on either side of the landscape feature. An assignment test identifies one individual (circle with solid outer line) as a recent immigrant from the other side of the landscape feature (arrow). (c) Current gene flow assessed by parentage analysis of offspring (squares). The offspring has one parent on the other side of the landscape feature (arrows). From Holderegger & Wagner (2008).

disease origin. This result suggested that landscape genetics can help predict populations at high risk of infection based on their genetic connectivity to infected host populations and might also help target geographic areas for disease surveillance and preventative measures such as increased harvest (e.g., along rivers to isolate or reduce contact rates between populations). To understand functional landscape connectivity, multiple sympatric species (e.g., hosts, parasites, predators, and prey) could be assessed using landscape genetic approaches.

Landscape genetics can be thought of as an extension of traditional metapopulation genetic models by allowing more complex modeling of connectivity in heterogeneous landscapes. Recent modeling approaches allow testing among many realistic landscape genetic models (e.g., with a range of barrier configurations and strengths). This can yield a more detailed understanding of interactions between landscapes, environment, and connectivity. Such detailed understanding is important in conservation for achieving goals such as accurate and precise localization of corridors and barriers. For example, many traditional metapopulation models yield a single migration rate parameter (m) between demes or habitat patches, and consider only a simple homogeneous matrix of inhospitable habitat between demes. More recent landscape genetic models assume a complex matrix of multiple habitat types and topography with different resistances to gene flow separating different pairs of demes or individuals (e.g., Nali et al. 2020).

The use of relatively complex landscape models allows testing of multiple hypotheses in realistic settings to better understand the relative importance of different features such as roads and vegetation cover on gene flow and landscape connectivity. Complex habitat models allow quantification of the relative resistance of different pathways and the identification of multiple potential pathways for dispersal or gene flow across the landscape. Each potential path can have a different cumulative (total) resistance to movement. The path with the least cumulative resistance is referred to as the **least cost path** (Epps et al. 2007). The least cost path is often longer in geographical distance than the direct straight-line path because the direct path often crosses poor habitat or partial barriers with high ecological cost to movement. Researchers calculate the least cost paths under many different hypotheses (**resistance models**) and then compare each matrix of least cost paths with a matrix of genetic relatedness (or genetic distance) to assess

the merit of each hypothesis. This approach is being widely used by ecologists and managers to understand habitat features that various wildlife and plants require for dispersal and gene flow.

An alternative and increasingly popular approach for modeling resistance across landscapes applies electrical **circuit theory** to landscape genetics (McRae 2006). With circuit theory analysis, landscapes are represented as nodes connected by resistors. The level of resistance among adjacent nodes is determined by the resistance surface. What distinguishes circuit theory from least cost path analysis is that, with circuit theory, pairwise resistances incorporate all possible pathways on the landscape, not just the least cost path. This can make circuit theory particularly appropriate for landscapes with many paths with similar resistance costs. As with least cost path analysis, researchers calculate pairwise resistance values under many different resistance model hypotheses and then compare each of these resistance surface matrices with a matrix of genetic distances to test which resistance model is best supported.

Studies that have compared least cost path with circuit theory approaches suggest that one approach is not consistently better than the other (Spear et al. 2016). For example, circuit theory modeled range-wide connectivity better than least cost paths (McRae & Beier 2007). In contrast, least cost paths modeled gene flow of Cope's giant salamander better than circuit theory, likely because this salamander is strongly associated with stream corridors (Trumbo et al. 2013). Thus, the most appropriate method will depend on the biology of the study species and the specific question the researcher is attempting to answer.

Another exciting advance in landscape genetics is the incorporation of genetic and environmental data collected across time, an approach that has been called **spatiotemporal landscape genetics** (Fenderson et al. 2020). Although the goal of many landscape genetic studies is to infer how landscape and environmental change affects gene flow and genetic variation, inferences are often limited by analyzing a single snapshot of genetic and environmental data. The ability to directly relate changes in genetic variation to environmental change is greatly improved by including historical genetic samples and environmental data from multiple time points. For example, Draheim et al. (2018) characterized spatial genetic structure of black bears in 2002, 2006, and 2010, and related changes in genetic structure to changes in the landscape. They found that changes in spatial genetic structure were associated with the magnitude of landscape change (e.g., number of patches changing from suitable to unsuitable) and changes in habitat fragmentation. Adding a temporal dimension to landscape genetic studies promises to greatly increase the power to relate changes in microevolutionary processes to landscape and environmental change.

A substantial challenge in landscape genetics—especially with a temporal dimension—is the lag times (of multiple generations) that often exist before the effect of a landscape modification on connectivity is detectable. For example, it can take 5–500 generations following new barrier establishment before a genetic signature of discontinuity is detectable genetically. The lag time depends on the number of loci, number and distribution of individuals sampled, the dispersal pattern of the focal species, and the modeling approach used to assess landscape connectivity (Landguth et al. 2010).

19.2.2 Corridor mapping

Corridor mapping is facilitated by landscape genetic approaches in which individuals are sampled widely across large landscapes. For example, Epps et al. (2007) applied a population-based landscape genetic approach using least cost path modeling to identify corridors and assess gene flow among bighorn sheep populations. The corridors predicted from the genetic modeling were largely consistent with known intermountain movements of bighorn sheep. In addition, the authors determined that gene flow was highest in landscapes with more than 10% sloping terrain; thus, gene flow occurred over longer distances when steep escape terrain was available, consistent with bighorn sheep biology and the use of cliffs to avoid predators. Their work also linked reduced genetic variation to the construction of interstate highways and canals that have reduced connectivity and heterozygosity by ~15% in only 40 years. This diminished connectivity

could reduce individual and population growth (Hogg et al. 2006) and threaten population persistence (Epps et al. 2005).

Understanding the effects of habitat and landscape features on functional connectivity is crucial in predicting the effects of habitat loss and climate change on the future viability of many species. Example 19.1 presents a landscape genetic analysis of wolverines in the US Rocky Mountains to develop corridor maps in order to: (1) evaluate current connectivity among extant populations; (2) evaluate the potential of colonization from extant populations to areas of recent historical extirpation; and (3) predict the effects of projected climate change on wolverine populations in the future.

19.2.3 Neutral landscape genomics

Genomic approaches are increasingly being applied to landscape genetics (Storfer et al. 2016; Forester et al. 2020). **Neutral landscape genomics** uses neutral loci presumably not under selection to characterize landscape connectivity and test the environmental factors that are related to gene flow and genetic variation. Neutral landscape genomics is conceptually similar to landscape genetics using traditional genetic markers such as microsatellite loci, but differs from landscape genetics in three important ways. First, in neutral landscape genomics, the first step is to remove loci under selection. This requires testing for loci under selection using an F_{ST} outlier test or a **genotype–environment association (GEA)** analysis (Section 19.2.4), and then removing any loci identified as potentially under selection, leaving presumably neutral loci for downstream analysis. This step is often ignored in landscape genetics, as it is relatively unlikely that any loci from a small subset of loci will be in or linked to loci under selection. However, even in these studies, some loci show a signature of divergent or stabilizing selection, which can bias estimates of genetic divergence (Luikart et al. 2003). A second way in which neutral landscape genomics differs from landscape genetics is that many more loci are available for estimating genetic distance among populations or relatedness among individuals, which should improve the precision of these estimates. A third way in which

Example 19.1 Landscape genetics and connectivity of wolverines in the US Rocky Mountains

The wolverine is a stocky and muscular carnivore that superficially resembles a small bear more than its fellow mustelids. It has a reputation for ferocity and strength out of proportion to its size, with the documented ability to kill prey many times its size. The wolverine is found in the northern boreal forests of North America, Europe, and Asia. Wolverines have experienced a steady decline in numbers for over 100 years because of trapping and habitat fragmentation, but now there is some evidence of population growth and range expansion.

Landscape features that influence wolverine population substructure and gene flow have been largely unknown. Schwartz et al. (2009a) examined 210 wolverines at 16 microsatellite loci to infer connectivity and movement patterns among wolverines in the Rocky Mountains in Montana, Idaho, and Wyoming, USA. They constructed a pairwise genetic distance (proportion of shared alleles) matrix among all individuals. Previous work indicated that the distribution of persistent spring snow was the key environmental feature affecting wolverine presence and movement. They built hypothetical resistance surfaces using different values for resistance to movement in areas not having persistent spring snow. They then estimated the correlation between matrices of pairwise genetic distances and movement costs (**cost distance**) among all individuals.

Significant positive correlations between genetic distance and cost distance were detected for all models based on spring snow cover. Models simulating large preferences for dispersing within areas characterized by persistent spring snow explained the data better than a model based on straight-line geographical distance. In all cases, cost models based on snow cover had significantly stronger correlations with observed genetic distances than straight-line (Euclidean) distances. For all cost models based on snow cover, there was no relationship between genetic distance and Euclidean distance once the spring snow-based correlation was removed. The least cost path analysis among all pairs of systematically gridded points suggested several important wolverine corridors connecting areas of spring snow cover (orange paths in Figure 19.7).

These results were used to derive empirically based least cost corridor maps. These corridor maps were concordant with previously published population subdivision patterns based on mitochondrial DNA and indicate that natural colonization of the southern Rocky Mountains by wolverines will be difficult but not impossible (Figure 19.7). In 2009, a male wolverine with a Global Positioning System (GPS) collar actually traveled from Wyoming to Colorado using a path nearly identical to the one predicted in this paper (M.K. Schwartz, personal communication). These results have been used to optimize corridors through the ranking of land parcels based on their cost and their contribution to connectivity (Dilkina et al. 2017).

neutral landscape genomics differs from landscape genetics is that the use of many loci can lead to non-independence among loci, which can bias results from analyses that assume independent loci. This issue is also often ignored in landscape genetics. One locus from each pair of loci in strong gametic disequilibrium should be removed before downstream analyses (Cesconeto et al. 2017).

Recent studies suggest that the higher precision of genetic distance estimates based on thousands of **single nucleotide polymorphisms (SNPs)** provides greater power to detect subtle effects of environmental and landscape features on gene flow than a small number of traditional genetic markers. For example, McCartney-Melstad et al. (2018) tested the effects of roads on genetic divergence in endangered eastern tiger salamanders in New York using SNPs from thousands of nuclear loci.

In contrast to a previous microsatellite study, they found highly structured populations over a small spatial scale and a significant effect of roads on genetic divergence. None of these patterns were found with microsatellite markers, suggesting that neutral landscape genomics is a much more powerful approach for detecting the effects of rapid urbanization on genetic structure. Power is important because it allows earlier detection of landscape fragmentation and the ability to detect more subtle effects of landscape features on connectivity, which gives managers more time and options for mitigating fragmentation (Selmoni et al. 2020).

Neutral landscape genomics may also be a powerful approach for detecting subtle landscape effects on connectivity in species with high dispersal rates and distances. Supporting this expectation, Trumbo et al. (2019) were able to detect the effects of natural

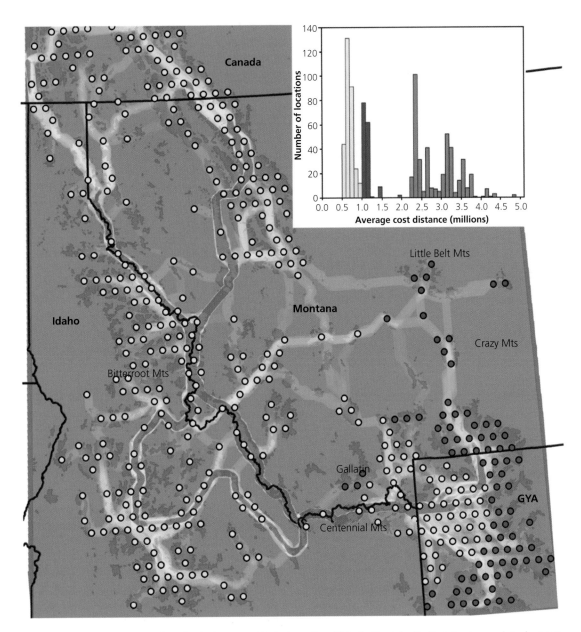

Figure 19.7 Landscape genetic analysis of wolverine connectivity in the northern US Rocky Mountains (Rockies) showing cumulative least cost paths between systematically placed locations (circles) in spring snow cover cells. Paths in orange are predicted to be used more often than those in cooler colors. The color of circles corresponds to the average resistance cost distance between that location and all other locations, based on the models. The graph was divided into four modes (three within the northern US Rockies, and one between the Greater Yellowstone Area (GYA) and Colorado (south of area shown)). The yellow mode has the lowest average cost distances (shown with yellow bars in inset graph), the blue mode the next lowest, and the pink mode (Crazy and Little Belt Mountains) the highest in the northern Rocky Mountains. The green bars show the cost distances between all points from Colorado to the GYA. Modified from Schwartz et al. (2009a).

and anthropogenic landscape features on gene flow in wide-ranging mountain lions over small spatial scales relative to the large dispersal distances traveled by mountain lions. These examples and others suggest that genomics will greatly improve the power of landscape genetics to characterize connectivity and identify landscape factors impeding and facilitating gene flow.

Characterizing genomic variation in multiple species across landscapes and environmental gradients can improve our understanding of the ecology, evolution, and conservation of natural populations (Hand et al. 2015). De Kort et al. (2018) studied the effects of climate warming, habitat fragmentation, genetic variation, and species interactions on the geographic distribution and persistence of the rare Alcon grassland butterfly in the Pyrenees Mountains in France. These butterfly populations are sensitive to grassland habitat configuration, and depend on both a rare marsh gentian plant and also an ant species. The study suggested that interactions between these species, climate warming, and livestock grazing will reduce habitat availability and reshuffle distributions of suitable habit patches, thereby reducing connectivity and population persistence of the Alcon butterfly. The study detected molecular genetic signatures of population bottlenecks and fragmentation in interacting species (plant and ant), suggesting their dispersal rates lag behind environmental changes. This bottlenecking, fragmentation, and reduced population connectivity could negatively affect the Alcon butterfly. Similar **community landscape genomics** studies could help us understand and conserve other species, including host–parasite, plant–pathogen, and predator–prey systems (Hand et al. 2015; Leo et al. 2016; Gamboa & Watanabe 2019).

19.2.4 Adaptive landscape genomics

In contrast to neutral landscape genomics, the goal of **adaptive landscape genomics** is to identify loci under selection that may be involved in local adaptation, and to test the environmental and landscape factors that drive and maintain divergence at these loci (Schwartz et al. 2009b; Storfer et al. 2016). Characterizing adaptive genetic variation using adaptive

landscape genomics is key for predicting the potential for adaptation to future environmental change (Grummer et al. 2019). For example, Razgour et al. (2019) used GEA analyses to quantify adaptive genetic variation related to climate and then incorporated this information into forecasts of range changes in response to climate change (Example 16.6). They found less predicted range loss based on models that include adaptive genetic variation compared with models that did not include adaptive variation. Moreover, as patterns of neutral genetic variation are sometimes a poor surrogate for patterns of adaptive genetic variation (Funk et al. 2012b; Prince et al. 2017), it is helpful to specifically identify loci involved in adaptation.

Two main approaches are commonly used to identify loci under selection in adaptive landscape genomics: F_{ST} outlier tests and GEA analyses. F_{ST} outlier tests are designed to identify loci with F_{ST} values that are higher or lower than expected based on neutrality that may be under divergent or stabilizing selection, respectively (Luikart et al. 2003; Beaumont & Balding 2004; Bonin 2008). Once loci under selection are identified, patterns at these loci can be compared with patterns at neutral loci to reveal adaptive differences and similarities among populations (Funk et al. 2012b). F_{ST} outlier tests should be interpreted with caution because a variety of factors other than selection can result in variation in F_{ST} values (Bierne et al. 2013; Lotterhos & Whitlock 2015; Hoban et al. 2016). GEA analyses identify loci involved in adaptation using associations between spatial distributions of alleles and environmental variables hypothesized to drive divergent selection (Rellstab et al. 2015; Forester et al. 2018). The inclusion of environmental predictor variables increases power compared with F_{ST} outlier tests, improves detection of contemporary selection, and allows identification of the environmental factors underlying adaptation (Rellstab et al. 2015; Forester et al. 2018). However, an advantage of F_{ST} outlier tests is that they can uncover loci under selection that are not related to environmental factors hypothesized to drive adaptation, thereby revealing unanticipated dimensions of adaptive divergence. Moreover, testing for outliers of other summary statistics (e.g., F_{IS}, heterozygosity, etc.) can help detect loci under selection (Luikart et al. 2003).

When available, adding phenotypic data to genotypic and environmental data can strengthen inferences about local adaptation (Sork et al. 2013; Forester et al. 2020). Local adaptation typically results from directional selection within populations for phenotypes that increase fitness in the local environment, which is manifest as divergent selection among populations. Thus, if a given environmental gradient (e.g., temperature, precipitation, canopy cover, etc.) drives local adaptation, we expect to see: (1) phenotypic differences across the environmental gradient that are hypothesized to increase fitness in the local environment (measured with phenotype–environment correlations); (2) associations between environmental variation and allele frequencies at some loci (detected with GEA analyses); (3) a genetic basis to phenotypic differences (quantified in a common garden experiment or with a genome-wide association analysis (GWAS); Chapter 11); and (4) ideally, overlap between some loci identified in the GEA and GWAS, showing that the loci underlying phenotypic differences among populations show evidence of divergent selection.

One of the few studies to meet this high bar for demonstrating local adaptation was in lodgepole pine and interior spruce (Yeaman et al. 2016). This study used a combination of phenotype–environment correlations, GWAS, and GEA to identify loci underlying local adaptation in these tree species. They found that adaptation to climate is highly polygenic in both species. Surprisingly, they also found that a large number of the same genes were involved in local adaptation in both species, despite 140 million years of separate evolution. As genomic data become cheaper and easier to collect, integrative approaches such as this that link environmental, genomic, and phenotypic data will provide deeper insights into adaptation in species of conservation concern.

19.3 Genetic effects of habitat fragmentation

Habitat loss and fragmentation are arguably the gravest threats to biodiversity globally (Haddad et al. 2015; Crooks et al. 2017). Habitat fragmentation is predicted to isolate populations and decrease population sizes, which can reduce gene flow and increase genetic drift (Templeton et al. 1990; Young et al. 1996). Reduced gene flow and increased genetic drift may in turn cause a loss of genetic variation, inbreeding depression, and reduced evolutionary potential (Chapter 18). Habitat fragmentation occurs in a variety of terrestrial and aquatic ecosystems, and includes forest fragmentation, grassland fragmentation, fragmentation caused by roads and urbanization, and fragmentation of rivers by dams and water diversions (Grill et al. 2019). The degree of isolation caused by fragmentation depends on barrier permeability and how inhospitable the habitat matrix is that surrounds remnant habitat patches (Revilla et al. 2004).

Habitat fragmentation has resulted in genetic isolation and a reduction in genetic variation for many species. For example, the black-capped vireo is a neotropical migrant songbird that breeds in oak–juniper savannah from south-central Texas through southern Oklahoma, USA. Due primarily to habitat loss and fragmentation caused by agriculture and urbanization, the species suffered dramatic declines and was listed as endangered under the US Endangered Species Act (ESA) in 1991. Using a combination of historical and contemporary sampling,

Figure 19.8 The riverine habitat of the Macquarie perch in southeastern Australia has been severely fragmented by barriers to movement, reducing genetic variation and effective population sizes in most remaining populations. Photo courtesy of the Victorian Fisheries Authority.

Athrey et al. (2012) found lower genetic diversity and increased genetic differentiation in contemporary compared with historical samples, which the authors attribute to increasing fragmentation of the species' habitat. Fragmentation has also affected genetic diversity in riverine species. The Macquarie perch has lost 95% of its habitat in its range in southeastern Australia and the rivers it occupies have become increasingly fragmented by barriers to movement (Figure 19.8). Pavlova et al. (2017) found that genetic diversity in most remaining populations of the Macquarie perch is low, and that effective population sizes are too small to maintain adaptive potential. The authors argue that supplementation of smaller populations will be required to avoid inbreeding depression and maintain adaptive potential (Section 19.5).

Some studies have found evidence that habitat fragmentation has not only reduced genetic variation, but has also resulted in inbreeding depression. A long-term study of a remnant population of greater prairie-chickens in Illinois, USA, documented declines driven in part by inbreeding depression (Westemeier et al. 1998). During 35 years of monitoring, this population declined from 2,000 to 50 individuals. Genetic variation and fitness (measured as fertility and hatching success) declined concurrently. Similarly, Florida panthers (puma) are a classic example of an isolated population suffering from inbreeding depression (Johnson et al. 2010b; Section 19.5). Evidence now indicates that some puma populations in southern California are similarly becoming isolated by urbanization, once again reducing genetic variation and causing inbreeding depression (Guest Box 19).

Recent work also demonstrates that habitat fragmentation has led to a loss of evolutionary potential in some species. Belasen et al. (2019) took advantage of a naturally fragmented island system to test the effects of isolation on diversity at the immunity locus **major histocompatability complex (MHC)** IIB in the frog *Thoropa taophora*. They found that MHC diversity was lower than observed in mainland populations. Given previous evidence for heterozygous advantage at MHC (Edwards & Hedrick 1998), reduced genetic diversity at this locus may decrease the ability of island populations to resist disease and evolve increased disease resistance.

The authors also found that populations with lower MHC diversity had greater eukaryotic parasite species richness, lending support to this concern.

Thompson et al. (2019) also provide evidence that fragmentation and degradation of a river system by dams and other anthropogenic activities have led to rapid loss of adaptive variation in Chinook salmon in the Rogue and Klamath rivers in southern Oregon and northern California, USA. In these rivers, Chinook salmon have spring-run and fall-run phenotypes that are determined by a single locus (Prince et al. 2017). However, spring-run Chinook salmon have declined precipitously in response to dam construction, logging, mining, and water diversions that have negatively impacted habitat quality and restricted movement to streams higher in these river systems that spring-run Chinook salmon prefer. As a result of this selection against the spring-run phenotype, the frequency of the spring-run allele has declined dramatically, threatening the potential of the spring-run phenotype to increase in frequency after planned dam removals (Section 18.4).

Fortunately, a variety of management strategies—including habitat corridors, underpasses/overpasses, and supplementation—have proven effective at mitigating the negative demographic and genetic effects of habitat fragmentation. Sharma et al. (2013) examined genetic connectivity using a landscape genetic approach in tigers in four tiger populations in central India to test whether the forest corridors in this landscape effectively facilitate connectivity. They found the highest rates of contemporary gene flow in populations connected by forest corridors, demonstrating their importance for the persistence of these tiger populations (Guest Box 9). An increasing body of evidence also indicates that supplementation of small, inbred populations can ameliorate inbreeding depression and increase population sizes (Section 19.5).

Genetic tools can also inform the effectiveness of management actions designed to increase connectivity. Sawaya et al. (2014) tested whether grizzly bears and black bears used crossing structures along the Trans-Canada Highway constructed to facilitate movement. They found that both bear species used crossing structures and that 47% of black bears and 27% of grizzly bears that used crossings successfully bred. They concluded that these structures successfully facilitated gene flow that otherwise would have been limited by this highway. Fragmentation will likely remain one of the most serious demographic and genetic threats to biodiversity for the foreseeable future, and its impacts will be exacerbated by climate change.

19.4 Genetic versus demographic connectivity

Dispersal can contribute significantly to population growth rates, gene flow, and, ultimately, species persistence. It also plays a major role in determining the rate at which populations or species can shift their range in response to changing environmental conditions (Chapter 16). Therefore, assessing the effects of dispersal is crucial for understanding population biology and evolution in natural systems (Wright 1951; Hanski & Gilpin 1997; Clobert et al. 2001). Likewise, effective protection of endangered species and management of economically important species often rely on estimates of **population connectivity** (Mills & Allendorf 1996; Drechsler et al. 2003), a concept based on the dispersal of individuals among discrete populations, but which can have very different meanings and implications depending on how it is measured.

Demographically connected populations are those in which population growth rates (λ, r) or specific vital rates (survival and birth rates) are affected by immigration or emigration (Lowe & Allendorf 2010a). Demographic connectivity is generally thought to promote population stability or growth ($\lambda \geq 1.0$), and this stabilizing effect can occur at two different scales. In individual populations, demographic connectivity can promote stability by providing an immigrant subsidy that compensates for low survival or birth rates of residents (i.e., low local recruitment). Demographic connectivity can also promote the stability of metapopulations by increasing colonization of unoccupied patches (i.e., discrete subpopulations), even when the extinction rate of occupied patches is high (Levins 1970; Hanski 1998).

Genetic data are often used to assess population connectivity because it is often difficult to measure dispersal directly, especially at large spatial

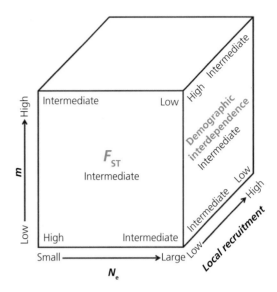

Figure 19.9 Combinations of migration rate (m), effective population size (N_e), and local recruitment resulting in different expected values of genetic divergence (F_{ST}) and demographic interdependence. Local recruitment is determined by births and deaths of resident individuals. Genetic divergence is primarily a function of mN, while demographic connectivity is primarily a function of m relative to local recruitment. From Lowe & Allendorf (2010b).

scales. As we saw in Chapter 9, however, genetic connectivity depends primarily on the absolute number (mN) of dispersers among populations (Equation 9.12). In contrast, demographic connectivity depends on the relative contributions to population growth rates of dispersal vs. local recruitment (i.e., survival and reproduction of residents; Hastings 1993). Therefore, estimates of genetic divergence alone provide little information on demographic connectivity (Figure 19.9; Lowe & Allendorf 2010a).

One way genetics can directly inform demographic connectivity is by providing dispersal estimates. Several genetic approaches have been developed for estimating dispersal (Cayuela et al. 2018). One genetic method for estimating dispersal is to use clustering and assignment methods that group individuals into discrete populations and assign individuals to one or more populations of origin (Sections 9.9.4 and 22.4). This allows identification of immigrants that are found in one population but assigned to another population. Population assignment methods can also identify offspring of immigrants (or later-generation descendants) to help estimate the amount of connectivity in current or recent generations (Rannala & Mountain 1997). Current dispersal rates and direction can be estimated using the assignment method of BayesAss (Wilson & Rannala 2003). Importantly, BayesAss was recently modified to use thousands of SNP loci, which increases power to identify demographically independent populations and management units (Section 20.5.3; Mussman et al. 2019).

Another genetic method for estimating dispersal uses parentage analysis and sibship reconstruction (Bode et al. 2018). By inferring parent–offspring relationships, it is possible to estimate natal dispersal distances by calculating the geographic distance between parent and offspring spatial positions. This method requires extensive sampling of all potential parents and offspring, and thus may be impractical for large populations. However, these and other genetic methods can complement field methods for estimating dispersal such as capture–mark–recapture analysis, which requires multiple years of capture–recapture data and high capture probabilities to estimate dispersal rates accurately.

19.5 Genetic rescue

Genetic rescue is a decrease in population extinction probability owing to gene flow, best measured as an increase in population growth rate (Thrall et al. 1998; Bell et al. 2019; Box 19.1). Genetic rescue is generally considered to occur when population growth rate or viability increases by more than can be attributed to just the demographic contribution of migrant individuals (Ingvarsson 2001). Genetic rescue can be caused by masking deleterious recessive alleles (reducing genetic load), heterozygous advantage, introduction of additive genetic variation on which selection can act, or a combination of these mechanisms. Genetic rescue may play a crucial role in the persistence of small natural populations and is an effective conservation tool under some circumstances (Tallmon et al. 2004b; Whiteley et al. 2015; Bell et al. 2019).

Box 19.1 Definitions of genetic rescue and related terms

Genetic rescue is a decrease in the extinction probability of a population caused by gene flow (Bell et al. 2019). It is best measured as an increase in population growth rate more than can be attributed to the demographic contribution of immigrants (Ingvarsson 2001; Box 19.2). Genetic rescue is typically attributed to the reduction of inbreeding depression by masking deleterious alleles. However, genetic rescue can also facilitate adaptation to changing environmental conditions by increasing additive genetic variation upon which selection can act. These mechanisms of genetic rescue will often co-occur in small populations that suffer both from inbreeding depression and maladaptation. Importantly, genetic rescue can occur "naturally" (e.g., when a puma moves from one population to another of its own accord; Guest Box 19) or it can be mediated by people as a management action.

Evolutionary rescue is defined as an adaptation-dependent reversal of demographic decline due to maladaptation to novel environmental conditions (Gonzalez et al.

2013; Carlson et al. 2014; Figure 16.1). Evolutionary rescue focuses on the beneficial demographic effects of adaptation to novel environmental conditions. Adaptation may occur due to selection on standing genetic variation or genetic variation introduced by mutation or gene flow. What distinguishes evolutionary rescue from genetic rescue is that: (1) evolutionary rescue emphasizes adaptation to new environmental conditions; and (2) evolutionary rescue can be caused by selection on standing genetic variation or new variation introduced by mutation or gene flow. In contrast, genetic rescue focuses on beneficial demographic effects caused by gene flow.

Another related term is **assisted gene flow**, which is the managed movement of individuals or gametes between populations within species ranges to facilitate adaptation to changing environmental conditions (Aitken & Whitlock 2013; Section 16.7.1). In small populations, assisted gene flow is useful for reducing inbreeding depression and becomes equivalent to genetic rescue.

Increasingly widespread evidence that genes from a pulse of immigrants into a local population often result in heterosis that increases population growth rates also has important implications for the study of evolution and metapopulation dynamics. However, the potential for **outbreeding depression** following heterosis in the first generation indicates that care is needed in considering the source of populations for rescue and evaluating potential risks (Byrne et al. 2011; Weeks et al. 2017). Genetic rescue is related to, but distinct from, other related terms (Box 19.1).

19.5.1 Evidence for genetic rescue

Multiple studies report positive fitness responses to gene flow into populations that have suffered recent demographic declines and suggest that natural selection can favor the offspring of immigrants (Box 19.2). Madsen et al. (1999) studied an isolated population of adders in Sweden that declined dramatically and has since suffered from severe inbreeding depression. The introduction of 20 males from a large and genetically variable population of adders resulted in a dramatic demographic recovery of this population. This recovery was brought about by increased survival rates, even though the number of litters produced by females per year actually declined during the initial phase of recovery.

Perhaps the most famous example of human-mediated genetic rescue is the case of the Florida panther (Johnson et al. 2010b). This subspecies is restricted to shrinking habitat between the urban centers of Miami and Naples, Florida, USA. The species was listed as endangered under the US ESA in 1967, and by the early 1990s, the population had declined to 20–25 adults and had reduced genetic variation compared with other puma populations. The population showed a variety of signs of inbreeding depression as well. Males had low sperm quality, low testosterone levels, and cryptorchidism (one or both testes fail to descend), and all individuals had a high incidence of heart defects, kinked tails, and high loads of parasites and infectious disease. The decision was made to attempt a human-mediated genetic rescue of Florida panthers. Eight female pumas from a different subspecies in Texas were translocated to the Florida panther population in 1995. The translocation was deemed a success, as Florida panther numbers increased threefold, heterozygosity doubled, survival and fitness increased, and inbreeding estimates decreased. Nonetheless, the Florida panther population remains small, isolated, and vulnerable to extinction. Thus, genetic management of this population will continue to be necessary for the foreseeable future (van de Kerk et al. 2019). Interestingly, natural dispersal of a single male puma into an inbred population of pumas in southern California also decreased inbreeding and increased heterozygosity, suggesting genetic rescue may play an important role in population persistence in small, isolated populations (Gustafson et al. 2017; Guest Box 19).

Several experimental studies have also provided evidence that gene flow can result in genetic rescue. In experimentally inbred populations of mustard, one immigrant per generation significantly increased fitness of four of six fitness traits in treatment populations compared with (no immigrant) control populations (Newman & Tallmon 2001). Interestingly, there was no fitness difference between one immigrant and an average of 2.5 immigrant treatments after six generations, but there was greater phenotypic divergence among populations in the one-migrant treatment, which could facilitate local adaptation in spatially structured populations subject to divergent selection pressures. Using flour beetles in a microcosm experiment, Hufbauer et al. (2015) found that genetic rescue increased population sizes and fitness substantially over the course of six generations. Gene flow into experimental populations of Trinidadian guppies in the wild and lab has also been shown to result in genetic rescue, even when immigrants are adaptively divergent from recipient populations (Kronenberger et al. 2018; Fitzpatrick et al. 2020; Example 19.2).

These and other examples suggest that genetic rescue may be of crucial importance to entire metapopulations by reducing local inbreeding depression and increasing the probability of local population persistence; in turn, this maintains a broad geographic range that buffers against metapopulation extinction and provides future immigrants for other populations. In long-established plant populations, it is conceivable that plants emerging from long-dormant seed banks could also provide an intergenerational genetic rescue.

19.5.2 Call for paradigm shift in use of genetic rescue

Recent calls have been made for a paradigm shift in the genetic management of small, isolated populations away from inaction and toward greater application of genetic rescue. Currently, widespread conservation practice is to manage populations in isolation to maintain genetic distinctiveness. Ralls et al. (2018) argue that if risk of outbreeding depression is low, then restoration of gene flow into small, inbred populations that have been isolated by human activities in the past 500 years using immigrants from large, noninbred source populations (Ralls et al. 2020) should be the default. Without supplementation, they argue, many fragmented populations are at high risk of extinction due to

inbreeding depression, maladaptation due to drift, and reduced ability to adapt to changing environmental conditions. They make the case that the benefits of restored gene flow usually far outweigh the risks of outbreeding depression and homogenization of locally adaptive genetic variation (Frankham et al. 2011). They also provide a framework for making evidence-based genetic management decisions for fragmented plant and animal populations that includes consideration of factors such as the degree of inbreeding, the availability of source populations, and the potential for outbreeding depression (Ralls et al. 2018).

Others have added that, although genetic rescue is not used as much as it should be as a management tool, many aspects of genetic rescue remain poorly understood (Bell et al. 2019). First, it is currently difficult to predict the magnitude of human-mediated genetic rescue, which will depend on many factors, including genetic load of the recipient population,

---

Box 19.2 Assessing the relative strength of different types of evidence for genetic rescue

When attempting a genetic rescue of a small, isolated population through translocation, it is important to monitor the population to assess whether the attempt was successful and to adjust management efforts as necessary. Different types of evidence vary in the strength of support they provide for genetic rescue (Bell et al. 2019). Since the ultimate goal of genetic rescue is to decrease a population's extinction risk, the best evidence for genetic rescue is an increase in population growth rate due to gene flow, as populations with higher growth rates are generally expected to have a lower probability of extinction. However, estimating population growth rates can be challenging.

The next best line of evidence for genetic rescue is an increase in lifetime reproductive success (number of offspring produced in an individual's lifetime) or a vital rate (e.g., survival, fecundity) to which population growth rate (λ) is highly sensitive in hybrids relative to residents (e.g., Hogg et al. 2006). Estimating lifetime reproductive success can also be challenging, as it will typically involve constructing a pedigree and inferring parentage for most individuals in the population. It is important to recognize that gene flow could increase relative fitness (of outbred individuals) but the population could still decline if environmental conditions

decline (e.g., drought, disease, habitat degradation). In this scenario, genetic rescue would be beneficial, but not sufficient, for population recovery, which can be important to forewarn or explain to wildlife managers.

The next strongest lines of evidence for genetic rescue are an increase in any component of fitness in hybrids relative to residents, or an increase in heterozygosity or migrant ancestry by more than is expected based on neutral expectations. Increased heterozygosity or migrant ancestry alone provides little evidence for genetic rescue, as it does not provide any information about the demographic consequences of gene flow. Gene flow will initially increase heterozygosity irrespective of whether genetic rescue or outbreeding depression occurs (Bell et al. 2019). Monitoring the effects of supplementation should occur over multiple generations and should focus on those metrics that provide the strongest evidence for genetic rescue. Robinson et al. (2020) showed that metrics providing the strongest evidence for rescue depend on the time since gene flow and if gene flow was beneficial or deleterious. Encouragingly, they found that using multiple metrics provides nonredundant information to improve power for evaluating effects of genetic rescue.

Example 19.2 Genetic rescue in experimental populations of Trinidadian guppies

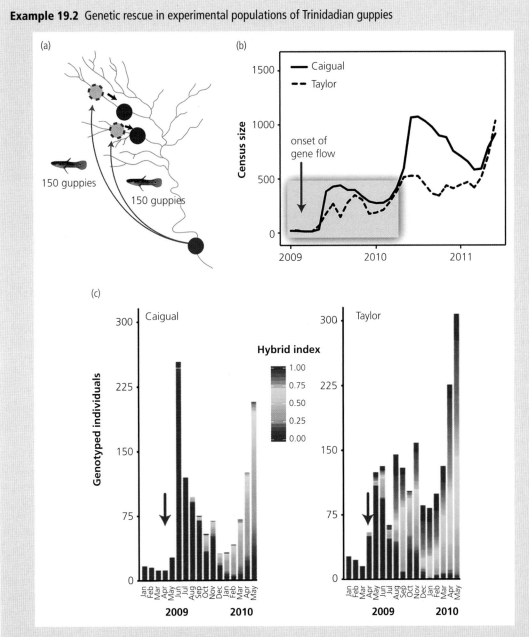

Figure 19.10 Gene flow manipulation experiment of Trinidadian guppies. (a) Map of the Guanapo River drainage, where experiments took place. In 2009, guppies were translocated from a downstream high-predation locality (red) into two headwater sites (dashed red) that were upstream of native recipient populations in low-predation environments (the Caigual and Taylor Rivers; dark blue). Downstream gene flow began shortly after the introduction (black arrows). (b) Population sizes in Caigual (solid line) and Taylor (dashed line) after the onset of gene flow from the upstream introduction sites. Gray box indicates the time period during which all captured individuals were genotyped at 12 microsatellite loci for parentage and hybrid index analyses. (c) Continuous hybrid index assignments based on microsatellite loci during the first 17 months of the study (approximately four to six guppy generations). Recipient populations prior to gene flow had a hybrid index = 0, and pure immigrant individuals had a hybrid index = 1. Red arrows indicate initiation of gene flow. From Fitzpatrick et al. (2020).

Continued

Example 19.2 *Continued*

A common problem in conservation biology is a lack of controlled, replicated experiments in which it is possible to attribute responses (e.g., changes in population growth rates) to treatments (e.g., gene flow) to conclusively link cause and effect (Tallmon 2017). Trinidadian guppies have been a powerful experimental system for testing the effects of gene flow on genetic variation, traits, and population dynamics (Fitzpatrick et al. 2016, 2020; Kronenberger et al. 2017, 2018). In Trinidad, translocation experiments allowed testing the effects of gene flow from a divergent source population into two small, isolated headwater guppy populations. Genetic rescue was documented in both streams, as demonstrated by sustained increases in population growth due to high hybrid fitness (Fitzpatrick et al. 2016; Figure 19.10). In addition, another study showed locally adaptive trait variation was maintained in several additional wild populations despite high levels of homogenizing gene flow (Fitzpatrick et al. 2015). Genomic data also showed that putative adaptive alleles were maintained despite gene flow from the divergent source population (Fitzpatrick et al. 2020).

Mesocosm experiments were also conducted mimicking the above-described field scenario, but also allowing additional replication, the capacity to manipulate the source of gene flow, and controls. In these experiments, immigrants from source populations with varying levels of neutral and adaptive differentiation were added to recipient populations with different amounts of genetic variation (Kronenberger et al. 2017, 2018). Gene flow increased population growth rates compared to controls without gene flow, particularly when the recipient population had low genetic variation.

These experimental studies with a model organism support the growing body of literature demonstrating that gene flow can increase fitness and population growth rates, particularly when recipient populations are small and inbred. Importantly, genetic rescue was observed even when immigrants were adaptively divergent from the recipient populations, suggesting that human-mediated genetic rescue should not automatically be ruled out if all potential source populations are adaptively differentiated from the recipient population. Nonetheless, when source populations are adaptively differentiated, it is particularly important to monitor at least to the second generation, which is when outbreeding depression is generally expected to become manifest if it occurs.

the degree to which increased hybrid fitness will translate into increased population growth rates, and the degree of divergence between the recipient and source populations. Second, the duration of human-mediated genetic rescue is a major outstanding question. Most genetic rescue studies have only monitored the fitness effects of gene flow for one to two generations when the beneficial effects of genetic rescue are expected to be maximal. A meta-analysis found that the beneficial effects of genetic rescue can extend into the third generation (Frankham 2016, 2018), but this study primarily relied on invertebrate laboratory studies, so the extent to which these results extend to other wild populations remains unknown. Third, when to expect outbreeding depression rather than genetic rescue is a lingering question.

Although a previous meta-analysis found limited evidence for outbreeding depression (Frankham et al. 2011), most of the studies only monitored fitness to the first or second generation. Thus, in summary, although evidence for the effectiveness of genetic rescue is increasing rapidly, these and other questions remain regarding under what circumstances and for how long genetic rescue will occur. Further research is needed to reduce uncertainty concerning genetic rescue. Researchers could collaborate with managers currently implementing genetic rescue to help improve the success of these attempts and to learn how to improve genetic rescue in the process, particularly including monitoring of multiple genetic and demographic metrics such as genetic variation (H_e), migrant ancestry (proportion of migrant alleles), abundance (census size), survival, and lifetime reproductive success, along with computer simulations to understand power to detect change in these metrics (Robinson et al. 2020).

19.5.3 Genomics and genetic rescue

Genomic approaches have tremendous potential to improve the planning and implementation stages of genetic rescue (Whiteley et al. 2015; Fitzpatrick & Funk 2020). One way in which genomics can help in the planning stages is by aiding in the identification of populations with high inbreeding coefficients and evidence of inbreeding depression that may benefit from genetic rescue. Importantly, individual inbreeding can now be estimated with high precision with or without a reference genome, as outlined in Chapter 17. The use of ~5,000 SNP loci provides more precise individual inbreeding estimates than 10 generations of complete pedigree information (Kardos et al. 2016b).

Another way in which genomics can assist in the planning stages is by helping identify appropriate source populations that are not too adaptively divergent from the recipient population and that have not been isolated from the recipient population for too long. Outbreeding depression is predicted to be most likely when crossing populations with many fixed chromosomal differences, populations that have been isolated for millions of years, or populations that are highly adaptively divergent.

Genomics can help test all of these factors (Fitzpatrick & Funk 2020). For example, chromosomal inversions can be mapped with a reference genome, and can even be detected when a full reference genome is not available (Kemppainen et al. 2015). Timing of divergence can also be estimated using identical by descent tracts (Pool et al. 2010; Gravel 2012) or based on the joint **site frequency spectrum**, which is the distribution of allele frequencies across polymorphic sites (Gutenkunst et al. 2009; Section 4.1.4). Finally, adaptive divergence can be tested for using adaptive landscape genomic approaches, as outlined in Section 19.2.4.

After augmenting gene flow, genomics can be used to monitor the outcome of genetic rescue and assess if and when an additional translocation is needed. Genomic monitoring (Chapter 23) can answer multiple questions about the outcome of a genetic rescue attempt (Fitzpatrick & Funk 2020). For example, genomics can increase power to test whether immigrants have interbred with residents, a necessary prerequisite for genetic rescue to occur. If interbreeding occurs between immigrants and residents, genomics can then be used to assess which immigrant individuals and alleles have the greatest effects on fitness, which can inform future genetic rescue attempts. For example, Miller et al. (2012) used a genomic approach to test the effects of supplementing an isolated population of bighorn sheep at the National Bison Range in Montana, USA. They found that individuals with a greater proportion of immigrant alleles had higher fitness. They also identified 30 loci that had effects on fitness above and beyond that predicted by overall levels of introgression. Lastly, genomics should be used to monitor changes in genetic variation and inbreeding after release of immigrants, given that eventually, genetic variation is likely to decrease after genetic rescue if the recipient population remains isolated or small. Genomic monitoring of adaptive alleles can also help ensure that locally adaptive alleles persist after translocation into a population (Fitzpatrick et al. 2020). Similarly, monitoring harmful alleles can be used to determine if they increase in frequency due to genetic drift (or linkage to beneficial alleles that are positively selected) following rescue, or if they are purged (e.g., Grossen et al. 2020). Appropriate application of genomics to planning and assessment of genetic rescue attempts promises to improve the success of this important management tool. Guidelines for evaluating outcomes of genetic rescue attempts, including use of computer simulations, are given by Robinson et al. (2020) and Ralls et al. (2020).

19.6 Long-term viability of metapopulations

There is sometimes confusion regarding when to address short- versus long-term genetic goals, and how they relate to the conservation of local populations versus entire species (Jamieson & Allendorf 2012; Section 18.8). Short-term goals are appropriate for the conservation of local populations. As indicated above, those goals are aimed at keeping the rate of inbreeding at a tolerable level. The effective population sizes at which this may be achieved (e.g., $N_e \geq 50$), however, are typically not large enough for new mutations to compensate for the loss of genetic variation through genetic drift. Some gene flow from neighboring populations is necessary to provide reasonable levels of genetic variation for

Example 19.3 Metapopulation structure and long-term productivity and persistence of sockeye salmon

Complex genetic population structure can play an important role in the long-term viability of populations and species. Sockeye salmon within major regions generally consist of hundreds of discrete or semi-isolated individual local demes (Hilborn et al. 2003). The amazing ability of sockeye salmon to return and spawn in their natal spawning sites results in substantial reproductive isolation among local demes. Local demes of sockeye salmon within major lake systems generally show pairwise F_{ST} values of 0.10–0.20, indicating relatively little gene flow.

These local demes occur in a variety of different habitats, which, combined with the low amount of gene flow, results in a complex of many locally adapted populations. Sockeye salmon spawning in tributaries to Bristol Bay, Alaska, display a wide variety of life history types associated with different breeding and rearing habitats. Bristol Bay sockeye salmon spawn in streams and rivers from 10 cm to several meters deep in substrate ranging from small gravel to cobble. Some streams have extremely clear water while others spawn in sediment-laden streams just downstream from melting glaciers. Sockeye salmon also spawn on the beaches in lakes with substantial groundwater. Different demes spawn at different times of the year. The date of spawning is associated with the long-term average thermal regime experienced by incubating eggs so that fry emerge in the spring in time to feed on zooplankton and aquatic insects. Fish from different demes have a variety of morphological, behavioral, and life history differences associated with this habitat complexity.

Up to 40 million fish are caught each year in the Bristol Bay sockeye fishery in several fishing areas associated with different major tributaries. There is large year-to-year variability in overall productivity, but the range of the productivity of this fishery has been generally consistent for nearly 100 years (Figure 19.11). However, the productivity of different demes and major drainage areas has changed dramatically over the years. The relative productivity of local demes has changed as the marine and freshwater climates change. Local reproductive units that are minor components of a mixed stock fishery during one climatic regime may dominate during others. Therefore, maintaining productivity over long time scales requires protection against the loss of local populations during certain environmental regimes (Schindler et al. 2010). US legislation in 2020 protected most of the Bristol Bay drainages from development, which is significant given the world's largest sockeye salmon fishery resides here.

The long-term stability of this complex system stands in stark contrast to the dramatic collapse and extirpation of a highly productive population of an introduced population of this species in the Flathead River drainage of Montana (Spencer et al. 1991). The life history form of this species that spends its entire life in freshwater is known as kokanee. Sockeye salmon were introduced into Flathead Lake in the early 20th century, and by the 1970s some 50,000–100,000 fish returned to spawn in one primary local population and supported a large recreational fishery. Opossum shrimp were introduced into Flathead Lake in 1983 and had a major effect on the food web in this ecosystem. A primary effect was the predation of opossum shrimp on the zooplankton that was the major food resource of the kokanee. The introduced shrimp also fed a population explosion of a voracious predator, lake trout. This productive single deme of kokanee went from over 100,000 spawners in 1985 to extirpation just 3 years later. This example illustrates the importance of maintaining multiple semi-isolated subpopulations with different life histories to help ensure long-term population and species persistence.

Continued

Example 19.3 *Continued*

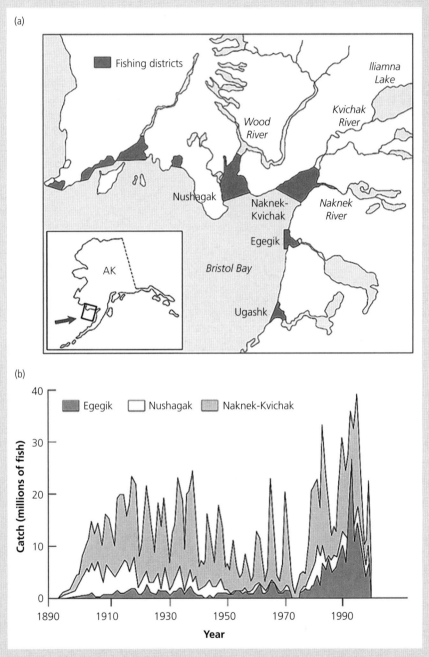

Figure 19.11 (a) Map of fishing districts (dark areas) around Bristol Bay, Alaska. (b) Catch history of the three major sockeye salmon fishing areas within Bristol Bay. The overall productivity of the system has been generally stable, but the relative contributions of the three major areas have changed greatly. For example, the Egegik district (see map) generally contributed less than 5% to the fishery until 1975, but has been a major contributor since then. From Hilborn et al. (2003).

quantitative traits to ensure long-term population persistence (Section 18.7).

The long-term viability of a metapopulation, or species, is influenced by the number, connectivity, and complexity of the subpopulations. Metapopulation viability can be increased by maintaining a number of populations across multiple, diverse, and semi-independent environments, as illustrated in Example 19.3. Local and metapopulation viability also can be increased by maintaining limited gene flow that prevents excessive loss of alleles without overwhelming natural selection for local adaptation (e.g., Figure 9.20).

The accumulation of mildly deleterious mutations as considered in Section 18.6 may also affect the long-term viability of metapopulations (Higgins & Lynch 2001). Under some circumstances, metapopulation dynamics can reduce the effective population size so that even mutations with a selection coefficient as high as $s = 0.1$ can behave as nearly neutral and cause the erosion of long-term metapopulation viability.

The long-term goal, where the loss of variation is balanced by new mutations, refers primarily to a global population, which may coincide with a species or subspecies that cannot rely on the input of novel genetic variation from neighboring populations. This global population may consist of one more or less panmictic unit, or it may be composed of multiple subpopulations that are connected by some gene flow, either naturally or through translocations (Mills & Allendorf 1996). It is the total assemblage of interconnected subpopulations forming a global population that must have an effective size meeting the criteria for long-term conservation (e.g., $N_e \geq 500$–1,000). The actual size of this global population will vary considerably from species to species depending on the number and size of the constituent subpopulations and on the pattern of gene flow between them (Waples 2002).

Guest Box 19 The eroding genomes of fragmented urban puma populations in California
Kyle D. Gustafson and Holly B. Ernest

Dense urbanization has had harmful effects on species in the southwestern USA, including an apex predator species, the puma (also commonly called cougars or mountain lions; Figure 19.12). Genetic research in 2003 suggested human development was a direct threat to puma population connectivity in California and called for puma management plans to address genetic threats faced by some populations (Ernest et al. 2003). It was not until 2014, however, that the effects of anthropogenic fragmentation on pumas became abundantly evident, particularly north of Los Angeles, in the Santa Monica Mountains (Riley et al. 2014) and between Los Angeles and San Diego, in the Santa Ana Mountains (Ernest et al. 2014). In both regions, freeway traffic was killing and isolating pumas (Riley et al. 2014; Vickers et al. 2015; Figure 19.12), and gene flow was limited to a single immigrant into each region, P12 (male puma 12) in the Santa Monica Mountains and M86 (male puma 86) in the Santa Ana Mountains. Although inbreeding depression is difficult to demonstrate, observation of kinked tails in the Santa Ana Mountains provided some phenotypic evidence of potential inbreeding depression in these populations (Ernest et al. 2014).

Detailed analyses of the impacts of immigrants P12 and M86 showed some degree of natural genetic rescue, including increased heterozygosity and reduced inbreeding coefficients, but little change in allelic richness (Riley et al. 2014; Gustafson et al. 2017; Figure 19.13). Unfortunately, the pool of alleles remains small in these two populations and without continued immigration, continued inbreeding, increased expression of deleterious alleles, and the progression of extinction vortices are predicted to ensue (Benson et al. 2016, 2019). A statewide analysis indicated that these two populations have the lowest genetic diversity in the state, but that other fragmented populations are also at risk (Gustafson et al. 2019).

The first puma genome sequence (PumCon 1.0) has now been published (Saremi et al. 2019). Until 2019, puma genetic research in California has relied entirely on traditional genetic markers (e.g., mitochondrial DNA and microsatellites). However, the puma reference genome will allow researchers to quantify inbreeding, inbreeding depression, and the effects of gene flow more accurately. Although only four pumas from California have had their genomes sequenced, the pumas from urban California that were sequenced have long runs of homozygosity, which are most likely explained by recent inbreeding (Saremi et al. 2019).

To further inform population conservation efforts, we are collaborating with the California Department of Fish and Wildlife and others on a project to genotype hundreds of individuals throughout the state of California at thousands of SNPs mapped to the puma reference genome. Our ultimate goal is to characterize genome sequence variation, chromosome structural variation, neutral variation, and functional variation for each puma population in California to guide their conservation. As genome sequencing continues to decrease in price, conservation geneticists will continue to expand their understanding of the relationships between genome-wide variation and associated inbreeding depression, fitness, and natural selection. Further, characterizing genomes will help identify source populations for genetic rescue efforts that minimize the likelihood of outbreeding depression.

(a)

(b)

Figure 19.12 (a) Aerial view of Interstate-15, which is the major barrier to puma movement to and from the Santa Ana Mountains in southern California (photo courtesy of Patrick Huber). (b) Male puma in the Santa Ana Mountains (photo courtesy of Courtney Aitken).

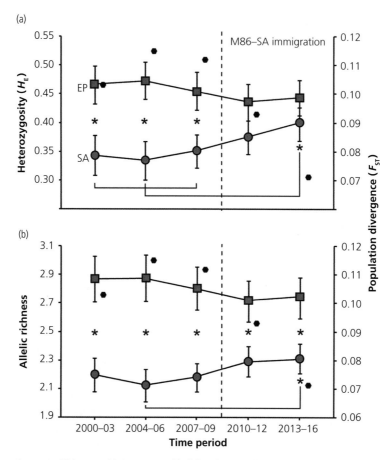

Figure 19.13 Temporal patterns of (a) expected heterozygosity, (b) allelic richness, and population divergence (F_{ST}: black hexagons; right y-axis) for the Eastern Peninsular Range (EP) and Santa Ana (SA) puma populations before (left of vertical dashed line) and after (right of the vertical dashed line) the migration of M86 into SA from EP. Asterisks indicate significant differences between the two populations. From Gustafson et al. (2017).

CHAPTER 20

Conservation Units

Flatwoods salamander, Section 20.7

The overriding purpose of defining ESUs is to ensure that evolutionary heritage is recognized and protected and that the evolutionary potential inherent across the set of ESUs is maintained.

(Craig Moritz 1994, p. 373)

At the end of the 20th Century the small rumblings associated with the use of DNA sequences in systematic studies turned into a major volcanic eruption that rapidly spread over the landscape.

(Vicki A. Funk 2018, p. 175)

The identification of appropriate taxonomic and population units for protection and management is essential for the conservation of biological diversity. For species identification and classification, genetic principles and methods are relatively well developed; nevertheless, species identification can be controversial. Within species, the identification and protection of genetically distinct local populations should be a major focus in conservation because the conservation of many distinct populations is crucial for maximizing evolutionary potential and minimizing extinction risks (Hughes et al. 1997; Hilborn et al. 2003; Luck et al. 2003). Furthermore, the local population is often considered the functional unit in ecosystems (Luck et al. 2003; Des Roches et al. 2021). For example, the US Marine Mammal Protection Act (MMPA) seeks to maintain populations as functioning elements of their ecosystem (MMPA Regulations, 50 CFR 216).

Identification of population units is necessary so that management and monitoring programs can be efficiently targeted toward distinct or independent populations. Biologists and managers must be able to identify populations and geographic boundaries between populations in order to effectively plan harvesting quotas, avoid overharvesting in a population or area, and to devise translocations and reintroductions of individuals to prevent mixing of adaptively differentiated populations (Banes et al. 2016). In addition, it is sometimes necessary to prioritize among population units (or taxa) to conserve, because limited financial resources preclude conservation of all units (Ryder 1986).

Finally, many governments and agencies have established legislation and policies to protect intraspecific population units. This requires the identification of population units. For example, the US Endangered Species Act (ESA) allows listing and full protection of **distinct population segments (DPSs)** of vertebrate species (Box 20.1). Other countries also have laws that depend upon the identification of distinct taxa and populations for the protection of species and habitats (Box 20.1). Species and subspecies identification is based upon traditional, established taxonomic criteria as well as genetic criteria, although the criteria for

Conservation and the Genomics of Populations, Third Edition. Fred W. Allendorf, *et al.*, Oxford University Press.

Box 20.1 The US ESA and conservation units

The US ESA is one of the most powerful pieces of conservation legislation ever enacted. It has been a major stimulus motivating biologists to develop criteria for identifying intraspecific population units for conservation. This is because the ESA provides legal protection for subspecies and DPSs of vertebrates, as if they were full species. According to the ESA:

The term "species" includes any subspecies of fish or wildlife and plants, and any distinct population segment of any species of vertebrate fish or wildlife which interbreeds when mature.

However, the ESA does not provide criteria or guidelines for delineating DPSs. The identification of intraspecific units for conservation is controversial. This is not surprising given that the definition of a "good species" is controversial (Section 20.5). Biologists have vigorously debated the criteria for identifying DPSs and other conservation units ever since the US Congress extended full protection of the ESA to "distinct" populations, but did not provide guidelines.

The IUCN Red List allows for the separate assessment of geographically distinct populations. These subpopulations are defined as "geographically or otherwise distinct groups in the population between which there is little demographic or genetic exchange" (typically one successful migrant individual or gamete per year or less; IUCN 2001, p. 10).

Legislation in other countries around the world has provisions that recognize and protect intraspecific units of conservation (Waples et al. 2013). For example, Canada passed the Species at Risk Act (SARA) in 2002 (https://laws-lois.justice.gc.ca/eng/acts/s-15.3/). The SARA aims to "prevent wildlife species from becoming extinct, and to secure the necessary actions for their recovery." Under the SARA, "wildlife species" means a "species, subspecies, variety or geographically or genetically distinct population of animal, plant or other organism, other than a bacterium or virus, which is wild by nature."

In Australia, the Environment Protection and Biodiversity Conservation Act of 1999 (EPBC Act) also allows protection for species, subspecies, and distinct populations. But, like the US ESA, there are challenges with defining and identifying intraspecific units (Woinarski & Fisher 1999). Unlike the ESA and SARA, the EPBC Act also recognizes and allows protection of ecological communities (an assemblage of native species that inhabits a particular area in nature). In South Africa, the National Environmental Management: Biodiversity Act (NEMBA) provides protection for "ecosystems, species, and 'categories of species' such as subspecies" (NEMBA 2004).

---

species identification are sometimes controversial. The criteria for delineating intraspecific units for conservation have been highly controversial.

New genomic approaches will improve our ability to delineate conservation units in two ways (Funk et al. 2012b). First, the thousands to millions of loci provided by genomics will increase statistical power to detect and quantify population structure. Second, genomics will allow characterization of adaptive differences among populations for nonmodel species. Nonetheless, many questions remain about how best to apply genomics to the delineation of conservation units, particularly information on adaptive variation (Waples & Lindley 2018; Section 20.6.1).

In this chapter, we examine the components of biodiversity and then consider methods to assess taxonomic and population relationships. We discuss the criteria, difficulties, and controversies in the identification of conservation units. We also consider the identification of appropriate population units for legal protection and for management actions such as supplemental translocations of individuals between geographic regions. Recall that in the previous chapter, we considered three spatial scales of genetic population structure for conservation: local population, metapopulation, and species.

20.1 What are we trying to protect?

Genes, species, and ecosystems are three primary levels of biodiversity (Figure 20.1) recognized by the Convention on Biological Diversity (CBD). There has been some controversy as to which level should receive priority for conservation efforts (e.g., Bowen 1999). However, it is clear that all three levels must be conserved for successful conservation of biodiversity. For example, it is as futile to

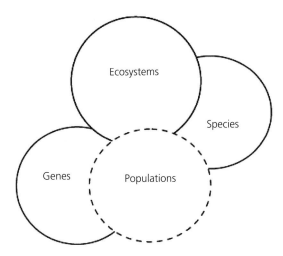

Figure 20.1 Primary levels of biodiversity recognized by the IUCN (solid circles), and a fourth level (populations)–recognized as perhaps most crucial for the long-term persistence of species. In reality, biodiversity exists across a continuum of many hierarchical levels of organization including genes, genomes (i.e., multilocus genotypes), local populations (dashed line), communities, ecosystems, and biomes. Additional levels of diversity include metapopulations, subspecies, genera, families, and so on.

conserve ecosystems without species, as it is to save species without large, healthy ecosystems. Nevertheless, an analysis found that genetic variation in wild animals and plants was generally not included in national plans to implement the CBD (Laikre et al. 2010a). In fact, fewer than 50% of the national plans that were reviewed included the goal of conserving genetic variation of wild populations. Other analyses suggest genetic factors are not sufficiently considered in conservation assessments or policy in general (Pierson et al. 2016; Garner et al. 2020).

An example of this kind of futility is that of the African rhinoceros, which is being protected, mainly in zoos and small nature reserves, but for which little habitat (free from poachers) is currently available. Without conserving vast habitats for future rhino populations, it seems pointless to protect rhinos in small nature reserves surrounded by armed guards and fences.

It is not too late for rhinos. Vast habitats do exist, and rhinos could be successful in these habitats if poaching is eliminated, which will require societal and policy changes in addition to biological

knowledge. In addition to conserving rhino species and their habitats, it is also important to conserve genetic variation within rhino species because variation is a prerequisite for long-term adaptive change and the avoidance of fitness decline through inbreeding depression. Clearly, it is important to recognize and conserve all levels of biodiversity: ecosystems, species, and genes.

The debate over whether to protect genes, species, or ecosystems is, in a way, a false trichotomy because each level is an important component of biodiversity as a whole. Nevertheless, considering each level separately can help us appreciate the interacting components of biodiversity, and the different ways that genetics can facilitate conservation at different levels. Appreciation of each level can also promote understanding and multidisciplinary collaborations across research domains. Finally, a fourth level of biodiversity—that of genetically distinct local populations—is arguably the most important level for focusing conservation efforts (Figure 20.1). The conservation of multiple genetically distinct populations is necessary to ensure long-term species survival and the stable functioning of ecosystems (Luck et al. 2003; Des Roches et al. 2021; Example 19.3).

We can also debate which temporal component of biodiversity to prioritize for conservation: past, present, or future biodiversity. All three components are important, although future biodiversity often warrants special concern (Box 20.2).

Another choice that is often debated is whether we should emphasize protecting the existing patterns of diversity or the processes that generate diversity (e.g., ecological and evolutionary processes themselves). Again the answer is, in general, both. It is clear that we should prioritize the preservation of the process of adaptation so that populations and species can continually adapt to future environmental changes. However, one important step toward preserving natural processes is to quantify, monitor, and maintain natural patterns of population subdivision and connectivity; for example, to identify intraspecific population units, boundaries, and corridors for dispersal in current and future environments. This would prevent extreme fragmentation and promote continued natural patterns of gene flow among populations.

Box 20.2 Temporal considerations in conservation: past, present, and future

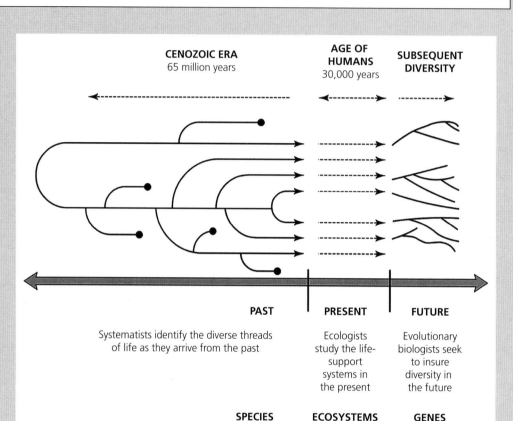

Figure 20.2 The temporal framework (past, present, and future), the corresponding disciplines (systematics, ecology, and evolutionary biology), and the levels of biodiversity (species, ecosystems, and genes) that are often considered when prioritizing biodiversity for conservation. Modified from Bowen (1999).

What temporal components of biodiversity do we wish to preserve? Do we want to conserve ancient isolated lineages, current patterns of diversity (ecological and genetic), or the diversity required for future adaptation and for novel diversity to evolve? Most would agree "all of the above." All three temporal components are interrelated and complementary (Figure 20.2). For example, conserving current diversity helps ensure future adaptive potential. Similarly, conserving and studying ancient lineages ("living fossils") can help us understand factors important for long-term persistence. Nevertheless, one can argue that the most impor-

tant temporal component to consider is future biodiversity: the ability of species and populations to adapt to future environments, for example, following global climate change. If populations do not adapt to future environments, then biodiversity will decline, leading to loss of ecosystem functioning and services. Figure 20.2 illustrates how different temporal components of biodiversity (past, present, and future) can be related to different scientific disciplines (systematics, ecology, and evolutionary biology, respectively). These components are also often related to different hierarchical levels of biodiversity: species, ecosystems, and genes, respectively.

How do we conserve the processes of evolution, including adaptive evolutionary change? We must first maintain healthy habitats and large wild populations, because only in large populations can natural selection proceed efficiently (Section 8.6). In small populations, genetic drift leads to random genetic change, which is generally nonadaptive. Drift can preclude selection from maintaining beneficial alleles and eliminating deleterious ones. To maintain evolutionary processes, we must also preserve multiple populations—ideally from different environments, so that selection pressures remain diverse and multilocus genotype diversity remains high. In this scenario, a wide range of local adaptations are preserved within species, as well as some possibility of adaptation to different future environmental challenges.

20.2 Systematics and taxonomy

The description and naming of distinct taxa are essential for most disciplines in biology. In conservation biology, the identification of taxa (taxonomy) and assessing their evolutionary relationships (systematics) is crucial for the design of efficient strategies for biodiversity management and conservation. For example, failing to recognize the existence of a distinct and threatened taxon can lead to insufficient protection and subsequent extinction. Identification of too many taxa (oversplitting) can waste limited conservation resources.

There are two fundamental aspects of evolution that we must consider: phenotypic change through time (**anagenesis**), and the branching pattern of reproductive relationships among taxa (**cladogenesis**). The two primary taxonomic approaches are based on these two aspects.

Historically, taxonomic classification was based primarily upon phenotypic similarity (**phenetics**), which reflects evolution via anagenesis; that is, groups of organisms that were phenotypically similar were grouped together. This classification is conducted using clustering algorithms that group organisms based exclusively on overall similarity. For example, populations that share similar morphological features or allele frequencies are grouped together into one species. In this case, the clustering by overall similarity of allele frequencies is phenetic. The resulting diagram (or tree) used to illustrate classification is called a **phenogram**, even if based upon genetic data such as allele frequencies.

A second approach is to classify organisms on the basis of their phylogenetic relationships (**cladistics**). Cladistic methods group together organisms that share **derived** traits (originating in a common ancestor), reflecting cladogenesis. Under cladistic classification, only **monophyletic** groups can be recognized, and only genealogical information is considered. The resulting diagram (or tree) used to illustrate relationships is called a **cladogram** (or sometimes, a **phylogeny**). Phylogenetics is discussed in Section 20.3.

Our current system of taxonomy combines cladistics and phenetics, and it is sometimes referred to as evolutionary classification (Mayr 1981). Under evolutionary classification, taxonomic groups are usually classified on the basis of phylogeny. However, groups that are extremely phenotypically divergent are sometimes recognized as separate taxa even though they are phylogenetically related. A good example of this is birds (Figure 20.3). Birds were derived from a dinosaur ancestor, as evidenced from the fossil record showing reptiles with feathers (bird–reptile intermediates; Prum 2003). Therefore, birds and dinosaurs are sister groups that should be classified together under a strictly cladistic classification scheme. However, birds underwent rapid evolutionary divergence associated with their development of flight. Therefore, birds are classified as a separate class while dinosaurs are classified as a reptile (class Reptilia). Sometimes in the literature dinosaurs are now referred to as nonavian dinosaurs, as a reminder of these relationships (Erickson et al. 2006).

There is a great deal of controversy associated with the correct method of classification. Effective classification makes use of all information available: morphology, physiology, behavior, life history, geography, parasite distributions, and genetics. The use of multiple lines of evidence is advocated in integrative taxonomy (Schlick-Steiner et al. 2010), although this can be challenging and using multiple lines of evidence in an iterative manner to test species boundaries in a hypothesis-driven

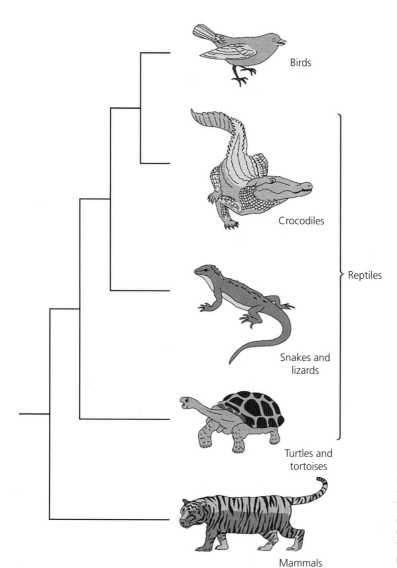

Figure 20.3 Phylogenetic relationships of birds, mammals, and reptiles. Note that crocodiles and birds are more closely related to each other than either is to other reptiles. That is, crocodiles share a more recent common ancestor with birds than they do with snakes, lizards, turtles, and tortoises. Therefore, the class Reptilia is not monophyletic.

context has also been proposed (Yeates et al. 2011).

20.3 Phylogeny reconstruction

A phylogenetic tree is a pictorial summary that illustrates the pattern and timing of branching events in the evolutionary history of taxa (Figure 20.4). A phylogenetic tree consists of **nodes** for the taxa being considered, and branches that connect taxa and show their relationships. Nodes are at the tips of branches and at branching points representing

extinct ancestral taxa (i.e., internal and ancestral nodes). A phylogenetic tree represents a hypothesis about relationships that is open to change as more taxa or characters are added. The same phylogeny can be drawn many different ways. Branches can be rotated at any internal node without changing the relationship between the taxa, as illustrated in Figure 20.5.

Branch lengths are often proportional to the amount of genetic divergence between taxa. If the amount of divergence is proportional to time, a phylogeny can show time since divergence between

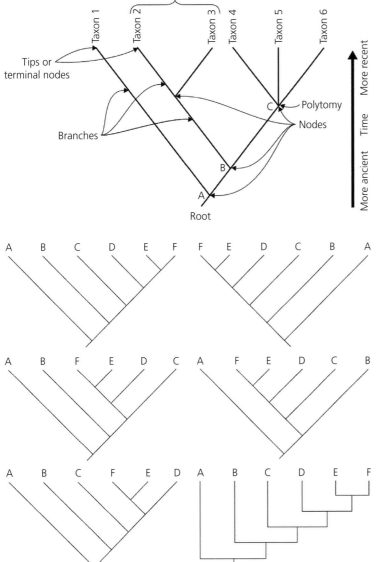

Figure 20.4 A rooted phylogenetic tree (phylogeny). A **polytomy** (node "C") occurs when more than two taxa are joined at the same node because data cannot resolve which shared a more recent common ancestor. From Freeman & Herron (1998).

Figure 20.5 Six rooted phylogenetic trees showing identical phylogenetic relationships among taxa. Branches can be rotated at nodes without changing the relationships represented on the trees. From Freeman & Herron (1998).

taxa. Molecular divergence (through mutation and drift) will be proportional to time if mutation accumulation is stochastically constant (like radioactive decay). The idea that molecular divergence can be constant is called the **molecular clock** concept. In conservation biology, the molecular clock and divergence estimates can help identify distinct populations and prioritize them based on their distinctiveness or divergence times. One serious problem with estimating divergence times is that extreme genetic drift, such as founder events and bottlenecks, can greatly inflate estimates of divergence times, leading to long branch lengths and misleading estimates of phylogenetic distinctiveness.

Phylogenetic trees can either be rooted or unrooted. **Rooted phylogenetic trees** show ancestral relationships and evolutionary history among taxa. Such trees are rooted by the common ancestor to all taxa (Figure 20.4). A rooted tree therefore shows the direction of evolutionary time. Accurate rooting of a phylogenetic tree is an important and crucial factor since inaccurate rooting can result in wrong interpretations of genetic changes between organisms and the directionality of evolution.

Unrooted phylogenetic trees only show the relatedness of organisms without showing ancestral relationships. This type of a tree does not indicate the origin of evolution of the groups of interest. It depicts only the relationship between organisms irrespective of the direction of the evolutionary timeline. Therefore, it is difficult to study the evolutionary relationships of the groups with respect to time using an unrooted tree. Figures 20.9, 20.10, 20.14, and 20.21 all show unrooted trees.

20.3.1 Methods

There are two basic steps in phylogeny reconstruction: (1) generate a matrix of character states (e.g., derived versus ancestral states); and (2) build a tree from the matrix. Cladistic methods use only shared derived traits, **synapomorphies**, to infer evolutionary relationships. Phenogram construction is based on overall similarity. Therefore, a phylogenetic tree may have a different topology from a phenogram using the same character state matrix (Example 20.1).

The actual construction of phylogenies is much more complicated than this simple example. It is sometimes difficult to determine the ancestral state of a character. Moreover, the number of possible evolutionary trees rises at an alarming rate with the number of taxa. For example, there are nearly 35 million possible rooted, bifurcating trees with just 10 taxa, and over 8×10^{21} possible trees with 20 taxa! In addition, there are a variety of other methods besides parsimony for inferring phylogenies (Hall 2004). The field of inferring phylogenies has been marked by more heated controversy than perhaps any other area of evolutionary biology (Felsenstein 2004).

20.3.2 Gene trees and species trees

It is important to recognize that different genes have different phylogenies, and that gene trees are often different from the true species phylogeny (Nichols 2001). Different gene phylogenies can arise due to four main phenomena: **lineage sorting** and associated genome sampling error, sampling error of individuals or populations, natural selection, or introgression following hybridization. Thus, many independent genes or DNA sequences should be used when assessing phylogenetic relationships (Wiens et al. 2010). Genomic methods now deliver large amounts of high-quality data for building phylogenies.

20.3.2.1 Lineage sorting and sampling error

Ancestral lineage sorting occurs when different DNA sequences from a mother taxon are sorted into different daughter species such that lineage divergence times do not reflect population divergence times. For example, two divergent lineages can be sorted into two recently isolated populations, where less-divergent lineages might become fixed in different ancient daughter populations. Lineage sorting makes it important to study many independent DNA sequences, to avoid sampling error associated with sampling too few or an unrepresentative set of genetic characters (loci).

Sampling error of individuals occurs when too few individuals or nonrepresentative sets of individuals are sampled from a species, such that the inferred gene tree differs from the true species tree. For example, many early studies using **mitochondrial DNA (mtDNA)** analysis included only a few individuals per geographic location, which could lead to erroneous phylogeny inference. Limited sampling is likely to detect only a subset of local lineages (i.e., alleles), especially when some lineages exist at low frequency.

We can use simple probability to estimate the sample size that we need to detect a rare lineage (haplotype) or allele. For example, how many individuals must we sample to have a greater than 95% chance of detecting an allele with frequency of 0.10 ($p = 0.10$)? Each time we examine one sample, we have a 0.90 chance ($1 - p$) of not detecting the allele

Example 20.1 Phenogram and cladogram of birds, crocodiles, and lizards

As we have seen, birds and crocodiles are sister taxa based upon phylogenetic analysis, but crocodiles are taxonomically classified as reptiles because of their phenetic similarity with snakes, lizards, and turtles. These conclusions are based on a large number of traits. Here we will consider five traits in Table 20.1 to demonstrate how a different phenogram and cladogram can result from the matrix of character states.

Table 20.1 Character states for five traits used to construct a phenogram and cladogram of lizards, crocodiles, and birds. Traits: 1, heart (three- or four-chambered); 2, inner ear bones (present or absent); 3, feathers (present or absent); 4, wings (present or absent); and 5, hollow bones (present or absent).

	Traits*				
Taxon	1	2	3	4	5
A Lizards	0	0	0	0	0
B Crocodiles	1	1	0	0	0
C Birds	1	1	1	1	1

* 0, ancestral; 1, derived.

Lizards and crocodiles are more phenotypically similar to each other than either is to birds because they share three out of five traits (0.60), while crocodiles and birds share just two out of five traits (0.40). Thus, the following phenotypic similarity matrix results (Table 20.2). We can construct a phenogram based upon clustering together the most phenotypically similar groups (Figure 20.6a). The phenotypic similarity of lizards and crocodiles results from their sharing ancestral character states because of the rapid phenotypic changes that occurred in birds associated with adaptation to flight.

Table 20.2 Phenotypic similarity matrix for lizards, crocodiles, and birds based upon proportion of shared characters states in Table 20.1.

	Lizards	Crocodiles	Birds
Lizards	1.0		
Crocodiles	0.6	1.0	
Birds	0.0	0.4	1.0

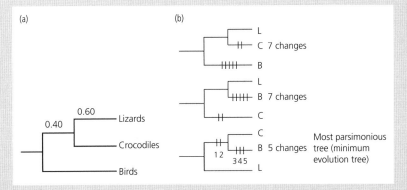

Figure 20.6 (a) Phenogram and (b) cladograms showing phenotypic and evolutionary relationships, respectively, among lizards, crocodiles, and birds. Numbers in (a) are proportion of shared phenotypic traits (e.g., a proportion of 0.60 phenotypic traits are shared between lizards and crocodiles). Vertical slashes in (b) on branches represent changes. Numbers below slashes on the bottom (most parsimonious) tree correspond to the traits (i.e., evolutionary change in traits) listed in Table 20.1.

Continued

Example 20.1 *Continued*

Parsimony methods were among the first to be used to infer phylogenies, and they are perhaps the easiest phylogenetic methods to explain and understand (Felsenstein 2004, p. 1). There are many possible phylogenies for any group of taxa. Parsimony is the principle that the most accurate phylogeny is the one that requires the minimum amount of evolution. To use parsimony, we must search all possible phylogenies and identify the one or ones that minimize the number of evolutionary changes.

in question and a 0.10 chance (p) of detecting it. Using the product rule (Box 5.1), the probability of not detecting an allele at $p = 0.1$ in a sample of size x is $(1 - p)^x$. Therefore, the sample size required to have a 95% chance of sampling an allele with frequency of 0.10 is 29 haploid individuals or 15 diploids for nuclear markers: $(1 - 0.1)^{29} = 0.047$.

20.3.2.2 Natural selection

Directional selection can cause gene trees to differ from species trees if a rare allele increases rapidly to fixation because of natural selection (**selective sweep**, Section 10.3.1). For example, a highly divergent (ancient) lineage may be swept to fixation in a recently derived species. Here the ancient age of the lineages would not match the recent age of the newly derived species. In another example, balancing selection could maintain the same lineages in each of two long-isolated species, and lead to erroneous estimation of species divergence, as well as a phylogeny discordant with the actual species phylogeny and with neutral genes (e.g., Bollmer et al. 2007). To avoid selection-induced errors in phylogeny reconstruction, many independent loci should be used. Analysis of many loci can help to identify a locus with unusual phylogenetic patterns due to selection (as in Section 9.7). For example, selection might cause rapid divergence at one locus that is not representative of the rest of the genome or of the true species tree.

20.3.2.3 Introgression

Introgression also causes gene trees to differ from species trees. For example, hybridization and subsequent backcrossing can cause an allele from species X to introgress into species Y. This has happened between wolves and coyotes that hybridize in the northeastern USA, where coyote mtDNA has introgressed into wolf populations. Here, female coyotes hybridize with male wolves, followed by the

F1 hybrids mating with wolves, such that coyote mtDNA introgresses into wolf populations (Roy et al. 1994). This kind of unidirectional introgression of maternally inherited mtDNA has been detected in deer, mice, fish, and many other species (Good et al. 2008). Introgression can also be unidirectional or asymmetric at nuclear loci, depending on demography and colonization history (Scascitelli et al. 2010).

20.3.2.4 mtDNA gene tree versus species tree

An example of a gene tree not being concordant with the species tree is illustrated in a study of polar bears and brown bears. Early work with mtDNA suggested that brown bears are **paraphyletic** with respect to polar bears (Talbot & Shields 1996). Subsequent phylogenetic analysis of the complete mtDNA genome, as well as geological and molecular age estimates of a 100,000-year-old **subfossil** bear specimen, indicated that polar bears adapted rapidly within ~20,000 years following their split from a brown bear precursor (Figure 20.7a; Lindqvist et al. 2010). However, we must remember that this conclusion is based on the phylogeny of mtDNA, which may not fully represent the phylogeny of the species themselves. A subsequent analysis of over 9,000 base pairs at 14 nuclear loci indicated that polar bears are monophyletic and diverged from brown bears some 600,000 years ago (Figure 20.7b; Hailer et al. 2012). The authors suggested that polar bears carry brown bear mtDNA because of past hybridization and introgression. These results show the importance of examining nuclear as well as mtDNA, in determining relationships among populations and species.

20.4 Genetic relationships within species

Identifying populations and describing population relationships are often difficult but crucial for conservation and management actions such as

(a)

Mitochondrial DNA

(b)

Nuclear DNA

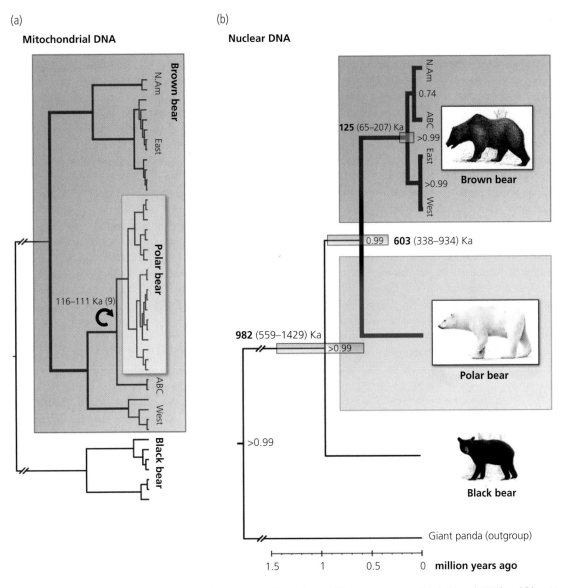

Figure 20.7 Phylogenetic trees of polar and brown bears based on (a) complete mtDNA genome sequences (Lindqvist et al. 2010), and (b) on 11 nuclear loci (Hailer et al. 2012). The circular arrow denotes mtDNA replacement in polar bears. Numbers next to nodes indicate statistical support, and gray bars are 95% highest credibility ranges for node ages (Ka = thousands of years ago).

monitoring population status, measuring gene flow, and planning translocation strategies (Waples & Gaggiotti 2006; Guest Box 20). Population relationships are generally assessed using multilocus allele frequency data and statistical approaches for clustering individuals or populations with a **dendrogram** or tree in order to identify genetically similar groups.

Population trees and phylogenetic trees look similar to each other, but they display fundamentally different types of information (Kalinowski 2009). Phylogenies show the time since the most recent common ancestor (TMRCA) between taxa. Phylogenies represent relationships among taxa that have been reproductively isolated for many generations. A phylogeny identifies monophyletic groups—isolated groups that shared a common ancestor. Phylogenetic trees can be used both for species and for genes (e.g., mtDNA; Nichols 2001). In the case of species, the branch points represent speciation events; in the case of genes, the branch points represent common ancestral genes.

Population trees, in contrast, generally identify groups that have similar allele frequencies because of ongoing genetic exchange (i.e., gene flow). The concept of TMRCA is not meaningful for populations with ongoing gene flow. Populations with high gene flow will have similar allele frequencies and cluster together in population trees.

The differences between population and phylogenetic trees described here are somewhat oversimplified to help explain the differences. In reality, there is a continuum in the degree of differentiation among populations in nature. Some populations within the same species may have been reproductively isolated for many generations. In this case, genealogical information and the phylogenetic approach can be used to infer population relationships (Section 20.4.3).

The description of genetic population structure is the most common topic for a conservation genetics paper in the literature. Individuals from several different geographic locations are genotyped at a number of loci to determine the patterns and amounts of gene flow among populations. This population-based approach assumes that all individuals sampled from one area were born there, and represent a local breeding population. However, powerful approaches have been developed that allow the description of population structure using an individual-based approach. That is, many individuals are sampled, generally over a wide geographic range, and then placed in population units on the basis of genotypic similarity.

20.4.1 Population-based approaches

There are a range of approaches that have been used to describe the genetic relationships among a series of populations. Here we will discuss several representative approaches.

Typically, the initial step in assessing population relationships using genotypic data from many individuals is to conduct statistical tests for differences in allele frequencies between sampling locations (Waples & Gaggiotti 2006; Morin et al. 2009). For example, a chi-square test is used to test for allele frequency differences between samples (e.g., Roff & Bentzen 1989). If two samples are not significantly different, they are often pooled together to represent one population. Before pooling, it is important to understand the statistical power to detect structure, given the sample size of loci and individuals, and the difference between statistical and biological significance (Waples 1998; Taylor & Dizon 1999; Ryman et al. 2006; Box 20.3). It can also be important to resample from the same geographic location in different years or seasons to test for sampling error and for stability of genetic composition through time. After distinct population samples have been identified, the genetic relationships (i.e., genetic similarity) among populations can be inferred.

20.4.1.1 Population dendrograms

Population relationships are often assessed by constructing a dendrogram based upon the genetic similarity of populations. The first step in dendrogram construction is to compute a genetic differentiation statistic (e.g., F_{ST} or Nei's D; Sections 9.1 and 9.8) between each pair of populations. A genetic distance can be computed using any kind of molecular marker (single nucleotide polymorphism (SNP) allele frequencies, DNA haplotypes) and a vast number of metrics (e.g., Cavalli-Sforza chord distance, Slatkin's R_{ST}, and Wright's F_{ST}; Section 9.8). This yields a **genetic distance matrix** (Table 20.3).

The second step is to use a clustering algorithm to group populations with similar allele frequencies (e.g., low F_{ST}). The most widely used clustering algorithms are UPGMA (unweighted pair group method with arithmetic averages) and neighbor-joining (Salemi & Van-Damme 2003). UPGMA clustering for dendrogram construction (Figure 20.8) is illustrated by a study assessing population relationships of Harper's beauty from Florida (Example 20.2).

Neighbor-joining is one of the most widely used algorithms for constructing dendrograms from a distance matrix (Salemi & VanDamme 2003). Neighbor-joining is different from UPGMA in that

Box 20.3 Statistical versus biological significance

It is critical to understand the difference between statistical significance and biological significance to avoid making incorrect inferences in conservation genomics and other fields of biology (Nakagawa & Cuthill 2007). In frequentist hypothesis testing (Appendix A4), **statistical significance** means that the null hypothesis of no effect (e.g., no differences in allele frequencies) is rejected. In contrast, **biological significance** means that the difference in values between groups (e.g., difference in allele frequencies), also known as the effect size, is biologically important. Biological importance is defined by the investigator. In the case of testing for genetic differentiation, it could be defined, for example, as a difference in allele frequencies that indicates demographic or evolutionary independence between populations.

Not detecting statistically significant genetic structure does not necessarily mean there is not biologically significant structure (McCormack & Maley 2015). A lack of statistical significance when the difference in allele frequencies is substantial can be caused by low power due to small sample sizes of loci or individuals. In particular, population differentiation is expected to be weak in species with high gene flow (Waples 1998). In these cases, large sample sizes are necessary to have the statistical power to detect any subtle genetic differentiation (Appendix A4).

Conversely, detecting statistically significant genetic structure does not necessarily mean that the structure is biologically significant. With the thousands of loci available with genomic data, statistical power is extremely high, such that statistical significance can be achieved even when there are relatively small, biologically unimportant differences in allele frequencies.

These examples highlight the importance of understanding the distinction between statistical and biological significance when interpreting statistical tests. To avoid making the common mistake of equating statistical significance with biological significance, investigators should define *a priori* what effect size they consider biologically significant and then design their study in such a way as to provide sufficient statistical power to detect this effect size (Taylor & Dizon 1999; Ryman et al. 2006).

---

the branch lengths for sister taxa (e.g., FL1 and FL2, Table 20.3) can be different, and thus can provide additional information on relationships between populations. For example, FL1 is more distant from FL3 than FL2 is from FL3 (Table 20.3). This is not evident in the UPGMA dendrogram (Figure 20.8), but would be in a neighbor-joining tree. It follows that neighbor-joining trees are especially useful when populations have substantial amounts of divergence. Other advantages include that neighbor-joining is fast and thus useful for large datasets and for **bootstrap analysis** (see next paragraph), which involves the construction of hundreds of replicate trees from resampling the data. It also permits correction for multiple character changes when computing distances between taxa. Disadvantages include that it gives only one possible tree and it depends on the model of evolution used.

Bootstrap analysis is a widely used sampling technique for assessing the statistical error when the underlying sampling distribution is unknown. In dendrogram construction, we can bootstrap-resample across loci from the original dataset, meaning that we sample with replacement from our set of loci until we obtain a new set of loci, called a "bootstrap replicate." For example, if we have genotyped 12 loci, we randomly draw 12 numbers from 1 to 12, and these numbers (loci) become our bootstrap replicate dataset. We repeat this procedure 1,000 times to obtain 1,000 datasets (and 1,000 dendrograms). The proportion of the random dendrograms with the same cluster (i.e., branch group) will be the **bootstrap support** for the cluster. For example, the cluster containing FL1 with FL2 has a bootstrap support of 85 (i.e., 85% of 1,000 bootstrap trees grouped FL1 with FL2).

20.4.1.2 Multidimensional representation of relationships among populations

Dendrograms cannot illustrate complex relationships among multiple populations because they consist of a one-dimensional branching diagram. Thus, dendrograms can oversimplify and obscure relationships among populations. Note that this

Example 20.2 Dendrogram construction via UPGMA clustering of lily populations

UPGMA clustering was used to assess relationships among five populations of Harper's beauty, an endangered perennial lily from northern Florida (Godt et al. 1997). Allele frequencies from 15 polymorphic allozyme loci were used to construct a genetic distance matrix (Table 20.3) and subsequently a dendrogram using the UPGMA algorithm.

Table 20.3 Genetic distance (D; Nei 1972) matrix based upon allele frequencies at 15 allozyme loci for 5 populations of Harper's beauty, an endangered perennial lily. Data from Godt et al. (1997).

	Population				
	FL1	**FL2**	**FL3**	**SC**	**NC**
FL1	—				
FL2	0.001	—			
FL3	0.003*	0.002*	—		
SC	0.029	0.032	0.030	—	
NC	0.059	0.055	0.060	0.062	—

The UPGMA algorithm starts by finding the two populations with the smallest inter-population distance in the matrix. It then joins the two populations together at an internal node. In our lily example here, populations "FL1" and "FL2" are grouped together first because the distance (0.001) is the smallest (underlined in Table 20.3). Next, the mean distance from FL1 (and from FL2) to each other population is used to cluster taxa. The next shortest distance is the mean of FL3 to FL1 and FL3 to FL2 (i.e., the mean of 0.002 and 0.003; see asterisks in Table 20.3); thus FL3 is clustered as the sister group of FL1 and FL2. Next, SC is clustered, followed by NC (Figure 20.8).

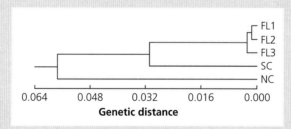

Figure 20.8 Dendrogram generated for Harper's beauty using the UPGMA clustering algorithm and the genetic distance matrix from Table 20.3. FL is Florida. SC and NC are South Carolina and North Carolina, respectively. From Godt et al. (1997).

In this example, the genetic distance is correlated with the geographic distance, in that South Carolina is geographically and genetically closer to Florida populations than is North Carolina.

one-dimensionality is not a limitation in using dendrograms to represent phylogenic relationships, as these can be represented by a one-dimensional branching diagram as long as there has not been secondary contact or hybridization following speciation. If secondary contact, hybridization, **horizontal gene transfer**, or multiple branching relationships exist among populations, this can potentially be represented by **phylogenetic networks** (Reeves & Richards 2007).

There are a variety of multivariate statistical techniques (e.g., principal component analysis, PCA) that summarize and can be used to visualize complex datasets with multiple dimensions (e.g., many loci and alleles) so that most of the variability in allele frequencies can be extracted and visualized on a two-dimensional or three-dimensional plot. Related multivariate statistical techniques include PCoA (principal coordinates analysis; Figure 20.9), FCA (frequency correspondence analysis), MDS (**multidimensional scaling**), and DAPC (discriminant analysis of principal components).

20.4.2 Individual-based approaches

Individual-based approaches are used to assess population relationships through first identifying populations by delineating genetically similar clusters of individuals. Clusters of genetically similar individuals are often identified by building a dendrogram in which each branch tip is an individual. Then, we quantify genetic relationships among the clusters (putative demes).

Individual-based methods for assessing population relationships make no *a priori* assumptions about how many populations exist or where boundaries between populations occur on the landscape. If individual-based methods are not used, we risk wrongly grouping individuals into populations based on somewhat arbitrary traits (e.g., color) or an assumed geographic barrier (a river) identified by humans subjectively (Section 19.2).

One example of erroneous *a priori* grouping would be migratory birds that we sample on migration routes or on overwintering grounds. Here, we might wrongly group together individuals from different breeding populations, only because we sampled them together at the same geographic location.

A similar error could be made in migratory butterflies, salmon, or whales if we sample mixtures containing individuals from different breeding groups originating from different geographic origins.

An individual-based approach was used by Langin et al. (2018) to assess relationships among populations of the white-tailed ptarmigan, an alpine-obligate species found in naturally fragmented habitats from Alaska to New Mexico that may be vulnerable to climate change. The authors built a tree of individuals based on pairwise genetic distance between individuals. Each individual was genotyped at 14,866 SNP loci. The number of base differences per site between each pair of individuals was computed for 14,693 presumably neutral loci, and then a clustering algorithm (neighbor-joining) was used to group similar individuals together on branches. The geographic location of origin of individuals was also plotted on the branch tips to help identify population units. The analysis provided strong support for two distinct intraspecific units in Colorado, USA, and neighboring states, and on Vancouver Island in British Columbia, Canada (Figure 20.10).

An individual-based and model-based approach that identifies populations as clusters of individuals was introduced by Pritchard et al. (2000). "Model-based" refers to the use of a model with k populations (demes) that are assumed to be in **Hardy–Weinberg (HW) proportions** and gametic equilibrium. This approach first tests whether our data fit a model with k = 1, 2, 3, or more populations. The method uses a computer algorithm to search for the set of individuals assigned to a given k groups that minimizes the amount of HW and gametic disequilibrium within groups (Section 10.1). Many possible sets of individuals are tested. Once k is inferred (step 1), the algorithm estimates, for each individual, the (posterior) probability (Q) of the individual's genotype originating from each population (step 2). If an individual is equally likely to have originated from population X and Y, then Q will be 0.50 for each population.

For example, Kozakiewicz et al. (2019) analyzed 13,520 SNPs using the *fastStructure* clustering algorithm to test the effects of major highways on the genetic structure of bobcats in an urban environment northwest of Los Angeles, California,

(a)

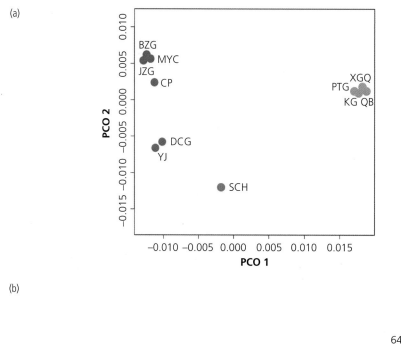

(b)

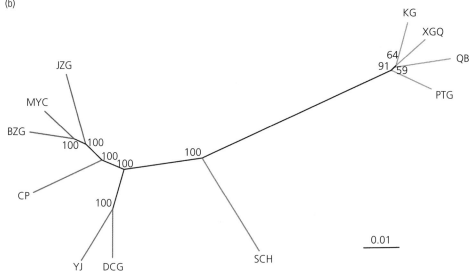

Figure 20.9 Patterns of population differentiation among 11 Andrew's toad populations at 15,777 SNPs shown using (a) a PCoA plot and (b) a neighbor-joining (NJ) tree. Populations from different regions across an altitudinal gradient from the Tibetan Plateau are shown in different colors. Values at nodes of the NJ tree refer to bootstrap values of 1,000 replicates. From Guo et al. (2016).

USA (Figure 20.11). They found that these highways blocked gene flow, resulting in detectable genetic differentiation among populations separated by highways. A strength of individual-based approaches is that they are not only useful for characterizing population structure, but also for identifying immigrants. In this example, Kozakiewicz et al. (2019) found several bobcats that originated in one population, but were sampled in another, indicating some contemporary dispersal of bobcats across these highways. These results are encouraging for the conservation of bobcats in this urban environment, as the data indicate ongoing dispersal and gene flow, which is crucial for maintenance of the long-term fitness and persistence of small, fragmented populations (Section 19.3).

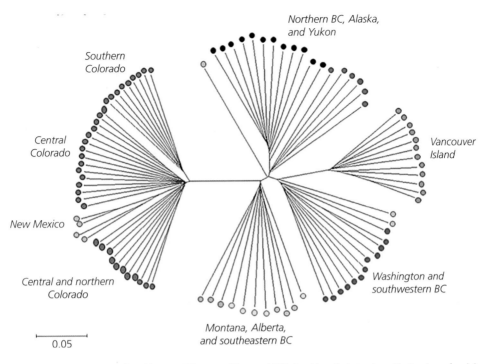

Figure 20.10 Neighbor-joining trees inferred from 14,693 presumably neutral SNPs in white-tailed ptarmigan. The terminus of each branch represents an individual and is color-coded by region. Branch lengths are proportional to the genetic distance between individuals (number of base pair differences per site). From Langin et al. (2018).

Individual-based analyses can also be conducted with many multivariate statistical methods (e.g., PCA) if individuals are used as the operational unit (instead of populations). These multivariate approaches make no prior assumptions about the population structure model (e.g., HW and gametic equilibrium are not assumed).

Individual-based methods are useful to identify cryptic subpopulations and localize population boundaries on the landscape. Once genetic boundaries (discontinuities) are located, we can test whether the boundaries are concordant with some environmental gradient, a phenotypic trait distribution, or some ecological or landscape feature (e.g., a river or temperature gradient). This approach of associating population genetic boundaries with landscape or environmental features is a **landscape genetics** approach (Section 19.2; Manel et al. 2003).

These approaches can provide misleading results if samples of individuals are not collected evenly across space. In a continuously distributed population (Figure 9.10), we might wrongly infer a genetic discontinuity (barrier) between sampling locations if clusters of individuals are sampled from distant locations (Frantz et al. 2009). Schwartz & McKelvey (2009) simulated a population of continuously distributed individuals, and then used common approaches to sample individuals from 10 sampling locations. An individual-based STRUC-TURE analysis was conducted, which suggested that many individuals could be assigned to the location from which they were sampled (Figure 20.12). This result could be interpreted to indicate the existence of discrete subpopulations, even though individual genotypes were from a continuously distributed population having no discrete subpopulations or genetic discontinuities. In addition, this analysis suggested that there were some long-distance migrants, which is incorrect because a simple neighbor-mating process was used to generate the genotypes. Schwartz & McKelvey (2009) provide useful guidance to avoid these types of misinterpretations.

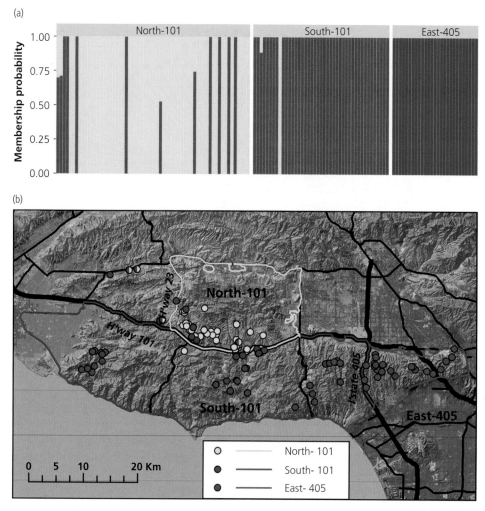

Figure 20.11 (a) *fastStructure* analysis of bobcats northwest of Los Angeles, California, USA, at 13,520 SNPs supports three spatially and genetically distinct populations separated by major highways. Each population is shown by a different color. Individuals are represented across the *x*-axis by a vertical bar that may be divided vertically into differently colored segments that represent the proportion of an individual's genome assigned to each population. Individuals that have a color different from the color of the local population are immigrants. (b) Map of study area showing the distribution of individuals assigned to different populations. Colored lines indicate the estimated spatial extent of each population's habitat, defined according to major roads and urban edges that form barriers to gene flow. Black lines denote major roads, while urban land cover density is indicated by red shading. Modified from Kozakiewicz et al. (2019).

20.4.3 Phylogeography

Phylogeography is the assessment of the correspondence between phylogenetic patterns and geographic patterns of distribution among taxa (Avise 2009). We expect to find phylogeographic structuring among populations with long-term geographic isolation. Substantial isolation for hundreds of

generations is generally required for new mutations to arise locally, and to preclude their spread beyond local populations. Phylogeographic structure is expected in species with limited dispersal capabilities, with **philopatry**, or with distributions that span strong barriers to gene flow (e.g., mountains, rivers, roads, and human development). In conservation biology, detecting phylogeographic

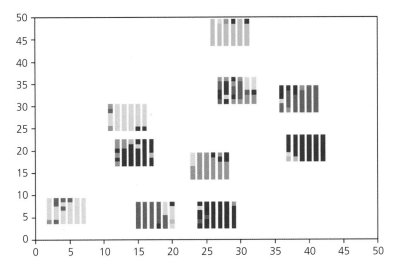

Figure 20.12 Potential problems when using individual-based methods to identify discrete subpopulations in the case of isolation by distance. Assignments using *STRUCTURE* (Pritchard et al. 2000) when 360 individuals are sampled in 10 clusters of individuals from simulations of a continuously distributed population. Each of the 10 sample locations consists of six vertical bars representing six individuals within each bar (36 individuals per sampling location). Each square represents an individual. Each square's shading indicates the cluster to which *STRUCTURE* assigned the individual. The values on the vertical and horizontal axes are references in the 50 × 50 grid containing 250 individuals total. Modified from Schwartz & McKelvey (2009).

structuring is important because it helps to identify long-isolated populations that might have distinct gene pools and local adaptations. Long-term reproductive isolation is one major criterion widely used to identify population units for conservation (Section 20.5.2).

Intraspecific phylogeography was pioneered by John Avise and colleagues (Avise et al. 1987). In a classic example, Avise et al. (1979b) analyzed mtDNA from 87 pocket gophers from across their range in the southeastern USA. The study revealed 23 different mtDNA genotypes, most of which were localized geographically (Figure 20.13). A major discontinuity in the maternal phylogeny clearly distinguished eastern and western populations. A potential conservation application of such results is that eastern and western populations of pocket gopher appear to be highly divergent with long-term isolation and thus potentially adaptive differences; this could warrant recognition as separate conservation units. However, additional data, including nuclear loci and nongenetic information, should be considered before making conservation management decisions (e.g., Section 20.6).

Phylogeographic studies can help to identify **biogeographic** provinces containing distinct flora and fauna worth conserving as separate geographic units in nature reserves. For example, multispecies phylogeographic studies in the southeastern USA (Avise 1992), northeastern Australia (Moritz & Faith 1998), and the northwestern USA (Soltis et al. 1997)

have revealed remarkably concordant phylogeographic patterns across multiple different species. Such multispecies concordance can be used to identify major biogeographic areas that can be prioritized as separate conservation units, and to identify locations to create nature reserves (Figure 20.14).

A formerly widely used but controversial phylogeographic approach is **nested clade phylogeographic analysis** (NCPA; Templeton 1998). There has been substantial debate over the usefulness of NCPA (Knowles & Maddison 2002; Knowles 2008; Templeton 2008; Garrick et al. 2008). A shortfall of NCPA is that it does not incorporate error or uncertainty. This is the same problem with many phylogeographic approaches. For example, NCPA does not consider interlocus variation, as do coalescent-based population genetic models (Section A10, Figure A12). Thus, NCPA might provide the correct inference about phylogeographic history, but we cannot easily quantify the probability of it being correct. Another limitation is that NCPA is somewhat ad hoc in using an inference key in order to distinguish between different historic processes, such as range expansion and population fragmentation. A study by Panchal & Beaumont (2007) evaluated NCPA by developing software to automate the NCPA procedure. Using simulations of random-mating populations, Panchal and Beaumont reported a high frequency of false positives for the detection of range expansions and for inferences of isolation by distance. The NCPA story

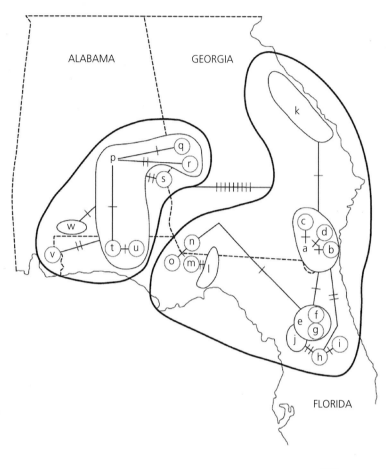

Figure 20.13 mtDNA phylogenetic network for 87 pocket gophers. mtDNA genotypes are represented by lower case letters and are connected by branches in a parsimony network. Slashes across branches are substitutions. Nine mutations separate the two major mtDNA clades encircled by heavy lines. From Avise (2004).

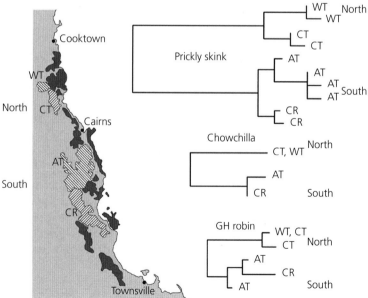

Figure 20.14 Neighbor-joining trees based on mtDNA sequence divergence for three species sampled from each of four areas (WT, CT, AT, and CR) from the tropical rainforests of northeastern Australia. Note the long branches separating the WT/CT in the north from the CR/AT populations in the south for all three species (prickly skink, chowchilla, and gray-headed robin). These results suggest long-term isolation for numerous species between the northern and southern rainforests. These regions merit conservation as separate units. From Moritz & Faith (1998).

illustrates the importance of conducting extensive performance evaluations (e.g., Section A9) before using any novel computational approach.

Fortunately, the field of **statistical phylogeography** combines the strengths of NCPA with formal model-based approaches and statistical tests to test alternative hypotheses to explain phylogeographic patterns (Knowles & Maddison 2002). As formal modeling and validated statistical phylogeography approaches are now available, it would be prudent not to use NCPA, or use it only in combination with well-evaluated model-based phylogeographic approaches (Turner et al. 2000; Beaumont et al. 2010).

Another exciting advance in comparative phylogeography is using trait-based hypotheses to explain variation in phylogeographic structure among species with different ecological and life history traits (Papadopoulou & Knowles 2016). Previously, comparative phylogeography was focused on trying to find concordant phylogeographic patterns among taxa in a given region. However, discordant patterns of genetic divergence may be due to taxon-specific trait differences that cause specific geographic features to be barriers to some species, but not others. For example, Paz et al. (2015) found that species traits, including body size, geographic range, biogeographic origin, and reproductive mode, were significant predictors of spatial genetic variation in frog communities in Panama. Similarly, Harvey et al. (2017) found that in Amazonian birds, habitat association predicted genetic divergence across the landscape. Specifically, upland forest bird species had greater genetic divergence and less gene flow among populations than closely related floodplain species. Comparative phylogeographic approaches such as these not only provide insights into interactions of taxa with their landscapes, but also can provide improved predictive frameworks for responses of forest species to climate change (Papadopoulou & Knowles 2016).

20.5 Units of conservation

It is crucial to identify species and units within species to help guide management, monitoring, other conservation efforts, and to apply laws to conserve taxa and their habitats. In this section, we consider the issues of identifying species and intraspecific conservation units.

20.5.1 Species

Identification of species is often problematic, even for some well-known taxa. One problem is that biologists cannot even agree on the appropriate criteria to define a species. In fact, more than two dozen species concepts have been proposed over the past few decades. Darwin (1859) wrote that species are simply highly differentiated varieties. He observed that there is often a continuum in the degree of divergence from between populations, to between varieties, species, and higher taxonomic classifications. In this view, the magnitude of differentiation that is required to merit species status can be somewhat arbitrary.

The **biological species concept** (BSC) of Mayr (1942, 1963) is the most widely used species definition, at least for animals. Ernst Mayr defined species as "groups of actual or potentially interbreeding natural populations that are reproductively isolated from other such groups" (Mayr 1942, p. 120). This concept emphasizes reproductive isolation and isolating mechanisms (e.g., pre- and postzygotic). Criticisms of this concept are that (1) it can be difficult to apply to allopatric organisms (because we cannot observe or test for natural reproductive barriers in disjunct populations); (2) it cannot easily accommodate asexual species; and (3) it has difficulties dealing with introgression between distinct forms, which is common in nature. Further, an emphasis on isolating mechanisms implies that selection counteracts gene flow. However, the BSC generally does not allow for interspecific gene flow, even for a few segments of the genome that can introgress between species (Wu 2001; Rieseberg 2011).

The **phylogenetic species concept** (PSC; Cracraft 1989) relies largely on monophyly such that all members of a species must share a single common ancestor. This concept has fewer problems dealing with asexual organisms (e.g., many plants, fish, etc.) and with allopatric forms. However, it does not readily accommodate hybridization and introgression. Some biologists have suggested it can lead to oversplitting, for example, as more and more characters are used, as with powerful DNA sequencing techniques, so more taxa might be identified. However, an iterative approach using multiple lines of evidence will minimize this risk.

A problem using PSC can arise if biologists interpret fixed DNA differences (monophyly) between populations as evidence for species status. For example, many species are becoming fragmented, and population fragments are becoming fixed (monophyletic) for different DNA polymorphisms. Under the PSC, this could cause the proliferation of new species if biologists strictly apply the PSC criterion of monophyly for species identification. This could result in oversplitting and the waste of limited conservation resources. This potential problem of fragmentation-induced oversplitting is described in a paper titled "Cladists in Wonderland" (Avise 2000). To avoid such oversplitting, multiple independent DNA sequences (i.e., not organelle DNA alone) should be used, along with nongenetic characters when possible.

The **ecological species concept** identifies species based on a distinct ecological niche (Van Valen 1976). The **evolutionary species concept** is used often by paleontologists to identify species based on change, but not necessarily splitting, within lineages (anagenesis; Simpson 1961).

The many different concepts overlap, but emphasize different types of information. Generally, it is important to consider many kinds of information or criteria when identifying and naming species. If most criteria or species concepts give the same classification or conclusion, then we are more confident in the conclusion (e.g., species status is warranted).

A **unified species concept**, or more general concept, has been proposed (De Queiroz & Weins 2007; Hart 2011; Hausdorf 2011). Generality or unification is achieved by treating "separately evolving metapopulation lineages" and isolation processes as the only necessary properties of species (Hart 2011). This author reported that a conceptual shift in focus is occurring away from species diagnoses based on published species definitions, and toward analyses of the processes acting on lineages of metapopulations that will likely lead to different recognizable species (De Queiroz & Weins 2007).

A possible advantage of this approach is that a species is recognized from **emergent properties** of measurable, ongoing population-level processes.

Similarly, a generalization of the genic concept (Wu 2001) has been proposed that defines species as groups of individuals that are characterized by features that would have negative fitness effects in other groups, if cross-breeding occurred between groups (Hausdorf 2011). This differential fitness concept has benefits, such as classifying groups that maintain differences and keep on differentiating despite occasional interbreeding, and it is not restricted to specific mutations or mechanisms causing speciation. In addition, it can be applied to the whole spectrum of organisms from biparentals to uniparentals (i.e., species with transmission of genotypes from only one parental type to all progeny).

African cichlid fishes illustrate some of the difficulties with the different species concepts. Approximately 1,500 species of African cichlids have recently evolved a diverse array of morphological differences (mouth structure, body color) and ecological differences (feeding and behaviors such as courtship). Morphological differences are pronounced among African cichlids. However, the degree of genetic differentiation among African cichlids is relatively low compared with other species, due to the recent radiation of species (less than 1–2 million years ago). Further complicating efforts to identify species using molecular markers is the fact that reproductive isolation can be transient. For example, some cichlid species are reproductively isolated due to mate choice based on fixed color differences between species. However, this isolation breaks down when murky water prevents visual color recognition and leads to temporary interspecific gene flow (Seehausen et al. 1997).

20.5.1.1 Cryptic species

Molecular genetic data can help to identify species, especially cryptic species that have similar phenotypes (Section 22.1). For example, an analysis of mtDNA, nuclear DNA, morphology, and calls revealed that in two different groups of Amazonian frogs, *Engystomops* toadlets and *Hypsiboas* treefrogs, species richness was severely underestimated (Funk et al. 2012a). In the toadlets, two recognized species were found to actually consist of five to seven species (a 150–250% increase in species richness). In the treefrogs, two recognized species represent six to nine species (a 200–350% increase). In both groups, morphological differences among cryptic species were subtle, but call differences were more pronounced.

Undiscovered cryptic species are not just restricted to remote tropical regions. A recent analysis combining microsatellites, mitochondrial DNA, SNPs, and morphology uncovered four distinct lineages of green salamanders, one of which was described as a new species restricted to a small geographical range in western North Carolina, USA (Patton et al. 2019). This new species faces pressing conservation threats due to rapid development in this region. The authors recommended treating the other three lineages as **evolutionarily significant units (ESUs;** Section 20.5.2).

Molecular data can also help to identify taxa that are relatively well studied. For example, a study of golden jackals and other canids using molecular genetic data detected previously unrecognized species. Golden jackals from Africa were long considered to be the same species as jackals found throughout Eurasia. However, Rueness et al. (2011) and Gaubert et al. (2012) found that mitochondrial haplotypes of some African golden jackals were more similar to haplotypes of gray wolves than Eurasian golden jackals, which was surprising given that gray wolves are not found in Africa, and that gray wolves are phenotypically divergent from golden jackals. Moreover, this result suggested that African golden jackals are a distinct species from Eurasian golden jackals.

Koepfli et al. (2015) followed up on this mitochondrial study to test the hypothesis that African golden jackals represent a distinct species from Eurasian golden jackals and to resolve the phylogenetic relationships among African golden jackals, Eurasian golden jackals, and other canids. The authors analyzed diverse types of data to address these questions, including mitochondrial genome sequences, sequences from 20 autosomal loci, microsatellite loci, X- and Y-linked zinc-finger protein genes, and whole genome nuclear sequences. They found robust and consistent support across all data types

that golden jackals from Africa and Eurasia are distinct monophyletic lineages that have been separated from each other for more than a million years (Figure 20.15). Based on these results, they recommended recognition of African and Eurasian golden jackals as distinct species: the African golden wolf and the Eurasian golden jackal. The authors also found that these two species are strikingly similar morphologically, which they argue suggests parallel evolution. Interestingly, more recent whole genome sequence data reveal that the African golden wolf originated through hybridization between gray wolves and Ethiopian wolves (Gopalakrishnan et al. 2018).

20.5.1.2 Oversplitting

Genetic data may show that currently recognized species are not genetically differentiated, indicating that they are not reproductively isolated. Some authors have recognized the black sea turtle (*Chelonia* spp.) as a distinct species on the basis of skull shape, body size, and color (Pritchard 1999). However, molecular analyses of mtDNA and three independent nuclear DNA fragments suggest that reproductive isolation does not exist between the black and green forms (Karl & Bowen 1999). Over the years, taxonomists have proposed more than a dozen species for different *Chelonia* populations—with oversplitting occurring in many other groups as well. Nevertheless, it is clear that black turtles are distinct and could merit recognition as an intraspecific conservation unit (e.g., Section 20.5.2) that possesses local adaptations. Unfortunately, populations are declining, and additional data on adaptive differences are needed (e.g., food sources and feeding behavior).

Another example of potential oversplitting is Holarctic redpoll finches. Some studies have recognized redpolls as three different species based on variation in plumage and morphology, but multiple genetic studies using restriction fragment length polymorphisms, mitochondrial control region sequences, and microsatellites have consistently failed to detect genetic differences among these species. Mason & Taylor (2015) followed up on these earlier genetic studies and used a large genomic dataset consisting of 20,712 anonymous

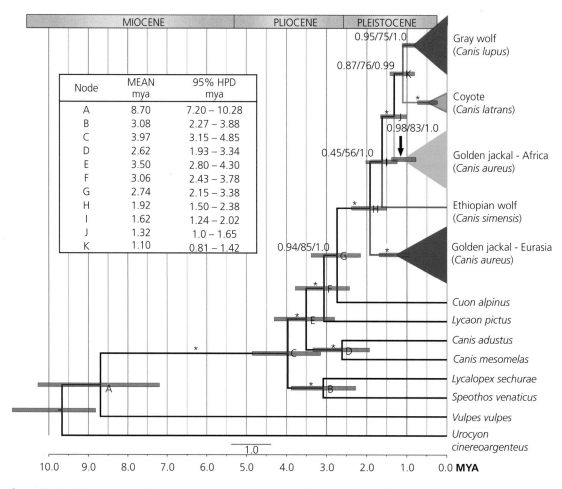

Figure 20.15 Phylogenetic tree with branch lengths scaled to time showing inferred evolutionary relationships of African golden jackals, Eurasian golden jackals, and other canids based on analyses of 20 nuclear gene segments. Values at nodes show nodal bootstrap support. HPD is the highest posterior density interval (see Section A3). Although African golden jackals are phenotypically similar to Eurasian golden jackals, these results support them being sister to the clade containing gray wolves and coyotes. From Koepfli et al. (2015).

SNPs and 215,825 SNPs within the redpoll transcriptome to test for genetic differences among the three redpoll species. Even with the greater statistical power and resolution of their dataset, they still found largely undifferentiated genomes among the currently recognized redpoll species. However, they did find differential gene expression related to observed phenotypic variation. Based on their genomic results, Mason & Taylor (2015) conclude that the patterns they observe are likely caused by ongoing gene flow among polymorphic populations or incomplete lineage sorting accompanying

recent or ongoing divergence. The taxonomy of redpolls remains unresolved.

20.5.2 Evolutionarily significant units

An ESU can be defined broadly as a population or group of populations that merit separate management or priority for conservation because of high distinctiveness both genetically and ecologically. The first use of the term ESU was by Ryder (1986). He used the example that five extant subspecies of tigers exist, but there is not space in zoos or captive breeding programs to maintain viable populations of all five. Thus sometimes we must choose which subspecies to prioritize for conservation action, and perhaps maintain only one or two global breeding populations (each perhaps consisting of more than one named subspecies). Since Ryder (1986), the term ESU has been used in a variety of frameworks for identifying conservation units (Box 20.4).

There is considerable confusion and controversy in the literature associated with the term ESU. For example, the ESA lacks any definition of a DPS (Box 20.1). Waples (1991) suggested that a population or group of populations of salmon would be a DPS if it is an ESU. This has led to some confusion because some biologists equate a DPS with an ESU. Here, we will use the term DPS when referring to officially recognized "species" under the ESA, and the term ESU in the more generally accepted sense.

It can be difficult to provide a single concise, detailed definition of the term ESU because of the controversy and different uses and definitions of the term in the literature. This ESU controversy is analogous to that surrounding the different species concepts mentioned in Section 20.5.1. The controversy is not surprising considering the problems surrounding the definition of species, and the fact that identifying intraspecific units is generally more difficult than identifying species (Waples 1991). It is also not surprising considering the different rates of evolution that often occur for different molecular markers and phenotypic traits used in ESU identification. Different evolutionary rates lead to problems analogous to those in the classification of

---

Box 20.4 Proposed definitions of ESUs

Ryder (1986): Populations that actually represent significant adaptive variation based on concordance between sets of data derived by different techniques. Ryder (1986) clearly argued that this subspecies problem is "considerably more than taxonomic esoterica". (Main focus: zoos for potential *ex situ* conservation of gene pools of threatened species.)

Waples (1991): Populations that are reproductively separate from other populations (e.g., as inferred from molecular markers) and that have distinct or different adaptations and that represent an important evolutionary legacy of a species. (Main focus: integrating different data types, and providing guidelines for identifying "distinct population segments" of Pacific salmon that are given "species" status for protection under the US ESA.)

Dizon et al. (1992): Populations that are distinctive based on morphology, geographic distribution, population parameters, and genetic data. (Main focus: concordance across some different data types, but always requiring some degree of genetic differentiation.)

Moritz (1994): Populations that are reciprocally monophyletic (Figure 20.16) for mtDNA haplotypes and that show significant divergence of allele frequencies at nuclear loci. (Main focus: defining practical criteria for recognizing ESUs based on population genetics theory, while considering that variants providing adaptation to recent or past environments may not be adaptive (or might even retard the response to natural selection) in future environments.)

USFWS & NOAA (1996b) (US policy for recognition of DPSs): (1) discreteness of the population segment in relation to the remainder of the species to which it belongs; and (2) the significance of the population segment to the species to which it belongs. This DPS policy is a further clarification of Waples' (1991) Pacific salmon ESU policy that applies to all species under the US ESA.

Crandall et al. (2000): Populations that lack (1) "ecological exchangeability" (i.e., they have different adaptations or selection pressures (e.g., life histories, morphology, quantitative trait locus variation, habitat, predators, etc.) and different ecological roles within a community); and (2) "genetic exchangeability" (e.g., they have had no recent gene flow, and show concordance between phylogenetic and geographic discontinuities). (Main focus: emphasizing adaptive variation and combining molecular and ecological criteria in an historical timeframe. Suggests returning to the more holistic or balanced two-part approach of Waples.)

Fraser & Bernatchez (2001): A lineage that demonstrates highly restricted gene flow from other such lineages within the higher organizational level (lineage) of the species. (Main focus: a context-based framework for delineating ESUs, which attempts to resolve conflicts among previous ESU definitions. Recognizes that different criteria will work better than others in some circumstances and can be used alone or in combination depending on the situation.)

birds as a taxonomic class separate from reptiles due to the rapid evolution of birds, when in fact the class Aves is monophyletic within the class Reptilia (Figure 20.3).

In practice, an understanding of the underlying principles and the criteria used in the different ESU frameworks will help when identifying ESUs. The main criteria for several different ESU concepts are listed in Box 20.4 (see also Fraser & Bernatchez 2001). Here we discuss some details about three widely used ESU frameworks, each with somewhat different criteria as follows: (1) reproductive isolation and adaptation (Waples 1991); (2) **reciprocal monophyly** (Moritz 1994); and (3) "exchangeability" of populations (Crandall et al. 2000). This will provide background on principles and concepts, as well as an historical perspective of the controversy surrounding the different frameworks for identifying units of conservation.

20.5.2.1 Isolation and adaptation

Waples (1991) was the first to provide a detailed framework for ESU identification. His framework included the following two main requirements for an ESU: (1) long-term reproductive isolation (generally hundreds of generations) so that an ESU represents a product of unique past evolutionary events that is unlikely to re-evolve, at least on an ecological timescale; and (2) ecological or adaptive distinctiveness such that the unit represents a reservoir of genetic and phenotypic variation likely important for future evolutionary potential. This second part requiring ecological and adaptive distinctiveness was termed the "evolutionary legacy" of a species by Waples (1991). This framework has become the official policy under the ESA (USFWS & NOAA 1996b).

Waples (2005b) has argued that ESU identification is often most helpful if an intermediate number of ESUs are recognized within each species, with the goal of preserving a number of genetically distinct populations. Waples (2005b) reviewed application of other published ESU concepts and criteria in Pacific salmon species based on the published criteria for these concepts, many of which are subjective or qualitative, and concluded that they often either identified only a single ESU or a large number (hundreds) of ESUs. There is a need for more empirical

examples such as this in which multiple ESU concepts are applied to common problems to evaluate their broader application in relation to reproductive isolation and adaptive uniqueness.

20.5.2.2 Reciprocal monophyly

Moritz (1994, p. 373) offered simple and thus readily applicable molecular criteria for recognizing an ESU: "ESUs should be reciprocally monophyletic for mtDNA (in animals) and show significant divergence of allele frequencies at nuclear loci." mtDNA is widely used in animals because it has a rapid rate of evolution, lacks recombination, and is uniparentally inherited, and thus facilitates phylogeny reconstruction. Chloroplast DNA markers are often used in plants as they also lack recombination and are uniparentally inherited, although have a much lower rate of evolution (Chapter 12) so are generally less useful at low taxonomic levels except for species with long evolutionary timeframes. Reciprocally monophyletic means that all DNA lineages within an ESU share a more recent common ancestor with each other than with lineages from other ESUs (Figure 20.16). These molecular criteria are relatively quick and easy to apply in most taxa because the necessary molecular markers (e.g., "universal" polymerase chain reaction (PCR) primers) and data analysis software have become widely available.

An occasionally cited advantage of the Moritz (1994) monophyly criterion is that it can employ population genetics theory to infer the time since population divergence. For example, it takes a mean of $4N_e$ generations for a newly isolated population to coalesce to a single gene copy and therefore become reciprocally monophyletic through drift and mutation at a nuclear locus (Neigel & Avise 1986). This means that if a population splits into two daughter populations of size $N_e = 1,000$, it would take an expected 1,000 generations to become reciprocally monophyletic for mtDNA. For mtDNA to become monophyletic it requires fewer generations because the effective population size is approximately four times smaller for mtDNA than for nuclear DNA; thus lineage sorting is faster (Section 9.6). Here it is important to recall that adaptive differentiation can occur in a much shorter time period than does monophyly.

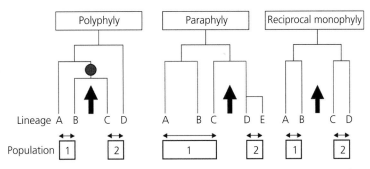

Figure 20.16 Development of phylogenetic relationships of alleles or lineages (A, B, C, D, E) in sister taxa (1 and 2). After a population splits into two because of the development of a barrier to reproduction (indicated by the large vertical arrows), the phylogenetic relationship of the alleles in the two sister populations usually proceeds from polyphyly through paraphyly to reciprocal monophyly. When two populations (1 and 2) first become isolated, they both will have some alleles that are more closely related to alleles in the other population (polyphyly). The filled circle at the root of the B and C branches indicates the most recent common ancestor between B and C (in the polyphyly example). After many generations of isolation, one population might become monophyletic for some alleles (D and E in population 2 in the paraphyly example). But the other population (1) might maintain an allele (C) that is more related to an allele in the other population. After approximately four N_e generations, both sister populations are expected to be monophyletic with respect to each other (reciprocal monophyly) at nuclear loci. Modified from Moritz (1994).

A disadvantage of the Moritz ESU concept is it generally ignores adaptive variation, unlike the two-step approaches that incorporate the "evolutionary legacy" of a species (Waples 1991). The framework of Moritz is based on a cladistic phylogenetic approach (Section 20.2) using neutral loci. Thus, unfortunately, the Moritz approach makes it likely that small populations (e.g., bottlenecked populations) could be identified as ESUs when in reality the populations are not evolutionarily distinct; small populations can quickly become monophyletic due to drift or lineage sorting. Worse perhaps, natural selection can lead to rapid adaptation, especially in large populations in which selection is efficient. Consequently, the strict Moritz framework could often fail to identify ESUs that have substantial adaptive differences.

One limitation of using only molecular information is that a phylogenetic tree might not equal the true population tree. This is analogous to the gene tree vs. species tree problem discussed in Section 20.3.2. This issue of population trees not equaling gene trees is more problematic at the intraspecific level because there is generally less time since reproductive isolation at the intraspecific level and thus more problems caused by lineage sorting and **paraphyly**. Consequently, problems of gene trees not matching population trees may be relatively common at the intraspecific level.

Unfortunately, in the conservation literature, mtDNA data alone have often been used to attempt to identify ESUs. This should occur less often now that nuclear DNA markers are more readily available.

20.5.2.3 Exchangeability

Crandall et al. (2000) suggested that ESU identification be based on the concepts of ecological and genetic "exchangeability." The idea of exchangeability is that individuals can be moved between populations and can occupy the same niche, and can perform the same ecological role as resident individuals, without any fitness reduction due to outbreeding depression. If we can reject the hypothesis of exchangeability between populations, then those populations represent ESUs. Ideally, exchangeability assessment would be based on heritable adaptive quantitative traits. Strengths of this approach are that it integrates genetic and ecological (adaptive) information, and that it is hypothesis-based.

Exchangeability can be tested using common garden experiments and reciprocal transplant experiments. For example, if two plant populations from different locations have no reduced fitness when transplanted between locations, they might be exchangeable and would not warrant separate ESU status (Section 2.7, Figure 2.9, and Figure 8.1).

The main problem with this approach is that it is not generally practical because it is difficult to test the hypothesis of exchangeability. For example, it is difficult to move a rhinoceros (or most endangered species) from one population to another and then to measure its fitness and the fitness of its offspring. Such studies are especially problematic in endangered species where experiments are often not feasible. Although difficult to test, exchangeability is a worthy concept to consider when identifying ESUs. Even when we cannot directly test for exchangeability, we might consider surrogate measures of exchangeability, such as life history differences, the degree of environmental differentiation, or the number of functional genes showing signatures of adaptive differentiation (e.g., Section 20.6.1). Surrogates are often used when applying Waples' ESU definition.

20.5.2.4 Synthesis

Substantial overlap in criteria exists among different ESU concepts. Several concepts promote a two-pronged approach involving isolation and adaptive divergence, but put different emphasis on the relative importance of these two criteria (Figure 20.17). The main common principles and criteria are the following: reproductive isolation (little gene flow), adaptive differentiation, and concordance across multiple data types (e.g., genetic, morphologic, behavioral, life history, and geographic). The longer the isolation and the more different the environment, the more likely populations are to represent distinct units that are worthy of preservation and separate management. We should not rely on any single criterion, such as reciprocal monophyly of mtDNA. In fact, the greater the number of different data types showing concordant differentiation between populations, the stronger the evidence for ESU status.

20.5.3 Management units

Management units (MUs) are populations that are demographically independent (Moritz 1994); that is, their population dynamics (e.g., growth rate) depend on local birth and death rates rather than on immigration. The identification of these units, similar to "stocks" recognized in fisheries biology,

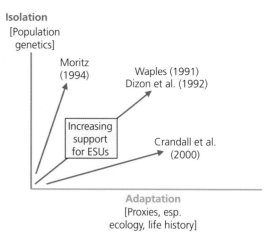

Figure 20.17 A general framework showing the relative importance assigned to intraspecific isolation vs. adaptation for evaluating the relative evidence in support of ESUs or other types of conservation units. Moritz's (1994) reciprocal monophyly of mitochondrial DNA concept focused almost entirely on isolation; Crandall et al.'s (2000) exchangeability concept placed more emphasis on adaptation; and Waples' (1991) and Dizon et al.'s (1992) frameworks placed roughly equivalent weight on both criteria. From Waples & Lindley (2018).

is useful for short-term management, such as delineating hunting or fishing areas, setting local harvest quotas, and monitoring habitat and population status (Palsbøll et al. 2007).

MUs often are subpopulations within a major metapopulation that represents an ESU. For example, fish populations are often structured on hierarchical levels, such as small streams (as MUs) that are nested within a major river drainage (ESU). MUs, unlike ESUs, generally do not show long-term independent evolution or strong adaptive differentiation. MUs should represent populations that are important for the long-term persistence of an ESU or species. The conservation of multiple populations, not just one or two, is critical for ensuring the long-term persistence of species (Hughes et al. 1997; Hobbs & Mooney 1998).

Waples & Gaggiotti (2006) have suggested that, based on work by Hastings (1993), the transition from demographic dependence to independence generally occurs when the fraction of immigrants in a subpopulation falls below 10%. Hastings (1993) identified the 10% threshold as the point where population dynamics in two patches transitions from behaving independently to behaving as a single population (Section 19.4).

Moritz (1994) originally defined the term "management unit" (MU) as a population that has substantially divergent allele frequencies at nuclear or mtDNA loci. This view was supported by Bentzen (1998), who said that significant genetic differences and departure from panmixia are strong evidence that the populations are demographically independent, and should be considered as separate MUs. However, Palsbøll et al. (2007) found that significant genetic differentiation can be detected consistently, even with migration rates greater than 20%, if many genetic markers are used.

We saw in Chapter 9 that genetic divergence is largely a function of the absolute number of migrants (mN). However, demographic independence is primarily a function of the proportion of migrants (m). Thus, allele frequency differentiation (e.g., F_{ST}) should not be used by itself to identify MUs (Figure 19.9). For example, large populations experience little drift (and little allele frequency differentiation) and thus can be demographically independent even if allele frequencies are similar. The same mN (and hence F_{ST}) can result in different migration rates (m) for different population sizes (N). As N goes up, m goes down for the same F_{ST} (Table 20.4). So in a large population, the proportion of migrants can be very small, and the population could be demographically independent, yet have a relatively low F_{ST}.

A related difficulty is determining whether migration rates would be sufficient for

Table 20.4 Inferring demographic independence of populations by using genetic differentiation data (F_{ST}) requires knowledge of the effective population size (N_e). Here, the island model of migration was assumed to compute mN_e (number of migrants) and m (proportion of migrants) from the F_{ST} (as in Figure 9.6). Recall that the effective population size is generally far less than the census size in natural populations (Section 7.10).

F_{ST}	N_e	m	mN_e	Demographic independence
0.06	50	0.080	4	Unlikely
0.06	100	0.040	4	Likely
0.06	1,000	0.004	4	Yes

recolonization on an ecological timescale; for example, if an MU became extinct or overharvested. Allele frequency data can be used to estimate the number of migrants (mN), but at moderate to high rates of migration ($m > 0.01–0.10$), genetic estimators are notoriously imprecise (Faubet et al. 2007), such that confidence intervals for the mN estimate might include infinity (Waples 1998).

20.5.3.1 Oversplitting and undersplitting

Two general errors can occur in MU diagnosis, as with ESU diagnosis. First, identification of too few units could lead to underprotection, which could lead to the reduction or loss of local populations. This problem could arise, for example, if statistical power is too low to detect genetic differentiation when differentiation is biologically significant (Box 20.3). For example, too few MUs (and underprotection) could result if only one MU is identified when the species is actually divided into five demographically independent units. Imagine that the sustainable harvest rate is 2% per year on the basis of total population, but that all the harvest comes from only one of the five MUs. Then the actual harvest rate for the single harvested MU is 10% (assuming equal sizes for the five MUs). This high harvest rate could result in overexploitation and perhaps extinction of the one harvested MU population. For example, if the harvested population's growth rate is only 4% per year and the harvest rate is 10%, overexploitation would be a problem (Taylor & Dizon 1999).

Here, undersplitting could result from either a lack of statistical power (e.g., due to too few data), or to the misidentification of population boundaries (e.g., due to cryptic population substructure). To help avoid misplacement of boundaries, researchers should sample many individuals that are widely distributed spatially, and use individual-based statistical methods (Section 20.4.2).

Second, diagnosing too many MUs (oversplitting) could lead to unnecessary waste of conservation management resources. This error could occur if, for example, populations are designated as MUs because they have statistically significant differences in allele frequencies, but this differentiation is not associated with important biological differences (Box 20.3). This becomes a potential problem as more and more molecular markers are used that are highly polymorphic and thus statistically powerful, although there is little evidence to date of this being a problem. For example, analysis of a large number of SNPs in redpolls did not lead to greater identification of genetic units, as described in Section 20.5.1.2. Similarly, in a study of Australian *Pelargonium* plants, analysis of SNPs in conjunction with other morphological and reproductive traits led to revised taxonomy through both splitting and lumping of taxa (Nicotra et al. 2016).

20.6 Integrating genetic, phenotypic, and environmental information

Many kinds of information should be integrated, including life history differences, environmental characteristics, phenotypic divergence, and patterns of gene flow for the identification of conservation units (Figure 20.18). For example, if two geographically distant populations (or sets of populations) show large molecular differences that are concordant with life history (e.g., flowering time) and morphologic (e.g., flower shape) differences, we would be relatively confident in designating them as two geographic or population units important for conservation. A recent study found strong support for two distinct ESUs in cycads based on combined analysis of 389 SNPs and morphological traits (Gutiérrez-Ortega et al. 2018; Example 20.3).

Researchers should always consider whether the environment or habitat type of different populations has been different for many generations, because this could lead to adaptations (even in the face of high gene flow) that are important for the long-term persistence of species. The more kinds of independent information that are concordant, the more certain one can be that a population merits recognition as a distinct conservation unit. The principle of considering multiple data types and testing for concordance is critical for identifying conservation units.

Difficulties arise when concordance is lacking among data types (Dussex et al. 2018a). For example, imagine that two populations show morphological differences in size or color of individuals, but show evidence of extensive recent gene flow. This scenario has arisen occasionally in studies that measure phenotypic traits from only small samples or nonrepresentative samples of individuals from each population (e.g., only 5–10 individuals of largely different sexes or ages from each population). In this example, taxonomic oversplitting results from biased or limited sampling, and conservation status as distinct units is not warranted. This hypothetical example relates to the green/black turtle species delineation dilemma described in Section 20.5.1, where more extensive sampling and studies of life history and adaptive traits would be helpful.

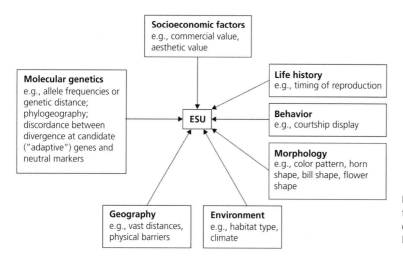

Figure 20.18 Sources of information that can help diagnose a population (or set of populations) as an ESU. Modified from Moritz et al. (1995).

Example 20.3 Integrating genetic and morphological data for delineating ESUs in cycads

Cycads are an ancient group of tropical and subtropical gymnosperms that are considered the most threatened group of plants in the world, with 88% of cycad species included on the International Union for Conservation of Nature (IUCN) Red List of Threatened species. The cycad species *Dioon sonorense* (Zamiaceae) is found along a climatic gradient in northwest Mexico, with the southern populations occurring in tropical forests and the northern populations in more xeric environments in the Sonoran Desert. As only the southern populations receive protection, Gutiérrez-Ortega et al. (2018) examined variation at 389 SNPs, 16 macromorphological, and nine epidermal (leaf) traits to test whether the northern populations harbor genetic or morphological variation not found in the southern populations that should also be conserved. They uncovered distinct northern and southern clusters that are consistent with observed morphological and epidermal trait variation. Southern and northern populations were clearly differentiated in all classes of data analyzed (Figure 20.19). Based on these results, the authors conclude that the northernmost populations of *D. sonorense* represent an ESU that also merits conservation. This study shows how integrating genetic and multiple phenotypic datasets is an effective strategy for conclusively delineating intraspecific units for conservation.

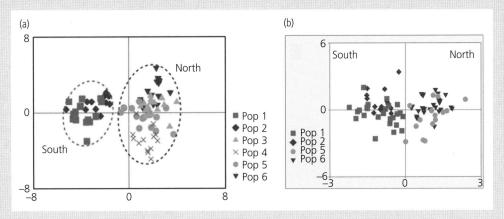

Figure 20.19 PCoA plots showing (a) variation among individuals at 389 SNPs and (b) 16 morphological traits. Southern and northern *Dioon sonorense* cycads are differentiated based on both classes of data. From Gutiérrez-Ortega et al. (2018).

20.6.1 Adaptive genetic variation

The incorporation of adaptive gene markers and gene expression studies can augment our understanding of conservation units (Funk et al. 2012b). For example, adaptive and neutral molecular variation can be integrated (Gebremedhin et al. 2009) by considering two separate axes in order to identify populations with high distinctiveness for both adaptive and neutral diversity (Figure 20.20).

Adaptive genetic markers often give similar patterns of population relationships as neutral markers. Langin et al. (2018) compared patterns of differentiation among populations of white-tailed ptarmigan at 14,693 presumably neutral SNPs and

77 presumably adaptive outlier loci (Figure 20.10). Both categories of loci identified ptarmigan from a disjunct southern population centered in Colorado, USA, as well as ptarmigan from Vancouver Island, British Columbia, Canada, as distinct groups, supporting their delineation as subspecies. This result suggests that neutral loci are adequate to predict patterns of divergence at adaptive loci.

In contrast, in other cases, adaptive loci identify patterns that are different from neutral loci or that are not detected with neutral loci, which can have important conservation ramifications. For example, Lamichhaney et al. (2012) compared Atlantic herring with herring from the Baltic Sea for several

Adaptive molecular or
phenotypic difference

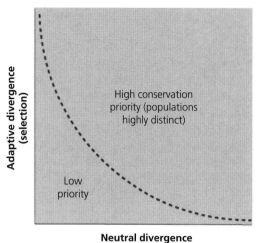

Figure 20.20 Adaptive information can be integrated with information from neutral markers and information on long-term isolation. Such an approach could help identify the most appropriate source population to translocate individuals into small, declining populations that might require supplementation to persist. This approach could also help rank or prioritize populations for conservation management. From Luikart et al. (2003).

hundred thousand SNPs. Previous studies with allozyme and microsatellite loci found little genetic differentiation, but a few loci showed significant genetic differentiation that was associated with a dramatic salinity gradient (Larsson et al. 2007; Gaggiotti et al. 2009; Limborg et al. 2012). The great majority of SNPs genotyped by Lamichhaney et al. (2012) showed little differences in allele frequency between the eight populations. However, several thousand SNPs showed striking differences in allele frequency, some approaching fixation for different alleles in the Atlantic and Baltic groups (Figure 20.21). The striking differences among loci apparently reflect adaptive differences between the Atlantic and Baltic population groups. Simulations confirmed that the distribution of the F_{ST} values deviated significantly from expectation for selectively neutral loci. These differences at adaptive loci are being used by managers to identify the origin of herring caught in a mixed-stock fishery (Bekkevold et al. 2015). This study illustrates the potential of adaptive markers to reveal patterns of differentiation not detectable with neutral loci. Other studies have also found that adaptive loci can yield substantially different inferences about population genetic relationships than neutral loci (Hancock et al. 2011; Funk et al. 2012b; Prince et al. 2017).

How should we identify and manage conservation units in these scenarios? Pitfalls arise when focusing on a set of adaptive loci rather than neutral patterns or genome-wide averages, largely because selection can be extremely complex and a complete understanding of adaptive divergence is unattainable (Luikart et al. 2003; Allendorf et al. 2010). Genes important for contemporary or past adaptations might not be those that will be crucial for adaptation in future environments. In addition, a focus on conserving variation at certain detectable adaptive genomic regions could result in loss of important genetic variation at other adaptive or neutral regions. Moreover, even when the same genomic regions are implicated in, for example, local adaptation across populations, the particular alleles involved may be different and perhaps even result in outbreeding depression when combined through admixture.

In summary, caution must be used when using adaptive loci for the identification of conservation units. Genetic patterns at neutral markers largely reflect the historic interaction of gene flow and genetic drift that are expected to affect the amount of genome-wide genetic variation within, and genetic divergence among, populations. These patterns are the foundation upon which natural selection

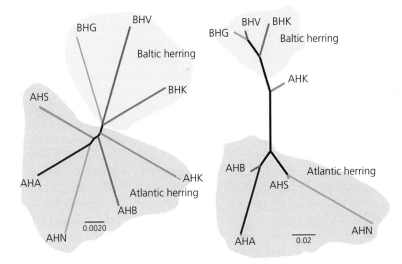

Figure 20.21 Neighbor-joining trees showing relationships among Atlantic and Baltic herring for (left) several hundred thousand SNPs and (right) only the 3,847 outlier SNPs that showed greater differentiation. The pattern with all SNPs is star-like with little differentiation between the Atlantic and Baltic population groups. The outlier SNPs show a clear pattern of differentiation between these two population groups. Note that the scales are different; the bars at the bottom represent 0.002 (left) and 0.02 (right) genetic distance units. From Lamichhaney et al. (2012).

operates to bring about adaptive differences among populations. Loci under selection generally should be used (1) as a supplement or complement to neutral loci, and (2) as one source of adaptive information that should be interpreted in combination with other data on phenotypes, life history, and the environment.

20.7 Communities

Identification of groups of multiple co-distributed species might help to delineate conservation units and the spatial boundaries between units. One example is the discovery of congruent genealogical patterns among many species distributed across the southeastern USA. Numerous species have phylogeographically concordant patterns across the Coastal Plain of the southeastern USA, such that a phylogeographic break (Avise 1996) occurs for many taxa near the Apalachicola River Basin in western Florida. This drainage system is the range limit and/or contact zone for reptiles, amphibians, fishes, mammals, spiders, trees, and estuarine macroinvertebrates (e.g., Engle & Summers 2000); it is also a phylogeographic barrier within many species (Figure 20.13; Avise 1992, 1996). Pauly et al. (2007) found that the flatwoods salamander has a phylogeographic break (species boundary) near the same river basin. The authors suggested that in

the absence of taxon-specific data, the existence of concordant multitaxa spatial distributions can provide strong predictions for locations of conservation units for unstudied species of conservation concern.

In contrast, analysis of >375 nuclear loci and mitochondrial genomes of four frog species in the southeastern USA found little correspondence of genetic patterns to putative biogeographic barriers and high discordance among species (Barrow et al. 2018). The authors suggest that variation in phylogeographic structure may be due to differences in the natural histories of these four species, as two species were habitat generalists with less genetic structure. This study highlights the need for caution in generalizing phylogeographic patterns beyond the species included in a given genetic study.

Similarly, if interacting species have concordant phylogeographic structure due to coevolution, then the phylogeographic distinctiveness of populations of one species could support the distinctiveness among populations within the other species. Criscione & Blouin (2007) found congruence in phylogeographic patterns between ESU boundaries of four Pacific salmonid species and of a trematode parasite. The authors suggested that the pattern from the trematode supports delineation of salmon ESU boundaries as biologically reasonable, as does the phylogeographic concordance among the four salmonid species.

Biek et al. (2006) used a rapidly evolving virus to help identify puma population boundaries and to assess population connectivity. Their study suggested that virus population genetic structure could help to identify puma MUs useful for conservation. Studies of multiple parasite species could further help to identify MUs or ESUs. For example, Whiteman et al. (2007) compared the phylogeographic and population genetic structure of three parasites of an endangered Galápagos hawk. This and other studies indicate that phylogeographic patterns of parasites can be predicted by the biogeographic history of their hosts, and that parasites can provide independent support for ESU boundary delineation. Landscape genomic approaches can be applied to multiple species including multiple host taxa and parasites to help identify MUs or ESUs (Hand et al. 2015).

Microbial community assemblages could also help to diagnose conservation units. **Community phylogeny**, **metagenomics**, and **metabarcoding** could identify distinct bacterial assemblages inferred by combining sequence data and ecological data to identify communities that represent both evolutionarily and ecologically distinct units (Cohan 2006). Microbial metagenomic spatial patterns could help to delineate distinct environments and geographic areas useful to help delineate conservation units for animal and plant species (e.g., Coleman & Chisholm 2010). Metagenomics is leading to new ways of thinking about microbial ecology and community ecology that supplant the concept of species or distinct populations in the sense of ESUs and MUs. In this paradigm, communities are becoming the units of evolutionary and ecological study among bacteria, archaea, and perhaps protists and fungi (Doolittle & Zhaxybayeva 2010). Metagenomics, metabarcoding, and phylogenetics also can be applied to metazoans to help identify communities and delineate MUs (e.g., using host–parasite metazoan–microbe assemblages, or environmental DNA sampled from many species in water).

Guest Box 20 Using genomics to reveal conservation units: The case of Haida Gwaii goshawks
Kenneth K. Askelson, Armando Geraldes, and Darren Irwin

Modern genomic analysis can be used to identify evolutionarily differentiated populations that may become conservation priorities. An example is provided by the northern goshawk, a charismatic bird of prey (Figure 20.22). Within North America, two subspecies have been recognized, *Accipiter gentilis atricapillus*, which is found across most of Canada and the USA, and *A. g. laingi*, which is usually described as inhabiting the coastal regions of British Columbia, southeast Alaska, and northwestern Washington. The *laingi* subspecies was first described by Taverner (1940) based on a small sample of individuals from the Haida Gwaii archipelago of British Columbia, which he described as being darker in plumage than *atricapillus*. This became the primary basis for the subspecies designation, but the coloration difference is subtle, leaving the ranges of the two forms uncertain. Given these challenges, habitat modeling has been used to define the range of *laingi* as coastal rainforest (COSEWIC 2000, 2013).

Due to small and declining population sizes combined with ongoing habitat destruction and alteration, the *laingi* subspecies has been formally listed as Threatened under SARA in Canada (COSEWIC 2000, 2013) and under the US ESA (USFWS 2012; COSEWIC 2013). Effective conservation policy and management has been complicated by uncertainty regarding the range of *laingi* and the difficulty of assigning individuals to this subspecies.

Working with government and industry wildlife biologists, we used a genomic approach to solve this problem (Geraldes et al. 2019). One challenge was obtaining high-quality genetic samples across a wide geographic range, as goshawks are difficult to sample. To make use of low-quality feather samples, we used a two-phased approach. First, we utilized high-quality samples and genotyping-by-sequencing to determine genotypes at tens of thousands of SNPs throughout the genomes of our sampled individuals. This allowed the identification of SNPs with strong geographic differentiation. Second, we developed genotyping assays for those highly differentiated SNPs and used those assays to genotype our low-quality samples.

The genome-wide data revealed a clear result (Geraldes et al. 2019): the strongest differentiation throughout North America was between Haida Gwaii and elsewhere. In contrast, other parts of the coast of British Columbia, Alaska, and Washington showed relatively subtle differentiation compared with elsewhere in North America. Genotypes at just 10 highly differentiated SNPs enabled confident assignment of individuals to the two major genetic groups (Figure 20.23).

The finding that the Haida Gwaii population is distinct is of much conservation importance, because this population has only an estimated 32–38 mature breeding individuals (COSEWIC 2013) and hence is at great risk of extinction. Furthermore, this result has prompted biologists and conservationists to debate whether the *laingi* name should now be used just for the birds from Haida Gwaii, or rather whether Haida Gwaii should be considered an important conservation unit (e.g., an ESU; Box 20.4) within the broader range of *laingi* as currently defined. This debate is complicated by there being no universally agreed-upon definition for subspecies. Despite these complexities, the finding that Haida Gwaii is the home of a genomically unique form of goshawk that is at extreme danger of extinction illustrates the importance of the use of genomic analysis as well as the potential for collaborations between academia, governments, and industry to advance effective conservation.

Figure 20.22 An adult goshawk. Photo courtesy of Adrian Rus.

Guest Box 20 Continued

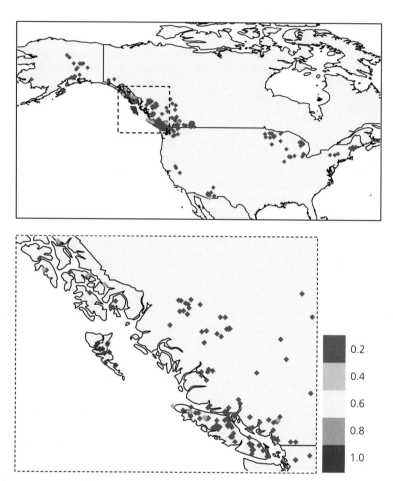

Figure 20.23 Map of North America (top) and the west coast of Canada (bottom; area indicated by the dotted lines in the upper map), with diamond symbols representing source localities of goshawks included in the study. The color of the symbols represents the estimated proportion of genetic ancestry from two genetic clusters (according to the scale bar on the lower right). Goshawks on Haida Gwaii (the island archipelago with red symbols) are distinctly different from those from all other sampling regions in the study. From Geraldes et al. (2019).

CHAPTER 21

Conservation Breeding and Restoration

Mellblom's spider orchid, Section 21.8

A major challenge of *ex situ* conservation will be to ensure that sexually propagated samples of rare plants do not become museum specimens incapable of surviving under natural conditions.

(Spencer Barrett & Joshua Kohn 1989, p. 32)

Although there are distinct differences between the high-tech approach of the age of genomics and the low-tech approach of pedigree analysis and animal breeding, the two approaches should be viewed as complementary rather than in conflict. The most effective genetic conservation will be achieved if we wisely combine the strengths of each.

(Robert C. Lacy 2009, p. 63)

Captive breeding represents the last chance of survival for many species faced with imminent extinction in the wild (Conde et al. 2011). For example, 28% of recovery plans for vertebrate species listed under the Environment Protection and Biodiversity Conservation Act (EPBC Act) of Australia include a captive breeding action (Harley et al. 2018). The Guam rail, black-footed ferret, and the kākāpō (Guest Box 6) would all almost certainly be extinct if the last few remaining individuals in the wild were not captured and brought into captivity where they have been bred successfully. Captive breeding has played a major role in the recovery of 17 of the 68 vertebrate species whose International Union for Conservation of Nature (IUCN) threat level has been reduced (Conway 2011). Less charismatic and well-known animal species have also avoided extinction by captive breeding programs. The white abalone became the first marine invertebrate to be listed under the US Endangered Species

Conservation and the Genomics of Populations, Third Edition. Fred W. Allendorf, *et al.*, Oxford University Press.
© Fred W. Allendorf, W. Chris Funk, Sally N. Aitken, Margaret Byrne, and Gordon Luikart (2022). DOI: 10.1093/oso/9780198856566.003.0021

Act (ESA) in 2001. A captive breeding program was begun in 1999 to bring this species back from the brink of extinction and establish a self-sustaining population in the wild (USGS 2002; Rogers-Bennett et al. 2016).

Several plant species also have been rescued from extinction by similar intervention. *Kokia cookei* is one of Hawai'i's most beautiful and endangered plants (Mehrhoff 1996). Four seeds were collected from the last remaining tree in 1915. Only one mature tree resulted from these four seeds and none of its progeny survived. In 1976, a branch from the last remaining tree was successfully grafted onto a closely related species. Twenty-eight grafted trees were transplanted back to Moloka'i in 1991 (USFWS 2008).

The IUCN has defined *ex situ* **conservation** as "the conservation of components of biological diversity outside their natural habitats" (IUCN 2002). There are various *ex situ* (or **offsite**) techniques that are potentially valuable tools in the conservation of a wide variety of taxa that are threatened with extinction (e.g., captive breeding and **germplasm** banking). Approximately 20% of all bird and mammal species recognized with **IUCN conservation status** are maintained in zoos (Conde et al. 2011). Unfortunately, there are fewer than 50 individuals in zoos for just over half of all these species (Conde et al. 2011). Moreover, there are at least 77 species of plants and animals that are extinct in the wild and only exist in captivity (Trask et al. 2020).

We use the more general term **conservation breeding** to include efforts to manage the breeding of plant and animal species that do not necessarily involve captivity. For example, kākāpō breeding is managed by moving groups of birds to predator-free islands, but they are not held in captivity (Guest Box 6).

Captive breeding has played a major role in the development of conservation biology. The first book on conservation biology (Soulé & Wilcox 1980) devoted five of 19 chapters to captive breeding. Modern conservation genetics had its beginnings in the use and application of genetic principles for the offsite preservation of plant genetic resources (Frankel 1974) and the development of genetically sound protocols for captive breeding programs

in zoos (Ralls et al. 1979). Some early conservation biologists equated conservation genetics with captive breeding.

The maintenance of genetic diversity and demographic security are the primary goals for management of conservation breeding programs. These two goals often are compatible. However, there are situations in which maintaining the genetic characteristics of a population can reduce the population growth rate so that a conflict arises (Guest Box 21). This is most likely to occur when a species with only a few remaining individuals is brought into captivity in a last-ditch effort for survival. Demographic security will best be achieved by rapidly increasing the census size of the captive population.

Maintenance of genetic diversity generally requires maximizing effective population size by reducing variation in reproductive success among individuals (Chapter 7). However, some individuals or pairs of individuals can be much more successful in captivity than others. Thus, maximizing the growth rate of a captive population can actually reduce the effective population size and result in more rapid erosion of genetic variation. In addition, allowing just a few founders to produce most of the captive population is expected to accelerate the rate of adaptation to captive conditions. Thus, maintaining the genetic characteristics of a captive population might come at the cost of reduced population growth rate.

Our goal in this chapter is to consider the genetic issues involved in conservation breeding and the introduction of individuals into the wild (Doremus 1999). We also provide an overview of the principles involved in actually genetically managing captive populations. Interested readers should consult other sources that provide detailed instructions for genetic management of conservation breeding programs (e.g., Ballou et al. 2010; Miller & Herbert 2010).

21.1 The role of conservation breeding

There are three primary roles of offsite conservation breeding as part of a management or recovery program to conserve a particular species: (1) provide demographic and genetic support for wild populations; (2) establish sources for founding new populations in the wild; and (3) prevent extinction of species that have no immediate chance of survival in the wild.

The genetic objectives of these three roles are very different. Captive individuals used to provide demographic and genetic support for wild populations should be genetically matched to the wild population and environment into which they will be introduced so that they do not reduce the fitness of the population by **outbreeding depression**. In contrast, introduced new populations should have enough genetic variation present so that they can become adapted to their new environment by natural selection. For the last role, the initial concern of a captive breeding program is to ensure that the species can be maintained in captivity (Midgley 1987). This might involve preferentially propagating individuals capable of reproducing in captivity and might result in adaptation to captivity.

Captive breeding can also make contributions to conservation through public education, research, and professional training, and entrance fees and donations to zoos can support captive breeding programs. Captive breeding populations in zoos and botanical gardens also contribute to public display of species, providing opportunities for the public to come into contact with a wide variety of species that might otherwise just be names or photos. The first author of this book became interested in biology because of visits to the Philadelphia Zoo as a child on class trips.

The goals of a display program are to establish an easily managed population that is well adapted to the captive environment (Frankham et al. 1986). These experiences provide an excellent opportunity for education and also provide the setting for the public to develop affection for and appreciation of a wide variety of species. Most people around the world will never have the opportunity to see a tiger, elephant, or a great ape in the wild. Zoos provide an important role in allowing the public to develop a first-hand connection to these species. People are more likely to support conservation efforts if they have knowledge, understanding, and appreciation of the species involved.

There is also a danger in this. Seeing elephants or tigers in the zoo can encourage the public and politicians to believe that these species are now protected from extinction. However, a species is not a collection of individuals that has been removed from the ecosystem in which they have survived and evolved for millions of years (Devall & Sessions 1984, p. 317):

A condor is 5 percent feathers, flesh, blood, and bone. All the rest is place. Condors are soaring manifestations of the place that built them and coded their genes.

The behavioral ecologist David Barash (1973, p. 214) has said this in a somewhat different fashion:

Thus, the bison cannot be separated from the prairie, or the epiphyte from its tropical perch. Any attempt to draw a line between these is clearly arbitrary, so the ecologist studies the bison-prairie, acacia-bromeliad units.

Thus, the display of charismatic species to the public should be accompanied with educational efforts that emphasize that long-term species existence can only occur within the complex web of connections and interactions in their native ecosystems.

21.1.1 When is conservation breeding an appropriate tool for conservation?

This is a crucial and difficult question. Conservation breeding should be used sparingly because it is difficult, expensive, and worldwide, resources are limited. In addition, directing resources to captive breeding and taking individuals into captivity might hamper efforts to recover species in the wild.

Captive breeding is perhaps too often promoted as a recovery technique. For example, Conservation Assessment and Management Plans under the Conservation Breeding Specialist Group of the IUCN have recommended captive breeding for 36% of the 3,314 taxa considered (Seal et al. 1993). In the USA, captive breeding was recommended in 64% of 314 approved recovery plans for species listed under the ESA (Tear et al. 1993). The resources are not available to include captive breeding in the recovery plans of such a high proportion of species (Conde et al. 2013). It should be used only for those species in which it can have the greatest effect.

Intensive field-based conservation may be the most effective and cost-efficient approach. Balmford et al. (1995) found that *in situ* management of well-protected reserves for large-bodied mammals resulted in comparable population growth rates and was consistently less expensive than captive propagation. These authors suggest that captive breeding is most cost-effective for smaller-bodied taxa and will only remain the best option for large mammals that are restricted to one or two vulnerable wild populations.

21.1.2 Priorities for conservation breeding

It is clear that only a relatively small proportion of the thousands of animal species that are threatened in the wild can be maintained in captivity because of constraints on space and other resources (Balmford et al. 1996; Snyder et al. 1996). It is generally assumed that a maximum of ~500 animal species could be maintained offsite in conservation breeding programs (IUDZG/CBSG 1993). As we have seen, however, captive breeding programs are often recommended for many taxa. Given this situation, what criteria should be used to determine which species should be maintained in conservation breeding programs?

Zoos have historically focused on large and charismatic species in breeding programs. Balmford et al. (1996) presented three general sets of criteria that should be considered in evaluating animal species for which captive breeding may be necessary:

1. Economic considerations. Which species can be conserved successfully in a captive breeding program most economically?
2. Biological suitability for captive breeding. Which species can be bred and raised successfully in captivity?
3. Likelihood of successful reintroduction. For which species is successful reintroduction to the wild a realistic option?

We suggest a fourth criterion: the potential effect on habitat preservation. Will development of a captive breeding program increase or decrease the likelihood of habitat protection?

Invertebrates are generally more cost-effective for captive breeding than are large and charismatic vertebrates for which enormous resources are required (Pearce-Kelly et al. 1998). Invertebrates have a relatively high probability of success for both the rearing and release phases. They also have small size

and require relatively little space and cost. They typically have high reproductive potential and population size increases relatively rapidly in captivity and after release. Finally, there is a wealth of knowledge and techniques for rearing numerous invertebrate species. For example, crickets, katydids, beetles, and butterflies have been widely and successfully raised in captivity.

A tragic accident with *Powelliphanta augusta*, a large, carnivorous land snail, demonstrates some of the dangers of captive breeding. This species was discovered in 1996 and was identified as a new species on the basis of molecular genetics and morphology (Walker et al. 2008). This species is only known from a small area on the South Island of New Zealand. The entire known habitat of this species was destroyed by coal mining by 2006. Some 2,300 individuals were translocated to two other areas, where their survival suggests that the populations will not persist. Some 1,600 snails were taken into captivity to develop a captive breeding program. However, a refrigerator malfunction resulted in 800 snails being frozen to death in November 2011 (Forest & Bird 2011).

Plants generally are better candidates for offsite breeding programs than animals for a variety of reasons (Li & Pritchard 2009). Many plants can be maintained for long periods as dormant seeds dried and often frozen in seed banks. This can be used to increase the generation interval and therefore reduce the rate of genetic change during offsite breeding. Other plants, such as trees, live a long time so that offsite breeding programs that might take hundreds of years might only represent a handful of generations. This again will minimize the rate of genetic change by genetic drift and selection. While many plant collections are maintained as seed banks, they can also be maintained through tissue culture, clonal lines, and in seed production areas. Other problems with offsite breeding can be reduced because of the variety of modes of reproduction that are possible for plants with short generation intervals (e.g., selfing, apomixis, and clonal reproduction; Figure 21.1).

Guidelines for undertaking **conservation collections** have been presented by the Center for Plant Conservation (1991). The decision to collect seeds for germplasm from a particular population or

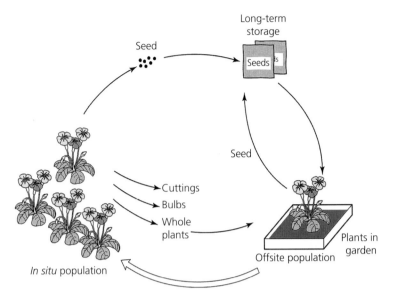

Figure 21.1 Possible modes of reproduction for offsite breeding of plants and possible interchange between offsite plants and *in situ* populations. Redrawn from Brown & Briggs (1991).

species must be made within a larger framework of conservation. The guidelines are based on a natural genetic hierarchy: species, populations (or ecotypes), individuals, and alleles. The goal is to address diversity at several levels of organization rather than sampling a particular species without regard to genetic variation and future long-term viability. This approach includes five sampling decisions:

1. Which species should be collected?
2. How many populations within a species should be sampled?
3. How many individuals should be sampled per populations?
4. How many propagules should be collected from each individual?
5. When should collections be made from multiple years?

21.1.3 Potential dangers of captive propagation

The use of captive breeding has been controversial. It is expensive, is sometimes ineffective, and can harm wild populations both indirectly and directly if not done correctly (Snyder et al. 1996). Perhaps the most serious criticism is that efforts directed toward captive breeding detract from grappling with the real problems (e.g., loss of habitat and protection). The dangers of captive breeding are clearly demonstrated by the use of fish hatcheries to maintain stocks of Pacific salmon on the west coast of North America (Box 21.1).

South Africa has recently declared that 24 species of native mammals (including lions, giraffes, and several zebra species) are landrace breeds, which will allow them to be genetically "improved" by artificial selection (Somers et al. 2020). Game ranching in South Africa occurs over 15% of the country. The escape of these "domesticated" varieties of wildlife could harm native species in a way that will be difficult or impossible to undo.

21.2 Reproductive technologies and genome banking

Reproductive technologies initially developed for agricultural species (e.g., cattle, sheep, and chickens) can be transferred to some related wild species to facilitate their conservation. These technologies include genome banking, cryopreservation, artificial insemination, and cloning.

Genome banking is the storage of sperm, ova, embryos, seeds, pollen, tissues, or DNA. Genome

Box 21.1 Who needs protection? We have hatcheries

Fish hatcheries have a long and generally unsuccessful history in conservation efforts to protect populations of fish. Pacific salmon began a rapid decline on the west coast of the lower USA in the late 1800s with the advent of the salmon-canning industry (Lichatowich 1999). The State of Oregon sought advice from the newly created US Commission on Fish and Fisheries, which was directed by Spencer Baird, a scientist with the Smithsonian Institution. In 1875, Baird (Lichatowich 1999, p. 112) recommended that:

> ... instead of protective laws, which cannot be enforced except at very great expense and with much ill feeling, measures be taken, either by the joint efforts of the States and Territories interested or by the United States, for the immediate establishment of a hatching establishment on the Columbia River, and the initiation during the present year of the method of artificial hatching of these fish.

Unfortunately, this recommendation from the leading fisheries scientist of the USA set in motion a paradigm for the conservation of salmon through hatcheries rather than facing the real problems of excessive fishing, dams that blocked spawning migrations, and habitat changes in the spawning rivers and streams. These efforts failed profoundly (Meffe 1992). Some 26 different groups of Pacific salmon and anadromous rainbow trout (steelhead) are listed as threatened or endangered under the US ESA at the time of writing this chapter. The role of hatcheries in salmon conservation continues to be controversial. There is disagreement if hatchery populations should be considered part of the distinct population segments that are listed and protected under the ESA (Myers et al. 2004).

A number of studies have been performed to assess the possible genetic effects on wild populations of releasing hatchery fish into the wild. The consensus is clear: hybridization with hatchery fish has a dramatic harmful effect on the fitness of wild populations of fish (Reisenbichler & Rubin 1999; Araki & Schmid 2010; O'Sullivan et al. 2020).

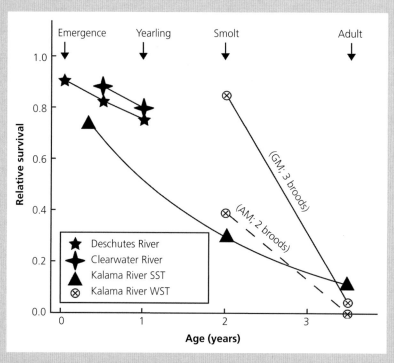

Figure 21.2 Results showing reduction in relative survival throughout the life cycle of progeny from hatchery steelhead spawning in the wild relative to the survival of progeny from wild fish. Data for the Kalama River winter steelhead (WST) are the geometric means (GM) for three year-classes, or the arithmetic means (AM) for two year-classes with an exceptional year-class omitted. From Reisenbichler & Rubin (1999).

Box 21.1 Continued

Reisenbichler and others have published several papers that compare the relative fitness of progeny of hatchery steelhead (the anadromous form of rainbow trout) with that of wild fish. Three primary results emerge from these studies. First, progeny from hatchery fish uniformly show reduced rates of survival. For example, Leider et al. (1990) found that the reproductive success of hatchery fish spawning in the wild relative to wild fish ranged from 5–15% in four successive year-classes. Second, progeny of hatchery fish have reduced survival at all life history stages between emer-

gence from the gravel until returning from the ocean as adults (Figure 21.2; Reisenbichler & Rubin 1999). Finally, the decline of fitness observed in hatchery fish is proportional to the number of generations that the hatchery stock has been maintained in captivity, but can occur in a single generation (Christie et al. 2012). Chilcote et al. (2011) found that intrinsic measures of population productivity of populations of steelhead, coho salmon, and Chinook salmon in Oregon declined as the number of hatchery fish in the spawning population increased.

---

resource banking can help move genetic material without moving individuals. It might, for example, allow for **genetic rescue** or **assisted gene flow** (Box 19.1) into isolated populations without the risks of translocating individuals. Genome banking also serves as an insurance against population or species extinction. It lengthens generation intervals and thereby reduces random genetic drift. It increases efficiency of captive breeding and reduces the number of individuals kept in captivity. Finally, banks are a source of tissue and DNA for basic and applied research.

Genome banking is widely used for agricultural crop and farm animal preservation to help ensure future agricultural productivity. Genome banking is also increasingly used for wild taxa (Chemnick et al. 2009; Breithoff & Harrison 2020). For example, for wild animals, there exists a genome bank at the Smithsonian Institution's National Zoo where there are more than 1,500 samples of frozen sperm or embryos from 69 species (including ~2% of mammalian species worldwide). Similarly, the San Diego Zoo maintains a "frozen zoo" with samples (including cell lines and tissues) from more than 7,000 species of endangered mammals, birds, and reptiles.

For wild plants, there is increasing interest in establishing seed banks. There are a large number of seed banking initiatives for threatened species, and common and agricultural plants throughout the world. For example, the Millennium Seed Bank at

Kew Gardens in the United Kingdom was established to store seeds from plants throughout the world and has over 2.3 billion seeds from over 39,000 different plant species. The FAO (2010) reported that 6 million accessions exist in over 1,300 seed banks around the world. However, less than 10% are from wild plants.

Cryopreservation is the freezing and storage (often in liquid nitrogen at −196°C) of sperm, ova, embryos, seeds, or tissues to manage and safeguard against loss of genetic variation in agricultural or wild populations. It is the principal storage method for animal material. In plants, seeds are often preserved dry at 15% relative humidity and at −20°C where they can remain viable for 50–200 years. But for some plant species recalcitrant under normal conditions or with short-lived seeds, cryostorage will increase longevity of seed storage (Streczynski et al. 2019). Tropical and rainforest species are often recalcitrant to normal seed banking storage conditions but can be stored in tissue culture. Tissue culture propagules are often maintained through cryopreservation.

Artificial insemination is a widely used and important technique for captive breeding. Artificial insemination allows animals to breed that would not breed naturally, perhaps due to behavior problems such as aggression toward mates. Further, a genetically important male can still be used within a breeding program long after his death. Finally, instead of moving animals, sperm can be collected and cryopreserved and shipped for artificial insemination. Artificial insemination has been used, for example, in breeding programs for the black-footed ferret and killer whales in the USA, the cheetah in Namibia, koalas in Australia, and gazelles in Spain and Saudi Arabia. Artificial insemination also has been used successfully with corn snakes at the Henry Doorly Zoo in Omaha, Nebraska, USA (Mattson 2007).

While propagating many plant species through cloning methods including rooting stem cuttings, grafting, and tissue culture is common and straightforward, the use of cloning for animal conservation is more difficult and controversial (Box 21.2). Cloning animals is generally conducted by (1) removing the nucleus from a donor egg cell of the animal that will carry the cloned embryo, and (2) injecting into the carrier's egg cell the nucleus from a cell of the animal to be cloned. For example, a nucleus from a tissue cell of a European wild sheep (mouflon) was injected into the nucleus-free cell of a close relative species, the domestic sheep. The resulting mouflon lamb was born and mothered by the domestic sheep. This is an example of cross-species cloning, which is more difficult than within-species cloning because of risks of incompatibility of mitochondrial genes from the donor egg and the nuclear genes from the animal to be cloned.

These technologies provide valuable opportunities for protecting species and increasing genetic variation within species on the brink of extinction. Nevertheless, it is essential that they be integrated so that they support ongoing conservation efforts rather than being used as alternatives.

21.3 Founding populations for conservation breeding programs

Developing a captive breeding program begins with the selection of the founding individuals. In many situations when a species is on the brink of extinction, there is no choice involved because all remaining individuals are brought into captivity. However, in other cases, captive breeding programs are established when long-term survival of a species in the wild is unlikely even though many individuals currently occur in the wild (e.g., tigers). In such cases, there are several questions to be decided. Which subspecies or populations should be the source of individuals to be brought into captivity? How many subspecies or populations should be maintained? Should subspecies and populations be maintained separately or mixed together? How many individuals should there be in the founder population?

Confirming the taxonomic identification of individuals used in a captive breeding program is essential. There are many examples of captive breeding programs in which molecular analysis detected that hybrid individuals were unknowingly used in the captive breeding program: Przewalski's horse (Oakenfull & Ryder 1998), greater rhea (Delsuc et al. 2007), and cutthroat trout (Metcalf et al. 2007).

We saw in Section 13.1 that morphological analysis is not a reliable way to detect hybrids and hybridization. Guest Box 13 showed pictures of four wild canids from remote regions of Australia and asked the reader to guess which ones are dingoes. Somewhat appropriately, the answer to this question is ambiguous. Most likely all of them are 100% dingo because of where they were photographed, even the ones on the right. Some of the bottom ones have dark muzzles, but it is unclear if this indicates hybridization. The bottom line is that they were all functionally dingoes, but their "purity" is unknown.

A study of 15 microsatellite loci in a breeding program for an endangered giant Galápagos tortoise from the island of Española discovered that one of 473 captive-bred tortoises released back to Española carried alleles at eight of 15 loci that were not present in any of the 15 founders (Milinkovich et al. 2007). Bayesian clustering analysis of microsatellite genotypes at nine loci suggested that this anomalous individual was a hybrid between an Española tortoise and a tortoise from the island of Pinzón (see Figure 21.3). It is thought that this individual was transported by humans to Española before the initiation of the captive breeding program. Efforts are currently underway to identify individuals carrying genes from Pinzón tortoises and remove them from the population.

Box 21.2 How useful is de-extinction of animals through cloning or gene editing?

The concept of de-extinction has attracted wide attention in the past few years (Sandler 2014; Shapiro 2015). The possibility of resurrecting individuals from extinct species is truly exciting. This has been proposed to be done by either cloning entire genomes or by inserting gene sequences from extinct species into living relatives using the **CRISPR-Cas9** (Clustered Regularly Interspaced Short Palindromic Repeats-CRISPR-associated system 9) technique (Section 14.5.4). For example, 14 loci from the extinct woolly mammoth have been used to replace these same 14 loci in the elephant genome (Shapiro 2015).

Although headline making, de-extinction is not currently a realistic conservation tool. Even if it does become possible to clone a few individuals of extinct species through partial or complete genome cloning, there are a host of problems that will make it impossible to recreate a genetically viable population.

Cloning animals is very expensive, and technologically feasible for only a few species that are related to model research organisms (e.g., mice) or important in agriculture (e.g., cattle and sheep). Further, the success rate is very low, less than 0.1 to 5% of re-nucleated embryos lead to a live birth (Holt et al. 2004). It is generally agreed that long-extinct species, such as the woolly mammoth from the frozen Siberian permafrost, cannot be cloned because their DNA is fragmented.

Cloning existing individuals to produce genetically iden-

tical individuals that can bolster population size (although not N_e) could help avoid extinction of critically endangered species such as the giant panda. However, the benefits of cloning compared with sexual reproduction in more traditional conservation breeding programs are questionable and the disadvantages are substantial. Cloning of plants for conservation or restoration can, however, be straightforward and useful, providing demographic support in very small populations.

Disadvantages of cloning are that cloned individuals are genetically identical and thus would be highly susceptible to the same infectious diseases and have low adaptive potential to environmental change. Further, the money spent on cloning would often be better spent preserving habitat and conducting less expensive breeding programs. Extensive healthy habitats are necessary to ensure long-term persistence of any species.

Cloning and genome editing should not be viewed as an alternative to habitat preservation and conservation breeding programs. In certain limited scenarios, perhaps cloning could be a last-resort approach in combination with habitat conservation and breeding programs to help ensure species persistence and even recover extinct taxa. There are a few programs dedicated to storing frozen tissues in hopes that one day these can be used to produce living animals, e.g., the Frozen Ark (http://frozenark.org) and the Frozen Zoo (http://www.sandiegozooglobal.org).

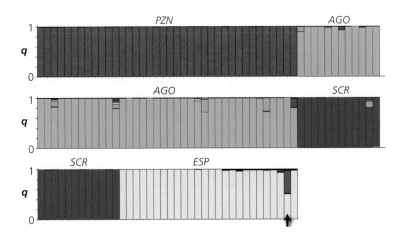

Figure 21.3 Inferred ancestry of giant Galápagos tortoises from the islands of Pinzón (*PZN*), Santiago (*AGO*), Santa Cruz (*SCR*), and Española (*ESP*) based upon nine microsatellite loci using the computer program *STRUCTURE* (Pritchard et al. 2000). The arrow indicates the hybrid individual found on Española that was apparently produced by a mating between Española and Pinzón parents. From Milinkovich et al. (2007).

Example 21.1 Assignment tests identify origins of captive tigers and identify valuable individuals for captive breeding

Tigers are disappearing from the wild; fewer than 3,000 remain (Luo et al. 2010). Three subspecies are extinct and one persists only in zoos. By contrast, captive tigers are flourishing, with 15,000–20,000 individuals worldwide, outnumbering their wild relatives many times over. Researchers have used genetic assignment tests to identify nonadmixed captive tigers that could be used for genetic management of subspecies in captivity and in the wild (Luo et al. 2008).

Subspecies-diagnostic genetic markers were developed by analyzing 134 reference voucher tigers from the five extant subspecies (Luo et al. 2004). Sequences of **mitochondrial DNA (mtDNA)** were used to assign maternal ancestry to a subspecies. The authors also genotyped 30 microsatellite loci and used Bayesian assignment tests implemented in *STRUCTURE* (Pritchard et al. 2000; Section 20.4.2) to calculate the probability (q) that a tiger could be assigned to a single subspecies or alternatively to quantify the extent of admixture among subspecies. The genetic differentiation among subspecies, based on the 134 voucher specimens, was high (microsatellite $R_{ST} = 0.314$, mtDNA $F_{ST} = 0.838$; Luo et al. 2004).

The assignment tests assigned 49 tigers with high certainty ($q > 0.90$) to 1 of 5 subspecies (Amur, Sumatran, Malayan, Bengal, and Indochinese). Fifty-two tigers had admixed subspecies origins (Figure 21.4). For example, 16 tigers had discordant mtDNA and microsatellite assignments and were therefore classified as admixed. Interestingly, 11 captive tigers previously thought to be "purebred" (nonadmixed) were hybrids. However, most assignments (80%) of captives were consistent with the origins provided by owners, including 42 named as a specific subspecies and 41 suspected admixed (Luo et al. 2008).

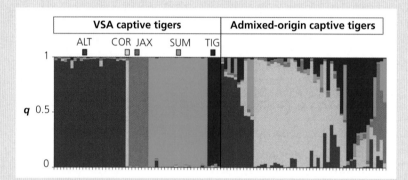

Figure 21.4 *STRUCTURE* population cluster analysis of 101 captive tigers. The left panel shows 49 tigers assigned to a single subspecies based on 134 reference tigers with verified subspecies ancestry (VSA). The right panel shows 52 tigers with apparent admixed origins. Each individual is represented by a thin vertical bar (defined by tick marks under colored regions on the *x*-axis) partitioned into five colored segments representing individual ancestry (q) to the five indicated tiger subspecies. ALT Amur, COR Indochinese, JAX Malayan, SUM Sumatran, and TIG Bengal. From Luo et al. (2008).

The tested captive tigers contain substantial genetic diversity that has not been observed in their wild counterparts. For example, 46 new microsatellite alleles (36 in nonhybridized individuals and 10 in admixed tigers) were observed that were not present in the 134 voucher specimens.

This study suggests that genetic assignment tests can help us identify captive individuals useful to supplement wild populations, and thereby provide insurance against extinction in the wild. Because captive and wild tigers today are consciously managed to avoid hybridization between subspecies, the discovery of 49 purebred tigers in a sample of 105 individuals (47%) has important conservation implications. Assignment of subspecies ancestry offers a powerful tool to considerably increase the number of nonadmixed tigers suitable for conservation management.

21.3.1 Source populations

Selecting the founding individuals and populations for a captive breeding program is an important and difficult problem for many species. Source populations should be selected in order to maximize genetic and ecological (adaptive) diversity, but also maintain local adaptation. For example, there are currently four remaining subspecies of tigers in the wild. Genetic results have indicated substantial genetic divergence among these subspecies(Luo et al. 2010; Example 21.1). A strong argument can be made that each of these subspecies represents a separate evolutionarily significant unit (ESU; Chapter 20) and separate captive breeding programs should be established for each. However, space and other resources for captive breeding of tigers are limited. There are currently ~1,000 spaces for tigers in captive breeding programs throughout the world. We then face a dilemma. How should we partition available captive breeding spaces among the four subspecies to enhance survival and retention of genetic variation?

Maguire & Lacy (1990) provided a useful consideration of this problem. They identified three conservation goals: (1) to maximize the number of surviving subspecies; (2) to maximize genetic variation at the species level; and (3) to maximize genetic diversity at the subspecies level. They choose a timeframe of 200 years (32 tiger generations) to achieve these recommendations for long-term conservation plans (Soulé et al. 1986). Their analysis also included consideration of the probabilities of persistence of the subspecies in the wild.

The two extreme options are to choose only one subspecies for captive breeding or to divide the 1,000 spaces equally among the four subspecies. They assume that the N_e/N_C ratio in captive tigers is 0.4 (Ballou & Seidensticker 1987). In the latter case, each of the four subspecies would have an N_e of ~100 tigers (250 × 0.40). Using Equation 6.7, we would expect to lose ~14% of the heterozygosity in each subspecies after 200 years ($t = 32$). General recommendations suggest a goal of retaining at least 90% of the heterozygosity after 200 years (Soulé et al. 1986). This would require an N_e of ~150, and an N_C of 375 for each subspecies. Maguire & Lacy

(1990) recommend devoting half of the available captive spaces to the Bengal subspecies and dividing the remainder equally among the other three subspecies.

21.3.2 Admixed founding populations

Another option for establishing a captive population is by hybridizing genetically divergent populations. For example, the State of Montana established a captive population of westslope cutthroat trout in 1985 to be used in a variety of restoration projects. Geographic populations of westslope cutthroat trout show substantial genetic divergence among populations, $F_{ST} = 0.32$ (Allendorf & Leary 1988). Space limitations required that only a single captive population could be maintained. The choice was to use a single representative population to establish the captive population or to create a hybrid captive population by crossing individuals from a wide spectrum of native westslope cutthroat trout populations.

Do we choose one population to be brought into captivity or do we create a captive population by hybridizing individuals from different populations? The genetic choice that we face here is between genes and genotypes. We can maximize the allelic diversity of westslope cutthroat trout in the captive population by including fish from many streams in our founding population. However, hybridizing these populations will cause the loss of the unique combination of alleles (genotypes) that exist in each population. These genotypes can be important for local adaptations. These combinations of genes, and the resulting locally adapted phenotypes, will be lost through hybridization. In addition, the hybridization of different populations could result in outbreeding depression (Section 13.4).

In some cases, genetically distinct populations have been brought into captivity and hybridized without realizing potential problems. For example, we saw in Table 3.2 that ~20% of orangutans born in captivity were hybrids between orangutans captured in Borneo and Sumatra. These two populations are fixed for chromosomal differences, and it has been proposed they should be considered separate species. Current conservation breeding plans avoid the production and use of hybrids between these taxa.

There are no simple prescriptive answers to the best strategy in establishing a captive population. In the case of westslope cutthroat trout, the captive population was established by mixing from some 20 natural populations. There was some concern in this case about possible outbreeding depression caused by mixing together so many local populations. However, the alternative of using just one local population that would contain such a small proportion of the total overall genetic variation was considered less desirable.

Some have suggested that individuals from interspecific hybrid swarms could be used as founders to recover extinct species. The giant tortoise from the island of Floreana in the Galápagos became extinct within decades after Charles Darwin's visit in which he described that massive numbers of these creatures were being removed to be stored in the hulls of ships to be used for food (Poulakakis et al. 2008). Examination of mtDNA and nuclear markers in historical museum specimens and extant tortoises from the nearby island of Isabela indicated the existence of a large number of individuals that are admixed and possess ancestry from both Floreana and Isabela ancestors. These admixed individuals could be used to found a captive breeding program in which screening for molecular markers could be used to increase the contribution from ancestors from Floreana.

21.3.3 Number of founder individuals

The number of founders recommended for establishing a captive population depends substantially on the proportion of rare alleles desired to be captured, and on the population growth rate expected in captivity. Approximately 30 diploid founders are required to have a 95% probability of sampling an allele at frequency 0.05. However, with 30 founders, there is only approximately a 45% probability of including an allele of frequency 0.01 (Equation 6.8 and Figure 6.8). This probability increases to 63% if 50 founders are used; ~150 founders are needed to have a 95% probability of including an allele at a frequency of 0.01. Thus, we recommend a

minimum of 30 founders and preferably at least 50. Fifty founders will maintain ~98% of the original heterozygosity (Equation 6.6). If the rate of population growth is low, additional founders or subsequent supplementation with additional individuals is recommended.

21.4 Genetic drift in captive populations

A primary genetic goal of captive breeding programs is to minimize genetic change in captivity (Lacy 2009). Genetic changes in captive populations might reduce the ability of captive populations to reproduce and survive when returned to the wild. There are two primary sources of genetic change in captivity: genetic drift and natural selection, which we consider in the next two sections.

21.4.1 Minimizing genetic drift

Genetic drift causes the loss of heterozygosity and allelic diversity. This reduced genetic diversity can have several consequences. First, inbreeding depression might limit population growth and lower the probability that the introduced population will persist. Second, reduced genetic diversity will limit the ability of introduced populations to evolve in their new or changing environments. In general, the effects of genetic drift can be minimized in captivity by managing the population to maximize the effective population size.

The primary method for minimizing genetic drift and maximizing effective population size is to equalize reproductive success among individuals. This is especially important for the founder individuals of a captive breeding program. We saw in Chapter 7 that the ideal population includes random variability in reproductive success. Under controlled captive conditions, it can be possible to reduce variance in reproductive success to near zero. In this case, the effective population size might actually be nearly twice as great as the census population size (Equation 7.5). The most effective method to reduce variance in reproductive success depends upon the type of breeding scheme used in captivity (Section 21.5).

Example 21.2 Chondrodystrophy in California condors

The captive population of California condors was founded with the last remaining 14 individuals in 1987 (Ralls et al. 2000). California condors have bred well in captivity and the first individuals were reintroduced into the wild in 1992. However, nearly 5% of birds born in captivity have suffered from chondrodystrophy, a lethal form of dwarfism. This defect is apparently caused by a recessive allele that occurs at a frequency of 0.09 in the captive population.

Such deleterious alleles are likely to occur in any captive population founded by a small number of founders (Laikre 1999). What should be done? Ralls et al. (2000) considered three management options for this allele: (1) reduce its frequency by selection; (2) minimize its phenotypic frequency by avoiding matings between possible heterozygotes; and (3) ignore it. Selective removal of this allele would require not using possible heterozygotes in the breeding program. It would be possible to eliminate all individuals possibly carrying this allele based upon pedigree relationships from the breeding program. This is a very high cost to pay for elimination of a trait that affects less than 5% of all birds. In addition, it is likely that other traits caused by deleterious recessive alleles occur in this population. Selective removal of relatively low frequency alleles at multiple loci is generally not worth the cost of reducing the effective population size and further eroding genetic variation genome-wide in the captive populations.

Ralls et al. (2000) recommended minimizing the phenotypic frequency of this trait by avoiding pairings between possible heterozygotes. They suggest that some selection would be feasible once the captive population has reached the carrying capacity in captivity. In addition, possible heterozygotes could be given a lower priority as candidates for introduction.

The location of the locus responsible for chondrodystrophy has been identified with a genomic approach (Norman et al. 2019). A simple molecular, diagnostic test has been developed that will be used to identify heterozygous individuals for this deleterious allele. This will allow avoiding matings between heterozygous individuals that produce affected progeny. In addition, it would be possible to reduce the frequency of this allele in the captive population by selecting against heterozygotes.

21.4.2 Accumulation of deleterious alleles

Deleterious alleles that are present at low frequencies in natural populations might drift to high frequencies in captive populations because of the founder effect combined with relaxed natural selection (Example 21.2). Joron & Brakefield (2003) have suggested that relaxed natural selection in captivity can mask reduced fitness due to inbreeding. For example, wolves bred for conservation purposes in Scandinavia were found to have a high frequency of hereditary blindness apparently caused by an autosomal recessive allele (Laikre et al. 1993). Only six founders were originally brought into captivity (Figure 17.9). At least one of these founders apparently was heterozygous for a recessive allele associated with blindness. It is also possible that partial blindness might actually have some advantage in captivity for a wild animal such as a wolf.

In addition, new mildly deleterious mutations will occur in captive populations; these mutations might drift to high frequency in populations with a small N_e because natural selection is not effective in small populations (Chapter 8). Many of these new mutations with mild deleterious effects could accumulate in small populations and lead to so-called **mutational meltdown** (Section 18.6).

21.4.3 Inbreeding or genetic drift?

It is crucial to understand the difference between the effects of inbreeding and those of genetic drift in captive populations. Some inbreeding (the mating of related individuals) is unavoidable in small captive populations; this is the so-called inbreeding effect of small populations (Chapter 6). In general, inbreeding should be avoided as much as possible in captive populations because the reduced fitness associated with inbreeding depression might threaten short-term persistence of the captive population.

However, the loss of genetic variation by genetic drift is more serious and lasting than inbreeding. The harmful effects of inbreeding last for a single generation and are undone by a single outcrossing.

That is, a mating between an inbred individual and an unrelated mate will produce a noninbred progeny. However, the loss of alleles in a species through genetic drift is permanent. The long-term genetic viability of a captive population is more affected by the unequal representation of founders and effective population size than by matings between related individuals.

Schemes of mating with maximum avoidance of inbreeding will minimize the initial rate of loss of heterozygosity. However, perhaps surprisingly, there are often other systems of mating that do a better job of retaining heterozygosity in the long term (Kimura & Crow 1963; Robertson 1964; Wright 1965b).

21.5 Natural selection and adaptation to captivity

Natural selection will occur in captivity and bring about adaptation to captive conditions. Such changes will almost inevitably reduce the adaptiveness of the captive population to wild or natural conditions. For example, tameness in response to contact with humans is generally advantageous in captivity, but can have serious harmful effects in the wild.

The emphasis of captive breeding protocols has been primarily to reduce genetic drift by maximizing effective population size. This emphasis is appropriate for captive breeding programs of mammals and birds in zoos that have a relatively small number of individuals that are managed using pedigrees (Ballou & Foose 1996). However, increasing effective population size for some captive species (e.g., fish and plants) can increase the rate of adaptation to captive conditions.

21.5.1 Adaptation to captivity

Adaptation to captivity is probably the greatest threat in species that produce many offspring (e.g., insects, fish, amphibians, etc.). For example, females of many fish species produce thousands of eggs. Extremely strong natural selection can occur in the first few generations when founding a captive hatchery population of fish. Christie et al. (2012) found evidence for adaptation to hatchery conditions after only a single generation in captivity for anadromous rainbow trout (steelhead). Williams & Hoffman (2009) have provided a valuable review of the literature for captive conservation breeding programs that reported a strategy to minimize adaptation to captivity.

Darwin (1896) was very interested in the genetic changes brought about by selection during the process of domestication of animals bred in captivity. He attributed such changes to three mechanisms:

1. systematic selection;
2. incidental (unintentional) selection;
3. natural selection.

Systematic selection occurs when purposeful selection occurs for some desirable characteristics. For example, many hatchery populations of fish are selected for rapid growth rate. Incidental selection occurs when captive management favors a particular phenotype without being aware of their preference. For example, hatchery personnel might unconsciously favor a particular phenotype (e.g., large, colorful, etc.) when choosing fish to be mated. Finally, natural selection will act to favor those individuals that have characteristics that are favored under captive conditions. For example, many wild fish will not feed when brought into captivity. Therefore, natural selection for behaviors that permit feeding and surviving in captivity will be very strong.

This issue was raised many years ago by A. Starker Leopold (1944) in his consideration of the effects of release of 14,000 hybrid (wild × domestic) turkeys on the wild population of turkeys in southern Missouri, USA. It was common practice throughout many parts of the USA to release such hybrid turkeys to enhance wild populations that were hunted. Hybrid stocks were used because of the great difficulty in raising wild turkeys in captivity. He found that the hybrid birds were unsuccessful in the wild because of their tranquility, early breeding, and inappropriate behavior of chicks in response to the warning note of the hen.

Systematic selection and incidental selection can be greatly reduced in captivity by intensive effort. However, genetic divergence between wild and captive populations because of natural selection cannot be eliminated. Efforts are currently under-way to reduce these effects in fish hatcheries by mimicking the natural environment. Nevertheless, it is impossible for a hatchery to simulate the complex and dynamic ecological heterogeneity of a natural habitat. In fact, any hatchery must create an environment that differs dramatically from the natural one to achieve its goal of producing more progeny per parent than occurs under natural circumstances. By definition then, a goal of reducing mortality while retaining natural environmental conditions cannot be achieved; it is impossible to synthetically create conditions that are both identical to the natural ones and at the same time provide a basis for increased survival.

21.5.2 Minimizing adaptation to captivity

Natural selection is most effective in large populations (Chapter 8). Thus, adaptation to captivity is expected to occur most rapidly in captive populations with a large N_e. Minimizing variance in reproductive success via pedigree management will also act to delay adaptation to captive conditions. However, pedigree management is not practical for many species kept in captivity.

In species with high fecundity (such as many fish, amphibians, and insects), rapid adaptation to captivity is most likely to occur because hundreds of progeny can be produced by single matings. Thus, natural selection can be very intense especially in the first few generations after being brought into captivity.

For example, the Apache trout, which is native to the southwestern USA, is currently listed as threatened under the US ESA (Daniel 2020). A single captive population, originating from individuals captured in the wild in 1983 and 1984, is the cornerstone of a recovery effort with the goal of establishing 30 discrete populations within the native range of this species. Advances in culture techniques and the high fecundity of these fish have resulted in a program that spawns hundreds of mature fish and produces hundreds of thousands of fry per year for reintroduction.

The large number of spawners suggests that the N_e of this population is very large so that the loss of genetic variation due to drift is not a concern.

Nevertheless, these circumstances are ideal for natural selection to bring about rapid adaptation to captive conditions that would reduce the probability of successful establishment of reintroduced populations.

21.5.3 Interaction of genetic drift and natural selection

In many regards, actions taken to reduce genetic drift will also reduce the potential for natural selection. For example, minimizing variability in reproductive success among individuals will both maximize N_e and reduce the effects of natural selection (Allendorf 1993). However, as we saw in Chapter 8, natural selection is most effective in very large populations. Therefore, intermediate size populations would be large enough to avoid rapid genetic drift, but not so large that even weak natural selection could bring about adaptation to captive conditions.

Woodworth et al. (2002) tested these predictions with experimental populations of *Drosophila* to mimic captive breeding. They evaluated adaptation to captivity under benign captive conditions for 50 generations using effective population sizes of 25, 50, 100, 250, and 500. The small populations demonstrated reduced fitness after 50 generations due to inbreeding depression. The large populations demonstrated the most rapid adaptation to captive conditions. The least genetic change in captivity was observed in intermediate size populations as measured by moving the populations to simulated wild conditions (Figure 21.5). These authors suggested that adaptation to captivity can be minimized by subdividing or fragmenting the captive population into a series of intermediate size populations. The effective population size of each population should be large enough to minimize the harmful effects of inbreeding and genetic drift, but small enough to minimize rapid adaptation to captive conditions.

21.6 Genetic management of conservation breeding programs

A primary genetic goal of captive breeding programs is to minimize genetic change caused by genetic drift and natural selection. Specific actions

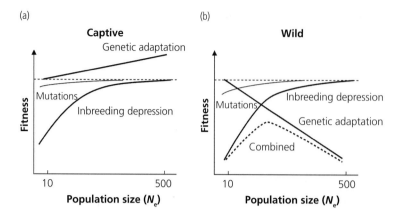

Figure 21.5 Expected relationship between fitness and population size (N_e) due to the inbreeding effect of small populations and genetic adaptation to captivity. The combined line represents the net effects of both factors. The effects are shown for populations maintained for ~50 generations under (a) benign captive conditions and (b) for these populations when introduced into the wild. Redrawn from Woodworth et al. (2002).

to achieve this goal depend upon the biology of the species. We first consider captive populations that are managed by keeping track of individual pedigrees (e.g., large mammals and birds). We then consider species for which large groups of individuals are held, for which it is difficult or impractical to keep track of individuals (e.g., fishes, amphibians, and insects). Most of our examples concern animals, but the same underlying genetic principles hold for plants. The wide variety of possible modes of clonal as well as sexual reproduction in plants (Figure 21.1) makes it harder to provide general guidelines to apply these genetic principles. Guerrant et al. (2004) provides an excellent review of maintaining offsite populations of plants for reintroduction.

21.6.1 Pedigreed populations

Genetic management by individual pedigrees is extremely powerful. It provides both maximum genetic information about the captive population and also maximum power to control the reproductive success of individuals chosen for mating. This approach is most appropriate for large mammals and birds. Much of the genetics literature dealing with management of captive populations deals with this situation.

Simply maximizing N_e might not be the best strategy for maintaining genetic variation in pedigreed populations (Ballou & Lacy 1995). Remember that genetic variation can be measured by either heterozygosity or allelic diversity. Maximizing N_e will minimize the loss of heterozygosity (by definition),

but it might not be the best approach to retain allelic diversity. A strategy that uses all of the information contained in a pedigree can be developed to minimize the loss of heterozygosity and allelic diversity.

Ballou & Lacy (1995) provide a lucid explanation of captive breeding strategies to maintain maximum genetic variation that is beyond the detail that we will consider here. This problem is difficult because the pedigrees of captive populations are often extremely complicated and genetic planning is often not initiated until after the first few generations of captivity.

Simple rules of thumb such as equalizing the genetic contributions of founders to the captive population are not valid. We can see this in the hypothetical example presented in Figure 21.6 in which there are four founders of a captive population. What would be the result of a breeding strategy that equalized the genetic contributions of the founders? We can be absolutely certain that we have lost one of the two alleles carried by founder 1 at every locus since this founder only contributed one offspring to the captive population. However, there is some possibility that both alleles from the three other founders have been retained because they have contributed multiple progeny. Thus, we can maximize the retention of allelic diversity in the captive population by weighting the desirable contribution of each founder by the expected proportion of a founder's alleles retained (founder genome equivalents; Lacy 1989).

Accurate calculations of kin relationships, inbreeding coefficients, and retention of founder

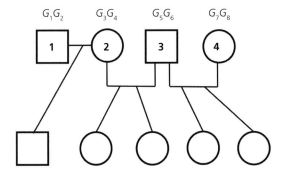

Figure 21.6 Hypothetical pedigree of a captive population founded by four individuals. We know that one allele (e.g., G_1 or G_2) at each locus has been lost from founder 1 because he left only one descendant in the captive population. Therefore, equalizing the contributions of these four founders in future generations would lead to an over-representation of alleles from founder 1.

alleles require complete knowledge of the pedigree. However, many pedigreed captive populations have some individuals with one or both parents unknown. Traditionally, such individuals have been treated as founders unrelated to all nondescendant animals. In some circumstances, this can cause substantial errors in estimating genetic parameters (Ballou & Lacy 1995). Genetic markers can often be used to resolve unknown relationships and can result in a substantially different view of captive populations (e.g., the whooping crane; Jones et al. 2002).

Genetic analysis of pedigreed populations can play a crucial role in verifying relationships (Ivy et al 2009), but pedigrees must be accurate. For example, Hammerly et al. (2016) used 10 microsatellite loci to verify the pedigree of a captive population of the critically endangered Attwater's prairie-chicken. A total of 933 individuals were genotyped, and the parentage assignment error rate was 4.1%. These authors concluded that sufficient errors exist in the Attwater's prairie-chicken pedigree to limit its ability to prevent inbreeding in the captive breeding population.

Sometimes a pedigree can be used to correct results from genetic marker analysis. Kathy Ralls provided us with the following experience in pedigree analysis with the California condor. She identified several problems with the relationships identified with molecular genetic analysis. For example,

according to the genomic data, condor-31 was closely related to three birds to which his full sibs were not at all closely related. Recognizing that this is not possible, Kathy explained this to the people doing the laboratory analysis. The lab people discovered that someone had mislabeled a sample, and that their "condor-31" was actually a mixture of condor-31 and condor-25. After they corrected the problem, they used the genetic analysis to identify a couple of previously unknown relationships among the founders. But if Kathy did not have confidence that the pedigree was correct, they would have reached quite a few incorrect conclusions.

Similarly, the founders of a captive population brought into captivity are generally assumed to be unrelated for pedigree analysis. However, this might often not be the case. Incorrect assignment of founder relatedness will result in erroneous estimates of inbreeding coefficients, effective population size, and population viability. For example, the last remaining individuals in the wild might consist of just a few groups of sibs. This information should be taken into account along with genetic marker analysis of relationships in order to maximize the retention of genetic variation in the captive population. Thus, correct classification of kin structure among founders is important for a captive breeding program.

21.6.2 Nonpedigreed populations

For many species held in captivity, it is difficult or impractical to keep track of individuals and pedigrees. For example, a single female of the endangered Colorado pikeminnow can produce as many as 20,000 eggs each year. Other procedures, therefore, need to be developed to achieve the goal of minimizing genetic change by genetic drift or selection.

The large census population sizes at which some species are maintained in captivity should not be taken to mean that genetic drift is not a concern. For example, Briscoe et al. (1992) studied the genetics of eight captive populations of *Drosophila* held in populations from 6 months to 23 years (8–365 generations) with thousands of individuals. All eight populations lost substantial heterozygosity at nine allozyme loci. Values of N_e estimated by the decline in heterozygosity were less than 5% of the census population size. Delpuech et al. (1993) reported similar results in their review of five species of insects held in captivity. Populations of all of these species had retained only ~20% or less of their original heterozygosity at allozyme loci.

These results demonstrate the importance of genetic monitoring of populations (Chapter 23). Regular examination of allele frequencies at genetic marker loci should be used to detect the effects of genetic drift in captive populations in which individual reproductive success is not being monitored.

Adaptation to captive conditions is an even greater concern for large populations held in captivity. For example, Frankham & Loebel (1992) found that the average fitness in captivity of *Drosophila* doubled after being maintained for eight generations in captivity. Many other studies have found evidence for rapid adaptation to captivity in a variety of organisms (Gilligan & Frankham 2003). The genes selected for in captivity are almost certain to decrease the fitness of individuals when they are returned to wild conditions. In addition, the strong selection in captivity will reduce the effective population size of the captive population. In fact, strong variance in reproductive success associated with this adaptation is the likely explanation of the small N_e/N_C ratios often found in captive populations.

A conceptual framework for minimizing the rate of adaptation to captivity (R) is provided by a modified form of the breeders' equation (Equation 11.9; Frankham & Loebel 1992):

$$R = \frac{H_N S(1 - m)}{G} \qquad (21.1)$$

where H_N is the narrow-sense heritability, S is the selection differential, m is the proportion of genes contributed from wild individuals, and G is the generation interval.

The goal is to minimize the rate of adaptation to captivity (R). Continued introduction of individuals from the wild (increased m) will slow the rate of adaptation to captivity. However, this will often not be possible. The generation interval (G) can be manipulated by increasing the average age of parents. For example, doubling the mean age of parents will double the generation interval and halve the rate of adaptation to captivity. However, increasing

the age of parents will also slow the rate of population growth so this approach is less feasible during the early stages of captivity before the population reaches carrying capacity.

Of course, reducing the intensity of selection (S) will slow adaptation to captivity. All efforts should be made to reduce differential survival and reproduction (fitness) in captivity. This can be done by minimizing mortality in captivity and by making the environmental conditions as close as possible to wild conditions.

Reducing differences in the number of progeny produced by individuals (family size) will also diminish the effects of selection in captivity (Allendorf 1993; Frankham et al. 2000). There will be no reproductive differences between individuals if all individuals produce the same number of progeny. In this situation, natural selection will only operate through differences in relative survival of genotypes within families of full- or half-sibs. In a random mating population, approximately one-half of the additive genetic variance is within families and half is between families. Therefore, the rate of adaptation will be reduced by ~50% by equalizing family size. Equalizing family size will also increase N_e.

Using genome-wide genetic information to infer relatedness among individuals has potential to minimize the loss of genetic variation in captive populations. de Cara et al. (2011) used simulations and concluded that using tens of thousands of **single nucleotide polymorphisms (SNPs)** spread throughout the genome generally performs better than using pedigree information to maintain heterozygosity and allelic diversity within captive populations.

21.7 Supportive breeding

Supportive breeding is the practice of bringing in a fraction of individuals from a wild population into captivity for reproduction and then returning their offspring into their native habitat where they mix with their wild counterparts (Ryman & Laikre 1991). The goal of these programs generally is to increase survival during key life stages in order to support the recovery of a wild population that is threatened with imminent extirpation. These programs would seem to pose a relatively small risk of causing genetic problems. Nevertheless, the favoring of only a segment of the wild population can also bring about changes in the wild population due to genetic drift and selection (Example 21.3).

21.7.1 Genetic drift and supportive breeding

Supportive breeding acts to increase the reproductive rate of one segment of the population (those brought into captivity). This will increase the variance in reproductive success (family size) among individuals and therefore potentially reduce effective population size. Demographic increases in population size can reduce the overall N_e and accelerate the loss of genetic variation. This effect is most likely to occur for species with high reproductive rates where large differentials in reproductive success

---

Example 21.3 Supportive breeding of the world's largest freshwater fish

The Mekong giant catfish is a spectacular example of the potential problem with supportive breeding (Hogan et al. 2004). This is perhaps the largest species of fish found in freshwater. It grows up to 3 meters long and weighs over 300 kg! A century ago, this species was found throughout the entire Mekong River from Vietnam to southern China. This species began disappearing from fish markets in the 1930s, and efforts to find individuals in fish markets have failed in the past few years. Very few individuals remain in the wild and the species is currently listed as Endangered on the IUCN "Red List."

The Department of Fisheries of Thailand began a captive breeding program in 1984. Over 300 adult fish have been captured in the wild and brought into captivity over the past 20 years. However, this program further threatens this species because of the removal of adult fish from the wild and the release of large numbers of young fish from very few parents. For example, over 20 wild adults were sacrificed in 1999 to supply eggs and milt for artificial propagation. More than 10,000 of these fingerlings were released back into the wild in 2001. However, genetic analysis of the progeny indicated that ~95% of these progeny were full sibs produced by just two parents (Hogan et al. 2004).

---

are possible (e.g., fishes, amphibians, reptiles, and insects).

Consider the situation where the breeding population consists of N_w effective parents that are reproducing in the wild and N_c effective parents that are breeding in captivity, and their progeny are then released into the wild to supplement the wild population (Figure 21.7). Figure 21.8 presents the overall N_e as a function of the progeny that are produced in captivity. The overall effective size can be substantially smaller than the effective number of parents reproducing in the wild when the contribution of the captive population is high. For example, consider the case where the wild population consists of 22 effective parents and that two of these parents are taken into captivity and then produce 50% of the total progeny. In this case, the total effective population size will be approximately six rather than the 22 that it would have been in the absence of a supportive breeding program.

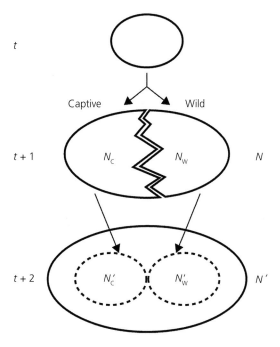

Figure 21.7 Schematic representation of supportive breeding. The total population of N individuals is divided into a captive and a wild group of size N_C and N_W that reproduce in captivity and in the wild, respectively. The N'_C and N'_W offspring are mixed before breeding in generation $t + 2$. From Ryman et al. (1995a).

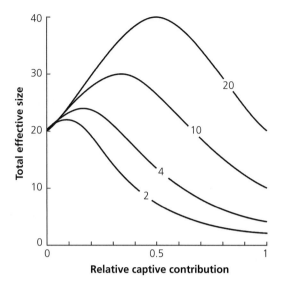

Figure 21.8 Total effective population size (wild + captive) when a natural population of 20 effective parents is supported by offspring from different numbers of captive parents, as indicated by the numbers on the different curves. The x-axis is the proportion of parents contributed by the captive parents. From Ryman & Laikre (1991).

Consideration of this problem has been extended to multiple generations (Wang & Ryman 2001; Duchesne & Bernatchez 2002). The effects of supportive breeding on N_e are complex. Moreover, the effects of supportive breeding on the inbreeding and variance effective population sizes (Section 7.6) can differ. Nevertheless, supportive breeding, when carried out successfully over multiple generations, might increase not only the census but also the effective size of the supported population as a whole. However, if supportive breeding does not result in a substantial and continuous increase of the census size of the breeding population, it can be genetically harmful because of elevated rates of inbreeding and genetic drift.

21.7.2 Natural selection and supportive breeding

Supportive breeding can also have important genetic effects on supplemented populations because alleles that are harmful in the wild but advantageous in captivity might rise to high frequencies in captive populations (Lynch & O'Hely 2001).

This genetic supplementation load will be especially severe when a captive population that is largely closed to import makes a contribution to the breeding pool of individuals in the wild.

Many papers have modeled possible harmful genetic effects of supportive breeding programs on natural populations (e.g., Duchesne & Bernatchez 2002; Ford 2002; Theodorou & Couvet 2004). These populations can be managed to increase rather than decrease effective population size. Nevertheless, the effects of supportive breeding on adaptation of wild populations are more difficult to predict. Selection in captivity can substantially reduce the fitness of a wild population during supportive breeding. The continual introduction of wild individuals into the captive population can reduce this effect, but is not expected to eliminate it. These programs can reduce the probability of local extirpations, but it is essential to carefully design the genetic aspects of these programs. Efforts are underway to improve the effectiveness of supportive breeding by integrating wild and captive populations (Redford et al. 2012; Canessa et al. 2016).

21.8 Reintroductions and translocations

To ensure a successful **reintroduction, introduction,** or supplemental **translocation,** we should consider several issues: (1) where to release the individuals, (2) how many populations to establish, (3) how many individuals to release, (4) age and sex of individuals to release, (5) which and how many source populations to use, and (6) how to monitor the population after the release of individuals. Genetics should play a role in all of these issues (Weeks et al. 2011; Jamieson & Lacy 2012; Maschinski & Albrecht 2017).

Monitoring after the release of individuals is crucially important to ensuring the success of reintroductions and translocations (Attard et al. 2016). Unfortunately, post-release studies of genetic contribution or of population status are seldom conducted. Molecular markers can help monitor genetic diversity, effective population size, and reproductive contributions of released individuals (Chapter 23). For example, if few founders actually reproduce, due to extreme polygamy, paternity analysis could detect the problem by identifying only a few males as fathers. If paternity analysis is not feasible, then monitoring for loss of alleles, rapid genetic change, and small effective population size could help determine if few founders reproduce (Examples 21.4 and 21.5).

21.8.1 Reintroduction of animals

Where to release individuals depends on habitat suitability and availability. To maximize chances of a successful reintroduction, the habitat should be similar to that which the individuals to be released are adapted. Obviously, sufficient food, water, breeding habitat, and shelter or escape terrain should be available. Furthermore, the habitat should be free from exotic predators or competitive invasive species. For example, when threatened marsupials are reintroduced in Australia, exotic foxes and feral cats should not be present because they are highly efficient at killing marsupials and preventing reestablishment of the population. In the case of the African rhino, there is abundant habitat, but little habitat free of human predators, that is, poachers (Section 20.1).

Reintroducing captive individuals into populations that are geographically and genetically distant from where their ancestors were originally removed can cause problems of outbreeding depression. For example, Banes et al. (2016) found that at least two orangutans reintroduced into Camp Leakey, an orangutan rehabilitation site on Borneo, were from a different subspecies than the orangutans in the area where they were introduced. These two reintroduced individuals produced at least 22 descendants, some of which have moved into the surrounding population from Camp Leakey. The effects of such hybridization between subspecies on fitness are unclear. However, zoos have not allowed hybridization between these subspecies because of concern about outbreeding depression. The data of Banes et al. (2016) are too limited to test the extent of possible outbreeding depression because of the slow reproductive rate of orangutans, which typically reproduce only once every 8 years.

How many populations should be reintroduced? More than one, and preferably several populations should be established and maintained. Populations should be independent demographically and

environmentally to avoid a catastrophic species-wide decline due to severe weather, floods, fire, or disease epidemics, for example. Two or more populations should be established within each different environment or for each divergent genetic lineage, whenever divergent environments or lineages exist within a species' range. The conservation of multiple populations across multiple diverse environments can help ensure long-term persistence of a species (Hilborn et al. 2003; Example 19.3). Genetics can help determine if populations are independent demographically through genetic mark-recapture to identify migrants. Molecular genetic markers are widely used to assess gene flow, an indicator of the degree of population independence or isolation.

How many individuals to release depends in part on the breeding system, effective population size, and population growth rate after the reintroduction. When feasible, at least 30–50 individuals should be reintroduced. More individuals will be required if the breeding system is strongly sex-biased (e.g., strong polygamy) or the effective population size is small compared with the census population size. Also, when population size does not increase above ~100–200 individuals within a few generations, more individuals should be released, when possible.

Some authors have suggested individuals should be chosen for breeding in captivity to increase genetic variation at certain loci that can be examined with molecular techniques that might have particular adaptive importance (Wayne et al. 1986; Hughes 1991). However, selecting for increased variation at a few detectable loci can reduce the effective population size and reduce genetic variation throughout the genome (Chapter 7).

The sex and age of individuals can influence the success of the reintroductions and translocations. For example, it is often important to release more females than males to maximize population growth, which limits demographic stochasticity and subsequent genetic drift. Many reintroductions of large game animals in western North America have used about 60–80% females. For supplemental translocations in polygynous species, it is better to reintroduce females than males if we want only limited gene flow, because a single male can potentially

Example 21.4 Recolonization by Pacific lamprey following dam removal

Pacific lampreys are a primitive fish with an unusual life history (Figure 21.9). They are anadromous and semelparous. They spend most of their life as filter-feeding larvae that live in freshwater. After the larval period, they metamorphose and move to the ocean where they attach their sucker mouth to fish or sperm whales and feed on blood and other body fluids. The adults live at least 1–2 years in the ocean and then return to freshwater to spawn.

Figure 21.9 Two of the first Pacific lamprey spawning in the Elwha River drainage after removal of two dams. These fish were identified as natural colonizers from nearby streams by genetic analysis. Photo courtesy of Amanda Anderson.

Pacific lampreys have declined dramatically along the northwest coast of the USA because of the construction of dams and other environmental changes (Close et al. 2002). They are an important ceremonial food for Native American tribes, harvested as a subsistence food by various tribes along the Pacific coast, and are highly regarded for their cultural value. In contrast, lampreys have been regarded as a nuisance by Euro-American management policies.

Removal of two hydropower dams in the Elwha River Basin in the State of Washington has been one of the largest river restoration projects ever attempted. The dams had eliminated Pacific lamprey populations upstream in the Elwha Basin. After the dam removal, 42 adult lampreys from nearby streams were translocated into the Elwha Basin and other lampreys from nearby streams naturally returned to the Elwha Basin to spawn. Hess et al. (2021) used 263 SNP loci to determine the source and number of successful spawners in order to determine the success of this restoration project.

Lamprey reproduction increased dramatically following removal of the dams. Genetic analysis revealed that production was primarily driven by natural colonization, rather than the translocated individuals. Given the high fecundity of lamprey, this reproduction could have been produced by a few spawners. Genetic analysis was used to determine the effective number of breeders (N_b; Section 7.5). The number captured in traps increased from 132 in 2014 to 1,805 in 2016. The effective number of spawners also increased from 14 for the 2013–2014 brood years, to ~130 for the 2015–2016 brood years, and to 160 for the 2017–2018 brood years. The effective number of breeders for the 2017–2018 brood years was similar to other nearby basins. The genetic analysis demonstrated that dam removal successfully increased Pacific lamprey productivity.

Example 21.5 Rapid genetic decline in a translocated plant

The Corrigin grevillea is one of the world's rarest plant species; only five plants were known in the wild in 2000 (Krauss et al. 2002). These plants occurred in degraded and isolated remnants of natural vegetation on road verges in Western Australia. In 1995, 10 plants were selected from the 47 plants known at the time to act as genetically representative founders for translocation into secure sites. Hundreds of ramets (tissue-cultured propagules of these 10 clones) were produced from these plants. By late 1998, 266 plants had been successfully translocated and were producing large numbers of seeds.

Krauss et al. (2002) used **amplified fragment length polymorphisms (AFLPs)** to determine the genetic contribution of the 10 founders to this translocated population and their first-generation progeny. They found that only eight clones, not 10, were present in the translocated population. In addition, 54% of all plants were a single clone. They also found that F_1s produced between founders were on average 22% more inbred and 20% less heterozygous than their founders, largely because 85% of all seeds were the product of only four clones. They estimated that the effective population size of the translocated population was approximately two. That is, the loss in heterozygosity from the founders to the next generation was what would be expected if two founders had been used.

These results demonstrate the importance of genetic monitoring of translocation programs (Chapter 23). Additional genotypes have been added to this population, and today there are hundreds of plants at this site. Continued genetic monitoring has shown that measures implemented to stabilize the erosion of genetic diversity have been successful, and heterozygosity has remained constant (Dixon & Krauss 2019).

breed with many females, thereby swamping a population with introduced genes. Further, a male in a polygynous population might never breed if he is not dominant, for example, making male-mediated gene flow highly variable and unpredictable. In territorial carnivores such as grizzly bears, it is often best to translocate females because males are more likely to fight for territory, sometimes to the death. Molecular genetic sexing can help determine sex before translocation in some species (birds and reptiles) where sex is cryptic.

Age can influence the likelihood that a translocated individual remains in the location of release and integrates socially into the new population. In large mammals, young juvenile or yearling individuals are often more likely than adults to integrate socially and/or not leave the release area.

Which and how many source populations should be used for reintroductions? For reintroductions and supplemental translocations, the source population generally should have high genetic diversity, genetic similarity, and environmental similarity when compared with the new or recipient population. Environmental similarity helps limit chances of maladaptation of the translocated individuals in the site of release, and environmental changes such as climate change may lead to matching contemporary rather than historic climates (Chapter 16). However, if populations have only recently become fragmented and differentiated, multiple differentiated source populations can help maximize genetic diversity in reintroductions or translocations, with little risk of outbreeding depression. For example, a source population with greater genetic divergence from the recipient population will result in a greater increase in heterozygosity in the recipient population.

If no individuals are available from a similar environment, then individuals from several source populations could be mixed upon release to maximize diversity for natural selection to act upon. Mixing of individuals from multiple sources is less desirable in supplemental translocations where some locally adapted individuals still persist because releasing many mixed individuals could swamp the local gene pool and lead to loss of locally adapted genotypes.

21.8.2 Restoration of plant communities

These same genetic principles apply to translocation of rare plants as well as to restoration of plant communities (Guest Box 21; Wood et al. 2020). Genetic considerations for plant translocations include confirming taxon or lineage identity, sourcing seed or producing propagules, numbers of founders, maximizing genetic diversity, mating system, and planting design. Maximizing genetic diversity in translocations is particularly important for rare plants where genetic diversity may already have been reduced. If seed is limiting, seed production areas can be established to increase availability of seed prior to translocations, and ensure a backup in case of limited survival. This may be particularly important for experimental translocations where the conditions for establishment and survival require research. A review of rare plant translocations in Australia has concluded that translocations have played an important role in the recovery of many species (Example 21.5; Silcock et al. 2019). For example, Mellblom's spider orchid, endangered under the EPBC Act, was down to just a few plants in a single location in the 1990s, but it now occurs in five separate subpopulations in southwestern Victoria.

Restoration is an important tool for the preservation of native plant communities (Kramer & Havens 2009; Byrne et al. 2011; Maschinski & Albrecht 2017). Restoration ecology is a synthesis of ecology and population genetics. In early restoration programs, native local plants were the preferred source because of the potential importance of local adaptations (Linhart & Grant 1996). A focus on maximizing genetic diversity in restoration programs has led to sourcing plants from multiple populations in composite or admixture sampling strategies (Broadhurst et al. 2008; Breed et al. 2013). More recently, restoration practitioners are concerned about persistence of restoration

under future climates and are considering a climate adjusted provenancing approach similar to assisted gene flow (Chapter 16; Prober et al. 2015). This approach focuses seed collections from multiple populations in the direction of climate change as this is generally known, and avoids matching populations with specific future climate conditions given the uncertainties of climate projections. Many studies are now using genomics to identify adaptations to climate along climate gradients (Chapter 16).

In addition, restoration projects often involve highly disturbed sites where local conditions have changed. In these cases, mixtures of genotypes possessing high levels of genetic variation are more likely to rapidly evolve genotypes adapted to the novel ecological challenges of severely disturbed sites.

Cultivated varieties (**cultivars**) of plants that have been selected for captive conditions may be a ready source of plants for restoration (Keller et al. 2000). Such cultivars are often readily available, and are much less expensive than acquiring progeny from wild seed sources. However, the widespread use of cultivars is likely to lead to the introduction of genes into the adjacent resident population through cross-pollination, although the degree of genetic introgression will depend on the breeding system. Thus, widespread introductions of cultivars could alter the resident neutral gene pool. For these reasons, the use of cultivars in restoration should be restricted.

Guest Box 21 Genetic management and reintroduction of Hawaiian silverswords
Robert H. Robichaux

The endemic Hawaiian silversword lineage (Asteraceae) is one of the world's premier examples of plant adaptive radiation. Although the lineage is a marvel of evolutionary diversification, it and the Hawaiian flora more broadly are confronted by a suite of threats, especially from alien species (Robichaux et al. 2017).

The iconic Mauna Kea silversword and Ka'ū silversword of Hawai'i Island exemplify the severity of the threats. Both taxa are long-lived, predominantly monocarpic (i.e., they flower only once before dying), and self-incompatible, and both have suffered precipitous declines following the introduction of alien ungulates to the island. Early sightings and collections indicate that the two taxa were formerly widely distributed on Mauna Kea and Mauna Loa volcanoes, respectively, with their larger populations containing many thousands of individuals. Because of ungulate browsing, the two taxa now total fewer than 25 and 365 remnant individuals, respectively.

The taxa also highlight the importance of incorporating genetic management into the beginning phases of plant reintroduction efforts. The reintroduction effort with the Mauna Kea silversword was initiated by the Hawai'i Division of Forestry and Wildlife (HDFW) in 1973. By 1996, HDFW had reintroduced more than 800 plants on Mauna Kea. Yet the available records indicated that the reintroduced plants were all first- or subsequent-generation offspring of only two maternal founders from the remnant population, with the maternal founders likely having been reciprocal pollen donors. Thus, the reintroduction effort itself had created a major population bottleneck, unintentionally compounding the impact of the ungulate-caused population crash. We partnered with HDFW to assess the genetic consequences of the bottleneck. Our analyses revealed that across eight microsatellite loci, the reintroduced population exhibited large reductions in the proportion of polymorphic loci, observed and effective numbers of alleles per locus, and expected heterozygosity compared with the remnant population (Friar et al. 2000). To ameliorate this genetic legacy, we have worked closely with HDFW, the Hawai'i Plant Extinction Prevention Program, and others to incorporate additional founders that have flowered intermittently in the remnant population since 1997 into the state's continuing reintroduction effort.

We also ensured that genetic management was central to our reintroduction effort with the Ka'ū silversword from the beginning (Robichaux et al. 2017). We initiated managed

breeding of the Ka'ū silversword in the late 1990s following our discovery of diminutive but old plants that had survived prior browsing on Mauna Loa. It proved feasible to extract the diminutive plants from the lava substrate and transport them to the Volcano Rare Plant Facility, including by helicopter where necessary. In greenhouses at the Facility, the plants grew to large size and flowered within 2–4 years. With managed breeding, we generated more than 21,000 seedlings from 169 founders. We reintroduced the seedlings into large protected sites in Hawai'i Volcanoes National Park, ensuring that no maternal founder accounted for more than 1.2% of the total seedlings (i.e., that no maternal founder was significantly over-represented). Flowering in the reintroduced populations in recent years has led to significant new seedling recruitment, thereby enhancing the prospects for recovery (Figure 21.10).

Figure 21.10 Ka'ū silverswords flowering in the reintroduced population in the Kīpukakulalio site in Hawai'i Volcanoes National Park. The large rosette of the plant in the foreground exceeds 70 cm in diameter and its massive compound inflorescence exceeds 2 meters in length. Photo courtesy of Robert Robichaux.

Chapter 22

Genetic Identification

Red-tailed black cockatoo, Example 22.3

Wildlife DNA forensics is essentially concerned with the identification of evidence items in order to determine the species, population, relationship or individual identity of a sample.

(Rob Ogden et al. 2009, p. 179)

DNA barcoding involves sequencing one or a few standard DNA regions to tell the world's species apart. Since its inception in 2003, DNA barcoding has grown into a global research programme involving thousands of researchers whose work has led to the production of millions of barcode sequences.

(Eric Coissac et al. 2016, p. 1423)

Genetic identification is the use of molecular genetic analyses to identify the species or population of origin of a sample, or to determine the identity of individuals or their relationships. Genetic identification of species, individuals, and their relationships is often a first step in many applications of molecular markers to conservation and management questions. Once individuals have been identified, we can estimate abundance of individuals (Section 18.1), monitor their movements, identify immigrants, estimate sex ratios, and monitor population genetic parameters relating to neutral and adaptive molecular variation.

Genetic identification has many applications in wildlife management and conservation, and assists in detection of invasive species, aids law enforcement, and informs biodiversity assessments. Poaching and trafficking are among the most serious threats to the persistence of many wild populations and species. Genetic identification has become a key tool in **wildlife forensics**, where genetic identification of species, individuals, or populations provides evidence to identify or verify illegal activities. Determination of the relatedness or parentage of individuals provides information on patterns of dispersal and migration and facilitates management of captive breeding populations.

Individual or species identification can also be used as an effective means of **genetic monitoring** of spatial or temporal changes in populations and species. Genetic monitoring using individual or species identification or identification of community composition has become more extensive in recent years, and is further discussed in Chapter 23.

While there are many challenges with applications of genetic identification from field sampling, laboratory methods, data analysis, and interpretation, there are substantial potential benefits to conservation in a wide range of applications in species management and forensics.

Conservation and the Genomics of Populations, Third Edition. Fred W. Allendorf, *et al.*, Oxford University Press.
© Fred W. Allendorf, W. Chris Funk, Sally N. Aitken, Margaret Byrne, and Gordon Luikart (2022). DOI: 10.1093/oso/9780198856566.003.0022

22.1 Species identification

Many kinds of molecular genetic markers have been used to identify species for both forensic and conservation applications. **Mitochondrial DNA (mtDNA)** analysis has been a widely used molecular approach for animal species identification because many species have distinctive mtDNA sequences, and because universal primers exist that work among taxa (mammals or even all vertebrates). Furthermore, mtDNA is relatively easy to extract from most tissues, including hair, elephant tusks, and old skins, because of its high copy number per cell (Section 3.2). **Chloroplast DNA (cpDNA)** is used for species identification in plants, but is less useful for identification as it has a low rate of evolution and consequently is less variable. Both mtDNA and cpDNA are generally maternally inherited and therefore are more limited in cases involving hybridization or where introgression is present (Chapter 13).

With the advent of genomics, nuclear DNA markers have proven more useful for species identification. Generation of a large number of **single nucleotide polymorphisms (SNPs)** provides power to determine individual genotypes and species identity of an individual. In addition, SNPs can be genotyped using small quantities of DNA or partially degraded DNA, making them especially useful in forensic applications and for noninvasive sampling (Morin et al. 2004).

22.1.1 DNA barcoding

DNA barcoding involves the analysis of short, standardized, and well-characterized DNA sequences for use as a tag for rapid and reliable species identification (Valentini et al. 2009; Bucklin et al. 2011; DeSalle & Goldstein 2019). The aim of barcoding is to identify the species of origin of an unknown individual sample using a known taxonomic classification, but it does not seek to determine evolutionary relationships.

Barcoding has become an ambitious international initiative (Box 22.1). The goal is to document the diversity of life on Earth through use of DNA sequences in species identification and species discovery. DNA barcoding has broad applications in biodiversity inventory, conservation and wildlife management, detection of invasive species, wildlife crime and illegal trade, and ecosystem monitoring. DNA barcoding is the basis of diet analysis (Section 22.1.3) and **environmental DNA (eDNA)** analysis (Section 22.1.4).

DNA barcoding is based on sequencing of **amplicons**, which have relatively low sequence variation within species, but high sequence divergence between species. In animals, mtDNA has a relatively high mutation rate and relatively high divergence between species, and thus has been useful for barcoding. The development of barcoding initiatives was based on early work using *cytochrome c oxidase I* (*COI*, a mtDNA gene) that demonstrated this low diversity within species and high differentiation among species. For example, Hebert et al. (2010) sequenced a 648-base pair (bp) region of *COI* in more than 1,300 lepidopteran species (butterflies and moths) from eastern North America and reported low intraspecific divergences, averaging only 0.4% within species, whereas congeneric species revealed 18-fold higher mean divergences (7.7%). This study reported that 99.3% of the 1,300 study species possessed diagnostic barcode sequences, with only nine cases of barcode-sharing among the 1,327 species.

cpDNA markers are commonly used in plants for species identification because mitochondrial genes in plants have a relatively low substitution rate. Universal primers for chloroplast noncoding regions are widely used (Taberlet et al. 2018). These primers work in a range of taxa from algae to gymnosperms and angiosperms. Hollingsworth et al. (2011) identified a standard two-locus DNA barcode for use in land plants. This combination of *rbcL* + *matK* was accepted by the CBOL Plant Working Group (2009) as providing a universal framework for the plant barcode. Sometimes a third locus is used, commonly *trnH-psbA* (Kress & Ericson 2012) or the internal transcribed spacer (ITS) region of the repetitive ribosomal genes (China Plant BOL Group et al. 2011).

Species identification via DNA barcoding or metabarcoding is increasingly used for conservation management. Sequencing of mtDNA is often used to detect the presence of endangered species in nature reserves and wildlife management areas. Use of barcoding of shed hair or scat to identify species

---

Box 22.1 International Barcode of Life

The International Barcode of Life initiative (iBOL; initially called the Consortium for the Barcode of Life, CBOL) has an ambitious goal of documenting the diversity of life on Earth (www.ibol.org). Initially proposed by Paul Hebert in 2003 at the University of Guelph (Hebert et al. 2003), it has grown to become an international initiative with substantial investment around the world (www.barcodeoflife.org). While initially focused on animals, it has grown to encompass plants, insects, and fungi, as well as protists and microbiota. Key features of iBOL are the systematic approach to generating the barcodes, the direct linking of the barcodes to specimens maintained in natural history collections, and making the barcode data available in a public data repository.

Direct linking of DNA barcodes with **vouchered specimens** maintained in natural history collections means that the identity of the barcode is linked to a reference specimen and therefore to any future taxonomic changes, thus maintaining the currency of the identification. The DNA samples are also kept in a secure repository, so that they will be available for study by future technologies (Vernooy et al. 2010). The linking to specimens and natural history collections also means that the samples are more likely to be managed in accordance with international agreements on access to genetic resources, such as the Nagoya protocol. Establishment and maintenance of the Barcode of Life Data Systems database (BOLD) provides a single point of curation of the data that is publicly accessible to all users. In mid-2020, BOLD held records for more than 8 million barcodes from over 300,000 described species.

iBOL has completed a major initial program, BARCODE 500K, that barcoded 500,000 species. This initiative involved researchers from 25 nations and attracted US$150 million in funding. A second major program, BIOSCAN, was launched in 2019 that aims to barcode 2.5 million species by 2026. These major programs are supported by many other initiatives and researchers across the world. By 2018, more than 12,000 studies using DNA barcoding had been published.

---

presence is common, particularly for species that are difficult to detect through traditional approaches such as trapping or spotting. For example, amplicon sequencing for 16S mtDNA was used to detect the presence of threatened wolverine, lynx, and fisher in snow tracks and hair samples from Montana and Idaho in the USA (Franklin et al. 2019).

DNA barcoding can also be used to detect the presence of invasive species, particularly when rare and at low densities. For example, microsatellite analysis of samples from hair snare traps was used to determine density of foxes that are a major predator of mammals in southwestern Australia (Berry et al. 2012). It demonstrated the effectiveness of noninvasive DNA analysis of hair samples compared with conventional trapping, and was used to determine the effectiveness of a baiting program to reduce fox density. Detection of invasive species in aquatic or terrestrial environments is particularly useful as it can confirm species presence at an early stage of invasion before the species becomes established and when eradication and/or control are more readily implemented.

Identification of infectious (invasive) and zoonotic disease pathogens is often only possible using DNA-based methods (Lam et al. 2020).

Recent advances in genomics have provided greater means of developing reference DNA sequence libraries using whole genome sequencing (Coissac et al. 2016). Sequences of the whole chloroplast genome and of the ITS region of ribosomal DNA (rDNA) are readily generated from **genome skimming** using low coverage shotgun sequencing. This approach provides greater power for identification than using two or three loci in plants. Combining cpDNA whole genome and ITS sequencing addresses the issue of limited differentiation among close relatives and can identify putative cases of introgression through differential identification in cpDNA and ITS sequences. A review of the level of discrimination in detecting species from the plot scale to the national flora of a small country showed 69–93% effectiveness (Hollingsworth et al. 2016).

DNA barcoding is particularly useful for identification of plant and insect samples that cannot be identified morphologically, for example, nonflowering samples or species with reduced diagnostic features. For example, identification of 45 test samples against a reference library of whole chloroplast and ITS sequences for 530 plant species of the Pilbara region of Australia demonstrated high levels of molecular identification of samples, with 100% match of DNA identity to morphological identity at genus level, and 73–80% to species level when using rDNA and cpDNA, respectively (Nevill et al. 2020). For five out of seven cases of mismatch, both cpDNA and rDNA did not match the morphological identification, suggesting the original identification was incorrect or that there is cryptic variation present within the five species.

Targeted sequence capture methods are also being developed to provide means of species identification using nuclear genes (Faircloth et al. 2012; Lemmon et al. 2012). This approach uses a target enrichment method to generate sequences for a specific set of genes and has been used in studies on a wide range of organisms (Section 4.5.2). Use of common sets of probes will facilitate a standardized approach across species. For example, a plant probe set recently developed has 353 nuclear genes suitable for amplification across the flowering plants (Johnson et al. 2019).

With developments in technology, portable next-generation sequencing methods exist for species identification from noninvasive samples (e.g., feces, hair) in the field or body parts (e.g., feathers) seized by customs officers (Seah et al. 2020). The development of new and better portable sequencers is an exciting frontier in DNA-based species identification.

22.1.2 DNA metabarcoding and metagenomics

While DNA barcoding focuses on identification of individual samples, **metabarcoding** uses a similar approach to identify multiple individuals from community samples (Cristescu 2014; Deiner et al. 2017). Metabarcoding generally involves **polymerase chain reaction (PCR)** amplification and then sequencing of one DNA locus to identify many species simultaneously. Similarly, **metagenomics** involves simultaneous identification of many species through the analysis of many genomes or DNA loci from individual samples with a focus on community composition through identification of species or taxonomic/genetic units. These community samples can be composites of specific samples, for example, bulk samples obtained from invertebrate traps, or from nonspecific community samples, usually from environmental samples such as water, soil, or snow. Analysis of these samples uses next-generation sequencing of the combined samples and matches sample sequences against reference libraries to either detect target species of interest or identify community composition. Natural history collections in herbaria and museums provide valuable repositories of identified and curated specimens from which to develop reference libraries for DNA metabarcoding applications (Dormontt et al. 2018).

While there are still many challenges in standardizing approaches in the field and in the laboratory, data processing, and availability of reference libraries (Deiner et al. 2017), metabarcoding and metagenomics have many applications in species identification for conservation in freshwater, marine, and terrestrial environments. The approach is useful for characterizing the diversity of microbial communities, invertebrate communities, or plant and animal communities when direct analysis of individuals or species is not feasible. Metagenomics can be applied to environmental samples, such as invertebrate malaise trap collections or pollen traps. For example, metagenomics has been used on water and soil samples for comparing bacterial communities, permafrost for assessing past animal and plant communities, ice for assessing past animal and plant communities during glaciations, rodents' middens for describing past plant communities, and feces samples for health assessments or diet analysis (Section 22.1.3).

One of the greatest advantages of metagenomics in environmental samples is that traditional bacterial or viral isolation or culture is not required to study microbes, many of which cannot be grown or detected using traditional culture or isolation approaches. This has significantly increased the feasibility of analysis of soil and water samples to identify microbial composition. Metagenomics has also improved our ability to detect and track parasites and pathogens (Lam et al. 2020).

22.1.3 Diet analysis

Identification of species that have been consumed by an endangered species can help to identify plant or animal resources important to protect or restore in order to conserve the species (Valentini et al. 2009). Diet analysis of predators can also identify the species they prey on, which can inform management of the predator and also any rare species that are being predated. Similarly, diet or gut analysis of parasites identifies host species, which can help search for rare species and may detect invasive species (Cutajar & Rawley 2020). The study of food webs is also fundamental to understanding how the feeding habits of different species can influence the community and ecosystems.

Diet analysis has long been conducted by morphological or microhistological analysis of plant or animal parts identified in fecal samples. Application of DNA metabarcoding or metagenomic analysis has greatly improved diet analysis from feces, and can also be used to analyze stomach contents (Pompanon et al. 2012). DNA analysis is especially useful when the food items cannot be identified using morphological traits of partially digested organisms, or by observing the animal's feeding behavior. Diet analysis in the giant wall gecko has provided information to understand the food sources of this rare species and the functions they provide in ecosystems (Example 22.1).

Several studies of diet in rare and threatened animals have shown that the species have much broader diversity of dietary composition than previously thought. For example, studies of caribou in Canada (Newmaster et al. 2013), lowland tapir in French Guiana (Hibert et al. 2013), golden-crowned sifaka in Madagascar (Quéméré et al. 2013), and Dorcas gazelle in Morocco (Baamrane et al. 2012) all provided information on broader diet to inform management of these species.

Diet analysis of predators can provide important information on the extent of predation on endangered species. For example, diet analysis of wild boar identified them as an important predator of the ground-nesting bird capercaillie grouse in Estonia (Oja et al. 2017). Introduced species can also compete for resources. The endangered Walia ibex has a highly restricted distribution in a mountainous region of Ethiopia and is being affected by changing land use in the area. Comparative analysis of diet in the Walia ibex and introduced goats in Simien Mountains National Park showed significant overlap in plant species consumed, including the five most common species in the diet of Walia ibex (Gebremedhin et al. 2016). This revealed competition for resources, especially during the dry season when resources are limited, and provides information for targeted management of this endangered species.

Until recently, the little brown bat was one of the most common and widespread bats in North America, although populations are now in decline due to white-nose syndrome (Frick et al. 2010) and the species is now endangered. Spatial and temporal diet analysis from fecal samples across Canada has shown a highly variable diet that changes over seasons (Clare et al. 2014). This analysis provides a means of identifying habitat quality based on prey species diversity to inform conservation strategies for this species.

Diet analysis has been much more commonly used in terrestrial compared with aquatic environments, most likely due to challenges in sampling in aquatic and marine species (de Sousa et al. 2019). Analysis of the feces of the endangered Australian sea lion showed a generalist diet that varied according to location, and included eels that had not previously been identified as a component of the diet (Berry et al. 2017). The study also showed that commercial species were not a dominant component of the diet, and that the little penguin was not being predated on by sea lions as previously thought (Berry et al. 2017).

22.1.4 Environmental DNA

eDNA is DNA in environmental samples, such as soil, sediment, water, snow, or air (Taberlet et al. 2018). Sampling eDNA is differentiated from noninvasive DNA sampling because noninvasive sampling directly targets collection of tissues, feathers, hair, feces, or urine spots (Beja-Pereira et al. 2009b), whereas eDNA involves using environmental samples containing mixtures of cells or free-floating DNA from multiple unknown organisms to identify species of interest.

Example 22.1 Diet analysis in the giant wall gecko

Diet analysis has been used to provide information to support management decisions for conservation of the endangered giant wall gecko on small islands in the Cabo Verde Archipelago off the coast of Africa. The species includes two subspecies that each occur on separate uninhabited islands, one on Branco Island and one on Raso Island. It has been extirpated from the larger island of Santa Luzia, where reintroductions are planned. To determine the diet of these animals, Pinho et al. (2018) analyzed feces from both subspecies using the *trnL* gene for plants, 16S for invertebrates, and 12S for vertebrates. They found a generalist diet of plants, invertebrates, and vertebrates in both subspecies, although the particular dietary species varied between the two according to the resources available on each island. Plants were the main component of the diet and a larger component than previously realized. The rare plant *Limonium brunneri* was found in the diet analysis, and this plant is declining on Santa Luzia where the giant wall gecko has gone extinct, suggesting that the giant wall gecko may play an important role in seed dispersal and pollination.

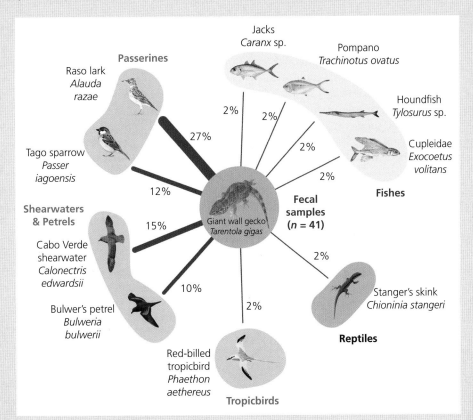

Figure 22.1 A network of vertebrate species found in the diet of the endangered giant wall gecko on Raso Island based on analysis of DNA in feces. The Raso lark is a critically endangered bird. From Lopes et al. (2019).

Identification of diet informs planning of reintroduction to Santa Luzia and also can be used to monitor the success of the reintroduction. Further analysis of the vertebrate component of the diet for the subspecies on Raso Island showed that birds made up the majority, with a small component of reptiles and fishes (Figure 22.1; Lopes et al. 2019). They also found that

Continued

Example 22.1 *Continued*

the critically endangered Raso lark was the main bird consumed both in the wet season (primary time for breeding) and in the dry season. The Raso lark only occurs on Raso Island and has been translocated to Santa Luzia Island. Identification of predation by giant wall geckos means that the translocation of the gecko to Santa Luzia requires reconsideration in the context of interactions between these two rare species.

Environmental DNA samples can contain live microorganisms, such as live bacteria, nematodes, larvae, and invertebrates, or sloughed cells from vertebrates in the sampling area from skins, scales, feathers, mucus, gametes, feces, blood, saliva, and intestinal matter. Early studies using environmental samples were focused on microbial communities, using species-diagnostic DNA sequences such as 16S or the ITS of rDNA (Herrera et al. 2007). More recently, studies have used specific primers designed for genera or species to target specific species, particularly using mitochondrial genes in animals. eDNA is often degraded into fragments of <200 bp, so methods need to be able to account for this. The use of eDNA for species identification has become an important tool due to sampling being efficient, nondestructive, and noninvasive, and delivering accurate identification of target species that can be applied across a broad taxonomic breadth (Beng & Corlet 2020).

The retention of DNA from local species within environments in which they occur allows detection of organisms without direct observation of the species themselves. Use of eDNA is particularly valuable for hard-to-detect, cryptic, rare, and elusive species, and during particular time periods or developmental stages. For these species, detection using eDNA can overcome potential bias from estimates of species diversity or distributions when using only traditional trapping or inventory methods. Determining their local distribution facilitates targeting of conservation and management actions, and identification and protection of critical habitat to enhance survival or reproductive success.

Use of eDNA has been particularly prevalent in marine and freshwater systems (Jerde et al. 2011). For example, there have been few detections of rare sturgeon fish in streams in Alabama, but eDNA can detect these species, enabling conservation measures to be implemented (Example 22.2). Similarly, the slackwater darter is a rare fish known from only a few locations in streams in Alabama and Tennessee. It spawns in freshwater habitats that are hard to access, making detection difficult with traditional methods. eDNA analysis of the *cytochrome b* mitochondrial gene in samples from streams and breeding habitat detected the slackwater darter at 23 breeding and nonbreeding sites where traditional methods had only found it at 1 site (Janosik & Johnston 2015). This included sites where the species had not been detected for some time, and one stream where it has not been found since the 1970s.

In addition to detecting rare and endangered species, eDNA can be used to detect invasive species, particularly at lower densities prior to large-scale establishment. The invasive bluegill sunfish is widely distributed in freshwater ponds in Japan, although little is known of their distribution in specific lakes and ponds or in island systems. Analysis of water samples using the *cytochrome b* mitochondrial gene detected bluegill sunfish in many ponds on the mainland, but at only one pond on each of four islands (Takahara et al. 2013). The low level of invasion in island ponds means that management actions can be focused on maintaining the conservation values of these island systems.

Similarly, Dougherty et. al. (2016) used eDNA analysis of the mitochondrial *COI* gene to detect presence of invasive rusty crayfish at low abundances in lakes of Wisconsin and Michigan, including in lakes where water quality was low and the species had not previously been recorded. eDNA analysis is increasingly used for detection of many aquatic invasive species, including zebra mussels, which cause substantial economic losses in many countries worldwide.

Example 22.2 Detecting rare sturgeon fish in Alabama

Two species of sturgeon fish occur in the Mobile River Basin of Alabama, USA. Alabama sturgeon is listed as critically endangered and Gulf sturgeon as vulnerable on the International Union for Conservation of Nature (IUCN) Red List. In particular, Alabama sturgeon is considered near extinct and very few specimens have been detected since 1997. Riverine sampling for these species is expensive and detection is low, making them good cases for use of an eDNA approach.

Analysis using the *cytochrome b* mitochondrial gene from eDNA sampled during the migration season showed positive detections of both species in streams of the Mobile River Basin (Pfleger et al. 2016). Gulf sturgeon were detected more frequently than Alabama sturgeon, which were detected at multiple locations in the river catchment, although at low levels (Figure 22.2). The detection of these species confirms their ongoing presence in the catchment and has identified priority locations for implementing management actions, while traditional sampling can focus on capturing fish for captive breeding.

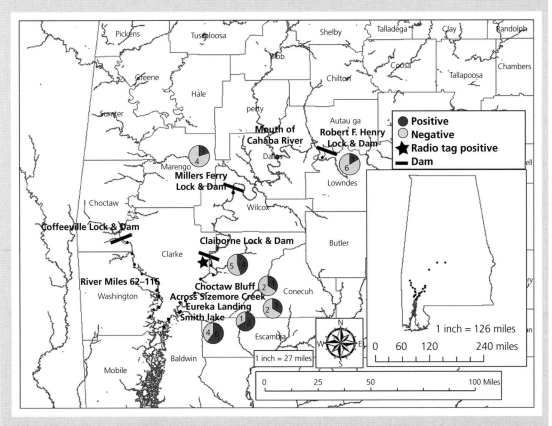

Figure 22.2 Locations of sampling for Alabama sturgeon in the Mobile River Basin, showing detection of the species using eDNA at multiple locations in 2015. Pie charts show the number of samples at locations that were positive detections (dark gray) versus negative detections (light gray). From Pfleger et al. (2016).

eDNA can also help distinguish between closely related native and exotic species where physical identification is difficult. For example, the Japanese giant salamander is a rare species that occurs in freshwater river systems in western Japan. Its close relative the Chinese giant salamander has been introduced and has spread in some rivers. Analysis of water samples using the *NADH1* mitochondrial gene detected the presence of both species, although hybrids between them could not be identified due to use of a mtDNA assay that only detects maternal inheritance (Fukumoto et al. 2015).

eDNA can also be used to improve surveys for biodiversity assessments and for rare species. Universal or pooled primers designed to amplify DNA from multiple species can be used to assay samples simultaneously. For example, a pool of multi-species primers has been developed to detect amphibian and reptile species in North America as a tool for expanded surveys (Lacoursière-Roussel et al. 2016). Similarly, a pooled set of primers has been developed for survey of freshwater turtles in Canada (Davy et al. 2015).

A range of factors can influence detection of target species in eDNA analysis. Determining abundance of species in eDNA analysis can be problematic and needs to be approached with caution. However, some studies have found correlations between eDNA concentrations and other indices of abundance (Section 23.2). For example, Lacoursière-Roussel et al. (2016) found a high correlation between eDNA concentration and visual survey results for the wood turtle, a threatened species restricted to rivers in the northeastern USA and southeastern Canada. Similarly, Kakuda et al. (2019) found that concentration of eDNA from the invasive red-eared sliders occurring in ponds in Japan were highly correlated with visual surveys, and that water quality factors, except for chlorophyll-a concentration, did not affect DNA concentration.

Another exciting advance in eDNA research is the estimation of allele frequencies from eDNA, allowing quantification of genetic variation within populations and genetic differentiation among populations for genetic monitoring. This approach has been used successfully to quantify genetic diversity of bowhead whales from "footprint" samples collected from the water surface after whales dived and additional samples collected along transect lines in Disko Bay, West Greenland (Székely et al. 2021).

22.1.5 Forensic genetics

Illegal "trade" of wildlife, including fish and plants, represents the world's third largest type of illegal trafficking, after drugs and weapons (Clynes 2011). Interpol estimates the illegal trade in wildlife to be worth more than US$20 billion per year (https://www.interpol.int/Crimes/Environmental-crime/Wildlife-crime). Poaching and illegal trade threaten taxa ranging from plants (e.g., orchids and hardwood trees) to insects (exotic tropical beetles and butterflies), reptiles (snakes, turtles, and lizards), fish (sturgeon for making caviar), birds (parrots and canaries), and mammals (especially trophy-horned ungulates, large carnivores, primates, elephants, rhinos, and cetaceans). Illegal logging is so extensive that in the year 2009, the harvest of contraband logs was estimated to be 100 million cubic meters globally (Clynes 2011).

The most important international treaty prohibiting the trade of endangered species is the **Convention on International Trade in Endangered Species of Wild Fauna and Flora (CITES)** that was established in 1973 in association with the United Nations Environmental Program (UNEP). The main international program for monitoring wildlife trade is TRAFFIC—a network of staff and researchers across 20 countries jointly sponsored by the World Wide Fund for Nature and IUCN. Other organizations that work to control illegal wildlife trade include the international organization WildAid, headquartered in San Francisco, California, USA, and PAW (Partnership for Action against Wildlife Crime) in the United Kingdom. Unfortunately, even with such programs, it is difficult to detect poaching and to enforce treaties and antipoaching laws.

Genetic identification of wild species has become prominent in the field of wildlife forensics and can provide critical evidence to enable prosecutions and act as a deterrent in illegal wildlife activities.

Developments and approaches in wildlife forensic genetics share much in common with human forensic genetics, although it has a broader focus. Wildlife forensics generally requires identification of species and often place of origin or captive-bred versus wild-caught, as well as identification of individuals, as is the approach in human forensics (Ogden 2011). In 1991, the Wildlife Forensics DNA Laboratory at Trent University in Ontario, Canada, was the first laboratory to produce DNA evidence to be used in a North American court. There are now more than 50 laboratories in North America, Europe, Australia, South Africa, and other countries that conduct forensics testing to help solve crimes such as illegal trafficking of plant and animal products, according to the Society for Wildlife Forensic Science.

Use of forensic genetics in law enforcement requires implementation of validated techniques and approaches that ensure quality (Ogden 2011). The Society for Wildlife Forensic Science has produced international standards and guidelines for validation and quality assurance in wildlife forensics. Validated sets of markers allow implementation of consistent approaches internationally. For example, Ewart et al. (2018) developed a standardized identification test for rhinoceroses based on the mtDNA cytochrome *b* region that can be used with low concentrations of DNA, such as obtained from horns, which are a major target of illegal poaching in these CITES protected species.

Identification of a sample to species is the most common use of forensic genetics. This has generally been through the use of a barcoding approach (Section 22.1.1) with a range of genes (Figure 22.3). It relies on the growing databases of reference sequences, including specific databases for validated forensic samples (e.g., Ahlers et al. 2017). The DNA Surveillance package (Ross et al. 2003) applies phylogenetic methods to the identification of species of whales, dolphins, and porpoises, and allows users to enter a genotype from a sample of unknown origin and then receive back an alignment, genetic distance estimates, and an evolutionary tree (Baker & Steel 2018).

Samples to be analyzed sometimes come from whole animals or plants but often come from a range of tissues, such as hairs, bone, ivory and horns, shells, feathers, scales, skin, claws, processed meat, wood, seeds, flower parts, or swabs from individuals or the environment. DNA from these samples is often limited and degraded, and mtDNA or cpDNA is the best approach for analysis of such material. Nuclear DNA markers such as microsatellites (SSRs) are generally used where individual identification is required to determine place of origin (particularly where populations are not well differentiated), in cases of identification of individuals from captive breeding versus wild populations, or to match samples with existing individuals (Figure 22.3). In these cases, assignment tests to determine probability of match or exclusion are required (Section 22.2.2), and also rely on sets of reference genotypes. Iyengar (2014) noted the development of reference allele frequency databases for a number of species targeted in illegal activities, including elephants, rhinoceros, mule deer, brown bears, wild boars, tigers, badgers, mouflon, birds of prey, and black cockatoos.

An early, widely publicized forensic application of wild population DNA analysis for conservation was the identification of illegally traded whale meat sold in Japanese and Korean markets (Baker et al. 1996). PCR-based analysis of mtDNA control region sequences revealed that about 50% of the whale meat sampled from markets had originated from protected species and not from the southern minke whale species that Japan is allowed to harvest under its scientific whaling program. For this study, the researchers were not allowed to export the tissue samples from Japan because many species of whale are protected by CITES, which forbids transportation of any tissues or DNA without a permit. Consequently, the researchers set up a portable PCR machine and amplified the mtDNA in a hotel in Japan. They subsequently transported the synthetic DNAs (not regulated by CITES) back to laboratories in the USA and New Zealand for sequence analysis.

Illegal poaching of ivory and rhinoceros horns is a major component of wildlife crime (Wyler & Sheikh 2013) and forensic genetics can assist in identifying hotspots of illegal activity. For example, analysis of a large seizure of ivory using an assignment

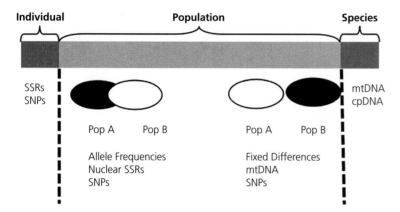

Figure 22.3 The identification of a sample to species or population, or matching to an individual, typically uses different genetic markers, from organelle DNA to nuclear DNA. SSRs are simple sequence repeats (microsatellites). From Ogden & Linacre (2015).

test of 16 microsatellite markers showed that they were from savannah elephants, and rather than originating from multiple locations, the ivory was most likely from a small area centered in Zambia (Wasser et al. 2007; Section 22.4.2).

Identification of wildlife poaching frequently relies on matching carcasses to evidence identified from other sites. Mouflon are a protected species that are similar to sheep. Lorenzini et al. (2011) used 16 microsatellite loci to obtain genotypes from the carcass of an animal being presented as a sheep, and blood and other trace samples obtained from the putative site of the killing. They showed that the carcass was a Sardinian mouflon and not a sheep, and the genotypes from the carcass and the blood at the site were a match with a 1 in 33 billion chance of being from unrelated individuals. Similarly, Barbanera et al. (2012) used mtDNA and 17 microsatellite loci to obtain genotypes from blood stains in the vehicle of a suspected poacher with animal carcasses believed to be Cypriot mouflon. Both of these studies required broader sampling of sheep and mouflon to provide the reference datasets to undertake the assignment analyses with confidence.

Birds, particularly parrots and cockatoos, are a target for smugglers, and eggs and nestlings of rare species are often harvested illegally from the wild. Forensics can be used to determine the species when eggs are seized at international borders. For example, mtDNA analysis of eggshell samples from seized eggs being smuggled through Sydney airport showed them to have come from cockatoos, macaws, and parrots that were subject to controlled trade by CITES (Johnson 2010). Analysis of identity and kinship can also provide evidence to identify local illegal harvest (Example 22.3).

Reptiles are also frequent targets for illegal harvest and smuggling operations. Kinship analysis can be used to demonstrate theft of putative captive animals that have been taken from the wild. For example, kinship analysis was used to demonstrate that six endangered Greek tortoises offered for sale were obtained through theft from a private breeder (Mucci et al. 2014). The analysis with 14 microsatellite loci showed that the tortoises offered for sale were the offspring and full sibs of the tortoises legally held by the breeder.

Illegal logging is a significant issue worldwide for protection and sustainable management of forests and the habitats they provide. Application of forensic genetics has been limited to date as DNA extraction from wood samples can be challenging. Early work by Ogden et al. (2008) produced a cpDNA SNP assay for identification of tropical hardwood in the genus *Gonystylus* (ramin), a CITES-listed genus subject to illegal trade from Asia. More recently, Ng et al. (2016) have developed microsatellite approaches to determine region and population of origin. Similarly, Dormontt et al. (2020) developed a nuclear SNP array for forensic analysis of bigleaf maple that can be implemented for wood samples that yield low DNA concentrations (Guest Box 22).

Example 22.3 Identifying illegal harvest of black cockatoos

Black cockatoos are a target for illegal harvest from the wild and forensic analysis has assisted in identification of illegal poaching activities. White et al. (2012) used 17–19 microsatellite markers to determine the identity of birds of two species of black cockatoo illegally harvested from the wild. In a case of suspected nest poaching, a member of the public alerted wildlife officers to a man climbing a ladder in a tree on a road verge (Figure 22.4). Wildlife officers went to the man's property and found a red-tailed black cockatoo nestling. Genetic analysis matched the genotype of the nestling to the genotype obtained from eggshell remains in the tree hollow from which it was suspected of having been removed, with a probability of a chance match in genotype being one in 389 billion. In another case, three white-tailed black cockatoos were found during a visit by wildlife officers to a property, and were claimed to have been captive-bred. However, genotyping identified them as having been removed from the wild, with one bird identified as being a full-sib match to another bird from a known wild population based on the authors' cockatoo population database.

(a)

(b)

(c)

Figure 22.4 Evidence in a case of nest poaching of endangered red-tailed black cockatoo in Western Australia. (a) A man climbing a tree on a road verge in Western Australia was photographed by a member of the public who suspected it was nest poaching. (b) Red-tailed black cockatoo nestling found at the suspect's property was taken into care. (c) Eggshell from the nest hollow was collected as evidence and used to DNA match the nestling to the hollow. From White et al. (2012).

22.2 Individual identification

Individual identification is one of the most widely used applications of molecular markers in conservation genetics, forensics, and molecular ecology. For example, in the lynx–bobcat hybridization study in Example 13.1, researchers had to analyze individual fecal samples to know the number of different individuals sampled. Another example is the matching of genotypes from blood samples from the kill site with the carcass in the wildlife poaching of mouflon (Lorenzini et al. 2011) described in Section 22.1.5.

To match individual samples, they must first be genotyped with either 10–20 highly polymorphic molecular markers such as microsatellites, or with a larger number of less polymorphic markers, such as SNPs. A **match probability** (or probability of identity) is then computed, by using allele frequencies estimated from the reference population, such as the National Park or the geographic location from which the sample at the "crime" scene originated. If allele frequencies from the reference population are not available, we can still estimate the match probability, but it requires additional markers to achieve reasonably high power to resolve individuals with high certainty (Menotte-Raymond et al. 1997; Box 22.2).

Microsatellites are widely used in forensics and genetic management because of several characteristics: (1) their short length (<300 bp) makes them relatively easy to PCR-amplify from partially degraded DNA; (2) they are generally highly polymorphic; and (3) alleles from the same locus can be easily identified. For human forensic investigations in the USA and United Kingdom, a standard set of 13 and 10 microsatellite loci are used, respectively (Watson 2000; Reilly 2001). These marker sets provide a chance of a match between two random people that is between about one in a million and one in a billion. Similar approaches of a common set of markers are used in wildlife forensics and management where there is sufficient demand for analysis of particular species. Microsatellites have not been replaced in human forensics by more powerful and reliable genomic approaches (e.g., SNPs and microhaplotypes) because of the millions of genotypes in the forensic databases that would have to be replaced (K.K. Kidd, personal communication).

SNPs can be used to determine individual identity and provide greater power in probability of identify, particularly in species where diversity may be reduced. SNPs may also provide greater genotyping success than microsatellites because SNPs can be PCR-amplified from smaller, more degraded DNA fragments (<100 bp) than microsatellites (von Thaden et al. 2017; Eriksson et al. 2020). However, the use of SNPs in forensics has been limited to date given the effectiveness of microsatellite markers to achieve identification. In addition, reference databases of individual genotypes have been established with microsatellites, and these would have to be replicated for use of SNPs.

An example of DNA-based individual identification for wildlife management is the identification of problem animals. For example, when a wolf or bear attacks a human, kills livestock, or steals a picnic basket, the problem animal will often be removed from the population. Removing the wrong bear could waste resources and eventually result in the removal of several animals before identifying the correct one. This could harm the population, especially if the individuals removed are reproductive females. Furthermore, knowing with certainty that the true problem individual was removed would reduce public concern. Thus, it is crucial to identify the correct individual before removing it. Here, matching DNA from the scene of the "crime" to an individual can help.

Other uses of individual identification in conservation management include **genetic tagging** for studying movements, particularly in species where dispersal or gene flow is difficult to detect, such as plants and marine species, and estimating population census size from mark–recapture methods (Section 18.1). Individual identification also helps to identify clonal individuals in plants and in some animals (corals, anemone, and fishes). The identification of clones is required for accurate estimation of patterns and rates of gene flow, geographic distributions of clones, and inbreeding versus outcrossing rates.

22.2.1 Probability of identity

The statistical power of molecular markers to identify all individuals from their multilocus genotype

---

Box 22.2 Computing the match probability (*MP*) for an individual sample (multilocus genotype)

Here we consider a scenario where we have one sample in hand (e.g., a bloodstain at a wildlife crime scene) and we want to compute the probability of sampling a different individual that has an identical multilocus genotype (in the same population). This is often called the match probability (*MP*). In this example, we compute the *MP* using two loci to demonstrate the method, recognizing that the greater the number of loci, the lower the *MP*.

Consider two loci that each has two alleles at the following frequencies: $p_1 = 0.50$, $q_1 = 0.50$; and $p_2 = 0.90$, $q_2 = 0.10$, respectively. A bloodstain from the scene of a wildlife crime (e.g., poaching in a National Park) has a genotype that is heterozygous at both of these loci. What is the probability that an individual sampled at random from the same population has the same genotype (as that of the individual whose bloodstain is "in hand")?

First we compute each single-locus *MP*:

Locus one: $2p_1q_1 = 2\,(0.50)\,(0.50) = 0.50$
Locus two: $2p_2q_2 = 2\,(0.90)\,(0.10) = 0.18$

Then, the multilocus *MP* is the product of the two single-locus probabilities: $0.50 \times 0.18 = 0.09$ (assuming independence between loci). We conclude there is a 9% chance of sampling a different individual with a double-heterozygous genotype identical to the one "in hand." Thus there is a 9% chance of matching the bloodstain to the wrong individual. Clearly, more loci and/or more polymorphic loci are

needed to have a reasonably low chance (e.g., <1/10,000) of a match to the wrong individual. Recall that here, we are assuming unrelated individuals, no substructure, and that the allele frequencies are known for the population considered.

What if the wildlife crime occurs in a population with no reference data (i.e., allele frequencies are unknown)? How can we estimate the probability that an individual sampled at random has the same double-heterozygous genotype as the individual "in hand"? Here we could assume that the frequency of the observed heterozygous genotype at each locus is high (e.g., 0.50). This is the highest frequency possible (assuming a bi-allelic locus and HW proportions), and gives the least power for individual identification. This assumption of a heterozygote genotype frequency of 0.50 is conservative and generally overestimates the true *MP* (Menotte-Raymond et al. 1997). This is especially true if a locus has >3 alleles because the two alleles in a heterozygote "in hand" could never have a population frequency as high as 0.50, which can only occur for a bi-allelic locus.

We then could compute the multilocus match probability as follows: $0.50 \times 0.50 = 0.25$. Here, the estimated 25% chance of sampling this multilocus genotype is much greater than the 9% chance estimated (above) by using the reference allele frequencies. This illustrates the advantage of having reference allele frequencies for the population.

---

is estimated as the average **probability of identity** (*PI*$_{av}$). *PI*$_{av}$ is the probability of randomly sampling two individuals that have the same genotype (for the loci being studied). If we use highly polymorphic molecular markers or a large number of less polymorphic markers, there is a low probability of two individuals sharing the same genotype at multiple loci. Thus, if we find two samples (e.g., bloodstains or tissues) with matching genotypes, we can determine with high probability that they come from the same animal or plant (e.g., found at a crime scene).

PI$_{av}$ is computed using the following equation:

$$PI_{av} = \sum_{i=1}^{n} p_i^4 + \sum_{i>j=1}^{n} (2p_ip_j)^2 \qquad (22.1)$$

where p_i and p_j are the frequencies of the ith and jth allele at the locus (Waits et al. 2001; Ayres & Overall 2004). Here, p_i^4 is simply the average probability of randomly sampling two homozygotes (e.g., aa), and $(2p_ip_j)^2$ is the average probability of randomly sampling two heterozygotes (e.g., Aa). This equation assumes **Hardy–Weinberg (HW) proportions** and that no substructure exists in the population. The multilocus *PI*$_{av}$ is computed by using the product rule (i.e., "multiplication rule," see Appendix Section A2), and multiplying together the single-locus probabilities (see match probability Box 22.2). This assumes use of independent loci, for example, on different chromosomes. A reasonably low multilocus *PI*$_{av}$ for forensics applications (e.g., matching blood from a wildlife crime scene to blood on a suspect's clothes) is ~1/10,000 to 1/100,000. Achieving

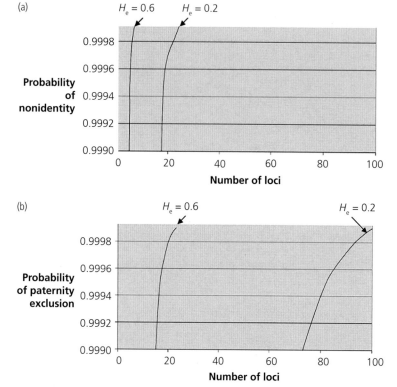

(a)

Figure 22.5 Relationship between the number of loci genotyped, and (a) the probability of nonidentity $(1 - PI_{av})$ and (b) the probability of paternity exclusion. The number of loci needed for paternity exclusion is approximately three- to fivefold higher than for individual identification. Also at low heterozygosity ($H_e = 0.2$; e.g., SNPs) three to five times more loci are needed compared with microsatellite loci ($H_e = 0.6$). The probability of identity (PI_{av}) was computed using Equation 22.1 and allele frequencies that give a heterozygosity (H_e) of 0.6, representing microsatellites, or 0.2, representing SNPs. For example, for a locus to have $H_e = 0.2$, two alleles would have frequencies of 0.885 and 0.115. The probability of paternity exclusion was computed using the expression from Jamieson & Taylor (1997) for the case where genotypes are known for the mother and the offspring when testing to exclude a randomly sampled male that is not the true father. Modified from Morin et al. (2004).

this low PI_{av} would require ~5–20 loci, depending on their polymorphism level (Figure 22.5).

PI_{av} is also often used to quantify the power of molecular markers for studies involving genetic tagging. A reasonably low PI_{av} for genetic tagging is ~1/100 (Waits et al. 2001). This is not as low as is needed for forensics, because it is less problematic to misidentify individuals in genetic tagging than in a law enforcement case where someone might be fined or imprisoned. To achieve a reasonably low PI_{av} for genetic tagging, ~5–10 highly polymorphic markers are often sufficient (Box 22.2). The number of SNPs obtained from genome sequencing or **restriction site-associated DNA sequencing (RADseq)** provides sufficient power for low PI_{av} (Section 4.5.1). Interestingly, the power of a set of markers is better predicted by heterozygosity than allelic richness; loci with the same heterozygosity ($H = 0.6$) but a different number of alleles (three or 10) have similar power to resolve individuals (Waits et al. 2001).

It is important to note that Equation 22.1 used to estimate PI_{av} assumes that individuals are unrelated

(e.g., no siblings), sampled randomly, and that no substructure or gametic disequilibrium (i.e., independent loci) exists. These assumptions are often violated in natural populations. The violation of assumptions could cause underestimation of the true PI_{av}. For example, in datasets from wolves and bears, PI_{av} was underestimated by up to three orders of magnitude (e.g., 1/100,000, which underestimates the true value of 1/100; Waits et al. 2001). To avoid problems with underestimation, other PI_{av}-related statistics such as PI_{av}-*sibs*, should also be used to compute the probability of identity. It is confusing that the "average probability of identity" (PI_{av}) is sometimes referred to as the "average match probability" (MP) in the literature, even though they are not equivalent terms.

22.2.2 Match probability

The match probability (MP) is a useful statistic related to PI_{av} (Equation 22.1). While PI_{av} is the average probability of randomly sampling two individuals consecutively that have the same geno-type, the *MP* is the actual probability of sampling one individual identical to the one already "in hand" (i.e., sampled previously). PI_{av} is for comput-ing the average power of a set of markers (consider-ing all genotypes, homozygotes and heterozygotes, in a given study), whereas *MP* gives a probabili-ty of sampling the individual genotype in question, that was previously detected or sampled (Box 22.2). *MP* requires the same assumptions (no substruc-ture, no gametic disequilibrium, and no siblings) as does PI_{av}, although more sophisticated *MP* statistics can correct for violations of these assumptions.

For example, Tnah et al. (2010) illustrated how DNA match probabilities can be computed for a tropical timber species across peninsular Malaysia. Match probabilities were assessed to help identify logs illegally removed from forest preserves. DNA typing was used to match stumps in forest preserves to logs sold or transported illegally. DNA match-ing of stumps to illegally trafficked logs can help to stop illegal deforestation. The authors genotyped 12 microsatellites from 30 populations and showed how effects of population substructure and inbreed-ing could be incorporated into estimates of match probabilities. This estimation procedure, along with a large database of genotypes from many popula-tions of the timber species, should help prosecute illegal loggers in Malaysia.

We have thus far considered only nuclear DNA markers for determining match probabilities and the average probability of identity. However, mtDNA can also be useful for individual identifi-cation. For example, the mtDNA control region in canids and felids has tandemly repeated sequences (Fridez et al. 1999; Savolainen et al. 2000). These repeats are highly polymorphic and heteroplasmic (i.e., multiple mtDNA clones with different repeat lengths are found within an individual). Thus, mtDNA analysis has occasionally been useful for individual differentiation because different individ-uals often have different mtDNA repeat profiles. Since mtDNA represents only one locus, it will pro-vide much less certainty than multilocus nuclear DNA methods. But it does have an advantage in some situations. It can be amplified from hair shafts, whereas unfragmented nuclear DNA is only found in the hair root bulb (Watson 2000), and animal hairs

are often found at poaching crime scenes and on people's clothing. Chloroplast markers are also used occasionally in forensics of plants.

22.3 Parentage and relatedness

Determination of the mother or father of an individ-ual can facilitate conservation in a variety of ways such as identifying mating patterns, quantifying reproductive fitness of parents, managing captive breeding programs, or verifying that individuals in pet stores originate from captive parents, as might be claimed by some pet-trade industry workers. Genetic analysis provides an effective way of iden-tifying parentage and relatedness and is particu-larly useful when this information is not available through other approaches.

22.3.1 Parentage

Parentage analysis involves comparing genotypes of offspring to potential parents in order to identify the actual parents. This analysis generally depends on the fact that an offspring will have one allele per locus from each parent.

Two computational approaches exist for genet-ic parentage analysis: **parentage exclusion** and **parentage assignment** (e.g., Marshall et al. 1998; Taberlet et al. 2001). Exclusion involves the deter-mination that both alleles at a locus in a candidate parent do not match either of the offspring's alleles, which leads to exclusion of that candidate parent as the true parent. With sufficient power, exclusion leads to confident identification of the true parent. Exclusion might be insufficient to resolve parent-age unambiguously in situations where multiple candidate parents cannot be excluded due to low diversity or relatedness, or when all potential par-ents have not been sampled. Another deficiency with exclusion is that genotyping errors can cause the true parent to be excluded erroneously.

In most cases parentage analysis involves identi-fication of the father, as the mother is often known. For example, an offspring often associates closely with its mother (e.g., in mammals and birds), or seed is collected from known plants. When some males cannot be sampled or excluded or when genotyping errors exist, a statistically based method can help infer paternity, particularly where the mother's genotype is known. **Paternity assignment** is statistically based and involves the use of probabilities and likelihood computation. Paternity assignment is widely used to estimate the probability that a given male is the father (Slate et al. 2000; Flanagan & Jones 2019). While statistical paternity inference is possible when not all potential fathers have been sampled, the power (or certainty) of assignment drops substantially when less than ~70–90% of males (potential fathers) are sampled (Marshall et al. 1998).

Parentage analysis requires more genetic markers than does individual identification (Section 22.2). For example, ~10–15 microsatellites (with heterozygosity 0.50–0.60) are required to achieve a high probability of paternity exclusion (>99.9%) (Figure 22.5b), whereas individual identification requires only 5–10 loci (Figure 22.5a). Parentage analysis when neither parent is known requires even more loci (e.g., >20 microsatellites) than paternity analysis where the mother is known. Parentage analysis also requires more loci when heterozygosity is low. For example, ~40–60 SNPs (heterozygosity > 0.2) are required for paternity exclusion when the mother is known (Figure 22.5b).

It is likely that parentage assignment with SNPs and unknown parents would often require ~100 SNP loci. However, most SNP analyses produce a large number of SNPs so the number of loci is not limiting. Comparisons of parentage assignment for SNPs versus microsatellites in birds have shown the effectiveness of SNPs. In the black-throated blue warbler, a panel of 97 SNPs had the same power of assignment as six microsatellites, giving 100% and 99% assignment at 95% confidence and **probabilities of nonexclusion** for the first parent of 1.9×10^{-2} and 1.9×10^{-3} for SNPs and microsatellites, respectively (Kaiser et al. 2017). Similarly, comparisons in the variegated fairy-wren showed nonexclusion probability for the first parent of 5.2×10^{-20} with 411 SNPs compared with 1.9×10^{-4} for 12 microsatellites (Thrasher et al. 2018).

To quantify the statistical power of molecular markers for parentage analysis (e.g., paternity exclusion), researchers can compute the expected **paternity exclusion probability** (*PE*), which is the probability of excluding a randomly chosen nonfather (e.g., Double et al. 1997). The power of a set of molecular markers for paternity exclusion (*PE*) with *n* alleles can be quantified using allele frequencies in the following equation (Jamieson & Taylor 1997):

$$PE = \sum_{i=1}^{n} p_i^2 (1 - p_i)^2 + \sum_{i>j=1}^{n} 2p_i p_j (1 - p_i - p_j)^2$$

(22.2)

where p_i and p_j are the frequency of the *i*th and *j*th allele, respectively. For one locus, this expression gives the average probability of excluding (as father) a randomly sampled nonfather, when the mother and offspring genotypes are both known. To compute the multilocus *PE*, we multiply together the *PE* for each locus, assuming independence among loci. Other expressions are available for estimating power for parent exclusion when neither parental genotype is known (Jamieson & Taylor 1997).

22.3.2 Mating systems and dispersal

Parentage analysis provides much insight into species' mating systems, particularly in estimating variance in reproductive success and detecting multiple paternities. Such information is helpful for population management. Variance in reproductive success influences the effective population size and thus the rate of loss of genetic variation, inbreeding, and efficiency of selection. For example, knowing that variance in reproductive success is high can help biologists to predict that the N_e is much smaller than the census population size (Section 7.10).

Parentage analysis can provide evidence of mating patterns in wild and reintroduced populations. Reintroduction from captive breeding is maintaining wild populations of the endangered green turtle in the Cayman Islands. Parentage analysis using 13 microsatellite loci has shown that 90% of the wild-caught breeding females are related to the captive breeding stock, demonstrating the significant contribution of captive reintroduction in maintaining the wild population (Barbanti et al. 2019). The founding stock for the reintroduction program was collected from a wide range of wild populations and genetic analysis of the population shows high levels of diversity. Paternity analysis in the threatened wood turtle in the Shawinigan River in Quebec used seven microsatellite loci to analyze clutches over 2 years (Bouchard et al. 2017). The mean number of mates per male was 1.24 and 1.78 across 2 years, although cases of polyandry and multiple paternity were observed for some males as is common in reptiles.

Knowledge of breeding and dispersal from sites can provide information to improve management of landscapes. For example, the corn crake is a migratory bird that nests in tall vegetation and is declining across Europe. It is hard to detect locations of nests and unfledged chicks, but females lay eggs close to the location of singing males. Parentage analysis using 15 microsatellite loci showed that young chicks <20 days old were within 150 meters of the singing males assigned parentage and independent unfledged chicks were up to 600 meters from the location of singing males (Green et al. 2019). This provides evidence to inform mowing requirements for broader management of the landscape to minimize impact to chicks.

Parentage analysis has been particularly useful in identifying and understanding dispersal patterns, particularly in species where dispersal is hard to determine, such as in plants and in marine systems. In these systems, paternity assignment where the mother is known or parentage analysis where both parents are unknown provide means of estimating mating patterns and population connectivity through dispersal of propagules.

Paternity assignment has been extensively used to understand pollen and seed dispersal, and the impacts of fragmentation on populations. Pollen dispersal, in particular, has been found to be extensive in both wind- and insect-dispersed species in temperate and tropical systems. For example, Gerber et al. (2014) genotyped stands of oak across Europe using eight microsatellite loci. They showed extensive dispersal of pollen and seed with up to 67% of pollen and 38% of seed dispersal from outside the stands.

Similar evaluation of pollen dispersal can be used to determine connectivity among populations that can be important in managing landscapes with rare plants. For example, Sampson et al. (2016) used seven microsatellite loci to genotype plants and seed crops of a rare insect-pollinated woody shrub restricted to ironstone soil in southwestern Australia. They found that 15% of seeds in a small population in this fragmented landscape had been fathered with pollen from a population at least 1 km away. They also found that pollinators moved over greater distances within a population where the vegetation remained intact compared with where it was degraded. This highlights the importance of habitat quality in maintaining effective pollination.

Analysis of seed dispersal is also being used to determine movement of the alpine tree line in response to climate change. Johnson et al. (2017) analyzed an alpine treeline ecotone of mountain hemlock in Alaska through genotyping of trees in the transect using 353 SNPs. Seeds are wind dispersed in this species and they found that the majority of 163 trees in the ecotone represented long distance seed dispersal over 450 meters.

Similarly, in marine environments dispersal can be challenging to determine and parentage analysis can identify dispersal routes and connectivity. For example, parentage analysis of the cinnamon anemonefish in the Keppel Islands of the Great Barrier Reef in Australia used 22 microsatellite loci to genotype adult and juvenile fish as well as egg clutches (Bonin et al. 2016). The analysis showed extensive connectivity among populations through larval dispersal, with 49% of larvae settling within 4 km and a maximum dispersal distance of 28 km. The analysis also showed that populations in marine reserves were a large source of larvae and juveniles, with one reserve population making the largest contribution to recruitment.

22.3.3 Relatedness

Relatedness estimation is useful in conservation of wild and captive populations to help understand animal social systems, to enable avoidance of inbreeding, and to optimize conservation breeding and translocation strategies (Gonçalves da Silva et al. 2010). Relatedness can be estimated using pedigrees (Section 17.1) or from genetic similarity between individuals estimated directly from genotyping. For example, founders of captive populations are often assumed to be unrelated, which

530 CONSERVATION AND THE GENOMICS OF POPULATIONS, THIRD EDITION

could bias breeding programs leading to mating between relatives, loss of genetic variation, and reduced fitness.

Computation of marker-based founder kinship coefficients can improve management and the maintenance of genetic variation in captive populations. Analysis can quantify **mean kinship** (*mk*), which is a measure of how closely related each animal is to the population. It is an important measure of just how rare an individual's combination of alleles is in the entire population. Animals with lower mean kinship values have relatively fewer alleles in common with the rest of the population, and are therefore more genetically valuable in a breeding program (Ballou & Lacy 1995; Section 21.6). For example, Gonçalves da Silva et al. (2010) used DNA markers to refine the captive breeding program for lowland tapirs. They used 10 microsatellite markers to genotype 49 captive individuals to evaluate relatedness and identified two individuals with low mean kinship (*mk* = 0.007), which were thus of high genetic value for the population. These individuals might otherwise have been excluded from the breeding program, in the absence of empirical genetic data, because of their unknown origins. Common assumptions of individuals of unknown origin being highly related would have led to overestimates of mean kinship and underestimates of future heterozygosity, when compared with values found when genetic markers were used to inform kinship and breeding.

The management of two critically endangered birds in New Zealand has included captive breeding programs, with the program for kakī (black stilt) in operation since the 1980s and for kākāriki karaka (orange-fronted parakeet) since 2003. Analysis of relatedness was undertaken using eight microsatellites, a large number of SNP markers, and estimates from pedigree records across eight and five generations for kakī and kākāriki karaka, respectively (Galla et al. 2020). Analysis based on SNPs provided high confidence in parent–offspring and full-sib relationships. This information is being used to inform future pairing between birds in the current program and those being brought into the program from the wild. The analysis has shown the power of SNPs in determining relatedness, particularly in species with low genetic diversity,

and the value of genetic analysis of captive breeding programs where detailed pedigrees are not available.

22.4 Population assignment and composition analysis

Genetic markers can help to identify the population of origin of individuals, groups of individuals, or products made from plants and animals, such as from endangered forest trees or horn or bone from threatened wildlife (Waser & Strobeck 1998; Manel et al. 2002). Determining the population or geographic region of origin of wildlife products can help to identify populations being poached, and the trade routes used by traffickers selling illegally harvested individuals. Such information could help law enforcement officials to target geographic regions with poaching problems. Genetic analysis of mixed aggregations enables identification of patterns of migration that provides information for management of wildlife programs, such as fish stocks, and landscape interactions such as for migratory birds.

22.4.1 Assignment of individuals

Population assignment tests work by assigning an individual genotype to the population in which its genotype has the highest expected frequency (Section 9.9.4; Manel et al. 2005). **Assignment tests** generally require samples and multilocus genotype data from each candidate population of origin. Assignment tests can potentially determine the population of origin of individuals, even when the F_{ST} is low, if many microsatellite loci are genotyped (Olsen et al. 2000) or SNP loci are used.

Researchers used assignment tests to identify the origin of a wolverine in California detected nearly 100 years after the last known sighting (Moriarty et al. 2009). The researchers used 16 microsatellite loci to assign the sample to populations identified from 261 current and historic genetic samples. The assignment tests suggested the animal of unknown origin was not a remnant of an historic California population, but was most closely related to populations from the western edge of the Rocky Mountains in Idaho. This is the first evidence of movement

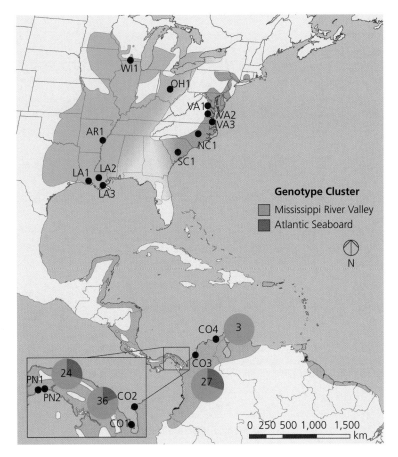

Figure 22.6 Analysis of the prothonotary warbler shows assignment of birds in their overwintering grounds in Central and South America to their breeding grounds in North America. Pie charts in Central and South American sites show number of sampled birds originating from the Mississippi River Valley (green) and the Atlantic Seaboard (purple). From DeSaix et al. (2019).

from wolverine populations of the Rocky Mountains to the Sierra Nevada Mountains in California.

Genetic analysis can be used to identify populations of origin and patterns of movement in species where migration occurs over large distances and individuals form mixed aggregations, such as in migratory birds or in marine systems. For example, DeSaix et al. (2019) used SNPs to investigate the populations of origin of the prothonotary warbler, a neotropical migratory bird that is of conservation concern. Although they identified low regional genetic structure in the Mississippi River Valley and the Atlantic Seaboard ($F_{ST} = 0.0055$), they were able to assign individuals from nonbreeding sites to their breeding region with 94% accuracy (Figure 22.6). They found mixing of populations in the nonbreeding grounds and low migratory connectivity. This information will assist in prioritization and management of the nonbreeding grounds.

Marine species can be highly connected but often return to natal sites to breed, and genetic information can be invaluable for sustainable fisheries management. Use of SNPs provides opportunities for population assignment to inform stock management, even in species with weak genetic structure. Genetic analysis of American and European lobsters showed high levels of accuracy of assignment to basin of origin in European lobsters (Jenkins et al. 2019) and to region of origin in American lobsters (Benestan et al. 2015, 2016). These analyses allow greater effectiveness in managing sustainable harvest and restocking practices, as well as in tracing of stock through the consumer chain. The analysis of Benestan et al. (2015, 2016) in American lobsters also demonstrates the power of assignment increased with the use of maximally differentiated SNPs up to the use of 3,000 SNPs, after which power declined. Power of assignment was also highest when over

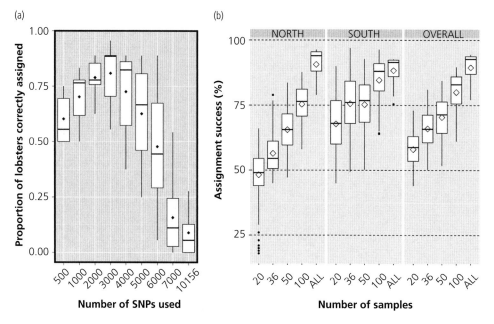

Figure 22.7 Assignment analysis of American lobsters with SNP loci shows power of assignment is greatest with 3,000 maximally differentiated SNPs and sampling of over 100 individuals per population. (a) The proportion of lobsters correctly identified based on number of SNPs, ordered from maximum to minimum differentiation among samples. (b) Assignment success as a function of the number of sampled individuals per population for north and south regions and overall. From Benestan et al. (2015, 2016).

100 individual reference samples were available per region (Figure 22.7).

Assignment tests have also been used along with genotypes from historical museum specimens to test whether individuals from a presumed extinct species might still exist in captivity. Analysis of Galápagos tortoises in a captive population found individuals that had been mistakenly identified and were actually from a species that was considered to be extinct (Example 22.4).

Another interesting example of assigning individuals for wildlife forensics comes from a fishing competition. A fisherman claimed to have caught a large salmon in Lake Saimaa, Finland. However, the organizers of the competition questioned the origin of the salmon because of its unusually large size (5.5 kg). A genetic analysis of seven microsatellite loci was conducted on the large fish and on 42 fish from the tournament lake, Saimaa. A statistical analysis was conducted using the exclusion–simulation assignment test (Primmer et al. 2000). The exclusion test suggested that the probability of finding the large fish's genotype in Lake Saimaa was <1 in 10,000. Thus, the competition organizers

excluded Lake Saimaa as the origin of the salmon. Subsequently, the fisherman confessed to having purchased the fish in a bait shop.

Population assignment in the fishing tournament example is based on **exclusion tests** and computer simulations to assess statistical confidence. In the exclusion–simulation approach, we "assign" an individual to one population only if all other populations can be excluded with high certainty (e.g., $P < 0.001$). We exclude a population if the genotype in question is unlikely to occur in the population ($P < 0.001$), assuming HW proportions and gametic equilibrium.

Similarly, population assignment based on exclusion has application in forensics. For example, illegal introduction of red deer in a forest in Luxembourg where deer are absent was detected using population assignment. Analysis of the deer populations from surrounding populations in Luxembourg, Belgium, and Germany using 13 microsatellite loci showed three genetic clusters (Frantz et al. 2006). The introduced animals were excluded from assignment to these genetic clusters, demonstrating that they had been introduced from elsewhere.

Example 22.4 Guiding conservation of Galápagos tortoises with assignment tests and parentage analysis

Galápagos tortoises are a complex of 11 species and are the subject of many *in situ* and *ex situ* conservation programs. Galápagos tortoises live up to nearly 200 years, and many in captivity have unknown origins. Russello et al. (2010) genotyped nine microsatellites and mtDNA from museum specimens of extinct Galápagos tortoise species to determine their genetic relationships. They then genotyped samples from animals held in captivity and assigned them to species from the reference samples. The Bayesian assignment testing identified nine Galápagos tortoises in a captive breeding facility on Santa Cruz Island that had some level of mixed ancestry but high genetic assignment ($q = 0.798–0.942$) to a species that had been considered extinct since the early 1900s. The findings led to captive breeding efforts to breed these nine animals and to reestablish this extinct species of tortoise in the wild on Floreana Island. Parentage analysis using 12 microsatellite loci is being undertaken to inform the breeding program of six females and three males. Eight of the nine animals are fertile, and analysis of 130 progeny produced over 3 years has shown breeding among six pairs, but with the majority of progeny arising from breeding between two males and three females. Nearly half of the progeny were produced by just one pair of animals (Miller et al. 2018; Figure 22.8).

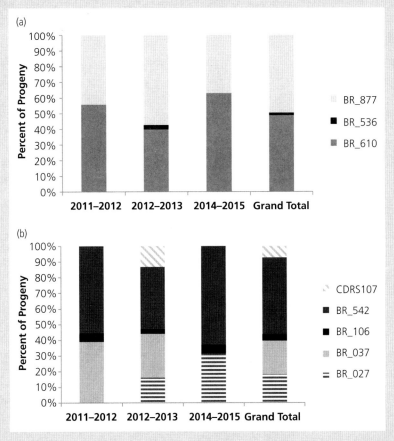

Figure 22.8 Parentage analysis in a species of Galápagos tortoise recently rediscovered in captivity, showing proportion of progeny arising from mating among eight fertile individuals. (a) Contribution to progeny from three sires, BR_877, BR_536, and BR_610. (b) Contribution to progeny from five dams, CDRS107, BR_542, BR_106, BR_037, and BR_027. From Miller et al. (2018).

An advantage of the exclusion–simulation approach is that it can be used when only one or a few populations have been sampled as the putative source population; it does not assume that the true population of origin has been sampled. Other approaches, such as Bayesian (Pritchard et al. 2000) and likelihood ratio tests (Banks & Eichert 2000), generally require samples from at least two populations, and assume that the true population has been sampled. If the true population of origin has not been sampled, the assignment probabilities from Bayesian and likelihood ratio approaches could be misleading. It seems prudent to apply both the exclusion-based and Bayesian assignment approaches (Manel et al. 2002). Recent assignment approaches for genome-scale datasets could potentially improve assignment (Mussmann et al. 2019).

It is estimated that fish fraud is a US$23 billion business worldwide. Fishing violations often involve harvest from areas that have been closed to fishing to allow the stocks to recover from overharvest. The FishPopTrace consortium was established to fight illegal fishing and fraudulent labeling of fish in European supermarkets (Stokstad 2010; Nielsen et al. 2012). Associated with this, a European fishing law was passed in 2009 that explicitly mentions genetic tests would be used as an enforcement tool to fight illegal harvest and fraudulent marketing. The FishPopTrace consortium created a fish SNP genotyping chip with 1,536 SNPs for each species, and found that the 20 most informative SNPs could correctly identify the population of origin for >95% of fish within each species, even though there is relatively high gene flow and low F_{ST} among marine fish populations. A key benefit of using SNPs for wildlife management in an international context is that it provides a consistent approach, reliable analysis across laboratories, and ease of data sharing.

Similarly, population assignment is used in fisheries for sustainable management of access to fishing areas, protection of breeding sites, and restocking from hatcheries. For example, Pacific cod has seasonal migration but restricted dispersal and shows a pattern of isolation by distance in the northwest and northeast Pacific Ocean. Population assignment using SNPs was able to assign samples

to populations of origin with an average of 85% success using 1,000 loci (Drinnan et al. 2018). Striped bass is a migratory fish in the North American Atlantic coast, with six regional management units identified based on genetic structure. Population assignment using SNPs assigns samples to regional unit with >99% success (LeBlanc et al. 2020). The analysis also showed greater variability in homing to their natal river than previously thought, with straying or colonization among rivers, particularly in the Chesapeake Bay and Delaware River.

22.4.2 Assignment of groups

Assignment of groups of individuals to a population (or geographic region) of origin is also feasible. Assigning a group of individuals can be done with a higher statistical certainty than assigning a single individual because more information is available in a group of genotypes than in a single individual's genotype. A user-friendly software program for assigning groups (as well as individuals) to a population of origin is available in the software *GeneClass 2.0* (Piry et al. 2004).

For example, the Alaska Department of Fish and Game confiscated a boatload of red king crab that they suspected was caught in an area closed to harvest near Bristol Bay in the Bering Sea (Seeb et al. 1989). The captain claimed that the crabs were caught near Adak Island in the Aleutian Islands, over 1,500 km away from the closed area. Comparison of allozyme allele frequencies in the confiscated crabs with those of 13 populations of king crab from Alaskan waters indicated that the confiscated crabs could not have been caught near Adak Island (in the Aleutian Islands), the only area open to harvest (Seeb et al. 1989). Allele frequencies at Adak Island significantly differed from the allele frequencies among the confiscated crab, but matched the samples from further north in the Bering Sea. Based upon these results, the vessel owner and captain agreed to pay the State of Alaska a US$565,000 penalty for fishing violations.

As mentioned above, this approach can be useful in forensic applications, particularly in combating the expanding wildlife trade (Section 22.1.5). Wasser et al. (2007) used a DNA assignment method to determine the geographic origin(s) of large sets of elephant ivory seizures. They showed that a joint analysis of multiple tusks performs better than sample-by-sample methods in assigning sample clusters to an origin. Authorities initially suspected that the largest ivory seizure since the 1989 ivory trade ban came from multiple locations across forest and savannah populations in Africa. However, DNA assignment analysis showed that the ivory was entirely from savannah elephants, probably originating from an area centered in Zambia. This finding allowed law enforcement to focus their investigation on a few trade routes and led to changes by the Zambian government to improve antipoaching efforts (Wasser et al. 2018).

22.4.3 Population composition analysis

Many species are harvested in mixed populations such as mixed-stock fisheries (salmon, marine mammals) and waterfowl. Other species also migrate in mixed groups (neotropical songbirds, butterflies, and others). Effective management of mixed-stock fisheries and mixed populations requires that the populations or stocks that compose the mixture be identified and the extent of their contribution determined (Pella & Milner 1987). Stocks are generally analogous to management units (e.g., demographically independent populations), which were discussed in Section 20.5.3.

Population composition analysis differs from individual-based assignment tests in that composition analysis estimates the percentage of the gene pool (or alleles) that originates from each local breeding population, whereas individual-based assignment methods estimate the actual number and identify individuals originating from each breeding population (Manel et al. 2005). Assignment methods compute the proportion of an individual's genome that originates from different breeding populations. The relative performance of the different assignment and composition analysis approaches depends on the question being addressed.

Mixed-stock analysis is a key component of management of fisheries where harvest occurs from mixed stocks that originate from discrete breeding locations. It uses a maximum likelihood approach (Manel et al. 2005) to estimate the proportion of alleles (and thus individuals) from each population in the pool of harvested fish. This analysis provides information on migration patterns, stock composition, and abundance that enables sustainable management of the harvest and informed management of the breeding grounds. These approaches are common in salmon that spawn in natal locations, often in river systems, but disperse to mixed stocks in ocean feeding grounds before returning to breeding grounds to spawn. The fundamental unit of replacement or recruitment for anadromous salmon is the local breeding population because of this homing (Rich 1939; Ricker 1972). That is, an adequate number of individuals for each local reproductive population is needed to ensure persistence of the many reproductive units that make up a fished stock of salmon. The homing of salmon to their natal streams produces a branching system of local reproductive populations that are demographically and genetically isolated. The demographic dynamics of a fish population are determined by the balance between reproductive potential (i.e., biological and physical limits to production) and losses due to natural death and fishing. Fishery scientists have focused on setting limits on fishing intensity so that adequate numbers of individuals escape fishing to provide sufficient recruitment to replace losses.

An important application of mixed population analysis in fisheries is illustrated by the management of Pacific salmon. The PacSNP database provides reference genotype data for management of chum salmon fish stocks harvest and bycatch by multiple nations and jurisdictions (Seeb et al. 2011). Analysis of 54 nuclear SNPs and three mitochondrial SNPs in samples from locations across the Pacific Rim showed genetic differentiation of stocks from regional areas representing local breeding grounds (Figure 22.9). Analysis of a mixed catch in the Bering Sea showed high admixture of populations from across the Pacific Rim. A time series analysis of catch from the Kamchatka Peninsula also demonstrated the variable composition of the metapopulation over months, with high composition from the Kamchatka/Anadyr region in May and June

(a)

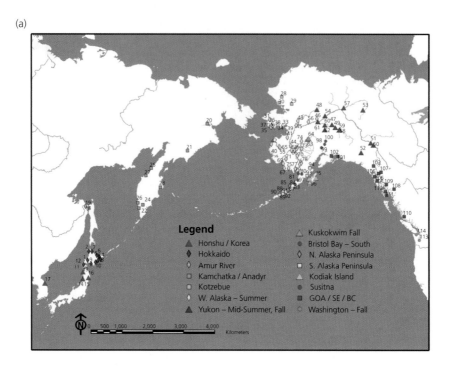

(b)

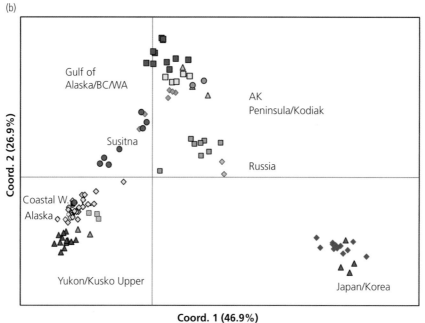

Figure 22.9 A large-scale reference database of chum salmon samples across the Pacific shows genetic structure that allows assignment testing in mixed-stock analysis to inform fishery management. (a) Sampling locations of chum salmon across the Pacific. (b) PCoA of genetic relationships among samples based on analysis of 54 SNPs and three mitochondrial genes. From Seeb et al. (2011).

transitioning to high composition of stocks from Japan/Korea by August. Further analysis of the fish from the British Columbia and Washington region of the eastern Pacific with 89 SNPs showed localized differentiation, with an average F_{ST} value of 0.058, and sufficient power to undertake mixed-stock analysis (Small et al. 2015).

This mixed-stock analysis allows real-time monitoring of harvest from each major population. It allows biologists to recommend closing the harvest if too many fish are harvested from any one major breeding population. This is critical to help prevent overfishing, longer-term closures of fishing, and the extinction of a major source population.

Guest Box 22 Tracking illegal logging using genomics
Eleanor E. Dormontt and Andrew J. Lowe

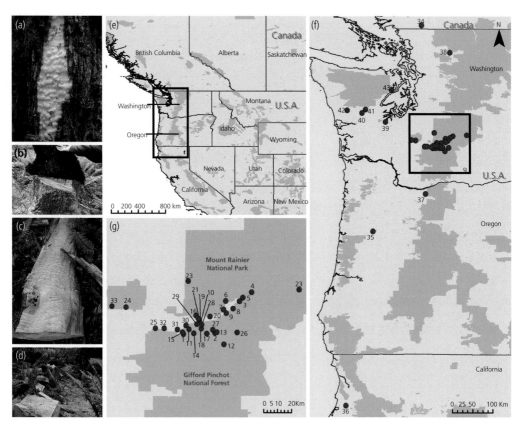

Figure 22.10 Bigleaf maple is subject to illegal logging in North America. (a) Trees are selected based on "figures" patterning observed through removal of the bark in a small area of "quilted" figuring shown here. (b) Once felled, the stump is often covered with moss to evade detection. (c) Bark is removed from the felled log, revealing the extent of figure patterning. The log is often then cut into blocks. (d) In some cases, little attempt is made to conceal the crime scene; only the most valuable (and portable) blocks are removed with the remaining timber left to decompose in the forest. (e) Western North America showing National Parks and Forests (green shaded areas). (f) Total study area with sampling locations (circles). (g) Area of most extensive sampling. From Dormontt et al. (2020). Photos courtesy of Anne Minden.

Globally, illegal logging is one of the largest illicit trades, worth over US$30 billion a year, with few prosecutions (Dormontt et al. 2015). While illegal logging is perhaps best known as a tropical forest issue, where in some countries it can account for up to 90% of all traded timber, there are also major problems with illegal logging in temperate countries (Lowe et al. 2016). Illegal logging activity in the USA has been estimated at US$1 billion annually.

One target of illegal logging in the Pacific Northwest of North America is bigleaf maple, a tree highly valued for

its grain and translucent pattern and popular for use in string instruments such as guitars. The removal of old growth maple trees from National Parks along the west coast of North America is a significant and increasing problem (Figure 22.10a–d). Motivated by the discovery of a cluster of illegally felled bigleaf maple trees in Washington State, the US Forest Service, with support through the World Resources Institute and Double Helix Tracking Technologies, worked with a group at the University of Adelaide to develop a DNA profiling tool to identify the timber in supply chains back to stumps from felled trees.

One hundred and twenty-eight SNPs were specifically selected for amplification to capture individual variation in bigleaf maple (Jardine et al. 2015). Using these markers, a DNA reference database was developed, consisting of 394 individuals from 43 sites across the species range (Figure 22.10e–g). The assay was subject to a strict forensic validation procedure based on the Scientific Working Group on DNA Analysis Methods validation guidelines to ensure reliability for forensic purposes. A range of sample types (leaf, cambium, timber) was analyzed to represent the variety of case-samples that may be encountered, and mother trees and seedlings compared to confirm Mendelian inheritance of the markers. The assay was demonstrated to work effectively at low DNA concentrations and to have high species-specificity.

Based on the reference data, the F_{ST}-corrected probability of identity was 1.785×10^{-25}, meaning that the chance of the assay returning a match between samples that did not originate from the same tree (or clone) was vanishingly small. The resulting publication was among the first to apply forensic validation criteria to an assay developed for individual identification of timber (Dormontt et al. 2020).

Using the developed assay, several pieces of wood seized from a sawmill were compared with samples from illegally felled stumps in the Gifford Pinchot National Forest, Washington, USA, and a match was detected. In a subsequent legal case in 2016, four defendants pleaded guilty to violations of the Lacey Act 2008, the first domestic prosecution under this legislation.

CHAPTER 23

Genetic Monitoring

Pinzón giant tortoise, Section 23.3

Genetic monitoring has been recognized in several international agreements and documents, and can be an important tool for the protection of biodiversity.

(Filippos A. Aravanopoulos 2011, p. 75)

Integrated sampling is required across multiple levels of biological organization (from genes to ecosystems) and across long temporal scales to properly understand the impacts of extreme events and predict subsequent adaptability of species to future climatic change.

(Carlos F. D. Gurgel et al. 2020, p. 1199)

Genetic monitoring is the use of molecular markers to track individuals, populations, species, or communities, or to quantify changes in population genetic metrics (such as effective population size, genetic diversity, and population divergence) or phenotypes over time (Schwartz et al. 2007). Monitoring has a temporal dimension that is different from assessment, which quantifies a population characteristic at only a single time point. Allendorf & Ryman (1987) presented one of the first uses of genetic monitoring in their consideration of maintaining hatchery stocks of fishes. Today, monitoring of wild populations is increasingly feasible thanks to continual improvements in molecular, statistical, and sampling techniques. Nevertheless, genetic monitoring is not widely conducted despite the fact that many national and international organizations have established principles and promoted strategies for monitoring biological diversity (Laikre et al. 2010b, 2020).

Three categories of genetic monitoring can be delineated (Figure 23.1). Category I is the use of genetic markers for traditional ecological population monitoring through the identification of individuals and species and the estimation of population census size, which is often conducted using noninvasive sampling (Section 22.1). Category II includes the use of genetic markers to monitor population genetic parameters such as N_e, allelic diversity, or population connectivity to detect potential population bottlenecks, fragmentation, or admixture. Category III is the use of DNA markers in or near adaptive genes to assess effects of environmental change, disease epizootics, or other stresses on the adaptive response of a population.

Recent advances in genomics will increase our power to detect genetic and demographic change over time. For example, using thousands of genomic markers will reduce the probability that multiple individuals will have the same genotype, improving

Conservation and the Genomics of Populations, Third Edition. Fred W. Allendorf, *et al.*, Oxford University Press.
© Fred W. Allendorf, W. Chris Funk, Sally N. Aitken, Margaret Byrne, and Gordon Luikart (2022). DOI: 10.1093/oso/9780198856566.003.0023

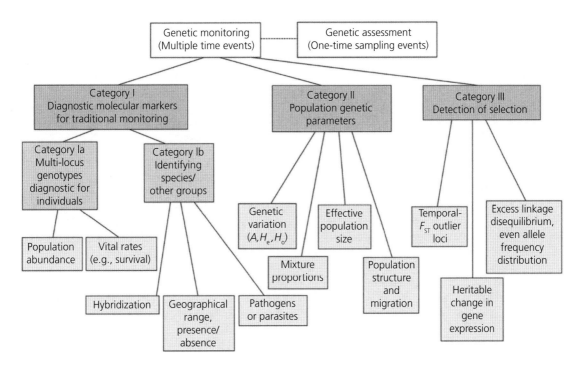

Figure 23.1 Categories of genetic monitoring. Category I includes the use of diagnostic molecular markers for traditional population monitoring through the identification of individuals and species, and the repeated (temporal) assessment of population size. Category II includes the use of genetic markers to monitor population genetic parameters. Category III involves use of DNA markers to detect change in frequency of adaptive alleles or gene expression associated with environmental change. Modified from Schwartz et al. (2007).

the precision of capture–mark–recapture estimates of N_C and vital rates like survival (Section 18.1.2). Similarly, the large number of markers available with genomics will increase the power to detect changes in genetic diversity as well as recent and historical bottlenecks (Hunter et al. 2018; Leroy et al. 2018; Santiago et al. 2020). Lastly, genomics greatly improves our ability to identify adaptive alleles and monitor changes in their frequencies in nonmodel species (Hansen et al. 2012; Flanagan et al. 2018).

23.1 Species presence

Genetic markers that are diagnostic of species facilitate monitoring changes in their presence, as well as changes in their geographic ranges. This falls under Category I genetic monitoring (Figure 23.1). For example, Hess et al. (2015) designed a genomic assay for Pacific lamprey that included two **single nucleotide polymorphisms (SNPs)** which were diagnostic of two other lamprey species that

are indistinguishable from the Pacific lamprey in the larval stage. This will allow them to monitor changes in the presence and geographic range of this declining species and plan management actions accordingly.

An increasingly important approach for monitoring species presence and distribution is environmental DNA (**eDNA**; Section 22.1.4). This approach allows detection of DNA shed by organisms in water, soil, air, snow, etc., as well as noninvasive characterization of individuals' diets and pathogens from their feces (Taberlet et al. 2018; Hohenlohe et al. 2019). A particularly creative example of using eDNA for monitoring is a study that screened mammal biodiversity in a Vietnamese rainforest by amplifying mammal DNA from hematophagous (blood-feeding) leeches (Schnell et al. 2012). The authors first demonstrated that mammalian blood DNA can be amplified using PCR from hematophagous leeches for at least 4 months after the leeches have fed. The authors then tested this approach using leeches caught in a tropical Vietnamese rainforest, which detected cryptic, rare, and even newly discovered mammalian species! Given that many species are rare, shy, and cryptic, the authors argue that this represents a quick, cost-effective, and standardized way to monitor mammalian biodiversity and occupancy (see also Lynggaard et al. 2019).

There are currently two main approaches for detecting species using eDNA. One uses **quantitative polymerase chain reaction (qPCR)** to amplify the DNA of a single, targeted species using species-specific primers (Goldberg et al. 2011). A newer approach combines eDNA and **metabarcoding** to detect whole communities using conserved primers rather than focusing on a single target species (Taberlet et al. 2012). Thus, eDNA metabarcoding is often a preferable option for monitoring entire species assemblages, including rare or invasive species. However, a potential concern is whether DNA metabarcoding is as sensitive of a technique as qPCR for detecting rare species. Harper et al. (2018) tested the relative sensitivity of these two approaches for detection of the great crested newt, which is a flagship pond species of international conservation concern in the United Kingdom and Europe. They found that eDNA metabarcoding was almost as effective as qPCR at detecting great crested newts. As a result, the authors recommend using eDNA metabarcoding to monitor entire freshwater communities. Other studies also found qPCR more sensitive than metabarcoding, and also that **digital droplet PCR** (ddPCR) is more sensitive than qPCR for detecting rare or invasive species (Wood et al. 2019).

Analysis of eDNA has become an important tool for monitoring invasive species, particularly aquatic invasive species. For example, Kamoroff et al. (2020) analyzed eDNA using qPCR to monitor invasive American bullfrogs in Yosemite Valley, Yosemite National Park, California, USA, after the National Park Service (NPS) began removing them. Starting in 2015, the NPS used a combination of eDNA, visual surveys, and audio recordings to test whether they had effectively eradicated bullfrogs from the valley. Using eDNA was an important component of their monitoring program, as in some cases, analysis of eDNA detected bullfrogs when visual surveys failed to detect them. Thanks to this effort, the NPS removed the last observed bullfrog in 2019. eDNA will likely continue to grow in importance for monitoring invasive species—including parasites and pathogens—especially when combined with new, cheaper, portable sequencing technology, which will allow rapid analysis (Sepulveda et al. 2020).

23.2 Population abundance

Another type of Category I genetic monitoring is monitoring of DNA tags (DNA fingerprints) in lieu of physical tags or natural markings in order to identify individuals and estimate population census size (N_C). For example, Stetz et al. (2010) used encounter data from 379 grizzly bears identified genetically through rub tree surveys of hair samples to parameterize a series of mark–recapture simulations to assess the ability of noninvasive genetic sampling to detect declines in population size (N_C). They concluded that annual tree rub surveys would provide >80% power to detect a 3% annual decline within 6 or fewer sampling years. Estimates of the true population size (starting at $N_C = 765$) were unbiased, and became very precise within about 4 years. Thus, annual tree rub surveys and DNA-based hair genotyping could provide a useful complement or alternative to traditional telemetry methods for monitoring trends in grizzly bear populations.

Another example of Category I monitoring involved noninvasive genetic monitoring of mortality and reproductive success of reintroduced otter populations in the Netherlands (Koelewijn et al. 2010). The study demonstrated that only a few dominant males successfully fathered offspring and thus the effective population size was small (Example 23.1).

An emerging frontier of eDNA monitoring is to infer changes in species density, abundance, and biomass. Chambert et al. (2018) developed a model relating eDNA concentration to animal count data from a subset of sites to allow inference of density from eDNA at other sites without animal count data. To test their model, they first conducted an experiment with carp at known densities in mesocosms. They then evaluated

Example 23.1 Genetic monitoring of an introduced otter population

Eurasian otters were extirpated from the Netherlands in 1989. From 2002–2008, 30 individuals were released into the northern Netherlands. Post-reintroduction success was monitored using noninvasive genetic analyses. The founding individuals were genotyped along with feces collected in the release area (Koelewijn et al. 2010). Researchers analyzed 1,265 fecal samples and anal secretions with 7–15 microsatellite loci. Of the 1,265 samples, 582 (46%) were successfully assigned to either a released or new genotype representing an offspring.

During the first three winters, seven microsatellites were sufficient for individual typing and parentage assessments. Subsequently, founder individuals died and relatedness increased, and 15 loci were required for parentage analyses, although the 7 loci were still sufficient for individual identification. For example, during the final 2007/8 season, the probability of identical genotypes in two random sibs (PI_{av}-sibs; Section 22.3) of the first set of seven loci was 2.1×10^{-3}, which is sufficient in small populations. When all 15 loci were used the PI_{av}-sibs decreased to 1.4×10^{-5}.

The researchers used genetic parentage assignment to identify 54 offspring (23 females and 31 males). The reproductive success among males was strongly skewed, with two dominant males fathering two-thirds of the offspring (Figure 23.2). One of the highly successful males was the son of the other. The effective population size was only about 30% of the detected number of individuals because of the large variance in reproductive success among males.

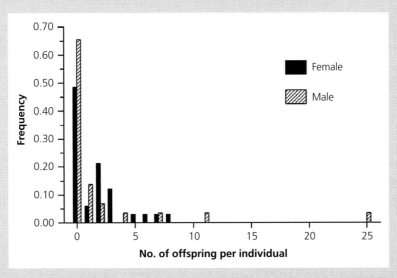

Figure 23.2 Reproductive success of introduced Eurasian otters during the period 2002–2008 in the Netherlands. The male with 11 offspring is the son of the male with 25 offspring. This high variance in male reproductive success greatly reduced the effective population size of this population. From Koelewijn et al. (2010).

Genetic sex identification revealed that males had a higher mortality rate (22 out of 41 males (54%) vs. nine out of 43 females (21%)), likely because most juvenile males dispersed to surrounding areas upon maturity. In contrast, juvenile females stayed inside the area next to the mother's territory. The main cause of mortality was traffic accidents.

This study demonstrates that noninvasive molecular methods can be used to monitor elusive species to reveal a comprehensive picture of population status. Nevertheless, the future of the Dutch otter population is unclear because it is new, small, and isolated, with a low effective population size. A connection with a second population in nearby wetlands is unlikely, and many animals are killed in traffic incidents when they move away from the current population. Furthermore, only a few males dominate the reproductive process, which lowers the effective population size and increases relatedness and inbreeding.

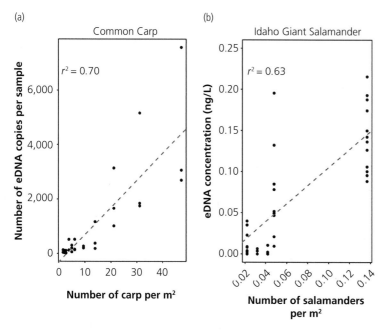

Figure 23.3 Linear regressions between eDNA concentration and estimates of (a) carp or (b) Idaho giant salamander density. Proportion of variance (r^2) values are shown on each graph. Environmental DNA measures on the y-axis are the number of eDNA copies quantified through droplet digital PCR for the carp, and concentration (ng/L) of eDNA quantified through qPCR for salamanders. From Chambert et al. (2018)

the model using field estimates of Idaho giant salamander densities. In both cases, eDNA abundance (copy number) explained a large proportion of the variation in density (Figure 23.3), providing proof of concept that eDNA can be used to estimate density of aquatic species. Carraro et al. (2018) demonstrated the great potential of integrating observed eDNA concentrations with information on the hydro-geomorphological features of rivers to monitor species distribution and abundance in river networks using two target species as a case study: a sessile invertebrate (a bryozoan) and its parasite (a myxozoan).

23.3 Genetic variation

One aspect of Category II genetic monitoring involves tracking changes in genetic variation (Figure 23.1). Most of these studies monitor changes in heterozygosity or allelic diversity. As explained in Chapter 7, allelic diversity is more sensitive to reductions in effective population size than heterozygosity, so a reduction in allelic diversity may be the first sign that a population is declining. Before starting a monitoring program to detect changes in genetic variation, it is important to design the study so that statistical power is sufficient to detect

a change in genetic diversity that is considered biologically significant (Box 20.3). Several papers provide guidelines on designing genetic monitoring programs (Hoban et al. 2014; Leroy et al. 2018; Luikart et al. 2021).

23.3.1 Changes in genetic variation in declining populations

One common goal of genetic monitoring studies is to test whether species of conservation concern have lost or are losing genetic variation in parallel with observed reductions in census population size. A powerful approach to test for changes in genetic variation is by analyzing **ancient DNA** (or historical DNA) extracted from museum specimens and comparing genetic variation in these historical samples with modern samples. D'Elia et al. (2016) used this approach to test whether genetic variation has been lost in endangered California condors, which experienced dramatic population declines after European colonization of the west coast of North America. California condors were originally found from southern British Columbia, Canada, south to Baja California, Mexico. By the middle of the 20th century, only 150 individuals

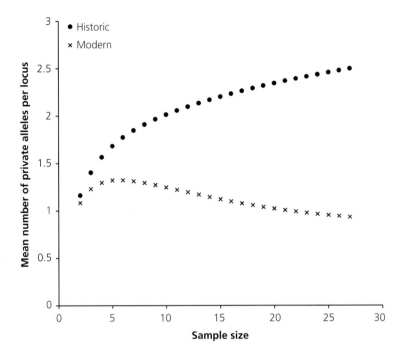

Figure 23.4 Mean number of private alleles per locus as a function of sample size for historical and modern samples of African lions at microsatellite loci using rarefaction. A private allele is an allele that is found in one sample (i.e., historical or modern) that is not found in the other sample. These results demonstrate that historic samples have more private alleles than modern samples, indicating a loss of allelic diversity across time. From Dures et al. (2019).

remained in the mountains of southern California. In the 1980s, the species was rescued from the brink of extinction, when all remaining 22 California condors were caught for captive breeding. California condors have since been reintroduced in their native range. D'Elia et al. (2016) examined genetic variation in California condors at the mitochondrial control region, and found a greater than 80% reduction in distinct haplotypes over the past two centuries, consistent with the hypothesis that condors were relatively abundant in the 19th century and then declined rapidly due to human-caused mortality.

Another example of research that harnessed the power of historical or ancient DNA for genetic monitoring is a study by Dures et al. (2019), which tested for a loss of genetic diversity in southern African lions. African lions have undergone dramatic reductions in their range and population size over the past century following European colonization. Dures et al. (2019) compared genetic variation at 16 microsatellite loci and the mitochondrial control region in museum samples collected in the late 19th century and early 20th century to samples from the modern extant population. They found a reduction in allelic diversity of ~15% between

these two time periods. They also found that the mean number of alleles only present in the historic sample (private alleles) was higher than the mean number of private alleles only present in the modern sample (Figure 23.4). Kotzé et al. (2019) have similarly found substantial reductions in genetic diversity, specifically in heterozygosity, in Cape mountain zebras in South Africa (Guest Box 23). These studies demonstrate the utility of genetic monitoring for confirming suspected population declines, and for showing the genetic effect of these declines.

23.3.2 Changes in genetic variation in response to environmental perturbations

Another important application of genetic monitoring is to test whether anthropogenic or naturally occurring environmental perturbations reduce genetic variation. Athrey et al. (2012) tested the effects of habitat loss and fragmentation on genetic variation in the endangered black-capped vireo. The black-capped vireo is a neotropical songbird that breeds in early-successional oak–juniper savannah in south-central Texas through southern Oklahoma, USA, and has experienced pronounced losses in its breeding and wintering habitat due to agriculture and urbanization over the past 50 years. To test whether this habitat loss has reduced genetic variation in this species, the authors compared genetic variation at microsatellite loci in historical samples collected in the early 20th century with modern samples collected in the early 21st century. As predicted, the authors found a significant reduction in allelic diversity and heterozygosity in modern samples compared with historical samples. Similarly, a comparison of microsatellite variation in historical versus contemporary harpy eagle samples from South America uncovered a reduction in allelic diversity and heterozygosity in those parts of its range most affected by deforestation (Banhos et al. 2016).

An increasing number of studies have tested the effects of extreme events (e.g., floods, fires, heatwaves, hurricanes, etc.) on genetic diversity. These studies require sampling of populations prior to, as well as following, the given extreme event, which typically only occurs by happenstance and is therefore rare. For example, a one in 500-year rainfall event happened to occur in the Colorado Front Range, USA, in September 2013 on top of the same 14 stream reaches where Poff et al. (2018) had sampled stream insects 2 years earlier. This allowed the research team to test the effects of the resulting extreme flooding on genetic variation in six stream insect species. They found that allelic richness declined after the event in two species. In contrast, allelic richness greatly expanded in a different species, suggesting resilience via recolonization from upstream populations. Another study found a striking loss of genetic diversity in two species of forest-forming seaweeds off the coast of Western Australia in response to a severe marine heatwave (Gurgel et al. 2020; Example 23.2).

23.3.3 Changes in genetic variation in response to management actions

Genetic monitoring can also be used to assess whether management actions have successfully staved off declines in genetic diversity, or have even increased genetic diversity (e.g., in response to genetic rescue; Section 19.5). Jensen et al. (2018) used a genomic approach to test whether an

ex situ head-start program developed for Pinzón giant tortoises was effective at maintaining genetic diversity in this island endemic species. The Pinzón giant tortoise is endemic to Pinzón Island in the Galápagos Islands and consists of a single population. Historically, this tortoise numbered in the thousands, but was reduced to 150–200 individuals due to a combination of over-harvesting for food by humans and introduced black rats, which ate all hatchling tortoises. To prevent mortality by rats, an *ex situ* head-start program was started in 1965 in which eggs or pre-emergent individuals were collected from natural nests and then reared in captivity until they were 4–5 years old, at which point rats are unable to depredate them. The program successfully repatriated 800 juvenile tortoises over a 50-year period, but the question remained as to whether the program also successfully retained genetic variation. To address this question, Jensen et al. (2018) compared genetic variation at 2,218 SNPs from historical samples collected in 1905–1906 with contemporary samples collected in 2014. They found that the amount and distribution of genetic variation in the historical and contemporary samples were very similar. Moreover, they found that relatedness among individuals in the historical sample was not significantly different from the relatedness among individuals in the contemporary samples, indicating that the head-start program collected eggs/individuals in a way that was not biased toward certain families (Figure 23.6). Jensen et al.'s (2018) study thus demonstrated that not only did the head-start program help rebound the population size of the Pinzón giant tortoise, but that it also successfully retained the genetic variation found in the population at the turn of the 20th century.

23.3.4 Meta-analyses of changes in genetic variation

Multiple research teams have conducted **meta-analyses** to test for overall trends in genetic variation over time. The general idea of a meta-analysis is to reanalyze multiple datasets in a single analysis to uncover overall effects. Leigh et al. (2019) reanalyzed a total of 99 studies that quantified changes in nuclear genetic variation (88 studies

Example 23.2 Loss of genetic variation in seaweeds caused by extreme heatwave

In 2011, an unprecedented heatwave affected ~2,000 km of coastline off Western Australia, during which sea temperatures increased 2.5–5°C above normal for several weeks. The heatwave caused widespread local extirpations and range shifts of whole marine communities. Seaweeds, which underpin biodiversity in this ecosystem, were most heavily affected. Ecological effects to seaweeds ranged from contraction of species' ranges to population expansions, but the effect of the heatwave on genetic diversity of seaweed was unknown.

Fortuitously, Gurgel et al. (2020) sampled two seaweed species—*Sargassum fallax* and *Scytothalia dorycarpa*—in this region prior to the heatwave, which put them in a position to resample these species and test the effects of the heatwave on genetic diversity. The authors estimated genetic variation in these species before and after the heatwave in a mitochondrial gene (*cox3*) and a chloroplast gene (*rbcL*).

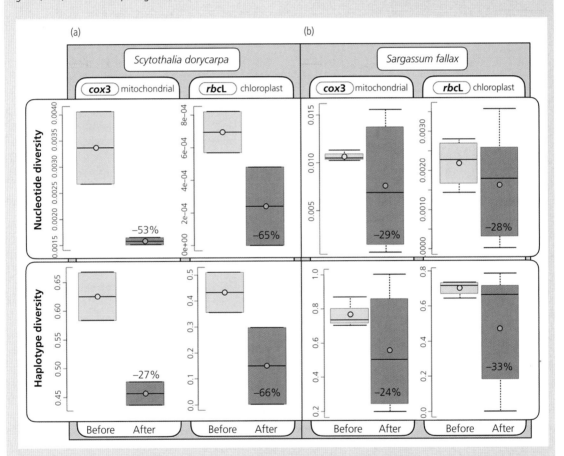

Figure 23.5 Boxplots showing a reduction in genetic diversity in two species of seaweed, *Scytothalia dorycarpa* (a) and *Sargassum fallax* (b), off the coast of Western Australia in response to an extreme heatwave. Nucleotide diversity and haplotype diversity averaged across localities before and after the heatwave for *Scytothalia* (a) and *Sargassum* (b) are shown for mitochondrial (*cox3*) and chloroplast (*rbcL*) genes. The percentage loss in genetic diversity is also presented. From Gurgel et al. (2020).

Continued

Example 23.2 *Continued*

The research team found a dramatic decrease in genetic variation following the heatwave. *Scytothalia dorycarpa* lost an average of 53–65% of nucleotide diversity and 27–66% of haplotype diversity (Figure 23.5). *Sargassum fallax* lost an average of 28–29% of nucleotide diversity and 24–33% of haplotype diversity (Figure 23.5). At some sites, 100% of diversity was lost. Another striking result was that these genetic effects of the heatwave were not reflected in measures of seaweed forest cover used to determine ecological effects. Thus, genetic monitoring was essential to detect the effects of the heatwave.

Given that heatwaves are increasing in frequency globally, these results suggest that heatwaves could severely affect seaweed forests. They also highlight the power of genetic monitoring to detect cryptic effects of extreme events that may not always be apparent from field surveys.

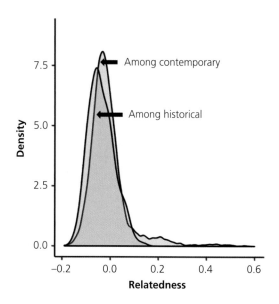

Figure 23.6 Frequency distributions of pairwise relatedness among individuals in the historical Pinzón giant tortoise sample and among individuals in the contemporary sample. Importantly, relatedness was not higher in the contemporary samples, indicating that the head-start program did not collect eggs/individuals in a way that was biased toward certain families. From Jensen et al. (2018).

using microsatellites; nine using SNPs; two using restriction enzyme-based markers) in 91 species over an average of 27 generations. They estimated a 5.4% global decrease in mean expected heterozygosity and a 6.5% decrease in allelic diversity. Island species showed a much larger decline in expected heterozygosity and allelic diversity of an average of 27.6% and 30.9%, respectively. The authors state that this was the first study to demonstrate a global decline in genetic diversity using a meta-analysis of genetic time series data.

In contrast to the study by Leigh et al. (2019), a meta-analysis of mitochondrial *cytochrome oxidase* I (*COI*) sequences did not find any consistent trends in intraspecific genetic diversity over time (Millette et al. 2020). In this second study, the authors conducted a meta-analysis of 175,247 *COI* sequences collected between 1980 and 2016 from 17,082 species of birds, fishes, insects, and mammals. Their time series analysis found approximately equal numbers of positive and negative trends in genetic diversity, resulting in no consistent overall trend in diversity over time.

However, a response to this study raised several issues with Millette et al.'s (2020) analysis (Paz-Vinas et al. 2021). One issue pointed out by Paz-Vinas et al. (2021) is that Millette et al. (2020) grouped sequences sampled at locations separated by ≤1,000 km. Given that most species have dispersal abilities far shorter than this distance, Paz-Vinas et al. (2021) argue that grouping samples collected across such large distances may conflate spatial and temporal effects, making it difficult to accurately quantify changes in genetic variation over time for single populations. Another issue Paz-Vinas et al. (2021) bring up is that the *COI* data that Millette et al. (2020) used for their meta-analysis come from public genetic databases in which sequence data are archived inconsistently. For example, many authors only archive novel sequences (rather than all sequences) from their study populations, which results in a biased estimate of genetic variation. Paz-Vinas et al. (2021) conclude that these and other issues compromise the soundness of Millette et al.'s (2020) analysis and results, but agree that carefully designed meta-analyses are essential for tracking and understanding trends in intraspecific genetic variation.

23.4 Effective population size

Another type of Category II genetic monitoring involves testing for changes in effective population size (N_e; Figure 23.1), which is one of the most important genetic parameters to monitor given that N_e determines the rate of loss of genetic variation, probability of inbreeding, and efficiency of natural selection (Chapter 7). Theory shows that a minimum N_e of 50 is necessary to keep inbreeding at a tolerable level, and an N_e of 500 is necessary to maintain sufficient genetic variation over the long term for adaptation to environmental change (Sections 18.8 and 19.6). Thus, it is important to monitor N_e over time to make sure populations do not drop below these targets. Two general strategies can be used for monitoring N_e: (1) estimate N_e at multiple time points to directly track how it changes; or (2) infer changes in N_e from a single contemporary sample based on analysis of patterns of genetic variation using population genetics theory.

23.4.1 Estimating effective population size at multiple time points

If samples over time are available, the most straightforward method of inferring changes in N_e is to estimate N_e at each time point and test for trends (Beaumont & Wang 2019). Lonsinger et al. (2018) used this approach to test whether N_e has declined in kit foxes in western Utah, USA, where their abundance has declined precipitously compared with historical levels, raising concern for their persistence. The authors estimated N_e using microsatellite data generated from museum samples collected from 1951–1969 and modern samples collected from 2013–2014 to test whether N_e has declined. They found that N_e decreased substantially between these time points (Figure 23.7; Chapter 7). Historical N_e estimates were 5.1–7.5 times higher than contemporary estimates. Moreover, although historical N_e estimates were close to the recommended N_e value of 500 to maintain adaptive potential, contemporary estimates were barely above an

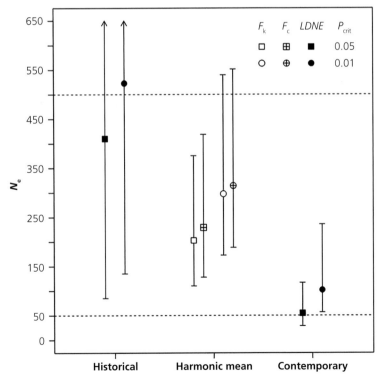

Figure 23.7 Estimates of effective population size (N_e) based on the single-sample gametic disequilibrium method using *LDNE* (Guest Box 10) for historical (1951–1969) and contemporary (2013–2014) kit foxes sampled from western Utah, USA. The harmonic mean N_e estimates were obtained using the two-sample temporal method (F_k and F_c), which is based on changes in allele frequencies between the historical and contemporary time periods. Rare alleles at frequencies below critical values of $P_{crit} = 0.05$ or $P_{crit} = 0.01$ were removed. Error bars represent 95% confidence intervals. The arrows on the upper bound of confidence intervals for the historical sample indicate that the N_e estimate was indistinguishable from infinity. Horizontal dashed lines show the N_e values necessary for minimizing the risk of inbreeding depression and maintaining adaptive potential under the 50/500 rule. From Lonsinger et al. (2018).

N_e of 50, which is recommended to prevent dangerous rates of inbreeding (Section 18.8). In contrast, the authors did not observe a concomitant loss of genetic diversity, which they hypothesize is due to ongoing immigration into this population from adjacent populations. This study provides yet another example of the utility of museum specimens for providing a baseline for comparison with modern samples to monitor changes in genetic diversity.

The effective number of breeders (N_b) per year (or reproductive cycle) is also important to monitor in age-structured species, in addition to N_e per generation (Section 7.5). For example, N_b is important when studying reproduction events, seasonal or annual processes, or sexual selection in age-structured species. Monitoring N_b can provide high power for early detection of population declines, reproductive failure, and recruitment decline. Luikart et al. (2021) provided an analysis of power to detect an N_b decline using the single-sample *LDNE* estimator of N_b (Waples et al. 2013). The method provided sufficient power to determine if N_b drops below 50 (or 500) when sampling reasonable numbers of loci and individuals. They also reported a linear regression method that provides high power to detect declines in N_b when sampling multiple cohorts through time. Software has been developed to allow biologists to evaluate the sensitivity of monitoring to detect changes in N_b (Antao et al. 2020).

23.4.2 Inferring changes in effective population size from contemporary samples

If historical samples are not available, we can still infer changes in N_e from genetic data collected from a single, contemporary sample using recent N_e estimation methods and population genetic theory. This is because genomes of organisms contain a record of past effective population sizes. We can detect historical population declines (bottlenecks) using DNA sequences and tests for coalescence events and shifts in the **site frequency spectrum** (SFS; Section 4.1.4), which is also referred to as the allele frequency distribution (e.g., Chen 2012). For example, the **pairwise sequentially Markovian coalescent** *(PSMC)* method uses coalescence events and linkage information along chromosomes to estimate the timing and duration of population size changes during the past thousands of years (Bunnefeld et al. 2015; Beeravolu et al. 2018). The rate of coalescence events at different historical time points informs us about population size because historical coalescence events (genome-wide) are more frequent when a population is small.

The *PSMC* approach was used by Ekblom et al. (2018) to test for changes in N_e in the Scandinavian wolverine population, which has recovered from a substantial population decline (Figure 23.8). The authors detected a long-term decline in N_e from 10,000 years before the last glaciation, to <500 after glaciation, and an even smaller contemporary N_e.

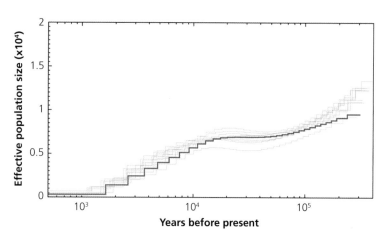

Figure 23.8 Plot of long-term demographic history of Scandinavian wolverines based on analysis of whole genome sequencing data from 11 individuals using the *PSMC* method. The bold line shows estimated changes in effective population size (N_e) for the reference individual, and the thin lines show changes in N_e for 10 additional resequenced individuals. From Ekblom et al. (2018).

This method uses long-range chromosomal linkage and recombination information along with genome assemblies now becoming available for nonmodel species. The *PSMC* method can be used on genome sequence data from a single diploid individual. Thus, it has been used on ancient or extinct species such as woolly mammoths to infer low genomic variation or inbreeding associated with population extirpation (Palkopoulou et al. 2015).

The *MSMC* approach (multiple sequentially Markovian coalescent) uses sequences from multiple individuals and generally outperforms *PSMC*. *MSMC* has better power than *PSMC* to detect more recent changes in effective population size, because using more alleles (from multiple individuals) increases the probability of detecting coalescence events in the relatively recent past (e.g., <2,000 years ago). *PSMC* cannot generally detect changes in effective population size within the past 10,000 years and is therefore less useful for conservation when information on recent population size is needed.

We can also detect population genetic bottlenecks using SFS information from shorter sequence blocks as obtained from short-read sequencing data (e.g., **restriction site-associated DNA sequencing (RADseq)**) or transcriptome data (Chapter 4). These short sequence block methods assume that intra-block recombination can be ignored, which is reasonable for short-read data like RADseq. Short- (and long-) sequence block tests incorporate information on runs of homozygosity, which are generated by historical bottlenecks and inbreeding. For example, 50% of blocks can be monomorphic (homozygous) for certain models of a recent strong bottleneck, whereas only 20% of blocks are monomorphic for populations of constant size (Bunnefeld et al. 2015).

Population bottlenecks (and expansions) can also be inferred from the distribution of the lengths of segments that are "identical-by-descent" (IBD) between individuals (Browning & Browning 2015; Kardos et al. 2017; Browning et al. 2018). A strength of this approach is that it estimates a time series of recent N_e (over the past few hundred generations) instead of simply testing for the presence of a recent bottleneck. However, vast numbers of SNPs are required (e.g., whole genome sequences), along with large sample sizes

(>50–100 individuals), a quality genome assembly, and recombination rate estimates across the genome. Kardos et al. (2017) inferred changes in historical N_e for collared flycatchers on a Baltic Sea island using this method and estimated IBD characterized from whole genome sequences (Figure 23.9).

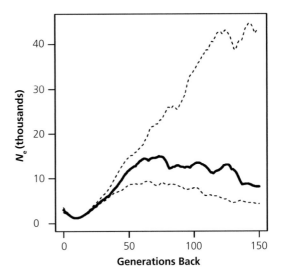

Figure 23.9 Estimates of recent historical effective population size (N_e) based on pairwise IBD segments in the Baltic collared flycatcher population. The solid black line represents the point estimates of N_e and the dashed lines represent the 95% confidence interval for N_e. From Kardos et al. (2017).

23.5 Population subdivision and gene flow

In addition to monitoring genetic variation and N_e within populations, another important type of Category II genetic monitoring focuses on tracking changes in population structure and gene flow. Habitat fragmentation, for example, can cause populations to become more isolated, which can increase population differentiation and reduce within population genetic variation (Section 19.3). Klinga et al. (2020) tested for changes in spatial genetic structure in the capercaillie, which is a ground-dwelling bird species of conservation concern found in mature forests in Europe and much of Russia. The researchers compared population structure at microsatellite loci in museum specimens collected in 1960–1990 with recent noninvasive samples collected in 2011–2015 in the Carpathian

Mountains of central and eastern Europe. They found greater genetic structure in recent samples than in historical samples, which they concluded indicates processes associated with loss of habitat are negatively affecting connectivity. Similar examples of increases in genetic structure and reductions in connectivity have been found in black-capped vireos (Athrey et al. 2012) and red-cockaded woodpeckers (Miller et al. 2019).

Genetic monitoring can also be used to assess whether management actions such as the establishment of corridors connecting habitat fragments or genetic rescue (Section 19.5) effectively restore connectivity. Montero et al. (2019) examined microsatellite and major histocompatibility complex (MHC) variation in fragmented mouse lemur populations in southeastern Madagascar before and after the establishment of corridors to test whether the corridors restored connectivity among forest remnants. They found an increase in the proportion of private alleles shared between fragment pairs after the establishment of the corridor, which led them to conclude that gene flow was augmented due to enhanced connectivity. Monitoring efforts such as this are essential for testing whether efforts to increase connectivity among habitat fragments are effective.

23.6 Adaptive variation

Monitoring to detect adaptive change—Category III genetic monitoring—can be accomplished by focusing on molecular markers or on quantitative traits or phenotypes using the tools of quantitative genetics (Chapter 11; Hansen et al. 2012; Flanagan et al. 2018). A challenge of monitoring quantitative traits is to distinguish between adaptive genetic responses and phenotypic plasticity (Gienapp et al. 2008). Monitoring molecular genetic markers can avoid this problem.

Genetic monitoring to assess adaptive change at the molecular level often involves the use of population genomics and genotype–environment associations (GEA) or outlier tests (Figure 9.14) to identify locus-specific effects (Sections 9.7.3, 10.3.1, and 19.2.4) including: (1) excessively high or low F_{ST}; (2) rapid loss of variation due to selective sweeps at a locus; (3) high linkage disequilibrium along a chromosome near a candidate adaptive locus; or (4) excessive deviation from mutation–drift equilibrium of the allele frequency distribution at a locus (Luikart et al. 2003). Tests for locus-specific outlier effects can be conducted in a monitoring framework, including the use of two samples separated by time from one population. The idea here is to test for locus-specific effects of a stress or challenge by genotyping a locus with an allele for which frequency is expected to change rapidly during the stress event. For example, a SNP locus near an immune system gene (MHC) could be monitored for excessive change in allele frequencies (compared with neutral loci) before and after a disease outbreak. In addition to this candidate gene approach, a genome-wide scan approach of genotyping thousands of loci could be conducted to monitor for genetic signatures associated with disease outbreaks (or other stresses or selection events).

Hansen et al. (2012) define criteria needed to convincingly demonstrate adaptive evolutionary change for DNA markers. These include requirements that the monitored genes are relevant to the specific environmental stress, that selection is tested for by comparison with observed or expected genetic drift, and that shifts in allele frequencies coincide with the changes expected in response to the environmental stress or change in question. Ideally, to further verify selection as the cause of evolutionary changes, researchers could conduct experiments such as reciprocal transplants or laboratory stress challenges to test whether locus-specific allele frequency changes occur as expected. For example, an allele for high-temperature tolerance would be expected to increase in frequency during a high-temperature challenge experiment. If experiments are not possible, it is helpful to test for the same allele frequency shift following the environmental change in each of multiple independent populations. Example 23.3 presents an excellent example that fulfills these criteria.

Although genomic approaches hold great promise for characterizing and monitoring adaptive genetic variation in nonmodel species, they may not always be a worthwhile use of resources, depending on conservation objectives and potential management actions. Flanagan et al. (2018) outline factors that conservation practitioners should consider to help them decide whether knowledge

Example 23.3 Genetic monitoring of adaptive variation in the Tasmanian devil

Devil facial tumor disease (DFTD) is a transmissible cancer in Tasmanian devils that has swept across most of the species' range, causing a decline in abundance of more than 80% in <20 years (Guest Box 4). Surprisingly, some populations have persisted in long-diseased areas, even though epidemiological models predicted extinction. Epstein et al. (2016) examined genetic variation at >90,000 SNPs before and after DFTD arrived to test for signatures of selection in regions of the genome that contain genes related to immune function or cancer risk in humans. They identified two genomic regions with concordant signatures of selection in three different geographical locations, suggesting that Tasmanian devils are evolving immune-modulated resistance that could aid the species' persistence.

The results for one of these genomic regions, which is on chromosome 2, is shown in Figure 23.10. The cereblon gene, a myeloma treatment target related to human cancer, is the possible target of selection in this region. Three separate locations in Tasmania were sampled from 1999–2014. The allele frequency changes at SNPs in this region were in the top 2.5% when comparing pre- and post-disease samples (Figure 23.10b). The authors identified a 100-kb window on either side of these SNPs based upon the presence of strong gametic disequilibrium (Figure 23.10c). They estimated *Rsb*, a statistic that is a measure of the increase in gametic disequilibrium before and after disease exposure. The observed *Rsb* values indicate increased haplotype homozygosity, which is indicative of a selective sweep (Figure 10.4). They fit a model of allele frequency change over time to estimate relative fitness advantages of SNPs. The point estimates for the mean fitness advantage per generation of favored alleles were 29% in Freycinet (18–43%), 28% in Narawntapu (0.5–101%), and 19% in West Pencil Pine (1–51%). These results suggest that there has been a substantial genomic response in very few generations (approximately four in Narawntapu and West Pencil Pine and approximately six in Freycinet) in these locations.

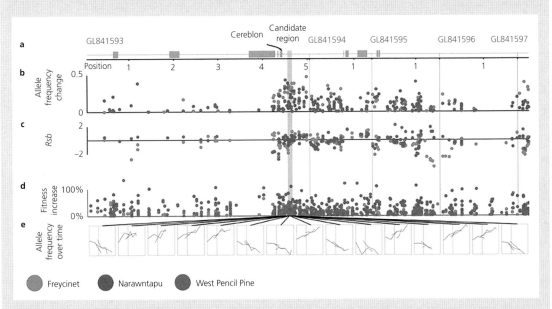

Figure 23.10 Tests for selection at SNPs in a candidate region on chromosome 2 in response to DFTD in three different populations of Tasmanian devils. (a) Scaffolds (GL841593, etc.), positions, and genes (gray boxes). The base pair position along each scaffold is given in Mb from the start of the scaffold (indicated with light gray vertical lines). (b) Allele frequency change between pre- and post-disease samples. (c) *Rsb* measures the increase in gametic disequilibrium over time which is expected with strong positive selection. (d) Estimates of the fitness advantage of the favored allele. (e) Trajectory of allele frequency change over time. From Epstein et al. (2016).

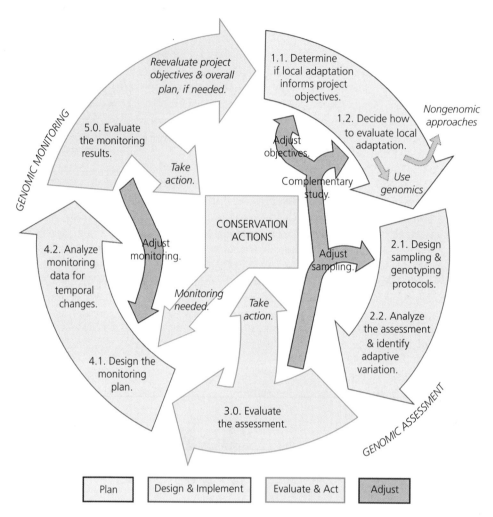

Figure 23.11 Adaptive management cycle for genomic assessment and monitoring of adaptive genetic variation. Stage 1 outlines the initial planning phase, stages 2 and 3 show the genomic assessment, and stages 4 and 5 are the genomic monitoring phase. Purple indicates planning steps; blue indicates design and implementation steps; green indicates evaluate and action steps; and red, unnumbered arrows highlight the importance of adjusting the plan throughout the entire adaptive management cycle. Modified from Flanagan et al. (2018).

of local adaptation will inform their decisions and, if so, when genomics is likely to be effective in detecting local adaptation. They also provide an **adaptive management** framework to plan an assessment and monitoring program to meet conservation objectives (Figure 23.11). The first step of this adaptive management framework is to determine if local adaptation informs the project objectives and whether genomics is the best approach for assessing adaptive variation. If so, the next step is to carefully design sampling and genotyping protocols

and conduct the study to identify adaptive variation. Based on these results, the next step is to make management decisions and take action. If it is decided that monitoring adaptive variation will advance conservation objectives, the next stage is to design and implement a monitoring program. The monitoring results are then evaluated, which may lead to new management actions. An important aspect of adaptive management is that throughout the entire cycle, the plan may need to be adjusted to improve both conservation outcomes and data collection to improve decision-making.

An example of Category III genetic monitoring is the detection of an adaptive response to ongoing climate change in wild populations of *Drosophila melanogaster* and *D. subobscura* fruit flies (Umina et al. 2005; Balanya et al. 2006). Each species displays a north–south cline in phenotypic traits influencing thermal adaptation and also in genetic markers involving well-characterized genes and chromosome inversions associated with temperature tolerance. Laboratory selection experiments independently suggest the clines reflect temperature-related selection (Umina et al. 2005). This provides compelling evidence that the traits and gene polymorphisms are under selection caused by a specific environmental factor: temperature.

Similarly, genomics was used in little brown myotis bats to detect immunogenetic variants related to white-nose syndrome (Donaldson et al. 2017). As genomic resources improve for species of conservation concern, Category III genetic monitoring of adaptive variation will become an increasingly important application of genomics to conservation and management.

23.7 Integrative genetic monitoring and the future

Adding demographic, phenotypic, and environmental data to genetic data can greatly strengthen inferences from genetic monitoring studies about any observed changes in genetic variation, as well as their causes. For example, if a genetic monitoring program reveals a reduction in genetic variation in a given population, this may raise a red flag that the population is declining, but by itself, would not necessarily inform an appropriate management response. However, if this reduction in genetic variation coincides with a change in the environment, such as the arrival of a predatory invasive species associated with a decline in abundance of the focal species, then this would suggest a cause of the reduction in genetic variation and a potential management action to stop or reverse it.

In another hypothetical example, observation of a relatively large change in the allele frequency at a locus hypothesized to be related to thermal performance might suggest an adaptive response to changes in temperature, but it could also be caused by genetic drift or be a sampling artifact. Yet if the given population experienced a heatwave and its thermal tolerance increased between sampling time points, then this would provide stronger evidence that the change in allele frequency at this locus was indeed caused by selection on thermal tolerance imposed by increasing temperatures.

Demographic data (on abundance, survival, fecundity, and population growth rates) can be particularly valuable for helping interpret genetic monitoring results. Nunziata et al. (2017) generated RADseq data for two species of salamanders, the marbled salamander and the mole salamander, at two time points (1993 and 2013 for the marbled salamander and 1984 and 2011 for the mole salamander). They then used a coalescent-based analysis to infer changes in N_e. In marbled salamanders, they observed an increase in N_e, but in mole salamanders, they observed a pronounced decline in N_e. Fortunately, 37 years of capture–mark–recapture (CMR) data for these same populations were available for estimating census population size and the effective number of breeders. The genetic estimates of changes in N_e matched the CMR results remarkably well (Figure 23.12), providing independent evidence that the inferred changes in N_e were real and demonstrating the utility of genetic monitoring for detecting changes in population size.

Environmental data, including information on changes in the abiotic (e.g., temperature) and biotic (e.g., habitat, interacting species) environment, can also be valuable for interpreting the causes of changes in genetic variation. As described in Section 23.3.2, Athrey et al. (2012) analyzed historical and contemporary samples of endangered black-capped vireos in Texas and Oklahoma, USA, using microsatellite data. They found a reduction in genetic diversity within populations and an increase in genetic differentiation among populations. In addition, evaluation of alternative demographic hypotheses using approximate Bayesian computation (Section A7) suggested that

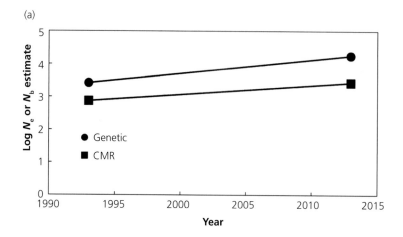

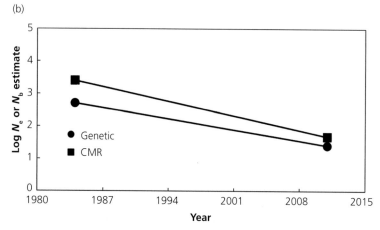

Figure 23.12 Comparisons of the log of effective number of breeders (N_b) based on analysis of capture–mark–recapture data (CMR; squares) and the log of N_e based on analysis of RADseq data using a coalescent maximum-likelihood model (circles) for two species of salamanders, (a) marbled salamanders and (b) mole salamanders. From Nunziata et al. (2017).

a bottleneck occurred in these populations at the turn of the 20th century. Alone, these results raise concern about the loss of genetic variation in this species of conservation concern, but they do not inform the cause of these genetic changes. However, environmental data suggest that the cause of these genetic changes is a reduction in the amount and increase in the fragmentation of the species' specialized oak–juniper savannah habitat. Once again, complementary nongenetic data increase confidence in the genetic monitoring results, and point to a potential cause of the observed changes.

In the future, other "omics" technologies, such as transcriptomics, proteomics, and metagenomics (Chapter 4), will likely be increasingly included in genetic monitoring programs. For example, metagenomics can be a useful tool for monitoring pathogens in wildlife species. Conceição-Neto et al. (2017) used metagenomics to monitor viral

pathogens in a small, threatened population of wolves and sympatric domestic dogs in central Portugal. Close contact of wolves with dogs is a concern for their health and survival given that wolves can contract pathogens from dogs, which are closely related to wolves (and actually considered a subspecies of the wolf). The authors screened diarrheic stools collected from wolves, feral dogs, and pet dogs using viral metagenomics and metabarcoding. They detected known viral pathogens, including canine distemper virus, a novel bocavirus, and canine minute virus. They then continued to screen wolves over time to monitor the presence of the above pathogenic viruses. Given that diseases can severely affect already imperiled wildlife populations, this approach can play an important role in the early detection of pathogens so that conservation practitioners can vaccinate or provide treatment to infected individuals.

Guest Box 23 African mammal conservation benefiting from genetic monitoring: The Cape mountain zebra in South Africa
Antoinette Kotzé and J. Paul Grobler

The Cape mountain zebra (Figure 23.13) is endemic to South Africa. The subspecies survived near extinction due to over-hunting and agricultural expansion, with the species suffering a 90% reduction in its distributional range. By the 1950s, the total number reached a low of 30 animals in three relict populations. The overall population size has since recovered to ~4,800 individuals found in ~75 localities. Populations are severely fragmented with risks of genetic drift and hybridization with the plains zebra. These factors could negatively impact the unique allelic diversity remaining in Cape mountain zebra. Three relic populations are currently managed in Nature Reserves in the Western Cape Province of South Africa as source populations.

Genetic diversity analysis as determined by Moodley & Harley (2005) found low diversity in the isolated relic populations, noticeable differentiation due to drift, and also increased incidence of tumors due to equine sarcoidosis. With the availability of samples collected for the previous study (1999–2001; $n = 36$) and more recent samples (2015–2016; $n = 106$), we decided to perform a follow-up study to determine temporal trends over 16 years (Kotzé et al. 2019).

Based on results from 14 microsatellite markers, all relic populations showed a reduction in allelic diversity and expected heterozygosity (up to 23%) between the initial and follow-up studies. For example, allelic diversity in the most diverse of the three populations declined from an average of 2.27 alleles per locus to 1.95

(despite a larger sample size being used for the later estimate). Expected heterozygosity in this population declined from 0.264 to 0.230 over this 16-year period. Significantly, a loss of private alleles was observed in all populations, which may result in negative consequences in terms of fitness and adaptability. A study of population structure using both principal component analysis and a *STRUCTURE*-based assignment test revealed strong population structure with clear separation between two subgroups. This structuring is notable considering the limited distribution, low numbers, and dwindling diversity of this species, and suggests significant drift among population fragments.

Based on these temporal genetic data, resumption of gene flow between isolated populations should be a priority to minimize further drift. To this end, the government implemented a Cape mountain zebra Biodiversity Management Plan in 2016 to identify accurately the complex challenges facing the species and to implement required conservation measures and monitoring. Longer-term priorities include the establishment of additional populations through the introduction of Cape mountain zebra into enlarged and new protected areas and to augment the genetic diversity in current populations. However, translocation and augmentation can create corridors for the spread of equine piroplasmosis. For the future, continued monitoring and management action is critical to maintain the fitness and adaptability of the species.

Figure 23.13 Cape mountain zebra mother with calf. Photo courtesy of CapeNature.

CHAPTER 24

Conservation Genetics in Practice

Primary guest chapter author: Helen R. Taylor

Deerhorn clarkia, Section 18.2

The ideal scientist thinks like a poet and only later works like a bookkeeper.

(E.O. Wilson 2013, p. 74)

The field of conservation genetics, when properly implemented, is a constant juggling act integrating molecular genetics, ecology, and demography with applied aspects concerning managing declining species or implementing conservation laws and policies.

(Susan M. Haig et al. 2016, p. 181)

Conservation genetics has become a well-established field of science over the past 50 years, and a dedicated journal, *Conservation Genetics*, began publishing in 2000. The field of genetics has come to play an important role by providing valuable information for application to conservation problems around the world. Genetics has played an especially crucial role in the application of the US Endangered Species Act (ESA), which has been the primary driver of conservation of imperiled species in the USA (Box 24.1). In this chapter, we consider the relationship between basic and applied science, discuss how genetics can be incorporated into conservation policy and management, and suggest how one can become an effective conservation geneticist. We invited Helen Taylor to help write this chapter; she is the primary author.

24.1 Basic and applied science

Science is often divided into basic and applied endeavors. According to the Glossary of the US National Science Foundation, the objective of basic research is to gain more complete knowledge or understanding of the fundamental aspects of phenomena and of observable facts, without specific applications toward processes or products in mind. In contrast, the objective of applied research is to gain knowledge or understanding necessary for determining the means by which a recognized need may be met.

The distinction between basic and applied science is somewhat artificial. The US National Academy of Sciences, one of the leading organizations of basic research scientists in the world, was founded in 1863 to solve the problem of using magnetic compasses on metal ships during the US Civil War. The National Science Foundation, the primary funder of basic research in the USA, was founded in 1950, and its first Director, Alan T. Waterman, was at the time the Chief Scientist of the Office of Naval Research.

Genetics has historically been both a basic and applied science. The first geneticists selected domesticated plants and animals for human use (Section 1.1). We saw in Section 5.1 that G.H. Hardy, who

Conservation and the Genomics of Populations, Third Edition. Fred W. Allendorf, *et al.*, Oxford University Press.
© Fred W. Allendorf, W. Chris Funk, Sally N. Aitken, Margaret Byrne, Gordon Luikart, and Helen R. Taylor (2022). DOI: 10.1093/oso/9780198856566.003.0024

Box 24.1 The role of genetics in the application of the US Endangered Species Act

The ESA has been the primary legal mechanism for conservation in the USA (Greenwald et al. 2019). It became law in 1973, far before similar legislation was passed elsewhere in the world (e.g., the Australian Environment Protection and Biodiversity Conservation Act of 1999 and the Canadian Species at Risk Act of 2002). Unfortunately, many key decisions under the ESA have been made by the courts in the process of litigation after a private citizen or NGO has filed a suit against the federal government. Genetics has often played a key role in these court decisions because genetics is easier to quantify and is more explicit in many ways than ecology. For example, the allowable proportion of admixture in hybrids between species, the population size needed for long-term viability, and the amount of connectivity needed between populations have all played crucial roles in litigation.

Genetics has been central in four major steps of the ESA process: (1) identifying units eligible to be listed; (2) determining whether species should be listed; (3) developing recovery plans; and (4) delisting. We explain these in the following text.

Species, subspecies, and **distinct population segments** (DPSs) of vertebrate species are all eligible to be listed under the ESA (Box 20.1). Species and subspecies are generally accepted taxonomic entities, but DPSs are legal entities that must be determined as part of the listing process (Waples 2006). The designation of DPSs has been based largely on population genetic data to determine the discreteness of individual populations within species (Kelly 2010). A substantial proportion of the vertebrate "species" listed under the ESA are actually DPSs. Approximately 445 vertebrate "species" that occur in the USA are listed under the ESA (https://ecos.fws.gov/ecp/report/boxscore,

accessed 10 September 2020); approximately 100 (20%) of these "species" are actually DPSs (https://ecos.fws.gov/ecp/report/dps, accessed 10 September 2020).

Population size has played a key role in determining whether species have been listed under the ESA (Wilcove et al. 1993). In many cases, the 50/500 rule for effective population sizes (Section 18.8) has been applied as a criterion to evaluate long-term viability (e.g., Kamath et al. 2015).

The ESA requires that recovery plans be developed for listed species. These plans describe actions for recovery and set criteria for when a species will be considered recovered. Pierson et al. (2016) found that 82% of plans included some mention of genetic considerations. They also found that plans for animal species were more likely to include genetics. M.C. Neel (personal communication) reviewed 469 recovery plans for plant species to 2010; she found that just over 70% included some genetic considerations (also see Zeigler et al. 2013).

Genetics has played a large role in delisting of plants by identifying entities that are not eligible for listing (e.g., hybrids or entities that are not distinct from a common species; M.C. Neel, personal communication). Many delistings of plant species have resulted from genetics detecting procedural errors rather than actual recovery.

Genetics has already played a key role in the implementation of the US ESA, and will likely play an even greater role in the future as genomics improves our ability to delineate conservation units and characterize evolutionary potential. This makes it more essential than ever that conservation geneticists in the USA learn to communicate the power and limits of genomics for informing conservation decisions (Section 24.3.3.4).

---

developed the practical **Hardy–Weinberg (HW) principle**, took pride in being a "pure" mathematician who never did anything useful (Hardy 1967). Sewall Wright, one of the founders of theoretical population genetics, worked for the US Department of Agriculture (USDA) for many years as the senior animal breeding manager before he took an academic position at the University of Chicago. We have encountered this close relationship between basic and applied science throughout this book. For example, we saw in Section 6.7.3 that Darwin applied his understanding of the relationship

between isolation and inbreeding depression to question the management of deer populations in isolated parks.

24.2 The role of science in the development of policy

Science and **policy** intersect across numerous disciplines, from medicine to economics and everything in-between. Ideally, all policy decisions would be evidence-based, informed by the most up-to-date research but, in practice, this is rarely the case and the relationship between science and policy is far from direct (Kasperson & Berberian 2011). Scientists frequently struggle with whether or not to try and inform policy, but in conservation science there is a very clear motivation to do so (Meffe 1998; Noss 2007).

24.2.1 The best available science

The US ESA includes a "best available science" mandate that requires that agencies must use the best scientific and commercial data to guide key decisions (Doremus 2004; Lowell & Kelly 2016). This mandate explicitly applies to three sections of the ESA: (1) listing decisions; (2) critical habitat designation; and (3) agency consultation. This mandate seeks to provide objectivity so that decisions are not based on the whims of decision-makers or on the basis of improper motivations. As a society, we look to science to provide that objectivity. However, the ESA does not define what is meant by the best available science. Moreover, science alone cannot answer all the relevant questions (Doremus 1997). For example, we might be able to use science to determine the risk of extinction of a particular species, but deciding what risk of extinction society should tolerate is a policy decision.

24.2.2 Advice versus advocacy

It has been pointed out that "much of what we do in conservation biology is essentially worthless if it is not translated into effective policy" (Meffe & Viederman 1995, p. 327). This is because effective conservation outcomes are contingent on how much policy and management strategies are supported by scientific evidence (Sutherland et al. 2004). The laws that protect species and ecosystems are shaped by policy-makers and, if the laws are not evidence-based, they are unlikely to be effective. However, the policy arena is very different from a typical scientific environment and many scientists may be uncomfortable there. Thus, to inform policy, scientists not only have to decide that it is something they should do—they also have to understand how policy works (Meffe & Viederman 1995).

In the first instance, it is important to distinguish between science advice and science advocacy. The former is evidence-based, while the latter reflects the interests of those providing information. As a result, they are viewed differently by government and by the public (Hutchings & Stenseth 2016). It is crucial that attempts to inform policy do not affect objective hypothesis testing (Meffe & Viederman 1995). It is also crucial to recognize that scientists are no longer viewed as assumed authority by either government or the public as they were historically (Bocking 2018). One only has to look at climate change skepticism and the anti-vaccination debate to see this attitude toward scientists in action. In the current sociopolitical climate, if conservation scientists want to inform policy, then they must learn to present their science in a format that is accessible to policy-makers and defensible to the public (Bocking 2018).

Policy is intimately tied to politics and is issue-driven (Haig et al. 2006) and works on a very different timescale from the majority of scientific disciplines. Scientists engaging in policy must be able to anticipate the evidence needs of policy-makers before those needs arise (Meffe & Viederman 1995) and deal with uncertainty due to evidence being demanded at a point when the full picture has not yet emerged. This is an uncomfortable position for conservation scientists, who are keenly aware of the potential consequences around the advice they are giving, but uncertainty is inherent in giving advice within policy timescales—therefore it is key to make the uncertainty explicit (Converse et al. 2013).

There are a number of steps scientists can take to ensure their interactions with policy-makers are more effective. Understanding the language of policy is important—compare the style in a journal like *Environmental Science and Policy* with that of a conservation journal to get an inkling of the different language use. Additionally, conservation scientists may have formed solid relationships with regional conservation managers, but these are not likely to be the people making policy decisions; understanding the structure of governance is also helpful (Sandström et al. 2019).

Decision-makers are typically busy, so gaining an understanding of how the policies you are interested in are made and supplying the relevant information will make you far more likely to be heard (Funk et al. 2019). You are also more likely to be heard, and viewed as credible, if you are presenting your case via a national academy, a legally responsible committee, or a chief science advisor, rather than as a lone scientist (Hutchings & Stenseth 2016). Finally, taking a broad view is key; understanding the ramifications of the decisions you are advising on outside the implications for your species or area of interest will help you understand the various motivations and tradeoffs the policy-makers you are advising are faced with (Haig et al. 2016). If this sounds like a lot of work, that's because it is, and the workload only increases when faced with governmental leaders who publicly question the value of science (Goldman et al. 2017).

Unfortunately, despite predictions in the late 1990s that science was about to become central to political decision-making (Meffe 1998), at the time of writing, in many parts of the world the political climate has swung in the opposite direction. Scientists in many countries have to fight harder than ever to make their voices heard. Conservation genetics is an example of a discipline that is still poorly integrated at the highest levels of conservation policy (Laikre et al. 2020; Hoban et al. 2020). In the next section, we examine this gap, the drivers behind it, the steps being taken to bridge it, and the challenges still to be faced in effectively integrating conservation genetics into conservation management and policy.

24.3 Integrating genetic data into conservation strategy

Research implementation gaps are a common challenge wherever there is an interface between science and practice. A disconnect between information generated by research and the information used to make policy and planning decisions has been noted in various fields including healthcare (Haines et al. 2004), marketing (Dibb & Simkin 2009), and education (Nagro et al. 2019), as well as conservation (Knight et al. 2008).

24.3.1 The conservation genetics gap

As the field of conservation genetics has become increasingly well established, the paucity of genetic data integrated into conservation planning has become increasingly obvious (Laikre et al. 2020).

The potential of genetics to inform conservation was first formally suggested in the early 1970s (Frankel 1974). Despite some historical debate regarding the importance of genetics to species persistence (e.g., Caro & Laurenson 1994), the risks of genetic factors to threatened species are now clearly acknowledged, and conservation genetics has become a recognized field. Despite this, genetic data and plans to collect them are often missing from conservation plans and policy at both national and international levels (Pierson et al. 2016; Cook & Sgrò 2017; Hoban et al. 2020). Where genetic factors are mentioned, it tends to be in quite vague terms such as "genetic diversity," but key concepts such as effective population size and inbreeding are usually missing (Cook & Sgrò 2017).

The disconnect between conservation genetics and management has been documented in numerous contexts, including in Europe (Hoban et al. 2013b; Pierson et al. 2016), the USA (Haig et al. 2016), Canada, South Africa, and Australia (Pierson et al. 2016; Cook & Sgrò 2017), New Zealand (Jamieson et al. 2006), the countries bordering the Baltic Sea (Laikre et al. 2016), and Latin America (Rodriguez-Clark et al. 2015). It has also been observed in documents produced by international organizations (Laikre 2010; Hoban et al. 2016). It is notable that the International Union for Conservation of Nature (IUCN) Red List, the most widely used international species threat status assignment tool, does not explicitly consider genetic threats to persistence (Laikre 2010; Garner et al. 2020).

Gathering more information from a wider range of contexts is important as the severity of the gap, and the drivers behind it, seem to vary by location. For example, only 52% of species recovery plans in Australia and 17% of plans across Europe include genetics (Pierson et al. 2016). Some of the reasons for this will be discussed in the following text.

In discussing the conservation genetics gap, it is important to define some key terms. Papers on this topic often refer to practitioners and scientists as mutually exclusively. This is misleading, as many practitioners are also trained scientists (some will even be conservation geneticists). To avoid any confusion, here we use "practitioners" to mean those working in conservation agencies and nongovernmental organizations (NGOs) as conservation managers and similar positions, and "academics" to refer to those working in conservation genetics research at universities and research institutes. In using these definitions, we acknowledge that practitioners can also contribute to research and co-author papers, but tend to sit outside academia.

24.3.2 What drives the gap and helps it persist?

The drivers behind the conservation genetics gap have been discussed extensively in the scientific literature. Suggestions of barriers to integrating genetics in conservation by practitioners include the following:

- Practitioners are unaware of the benefits of genetics to conservation (Pierson 2016).
- Genetic factors are perceived as long-term threats and given lower priority than more pressing conservation concerns (Cook & Sgrò 2017).
- Practitioners not having access to scientific publications (Fabian et al. 2019).
- A belief among practitioners that genetics data are expensive to produce (Vernesi & Bruford 2009).
- Poor communication between academics and practitioners coupled with overuse of jargon (Hoban et al. 2013a).
- The complexity of linking demographic data with genetic data (Waples et al. 2008).
- A lack of funding or incentives for academics to make any efforts to bridge the gap (Haig et al. 2016).
- Geneticists have sometimes over-promised what genetics can deliver to managers (Ovenden & Moore 2017).

These are all sensible, well-reasoned suggestions, but they are primarily from academics, with little input from practitioners. Recently, several studies have sought information from practitioners on why genetics is not better integrated into conservation planning (Taylor et al. 2017b; Cook & Sgrò 2019b; Kadykalo et al. 2020). In contrast to previous suggestions that practitioners did not see the value of genetics, several of these studies found that practitioners were very positive regarding the usefulness

of genetics for conservation and were interested to see genetics better integrated into conservation practice. In these same studies, most practitioners believed they did not have sufficient knowledge regarding conservation genetics and wanted more. A test study on Australian practitioners found that they had a good understanding of how best to manage genetic diversity in practical scenarios, but were unfamiliar with the technical terms associated with the topic (Cook & Sgrò 2019a). This highlights the importance of reducing the use of jargon when communicating with practitioners.

Although practitioners are often keen to include more genetics in their work, they are often unsure who to approach to get assistance, and they have a lack of understanding of the funding issues that surround many conservation genetics research projects (Taylor et al. 2017b). Both anecdotal and survey evidence suggest that some practitioners are not aware of all the issues to which genetics might be applied (Taylor et al. 2017b; Holderegger et al. 2019).

Additional problems have also been identified via direct experience working alongside practitioners. The lack of access to scientific literature in general is problematic but, outside English-speaking countries, a lack of access in the local language of the practitioners is especially problematic (Holderegger et al. 2019). Practitioners also experience issues that academics may not be aware of regarding navigating various levels and branches of governance. Regional conservation managers often attempt to resolve priorities in implementation of management actions (Sandström et al. 2019). Differences in integration of genetics at different levels of management can also cause difficulties for practitioners (Cook & Sgrò 2017). By identifying these varied drivers of the conservation gap, we can also start to identify ways to bridge it.

24.3.3 Bridging the gap

There are various approaches that can be used to close the gap between the science of conservation genetics and its application to problems in conservation (e.g., Bernos et al. 2020; Hohenlohe et al. 2021). In the following text, we discuss some of the most widely suggested approaches in more detail.

24.3.3.1 In-house geneticists

One approach to integrating genetics into conservation management is to set up genetics departments within government agencies. This is a tactic used widely in the USA, with agencies such as the US Forest Service (USFS), the US Geological Survey (USGS), the US Fish and Wildlife Service (USFWS), and the National Oceanic and Atmospheric Administration (NOAA) (as well as some state agencies) all having their own in-house genetics teams (Guest Box 24). In view of the much higher rates of integration of genetics into species planning in the USA, this integrative strategy seems to work (Pierson et al. 2016). There have been several calls for agencies to foster evidence-based conservation by hiring more in-house researchers across a variety of disciplines (Cook et al. 2013; Roux et al. 2019). However, practitioners in some studies preferred to increase genetic capabilities by collaborating with academics outside their organizations (Taylor et al. 2017b; Taft et al. 2020). This desire for collaboration means that there are many opportunities for academics to form conservation genetics research partnerships. However, the lack of knowledge among practitioners regarding who to talk to about genetic work means that there is a need for improved communication.

24.3.3.2 Boundary organizations

The need to deliver scientific information to conservation agencies could be met by boundary organizations. Such organizations act as a bridge between science, policy, and practice, facilitating communication between the three. These organizations have been suggested as a potential solution to bridge research implementation gaps in conservation in general (Cook et al. 2013).

Forming multi-institutional organizations with the goal of bridging the conservation genetics gap is certainly one way to spread the workload involved in providing networking and training opportunities. A variety of organizations have been formed with this purpose in mind. Some, like the Genomic Biodiversity Knowledge for Resilient Ecosystems initiative (G-BiKE), come with funding attached, while others, such as the IUCN Conservation Genetics Specialist Group (CGSG), do not. A lack of funding can make life difficult, even for large groups of interested parties, as organizing meetings

across several countries or arranging funding for production of materials can be challenging, and such groups often still demand a fair amount of time in kind from their already busy members. Even funded organizations such as Baltic Sea Marine Biodiversity (BAMBI) and Conservation Genetic Resources for Effective Species Survival (CONGRESS) usually only have fixed-term funding, and both these initiatives have, indeed, come to the end of their funded period.

While such organizations may make excellent progress during their tenure, there is always the question of what comes next? In the case of CONGRESS, more funding was obtained, and G-BiKE was formed as a next step, but this is not always the case. To our knowledge, a solution to the funding conundrum does not yet exist. However, in a post-COVID-19 pandemic and, hopefully, more carbon-neutral world, it could be that greater acceptance of video conferencing puts less financial and time pressure on both academics and practitioners wishing to take part in gap-bridging initiatives.

Zoo-based organizations with their own conservation genetics teams are often already bridging the gap between conservation genetics and practice and acting as boundary organizations. These teams sit at a natural interface between captive breeding and reintroduction programs and thus have to liaise with multiple practitioners and other stakeholders to make their projects a success (Conde et al. 2013). Zoo-based research teams have produced some well-known conservation genetics software such as *VORTEX* (Chapter 18) and *PmX* (Lacy & Pollack 2020; Ballou et al. 2020). Additionally, thanks to their international connections, zoos are well placed to assist with capacity-building, such as the Royal Zoological Society of Scotland's (RZSS) WildGenes team's work assisting with setting up the first conservation genetics laboratory in Cambodia and collaborating with the Centre for Molecular Dynamics in Nepal. In contrast to specially formed boundary organizations, zoo-based teams tend to have long-term core funding via zoo visitor entrance fees, members, and patrons. As a result, partnering with a zoo-based conservation genetics team can be an excellent way to make use of bridges across the conservation genetics gap that have already been built (Example 24.1).

Example 24.1 Royal Zoological Society of Scotland's WildGenes team—a lesson in effective gap-bridging

RZSS's WildGenes program was founded in 2010 and has a team of laboratory and population genetics bioinformatics specialists based in the RZSS Conservation Department at Edinburgh Zoo. The team is involved in conservation management of both UK native and overseas species, working with animals as diverse as wildcats (Senn et al. 2019), dama gazelles (Senn et al. 2014a), Siamese crocodiles, northern rockhopper penguins, beavers (Senn et al. 2014b), and Himalayan wolves (Werhahn et al. 2018).

The team's work ranges from providing genetic data for reintroduction management (e.g., Senn et al. 2014b), to characterizing wild animal diets, and identifying and tracking illegally traded wildlife products (Bourgeois et al. 2018). RZSS is also a BioBank hub for the European Association of Zoos and Aquaria and, in collaboration with the National Museum of Scotland, is a partner on the CryoArks initiative.

In addition to running their own research programs, WildGenes staff also work to build conservation genetics capacity in low-income countries. Team members facilitated the setup and running of the conservation genetics laboratory at the Royal University of Phnom Penh in Cambodia, and have developed protocols in tandem with the Centre for Molecular Dynamics in Nepal.

The projects WildGenes participates in involve liaising with universities, laboratories, government agencies, funders, and other NGOs. Every project the team undertakes tends to be multistakeholder and thus is underpinned by clear contracts, memoranda of understanding (MOUs), and partnership agreements as necessary. There are careful discussions around publication authorship, sample storage, and guardianship, and the team also contributes to management plans (e.g., RZSS et al. 2019). Being based in a zoo provides the team with close links to an expert veterinary and keeping team, who can provide samples from captive animals for the development of genetic tools. The zoo setting also brings excellent opportunities for public communication of the conservation work being undertaken on site. There are still challenges regarding how best to communicate conservation genetics work to zoo visitors, but collaboration with education and visitor experience teams can help with this process.

At the heart of the success of WildGenes is a clear understanding of how to build successful, mutually beneficial partnerships, communicate and collaborate with diverse groups of people, and deliver data that are meaningful in a conservation context and driven by species management requirements.

24.3.3.3 Integration into threat classification systems

Arguably, one of the most effective ways to get genetics firmly embedded in the conservation agenda is to include genetic metrics in threat ranking programs such as the IUCN Red List. Despite clear connections between genetic factors and extinction risk (Chapter 18), identifying universal metrics that can be used across species to assess endangerment status has proved challenging. Several metrics have been suggested as suitable, including effective population size, various measures connected to inbreeding, inbreeding depression, loss of function mutations, and measures of fragmentation. Similar suggestions have been made for more powerful genetic targets to integrate into the Convention on Biological Diversity's post-2020 biodiversity framework:

effective population size and loss of genetically distinct units within species (Hoban et al. 2020).

None of these proposed metrics is without its flaws. Not least is the fact that many species on the IUCN Red List may lack genetic data precisely because collecting them is so infrequently included in species recovery planning (Pierson et al. 2016). It is possible to use 10% of census size as a proxy for effective population size (Hoban et al. 2020), but many species are missing even that basic information, and the IUCN Red List and others do not strongly promote reporting such data (Kindsvater et al. 2018). Additionally, many of the metrics proposed can be calculated in several different ways, and will be affected by the genetic methods and markers used to estimate them, making standardized comparisons across species extremely challenging. The IUCN CGSG and the Group on Earth

Observations Biodiversity Observation Networks (GEOBON) Genetic Composition Working Group are both working on defining genetic metrics that can be integrated into threat listings and made useful to both policy-makers and practitioners.

24.3.3.4 Improved communication

There have been repeated calls for improved communication between academics and practitioners (Shafer et al. 2015; Holderegger et al. 2019). Employing trained science communicators to act as "knowledge-brokers" and explain scientific results in lay terms is one possible solution (Meyer 2010; Cvitanovic et al. 2017), yet this is unlikely to be feasible except in specific well-funded initiatives or large agencies. It seems more likely that academics themselves will have to be the knowledge-brokers.

Less jargon and better applied framing of studies are important. But so are opportunities for practitioners and academics to network and communicate. Multidisciplinary conservation conferences can be good venues for networking and ideas exchange. However, at large conferences, talks featuring conservation genetics are often sidelined into specific genetics-themed sessions, which may be unattractive for practitioners (and, indeed other nongeneticists) to attend (Taylor & Soanes 2016). Further, funding is not always available for those outside academia to attend large overseas conferences. A better approach might be to organize smaller, local events specifically targeted at academics and other scientists showcasing genetic research to practitioners in a digestible format while conversing with them about their research needs (Holderegger et al. 2019). At these same events, conservation practitioners could have the opportunity to present conservation problems they are currently grappling with, so that academics can better appreciate the challenges and questions practitioners face.

There are more general ways to improve communication with practitioners. Putting in the effort to build trust, and establishing long-term collaborations around projects with mutual benefits are effective ways to bridge the gap (Box 24.2). For example, BioPlatforms Australia, a genomics infrastructure facility, has established four framework initiatives focused on applying genomics to conservation, and engagement between scientists and practitioners is

a core component of the approach. Another way to build trust and facilitate collaboration is to establish an MOU between an academic lab and a management agency. This allows both sides to understand expectations and to agree who releases sensitive information such as discovery of invasive species, disease pathogens, or endangered species on public or private lands (e.g., Sepulveda et al. 2020).

Publishing alternate versions of textbooks in language geared specifically toward practitioners is also a helpful step (e.g., Frankham et al. 2019), as is making pay-walled publications available to those outside of academia, and publishing pieces in club bulletins, professional society newsletters, and magazines (Holderegger et al. 2019). Jargon-free case studies such as those provided by the USFWS (https://training.fws.gov/courses/csp/csp3157/content/index.html) are a good way to communicate the benefits of genetic approaches in a relatable context. In non-English-speaking countries, producing textbooks and information in local languages is also a potentially useful approach (Holderegger et al. 2019). Incorporating genetic information and recommendations into practitioner guidelines provides information in accessible formats directly relevant to practitioners (e.g., Commander et al. 2018). Finally, many practitioners are not experienced coders. Therefore, when academics are designing software in computer languages such as R that they hope might be accessible and useful to nonacademics (e.g., Weiser et al. 2012), building in time and resources to produce either extremely user-friendly code or a graphical user interface will make it far more likely the tool will be adopted by practitioners. Videos explaining how to run and interpret results from such software are also useful.

24.3.3.5 Practitioner training

Many practitioners surveyed in the studies discussed in Section 24.3.2 felt that more training would be useful to help improve understanding of conservation genetics concepts. This requires careful thought regarding training methods. Lectures rather than group-based discussions have been found to be more effective in changing beliefs on the importance of genetic diversity and improving self-reported knowledge of the topic among practitioners (Lundmark et al. 2017). However, several

Box 24.2 Devil Tools and Tech: working across the gap to conserve Tasmanian devils

Bringing together researchers, practitioners, and policy-makers on one project has proved a successful way to integrate conservation genetics into the management of the Endangered Tasmanian devil (Guest Box 4). The resulting Devil Tools and Tech program has resulted in useful outcomes for both science and conservation management.

Tasmanian devil numbers have declined by almost 80% across their range since the late 1990s due to the advent of devil facial tumor disease (DFTD), a transmissible cancer that can lead to death (Figure 24.1; Lazenby et al. 2018). Due to previous population bottlenecks, devils already exhibited relatively low genetic diversity (Jones, M.E. et al. 2004; Morris et al. 2013). When an insurance population of DFTD-free animals was established in 2006, there were concerns that there might be further loss of genetic diversity and risks of inbreeding due to founders having been caught in close geographical proximity (Hogg et al. 2015). Genetic diversity in devils was too low for the existing microsatellite markers available for the species to be useful in establishing relatedness (Hogg et al. 2015), and thus a research need, driven by management requirements, was identified.

Described by its leaders as "bridging the gap between the lab bench and the forest floor" (Hogg et al. 2017, p. 134), the Devil Tools and Tech project brought together teams from government, industry, and academia. The Save the Tasmanian Devil Programme is a government initiative that contracted out management of the devil insurance population to the Zoo and Aquarium Association Australasia. They, in turn, invited the University of Sydney to bring its genetics expertise to bear on the project and, finally, San Diego Zoo Global joined the partnership by funding a postdoctoral research position (Hogg et al. 2017).

The genetic research conducted by the Devil Tools and Tech team has been explicitly guided by management questions, and the management team was consulted while research was being conducted. Regular communication between leads from each organization ensured everyone's needs were articulated and being met. The egalitarian nature of the project meant that all partners invested financially and intellectually in the work. Managers were also given access to raw data so that they could be rapidly integrated into decision-making, rather than waiting for publication. The data produced have led to further questions and the project continues to evolve. The result of Devil Tools and Tech has been more than 40 published research papers containing data that are being used for effective species management

(e.g., Wright et al. 2015; Grueber et al. 2015b; McLennan et al. 2020).

(a)

(b)

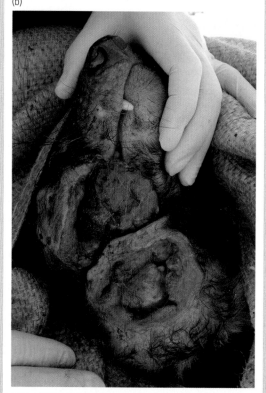

Figure 24.1 (a) Healthy Tasmanian devil without DFTD. (b) Tasmanian devil with advanced DFTD. Photos courtesy of Carolyn Hogg and Samantha Fox.

months later, while the positive effects of training on self-assessed knowledge persist, the difference between the two training methods disappears, as do changes in beliefs around policy, suggesting that it is important to continue knowledge communication with practitioners over time (Lundmark et al. 2019). As mentioned in Section 24.3.2, it is possible that practical, scenario-based training may be more effective for those without a genetics background (Cook & Sgrò 2019b). Either way, training requires a commitment from both sides in terms of time and, for academics, development of materials.

24.3.3.6 Inclusivity

A key point to remember is that it is not just practitioners who might form mutually beneficial relationships with academics. Investment in communication and training for other stakeholder groups and project partners, especially in countries with Indigenous Peoples, is also important. The UN Declaration on the Rights of Indigenous Peoples (UNDRIP) recognizes the broad intellectual rights of Indigenous People to genetic resources as well as traditional knowledge of plants and animals. Indigenous groups are playing increasing roles as partners in conservation research and management broadly. These partnerships need to be nurtured and grown through sustained communication and respect for traditional knowledge. Kadykalo et al. (2020) broadened their 2020 survey on genomics in Canadian freshwater fisheries out to a wide variety of potential "knowledge users," including First Nations, NGOs, professional societies, private environmental consultants, and retired government employees, in addition to practitioners in the sense used in this chapter. Again, there was positivity toward genomics, but a lack of familiarity with the topic (Kadykalo et al. 2020). There are few templates for how to conduct conservation genetic or genomic studies to include a diversity of partners and stakeholders, but those that are emerging show clear benefits to these partnerships, especially with Indigenous rights holders (for example, Gemmell et al. 2020).

24.3.4 Could genomics widen the gap?

Genomics brings obvious advantages to conservation genetic research, but also has the potential to complicate discussions with practitioners and raise additional barriers. Concerns around the additional jargon that comes with genomics, the workflow bottlenecks that can result from massive datasets, and the different skill sets required to work on these projects have already been raised in terms of widening the gap. There was already an established perception among practitioners that genetic research is expensive (Taylor et al. 2017b), and genomics could exacerbate this, with the main expense at this point being data analyses rather than laboratory costs (Holderegger et al. 2019).

When questioned, practitioners themselves have several concerns around shifting to conservation genomics. Qualms among practitioners regarding genomics range from requirements for high-quality samples leading to increased handling time for sensitive/threatened species, to difficulties integrating and prioritizing such large amounts of data, and occasional concerns that some stakeholders (including the general public) might be confused between genomics and genetic modification (Kadykalo et al. 2020). This last point is particularly pertinent in countries like New Zealand, which has an extremely strong anti-GMO stance (Taylor et al. 2017b). Finally, there is an awareness that genomic studies were, at least initially, focused on commercially important species and that this bias might continue at the expense of less commercially viable, but more severely threatened species (Kadykalo et al. 2020).

Genomics might also exacerbate the frequent and unhelpful disparity between the questions that practitioners need answers for and the questions that researchers are interested in and can publish (Haig et al. 2016). This is in part driven by the way academics are rewarded professionally, the pressure to publish in high impact factor journals, and funders prioritizing "novel" research. Often the basic questions that need addressing for conservation management do not fit the bill. There is an awareness of this among at least some practitioners (Smith et al. 2016), but that does not solve the issue. Genomics allows academics to delve into a wide range of functional questions, but, outside of adaptive potential and improved estimates of inbreeding, these are often not directly applicable to conservation. Building in useful applied conservation questions to more far-reaching and esoteric genomics projects will be key to avoiding this problem.

24.3.5 The genetics gap within conservation science

It is worth briefly noting that there is not just a gap between academics and practitioners, but between conservation genetics researchers and researchers in other conservation fields. The potential positive impacts of conservation genetics are often underestimated or completely overlooked, even by extremely high-profile conservationists (Taylor & Gemmell 2016). As mentioned in Section 24.3.3.4, talks featuring genetics are frequently siloed at multidisciplinary conservation conferences, creating an echo chamber effect where conservation geneticists discuss the merits of their field among themselves (Taylor & Soanes 2016). There are even gaps between conservation genetics and the broader fields of genetics and genomics; those working on cutting-edge genomics analyses are often isolated from those applying the techniques to conservation problems, resulting in a lack of knowledge exchange (Galla et al. 2016). Conservation problems require interdisciplinary, holistic solutions (Liu et al. 2015); clearly there is a need for conservation geneticists to improve their communication of the benefits of their chosen field across the board to enable this.

24.4 How do I become a conservation geneticist?

There is no straightforward course of study available to become a conservation geneticist (Box 24.3). In this section, we consider what type of training one should consider to become a conservation geneticist.

24.4.1 What is a conservation geneticist?

First and foremost, a good conservation geneticist is a population geneticist. To be effective in conservation genetics requires numerous other skills, but underlying all of those has to be a firm knowledge and understanding of genetic theory. This is non-negotiable and means that if you have an interest in applying genetics to conservation, it would be helpful to seek out a graduate training program featuring modules in population genetics. As a graduate student, you need at least one supervisor/advisor who specializes in the subject. Graduate students seeking to "do some genetics" on their study species with no prior knowledge of the topic without qualified supervision often end up not achieving effective outcomes from their work or for their own development.

Fundamental concepts in population genetics, such as HW proportions and gametic disequilibrium, are frequently misunderstood or overlooked in population genetic studies (Waples 2015). This suggests a lack of understanding of population genetic theory among authors, reviewers, and journal editors. We saw in Figure P.2 in the Preface that empirical population genetics requires three primary activities: collecting data, analyzing data, and applying appropriate theory. Generating population genetics data has become increasingly automated and can be done with relatively little training or by fee-for-service laboratories. Powerful and relatively easy-to-use software exists for analyzing data, but considerable knowledge is needed to use them correctly, and understand the assumptions and limitations of approaches. Finally, understanding the necessary theory can only be accomplished through years of education and training.

Understanding genetic theory has become even more important in the genomic era. There is a current tendency in population genomic work to focus on the challenges around coding, and getting software to run through enormous datasets. Again, this frequently comes at the expense of understanding the theory and assumptions that underlie the software being used (Allendorf 2017), rendering such software a black box. Population genomics software

Box 24.3 The career of one conservation geneticist

The career of the first author of this book demonstrates the interrelationship between basic and applied science in conservation genetics. As a graduate student, he was supported by the State of Washington Department of Game to study the population genetics of steelhead (anadromous rainbow trout). He provided the first study of the genetic relationships among populations of steelhead in the State of Washington, which was used to guide a number of management programs. He also helped develop an experimental design that used an allozyme locus to genetically mark fish to test the relative fitness of hatchery and wild fish in natural conditions (Chilcote et al. 1986).

At the same time, much of his dissertation research involved esoteric work in the inheritance of duplicated genes in salmonid fishes, which are derived from a tetraploid ancestor. This work was published in a number of genetic journals devoted to basic science (*Genetics*, *Heredity*,

and *Journal of Heredity*). Every scientist working in the field of conservation genetics has an analogous story of working at the interface between basic and applied science.

His academic position, which began in 1976, required the annual teaching of a course in population and ecological genetics. In the early 1980s, this course morphed into conservation genetics, and eventually this book. The material in the course remained largely unchanged, but there was an emphasis on using examples with relevance for conservation. He has long believed that many of the most interesting questions in basic population genetics are also of important relevance in conservation (e.g., the genetic basis of local adaptation, the genetic basis of inbreeding depression, etc.). As well as teaching a university course in conservation genetics, he has also taught workshops in conservation genetics throughout the world (Figure 24.2)

Figure 24.2 Students in a conservation genetics workshop taught by the first author at the National Zoological Garden, Pretoria, South Africa, in November 2016. Photo courtesy of Diane V. Haddon.

will take the data it is given and give an answer in exchange, but without an understanding of what assumptions the program is making, the answer may well be meaningless and certainly should not be used to advise conservation strategy.

The large numbers of markers available with genomics underlie the importance of understanding basic theory, and also of understanding the difference between statistical and biological significance (Box 20.3). For example, testing for differences in allele frequencies between two population samples is one of the first steps in data analysis. The null hypothesis, in this case, is that the two samples are drawn from a population with the same allele frequencies (i.e., panmixia). However, no two populations will have exactly the same allele frequencies. Therefore, if we look at many markers using large sample sizes, we will almost always reject the statistical null hypothesis of no difference in allele frequencies. This statistical significance should not be interpreted to mean that these two samples are drawn from populations with biologically meaningful differences in allele frequencies (Palsbøll et al. 2007).

Overall, if the intention is to give advice on species management based on genetic data, it is imperative that the scientists interpreting those data have a thorough understanding of the theory underpinning their assertions. When it comes to threatened species management, the stakes are too high for misdirection based on poor understanding of theory. We cannot stress this enough.

24.4.2 Diverse skill sets

Other than a strong grasp of population genetics theory, conservation geneticists need to embody a wide range of different skill sets. The journey from identifying a conservation question that needs addressing, through obtaining funding, forming collaborations, sample collection, lab work, analysis, dissemination of results, and informing policy is unlikely to be completed by one person working in isolation. Conservation genetics is a discipline that simultaneously advances scientific understanding and (when executed effectively) contributes to decision-making. Thus, there is room for a variety of people and skills on any conservation genetics project and space within the field to tailor your career to your particular skills and interests.

It is also important to note that conservation geneticists need not be based in universities (although they will almost certainly begin their careers there). Conservation geneticists find roles at research institutes, government agencies, and conservation NGOs, among other organizations, and may move through several kinds of organization over the course of their career. As stated in Section 24.3.1, one does not cease to be a conservation geneticist by dint of no longer working in academic research, although an unfortunate snobbery around leaving academia stubbornly persists in some quarters (among a subset of academics). A growing number of academics recognize the need to translate research to real-world applications, although many won't have much experience with this. You may find that career academics are not always well equipped to discuss alternative conservation genetic career paths with you, as some have no direct experience of these paths themselves. This makes networking with those already working at nonacademic organizations important in terms of understanding what other opportunities might be available and how to navigate an alternative path.

The skills you need to cultivate for a career in conservation genetics will depend very much on where your interests lie. At this point, almost everyone working in genetics will need to learn some kind of coding language in order to analyze their data and run basic analyses. However, those who wish to dive into large-scale genomic analyses and the development of tools to facilitate these analyses will need to expand their bioinformatics skills much further. Alternatively, you may be more inclined to the laboratory side of conservation genetic work and thus need to develop your understating of sample curation, the Nagoya protocol (Secretariat of the Convention on Biological Diversity 2011), and laboratory genetic chemistry. Maybe on-the-ground conservation management will appeal to you, in which case, building your field experience and logistical skills in tandem with bringing your conservation genetics knowledge to bear in devising management plans will be key. Other conservation geneticists may be drawn more to the policy and implementation side of the process, requiring them to grow their understanding of political and legal processes. The paths outlined here are examples. These examples are neither exhaustive nor mutually exclusive; there will be multiple skill sets existing within various people and a need for collaboration between people working across all these areas (and more) that demands at least some understanding of the workings and challenges of each component part. We assume, though, that a common goal for anyone engaged in a conservation genetics career is to use genetic theory and data to improve the chances of persistence for biodiversity. We suggest some key skills that we believe will be helpful in pursuing that goal.

24.4.3 Communication and collaboration

In our experience, there are two main skills, in addition to genetics itself, that will help you be an effective conservation geneticist, in whatever form that takes: the ability to communicate and the ability to collaborate. There is some overlap in these skill sets (communication is an essential part of collaboration) and both are commonly overlooked in scientific training.

Depending on their role, conservation geneticists may be required to communicate their specific research or population genetic concepts in general to a variety of audiences, both verbally and in writing. These audiences could include other conservation scientists, conservation practitioners, politicians and policy-makers, and the general public. Unfortunately, conservation genetics and its associated theory are couched in a large amount of jargon (Hoban et al. 2013a). Words like "allele" and "heterozygosity" mean little to those outside of genetics, and even commonplace words like "gene" are surprisingly difficult to define (Pearson 2006).

Thus, as a conservation geneticist, it is important to learn new ways of talking about our work that avoid intimidating or confusing language, and tell clear stories that help people of all backgrounds understand how this work contributes to successful conservation. The bottom line for any kind of science communication should be "why you should care." Storytelling that puts science into context typically helps audiences engage with a given message and make a more meaningful connection with a topic (Dahlstrom 2014; Sundin et al. 2018). Messages have to be tailored to each given audience—there is no one-size-fits-all solution and people will care about an issue for different reasons. Similarly, we have to find more intuitive ways to communicate data; an F_{ST} table will mean very little to anyone in conservation not familiar with population genetics.

Effective communication is also necessary for forming effective collaborations. Again, conservation geneticists need to collaborate on various levels. As mentioned earlier, it is unlikely that any single conservation geneticist will have all the skills required for a given project so, at the very least, conservation geneticists will need to collaborate with other scientists. Increasingly, conservation genetics labs employ bioinformaticians to help with the enormous amounts of data generated by genomics projects. Bioinformaticians do not always have a population genetics background and they almost certainly won't have a background in whatever natural system is being analyzed. Thus, even within lab groups, effective communication is essential.

At a broader scale, conservation geneticists will potentially need to collaborate with and form partnerships with scientists from other disciplines, Indigenous groups, local communities, and a variety of other stakeholders. Although there is no one formula for successful collaborations, self-awareness, empathy, and the ability to listen are all key skills for working in a group, especially a diverse one with many contrasting points of view. Collaboration also requires strong leaders, which, again, is not something featured in scientific training.

The training required to obtain these soft skills may not be available in your department at your academic institution. This is an issue across conservation science, not just in conservation genetics (Blickley et al. 2013; Andrade et al. 2014). Three key areas have been identified where capacity-building is required for conservation scientists: communication skills, interpersonal skills, and boundary-crossing skills (Elliott et al. 2018).

Unfortunately, it is currently incumbent on conservation geneticists to develop their communication and collaboration skills on their own.

However, resources are out there. A growing number of universities have their own science communication or science and society departments and courses. There are also some external bodies offering media training for scientists (e.g., New Zealand's Royal Society-funding Science Media Centre https://www.sciencemediacentre.co.nz, and the Liber Ero Fellowship Program in Canada for postdoctoral scientists providing training in conservation research and communication http://liberero.ca). The IUCN's Conservation Planning Specialist Group (CPSG) runs courses on facilitation both online and in person with a strong focus on effective group decision-making (http://www.cpsg.org/our-approach/training). If there are colleagues in your lab group who sit on advisory boards or planning groups for conservation projects, you may be able to attend as an observer to increase your understanding of the process. Elective study on tools such as structured decision-making (Martin et al. 2009) could also be useful; the IUCN Conservation Translocation Specialist Group (CTSG) offers annual training in this topic (https://iucn-ctsg.org). Finally, pay attention to current affairs and political news to better understand the context you are working in.

The conservation genetics job market is, like all other fields of conservation, and indeed science in general, extremely competitive. Everyone you are competing with will have been to university and have some form of postgraduate degree. They will likely also have published papers. These are stan-dard achievements among your peers. However, if you can demonstrate an excellent understanding of population genetic theory, coupled with something different such as science communication training or experience, participation in multistakeholder workshops, or experience in different work environments, you will be ahead of the game. What is more, you are far more likely to be effective in improving biodiversity conservation outcomes using genetic tools and data.

24.5 The future

We are currently facing a crisis in the loss of biodiversity and ecosystem function throughout the world. Genetics itself will not provide an answer to these problems. The primary issues that society must address are habitat loss, climate change, translocation of species, pollution, overexploitation, etc. Nevertheless, genetics and genomics are essential ingredients to help manage this crisis. We will not be able to make well-informed decisions without all aspects of conservation science.

The application of genomics to conservation has just begun. Genomics is now being applied in ways that we could not even imagine when we wrote the previous edition of this book some 10 years ago. We are sure that similar unimagined applications of genomics will be developed in the near future. We hope that some of the students and practitioners using this book will be involved in developing these applications.

Guest Box 24 Making genetics applicable to managers
Michael K. Schwartz

One commonality among nearly all conservation geneticists is the desire for their work to be used to affect the conservation and management of species. Thus, the natural question arises: "How can one be effective at having their conservation genetics research used in the decision-making and conservation process?" Below I provide my perspective as a research scientist embedded in a federal land management agency. The answer I give is not unique to conservation genetics, but is universal to any scientist who wants to see their work used in a management context.

To address this question we need to first ask: "What is a scientist?" or "What do we mean by science?" Is a scientist someone who systematically collects data relevant to a pressing question? Or someone who uses the scientific method to test a hypothesis? What if that hypothesis has been tested over and over with the same species but in different locations? Author Gary Zukav (1979, p. 9) once noted that there are scientists and technicians of science. Zukav wrote: "When most people say 'scientist', they mean 'technician'. A technician is a highly trained person whose job is to apply known techniques and principles. A scientist is a person who seeks to know the true nature of physical reality."

Many management agencies desire to be science-based. However, what they mean is they need highly trained people to apply known and defensible techniques and principles to learn about their pressing issues. They may want to understand an ecosystem better and its true nature, but given their sense of urgency and the need to triage issues due to limited funding, they don't prioritize funding scientific discovery. This is the heart of the mismatch between the questions practitioners have and the questions researchers are interested in and can publish.

Recognizing this mismatch is the start of being more effective. Too many times conservation geneticists will answer a question that is of interest to them, but of little use to the manager. Or worse, they will sell a manager a research project that they know will produce an answer that cannot be translated to a management product. There are three things that I have seen successfully eliminate this gulf between scientists and managers:

1. Empathize with the manager: Understand what the manager's specific information need is and consider if the results from a proposed project meet that need. If not, maybe your interest is more about basic science than conservation. While your interests may be useful in the long run and may provide a major breakthrough, a manager may prefer an answer to an immediately pressing problem. The best-designed studies can accomplish both sets of needs (advancing science and providing scientific information) with the same dataset. Most managers I know will seek to provide additional funding for basic research if the project also provides the immediate and direct technical information they desperately need.

2. Project results: When we test hypotheses and collect appropriate data, we know what the data stream will look like and ways the study might resolve. Show the manager what the possible outcomes will be and ask how they will use this information. Create mock-ups of what key figures may look like once the results are generated. Then, work together to target the right study to meet everyone's needs.

3. Communicate consistently: Obtaining funding from a management agency for research is just the beginning of the process. At this point, the worst thing to do is to walk away, only to reappear a year or two later with an answer. It will probably be the answer to the wrong question. Have both formal and informal communication about the work throughout the entire process. This gives deeper understanding and buy-in to the study and results. When managers engage with research throughout the process, I have found them more likely to provide additional support if needed and more effectively communicate results to the broader management community.

Overall, the best scientists I know working in management agencies have been successful in their careers by finding ways to leverage data collected in the applied realm to test basic science questions. However, they all dependably focus on being dedicated partners to co-produce knowledge with their management colleagues first. Each one spends a lot of time listening to management needs early on in the project development phase with an open mind, is honest in what their research can and cannot answer, and consistently communicates from inception of the project to dissemination of the final results. In the best of cases, the managers are so engaged with the research that they are the ones that disseminate findings to their colleagues as they are more effective messengers.

Glossary

acrocentric Chromosomes and chromatids with a centromere near one end.

adaptation Evolutionary change resulting from natural selection that increases fitness, or a trait that increases fitness.

adaptational lag A population in a changing environment is sometimes unable to adapt quickly enough to keep up with the necessary rate of change to maintain fitness.

adaptive introgression When the introduction of genetic material from another population or species increases the fitness of the population.

adaptive landscape genomics The study of the relationships between spatial environmental variation and the genomics of adaptation of organisms to those environmental factors.

adaptive management An iterative process of decision-making, whereby as knowledge improves, this information is used to inform and modify management approaches over time.

adaptors Short single- or double-stranded oligonucleotides that can be ligated to DNA or RNA fragments to barcode or amplify those fragments. Also spelled adapters.

addition rule *See* **sum rule.**

additive coefficient of variation Additive genetic variation standardized by the trait mean.

additive genetic variation The portion of total genetic variation that is the average effect of substituting one allele, responsible for a phenotypic trait, for another. The proportion of variation that responds to natural selection.

admixture The formation of novel genetic combinations through hybridization of genetically distinct groups.

agamospermy The asexual formation of seeds without fertilization in which mitotic division is sometimes stimulated by male gametes. Also called **apomixis.**

alien species A nonnative species or any propagule of that species such as eggs, spores, or tissue present in an ecosystem.

allele Alternative form of a gene.

allelic diversity A measure of genetic diversity based on the average number of alleles per locus present in a population.

allelic dropout Occurs when an allele is not amplified in the polymerase chain reaction, resulting in apparent homozygosity at a locus.

allelic richness A measure of the number of alleles per locus; allows comparison between samples of different sizes by using various statistical techniques (e.g., rarefaction).

allopatric Species or populations that occur in geographically separate areas.

allopolyploidy Polyploidy originating through the addition of unlike chromosome sets, often in conjunction with hybridization between species.

allozygous An individual whose alleles at a locus are descended from different ancestral alleles in the base population. Allozygotes may be either homozygous or heterozygous in state at this locus.

allozyme An allelic form of an enzyme detected through protein electrophoresis.

alpha value The chosen significance level of a statistical test that is the probability of rejecting the null hypothesis when it is true, e.g., $\alpha = 0.05$ represents a 5% risk of a false positive.

amplicon A segment of DNA formed by natural or artificial amplification events; for example, by polymerase chain reactions.

amplicon sequencing Amplification and sequencing of a selected piece of DNA or RNA.

amplified fragment length polymorphism (AFLP) A technique that uses PCR to amplify genomic DNA, cleaved by restriction enzymes, in order to generate DNA fingerprints. Of note, the original description said that AFLP is not an acronym for amplified fragment length polymorphism because it does not detect length polymorphisms (Vos et al. 1995).

amplify To use PCR to make many copies of a segment of DNA.

anagenesis Evolutionary changes that occur within a single lineage through time. *See* **cladogenesis.**

analysis of molecular variation (AMOVA) A statistical approach to partition the total genetic variation in a species into components within and among populations or groups at different levels of hierarchical subdivision. Analogous to ANOVA in statistics.

ancient DNA (aDNA) DNA isolated from ancient samples.

aneuploid A chromosomal condition resulting from either an excess or deficit of a chromosome or chromosomes so that the chromosome number is not an exact multiple of the typical haploid set in the species.

animal model A statistical model in which phenotypic variance is compartmentalized into environmental, genetic, and other causes, as in mixed models.

anneal The joining of single strands of DNA because of the pairing of complementary bases. In PCR, primers anneal to complementary target DNA sequences during cooling of the DNA (after DNA is made single-stranded by heating).

annotation Genome annotation is the process of identifying genes and coding regions in a genome and determining the function of those genes.

annual plant Those that germinate, flower, and mature seed in the same year, and survive winter or drought periods as seeds.

antagonistic genetic correlations Genetic tradeoffs between two traits such that an increase in one trait in a favorable direction (i.e., increasing fitness) results in a corresponding change in another trait that reduces fitness.

apomixis *See* **agamospermy**.

approximate Bayesian computation (ABC) A statistical framework using simulation modeling to approximate the Bayesian posterior distribution of parameters of interest (e.g., N_e, mN), often by using multiple summary statistics (H_e, number of alleles, F_{ST}).

arithmetic mean The average value of a set of numbers calculated by summing the numbers and dividing by the number of observations.

artificial selection Anthropogenic selection of phenotypes with a heritable genetic basis to elicit a desired phenotypic change in succeeding generations.

ascertainment bias Selection of loci for marker development (e.g., single nucleotide polymorphisms or microsatellites) from an unrepresentative sample of individuals, or using a particular method, which yields loci that are not representative of the spectrum of allele frequencies in a population.

assembly (genome) The computational process of combining many short DNA or RNA sequence reads into longer **contigs** and **scaffolds**, often with the ultimate goal of generating sequences for whole chromosomes.

assignment tests A statistical method using multilocus genotypes to assign individuals to the population from which they most likely originated.

assisted colonization Anthropogenic movement of individuals to sites where the species does not occur or has not been known to occur in recent history to increase the range of a species.

assisted gene flow Anthropogenic movement of individuals or gametes within the range of a species to facilitate adaptation to changing environmental conditions. In small populations, assisted gene flow is useful for reducing inbreeding depression and becomes equivalent to **genetic rescue.**

assisted migration Anthropogenic movement of individuals to new locations within or outside of their species range, often to match organisms with their historic climates as climate change occurs.

assisted range expansion Anthropogenic movement of individuals to regions adjacent to existing species ranges on the leading edge of range expansion.

association mapping A method using gametic disequilibrium to associate phenotypes to genotypes in order to map quantitative trait loci. Also called linkage disequilibrium mapping. *See* **genome-wide association mapping**.

associative overdominance An increase in fitness of heterozygotes at a neutral locus because it is in gametic disequilibrium at a locus that is under selection. Also known as **pseudo-overdominance**.

assortative mating Preferential mating between individuals with a similar (or a different) phenotype is referred to as positive (or negative) assortative mating.

autosomal A locus that is located on an autosome (i.e., not on a sex chromosome).

autosomes Chromosomes that do not differ between sexes.

autozygosity A measure of the expected homozygosity where alleles are identical by descent.

autozygous Individuals whose alleles at a locus are identical by descent from the same ancestral allele.

balancing selection Diversifying selection that maintains a polymorphism, resulting from frequency-dependent selection, spatially heterogeneous selection, or heterozygous advantage.

barcodes DNA sequences that identify individuals in multiplexed DNA libraries, or that identify species from unknown samples (e.g., eDNA).

Barr bodies Inactivated X-chromosomes in many female mammals that condense to form a darkly colored structure in the nuclei of somatic cells.

Bayes factor The ratio of the likelihood of one hypothesis to the likelihood of another. It provides a Bayesian alternative to classical hypothesis testing.

Bayesian Statistical approaches based on Bayes' theorem that include prior knowledge in assessing the probability of an event.

B chromosome *See* **supernumerary chromosome**.

bet-hedging A long-term strategy that maximizes geometric mean fitness of a genotype in variable environments by decreasing individual fitness variance, despite lowering the arithmetic mean fitness.

binning Combining alleles into groups. Binning is often used to combine all alleles except the one with the highest frequency into one group to calculate minor allele frequency and allow for comparisons with other bi-allelic loci.

binomial proportion A population will be in binomial proportions when it conforms to the binomial distribution so that the occurrence of a given event X, r_i times with a probability (p_i) of success, in a population of n total events is not significantly different than what would be expected based on random chance alone.

bioclimatic envelope model Describes the multivariate climatic niche space occupied by a species, usually developed through correlating species occurrence with climatic variables. Also called a **species distribution model (SDM)**.

biogeography The study of the geographic distribution of species and the principles and factors that influence these distributions.

bioinformatics The application of code and script writing and informatics techniques used in statistical, mathematical, or computer science to understand and organize biological data such as DNA sequence data.

biological significance A statistical difference between groups or effect size that is biologically as opposed to statistically important.

biological species concept (BSC) Groups of naturally occurring, interbreeding populations that are reproductively isolated from other such groups or species.

biotinylated A DNA, RNA, or protein molecule that has a covalently attached biotin that allows for the isolation and purification of selected fragments from a mixed sample (e.g., in targeted sequence capture).

biparental inbreeding Inbreeding due to matings between related individuals (i.e., not selfing).

BLAST Basic Local Alignment Search Tool; software to search for a DNA sequence similar to the sequence in hand.

Bonferroni correction A correction used when several independent statistical tests are being performed simultaneously (since while a given α-value may be appropriate for each individual comparison, it is not for the set of all comparisons). In order to avoid a lot of spurious positives, the α-value needs to be adjusted to account for the number of comparisons being performed.

bootstrap analysis A nonparametric statistical analysis for computing confidence intervals for a phylogeny or a point estimate (e.g., of F_{ST}). Resampling of a dataset with replacement to estimate the proportion of times an event (such as the positioning of a node on a phylogenetic tree) appears.

bootstrap support The proportion of bootstrapped samples that produce the same result, for example, the same phylogenetic tree.

bottleneck A special case of strong genetic drift where a population experiences a loss of genetic variation by temporarily going through a marked reduction in effective population size. In demography, a severe transient reduction in population size.

branch length Length of branches on a phylogenetic tree. Generally proportional to the amount of genetic divergence between populations or species.

breeding value A measure of the value of an individual for breeding purposes, as assessed by the mean performance of its progeny.

broad-sense heritability (H_B) The proportion of phenotypic variation within a population that is due to genetic differences among individuals.

burn-in The initial steps in a Markov Chain Monte Carlo (MCMC) simulation that allow the run to reach a near equilibrium state before parameters are estimated.

candidate gene A gene that is thought to be more likely to affect a trait based on known functions compared with a random gene from the genome.

census population size (N_C) The number of adult individuals in a population.

centimorgan (cM) The genetic distance between genetic markers or loci. Two markers that are 1 cM apart on a chromosome have a probability of having a crossing over event between them in a single generation of 1%. If two markers are 50 cM apart, they are inherited independently, and may be on different parts of the same chromosome or on different chromosomes.

central limit theorem The distribution of a sample mean will approach a normal distribution as the number of observations increases.

centromere A constricted region of a chromosome containing spindle microtubules, responsible for chromosomal movement during mitosis and meiosis.

chi-square test A test of statistical significance based on the chi-squared statistic, which determines how closely experimental observed values fit theoretical expected values.

chloroplast DNA (cpDNA) A circular DNA molecule located in the chloroplast.

chromosomal satellites Small chromosomal segments separated by secondary constrictions from the main body of a chromosome.

chromosomal tract Contiguous segment of a chromosome.

chromosome A molecule of DNA in association with proteins (histones and nonhistones) constituting a linear array of genes. In bacteria and archaea, the circular DNA molecule contains the set of instructions necessary for the cell.

circuit theory An application of electrical circuit theory to landscape genetics, where landscapes consist of nodes (populations or individuals) connected by resistors (the intervening landscape). Gene flow is modeled as electric current flowing between nodes through the resistors, using a landscape resistance surface.

clade A species, or group of species, that has originated from and includes all the descendants from a common ancestor. A monophyletic group.

cladistics The classification of organisms based on phylogenies.

cladogenesis The splitting of a single evolutionary lineage into multiple lineages.

cladogram *See* **phylogeny**.

climate-adjusted provenancing Seeds are sourced from multiple locations with warmer, drier climates than a restoration site in response to climate change. A type of **assisted gene flow**.

cline A gradual directional change in a character or allele frequency across a geographic or environmental gradient.

close-kin mark–recapture (CKMR) A mark–recapture method based on identifying the number of juveniles whose parents can be identified. Each juvenile captured "marks" two adults.

coadapted gene complex A multilocus genotype of alleles at several loci that results in high fitness and is commonly inherited as a unit because the loci are closely linked.

coalescent The point at which the ancestry of two alleles converges (coalesces) at a common ancestral sequence.

coancestry Degree of relationship between two parents of a diploid individual.

coding Using a programming language to write commands for a computer to follow. It allows creation of a script, application, software, or operating system. Coding of a few lines produces a script. Coding multiples sets of scripts or commands produces software.

codominant Alleles are codominant when one allele at a locus cannot mask or obscure the presence of another. Codominance can refer to alleles for genetic markers, or for the phenotypic effect of alleles.

codon A three-nucleotide sequence on a strand of mRNA that gets translated into a specific amino acid within a protein.

common garden An experiment where individuals from different populations are reared in a common environment to facilitate the study of genetic differences in phenotypes. The term applies to both plants and animals.

community genetics The study of how genetic variation within a species affects interactions among species, community structure, and diversity.

community landscape genomics The study of genomic variation in multiple species simultaneously across landscapes.

community phylogeny Application of phylogenetics to inform studies of community ecology, including understanding the processes governing species assemblages on the basis of evolutionary relationships among co-existing species.

comparative genomics The comparative study of genomic characteristics of different species, including DNA sequences, genes, regulatory regions, gene order, signals of local adaptation, and structural rearrangements.

complementary DNA (cDNA) Synthetic DNA which is reverse transcribed from RNA by the enzyme reverse transcriptase. The sequence is complementary to the original DNA template for the RNA.

conditional neutrality The situation where genetic variants affect fitness in one environment but not in another.

conservation breeding The breeding of plant and animal species for conservation. Also known as captive breeding.

conservation collections Living collections of rare or endangered organisms established for the purpose of contributing to the survival and recovery of a species.

conservation genomics The use of new genomic techniques to solve problems in conservation biology.

conservation unit A population or group of populations that is considered distinct for purposes of conservation, such as a management unit (MU), distinct population segment (DPS), or evolutionarily significant unit (ESU).

conspecific A member of the same species.

contig A DNA sequence assembled from a series of short, overlapping sequences, such as those obtained from next-generation sequences.

continuous characters Phenotypic traits that are distributed continuously throughout the population (e.g., height or weight).

Convention on International Trade in Endangered Species of Wild Fauna and Flora (CITES) An international agreement that bans international trade or shipment of an agreed-upon list of endangered species, and that regulates and monitors trade in others that might become endangered.

converged The point where an MCMC simulation has become independent of starting parameter biases, or has been "burnt in." Typically, thousands of simulation steps are required (and discarded) before the MCMC simulation is used to estimate a parameter (e.g., N_e, mN, etc.).

cost distance The sum of the resistance (cost) of each cell along a given movement route between two points in a landscape.

countergradient variation Occurs when genetic effects on a trait oppose or compensate for environmental effects so that phenotypic differences across an environmental gradient among populations are minimized.

CRISPR-Cas9 A genome editing technology that uses elements of a prokaryotic immune system to insert, delete, or modify DNA sequences.

crossover interference The existence of a crossover during meiosis in one chromosomal interval reduces the probability of another crossover. As a result, there is often only one crossover event per chromosome pair per meiosis in many species.

cryptic species Closely related species that are morphologically difficult to distinguish, but can be identified with genetic data.

cultivars A human cultivated plant that was derived through anthropogenic selection and breeding.

cytoplasmic genes Genes located in cellular organelles such as mitochondria and chloroplasts.

cytoplasmic male sterility (CMS) Total or partial male sterility in plants caused by interactions of nuclear and mitochondrial genes.

data scraping A technique in which a computer program extracts data from human-readable output coming from another program.

de novo Latin for from the beginning or anew.

de novo sequence assembly The assembly of DNA or RNA sequence data from scratch, without a reference genome or transcriptome.

de-extinction The resurrection of species that have gone extinct by inserting gene sequences from extinct species into living relatives.

degrees of freedom (df) The total number of items in a dataset that are free to vary independently of each other. For example, assume there are four cookies and four children. The first three children get to choose a cookie, but the fourth does not. Therefore, there are three degrees of freedom.

deleterious Having harmful effects, for example, associated with lower fitness.

deme A local group of individuals that mate at random.

demographic Topics relating to the structure and dynamics of populations, such as birth, death, and dispersal rates.

demographic rescue The decrease in the probability of extinction of an isolated population brought about by immigration that increases population size.

demographic stochasticity Differences in the dynamics of a population that are the effects of random events on individuals in the population.

demographic swamping Loss of reproductive potential and reduction in population growth rate due to hybridization.

dendrogram A tree diagram that serves as a visual representation of the relationships between populations within a species.

derived A derived character is one found only in a particular lineage within a larger group. For example, feathers are derived characters that distinguish birds from their reptile ancestors.

descriptive statistics Statistics used to summarize sample data, usually early in data exploration and analysis, that do not involve fitting data to a probability distribution or model or testing hypotheses.

deterministic Events that have no random or probabilistic aspects but rather occur in a completely predictable fashion.

deterministic models Models that will always produce the same result if the same initial conditions and parameter values are used.

diagnostic locus A locus that is fixed or nearly fixed for different alleles in different genetic groups.

digital droplet polymerase chain reaction (ddPCR) A method of digital PCR based on fractionating DNA samples into thousands of water–oil emulsion droplets and conducting PCR within those droplets.

dioecious Varieties or species of plants that have separate male and female reproductive organs on unisexual individuals.

diploid The condition in which a cell or individual has two copies of every chromosome.

directional selection The increase in the frequency of a selectively advantageous allele, gene, or phenotypic trait in a population.

discrete generations Generations that can be defined by whole integers and in which all individuals will breed only with individuals in their generation (e.g., pink salmon, or annual flowers without a seed bank).

dispersal In ecological literature, dispersal is the movement of individuals from one genetic population (or birthplace) into another. Dispersal is also known as **migration** in genetics literature.

dispersal lags A lag between when habitat becomes suitable for a species, for example, with a changing climate, and when that species fully occupies it,

distinct population segment (DPS) A level of classification under the US Endangered Species Act that allows for legal protection of populations that are distinct, relatively reproductively isolated, and represent a significant evolutionary lineage to the species.

DNA barcoding The use of a short DNA sequence from a standardized region of the genome to help discover, characterize, and distinguish species, and to assign unidentified individuals to species.

DNA fingerprinting Individual identification through the use of multilocus genotyping.

DNA methylation The addition of methyl groups to DNA. Methylation can change gene expression and is one of the mechanisms of epigenetic effects.

Dobzhansky–Muller incompatibilities Genic interactions between alleles at multiple loci in which alleles that enhance fitness within their parental genetic backgrounds may reduce fitness in the novel genetic background produced by hybridization.

dominance genetic variation The proportion of total genetic variation that can be attributed to the interactions of alleles at a locus in heterozygotes.

dominant An allele whose phenotypic effect is expressed in both homozygotes (AA) and heterozygotes (Aa).

double digest RAD (ddRAD) One of the **RADseq** family of reduced representation sequencing, involving initial digestion with two restriction enzymes.

drift load Reduction in average fitness of a population because of the increase in frequency of deleterious alleles caused by genetic drift in small populations.

ecological crunches Episodes of extreme environmental conditions.

ecological species concept Defines species as a set of organisms that is adapted to a particular ecological niche and the resources it contains.

ecosystem A community of organisms and its environment.

ecosystem services The products and services humans receive from functioning ecosystems.

ecotone The region that encompasses the shift between two biological communities.

effect size The size of the difference between two groups of observations. For example, the average phenotypic difference between two alleles at a locus.

effective number of alleles The number of equally frequent alleles that would create the same heterozygosity as observed in the population.

effective population size (N_e) The size of the ideal population that would experience the same amount of genetic drift as the observed population.

electrophoresis The movement of protein or DNA molecules through a medium across an electric field.

emergent properties When several simple entities interact to form complex behaviors or structures as a collective. The complex structures or behaviors are not a property of any single entity and cannot be predicted from the behavior of a subset of entities.

empirical probability The ratio of the number of "favorable" outcomes to the total number of trials in an actual sequence of experiments.

endemic Present exclusively in one particular area, for example, the tuatara is endemic to New Zealand.

endonuclease *See* **restriction enzyme**.

ensemble models A group of climate model simulations used for climate change projections.

environmental DNA (eDNA) DNA collected from an environmental sample such as water or soil.

environmental stochasticity Random variation in environmental factors that influence population parameters affecting all individuals in that population.

epigenetic marks Modifications of DNA and associated histones including DNA methylation and histone modification that alter gene activity and expression but do not change DNA sequence.

epigenetics Changes in gene expression that are stable through cell division but do not involve changes in the underlying DNA sequence.

epigenome The entire collection of epigenetic marks on a genome including DNA methylation and histone modification.

epistasis In statistical genetics, this term refers to an interaction of different loci such that the multiple locus phenotype is different than predicted by simply combining the effects of each individual locus (i.e., there is a significant gene–gene interaction).

epistatic genetic variation The proportion of total genetic variation that can be attributed to the interaction between loci producing a combined effect different from the sum of the effects of the individual loci.

establishment lag The delay between the introduction of a species outside of its native range and the establishment of the species in that location.

evolutionarily significant unit (ESU) A classification of populations that have substantial reproductive isolation which has led to adaptive differences so that the population represents a significant evolutionary component of the species. Both evolutionarily and evolutionary are used in the literature, but the original term was "evolutionarily" (Ryder 1986).

evolutionary potential The capacity for a species to evolve and adapt to new conditions.

evolutionary rescue An adaptation-dependent reversal of demographic decline due to maladaptation to novel environmental conditions that results from selection on standing genetic variation that is already present in a population or new variation introduced through gene flow or mutation.

evolutionary species concept A species that encompasses a single evolutionary lineage of organisms that is distinct from all other such lineages.

evolvability *See* **evolutionary potential**.

exact tests An approach to compute the exact *P*-value for a statistical test rather than using an approximation such as the chi-square distribution.

exclusion tests A statistical genetic test to identify likely immigrants by excluding individuals as residents because their multilocus genotype is unlikely to occur in the focal population.

exome capture A technique for sequencing all of the protein-coding regions of a genome through targeted sequence capture. DNA is fragmented, and then DNA probes are used to capture all exon sequences, allowing them to be separated from nontargeted sequences. The targeted fragments are then amplified and sequenced.

exome sequencing *See* **exome capture**.

exon A coding portion of a gene that produces a functional gene product (e.g., a peptide).

expectation maximization (EM) algorithm An iterative mathematical algorithm used to estimate allele or gamete frequencies. EM alternates between the expectation (E) step, which computes the log-likelihood evaluated using the current estimates for the variables, and the maximization (M) step, which computes parameters maximizing the expected log-likelihood found in the E step. These parameter estimates are then used to determine the distribution of the variables in the next E step, until the values converge.

experimental evolution Studies that expose experimental populations to different conditions and evaluate the capacity of the population to adapt to those conditions.

exploitation Use of a natural resource.

expressed sequence tags (ESTs) A cDNA sequence (i.e., a transcribed mRNA sequence) of 200–500 nucleotides that is usually generated from the $3'$ end of cDNA clones. ESTs are used to identify full length genes, to study gene expression via transcript profiling, and for phylogenetic studies.

ex *situ* conservation The conservation of important evolutionary lineages of species outside the species natural habitat.

extant Currently living; not extinct.

extinction The loss of a species or other taxon so that it no longer exists anywhere.

extinction lag The time between an event or events that will ultimately result in extinction of a species and the final disappearance of that species.

extinction vortex The mutual reinforcement among biotic and abiotic processes that can drive small populations to extinction. For example, small population size causes increased inbreeding, which reduces individual fitness, which can further reduce the population size, leading to further increased inbreeding, and so on.

extirpation The loss of a species or subspecies from a particular area but not from its entire range, also known as local extinction.

extrinsic outbreeding depression Outbreeding depression that results from the reduced fitness of hybrids under local environmental conditions rather than from intrinsic genetic incompatibilities.

false negative Incorrectly failing to reject the null hypothesis when it is false, for example, concluding no difference exists between tested groups when it does. Also called Type II error.

false positive Incorrectly rejecting a true null hypothesis, for example, concluding a difference exists between tested groups when it doesn't. Also called Type I error.

fecundity The reproductive capacity of an individual or population (e.g., the number of eggs or young produced by an individual per unit time).

fertility The ability to conceive and have offspring.

filtering The use of quality control steps in analyses of next-generation sequences to remove errors and eliminate poor quality sequence reads, genotypes, or samples from a dataset.

Fisher–Wright model *See* **Wright–Fisher model**.

fitness The ability of an individual or genotype to survive and produce viable offspring. Quantified as the number or relative proportion of offspring contributed to the next generation.

fitness rebound Following an episode of inbreeding depression, successive generations of breeding may result in a rebound in fitness due to the selective decrease in frequency of deleterious alleles (purging). If inbreeding depression is due to deleterious recessive alleles (with negative fitness effects in a homozygous state), then successive generations of inbreeding may result in a rebound in fitness due to the selective decrease in frequency of deleterious alleles.

fitness sets The set of fitness estimates associated with a set of genotypes at a locus.

fixation The shift from a locus being polymorphic to being monomorphic within a population; that is, one allele becoming fixed in the population ($p = 1.0$) and the other allele(s) being lost ($p = 0$).

fixation index The proportional increase of homozygosity through population subdivision. F_{ST} is sometimes referred to as the fixation index.

fluctuating asymmetry (FA) Asymmetry in which deviations from symmetry are randomly distributed about a mean of zero. FA provides a simple measure of developmental precision or stability.

forensics The use of scientific methods and techniques, such as genetic fingerprinting, to solve crimes.

foundation species A dominant primary producer in an ecosystem both in terms of abundance and influence.

founder effect A loss of genetic variation in a population established by a small number of individuals that carry only a fraction of the original genetic diversity from a larger population.

founder-specific inbreeding coefficient The probability that an individual is homozygous (identical by descent) for an allele descended from a specific founder. The sum, across all founders, of the founder-specific partial inbreeding coefficients for an individual is equal to the inbreeding coefficient for that individual.

frequency-dependent selection Natural selection in which fitness varies as a function of the frequency of a phenotype.

frequentist A framework for making inferences from statistical samples. The result of a frequentist approach is either to accept or reject a hypothesis following a significance test or a conclusion if a confidence interval includes the true value.

functional connectivity The degree to which individuals or gametes move through a landscape, which is affected by landscape features and environmental heterogeneity.

F_1 First filial generation; that is, progeny of a focal controlled cross.

F_2 Second filial generation; that is, progeny of an F_1 cross.

gametes Haploid reproductive cells ready for fertilization; that is, eggs (female) and sperm (male).

gametic disequilibrium Nonrandom association of alleles at different loci within a population. Also known as **linkage disequilibrium**.

gametic equilibrium Random association of alleles at different loci within a population. Also known as **linkage equilibrium**.

gene A segment of DNA whose nucleotide sequence codes for protein or RNA, or that regulates other genes.

gene drive A mechanism, either natural or genetically engineered, that propagates a particular allele to spread throughout a population. Gene drives are driven by selfish gene elements.

gene editing The intentional alteration of a DNA sequence through deleting, inserting, or replacing a sequence. *See* **CRISPR-Cas9**.

gene flow Exchange of genetic information between demes through migration.

gene genealogies The tracing of the history of inheritance of the genes in an individual. Gene genealogies are most easily constructed using nonrecombining DNA such as mtDNA or the mammalian Y-chromosome.

Gene Ontology (GO) A major bioinformatics initiative to unify the representation of gene and gene product attributes across all species.

generation interval The average age of all parents within a population when their progeny are born.

genetic architecture The genetic basis of a phenotypic trait and its properties; for example, the number of genes affecting that trait and their corresponding phenotypic effect sizes.

genetic assimilation A process in which phenotypically plastic characters that were originally "acquired" become converted into inherited characters by natural selection. This term has also been applied to the situation in which hybrids are fertile and displace one or both parental taxa through the production of hybrid swarms (i.e., genomic extinction).

genetic correlation The genetic relationship between two quantitative traits; that is, the proportion of variance they share due to genetic causes.

genetic distance matrix A pairwise matrix composed of distances between population (or individual) pairs that is calculated using a measure of genetic divergence, such as F_{ST}.

genetic divergence The evolutionary change in allele frequencies between populations.

genetic draft A stochastic process in which selective substitutions at one locus will reduce genetic diversity at neutral linked loci through hitchhiking.

genetic drift Random changes in allele frequencies in a population between generations due to sampling individuals that become parents and binomial sampling of alleles during meiosis.

genetic engineering A process in which an organism's genes are artificially modified, often through splicing DNA fragments from different chromosomes or species, to achieve a desired result. Also called **genetic modification**.

genetic exchange *See* **gene flow**.

genetic hitchhiking The increase in frequency of a neutral or weakly selected mutation due to linkage with a positively selected mutation.

genetic linkage map A map of the relative locations of loci based on the amount of recombination that occurs between the loci.

genetic load The decrease in the mean fitness of individuals in a population compared with the theoretical mean fitness if all individuals had the most favored genotype.

genetic map *See* **linkage map**.

genetic modification *See* **genetic engineering**.

genetic monitoring The quantification of temporal change in population genetic metrics or other population data generated using genetic markers.

genetic offset The predicted difference between the current genetic makeup of a population and the genetic composition that will maintain the existing adaptive gene–environment relationship.

genetic rescue A decrease in population extinction probability because of gene flow, occurring naturally or human-mediated.

genetic stochasticity Random genetic changes in a population, for example, due to genetic drift and inbreeding.

genetic swamping The loss of locally adapted alleles or genotypes caused by constant immigration and gene flow.

genetic tagging Sampling and genotyping an individual so it can be identified again, for example, in a mark–recapture study.

genetic variance In quantitative genetics, the variance of phenotypic values caused by genetic differences among individuals.

genetics The study of how genes are transmitted from one generation to the next and how those genes affect the phenotypes of the progeny.

genome skimming A sequencing approach using low-coverage, shallow sequencing that results in complete information for parts of the genome with high copy number such as ribosomal and mitochondrial DNA and microsatellites, but incomplete sequencing of other regions.

genome-wide association studies (GWAS) The genotyping of many loci in different individuals to see whether any alleles or loci are associated with a particular trait.

genomic architecture The arrangement of structural and functional elements (e.g., genes, regulatory regions) within a genome.

genomic extinction The situation in which hybrids are fertile and displace one or both parental taxa through the production of hybrid swarms so that the parental genomes no longer exist, even though the parental alleles are still present.

genomic ratchet A process where hybridization producing fertile offspring will result in a hybrid swarm over time, even in the presence of outbreeding depression.

genomic resources Sequences and associated genetic knowledge about a species that facilitate genomic analysis.

genomic vulnerability *See* **genetic offset**.

genomics The study of the structure, function, evolution and mapping of genomes undertaken through analysis of large numbers of genes or markers.

genotype An organism's genetic composition.

genotype-by-environment (G × E) interactions The phenotypic variation that results from interactions between genes and environments.

genotype–environment associations (GEAs) Statistical associations between alleles at a locus in populations and environmental variables for population origins.

genotyping-by-sequencing One of the restriction enzyme-associated reduced representation sequencing methods (*See* **RADseq**).

geometric mean The nth root of the product of n numbers. For example, the geometric mean of 2, 3, and 4 is $(2 \times 3 \times 4)^{1/3} = 2.884$.

germplasm Living tissues (e.g., seeds or tissue culture) from which plants can be grown.

global climate model (GCM) Complex atmospheric and oceanic models based on fluid dynamics and thermodynamics, widely used for predicting climate change.

global effective population size The effective population size of an entire species.

gnomics The study of the complete genetic information of small creatures that live in the depths of the earth and guard buried treasure.

gynodioecy The occurrence of female and hermaphroditic individuals in a population of plants.

gynogenesis A form of parthenogenesis where sperm is required to be present for reproduction but does not fertilize the egg.

haldane A unit of evolutionary change equal to one phenotypic standard deviation per generation, named for J.B.S. Haldane.

haplodiploid A genetic system whereby females develop from fertilized diploid eggs and males develop from unfertilized haploid eggs, for example, in bees and wasps.

haplogroup A group of similar haplotypes that share a common ancestor.

haploid The condition in which a cell or individual has one copy of every chromosome.

haplotype The combination of alleles at loci that are found in close proximity on a single chromosome or within an mtDNA molecule.

Hardy–Weinberg principle The principle that allele and genotype frequencies will reach equilibrium in binomial proportions after one generation and remain constant in large random mating populations that experience no migration, selection, mutation, or nonrandom mating.

Hardy–Weinberg proportion A state in which a population's genotypic proportions equal those expected with the binomial distribution.

harmonic mean The reciprocal of the arithmetic mean of reciprocals in a series of n numbers. For example, the harmonic mean of 2, 3, and 4 is $[3/(1/2 + 1/3 + 1/4) = 2.769]$.

harvest To take or kill individuals (e.g, fish, deer, or trees) for food, sport, or other uses.

hemizygous A term used to denote the presence of only one copy of an allele due to a locus being in a haploid genome, on a sex chromosome, or only one copy of the locus being present in an aneuploid organism.

heritability The proportion of total phenotypic variation within a population that is due to individual genetic variation (H_B; broad-sense heritability). Heritability is more commonly referred to as the proportion of phenotypic variation within a population that is due to additive genetic variation (H_N; narrow-sense heritability).

hermaphrodite An individual that produces both female and male gametes.

heterochromatin Highly folded chromosomal regions that contain few functional genes. When these traits are characteristic of an entire chromosome, it is a heterochromosome or supernumerary chromosome.

heterogametic The sex with different sex chromosomes (e.g., the male in mammals [XY] and female in birds [ZW]).

heteromorphic Having different forms.

heteroplasmy The presence of more than one mtDNA haplotype in a cell or tissue.

heterosis The case when hybrid progeny have higher fitness than either of the parental organisms. Also called **hybrid vigor.**

heterozygosity (*H*) A measure of genetic variation that estimates either the observed or expected proportion of individuals in a population that are heterozygotes.

heterozygosity–fitness correlations (HFC) The observation that individuals with greater heterozygosity at marker loci have greater fitness.

heterozygote An organism that has different alleles at a locus (e.g., *Aa*).

heterozygous advantage A situation where heterozygous genotypes are more fit than homozygous genotypes. This fitness advantage can create a stable polymorphism. Also called **overdominance.**

heterozygous disadvantage A situation where heterozygous genotypes are less fit than homozygous genotypes. Also called **underdominance.**

Hill–Robertson effect The case where selection at one locus will reduce the effective population size of linked loci, increasing the chance of genetic drift forming negative genetic associations that reduce the ability of associating loci to respond to selection. *See* **genetic draft**.

histone modification Histones are the main proteins around which DNA spools in chromatin. Histone modification including different types of methylation and acetylation affect gene expression.

hitchhiking The increase in frequency of a selectively neutral allele through gametic disequilibrium with a beneficial allele that selection increases in frequency in a population.

homogametic The sex that possesses the same sex chromosomes (e.g., the female in mammals [XX] and the male in birds [ZZ]).

homology Similarity in structural features such as genes or morphology that are derived from a shared ancestor by common descent.

homoplasmy The presence of a single mitochondrial DNA haplotype within a cell.

homoplasy Independent evolution or origin of similar traits, or gene sequences. At a locus, homoplasy can result from back-mutation or mutation to an existing allelic state.

homozygosity A measure of the proportion of individuals in a population that are homozygous, and is the reciprocal of heterozygosity.

homozygote An organism that has two or more copies of the same allele at a locus (e.g., *AA* or *aa*).

horizontal gene transfer Incorporation of genetic material into an individual without vertical (parent–offspring) transmission. Common among single-celled organisms (e.g., transfer of genes for antibiotic resistance in bacteria).

hybrid sink The situation where immigration of locally unfit genotypes produces hybrids with low fitness that reduces local density and thereby increases the immigration rate.

hybrid swarm A population of individuals that are all hybrids by varying numbers of generations of backcrossing with parental types and matings among hybrids.

hybrid vigor *See* **heterosis**.

hybrid zone An area of sympatry between two genetically distinct populations or species where hybridization occurs without forming a hybrid swarm in either parental population beyond the area of co-occurrence.

hybridity Genomic mixture of two taxa due to hybridization.

hybridization Mating between individuals of two genetically distinct populations or species.

ideal population *See* **Wright–Fisher population model**.

identical by descent Alleles that are copies of the same allele from a common ancestor.

inbreeding The mating between related individuals which results in an increase of homozygosity in the progeny because they possess alleles that are identical by descent.

inbreeding coefficient A measure of the level of inbreeding in a population, developed by Sewall Wright, that determines the probability that an individual possesses two alleles at a locus that are identical by descent.

inbreeding depression The relative reduction in fitness of progeny from matings between related individuals compared with progeny from unrelated individuals.

inbreeding effect of small populations The inbreeding that develops over time in a small population even if individuals are mating at random because after a few generations, all individuals will be related to one another.

inbreeding effective number (N_{el}) The size of the ideal population that loses heterozygosity at the same rate as the observed population.

inbreeding load The reduction in fitness of inbred individuals compared to noninbred individuals.

insertion–deletion (indel) A polymorphism where a specific DNA sequence is either present (insertion) or absent (deletion).

***in situ* conservation** The conservation of a population or species in its natural habitat.

interdemic genetic drift The reduction in effective population size of a metapopulation caused by random differences in the productivity of subpopulations.

interlocus allelic love The nonrandom association of alleles (and therefore genotypes) at different loci within a population. Also known as **gametic disequilibrium** or **linkage disequilibrium**.

intrinsic outbreeding depression The reduction in fitness in hybrids due to genetic incompatibilities between the taxa hybridizing, whether between species or long-diverged populations within species.

introduction The placement or escape of a species or individual into a novel habitat.

introgression The incorporation of genes from one population to another through hybridization that results in fertile offspring that further hybridize and backcross to parental populations.

intron A portion of a gene that produces a section of mRNA strand that is cleaved prior to translation; a noncoding region between the exons.

invasive species An introduced alien species that is likely to cause harm to the natural ecosystem, the economy, or human health.

island model of migration A model of migration in which a population is subdivided into a series of demes, of equal size N, that randomly exchange migrants at a given rate, m.

isolates Genetic samples of disease-causing organisms such as fungi or bacteria.

isolation by distance (IBD) The case where genetic differentiation is greater the further that individuals (or populations) are from each other because gene flow decreases as geographic distance increases. Originally used when individuals are distributed continuously on the landscape and are not subdivided into discrete local populations by barriers to gene flow (Wright 1943).

isozymes Alternative forms of an enzyme that may be alleles produced by the same locus (allozymes), or by paralogous loci.

ISSR Inter-simple sequence repeat markers that use PCR with primers based on simple sequence repeats of microsatellites.

IUCN conservation status The global ranking of the population health and risk of extinction of species by the International Union for the Conservation of Nature.

jackknife A resampling technique used to estimate variance and bias. The jackknife estimator of a parameter is found by systematically leaving out each observation from a dataset and calculating the estimate and then finding the average of these calculations.

karyotype The composition of the chromosomal complement of a cell, individual, or species.

landscape genetics An interdisciplinary field that assesses the influence of landscape and environmental variables on dispersal, gene flow, and genetic variation.

landscape genomics Landscape genetics with many markers, often in adaptive genes and with samples of individuals across environmental gradients, to study both neutral and adaptive patterns and processes (e.g., adaptive differentiation and gene flow).

landscape resistance barrier A feature of the landscape (e.g., a river or land-cover type) that impedes movement and therefore reduces gene flow.

leading edge population A population at the expanding margin of a species range during a range shift.

least cost path Path with the least cumulative resistance to movement in landscape genetics. The least cost path is often longer in geographic distance than the direct straight-line path because the direct path often crosses poor habitat with high ecological cost to movement.

lek Area where the males of a population congregate and display for females.

lethal equivalent The number of deleterious alleles in an individual whose cumulative effect is the same as that of a single lethal allele.

library *See* **sequencing library**.

likelihood The probability of the data given the hypothesis.

lineage sorting A process where different gene lineages within an ancestral taxon are lost by drift or replaced by unique lineages evolving in a different derived taxon.

linkage The nonrandom segregation of two loci on the same chromosome. Linkage is measured by the rate of recombination (r) between loci; $r < 0.5$ for linked loci, and 0.5 for unlinked loci.

linkage disequilibrium (LD) Nonrandom association of alleles at different loci within a population. Also known as **gametic disequilibrium**.

linkage equilibrium Random association of alleles at different loci within a population. Also known as **gametic equilibrium**.

linkage map A table or diagram that indicates the position of genes or markers relative to one another, based on the frequency of recombination rather than the physical distance between them. Also known as a **genetic map**.

local adaptation Greater fitness of individuals in their local habitats compared with nonlocal individuals due to natural selection.

local scale The spatial scale at which individuals routinely interact with their environment.

locus The location of a gene or marker on a chromosome.

log of odds ratio (LOD) The odds ratio is the odds of an event occurring in one group to the odds of it occurring in another group. For example, if 80% of the individuals in a population are *Aa* and 20% are *AA*, then the odds of *Aa* over *AA* is four; there are four times as many *Aa* as *AA* genotypes. The natural log of this ratio is often computed because it is convenient to work with statistically.

major histocompatibility complex (MHC) A group of highly polymorphic genes in vertebrates that code for immune system proteins on cell surfaces that help recognize foreign substances.

management unit (MU) A local population that is managed as a distinct unit because of its demographic independence.

Manhattan plot A type of scatterplot often used to display GWAS results, with genomic position along chromosomes or linkage groups on the *x*-axis and statistical values for marker–trait associations on the *y*-axis. They are named Manhattan plots because they resemble a city skyline with many low values and a few genomic peaks with higher values.

marginal overdominance Greater overall fitness associated with heterozygous genotypes, which are not the most fit in any single environment, due to an organism's interactions with multiple environments that each favor different alleles.

Markov chain A mathematical system that undergoes transitions from one state to another, as a random process in which the next state depends only on the current state.

Markov chain Monte Carlo (MCMC) A tool or algorithm for sampling from probability distributions based on constructing a Markov chain. The state of the chain after many steps is then used as a sample from the desired distribution. Sometimes called a random walk Monte Carlo method.

match probability (MP) The probability of sampling an individual that has an identical multilocus genotype to the one already sampled ("in hand").

maternal effects The influence of the genotype or phenotype of the mother on the phenotype of the offspring. Maternal effects are not heritable because they have no genetic basis in the offspring.

maximum likelihood A statistical method of determining which of two or more competing alternative hypotheses (such as alternative phylogenetic trees) yields the best fit to the data.

maximum likelihood estimate (MLE) A method of parameter estimation that obtains the parameter value that maximizes the likelihood of the observed data.

mean kinship (*mk*) A measure of how closely related each individual is to the entire population. Animals with lower *mk* have relatively fewer genes in common with the rest of the population, and are therefore more genetically valuable in a breeding program.

Mendelian segregation The random separation of paired alleles (or chromosomes) into different gametes.

meristic character A trait of an organism that can be counted using integers (e.g., fin rays or ribs).

meta-analyses Statistical analyses that combine and analyze the results from multiple similar studies.

metabarcoding Genetic tests, often involving PCR amplification and sequencing of a single DNA locus, to identify many species simultaneously.

metabolome The entire set of small-molecule metabolites found in a cell, tissue, or organism.

metacentric A chromosome in which the centromere is centrally located.

metagenomics The analysis of DNA from the many species contained in an environmental sample, facilitated by high-throughput sequencing.

metapopulation A collection of spatially divided subpopulations with some probability of extinction that are connected by gene flow among them.

metapopulation scale The spatial scale at which individuals migrate between local subpopulations, often across habitat that is unsuitable for colonization.

metatranscriptomics The analysis of all of the RNA in an environmental sample or tissue.

methylated Having a methyl group added to a molecule. Two of the four bases in DNA, cytosine and adenine, can be methylated, a state that represses gene activity.

methylation-sensitive Restriction enzymes that can only cleave DNA at restriction sites if the DNA is not methylated.

Metropolis–Hastings methods A popular MCMC technique used to simulate observations from a probability distribution.

microbiome The aggregate community of all microorganisms found in an environment, for example, an animal gut, soil sample, or water sample.

microchromosomes Small chromosomes which, unlike heterochromosomes, carry functional genes.

microhaplotype A locus with two or more single nucleotide polymorphisms that occur within a short segment of DNA (e.g., 200 bp) that can be covered by a single sequence read and collectively define a multiallelic locus.

microsatellite Tandemly repeated DNA consisting of short sequences of 1–6 nucleotides repeated ~5–100 times. Also known as **VNTR** or **SSR**.

migration The movement of individuals from one genetically distinct population to another, resulting in gene flow.

migration–drift equilibrium A dynamic balance between the loss of genetic variation due to genetic drift and gain via migration within populations.

migration load The reduction in fitness of a population caused by the immigration of individuals that are less adapted to the local environment than nonimmigrants.

minimum viable population (MVP) The minimum population size at which a population is likely to persist over some defined period of time.

minisatellite Regions of DNA in which repeat units of 10–50 bp are tandemly arranged in arrays of 0.5–30 kb in length.

missing heritability problem In most genome-wide association studies (GWAS), the effects of all individual loci statistically associated with a trait do not explain most or all of the heritability of the trait.

mitochondrial DNA (mtDNA) A small, circular, haploid DNA molecule found in the mitochondria cellular organelle of eukaryotes.

molecular clock The observation that substitutions accumulate at relatively constant rates, thereby allowing researchers to estimate the time since two species shared a common ancestor.

molecular epidemiology The use of molecular genetic analysis to infer patterns of pathogen transmission.

molecular genetics The branch of genetics that studies the molecular structure and function of genes, or that (more generally) uses molecular markers to test hypotheses.

molecular mutations Changes to the genetic material of a cell, including single nucleotide changes, deletions, and insertions of nucleotides, as well as recombination and inversion of DNA sequences.

moments Quantitative measures of the shape of a distribution. The first moment of the distribution of the random variable X is the population mean. The second and third moments are the variance and skewness, respectively.

monoecious A plant in which male and female organs are found on the same plant.

monoecy Individuals can produce both male and female gametes (as opposed to dioecy, where individuals are either male or female).

monomorphic The presence of only one allele at a locus, or the presence of a common allele at a high frequency (>95% or 99%) in a population.

monophyletic A group of taxa that includes all species, ancestral and derived, from a common ancestor.

monotypic A taxonomic group that encompasses only one taxonomic representative.

Monte Carlo process A process of drawing a random number at each step of a random walk, named for a city famous for gambling.

morphology The study of the physical structures of an organism, including the evolution and development of these structures.

multidimensional scaling A statistical graphing technique used to represent genetic distances between samples in two or three dimensions, and thereby visualizing similarities and differences between different groups or samples.

multiplexed Short, unique DNA barcodes are added to DNA fragments from individual samples before combining samples for library preparation and sequencing so that sequence data can then be sorted and analyzed by individual.

multivariate exploratory techniques The study of three or more variables as a first step in analyzing data for identifying patterns, detecting errors, or selecting appropriate models or statistical tests for further analyses.

mutagenesis The natural or intentional formation of mutations in a genome.

mutation A change in the DNA sequence or chromosome in the transmission of genetic information from parent to progeny. *See* **molecular mutations**.

mutation load The decrease in fitness caused by the accumulation of deleterious alleles.

mutational meltdown The process by which a small population accumulates new deleterious mutations, which leads to loss of fitness and decline of the population size, which leads to further accumulation of deleterious mutations because of the ineffectiveness of selection in small populations.

mutualist species A species in a symbiotic relationship in which both species benefit.

narrow-sense heritability (H_N) The proportion of individual phenotypic variation that is due to additive genetic variation.

native species A species that historically or currently occurs in a given geographical region.

natural catastrophes Natural events causing great damage to populations that decrease their population size and increases their probability of extinction.

natural selection Differential contribution of genotypes to the next generation due to differences in survival or reproduction.

neighborhood The spatial area in a continuously distributed population that can be considered panmictic.

nested clade analysis (NCA) *See* **nested clade phylogeographic analysis**.

nested clade phylogeographic analysis (NCPA) A statistical approach to describe how genetic variation is distributed spatially within a species' geographic range. This method uses a haplotype tree to define a nested series of branches (clades), thereby allowing a nested analysis of the spatial distribution of genetic variation, often with the goal of resolving between past fragmentation, colonization, or range expansion events.

neutral landscape genomics Studies using selectively neutral loci to characterize landscape connectivity of populations and genetic variation.

next-generation sequencing (NGS) DNA sequencing technologies that produce millions or billions of short reads (from 25–500 bp) in a few days.

node A branching point or end point on a phylogenic tree that represents either an ancestral taxon (internal node) or an extant taxon (external node).

nonindigenous species Species present in a geographical area that were introduced and did not historically occur.

noninvasive Sampling DNA with no contact or skin break of a target organism. For example, collection of sloughed skin, shed feathers, feces, urine, or hair from a tree's bark.

nonparametric Statistical approaches that make few or no assumptions about the distribution of data or test statistic. These include tests that involve ranking observations, such as Wilcoxon's signed-ranks test.

nuclear DNA (nDNA) DNA in the cell nucleus of eukaryotes.

nuclear gene A gene located in the nucleus of a eukaryotic cell.

nucleotide diversity (π) The average number of nucleotide differences per site between two DNA sequences in all possible pairs in the sample population.

nucleotides The building blocks of DNA and RNA made up of a nitrogen-containing purine or pyrimidine base linked to a sugar (ribose or deoxyribose) and a phosphate group.

null allele An allele that is not detectable either due to a failure to produce a functional product (e.g., allozymes) or a mutation in a primer site that precludes amplification during PCR analysis (e.g. for microsatellites).

Ockham's razor The principle that the least complicated explanation (most parsimonious hypothesis) generally should be accepted to explain the data at hand.

offsite conservation *See ex situ* conservation.

operational taxonomic units (OTUs) Groups of similar, often unknown, microbes, clustered based on DNA sequence similarity, that can serve as proxies for classic taxonomic levels (e.g., species).

orthologous Homologous genes found in different species that are derived from a common ancestral gene.

otoliths A bony structure found in the inner ear of vertebrates, often used to estimate the age of fish as they form annual growth rings.

outbreeding depression The reduction in fitness of hybrids compared with parental types.

outlier loci Loci that fall outside the range of the expected distribution for some summary statistic compared with that of neutral loci in a sample (e.g., extremely high values of F_{ST}).

overdominance *See* **heterozygous advantage**.

overexploitation Use of a resource to the point of diminishing returns or destruction of the resource.

overlapping generations A breeding system where sexual maturity does not occur at a specific age, or where individuals breed more than once, causing individuals of different ages to interbreed in a given year.

pairwise sequentially Markovian coalescent (PSMC) An analytical approach that uses coalescence events and linkage information along chromosomes to estimate the timing and duration of population size changes over past millennia.

panmictic Randomly mating.

paracentric inversion A chromosomal inversion that does not include the centromere because both breaks were on the same chromosomal arm.

paralogous Homologous genes or sequences within a species that arose from an ancestral gene that was duplicated, where one copy may have evolved a new function.

paralogs Paralogous genes or sequences.

parameter Numerical characteristic of a statistical population or model, often referring to the population parametric value computed from the total population of observations (not a sample).

parametric Statistical approaches that assume data come from a distribution that can be adequately modeled by a probability distribution with a fixed set of parameters.

paraphyletic A clade that does not include all of the descendants from the most recent common ancestor taxon.

parentage analysis The assessment of the maternity or paternity of a given individual.

parentage assignment Likelihood approaches for identifying parents that estimate the probability of each candidate parent being the actual parent based on parent and offspring genotypes as well as population allele frequencies.

parentage exclusion Computational approaches that assign parentage to an offspring based on excluding possible parents based on the principle that the offspring should should share at least one allele at each locus with the parent.

parsimony The principle that the preferred phylogeny of an organism is the one that requires the fewest evolutionary changes; the simplest explanation.

parthenogenesis The development of an embryo without fertilization. In animals, it is development of an embryo from an unfertilized egg cell. In plants, apomixis is a form of parthenogenesis that involves no meiosis.

parthenogenetic A form of asexual reproduction without fertilization.

paternity assignment Likelihood approaches to identify the paternal parent that estimate the probability of each candidate paternal parent being the actual parent based on offspring and candidate paternal genotypes, where the maternal genotype is known. *See* **parentage assignment**.

paternity exclusion probability The probability that an individual is not the male parent of an offspring based on their genotype. *See* **parentage exclusion**.

PCR duplication A process during PCR amplification of DNA where one DNA fragment gets amplified much more than expected, producing a high frequency of identical sequence reads. When this happens to one allele at a locus, it will create bias in allele frequency estimates.

performance evaluation An evaluation of the performance of an analysis by quantifying the accuracy, precision, power, or robustness of a statistical estimator or test.

pericentric inversion A chromosomal inversion that includes the centromere because the breaks were on opposite chromosomal arms.

permanent translocation heterozygote Occurs when two pairs of homologous chromosomes have reciprocally exchanged nonhomologous segments between one member of each pair. Therefore, each of the affected chromosome pairs contain both homologous and nonhomologous segments. Put another way, each such pair has one translocated chromosome, and one normal (untranslocated) chromosome.

permutation test A test in which the distribution of the test statistic under the null hypothesis is obtained by calculating all possible values of the test statistic under rearrangements of the labels on the observed data points. Also called a randomization test.

phenetics Taxonomic classification solely based on overall similarity of phenotypic traits regardless of phylogeny.

phenogram A branching diagram or tree that is based on estimates of overall similarity between taxa derived from a suite of characters.

phenology The study of periodic plant and animal life-cycle events and how these are influenced by seasonal and interannual variations in climate.

phenotype The observable characteristics of an organism that are the product of the organism's genotype and environment.

phenotypic plasticity Phenotypic differences between individuals with similar genotypes or changes in the phenotype of an individual genotype over time due to differences in environmental factors during development. Often results from changes in gene expression.

philopatry A characteristic of reproduction of organisms where individuals faithfully return home to natal sites. Individuals exhibiting philopatry are philopatric.

phylogenetic networks A graph used to depict evolutionary relationships among sequences, genes, chromosomes, or genomes. These graphs can take many forms, and can illustrate events including hybridization, gene duplication, or horizontal gene transfer.

phylogenetic species concept (PSC) States that a species is a discrete lineage or recognizable monophyletic group.

phylogeny A diagram illustrating the evolutionary relationships among taxa or genes.

phylogeography The assessment of the geographic distributions of the taxa of a phylogeny to understand the evolutionary history (e.g., origin and spread) of a given taxon.

physical genome maps These maps estimate the physical distance between genes or markers in base pairs, based on genome sequencing and assembly, in contrast to **genetic maps** that are based on recombination rates between loci.

pleiotropy The case where a single gene affects more than one trait.

Poisson distribution A probability distribution, with identical mean and variance, that characterizes discrete events occurring independently of one another in time, where the mean probability of the events is very small.

policy A set of principles to guide decisions and achieve desired outcomes. A policy is a statement of intent, and is implemented as a procedure or protocol.

polygenic A phenotype affected by many genes.

polymerase A molecule that catalyzes the synthesis of DNA or RNA from a single-stranded template and free deoxynucleotides (e.g., during PCR).

polymerase chain reaction (PCR) A technique to replicate a desired segment of DNA.

polymorphic The presence of more than one allele at a locus. Generally defined as when the most common allele is at a frequency less than 95% or 99%.

polymorphism The presence of more than one allele at a locus. Polymorphism is also used as a measure of the proportion of loci in a population that are genetically variable or polymorphic (P).

polyphyletic A group of taxa classified together that have descended from different ancestral taxa (i.e., taxa that do not all share the same recent common ancestor).

polyploid Individuals whose genome consists of more than two sets of chromosomes (e.g., tetraploids).

polytomy A node on a phylogenetic tree from which more than two branches emerge. Sometimes used to indicate either radiation events or ambiguities in knowledge about relationships.

pool-seq An approach involving pooling equal amounts of DNA from multiple individuals per population or treatment prior to sequencing. This approach reduces costs and allows for estimation of allele frequencies but does not provide genotypes of individuals.

population connectivity Measure of the extent of dispersal of individuals or gametes among discrete populations that can have different meanings and be estimated in different ways.

population genomics The use of genome-wide data at thousands to hundreds of thousands or even millions of variable loci to infer population demographics and

evolutionary processes, including selection, genetic drift, gene flow, and mutation.

population parametric Parametric estimates of parameters (e.g., mean, variance) for a population.

population-scaled mutation rate (θ) An important parameter in population genetics that scales with population genetic diversity, estimated as $\theta = 4N_e\mu$, where μ is the neutral mutation rate.

population viability analysis (PVA) The general term for the application of models that account for multiple threats facing the persistence of a population, to access the likelihood of the population's persistence over a given period of time.

portfolio effect The effect of genetic and life history diversity on ecological stability analogous to financial stability derived from a diversified investment portfolio.

positive assortative mating When individuals tend to mate with phenotypically similar individuals.

posterior probability A Bayesian estimate of the probability that a hypothesis is true in the light of relevant observations.

prevalence The proportion of individuals in a population infected with a pathogen or exhibiting a particular disease, or proportion of a disease vector (e.g., a mosquito or tick) carrying a pathogen.

primer A small oligonucleotide (typically 18–22 bp long) that anneals to a specific single-stranded DNA sequence to serve as a starting point for DNA replication (e.g., extension by polymerase during PCR).

prior probability The prior probability distribution (i.e., "prior") of an uncertain quantity k (e.g., the number of populations represented in a sample of individuals) is the probability distribution that would express one's uncertainty about k before the data (e.g., genotypes) are taken into account. It attributes uncertainty rather than randomness to the quantity.

private allele An allele present in only one of many populations sampled.

probabilistic maturation reaction norm (PMRN) A reaction norm is the set of phenotypes expressed by a single genotype across a range of environments reflecting phenotypic plasticity. A probabilistic maturation reaction norm describes an individual's probability of maturing at a given age as a function of size and other relevant phenotypic traits.

probability The certainty of an event occurring. The observed probability of an event, r, will approach the true probability as the number of trials, n, approaches infinity.

probability density function (PDF) A function for a continuous, random variable for which the value at any point on the function provides an estimate of the probability of that value. For example, the birth weights of sheep follow a normal distribution, which is a probability density.

probability distribution function A function used to define the distribution of a probability. A discrete probability distribution function, such as the Poisson or binomial, informs how the total probability of one is distributed over all possible outcomes.

probability of identity (*PI*) The probability that two unrelated (randomly sampled) individuals would have an identical genotype. This probability becomes very small if many highly polymorphic loci are considered.

probability of nonexclusion The probability that a candidate parent cannot be excluded as the actual parent of an offspring based on the genotypes of the candidate parent and offspring. *See* **parentage exclusion**.

product rule A statistical rule that states that the probability of n_i independent events occurring is equal to the product of the probabilities of each n independent event.

propagule A dispersal vector. Any disseminative unit or part of an organism capable of independent growth (e.g., a seed, spore, mycelial fragment, sclerotium bud, tuber, root, or shoot).

propagule pressure A measure of the introduction of nonindigenous individuals that includes the number of individuals (or propagules) introduced and the number of introductions.

proportion of individual admixture The proportion of alleles in an individual's genotype originating from a species or populations that are genetically differentiated.

protein A polypeptide molecule.

proteome The complete set of proteins found within a cell, tissue, or organism.

provenance trial A field common garden experiment with plants, most often trees, from different populations planted together in one or more locations. Often a partial **reciprocal transplant experiment**.

pseudoautosomal Regions or genes of sex chromosomes that are present on both the X and Y, or the Z and W, chromosomes.

pseudo-overdominance *See* **associative overdominance**.

purging The removal of deleterious recessive alleles from a population through inbreeding, which increases homozygosity, which in turn increases the ability of selection to act on recessive alleles.

purging selection Selection against recessive deleterious alleles that prevents them from reaching high frequencies.

quantitative polymerase chain reaction (qPCR) PCR that amplifies and simultaneously quantifies the number of copies of the targeted DNA molecule.

quantitative trait A phenotype (characteristic) that varies in degree due to polygenic effects (of two or more loci) and the environment.

quantitative trait loci Genetic loci that affect phenotypic variation (and potentially fitness), which are identified by a statistically significant association between genetic markers and measurable phenotypes. Quantitative traits are often influenced by multiple loci as well as environmental factors.

quasi-extinction When population size falls below some predefined threshold, below which the population is unlikely to persist.

RAD sequencing (RADseq) Sequencing of the DNA segment that immediately flanks each side of a restriction enzyme site throughout the genome to discover or genotype single nucleotide polymorphisms.

RAD tags Anonymous markers genotyped through RADseq.

random forest A machine-learning ensemble method used for classification, regression, and other purposes that builds a "forest" of decision trees based on repeated subsampling of data and variables, and assigns importance values to variables.

randomization test *See* **permutation test**.

RAPD Randomly amplified polymorphic DNA. A method of analysis in which PCR amplification using arbitrary oligonucleotide primers is used to create a multilocus band profile. RAPDs are no longer used because of their poor repeatability.

Rapture A reduced representation sequencing approach that adds a targeted DNA capture step to RADseq, reducing the extent of missing data.

rarefaction A technique to compare allelic diversity in samples of different sizes. Rarefaction estimates the expected total number of alleles in a smaller sample drawn at random from a large pool. Rarefaction allows the comparison of the diversity in samples of different sizes.

rarefaction curves Plots of sampling effort (*x*-axis) versus diversity captured (e.g., species richness, number of OTUs, or number of alleles). They can be used to compare diversity estimates at a common sample size.

read depth The number of separate reads from NGS for a specific genomic region.

rear edge population A population at the contracting or trailing edge of a species range during range migration.

reciprocal monophyly A genetic lineage is reciprocally monophyletic when all members of the lineage share a more recent common ancestor with each other than with any other lineage on a phylogenetic tree.

reciprocal transplant experiment A field common garden experiment containing multiple populations, where all populations are planted on sites in every population's place of origin for evaluating the degree of local adaptation.

recombinant inbred lines (RILs) Homozygous lineages produced through repeated inbreeding (or self-pollination in some plants).

recombination The process that generates a haploid product of meiosis with a genotype differing from both haploid genotypes that originally combined to form the diploid zygote.

reduced representation sequencing Partial sequencing of genomes of multiple individuals for population genomic analyses targeting a subset of regions, either anonymous or targeted, using methods including **RADseq**, **exome capture**, **Rapture**, and other **targeted sequence capture** approaches.

reference genome A fully sequenced, assembled, and preferably annotated genome of one individual of a species against which other genomic data can be aligned and interpreted. Highly repetitive regions are often excluded.

reference transcriptome An assembled set of all RNA transcripts of an individual or species derived from sequencing cDNA reverse-transcribed from mRNA. A reference transcriptome may be assembled from a single individual, tissue, or treatment, or may combine transcribed sequences from a variety of samples. For species lacking a reference genome assembly, sequence reads for coding regions can be aligned using a reference transcriptome.

reintroduction The introduction of a species or population into a historic habitat from which it had previously been extirpated.

relative fitness A measure of fitness that is the ratio of a given genotype's absolute fitness to the genotype with the highest absolute fitness. Relative fitness is used to model the effects of natural selection on allele frequencies.

Representative Concentration Pathways (RCPs) Four atmospheric greenhouse gas concentration trajectories (RCP 2.6, 4.5, 6.0, and 8.5) developed and used by the Intergovernmental Panel on Climate Change (IPCC) to describe different future climate scenarios.

rescue effect When immigration into an isolated deme reduces the probability of the extinction of that deme either because of genetic or demographic effects. *See* **demographic rescue** and **genetic rescue**.

resistance models Spatial models in which each location across a landscape is given a weight or "resistance value" reflecting the influence of variables (e.g., land cover, slope, elevation) on movement by the species in question.

restricted (or residual) maximum likelihood (REML) A maximum likelihood estimation approach that can be used to estimate variance components (e.g., in quantitative genetics) where maximum likelihood is not estimated directly from all data, but instead uses a likelihood function derived from a transformed dataset.

restriction endonuclease An enzyme (*see* **endonuclease**), isolated from bacteria, that cleaves DNA at a specific nucleotide sequence. Over 3,000 such enzymes exist that recognize and cut hundreds of different DNA sequences; used in RFLP, RAD, and AFLP analysis and to construct recombinant DNA (in genetic engineering).

restriction enzyme *See* restriction endonuclease.

restriction fragment length polymorphism (RFLP) A method of genetic analysis that examines polymorphisms based on differences in the number of fragments produced by the digestion of DNA with specific endonucleases. The variation in the number of fragments is created by substitutions within restriction sites for a given endonuclease.

restriction site-associated DNA sequencing (RADseq) A family of reduced-representation sequencing methods that digest DNA with one or more restriction enzymes and sequence DNA adjacent to those restriction sites.

resurrection studies Experiments that revive ancestral populations from stored propagules (i.e., seed, eggs, or spores) and compare them with contemporary individuals, allowing for the detection and characterization of evolution.

retrotransposon A mobile DNA sequence that can move to new locations through an RNA intermediate.

reverse mutation rate Back mutation rate. The rate at which a nonfunctional gene's ability to produce a functional product is restored.

reverse speciation The loss of differences between species brought about by hybridization and admixture of two isolated gene pools into a single admixed population.

ribonucleic acid (RNA) A polynucleotide similar to DNA that contains ribose in place of deoxyribose, and uracil in place of thymine. RNA is involved in the transfer of information from DNA, programming protein synthesis, regulating gene expression, and maintaining ribosome structure.

risk of nonadaptedness A projection of the extent of maladaptation due to climate change based on the difference in predicted allele frequencies locus by locus between current and future conditions.

RNA sequencing (RNAseq) The use of next-generation sequencing to sequence and quantify the total amount of RNA is a sample. RNA is reverse-transcribed to more stable cDNA prior to sequencing. The sequence read depth for a given exon can be used as an estimate of gene expression levels.

Robertsonian fission An event where a metacentric chromosome breaks near the centromere to form two acrocentric chromosomes.

Robertsonian fusion An event where two acrocentric chromosomes fuse to form one metacentric chromosome.

Robertsonian translocation A special type of translocation where the break occurs near the centromere or telomere and involves the whole chromosomal arm so balanced gametes are usually produced.

rooted phylogenetic tree A type of phylogenetic tree that shows evolutionary relationships among taxa by rooting the tree to a common ancestor to all taxa, usually represented by the point that an outgroup splits from the taxa of interest.

runs of homozygosity Genomic regions where an individual is autozygous; that is, has DNA sequences on homologous chromosomes that are identical by descent.

sampling distribution The sampling distribution of a statistic is the distribution of that statistic, considered as a random variable, when derived from a random sample.

sampling error Error associated with sampling, which can be large if too few or nonrepresentative samples are used.

scaffolds A portion of a genome sequence assembled from whole-genome shotgun sequences, containing fully sequenced and assembled regions called contigs bridged by paired-end sequences that have some gaps.

script A series of commands or statements executed or interpreted by a shell or interpreter without user interaction or the need for compiling as required by certain programming languages. Examples are batch files or macros.

secondary contact zone Contact between populations that had previously been geographically separate (i.e., allopatric).

segregation load The decrease in mean fitness in a population caused by heterozygous advantage; that is, homozygotes will have lower fitness than heterozygotes, reducing mean fitness.

selection coefficient The reduction in relative fitness, and therefore genetic contribution to future generations, of one genotype compared with others.

selection differential The difference between the mean value of a quantitative trait found in a population as a whole compared with the mean value of the trait in the selected breeding population.

selective sweep The rapid increase in frequency by natural selection of an initially rare allele that also fixes (or nearly fixes) alleles at closely linked loci and thus reduces the genetic variation in a region of a chromosome.

selectively neutral Variation in DNA sequence or among alleles that has no phenotypic effect and therefore has no effect on fitness.

selfing Self-pollination in plants; the most extreme form of inbreeding.

sensitivity testing A method used in population viability analyses where the effects of parameters on the persistence of populations are determined by testing a range of possible values for each parameter.

sequencing library Collection of DNA fragments from a given organism, ligated to short DNA sequences (adaptors) or "stored" in a virus or bacteria, used for next-generation sequencing.

sequential Bonferroni correction A method, similar to the Bonferroni correction, that is used to reduce the probability of a false positive statistical error when conducting multiple simultaneous tests.

sex chromosomes Chromosomes that pair during meiosis but differ in the heterogametic sex.

sex-linked locus A locus that is located on a sex chromosome.

sexual selection Selection due to differential mating success, either through competition for mates or mate choice.

shadow effect A case in genetic mark–recapture studies in which a novel capture is labeled as a recapture due to identical genotypes at the loci studied.

shell (UNIX) A layer of software that runs on top of the operating system. The shell interprets user commands and dispatches them to the operating system.

simulations Computer models that simulate population processes over generations to explore the effects of various factors (e.g., population size, spatial distribution, connectivity) on the genetics of populations (e.g., genetic variation, inbreeding).

single nucleotide polymorphism (SNP) A nucleotide site (base pair) in a DNA sequence that is polymorphic in a population and can be used as a marker to assess genetic variation within and among populations.

singleton An allele that is present exactly once in a sample. Also, a sequencing read that is present exactly once.

site frequency spectrum (SFS) The frequency distribution of alleles that are segregating within DNA sequences. The SFS provides evidence of the demographic history of populations.

small RNAs Small (<200 bp) noncoding RNA molecules that are often involved in RNA silencing.

sonication Fragmentation of DNA with sound energy at ultrasonic frequencies.

space-for-time substitutions The use of spatial environmental (e.g., climatic) variation in common garden experiments as a surrogate for the effects of environmental changes on populations in one location over time.

spatial genetic structure (SGS) The nonrandom distribution of genetic variation spatially within a population, which may be due to mating system, gene flow, or selection.

spatial sorting An evolutionary process by which genes change in frequency as a function of their effects on dispersal, rather than their effects on survival and reproduction.

spatiotemporal landscape genetics The incorporation of genetic and environmental data collected both spatially and temporally in landscape genetics.

species A group of individuals that naturally interbreed among themselves. *See* **species concepts**.

species concepts The ideas of what constitutes a species, such as reproductive isolation (**Biological Species Concept**, BSC), or monophyly of a lineage (**Phylogenetic Species Concept**).

species distribution models (SDMs) Models based on the statistical relationship between species occurrence and local environmental conditions, often climate variables.

species scale The spatial scale encompassing an entire species' distribution.

stabilizing selection Selection favoring intermediate phenotypes over extreme states of the phenotype.

stable equilibrium An equilibrium of allele frequencies to which the population returns if allele frequencies are perturbed away from the equilibrium.

stable polymorphism A polymorphism that is maintained at a locus through natural selection.

standard error The standard deviation of the sampling distribution of a statistic.

standing genetic variation Allelic variation present in a population at one point in time.

statistic A numerical measure of some attribute of a sample such as its mean or median.

statistical estimators A statistic or algorithm that creates an estimate of a population parameter from a sample. For example, the sample mean ($\bar{x}$) is an estimator for the population mean, μ.

statistical inference The process of drawing conclusions from data that are subject to random variation such as sampling variation.

statistical phylogeography Model-based phylogeographic inference that tests alternative hypotheses by taking into account the stochasticity of the genetic process.

statistical significance Rejection of the null hypothesis.

stepping-stone model of migration A model of migration in which exchange between nearby populations is more likely to occur than exchange between more distant populations.

stepwise mutation model (SMM) A model of mutation in which the microsatellite allele length has an equal probability of either increasing or decreasing (usually by a single repeat unit, as in the strict one-step SMM).

stochastic The presence of a random variable in determining the outcome of an event.

stock A term used in fisheries management that refers to a population that is demographically independent (e.g., an MU).

subfossil Remains not ancient enough to be considered true fossils, but not recent enough to be considered modern.

subjective probability A probability based on a person's judgment, belief, or expert opinion about how likely a specific outcome is to occur.

subpopulations Groups within a population delineated by reduced levels of gene flow with other groups.

subspecies A taxonomically defined subdivision within a species that is physically or genetically distinct, and often geographically separated.

sum rule A statistical rule that states that the probability of n_i mutually exclusive, independent events occurring is equal to the sum of the probabilities of each n event.

supergene Allelic combinations found at closely linked loci that affect related traits and are inherited together.

supernumerary chromosome A chromosome, often present in varying numbers, that is not needed for normal development, lacks functional genes, and does not segregate during meiosis. Also called B chromosomes.

supportive breeding The practice of removing a subset of individuals from a wild population for captive breeding, and then releasing the captive-born offspring back into their native habitat to intermix with wild-born individuals and increase population size or persistence.

sympatric Populations or species that occupy the same geographic area.

synapomorphy A shared derived trait between evolutionary lineages. A homology that evolved in an ancestor common to all species on one branch of a phylogeny, but not common to species on other branches.

synonymous A nucleotide substitution within the coding region of a gene that changes a codon sequence but does not change the amino acid coded for. Also called silent mutations.

syntenic The location of two or more loci on the same chromosome. In recent literature, syntenic sometimes refers to the conservation of gene order on a chromosome between individuals or species that are being compared.

Taq The bacterium *Thermus aquaticus* from which a heat-stable DNA polymerase used in PCR was isolated.

targeted sequence capture A method for sequencing selected rather than anonymous genomic regions, concentrating resources on loci of interest, for example, **exome capture**.

telomere A region of tandemly repeated segments of a short DNA sequence, one strand of which is G-rich and the other strand is C-rich, that forms the ends of linear eukaryotic chromosomes.

tests for differences Statistical tests of differences between groups (e.g., populations) where the null hypothesis is that there is no difference.

tests for relationship Statistical tests of associations between two variables. The null hypothesis is that there is no relationship.

third-generation sequencing Sequencing technologies that can produce individual sequence reads that are thousands of base pairs long or longer.

threshold character A phenotypic character that contains a few discrete states controlled by many genes underlying continuous variation, which affects a character phenotypically only when a certain physiological threshold is exceeded.

transcriptome The set of all of the messenger RNA transcripts of a tissue, individual, or species.

transgenic An individual or species with genes inserted from another species (e.g., a genetically engineered organism), also called genetically modified.

transgressive segregation Hybridization events that produce progeny that express phenotypic values outside the range of parental phenotypic values. These differences are usually due to the disruption of polygenic traits.

transition A point mutation in which a purine base (A or G) is substituted for a different purine base, or a pyrimidine base (C or T) is substituted for a different pyrimidine base; for example, an A:T to G:C transition.

translocation (1) Anthropogenic movement of individuals from one location to another that is often intended to establish a new population, to achieve genetic or demographic rescue, or to facilitate adaptation to a rapidly changing climate.

translocation (2) A chromosomal rearrangement occurring when a piece of a chromosome is broken off and joined to another location.

transposable element Any DNA sequence that can insert into a chromosome, exit, and relocate; includes insertion sequences, transposons, some bacteriophages, and controlling elements. A region of the genome flanked by inverted repeats, a copy of which can be inserted at another place; also called a transposon or a jumping gene.

transposon A mobile element of DNA that jumps to new genomic locations through a DNA intermediate, and which usually carries genes other than those that encode for transposase proteins used to catalyze movement.

transversion A point mutation in which a purine base is substituted for a pyrimidine base or vice versa; for example, an A: T to C:G transversion. *See* **transition**.

Type I statistical error *See* **false positive**.

Type II statistical error *See* **false negative**.

underdominance The opposite of overdominance at a locus, where the heterozygous genotype has lower fitness than either homozygotes. Also called heterozygous disadvantage.

unified species concept Defines species as separately evolving lineages of metapopulations.

unigene A bioinformatically determined gene identified from RNA transcript sequencing.

unrooted phylogenetic trees A type of phylogenetic tree that indicates the relative relatedness of taxa without showing their ancestral relationships.

unstable equilibrium An equilibrium of allele frequencies in which allele frequencies will move away from the equilibrium if they are perturbed.

variance A statistical measure of variation calculated as the mean of the squared deviations from the arithmetic mean (the standard deviation squared). In quantitative genetics, phenotypic variance is partitioned into genetic, environmental, and interaction components.

variance effective number (N_{eV}) The size of the ideal population that experiences changes in allele frequency at the same rate as the observed population.

viability The probability of survival of a given genotype to reproductive maturity (or of a population to persist through a certain time interval).

vouchered specimens A preserved specimen that provides a verified and permanent record of that species (for example, in an herbarium or natural history collection).

Wahlund principle The deficit of heterozygotes compared with expected HW proportions because of the presence of two or more panmictic demes with different allele frequencies.

whole genome resequencing Sequencing genomic DNA from multiple individuals at lower read depth than would be used for assembling a reference genome for population genomic analyses.

wildlife forensics The genetic identification of species, populations, or individuals to detect and provide evidence of illegal activities.

Wright–Fisher population model A constant-size population of size N in which the next generation is produced by drawing $2N$ genes at random from a large gamete pool to which all individuals contribute equally.

zoonoses Any disease of nonhuman animals that can be transmitted to humans.

zygote A fertilized egg cell resulting from the fusion of an egg and sperm.

Probability, Statistics, and Coding

Appendix (pages 596–628) is only available online at www.oup.com/companion/AllendorfCGP3e

References

References (pages 629–711) are only available online at www.oup.com/companion/AllendorfCGP3e

Index

Page numbers in *italics* refer to figures and tables. Page numbers in **bold** refer to the Appendix which is available only online at www.oup.com/companion/AllendorfCGP3e.